Survey Measurement
and Process Quality

WILEY SERIES IN PROBABILITY AND STATISTICS

Established by WALTER A. SHEWHART and SAMUEL S. WILKS

Editors: *Vic Barnett, Ralph A. Bradley, Nicholas I. Fisher, J. Stuart Hunter, J. B. Kadane, David G. Kendall, David W. Scott, Adrian F. M. Smith, Jozef L. Teugels, Geoffrey S. Watson*

A complete list of the titles in this series appears at the end of this volume.

Survey Measurement and Process Quality

Edited by

LARS LYBERG

PAUL BIEMER

MARTIN COLLINS

EDITH DE LEEUW

CATHRYN DIPPO

NORBERT SCHWARZ

DENNIS TREWIN

A Wiley-Interscience Publication

JOHN WILEY & SONS, INC.

New York • Chichester • Weinheim • Brisbane • Singapore • Toronto

Copyright © 1997 by John Wiley & Sons, Inc.

All rights reserved. Published simultaneously in Canada.

Library of Congress Cataloging in Publication Data:

Survey measurement and process quality / Lars Lyberg . . . [et al.]
    p.   cm.—(Wiley series in probability and statistics.
  Applied probability and statistics)
    Includes index.
    ISBN 0-471-16559-X (cloth : alk. paper)
    1. Social sciences—Statistical methods. 2. Surveys—Methodology.
  3. Social sciences—Research—Evaluation. 4. Surveys—Evaluation.
I. Lyberg, Lars. II. Series.
  HA29.S843   1996
  300′.723—dc20

96-44720
CIP

Contents

SECTION E. ERROR EFFECTS ON ESTIMATION, ANALYSES, AND INTERPRETATION

Contributors

Reginald P. Baker, National Opinion Research Center, Chicago, Illinois, U.S.A.

Alison K. Baldwin, National Opinion Research Center, Chicago, Illinois, U.S.A.

Francesca Bassi, University of Padua, Padua, Italy

Mary K. Batcher, Internal Revenue Service, Washington, DC, U.S.A.

Jelke G. Bethlehem, Statistics Netherlands, Voorburg, The Netherlands

Paul P. Biemer, Research Triangle Institute, Research Triangle Park, North Carolina, U.S.A.

Steven Blixt, MBNA-America, Wilmington, Delaware, U.S.A.

Bill Blyth, Taylor Nelson AGB, London, United Kingdom

Pamela Campanelli, Social Community Planning Research, London, United Kingdom

Noel Chavez, University of Illinois, Chicago, Illinois, U.S.A.

Michael Colledge, Australian Bureau of Statistics, Belconnen, Australia

Martin Collins, City University Business School, London, United Kingdom

Frederick Conrad, Bureau of Labor Statistics, Washington, DC, U.S.A.

Mick P. Couper, University of Michigan, Ann Arbor, Michigan, U.S.A.

Edith de Leeuw, Vrije Universiteit, Amsterdam, The Netherlands

Don A. Dillman, Washington State University, Pullman, Washington, U.S.A.

Cathryn S. Dippo, Bureau of Labor Statistics, Washington, DC, U.S.A.

Jennifer Dykema, University of Wisconsin, Madison, Wisconsin, U.S.A.

James L. Esposito, Bureau of Labor Statistics, Washington, DC, U.S.A.

Luigi Fabbris, University of Padua, Padua, Italy

Leandre R. Fabrigar, Queen's University at Kingston, Ontario, Canada

Wayne Fuller, Iowa State University, Ames, Iowa, U.S.A.

Leopold Granquist, Statistics Sweden, Stockholm, Sweden

Bill Gross, Australian Bureau of Statistics, Belconnen, Australia

Patricia M. Guenther, Department of Agriculture, Washington, DC, U.S.A.

Sue Ellen Hansen, University of Michigan, Ann Arbor, Michigan, U.S.A.

K. Piyasena Hapuarachchi, Statistics Canada, Ottawa, Canada

Anthony Heath, Nuffield College, Oxford, United Kingdom

John Horm, National Center for Health Statistics, Hyattsville, Maryland, U.S.A.

Joop Hox, University of Amsterdam, Amsterdam, The Netherlands

Cleo R. Jenkins, Bureau of the Census, Washington, DC, U.S.A.

Jared B. Jobe, National Center for Health Statistics, Hyattsville, Maryland, U.S.A.

Timothy Johnson, University of Illinois, Chicago, Illinois, U.S.A.

Daniel Kasprzyk, Department of Education, Washington, DC, U.S.A.

John G. Kovar, Statistics Canada, Ottawa, Canada

Jon A. Krosnick, Ohio State University, Columbus, Ohio, U.S.A.

Jouni Kuha, Nuffield College, Oxford, United Kingdom

Loretta Lacey, University of Illinois, Chicago, Illinois, U.S.A.

Rolf Langeheine, University of Kiel, Kiel, Germany

James M. Lepkowski, University of Michigan, Ann Arbor, Michigan, U.S.A.

Susan Linacre, Australian Bureau of Statistics, Belconnen, Australia

Lars E. Lyberg, Statistics Sweden, Stockholm, Sweden

Mary March, Statistics Canada, Ottawa, Canada

David A. Marker, Westat Inc., Rockville, Maryland, U.S.A.

Jean S. Martin, Office for National Statistics, London, United Kingdom

Nick Moon, NOP Social and Political, London, United Kingdom

David R. Morganstein, Westat Inc., Rockville, Maryland, U.S.A.

Colm O'Muircheartaigh, London School of Economics and Political Science, London, United Kingdom

William L. Nicholls II, Bureau of the Census, Washington, DC, U.S.A.

Diane O'Rourke, University of Illinois, Urbana, Illinois, U.S.A.

Sarah M. Nusser, Iowa State University, Ames, Iowa, U.S.A.

William F. Pratt, National Center for Health Statistics, Hyattsville, Maryland, U.S.A.

Thomas W. Pullum, University of Texas, Austin, Texas, U.S.A.

J. N. K. Rao, Carleton University, Ottawa, Canada

Kenneth A. Rasinski, National Opinion Research Center, Chicago, Illinois, U.S.A.

Jennifer M. Rothgeb, Bureau of the Census, Washington, DC, U.S.A.

Sally Sadosky, University of Michigan, Ann Arbor, Michigan, U.S.A.

Fritz J. Scheuren, George Washington University, Washington, DC, U.S.A.

Norbert Schwarz, University of Michigan, Ann Arbor, Michigan, U.S.A.

Jacqueline Scott, Queen's College, Cambridge, United Kingdom

Randy R. Sitter, Carleton University, Ottawa, Canada

Chris Skinner, University of Southampton, Southampton, United Kingdom

Tessa Staples, Office for National Statistics, London, United Kingdom

S. Lynne Stokes, University of Texas, Austin, Texas, U.S.A.

Seymour Sudman, University of Illinois, Urbana, Illinois, U.S.A.

Katarina Thomson, Social Community Planning Research, London, United Kingdom

Roger Tourangeau, National Opinion Research Center, Chicago, Illinois, U.S.A.

Dennis Trewin, Australian Bureau of Statistics, Belconnen, Australia

Frank van de Pol, Statistics Netherlands, Voorburg, The Netherlands

Richard Warnecke, University of Illinois, Chicago, Illinois, U.S.A.

Adam Wronski, Statistics Canada, Ottawa, Canada

Michaela Wänke, University of Heidelberg, Heidelberg, Germany

Preface

Survey quality is directly related to survey errors. Survey errors can be decomposed in two broad categories: sampling and nonsampling errors. Comprehensive theories exist for the treatment of sampling errors. As for nonsampling errors, no such theory exists. Indeed, there has been extensive interest in nonsampling error research over the last 50 years which has resulted in an abundance of literature describing the treatment of various specific error sources and attempts at an integrated treatment or simultaneous modeling of several specific error sources. In some ways this research has been very successful and has given rise to efficient methods of dealing with some error sources. On the other hand, many types of errors have exhibited a resistance to the suggested solutions. One reason, of course, is that errors are generated by a vast number of sources, which is then further complicated by the great complexity of some survey designs.

Another reason is that nonsampling errors usually are not additive. Reducing the level of one type of error might very well increase some other type of error. Indeed, some survey quality goals are conflicting, which introduces an irrational element into the decision making. For instance, attempts at reducing non-response rates by intense follow-ups might sabotage timeliness; wealth of detail regarding survey data might violate confidentiality safeguards; and reducing processing errors might call for questionnaires that are less user-friendly. Admittedly, some error sources can be dealt with, but others are so elusive that they defy expression, not to mention estimation.

Left uncontrolled, nonsampling errors can render the resulting survey data useless for many important survey objectives. Post-survey quality assessment measures such as reliability, validity, and bias estimates are very important indicators of data accuracy, but, except for repeated surveys, may be of little value for improving the survey data. Rather, interest must shift from post-survey quality evaluation and possible correction to controlling the survey processes such as questionnaire construction, interviewing and other data collection activities, coding, data capture, editing, and analysis. Process quality generates product quality.

Many survey organizations throughout the world are now working with the concepts of Total Quality Management (TQM) in the context of survey design, data collection, and data processing. Methods for monitoring and ensuring process quality such as process control, quality teams, customer focus, decisions based on scientific methods, and so on which have been developed in industrial settings are being successfully applied in survey work. Very little of this work is reported in the literature yet the potential of these methods is enormous.

Given the importance of the topic of survey measurement and process quality, the Survey Research Methods Section (SRM) of the American Statistical Association (ASA) in 1992 determined that survey measurement and process quality should be the topic of an SRM-sponsored conference and edited monograph and approached Lars Lyberg to develop the idea. It was decided that the conference should seek the participation of researchers worldwide and that it should take place in a European country to ensure a wider dissemination of research findings and exploit the great interest emerging in countries outside the U.S. which, geographically, has been the locus of interest in this topic. Also, the SRM emphasized the need for interdisciplinary and cross-cultural research.

By early 1993, an organizing/editing committee was formed consisting of: Lars Lyberg (Chair), Paul P. Biemer, Martin Collins, Edith de Leeuw, Cathryn Dippo, Norbert Schwarz, and Dennis Trewin. Lee Decker was enlisted to be in charge of the conference logistics and most administrative issues related to the conference. Patricia Dean was enlisted as consulting editor for the monograph. Lilli Japec was enlisted to be in charge of contributed papers and other planning tasks.

The committee contacted numerous research organizations for financial contributions. The committee also developed the monograph outline and began to identify and contact researchers throughout the world as potential authors. Abstracts were requested and 133 abstracts were received from researchers interested in writing for the monograph. From these, the committee selected 34 which would be the chapters in this monograph and developed the conference program. Steve Quigley orchestrated John Wiley & Sons' role in publishing this monograph.

Five professional organizations were asked to sponsor the conference: the American Statistical Association (ASA), the International Association of Survey Statisticians (IASS), the Market Research Society (MRS), the Royal Statistical Society (RSS), and the World Association for Public Opinion Research (WAPOR). All five organizations enthusiastically agreed and also contributed funds to support the project. In addition, the following research organizations contributed funds:

Australian Bureau of Statistics
Central Statistical Office, Dublin
Economic & Social Research Council, London

International Labour Office, Geneva
National Opinion Resarch Center, Chicago
NSS, The Hague
National Science Foundation, Washington, DC
Office for National Statistics, London
Research Triangle Institute, Research Triangle Park
SPSS, London
Statistics Denmark
Statistics Finland
Statistics Netherlands
Statistics New Zealand
Statistics Sweden
Survey Research Center, Institute for Social Research, University of Michigan
U.S. Bureau of Labor Statistics
U.S. Bureau of the Census
U.S. Department of Agriculture/NASS
U.S. National Center for Health Statistics
Westat, Inc., Rockville
ZUMA, Mannheim

Without the financial support of these organizations, the conference and edited monograph would not have been possible.

The International Conference on Survey Measurement and Process Quality was held April 1–4, 1995 in Bristol, United Kingdom. It drew 291 attendees from 27 countries. The program consisted of the 34 invited papers chosen for the present monograph and 113 contributed papers. Authors of contributed papers were offered the opportunity to submit their papers to a proceedings volume published by the American Statistical Association. The volume contains 60 of the contributed papers. Additionally, two short courses were offered, one on multilevel analysis for survey research and one on TQM in statistical organizations. The organizing committee was also very pleased by the fact that the Sir Ronald Fisher Memorial Committee of Great Britain chose our conference for its XIXth Fisher Memorial Lecture. This lecture is included in the proceedings volume.

In designing this book, the aim has been to discuss the most important issues in the field of survey measurement and process quality, attempting whenever possible to integrate various perspectives. Thus, each chapter has undergone extensive editing, review, and revision. The book is organized into five sections. The section titles and their editors are:

Section A: Questionnaire Design (Norbert Schwarz)
Section B: Data Collection (Edith de Leeuw and Martin Collins)
Section C: Post Survey Processing and Operations (Lars Lyberg)
Section D: Quality Assessment and Control (Cathryn Dippo)

Section E: Error Effects on Estimation, Analyses, and Interpretation (Paul Biemer and Dennis Trewin).

The diversity of orientations of the authors for the monograph made it impossible to impose a unified terminology and set of notation across all chapters. Except for Section E, the statistical level of the monograph is quite accessible by graduate students in sociology, psychology, communication, education, or marketing research. Section E, however, requires a fairly thorough foundation in survey sampling and mathematical statistics.

Although the present book can serve as a course text, its primary audience is researchers having some prior knowledge in survey research. Since it contains a number of review articles on survey measurement and process quality in several disciplines, it will be useful to researchers actively involved in this field who want a discussion from different theoretical perspectives. The book will also be useful to methodologists who want to learn more about improving the quality of surveys through better design, data collection, and analytical techniques and by focusing on processes. The book reflects current knowledge in 1995, to the best of our editorial judgment. As a group, we hope that its publication will stimulate future research in this exciting field.

Most section editors had responsibilities as the secondary editor for one other section as well. The authors of the chapters, in addition to their extensive writing and revising activities, were also involved in the review of other monograph chapters. They were encouraged to seek outside reviews for their chapters on their own. Thus, the monograph reflects the efforts and contributions of scores of writers, editors, and reviewers. The committee would like to sincerely thank Patricia Dean who performed the final editing of all manuscripts. She served as consulting editor and her efforts went far beyond regular language editing in providing authors with suggestions regarding style and organization of all chapters.

We are again grateful to Lee Decker of the ASA, who handled all the logistical details associated with the conference and the proceedings volume with great care and efficiency. Lilli Japec deserves great appreciation for all the activities she performed so ably for the conference and monograph.

Sincere thanks go to Joke Hoogenboom and Truus Kantebeen at the SRM Documentation Centre at the Erasmus University in Rotterdam, who compiled a booklet "SMPQ, A Selected Bibliography" that was available at the conference. They also did literature searches for all sections of the monograph.

Thanks are due to Barbara Bailar, Seymour Sudman, and Judith Lessler, who while serving in various ASA positions promoted the idea of an international conference on this topic to be held outside the U.S., which is the usual locus for these activities. We are appreciative of the efforts of Dan Kasprzyk, Jun Liu, Joop Hox, John Bosley, Fred Conrad, Sylvia Fisher, Roberta Sangster, Linda Stinson, and Clyde Tucker who assisted the committee in the review of a number of chapters. Sincere thanks go to Eva von Brömssen who compiled the initial list of index entries. Our employing organizations also deserve great appreciation for supporting our activities in conducting the conference and assembling the monograph: Statistics Sweden (Lyberg); University of Michigan (Schwarz); Vrije Universiteit, Amsterdam (de Leeuw); City University (Collins); U.S. Bureau of Labor Statistics (Dippo); Research Triangle Institute (Biemer); and Statistics New Zealand and Australian Bureau of Statistics (Trewin).

LARS LYBERG
PAUL BIEMER
MARTIN COLLINS
EDITH DE LEEUW
CATHRYN DIPPO
NORBERT SCHWARZ
DENNIS TREWIN

Survey Measurement
and Process Quality

INTRODUCTION

Measurement Error in Surveys: A Historical Perspective

Colm O'Muircheartaigh
London School of Economics and Political Science

1 ORIGINS OF SURVEY ASSESSMENT

In considering the different contexts in which early users of surveys operated, and the different perspectives they brought to their operations, it is difficult to find common criteria against which to measure the success of their endeavors. The history of surveys (in their modern sense) goes back only 100 years, but from the outset there was a great diversity in the settings, topics, philosophies, and executing agencies involved. In the initial stages there was no particular distinction drawn between the issues of survey design and execution and the issues of error in surveys.

The concept of quality, and indeed the concept of error, can only be defined satisfactorily in the same context as that in which the work is conducted. To the extent that the context varies, and the objectives vary, the meaning of error will also vary. I propose that as a definition of error we adopt the following: *work purporting to do what it does not do.* Rather than specify an arbitrary (pseudo-objective) criterion, this redefines the problem in terms of the aims and frame of reference of the researcher. It immediately removes the need to consider *true value* concepts in any absolute sense, and forces a consideration of the needs for which the data are being collected. Broadly speaking, every survey operation has an objective, an outcome, and a description of that outcome. Errors (quality failures) will be found in the mismatches among these elements.

There are three distinct strands in the historical development of survey research: governmental/official statistics; academic/social research; and

Survey Measurement and Process Quality, Edited by Lyberg, Biemer, Collins, de Leeuw, Dippo, Schwarz, Trewin.
ISBN 0-471-16559-X © 1997 John Wiley & Sons, Inc.

commercial/advertising/market research. Each of these brought with it to the survey its own intellectual baggage, its own disciplinary perspective, and its own criteria for evaluating success and failure.

The International Statistical Institute (ISI) was the locus of debate for official statisticians at the end of the 19th century when Kiaer, director of the Norwegian Bureau of Statistics, presented a report of his experience with "representative investigations" and advocated further investigation of the field. In this context the evaluation of surveys was largely statistical and the survey was seen as a substitute for complete enumeration of the population. Bowley— the first professor of statistics at the London School of Economics and Political Science—through his work on sampling (1906 and 1926) and on measurement (1915) was one of the principal figures in the development of the scientific sample survey. This became and has remained the dominant methodology in the collection of data for government, and the government sample survey agency became an important purveyor of data both to politicians and to statesmen. Symptomatic of their genesis, these agencies tended to be located in the national statistical office, and their professional staff tended to be trained in mathematics or in statistics. Here the concept of error became synonymous with the variance of the estimator (essentially the variance of the sampling distribution following Neyman's influential paper in 1934 (Neyman, 1934)). This equivalence of quality and variance and its measurement by repeated sampling, with some acknowledgment of bias, was confirmed by the work of Mahalanobis in India, reported in the mid-1940s (see Mahalanobis, 1944, 1946), and in particular by his design of interpenetrating samples for the estimation of fluctuations or variability introduced by fieldworkers and others. The influence of statisticians on the conceptualization of error and its measurement has continued in this tradition, and can be found in all the classic texts of survey sampling (Yates, 1949; Cochran, 1953; Hansen et al., 1953; Kish, 1965). In this tradition the term "error" has more than one meaning (see, for example, Groves (1989)) but it is used loosely to describe any source of variation in the results or output or estimates from a survey.

While recognizing the powerful position occupied by the scientific sample survey in social research, it is worth noting that Kiaer's proposal to the ISI in 1895 was not universally welcomed, and would almost certainly have been rejected had it not been for the support of the monographers whose work consisted of the detailed examination of one or a small number of cases (what might today be called the *case study* approach).

The involvement of the monographers in the debate at the ISI is interesting particularly because it provides a link to the second major strand in the development of surveys. This was the *Social Policy* and *Social Research* movements, whose beginnings are perhaps best represented by Booth's study, from 1889 to 1903, of poverty in London, and the Hull House papers in the U.S.A. in 1892. Though not in any way a formal or organized movement, there were certain commonalities of approach and objectives across a wide range of activities. The goal of this movement was social reform, and the mechanism was community description. Here the success or failure of the activity was the effect the findings had on decision makers and politicians.

The principal influences on this group were the social reform movement and the emerging sociology discipline. Some of the pioneers of sample surveys spanned both official statistics and the social survey; in particular, Bowley (who made a substantial contribution to the development of survey sampling) produced a seminal work on social measurement in 1915 which helped define the parameters of data quality and error for factual or behavioral information. Bogardus (1925), Thurstone (1928), and Likert (1932) provided scientific approaches to the measurement of attitudes. In this field the disciplinary orientation was that of sociology and social psychology, with some influence from social statistics and psychometrics. Likert, who was subsequently the founding director of the Institute for Social Research at the University of Michigan in 1946, reflected the same practical orientation as the early pioneers in the Social Research movement in his later work on organizations (though with extensive use of attitude measurement).

The third strand arose from the expansion of means of communication and growth in the marketplace. From modest beginnings in the 1890s (Gale and others), there was a steady increase in the extent of advertising and a development and formalization of its companion, market research. The emphasis was on commercial information, originally in the hands of producers of goods and services and collected and evaluated informally (Parlin, 1915); market research, however, developed gradually into a specialized activity in its own right.

Here the effect of psychologists was particularly strong. The work of Link and others in the Psychological Corporation was influential in providing an apparently scientific basis for measurement in the market research area. For those psychologists, experimental psychology took precedence over social psychology. The terminology and the approach were redolent of science and technology. The term "error" was not used explicitly; rather there was a description of *reliability* and *validity* of instruments. This contrasts particularly with the "error" orientation of the statisticians.

Thus the field of survey research as it became established in the 1940s and 1950s involved three different sectors—government, the academic community, and business; it had three different disciplinary bases—statistics, sociology and experimental psychology; and it had developed different frameworks and terminologies in each of these areas.

2 FRAMEWORK

In general in describing data quality or errors in surveys, models concentrate on the survey operation itself, in particular on the data collection operation. The models may be either mathematical (presenting departures from the ideal as disturbance terms in an algebraic equation) or schematic (conceptual models describing the operational components of the data collection process). The conceptual models focus on the interview as the core of the process. Building on the work of Hyman (1954), Kahn and Cannell (1957), Scheuch (1967) and others, Sudman and Bradburn present one of the more useful of these in their book on response effects in surveys (Sudman and Bradburn, 1974). This (schematic) model presents the relationship among the interviewer, the respondents and the task in determining the outcome of the survey interview. The elaborated model identifies the potential contribution of a number of the key elements in each of these to the overall quality of the survey response.

- The *interviewer*, as the agent of the researcher, is seen to carry the lion's share of responsibility for the outcome of the data collection process. Sudman and Bradburn distinguish three elements: the formal constraints placed on the interviewer; the actual behavior of the interviewer; and the extra-role characteristics of the interviewer.
- The *respondent* has not generally been examined as a source of error (apart from a general complaint about poor performance of his/her task). Survey research has tended to take the respondent for granted, though many of the early writers referred to the need to motivate the respondent. The overall approach has, however, been to consider the respondent an obstacle to be overcome rather than an active participant in the process.
- In general, models of response errors in surveys focus on the *task*, which is constrained and structured to accomplish the research goals—in particular to provide the data necessary for analysis. The task includes the length and location of the interview, the question wording, questionnaire construction, the types of data sought, and their implications in terms of memory and social desirability.
- The *responses* are the outcome of the data collection exercise, and the raw material for survey analysis. Most survey analyses treat these as free from error; the statistical approach to measurement error considers the response to be a combination of the *true value* of the data for the individual plus a disturbance described as a *response deviation* or *response effect*.

It is clear that any model of the survey process, and therefore any general model of survey error, will have to include these elements. It is not, however, sufficient to consider these elements, as they do not take into account the context of a survey nor can they distinguish among different survey objectives. To compare the different approaches to survey research described in Section 1, however, it is necessary to provide an over-arching framework that encompasses the concerns of all three major sectors.

One possible framework draws on some ideas presented by Kish in his book on statistical design for research (Kish, 1987). Kish suggests that there are three issues in relation to which a researcher needs to locate a research design; I propose that a similar typology could be used to classify the dimensions that would encompass most sources of error. Each of these "dimensions" is itself multi-dimensional; they are *representation, randomization,* and *realism.*

As survey research deals with applied social science, our understanding of measurement in surveys must also be grounded in actual measures on population elements. Social theory does not have this requirement, nor indeed does statistical theory. At this empirical level, however, the strength and even the direction of relationships between variables are always conditional on the elements, and thus it is critical that any conclusions from a survey should be based on a suitable set of elements from the population, and that comparisons between subclasses of elements should be based on comparable subsets of elements. *Representation* involves issues such as the use of probability sampling, stratification, the avoidance of selection bias, and a consideration of non-response. In general we do not believe that any finding in social science will apply uniformly to all situations, in all circumstances, for all elements of the population. Indeed a good deal of social science is dedicated to understanding the ways in which differences occur across subgroups of populations or between populations. *Representation* reflects this set of concerns with regard to the elements included in the investigation. In particular it refers to the extent to which the target population is adequately mirrored in the sample of elements. In a perfectly specified model, there would be no need to be concerned about which elements from the population appeared in the sample. In the absence of complete and perfect specification of a model (with all variables with potential to influence the variables or relationship under consideration being included), the notion of representation specifically covers the appropriate representation of domains (or subclasses), the avoidance of selection bias, and the minimization of differential nonresponse. The term representative sampling has a chequered history in statistics (see, for instance, Kruskal and Mosteller, 1980; O'Muircheartaigh and Soon, 1981). It carries with it an aura of general (possibly) unjustified respectability; it can be taken to mean the absence of selective forces (that could lead to selection biases); its original connotation was that of a miniature or mirror of the population; it has sometimes been seen as a typical or ideal case; it can imply adequate coverage of the population (cf. stratification); its highest manifestation is in probability sampling, which is the approach in academic and (most) governmental research.

Randomization (and its converse in this context, *control*) covers issues of experimentation and control of confounding variables. Though surveys rarely involve the use of formal experiments for substantive purposes, the identification of sources of measurement error (distortion in the data) and the estimation of the magnitudes of these "errors" frequently do. Randomization is used to avoid, or reduce the probability of, spurious correlations or mis-identification of effects. (It may be worth pointing out that randomization (or at least random selection) is also used to achieve representation in probability sampling.)

Realism arises as an issue in this context in two ways. *Realism in variables* concerns the extent to which the measured or manifest variables relate to the constructs they are meant to describe; *realism in environment* concerns the degree to which the setting of the data collection or experiment is similar to the real-life context with which the researcher is concerned. The survey context may be contrasted with observational studies in which both the variables and the environment are closer to the reality we would like to measure. These dimensions are related to the ideas of *internal validity* and *external validity* used by Campbell and Stanley (1963) and others in describing the evaluation of social research. The validity of a comparison within the context of a particular survey is the realm of internal validity; the extent to which an internally valid conclusion can be generalized outside that particular context is the realm of external validity.

In the following sections the different components of the response process are presented. Each of them concentrates on a different element of the basic model. Section 3 presents the perspective of official (government) statistics and concentrates on the *responses*; this tradition is still followed by the *hard science* school of survey research. Section 4 considers the elements of the *task*. Section 5 takes as its focus first the *interviewer*, then the *respondent* and the interrelationship between them; in Sections 4 and 5 most of the contributions to progress have been made by either the psychologists involved in market and opinion research, or by the sociologists and social psychologists involved in social and policy research. In Section 6 some recent developments are used to illustrate how measurement error in surveys is being reconsidered. Section 7 presents some tentative conclusions.

3 THE EFFECT OF THE SAMPLING PERSPECTIVE— STATISTICAL ANALYSIS OF THE RESPONSES: OFFICIAL STATISTICS AND SURVEY STATISTICS

The sample survey was seen by Kiaer (1897) and its other originators in government work as an alternative to complete enumeration necessitated by the demand for more detail and more careful measurement. In 1897 Kiaer wrote "In order to arrive at a deeper understanding of the social phenomena ... it is necessary to ... formulate a whole series of special questions ... prohibitive to conduct a complete survey of the population of a country, indeed even one for all the inhabitants of a large town" (p. 38). It was the necessity to *sample* that brought about the difference between the survey and the usual government enquiry, and it was the errors that might contaminate the results because of this that became the focus of concern for the first generation of statisticians and others involved with government surveys. Kiaer suggested *replication*— simply constructing a set of comparable subsamples (in essence repeating the sampling operation)—as the means of evaluating the survey results (p. 51); this was, as far as I know, the first *total variance model*.

This approach was taken on board by Bowley and other statisticians and culminated in the classic 1934 paper by Neyman to the Royal Statistical Society "On the Two Different Aspects of the Representative Method" which crystallized the ideas in his concept of the *sampling distribution*—the set of all possible outcomes of the sample design and sampling operation. The quality of a sample design, and thus a sample survey, could be encapsulated in the *sampling error* of the estimates; it is worth noting that though the general term "error" was used, this was a measure purely of variance or variability, and not necessarily a measure of error in any general sense. In particular, *bias* (or systematic error) was not a primary consideration, except as a technical issue in the choice among statistical estimators.

The Kiaer–Bowley–Neyman approach produced the sequence of texts on sampling which have defined the social survey field for statisticians ever since. The sequence of classic sampling texts began with Yates (1949, prepared at the request of the United Nations Sub-Commission on Statistical Sampling for a manual to assist in the execution of the projected 1950 World Census of Population), and Deming (1950), followed by Hansen *et al.*, (1953), Cochran (1953), and Sukhatme (1953), and concluded with Kish (1965). With these texts survey statisticians defined their field as that of measuring the effect of sample design on the imprecision of survey estimates. Where other considerations were included they tended to be relegated to a subsidiary role, or confined to chapters towards the end of the book. The texts vary a good deal in terms of the relative weight given to mathematical statistics; the most successful as a textbook, however, Cochran (2nd edition 1963, 3rd edition 1977) was the most mathematical and the least influenced by nonsampling and nonstatistical concerns.

A second strand was present in the work of Mahalanobis in India. He, like Kiaer, advocated the use of replication, using what he called *interpenetrating samples*, to estimate the precision of estimates derived from a survey. He defined these as "independent replicated networks of sampling units." He was, moreover, the first statistician to emphasize the *human agency* in surveys (1946, p. 329); he classified errors as those of sampling, recording, and physical fluctuations (instability). To estimate variance, he advocated that different replicates should be dealt with by "different parties of field investigators" so that human error as well as sampling errors would be included in the estimates of precision; he also carried out tests of significance among replicates. Mahalanobis may also be credited with perceiving that an additional advantage of partial investigations (using his interpenetrating samples) was that they facilitated the estimation of error, something previously not a part of the reports of government agencies. Indeed one of his early evaluations, based on sampling by the Indian Civil Service between 1923 and 1925, showed a bias in the estimation of crop yields (1946, p. 337).

Replication remains the primary instrument of statisticians when dealing with error. The traditional division between *variance*—the variability across replications, however defined—and *bias*—systematic deviation from some correct or true value—still informs the statistician's approach to error. Replication (or Mahalanobis's interpenetration) is the method of producing *measurability* in the sense of being able to estimate the variability of an estimate from within the process itself. In sampling, this was brought about by selecting a number of sampling units independently and using the differences among them as a guide to the inherent stability or instability of the overall estimates. For simple response variance, the statisticians simply repeated the observations on a subset of respondents and compared the outcomes; this is usually called a reinterview program; this gives replication in the sense of repetition. In the context of interviewer effect, the replication is within the survey operation and is brought about by constructing comparable subsets of cases and comparing them. To measure interviewer effect, respondents are allocated at random to different interviewers and the responses obtained are compared. Statistical theory tells us how much variability we could expect among these interviewer workloads if there is no systematic interviewer effect on the responses. To the extent that the variation is larger than that predicted by the null model, we attribute the effect to the interviewers.

The early 1960s saw the next step forward in statisticians' consideration of survey error. Hansen *et al.* (1961) in a seminal paper presented what became known as the "U.S. Census Bureau" model of survey error. They defined the *essential survey conditions* as the stable characteristics of the survey process and the survey organization carrying it out; variance was defined relative to those essential survey conditions. The observation is seen as being composed of two parts, its *true value*, and a deviation from that value—the *response deviation*. Though Hansen and his colleagues were well aware that the notion of a "true value" is problematic, they proposed it as a useful basis for the definition and then estimation of error. Their model is essentially a variance-covariance model and permits considerable generalization (see, for example, Fellegi, 1964, 1974) and has been extremely influential among survey statisticians. In particular it allows the incorporation of the effects of interviewers and supervisors, and the possibility of correlated errors within households.

About the same time, Kish (1962) presented findings using an alternative technical approach using analysis of variance (ANOVA) models in dealing with interviewer error; this was the approach favored by Yates, among others, and derived from the experimental design perspective of agricultural statisticians. Again the statistician simplifies reality so that it can be accommodated within the structure of his/her models; the effect of the interviewer is seen as an additive effect to the response of each respondent interviewed by that interviewer. The approach is easily generalizable to the effects of other agents in the survey (coders, supervisors, editors, etc.) (see, for instance, Hartley and Rao, 1978); one drawback is that the ANOVA models do not easily lend themselves to the analysis of categorical data.

These two approaches have in common the objective of estimating the variance of the mean of a single variable (or proportion). The focus of a survey is seen as the estimation of some important descriptive feature of the population. Thus the total variance is seen as the sum of the various sources of error affecting the estimate. Starting with the variance of the sample mean of a variable measured without error and based on a simple random sample (SRS), a set of additional components may easily be added to the variance, each representing a separate source. Thus processing, nonresponse, noncoverage, and measurement errors can all be incorporated in the approach. For generality any biases—whatever their sources—may also be added in, giving the mean squared error as the total error of the estimate. This concentration on estimates of a single descriptive parameter arose partly from the government statistics orientation (which was frequently concerned with estimating a single important characteristic of the population, such as the unemployment rate) and partly from the general statistics tradition of interval estimation of the parameters of a distribution.

The survey statistics approach remained for two decades directed at such descriptive parameters. In the 1980s, however, the use of statistical models in data analysis and the controversy in survey sampling between the proponents of design-based and model-based inference led to the incorporation of response errors directly into statistical models in surveys. O'Muircheartaigh and Wiggins (1981) modelled interviewer effects explicitly in a log–linear analysis of the effects of aircraft noise on annoyance; Aitkin, Anderson and Hinde (1981) and Anderson and Aitkin (1985) used variance component models (also known as multi-level models) to investigate interviewer variability; Hox, de Leeuw, and Kreft (1991), Pannekoek (1991), and Wiggins et al., (1992) provide applications and extensions of these methods.

These more recent developments mark a change in the statistical orientation from the original survey/statistical view of measurement error as a component of the "accuracy" of an estimate to a broader concern with the way in which measurement error can have an effect on substantive analysis of survey data. This latter view was of course always present in the case of the other survey traditions.

4 THE TASK

The task represents the core of the inquiry and the circumstances under which it is conducted. At its center is the *questionnaire* or *interview schedule*. The exact nature and function of the questionnaire was not by any means universally agreed upon in the early years of survey research (nor indeed is it now).

Among the early exponents, Galton—not best known for his views on social surveys—set out in *Inquiries into the Human Faculty* (1883) his four requirements for questionnaires (see Ruckmick, 1930); they should "... (*a*) *be quickly and easily understood*; (*b*) *admit of easy reply*; (*c*) *cover the ground of enquiry*; (*d*) *tempt the co-respondents to write freely in fuller explanation of their replies and cognate topics as well. ... These separate letters have proved more instructing and interesting by far than the replies to the set questions.*" Booth (1889), a social reformer, took an instrumental view of the problem of data collection. "*The root idea ... every fact I needed was known to someone, and that the information had simply to be collected and put together.*"

In due course, however, social researchers began to question the reliability and validity of their data, and in some cases began to carry out experiments or other investigations to test and evaluate their methods. Psychologists had of course been aware of the possible effects of changing question wording; Muscio (1917) who describes experiments on question form refers to research by Lipmann published in 1907. One of the earliest directly related to the social survey was reported by Hobson in the *Journal of the American Statistical Association* in 1916 and contrasted different questionnaire types and wording in mail surveys. The bulk of research on this area did not emerge until the 1930s, with Link and Gallup to the fore, and the 1940s, which saw a dramatic increase in published work on methodology.

In this area of endeavor also there was evidence of the different perspectives of practitioners. From the psychologists came scaling methods, in particular the development of formal attitude scales (Thurstone, Likert, Guttman) spanning the period from 1920 to 1950. They brought with them their own terminology, using the terms reliability and validity to describe the characteristics of their measures. These terms have reassuring connotations of science about them, and they also emphasize the positive rather than the negative aspects of the measures. They may be contrasted with the statistician's *variance* and *bias*— terms with broadly similar meanings but emphasizing the imperfections rather than the strengths of estimators. The psychological tradition also stressed the possibility of experimentation and experimental comparisons, often—as in the construction of attitude scales—using internal consistency, split-half experiments, and test-retest comparisons to examine their instruments. The Psychological Corporation was founded by Cattell to put applied psychology on a business footing; by 1923 nearly half the members of the American Psychological Association were stockholders (see Converse, 1986, p. 107). One of the foremost practitioners in methodological (and applied) research was Henry Link, who joined the Psychological Corporation in 1930, and produced a succession of insightful papers on survey and scaling methodology over the next twenty years.

On the academic front, the Office of Public Opinion Research at Princeton University was established in 1940 for the purpose of "(1) *studying techniques of public opinion research*; (2) *gaining some insight into the psychological problems of public opinion motivation; and (3) building up archives of public opinion data for the use of qualified students*" (Cantril, 1944 p. x). Among the task-related topics considered by Cantril and his colleagues were question meaning, question wording, the measurement of intensity (as distinct from direction) of opinion, and the use of batteries of questions.

On the whole, commercial market research practitioners were serious researchers into their own techniques. Cantril had forged a valuable alliance with Gallup—considered by many the father of the opinion poll—and his American and British Institutes of Public Opinion. During the period 1936 to 1949 Gallup conducted almost 400 split-ballot experiments. Many of these were designed by Cantril, to whom he gave access "*without restrictions or stipulations concerning the publication of results.*" The major fruit of this effort was *Gauging*

Public Opinion (1944) a compendium of studies of the methods of survey design, execution and analysis. It is a pity that there have not been more examples of such cross-sector collaboration in the fifty years since.

The other major work of that period arose from another exceptional (in this cased forced) cross-sector partnership. During World War II a Research Branch had been set up in the Division of Morale, U.S. Army, directed by Stouffer from 1941–45 (see Converse, p. 165 *passim* for an excellent description). The war brought about a sustained program of opinion research on matters of interest to the military establishment. The research was not conducted with any intention of furthering social science; the branch was set up to do a practical job by providing information to the authorities about the attitudes and views of military personnel. Staff and consultants of the Research Branch would produce a few years after the war the four volume work *Studies in Social Psychology in World War II* (1949–50) after a considerable amount of additional effort had been devoted to the material collected during the war. Volumes 1 and 2, together entitled *The American Soldier*, are a shining example of how findings of methodological and theoretical interest can be found through appropriate analysis of routine applied work.

Throughout the 1950s and 1960s studies were published which illustrated various strengths and weaknesses of survey instruments (see for instance Parry and Crossley, 1950; Cannell and Fowler, 1963 as examples of validation studies). Sudman and Bradburn (1974) provided an overview of research to date and formulated an explicit model of the survey response process. Schuman and Presser (1981) published a valuable book describing their own and others' research on question wording.

Studies bearing on the survey task can be classified according to whether they used *validation information* or *internal consistency analysis*. In the former case—much the less frequent—information was available external to the survey that permitted checking the validity of the individual (or group) responses to the survey questions. Such information is hardly ever available in a substantive survey, and is in any case restricted to behavioral information. Methodological validation studies may themselves be considered as falling in two categories— *identification* studies where the objective is to identify and quantify errors of particular kinds, and *diagnostic* studies where the objective is to discover the factors generating a particular error and to devise an improved procedure to eliminate or reduce it.

The principal characteristics of the task that have over the years been identified as having a potential effect on the magnitude of the measurement errors include: the location of the interview and method of administration; the designated respondent; the length of the questionnaire; the position and structure of the questions; the question wording; the length and difficulty of the questions; the saliency of the questions; and the level of threat and possibility of socially desirable response. Other factors examined or postulated to have an effect were respondent burden, memory effects, the classic open versus closed question debate, the mode of data collection, the explicit *don't know* category, the number of points on a scale, and many more.

The difficulty with the literature on task variables is that no clear pattern was found to explain the many and various effects that were demonstrated to exist. The lack of a clear theoretical framework made it difficult to classify effects in any parsimonious way and hindered attempts to formulate a general theory of survey methods. While statisticians were content simply to quantify errors (and add their effects to the total variance or mean squared error), no satisfactory psychological, sociological, or practical principles were found that underpinned the miscellany of effects observed. Recognizing the extent to which responses, and hence response effects, could be context dependent, it was suggested (Cantril, 1944, p. 49) that any result should be replicated across a variety of contexts: "*Since any single opinion datum is meaningful only in so far as it is related to a larger personal and social context, it is essential that responses to many single questions asked by the polls be compared with responses to other questions which place the same issue in different contingencies*". This recognition of the context-dependent nature of survey responses may be seen as a precursor of later attempts to systematize our understanding of question-related effects.

5 THE INTERVIEWER AND THE RESPONDENT

It is impossible to separate entirely the function and behavior of the interviewer from the function and behavior of the respondent in social surveys. This section describes how their roles developed and changed as the social survey and, more particularly, the way we think about the social survey changed and changed again.

The nature of data collection in surveys was by no means standardized in the early days of social investigations. In the case of many of the classical poverty studies there was no direct communication with the individuals about whom data were being collected. For Booth, for instance, "*the facts existed*" and all that remained was finding an efficient method of collecting them. He consequently used "expert" informants, such as School Attendance Officers, to provide detailed knowledge about children, and their parents, and their living conditions; Beatrice Webb later termed this procedure "*the method of wholesale interviewing*"; in general there was no concern about the performance of the interviewers. The practice of having the interviewer record information without reference to the respondent continued, and in a limited context still continues for some kinds of information; Yates (1949 and subsequent editions) states: "*Thus it is better to inspect a house to see if it shows signs of damp than to ask the occupant if it is damp.*" DuBois, in Philadelphia in 1896, used a variety of methods, including a house to house canvass, to collect data on the black population of the city (DuBois, 1899). Rowntree (1902), a decade later than Booth, obtained his information directly from families by using interviewers.

There was also considerable variation in *how* interviewers were used. There were two poles to the interviewing role. One pole was represented by the expert interviewer who obtained information by having a "conversation" with the respondent, usually without taking notes at the time. The other was the "questionnaire" interviewer, who had a blank form and a prepared set of questions. Among the social policy reformers there tended to be some formal instruments in all cases, either a schedule of questions (leading to the term *interview schedule* for the questionnaire and accompanying instructions used by an interviewer) or a scorecard or tabulation card (subsequently becoming the *questionnaire*). Among market and opinion researchers there was a similar dichotomy: expert interviewers tended to be used when dealing with business executives or merchants, whereas "questionnaire" interviewers dealt with the general public. The questionnaire interviewers were given a short set of preprinted questions and wrote down the responses—in the presence of the respondent—on blanks provided for this purpose.

Partly because of the vagueness of the definition of a social survey and the absence of a generally recognized set of standards, the social survey was by no means held in universally high regard in its early days. Thomas (1912) in the American Journal of Sociology, opined that "... *interviews in the main may be treated as a body of error to be used for purposes of comparison in future observations.*" Gillin (1915) in the *Journal of the American Statistical Association*, decried the tendency towards lack of quality control: "... *the survey is in danger of becoming a by-word and degenerating into a pleasant pastime for otherwise unoccupied people.*" The interviewer and his/her behavior was always seen as central to the quality of the survey. As early as the 1920s investigations were carried out into possible contaminating influences of the interviewers (see, for example, Rice, 1929). By the 1930s momentum was gathering behind attempts to standardize the interviewer's behavior, drawing in many cases on criticisms of interviewing in contexts such as job interviews (see Hovland and Wonderlic, 1939; Bingham and Moore, 1934). In the 1940s a more systematic examination of the effect of interviewers was undertaken. Katz (1942) compared white-collar and working class interviewers and found a conservative tendency in results from white-collar interviewers; Cantril and his colleagues considered issues of interviewer bias, reliability of interviewers' ratings, interviewers' training, and rapport (Cantril, 1944); Wechsler (1940) drew attention to the manner of delivery of interviewers: "*An interviewer may ask a question in belligerent, positive tones with the obvious inference that he won't take no for an answer. He (or she) may use cadences so gentle that the person being interviewed will hesitate to voice what may appear to be a dissenting voice. ... Individually such cases may seem trifling. Add them up.*"

Gradually considerable pressure grew to standardize interviewers' behavior and the leading survey organizations responded. In *Public Opinion Quarterly* (POQ) in 1942, Williams described the *Basic Instructions for Interviewers* for the National Opinion Research Center (NORC). These instructions are a model of clarity in their intent and describe a very carefully standardized practice. The Central City Conference in 1946 included a discussion of interviewing in which

Moloney described the ideal interviewer ("*a married woman, 37 years old, neither adverse to nor steamed up about politics, and able to understand and follow instructions*"); this stereotype has persisted a long time. Tamulonis in the same session described experiments she had conducted that showed that middle of the road interviewers were less likely to bias their respondents than interviewers holding an extreme view on a question. Sheatsley (1947–48) adds advice on the use of interviewer report forms which presages the method of interaction coding proposed by Cannell in the 1980s.

The 1950s saw the publication of two major books on interviewing, by Hyman (1954) and Kahn and Cannell (1957). By this time the fundamentals of survey interviewing had been consolidated; interviewers were expected to behave as neutral instruments—professionals who read their questions as written without implying any favoritism towards any particular answer or expressing surprise or pleasure at a particular response. The ideal interviewer would have a highly structured role, reinforced by rigorous training, and with no extra-role characteristics that might be salient for the questions being asked.

The role of the research subject (later the respondent) during this time had ranged from unimportant to insignificant. There was always concern about how information could be obtained. Booth, with his method of "wholesale interviewing" did not consult the individuals about whom he desired information. Though Rowntree and DuBois did ask the questions directly, there remained a general feeling that the respondent was something of an obstacle to effective research. There were seen to be problems with collecting data, but these problems were not *centered* on the respondent. Occasionally there would be concern that a task was simply too long and tedious. White, in 1936, wrote of the Federal Survey of Consumer Purchases that "*the respondent is supposed to know the number of quarts of oil which he has bought during the year and have an account of the money he has spent for bridge and ferry tolls*"; White felt that anyone who actually kept such records "*would be so abnormal that their reports might not be representative of the public in general.*" On the whole, however, neither the respondent nor respondent fatigue was seen as a problem; the proceedings of the Central City Conference in 1946 sums up the lack of concern about respondents (and possibly sexism) in the observation "*They'd rather give their opinions than do their washing*"!

Even among thoughtful methodologists, the general view of the respondent was of a relatively passive actor in the research process; see, for instance, Sudman and Bradburn (1974): "*The primary demand of the respondent's role is that he answer the interviewer's questions.*" The main concern was with motivating the respondent to do so. Gates and Rissland (1923) were among the first to formalize this issue. Cannell *et al.* (1975) reinforces the importance of motivation.

The influences that dictated the view of the interviewer and respondent roles described above were primarily the production requirements of large scale surveys. Survey methodologists had, however, been concerned about the possible shortcomings of the method. Cannell and Kahn (1968), in a review article, point out that the *interaction* between the respondent and interviewer is central to the interview. Since then a series of studies has revealed an increasingly complex view of the interaction that takes place in an interview.

The studies by Cannell and his colleagues (summarized in Cannell *et al.*, 1981) led to an explicit involvement of the respondent in the interview (through, for example, commitment procedures) and the devising of feedback from interviewers that reinforced appropriate role behavior by respondents. The group also developed behavior coding schemes for interactions in the interview, leading eventually to a sophisticated instrument for pretesting survey questions and questionnaires (Cannell *et al.*, 1989).

Rapport had always been seen as a key feature of a successful interview; the term encompassed a variety of qualities that implied success on the part of the interviewer in generating satisfactory motivation for the respondent. In the recent literature the concept of rapport has been replaced with "interviewing style," with the distinction made between formal or professional style on the one hand, and informal, interpersonal, socio-emotional, or personal styles on the other (see Schaeffer, 1991). Dijkstra and van der Zouwen (1982) argue that *"researchers need a theory of the processes which affect the response behaviour"*; the detailed study of the interactions that take place during the interview is one way of developing such a theory.

There had, of course, always been critics of the standardized interview; the criticisms were similar to those initially directed at the self-completion questionnaire. Even as, or perhaps because, surveys became increasingly used in social research, these criticisms surfaced again. Cicourel (1964) argued that there was no necessary relation between the researcher's agenda and the lives of the respondents. This point of view emerges again in the literature that considers the survey interview from the respondent's perspective.

The more qualitative approaches to the analysis of the survey interview tend to concentrate on the role of the respondent, or at least give equal status to the respondent in considering the issue. Suchman and Jordan (1990), in an analysis of a small number of interviews, describe the various and very different images that respondents may have of the survey interview, ranging from an interrogation or test through to conversation and even therapy. The contribution of cognitive psychology can be seen in Tourangeau and Rasinski (1988) who provide a model of the response process which develops earlier descriptions in the survey literature, particularly Cannell's work. This model of the process makes explicit some of the ways in which the active participation of the respondent is necessary if there is to be any chance of obtaining reliable and valid information. Even the rather esoteric field of "conversational analysis" has shown potential to contribute to our understanding of the survey response process (see Schaeffer, 1991).

There is a common thread that connects the contribution of all these sources. Overall the emphasis has shifted from the interviewer as the sole focus of attention, acting as a neutral agent, following instructions uniformly regardless of the particular respondent, to a joint consideration of the interviewer and the respondent taking part in a "speech event," a communication, where each has expectations as well as responsibilities. One pole of current thought is that of empowering the respondent (at its most extreme allowing the respondent to set the research agenda); the other pole is the ever more detailed specification of permissible interactions, so that while acknowledging the importance of respondent behavior, the intention is to control it rather than to liberate it. These two poles are strangely similar to the original situation one hundred years ago, when the expert interviewer and the questionnaire interviewer represented the two extremes of data collection.

6 RECENT DEVELOPMENTS

In this section I have chosen four examples of ways in which modern research in the methodology of surveys has begun to change our view of error and our treatment of it. They are chosen not to be exhaustive but to illustrate the breadth of the issues involved.

6.1 Statistical Modelling

A relatively theoretical paper in survey sampling in 1955 was instrumental (at least in part) in setting in train a profound change in the survey statistician's view of his/her role and the most appropriate ways to accommodate error in the statistical analysis of survey data. Godambe (1955) demonstrated the fact (neither surprising nor remarkable in itself) that there is no uniformly best estimator (even for the population mean) for *all* finite populations. This led, starting in the mid-1960s and lasting some twenty years, to a burst of interest in what became known as model-based estimation, i.e., analysis based not on the sample design (the joint probabilities of selection of the elements) but on some formal model that related the different variables in the analysis to each other, regardless of the configuration of the sample.

This tendency was strengthened by the substantive demands of economics, psychology, and sociology, which lead to relatively complex analyses. The statistical methods used for these analyses are usually (or at least frequently) based on a *generalized linear model*; such models almost always assume *i.i.d.* (independent identically distributed) observations and often assume normality for the distribution of errors. The emphasis is on model specification and testing.

In contrast, the survey statistician's approach has traditionally been design-based. As Section 3 illustrated, the primary concern is with inference to the particular finite population from which the sample was selected; response errors are conceptualized as another distortion of complete enumeration. In consequence weights are used to adjust the estimates in the light of the sample design and nonresponse, and standard errors are computed taking into account the effect of clustering of the sample and the correlated effect of interviewers and other agents. These adjusted standard errors are then used mainly to present confidence intervals for descriptive statistics or to test hypotheses about such statistics.

This difference of approaches illustrates the different emphasis placed on randomization and control by survey statisticians. For the survey statistician the overriding criterion has been randomization for *representation* through the sample design; for experimental statisticians and substantive analysts the emphasis has been on randomization and modelling for *control* of extraneous variables.

Recent developments in statistical methodology, allied to the dramatic increase in the power and accessibility of computers, have provided hope that these conflicting approaches may be reconciled. *Multilevel modelling* (a development of analysis of variance models) (Bryk and Raudenbush, 1992; Goldstein, 1995; Longford, 1993), were originally designed to analyze hierarchical data structures such as education systems—schools, teachers, classes, pupils. This framework has been adapted to deal with the clustering imposed on survey observations by, *inter alia*, the sample design, response error, and repeated observations on the same individuals. The analysis incorporates these elements explicitly in the analysis rather than ignoring them—as would be the case for standard statistical analysis—or adding them on as adjustments—which would be the effect of calculating design effects or interviewer effects for the standard errors. *Latent variable modelling* (Joreskog, 1973; Joreskog and Sorbom, 1989; Bartholomew, 1987) has its origins in the scaling literature in psychometrics. In such analyses the variance–covariance structure of the data and the errors is modelled explicitly and simultaneously. There are nontrivial challenges in formulating appropriate assumptions for each context, but there is no doubt that the approach has considerable potential.

6.2 Cognitive Approaches

The context-dependent nature of survey responses and the errors associated with them constituted a significant obstacle to any coherent approach to task-related, interviewer-related, and respondent-related effects. Though much of the early work in surveys was carried out by psychologists, there was very little formal theoretical consideration of survey data collection. One of the first discussions by a psychologist of memory issues in surveys was by Baddeley at a conference organized in Britain by the Royal Statistical Society and the Statistics Committee of the (British) Social Science Research Council in 1978 (Moss and Goldstein, 1979).

Subsequently, a conference organized by the U.S. Social Science Research Council brought together a selection of survey researchers, statisticians, and psychologists to explore the possible advances that might be made by combining the techniques of the different disciplines, in particular by looking at how ideas in cognitive and social psychology might contribute to an improvement in survey methods (see Jabine *et al.*, 1984). This led, directly or indirectly, to considerable advances in the methodology of question design, in particular in the development of a broader framework within which instrument design could be considered. A valuable set of papers is to be found in Tanur (1992).

The aim of the *cognitive approaches to survey questions* is to understand the process of response generation and to produce "stable" predictors of effects, solutions to practical questions in questionnaire design, and "rules" of question formulation. An example of progress in this direction is the work by Schwarz and his colleagues on assimilation and contrast effects (Schwarz *et al.*, 1991).

6.3 Computer Assisted Methods

The development of computer assisted methods in survey data collection arose from two different stimuli. On the one hand the movement towards telephone interviewing, especially in North America, led to the setting up of centralized telephone facilities where many interviewers worked simultaneously on a particular survey. This provided an opportunity for utilizing the power of a mainframe computer at the same location for data entry and for data analysis, and facilitated the development of computer assisted interviewing in which the questions themselves, the routing through the questionnaire, and a number of checks on the data were incorporated in the system. On the other hand, difficulties and bottlenecks in survey processing led to the development of computer assisted data entry systems, especially in Europe, in which the editing of the data was carried out at the time of the data entry. Together these led to the development of ever more powerful and flexible computer assisted systems for face to face, telephone, and self-completion surveys.

The great potential of computer assisted methods arises from three features. First, the systems are already *flexible*, and are becoming consistently more so. The survey instrument can be hierarchically structured, with lengthy material being called upon only when needed. Second, the systems are also inherently *portable*; there are in practice very few constraints on the size of the instrument and the size of the instrument is not related to the physical interviewer burden of transporting it. Third, the systems are *interactive*; the specific interview schedule can be tailored to the specific needs of the respondent, and can be modified depending on the preferences stated by the respondent (or modified by the respondent during the course of the interview).

Two examples may help to illustrate these strengths. In consumer expenditure surveys, not only can long coding frames be incorporated into the instrument, but the order of completion of the questionnaire (by person or by activity, for instance) can be determined by the respondent during the interview, and may vary for different topics. This removes the necessity to have a single structure for the questionnaire, which would undoubtedly lead to a "respondent-unfriendly" structure for at least some respondents. In panel surveys, there has always been a problem in deciding whether independent or dependent reinterviewing should be used. With an appropriate computer assisted system the prior information can be stored in the system, but released to the interviewer only when the new response has been entered, thereby combining the advantages of both approaches.

Computer assisted methods have one other great advantage in the context

of measurement error—the capacity to introduce randomization in the allocation of treatments to respondents. We return to this advantage in the next
section.

6.4 Error Profiles

Traditionally error indicators, when presented at all, tend to be restricted to a
particular perspective. Thus there might be an appendix presenting some
computation of sampling error or nonresponse, or there might be in the text a
discussion of difficulties in interpretation or administration for a set of
questions. There has been some movement in recent years towards the idea of
error profiles for surveys, where this would be a multi-dimensional assessment
of the whole survey and incorporate both quantitative and qualitative measures.
The Italian Labour Force Survey and the Survey of Income and Program
Participation in the U.S.A. are cases where this has been attempted.

7 CONCLUSION

The broad tenor of this chapter has been to emphasize that the social survey
should not be considered without its broader context. In the framework
presented in Section 2, this means that any assessment of social data should
take into account the positioning of the inquiry in relation to the three
dimensions of *representation, randomization,* and *realism,* as well as considering
the more traditional elements of interviewer, respondent, and task. If we see
the origin of the three-dimensional space as the point where we have optimized
on all three dimensions, then the most important lesson to learn is that in
general a shift towards the origin on one dimension is likely to be associated
with a movement away from the origin on another. Three examples may
illuminate this.

1. In consumer panels in market research respondents are asked to complete
   in exhaustive detail all the transactions of a particular kind they carry
   out over a period of time, sometimes extending over many weeks. In order
   to obtain a sufficient number of respondents, it is necessary to approach
   a very much larger number of individuals, perhaps as many as ten for
   every successful recruit. The characteristics of an individual that are
   associated with a willingness to make the commitment to undertake the
   task probably mean that the eventual recruits are very different from the
   rest of the population in terms of characteristics that are relevant to
   the variables being studied. Remember that White, as early as 1936, had
   voiced this concern (see Section 5). This is an example of a situation in
   which the *representation* dimension is sacrificed in order to make the *task*
   feasible.

2. The redesign of the U.S. Current Population Survey (CPS) was a major methodological exercise undertaken jointly by the U.S. Bureau of Labor Statistics and the U.S. Bureau of the Census. During the redesign process a number of field and laboratory experiments were carried out to test various components of the new procedures. At the penultimate stage the new CPS was run in parallel with the old CPS for a period to make it possible to splice the time series together when the new version was put officially in place. This parallel operation indicated that the new version would lead to a 0.5% increase in the reported level of unemployment. At the time of the official change-over, the old version was continued in parallel as a further check on the splicing of the two series. To everyone's surprise, after the change-over it was the old CPS that showed the higher estimates of unemployment.

   The implementation of the new version and the decision to run the two versions in parallel is an example of good methodological practice. The probable explanation of the surprising outcome lies in the fact that the procedures concentrated on *representation* and *control* (*randomization*) in the final experiment but did not consider the issues of *realism*. It was known to all the participants which was the *actual* CPS and which was the parallel version; the procedures were necessarily different, and certainly the perception of the surveys was different. An alternative strategy that would have addressed this issue would have been to implement the new CPS in a number of states while continuing to run the old CPS in others; this might have brought to light the importance of the unmeasured factors implicit in the treatment of the real policy-relevant survey.

3. The third example is in a sense a counterexample. One of the great difficulties of methodological work in social surveys has been the impracticability of incorporating experiments into survey fieldwork. The bulkiness of interview materials if interviewers were to be able to apply a number of different options within an interview schedule is one problem. The difficulty of devising methods to randomize treatments (or versions of questionnaires) in the field is another. The problem of training interviewers to carry out their instructions is a third. Concealing from the respondent that an experiment was being carried out is yet another.

   Computer assisted methods of data collection have made it possible to incorporate experiments into survey fieldwork with a minimum of disruption of the survey operation. The system can carry out the randomization, access the appropriate part of the schedule, and record the results, without any direct intervention from either the interviewer or the respondent. This has made it possible to embed controlled (*randomized*) experiments in *real* and *representative* surveys.

Bowley expressed, in his 1915 book on measurement, a suitably broad perspective of the issues:

"The main task . . . is to discover exactly what is the critical thing to examine, and to devise the most perfect machinery for examining it with the minimum of effort. . . . In conclusion, we ought to realise that measurement is a means to an end; it is only the childish mind that delights in numbers for their own sake. On the one side, measurement should result in accurate and comprehensible description, that makes possible the visualization of complex phenomena; on the other, it is necessary to the practical reformer, that he may know the magnitude of the problem before him, and make his plans on an adequate scale."

This inclusion in the definition of measurement of both the substantive decisions about what to examine and the technical issues of how to examine it is a useful reminder that there is a danger that those directly involved in the measurement process can at times become too restricted in their vision by the technical fascination of their own discipline. We need to step outside those confines and acknowledge that others also have something to offer. Survey methodologists should become more *externalist* and less *internalist* (see Converse, 1986) and accept that they should respect others' criteria as well as their own. The rest of this book provides an ideal opportunity to do this. I would urge everyone to read not only the particular chapters close to their own concerns but a selection of those in which they feel they have no direct professional interest. The increasing specialization of survey research will otherwise mean that the raison d'être of surveys will disappear from the vision of those most intimately involved in their development.

REFERENCES

Aitkin, M., Anderson, D., and Hinde, J. (1981), "Statistical Modelling of Data on Teaching Styles," *Journal of the Royal Statistical Society, A*, Vol. 144, pp. 419–461.

Anderson, D., and Aitkin, M. (1985), "Variance Component Models with Binary Response," *Journal of the Royal Statistical Society, B*, Vol. 47, pp. 203–210.

Bartholomew, D.J. (1987), *Latent Variable Models and Factor Analysis*, London: Griffin.

Bingham, W.V., and Moore, B.V. (1934), *How to Interview*, revised edition, New York: Harper.

Bogardus, E.S. (1925), "Measuring Social Distances," *Journal of Applied Sociology*, Vol. 9, pp. 299–308.

Booth, C., (ed.) (1889), *Labour and the Life of the People of London*, London: Macmillan.

Bowley, A.L. (1906), "Presidential Address to the Economic Section of the British Association for the Advancement of Science," *Journal of the Royal Statistical Society*, Vol. 69, pp. 540–558.

Bowley, A.L. (1915), *The Nature and Purpose of the Measurement of Social Phenomena*, London: P.S. King and Son, Ltd.

Bowley, A.L. (1926), "Measurement of the Precision Attained in Sampling," *Bulletin of* the *International Statistical Institute*, Vol. 22, No. 1, pp. 1–62 of special annex following p. 451.

Bryk, A.S., and Raudenbush, S.W. (1992), *Hierarchical Linear Models*, Newbury Park: Sage.

Campbell, D.T., and Stanley, J.C. (1963), *Experimental and Quasi-experimental Designs for Research*, Chicago: Rand McNally.

Cannell, C.F., and Fowler, F.J. (1963), "Comparison of a Self-enumerative Procedure and a Personal Interview: A Validity Study," *Public Opinion Quarterly*, Vol. 27, pp. 250–264.

Cannell, C.F., and Kahn, R. (1968), "Interviewing," in G. Lindzey, and E. Aronson (eds.), *The Handbook of Social Psychology*, Vol. 2, pp. 526–595, Reading, Mass.: Addison-Wesley.

Cannell, C.F., Kalton, G., Oksenberg, L., Bischoping, K., and Fowler, F. (1989), "New Techniques for Pretesting Survey Questions," Final Report for grant HS05616, National Center for Health Services Research, Ann Arbor: University of Michigan.

Cannell, C.F., Lawson, S.A., and Hausser, D.L. (1975), *A Technique for Evaluating Interviewer Performance*, Ann Arbor: Institute for Social Research, University of Michigan.

Cannell, C.F., Miller, P.V., and Oksenberg, L.F. (1981), "Research on Interviewing Techniques," in S. Leinhardt (ed.), *Sociological Methodology*, pp. 389–437, San Francisco: Jossey-Bass.

Cantril, H., and Associates (1944), *Gauging Public Opinion*, Princeton: Princeton University Press.

Cicourel, A. (1964), *Method and Measurement in Sociology*, New York: Free Press.

Cochran, W.G. (1953), *Sampling Techniques*, New York: John Wiley and Sons; 2nd edition 1963; 3rd edition 1977.

Converse, J.M. (1986), *Survey Research in the United States: Roots and Emergence 1890–1960*, Berkeley: University of California Press.

Deming, W.E. (1950), *Some Theory of Sampling*, New York: Wiley.

Dijkstra, W., and van der Zouwen, J., (eds.) (1982), *Response Behaviour in the Survey-Interview*, London: Academic Press.

DuBois, W.E.B. (1899), *The Philadelphia Negro: A Social Study*, Philadelphia: Ginn.

Fellegi, I.P. (1964), "Response Variance and Its Estimation," *Journal of the American Statistical Association*, Vol. 59, pp. 1016–1041.

Fellegi, I.P. (1974), "An Improved Method of Estimating the Correlated Response Variance," *Journal of the American Statistical Association*, Vol. 69, pp. 496–501.

Galton, F. (1883), *Inquiries into the Human Faculty* quoted in Ruckmick (1930) *op. cit.*

Gates, G.S., and Rissland, L.Q. (1923), "The Effect of Encouragement and of Discouragement Upon Performance," *Journal of Educational Psychology*, Vol. 14, pp. 21–26.

Gillin, J.L. (1915), "The Social Survey and Its Further Development," *Journal of the American Statistical Association*, Vol. 14, pp. 603–610.

Godambe, V.P. (1955), "A Unified Theory of Sampling from Finite Populations," *Journal of the Royal Statistical Society*, Series B, Vol. 17, pp. 269–278.

Goldstein, H. (1995), *Multilevel Statistical Models*, 2nd edition, London: Edward Arnold.

Groves, R.M. (1989), *Survey Errors and Survey Costs*, New York: John Wiley.

Hansen, M.H., Hurwitz, W.N., and Bershad, M.A. (1961), "Measurement Errors in Censuses and Surveys," *Bulletin of the International Statistical Institute*, Vol. 38(2), pp. 359–374.

Hansen, M.H., Hurwitz, W.N., and Madow, W.G. (1953), *Sample Survey Methods and Theory. Volume I: Methods and Applications. Volume II: Theory*, New York: Wiley.

Hartley, H.O., and Rao, J.N.K. (1978), "Estimation of Nonsampling Variance Components in Sample Surveys," in N.K. Namboodiri (ed.), *Survey Sampling and Measurement*, pp. 35–43, New York: Academic Press.

Hovland, C.I., and Wonderlic, E.F. (1939), "Prediction of Industrial Success from a Standardized Interview," *Journal of Applied Psychology*, Vol. 23, pp. 537–546.

Hox, J.J., de Leeuw, E.D., and Kreft, I.G.G. (1991), "The Effect of Interviewer and Respondent Characteristics on the Quality of Survey Data: A Multilevel Model," in P.P. Biemer, R.M. Groves, L.E. Lyberg, N.A. Mathiowetz, and S. Sudman (eds.), *Measurement Errors in Surveys*, pp. 439–461, New York: John Wiley & Sons, Inc.

Hyman, H.H. (1954), *Interviewing in Social Research*, Chicago: Chicago University Press.

Jabine, T.B., Loftus, E., Straf, M.L., Tanur, J.M., and Tourangeau, R., (eds.) (1984), *Cognitive Aspects of Survey Methodology: Building a Bridge Between Disciplines*, Washington, D.C.: National Academic Press.

Joreskog, K.G. (1973), "Analyzing Psychological Data by Structural Analysis of Covariance Matrices," in C. Atkinson, H. Krantz, R.D. Luce, and P. Suppes (eds.), *Contemporary Developments in Mathematical Psychology*, pp. 1–56, San Francisco: Freeman.

Joreskog, K.G., and Sorbom, D. (1989), *LISREL 7: A Guide to the Program and Applications*, The Netherlands: SPSS International B.V.

Kahn, R.L., and Cannell, C.F. (1957), *The Dynamics of Interviewing: Theory Techniques and Cases*, New York: John Wiley and Sons.

Katz, D. (1942), "Do Interviewers Bias Poll Results?" *Public Opinion Quarterly*, Vol. 6, pp. 248–268.

Kiaer, A.N. (1897), *The Representative Method of Statistical Surveys*, Oslo: Kristiania.

Kish, L. (1962), "Studies of Interviewer Variance for Attitudinal Variables," *Journal of the American Statistical Association*, Vol. 57, pp. 92–115.

Kish, L. (1965), *Survey Sampling*, New York: John Wiley and Sons.

Kish, L. (1987), *Statistical Design for Research*, New York: John Wiley and Sons.

Kruskal, W., and Mosteller, F. (1980), "Representative Sampling IV: The History of the Concept in Statistics," *International Statistical Review*, Vol. 48, pp. 169–195.

Likert, R. (1932), *A Technique for the Measurement of Attitudes* (Archives of Psychology, no. 140), New York: Columbia University Press.

Longford, N.T. (1993), *Random Coefficient Models*, Oxford: Clarendon Press.

Mahalanobis, P.C. (1944), "On Large Scale Sample Surveys," *Roy. Soc. Phil. Trans.*, B, Vol. 231, pp. 329–451.

Mahalanobis, P.C. (1946), "Recent Experiments in Statistical Sampling in the Indian Statistical Institute," *Journal of the Royal Statistical Society*, Vol. 109, pp. 326–378.

Moss, L., and Goldstein, H., (eds.) (1979), *The Recall Method in Social Surveys*, London: Institute of Education, University of London.

Muscio, B. (1917), "The Influence of the Form of a Question," *The British Journal of Psychology*, Vol. 8, pp. 351–389.

Neyman, J. (1934), "On the Two Different Aspects of the Representative Method: The Method of Stratified Sampling and the Method of Purposive Selection," *Journal of the Royal Statistical Society*, Vol. 97, pp. 558–606.

O'Muircheartaigh, C.A., and Soon, T.W. (1981), "The Impact of Sampling Theory on Survey Sampling Practice: A Review," *Bulletin of the International Statistical Institute*, Vol. 49, Book 1, pp. 465–493.

O'Muircheartaigh, C.A., and Wiggins, R.D. (1981), "The Impact of Interviewer Variability in an Epidemiological Survey," *Psychological Medicine*, Vol. 11, pp. 817–824

Parlin, C.C. (1915), *The Merchandising of Automobiles, An Address to Retailers*, Philadelphia: Curtis Publishing Co.

Parry, H.J., and Crossley, H.M. (1950), "Validity of Responses to Survey Questions," *Public Opinion Quarterly*, Vol. 14, pp. 61–80.

Pannekoek, J. (1991), "A Mixed Model for Analyzing Measurement Errors for Dichotomous Variables," in P.P. Biemer, R.M. Groves, L.E. Lyberg, N.A. Mathiowetz, and S. Sudman (eds.), *Measurement Errors in Surveys*, pp. 517–530, New York: John Wiley & Sons, Inc.

Rice, S.A. (1929), "Contagious Bias in the Interview," *American Journal of Sociology*, Vol. 35, pp. 420–423.

Rowntree, B.S. (1902), *Poverty: A Study of Town Life*, London: Longmans.

Ruckmick, C.A. (1930), "The Uses and Abuses of the Questionnaire Procedure," *Journal of Applied Psychology*, Vol. 14, pp. 32–41.

Schaeffer, N.C. (1991), "Conversation with a Purpose—or Conversation? Interaction in the Standardized Interview," in P.P. Biemer, R.M. Groves, L.E. Lyberg, N.A. Mathiowetz, and S. Sudman (eds.), *Measurement Errors in Surveys*, pp. 367–391, New York: John Wiley & Sons, Inc.

Scheuch, E.K. (1967), "Das Interview in der Sozialforschung," in R. Konig (ed.), *Handbuch der Empirischen Sozialforschung*, Vol. 1, 1360196, Stuttgart: F. Enke.

Schuman, H., and Presser, S. (1981), *Questions and Answers in Attitude Surveys*, New York: Academic Press.

Schwarz, N., Strack, F., and Mai, H. (1991), "Assimilation and Contrast Effects in Part-Whole Question Sequences: a Conversational Analysis," *Public Opinion Quarterly*, Vol. 55, pp. 3–23.

Sheatsley, P.B. (1947–48), "Some Uses of Interviewer-Report Forms," *Public Opinion Quarterly*, Vol. 11, pp. 601–611.

Stouffer, S., and Associates (1949), *The American Soldier, Vol. 1, Adjustment During Army Life*, Princeton: Princeton University Press.

Stouffer, S., and Associates (1949), *The American Soldier, Vol. 2, Combat and Aftermath*, Princeton: Princeton University Press.

Suchman, L., and Jordan, B. (1990), "Interactional Troubles in Face-to-Face Survey Interviews," *Journal of the American Statistical Association*, Vol. 85, pp. 232–241.

Sudman, S., and Bradburn, N. (1974), *Response Effects in Surveys: a Review and Synthesis*, Chicago: Aldine Publishing Co.

Sukhatme, P.V. (1953), *Sampling Theory of Surveys with Applications*, New Delhi: The Indian Society of Agricultural Statistics.

Tanur, J. (ed.) (1992), *Questions About Questions: Inquiries into the Cognitive Bases for Surveys*, New York: Russell Sage Foundation.

Thomas, W.I. (1912), "Race Psychology: Standpoint and Questionnaire, with Particular Reference to the Immigrant and the Negro," *American Journal of Sociology*, Vol. 17, pp. 725–775.

Thurstone, L.L. (1928), "Attitudes Can Be Measured," *American Journal of Sociology*, Vol. 33, pp. 529–544.

Tourangeau, R., and Rasinski, K.A. (1988), "Cognitive Processes Underlying Context Effects in Attitude Measurement," *Psychological Bulletin*, Vol. 103, pp. 299–314.

Wechsler, J. (1940), "Interviews and Interviewers," *Public Opinion Quarterly*, Vol. 4, pp. 258–260.

White, P. (1936), "New Deal in Questionnaires," *Market Research*, June, 4, 23.

Wiggins, R.D., Longford, N., and O'Muircheartaigh, C.A. (1992), "A Variance Components Approach to Interviewer Effects" in A. Westlake, R. Banks, C. Payne, and T. Orchard (eds.), *Survey and Statistical Computing*, Amsterdam: Elsevier Science Publishers.

Williams, D. (1942), "Basic Instructions for Interviewers," *Public Opinion Quarterly*, Vol. 6, pp. 634–641.

Yates, F. (1949), *Sampling Techniques for Censuses and Surveys*, London: Griffin.

SECTION A

Questionnaire Design

CHAPTER 1

Questionnaire Design: The Rocky Road from Concepts to Answers

Norbert Schwarz
University of Michigan, Ann Arbor

1.1 INTRODUCTION

According to textbook knowledge, research is conducted to answer theoretical or applied questions that the investigators—or their clients or funding agencies—consider of interest. Moreover, the specifics of the research design are tailored to meet the theoretical or applied objectives. Depending on one's research environment, these assumptions reflect a "trivial truism" (as one colleague put it) or the "lofty illusions of academic theoreticians" (as another colleague put it). This chapter, and the contributions to the section on "Questionnaire Design" address different aspects of the long and rocky road from conceptual issues to the answers provided by survey respondents. The first part of this chapter reviews key elements of survey design, which are elaborated upon in more detail in subsequent sections of this volume.

1.2 ELEMENTS OF SURVEY DESIGN

Schuman and Kalton (1985) delineate and discuss the major components of a survey. Starting from a set of research objectives, researchers specify the population of interest and draw an appropriate sample. The research objectives further determine the concepts to be investigated, which need to be translated into appropriate questions. As Schuman and Kalton (1985, p. 640) observed, "Ideally, sampling design and question construction should proceed hand in

Survey Measurement and Process Quality, Edited by Lyberg, Biemer, Collins, de Leeuw, Dippo, Schwarz, Trewin.
ISBN 0-471-16559-X © 1997 John Wiley & Sons, Inc.

hand, both guided by the problem to be investigated. When these stages are not well integrated—a rather common failing—one ends up with questions that do not fit part of the sample or with a sample that provides too few cases for a key analysis." Note that neither the sample nor the specific question asked embodies the researcher's primary interest. Rather, "investigators use one (sample, question) to make inferences about the other (population, concept), with the latter being what one is primarily interested in. Sampling populations and operationalizing concepts are each intended to allow us to go from the observed to the unobserved" (Schuman and Kalton, 1985, pp. 640–641).

The nature of the sample and of the questions to be asked, as well as the budget available, further determine the choice of administration mode, which, in turn, may require adjustments in question wording. Following pretesting, the survey is administered to collect relevant data. At this stage, the questions written by the researcher need to be appropriately delivered by the interviewer, unless the survey is self-administered. To answer the question, respondents have to understand its meaning, which may or may not match the meaning that the researcher had in mind. Next, they have to retrieve relevant information from memory to form a judgment, which they need to format to fit the response alternatives provided. Moreover, they may want to edit their answer before they convey it to the interviewer, due to reasons of social desirability and self-presentation concerns. Finally, the interviewer needs to understand the respondent's answer and needs to record it for subsequent processing.

At the post-survey stage, the interviewer's protocol may need editing and coding prior to data processing. Finally, data analysis, interpretation and dissemination of the findings complete the research process.

1.3   RESEARCH OBJECTIVES, CONCEPTS, AND OPERATIONALIZATIONS

As Hox (Chapter 2) notes, survey methodologists have paid much attention to issues of sampling and questionnaire construction. In contrast, considerably less effort has been devoted to the steps that precede questionnaire construction, most notably the clarification of the research objectives and the elaboration of the theoretical concepts that are to be translated into specific questions. In fact, most textbooks on research methodology do not cover these issues, aside from offering the global advice that the methods used should be tailored to meet the research objectives.

As anyone involved in methodological consulting can testify, the research objectives of many studies are surprisingly ill-defined. Asking a researcher what exactly should be measured by a question for which purpose frequently elicits vague answers—if not different answers from different researchers involved in the same project. This problem is compounded by large-scale survey programs involving different groups of researchers, a heterogeneous board of directors, and often an even more heterogeneous external clientele of data users (cf. Davis *et al.*, 1994). In this case, the set of questions that is finally agreed upon reflects the outcome of complex negotiations that run the risk of favoring ill-defined concepts. Apparently, vaguely defined concepts allow different groups of researchers to relate them to different research objectives, which makes these concepts likely candidates for adoption by compromise—at the expense of being poorly targeted towards any one of the specific objectives addressed. Moreover, researchers frequently hesitate to change previously asked questions in the interest of continuing an existing time series, even though the previously asked question may fall short of current research objectives. Systematic analyses of the negotiations involved in determining a question program would provide an exciting topic for the sociology of science, with potentially important implications for the organization of large-scale social science research.

Having agreed upon the research objectives, researchers need to develop an appropriate set of concepts to be translated into survey questions. That issues of conceptual development are rarely covered in methodology textbooks reflects that the development of theoretical concepts is usually assigned to the context of discovery, rather than the context of verification (e.g., Popper, 1968). As Hox (Chapter 2) notes, "concepts are not verified or falsified, they are judged on the basis of their fruitfulness for the research process." It is useful, however, to distinguish between the creative act of theoretical discovery, for which specified rules are likely to be unduly restrictive (though advice is possible, see Root-Bernstein, 1989), and the logical development of theoretical concepts.

The latter involves the elaboration of the nomological network of the concept, the definition of subdomains of its meaning, and the identification of appropriate empirical indicators. In the case of surveys, these empirical indicators need to be translated into a specific question, or set of questions, to be asked. Hox (Chapter 2) reviews a range of different strategies of concept development and highlights their theoretical bases in the philosophy of science.

1.4 WRITING QUESTIONS AND DESIGNING QUESTIONNAIRES

Having agreed on the desired empirical indicators, researchers face the task of writing questions that elicit the intended information. In the history of survey methodology, question writing has typically been considered an "art" that is to be acquired by experience. This has changed only recently as researchers began to explore the cognitive and communicative processes underlying survey responses. Drawing on psychological theories of language comprehension, memory, and judgment, psychologists and survey methodologists have begun to formulate explicit models of the question answering process and have tested these models in tightly controlled laboratory experiments and split-ballot surveys. Several edited volumes (Jabine *et al.*, 1984; Jobe and Loftus, 1991; Hippler *et al.*, 1987; Schwarz and Sudman, 1992, 1994, 1996; Tanur, 1992) and a comprehensive monograph (Sudman *et al.*, 1996) summarize the rapid progress made since the early 1980s.

This development is also reflected in the rapid institutionalization of cognitive laboratories at major survey centers and government agencies, in the U.S. as well as in Europe (see Dippo and Norwood, 1992). These laboratories draw on a wide range of different methods (reviewed in Forsyth and Lessler, 1991, and the contributions to Schwarz and Sudman, 1996) to investigate the survey response process and to identify problems with individual questions. The application of these methods has led to major changes in the pretesting of survey questionnaires, involving procedures that are more focused and less expensive than global field pretests. The fruitfulness of these approaches is illustrated by the way Johnson *et al.* (Chapter 4) use think-aloud procedures to explore the question answering strategies used in a culturally diverse sample.

In addition to Johnson *et al.*, several other chapters in the present volume address issues of question writing and questionnaire design, explicitly drawing on cognitive theories. To set the stage for these chapters, it is useful to review key aspects of the question answering process.

1.4.1   Asking and Answering Questions: Cognitive and Communicative Processes

Answering a survey question requires that respondents perform several tasks (see Cannell *et al.*, 1981; Strack and Martin, 1987; Tourangeau, 1984, for closely related versions of the same basic assumptions). Not surprisingly, the respondents' first task consists in interpreting the question to understand what is meant. If the question is an opinion question, they may either retrieve a previously formed opinion from memory, or they may "compute" an opinion on the spot. To do so, they need to retrieve relevant information from memory to form a mental representation of the target that they are to evaluate. In most cases, they will also need to retrieve or construct some standard against which the target is evaluated.

If the question is a behavioral question, respondents need to recall or reconstruct relevant instances of this behavior from memory. If the question specifies a reference period (such as "last week" or "last month"), they must also determine whether these instances occurred during the reference period or not. Similarly, if the question refers to their "usual" behavior, respondents have to determine whether the recalled or reconstructed instances are reasonably representative or whether they reflect a deviation from their usual behavior. If they cannot recall or reconstruct specific instances of the behavior, or are not sufficiently motivated to engage in this effort, respondents may rely on their general knowledge or other salient information that may bear on their task in computing an estimate.

Once a "private" judgment is formed in respondents' minds, respondents have to communicate it to the researcher. To do so, they may need to format their judgment to fit the response alternatives provided as part of the question. Moreover, respondents may wish to edit their responses before they communicate them, due to influences of social desirability and situational adequacy.

Accordingly, interpreting the question, generating an opinion or a representation of the relevant behavior, formatting the response, and editing the answer are the main psychological components of a process that starts with respondents' exposure to a survey question and ends with their overt report (Strack and Martin, 1987; Tourangeau, 1984).

1.4.1.1 Understanding the Question

The key issue at the question comprehension stage is whether the respondent's understanding of the question does or does not match what the researcher had in mind. From a psychological point of view, question comprehension reflects the operation of two intertwined processes (see Clark and Schober, 1992; Strack and Schwarz, 1992; Sudman et al., 1996, chapter 3). The first refers to the semantic understanding of the utterance. Comprehending the *literal meaning* of a sentence involves the identification of words, the recall of lexical information from semantic memory, and the construction of a meaning of the utterance, which is constrained by its context. Not surprisingly, survey textbooks urge researchers to write simple questions and to avoid unfamiliar or ambiguous terms (cf. Sudman and Bradburn, 1983).

As Johnson et al. (Chapter 4) note, semantic comprehension problems are severely compounded when the sample is culturally diverse. Different respondents may associate different meanings with the same term, potentially requiring that one asks different questions to convey the same meaning. In this regard, it is important to keep in mind that respondents answer the question as they understand it. If so, any effort at standardization has to focus on the standardization of conveyed meaning, rather than the standardization of surface characteristics of the question asked. This is particularly true when a question is translated into different languages, where attempts to maintain surface characteristics of question wording may strongly interfere with conveyed meaning (see Alwin et al., 1994, for a discussion of measurement issues in multi-national/multi-lingual surveys). These issues provide a challenging agenda for future cognitive research and may require a rethinking of standardized question wording.

Note, however, that the issue of standardized question wording is distinct from the issue of standardized measures addressed by Heath and Martin (Chapter 3). Heath and Martin ask why social scientists rarely use standardized measures of key theoretical constructs and emphasize the benefits that such measures may have for comparisons across studies. A crucial requirement for the use of standardized measures, however, is that the questions that comprise a given scale convey the same meaning to all (or, at least, most) respondents. To the extent that we may need to ask (somewhat) different questions to convey the same meaning to different respondents, standardizing the meaning of a scale may, ironically, require variation in the specific wordings used, in particular when the scale is to be used in cross-national or cross-cultural research.

Complicating things further, understanding the words is not sufficient to answer a question, in contrast to what one may conclude from a perusal of books on questionnaire design. For example, if respondents are asked "What have you done today?" they are likely to understand the meaning of the words. Yet, they still need to determine what kind of activities the researcher is interested in. Should they report, for example, that they took a shower, or not? Hence, understanding a question in a way that allows an appropriate answer requires not only an understanding of the *literal meaning* of the question, but involves inferences about the questioner's intention to determine the *pragmatic meaning* of the question (see Clark and Schober, 1992; Strack and Schwarz, 1992; Sudman *et al.*, 1996, chapter 3, for more detailed discussions). To determine the questioner's intentions, respondents draw on the content of related questions and on the response alternatives provided to them. Wänke and Schwarz (Chapter 5) address the impact of pragmatic processes on question comprehension and the emergence of context effects in more detail, and Schwarz and Hippler (1991) review the role of response alternatives in question comprehension.

1.4.1.2   *Recalling or Computing a Judgment*

After respondents determine what the researcher is interested in, they need to recall relevant information from memory. In some cases, respondents may have direct access to a previously formed relevant judgment, pertaining to their attitude or behavior, that they can offer as an answer. In most cases, however, they will not find an appropriate answer readily stored in memory and will need to compute a judgment on the spot. This judgment will be based on the information that is most accessible at this point in time, which is often the information that has been used to answer related preceding questions (see Bodenhausen and Wyer, 1987; Higgins, 1989, for a discussion of information accessibility).

The impact of preceding questions on the accessibility of potentially relevant information accounts for many context effects in attitude measurement (see Sudman *et al.*, 1996, chapters 4 and 5; Tourangeau and Rasinski, 1988, for reviews). The specific outcome of this process, however, depends not only on *what* comes to mind, but also on how this information is *used*. Wänke and Schwarz (Chapter 5) discuss theoretical issues of information accessibility and use in attitude measurement in more detail, focusing on the emergence of context effects and the role of buffer items.

If the question is a behavioral question, the obtained responses depend on the recall and estimation strategies that respondents use (see Bradburn *et al.*, 1987; Pearson *et al.*, 1992; Schwarz, 1990; Sudman *et al.*, 1996, chapters 7 to 9, for reviews, and the contributions to Schwarz and Sudman, 1994, for recent research examples). Johnson *et al.* (Chapter 4) discuss some of these strategies in more detail. Their findings indicate that the use of these strategies differs as a function of respondents' cultural background. This raises the possibility that differences in the behavioral reports provided by different cultural groups may reflect systematic differences in the recall strategies chosen rather than, or in addition to, differences in actual behavior.

1.4.1.3 Formatting the Response

Once respondents have formed a judgment, they cannot typically report it in their own words. Rather, they are supposed to report it by endorsing one of the response alternatives provided by the researcher. This requires that they format their response in line with the options given. Accordingly, the researcher's choice of response alternatives may strongly affect survey results (see Schwarz and Hippler, 1991, for a review). Note, however, that the influence of response alternatives is not limited to the formatting stage. Rather, response alternatives may influence other steps of the question answering sequence as well, including question comprehension, recall strategies, and editing of the public answer.

The only effects that seem to occur unequivocally at the formatting stage pertain to the anchoring of rating scales (e.g., Ostrom and Upshaw, 1968; Parducci, 1983). Krosnick and Fabrigar (Chapter 6) address this and many other issues related to the use of rating scales. Their chapter provides a comprehensive meta-analytic review of the extensive literature on the design and use of rating scales and answers many questions of applied importance.

1.4.1.4 Editing the Response

Finally, respondents may want to edit their responses before they communicate them, reflecting considerations of social desirability and self-presentation. Not surprisingly, the impact of these considerations is more pronounced in face-to-face than in telephone interviews and is largely reduced in self-administered questionnaires. DeMaio (1984) reviews the survey literature on social desirability, and van der Zouwen et al. (1991) as well as Hox et al. (1991) explore the role of the interviewer in this regard. These issues are addressed in more detail in Section B of the present volume.

1.4.2 Pretesting Questionnaires: Recent Developments

An area of survey practice that has seen rapid improvements as a result of cognitive research is the pretesting of questionnaires, where cognitive methods—initially employed to gain insight into respondents' thought processes—are increasingly used to supplement traditional field pretesting (see Sudman et al., 1996, chapter 2, and the contributions to Schwarz and Sudman, 1996, for reviews). These methods have the potential to uncover many problems that are likely to go unnoticed in field pretesting. This is particularly true for question comprehension problems, which are only discovered in field pretesting when respondents ask for clarification or give obviously meaningless answers. In contrast, asking pretest respondents to paraphrase the question or to think aloud while answering it provides insights into comprehension problems that may not result in explicit queries or recognizably meaningless answers. Moreover, these procedures reduce the number of respondents needed for pretesting, rendering pretests more cost efficient—although this assertion may reflect psychologists' potentially misleading assumption that cognitive processes show little variation as a function of sociodemographic characteristics (see Groves,

1996, for a critical discussion). At present, a variety of different procedures is routinely used by major survey organizations.

The most widely used method is the collection of *verbal protocols*, in the form of concurrent or retrospective think-aloud procedures (see Groves *et al.*, 1992; Willis *et al.*, 1991, for examples). Whereas concurrent think-aloud procedures require respondents to articulate their thoughts as they answer the question, retrospective think-aloud procedures require respondents to describe how they arrived at an answer after they provided it. The latter procedure is experienced as less burdensome by respondents, but carries a greater risk that the obtained data are based on respondents' subjective theories of how they would arrive at an answer. In contrast, concurrent think-alouds are more likely to reflect respondents' actual thought processes as they unfold in real time (see Ericsson and Simon, 1980, 1984; Nisbett and Wilson, 1977, for a more detailed discussion). Related to these more elaborate techniques, respondents are often asked to paraphrase the question, thus providing insight into their interpretation of question meaning. Johnson *et al.* (Chapter 4) illustrate the fruitfulness of these procedures.

Although the use of extensive think-aloud procedures is a recent development that is largely restricted to laboratory settings, it is worth noting that asking respondents to report on their thought processes after they have answered a question has a long tradition in survey research. For example, Cantril (1944) asked respondents what they actually thought of after they answered a question and Nuckols (1953) prompted respondents to paraphrase questions in their own words to check on question comprehension. Extending this work, Schuman (1966) suggested that closed questions may be followed by a "random probe" (administered to a random subsample), inviting respondents to elaborate on their answers. Although Schuman's random probes were not intended to check question interpretation or to explore respondents' thought processes, probes that address these issues may be a fruitful way to extend research into question comprehension beyond the small set of respondents used in laboratory settings.

A procedure that is frequently applied in production interviews rather than laboratory settings is known as *behavior coding* (e.g., Cannell *et al.*, 1981; see Fowler and Cannell, 1996, for a review). Although initially developed for the assessment of interviewer behavior, this procedure has proved efficient in identifying problems with questionnaires. It involves tape recording interviews and coding behaviors such as respondent requests for clarification, interruptions, or inadequate answers. Much as regular field pretesting, however, this method fails to identify some problems that surface in verbal protocols, such as misunderstanding what a question means. Recent extensions of this approach include automated coding of interview transcripts, for which standardized coding schemes have been developed (e.g., Bolton and Bronkhorst, 1996).

Based on insights from verbal protocols and behavior coding, Forsyth *et al.* (1992; see also Lessler and Forsyth, 1996) have developed a detailed coding scheme that allows cognitive experts to identify likely question problems in advance. At present, this development represents one of the most routinely applicable outcomes of the knowledge accumulating from cognitive research. Other techniques, such as the use of sorting procedures (e.g., Brewer *et al.*, 1989; Brewer and Lui, 1996) or response latency measurement (see Bassili, 1996, for a review), are more limited in scope and have not yet been routinely employed in pretesting.

1.4.3 Summary

The recent collaboration of cognitive and social psychologists and survey methodologists has resulted in an improved understanding of the cognitive and communicative processes that underlie survey responding and has had a considerable effect on questionnaire pretesting. Whereas much remains to be learned, cognitive research has identified strategies that improve retrospective behavioral reports (see the contributions in Schwarz and Sudman, 1994; Tanur, 1992) and has contributed to an understanding of the emergence of context effects in attitude measurement (see the contributions in Schwarz and Sudman, 1992; Tanur, 1992). To what extent these theoretical insights can be successfully translated into standard survey practice, on the other hand, remains to be seen.

At present, the theoretical frameworks developed in this collaboration have received considerable empirical support under conditions explicitly designed to test key predictions of the respective models. Thus, researchers working on context effects in attitude measurement, for example, can reliably produce assimilation and contrast effects—provided that the questions are written to achieve a clear operationalization of the theoretically relevant variables. Unfortunately, this success does not imply that the same researchers can predict the behavior of any given question—in many cases, it is simply unclear to what extent a given question reflects the key theoretical variables, thus limiting the applied use of the theoretical models. Research into retrospective reports of behavior, on the other hand, has proved more directly applicable. This reflects, in part, that the recall tasks posed most frequently by retrospective behavioral questions are less variable in nature than the tasks posed by attitude questions. Nevertheless, the facet of cognitive work that has most quickly found its way into survey practice is methods initially developed to gain insight into individuals' thought processes. These methods have been adopted for questionnaire pretesting and have proved fruitful in identifying potential problems at an early stage.

1.5 MODES OF DATA COLLECTION: IMPLICATIONS FOR RESPONDENTS' TASKS AND QUESTIONNAIRE CONSTRUCTION

The choice of administration mode is determined by many factors, with cost and sampling considerations usually being the dominant ones (see Groves, 1989, for a discussion). Not surprisingly, the decision to rely on face-to-face interviews, telephone interviews or self-administered questionnaires has important implications for questionnaire construction. Below, I summarize some of the key differences between face-to-face interviews, telephone interviews and self-administered questionnaires from a psychological perspective (for a more extended discussion see Schwarz et al., 1991). Many of the issues arising from these differences are addressed in Section B on data collection.

1.5.1   Cognitive Variables

The most obvious difference between these modes of data collection is the sensory channel in which the material is presented. In self-administered questionnaires, the items are visually displayed to the respondent who has to read the material, rendering research on visual perception and reading highly relevant to questionnaire design, as Jenkins and Dillman (Chapter 7) note. In telephone interviews, as the other extreme, the items and the response alternatives are read to respondents, whereas both modes of presentation may occur in face-to-face interviews.

Closely related to this distinction is the temporal order in which the material is presented. Telephone and face-to-face interviews have a strict sequential organization. Hence, respondents have to process the information in the temporal succession and the pace in which it is presented by the interviewer. They usually cannot go back and forth or spend relatively more or less time on some particular item. And even if respondents are allowed to return to previous items should they want to correct their responses, they rarely do so, in part because tracking one's previous responses presents a difficult memory task under telephone and face-to-face conditions. In contrast, keeping track of one's responses, and going back and forth between items, poses no difficulties under self-administered questionnaire conditions. As a result, self-administered questionnaires render the sequential organization of questions less influential and subsequent questions have been found to influence the responses given to *preceding* ones (e.g., Bishop et al. 1988; Schwarz and Hippler, 1995). Accordingly, the emergence of context effects is order-dependent under face-to-face and telephone interview conditions, but not under self-administered questionnaire conditions.

In addition, different administration modes differ in the time pressure they impose on respondents. Time pressure interferes with extensive recall processes and increases reliance on simplifying processing strategies (Krosnick, 1991; Kruglanski, 1980). The greatest time pressure can be expected under telephone interview conditions, where moments of silent reflection cannot be bridged by nonverbal communication that indicates that the respondent is still paying attention to the task (Ball, 1968; Groves and Kahn, 1979). If the question poses challenging recall or judgment tasks, it is therefore particularly important to encourage respondents to take the time they may need under telephone interview conditions. The least degree of time pressure is induced by self-administered questionnaires that allow respondents to work at their own pace.

Face-to-face interviews create intermediate time pressure, due to the possibility of bridging pauses by nonverbal communication.

1.5.2 Social Interaction Variables

While social interaction is severely constrained in all standardized survey interviews, the modes of data collection differ in the degree to which they restrict nonverbal communication. While face-to-face interviews provide full access to the nonverbal cues of the participants, participants in telephone interviews are restricted to paraverbal cues, whereas social interaction is largely absent under self-administered conditions. Psychological research has identified various functions of nonverbal cues during face-to-face interaction (see Argyle, 1969 for a review). Most importantly, nonverbal cues serve to indicate mutual attention and responsiveness and provide feedback as well as illustrations for what is being said (in the form of gestures). Given the absence of these helpful cues under self-administered questionnaire conditions, particular care needs to be taken to maintain respondent motivation in the absence of interviewer input and to render the questionnaire self-explanatory, as Jenkins and Dillman (Chapter 7) emphasize. Their chapter provides much useful advice in this regard.

Further contributing to the need of self-explanatory questionnaires, respondents have no opportunity to elicit additional explanations from the interviewer under self-administered conditions. In contrast, interviewer support is easiest under face-to-face conditions, where the interviewer can monitor the respondent's nonverbal expressions. Telephone interview conditions fall in between these extremes, reflecting that the interviewer is limited to monitoring the respondent's verbal utterances. Even though any additional information provided by the interviewer is usually restricted to prescribed feedback, it may help respondents to determine the meaning of the questions. Even the uninformative—but not unusual—clarification, "whatever it means to you," may be likely to shortcut further attempts to screen question context in search for an appropriate interpretation. Under self-administered questionnaire conditions, on the other hand, respondents have to depend on the context that is explicitly provided by the questionnaire to draw inferences about the intended meaning of questions— and they have the time and opportunity to do so. This is likely to increase the impact of related questions and response alternatives on question interpretation (Strack and Schwarz, 1992).

Whereas self-administered questionnaires are therefore likely to increase reliance on contextual information (although potentially independent of the order in which it is presented), they have the advantage to be free of interviewer effects. In general, interviewer characteristics are more likely to be noticed by respondents when they have face-to-face contact than when the interviewer cannot be seen. This is the case with telephone interviews, where the identification of interviewer characteristics is limited to characteristics that may be inferred from paralinguistic cues and speech styles (such as sex, age, or race). Under self-administered questionnaire conditions, of course, no interviewer is required, although respondents may pick up characteristics of the researcher from the cover letter, the person who dropped off the questionnaire, and so on. While respondents' perception of interviewer characteristics may increase socially desirable responses, it may also serve to increase rapport with the interviewer, rendering the potential influence of interviewer characteristics on response behavior ambivalent.

1.5.3 Other Differences

In addition, the modes of data collection differ in their degree of perceived confidentiality and in the extent to which they allow the researcher to control external distractions, e.g., from other household members. Moreover, different administration modes may result in differential self-selection of respondents with different characteristics. In general, respondents with a low level of education are assumed to be underrepresented in mail surveys relative to face-to-face and telephone interviews (e.g., Dillman, 1978). More important, however, respondents in mail surveys have the opportunity to preview the questionnaire before they decide to participate (Dillman, 1978). In contrast, respondents in face-to-face or telephone surveys have to make this decision on the basis of the information provided in the introduction (and are unlikely to revise their decision once the interview proceeds). As a result, mail surveys run a considerably higher risk of topic-related self-selection than face-to-face or telephone surveys.

Although topic-related self-selection becomes less problematic as the response rate of mail surveys increases, self-selection problems remain even at high response rates. As an illustration, suppose that a mail and a telephone survey both obtain an 80% response rate. In the telephone survey, the 20% nonresponse includes participants who refuse because they are called at a bad time, are on vacation, or whatever. However, it does not include respondents who thought about the issue and decided it was not worth their time. In contrast, mail respondents can work on the questionnaire at a time of their choice, thus potentially reducing nonresponse due to timing problems. On the other hand, however, they have the possibility to screen the questionnaire and are more likely to participate in the survey if they find the issue of interest. As a result, an identical nonresponse rate of 20% under both modes is likely to be unrelated to the topic under interview conditions, but not under mail conditions. Hence, similar response rates under different modes do not necessarily indicate comparable samples.

Moreover, the variable that drives self-selection under mail conditions is respondents' interest in the topic, which may be only weakly related to sociodemographic variables. Accordingly, topic-driven self-selection may be present even if the completed sample seems representative with regard to sociodemographic variables like age, sex, income, and so on. To assess the potential problem of topic-related self-selection, future research will need to assess respondents' interest in the topic and its relationship to substantive responses under different administration modes.

1.6 CONCLUSIONS

This introduction to the section on Questionnaire Design reviewed some of the key elements of the survey process, focusing primarily on cognitive and communicative aspects of the response process. Other components of the survey process will be addressed in subsequent sections of this volume. As this selective review illustrates, our theoretical understanding of how respondents make sense of a question and arrive at an answer has improved during recent years. Moreover, researchers have begun to translate some of the recent theoretical insights into survey practice, as the contributions to the present volume illustrate, and their findings document the potential fruitfulness of cognitive approaches to questionnaire construction. Nevertheless, much remains to be learned, and after a decade of cognitive research, its influence on standard survey practice is still more limited than optimists may have hoped for. Disappointing as this may be in some respects, it is worth emphasizing that without the systematic development and testing of guiding theoretical principles, our knowledge about survey measurement is "likely to remain a set of scattered findings, with repeated failure at replication of results," as Groves and Lyberg (1988, p. 210) noted in a related discussion.

REFERENCES

Alwin, D.F., Braun, M., Harkness, J., and Scott, J. (1994), "Measurement in Multinational Surveys," in I. Borg and P.P. Mohler (eds.), *Trends and Perspectives in Empirical Social Research*, Berlin, FRG: de Gruyter, pp. 26–39.

Argyle, M. (1969), *Social Interaction*, London: Methuen.

Ball, D.W. (1968), "Toward a Sociology of Telephones and Telephoners", in M. Truzzi (ed.), *Sociology and Everyday Life*, Englewood Cliffs, NJ: Prentice-Hall, pp. 59–75.

Bassili, J.N. (1996), "The How and Why of Response Latency Measurement in Telephone Surveys," in N. Schwarz, and S. Sudman (eds.), *Answering Questions: Methodology for Determining Cognitive and Communicative Processes in Survey Research*. San Francisco: Jossey-Bass, pp. 319–346.

Bishop, G.F., Hippler, H.J., Schwarz, N., and Strack, F. (1988), "A Comparison of Response Effects in Self-administered and Telephone Surveys," in R.M. Groves, P. Biemer, L. Lyberg, J. Massey, W. Nicholls, and J. Waksberg (eds.), *Telephone Survey Methodology*, New York: Wiley, pp. 321–340.

Bodenhausen, G.V., and Wyer, R.S. (1987), "Social Cognition and Social Reality: Information Acquisition and Use in the Laboratory and the Real World, in H.J. Hippler, N. Schwarz, and S. Sudman (eds.), Social Information Processing and Survey Methodology. New York: Springer Verlag, pp. 6–41.

Bolton, R.N., and Bronkhorst, T.M. (1996), "Questionnaire Pretesting: Computer Assisted Coding of Concurrent Protocols," in N. Schwarz and S. Sudman (eds.), *Answering Questions: Methodology for Determining Cognitive and Communicative Processes in Survey Research*, San Francisco: Jossey-Bass, pp. 37–64.

Bradburn, N.M., Rips, L.J., and Shevell, S.K. (1987), "Answering Autobiographical Questions: The Impact of Memory and Inference on Surveys," *Science*, 236, pp. 157–161.

Brewer, M.B., Dull, V.T., and Jobe, J.B. (1989), "A Social Cognition Approach to Reporting Chronic Conditions in Health Surveys," National Center for Health Statistics, *Vital Health Statistics*, 6(3).

Brewer, M.L., and Lui, L.N. (1996), "Use of Sorting Tasks to Assess Cognitive Structures," in N. Schwarz, and S. Sudman (eds.), *Answering Questions: Methodology for Determining Cognitive and Communicative Processes in Survey Research*, San Francisco: Jossey-Bass, pp. 373–387.

Cannell, C.F., Miller, P.V., and Oksenberg, L. (1981), "Research on Interviewing Techniques," in S. Leinhardt (ed.), *Sociological Methodology*, San Francisco: Jossey-Bass.

Cantril, H. (1944), *Gauging Public Opinion*, Princeton, NJ: Princeton University Press.

Clark, H.H., and Schober, M.F. (1992), "Asking Questions and Influencing Answers," in J.M. Tanur (ed.), *Questions About Questions*, New York: Russell Sage, pp. 15–48.

Davis, J.A., Mohler, P.P., and Smith, T. W. (1994), "Nationwide General Social Surveys," in I. Borg and P.P. Mohler (eds.), *Trends and Perspectives in Empirical Social Research*, Berlin, FRG: de Gruyter, pp. 15–25.

DeMaio, T.J. (1984), "Social Desirability and Survey Measurement: A Review," in C.F. Turner and E. Martin (eds.), *Surveying Subjective Phenomena*, New York: Russell Sage, Vol. 2, pp. 257–281.

Dillman, D.A. (1978), *Mail and Telephone Surveys: The Total Design Method*, New York: Wiley.

Dippo, C.S., and Norwood, J. L. (1992), "A Review of Research at the Bureau of Labor Statistics," in J.M. Tanur (ed.), *Questions About Questions*, New York: Russell Sage, pp. 271–290.

Ericsson, K.A., and Simon, H. (1980), "Verbal Reports as Data," *Psychological Review*, 8, pp. 215–251.

Ericsson, K.A., and Simon, H.A. (1984), *Protocol Analysis: Verbal Reports as Data*, Cambridge, MA: MIT Press.

Forsyth, B.H., and Lessler, J.T. (1991), "Cognitive Laboratory Methods: A Taxonomy," in P. Biemer, R. Groves, L. Lyberg, N. Mathiowetz, and S. Sudman (eds.), *Measurement Errors in Surveys*, Chichester: Wiley, pp. 393–418.

Forsyth, B.H., Lessler, J.L., and Hubbard, M.L. (1992), "Cognitive Evaluation of the Questionnaire," in C.F. Turner, J.T. Lessler, and J.C. Gfroerer (eds.), *Survey Measurement of Drug Use: Methodological Studies*, Washington, DC: DHHS Publication No. 92-1929.

Fowler, F.J., and Cannell, C.F. (1996), "Using Behavioral Coding to Identify Cognitive Problems with Survey Questions," in N. Schwarz, and S. Sudman (eds.), *Answering Questions: Methodology for Determining Cognitive and Communicative Processes in Survey Research*, San Francisco: Jossey-Bass, pp. 15–36.

Groves, R.M. (1989), *Survey Errors and Survey Costs*, New York: Wiley.

Groves, R.M. (1996), "How Do We Know that What We Think They Think Is Really What They Think?," in N. Schwarz, and S. Sudman (eds.), *Answering Questions: Methodology for Determining Cognitive and Communicative Processes in Survey Research*, San Francisco: Jossey-Bass, pp. 389–402.

Groves, R.M., Fultz, N.H., and Martin, E. (1992), "Direct Questioning About Comprehension in a Survey Setting," in J.M. Tanur (ed.), *Questions About Questions*, New York: Russell Sage, pp. 49–61.

Groves, R.M., and Kahn, R.L. (1979), *Surveys by Telephone: A National Comparison with Personal Interviews*, New York: Academic Press.

Groves, R.M., and Lyberg, L.E. (1988), "An Overview of Nonresponse Issues in Telephone Surveys," in R.M. Groves, P. Biemer, L. Lyberg, J. Massey, W. Nicholls, and J. Waksberg (eds.), *Telephone Survey Methodology*, New York: Wiley, pp. 191–211.

Higgins, E.T. (1989), "Knowledge Accessibility and Activation: Subjectivity and Suffering from Unconscious Sources," in J.S. Uleman, and J.A. Bargh (eds.), *Unintended Thought*, New York: Guilford Press, pp. 75–123.

Hippler, H.J., Schwarz, N., and Sudman, S. (eds.) (1987), *Social Information Processing and Survey Methodology*, New York: Springer Verlag.

Hox, J.J., de Leeuw, E.D., and Kreft, I.G.G. (1991), "The Effect of Interviewer and Respondent Characteristics on the Quality of Survey Data: A Multilevel Model," in P. Biemer, R. Groves, L. Lyberg, N. Mathiowetz, and S. Sudman (eds.), *Measurement Errors in Surveys*, Chichester: Wiley, pp. 393–418.

Jabine, T.B., Straf, M.L., Tanur, J.M., and Tourangeau, R. (eds.) (1984), *Cognitive Aspects of Survey Methodology: Building a Bridge Between Disciplines*, Washington, DC: National Academy Press.

Jobe, J., and Loftus, E. (eds.) (1991), "Cognitive Aspects of Survey Methodology," special issue of *Applied Cognitive Psychology*, 5.

Krosnick, J.A. (1991), "Response Strategies for Coping with the Cognitive Demands of Attitude Measures in Surveys," *Applied Cognitive Psychology*, 5, pp. 213–236.

Kruglanski, A.W. (1980), "Lay Epistemologic Process and Contents," *Psychological Review*, 87, pp. 70–87.

Lessler, J.T., and Forsyth, B. H. (1996), "A Coding System for Appraising Questionnaires," in N. Schwarz, and S. Sudman (eds.), *Answering Questions: Methodology for Determining Cognitive and Communicative Processes in Survey Research*, San Francisco: Jossey-Bass, pp. 259–292.

Nisbett, R.E., and Wilson, T.D. (1977), "Telling More than We Know: Verbal Reports on Mental Processes," *Psychological Review*, 84, pp. 231–259.

Nuckols, R. (1953), "A Note on Pre-testing Public Opinion Questions," *Journal of Applied Psychology*, 37, pp. 119–120.

Ostrom, T.M., and Upshaw, H.S. (1968), "Psychological Perspective and Attitude Change," in A.C. Greenwald, T.C. Brock, and T.M. Ostrom, (eds.), *Psychological Foundations of Attitudes*, New York: Academic Press.

Parducci, A. (1983), "Category Ratings and the Relational Character of Judgment," in H.G. Geissler, H.F.J.M. Bulfart, E.L.H. Leeuwenberg, and V. Sarris (eds.),

Modern Issues in Perception, Berlin: VEB Deutscher Verlag der Wissenschaften, pp. 262–282.

Payne, S.L. (1951), *The Art of Asking Questions*, Princeton: Princeton University Press.

Pearson, R.W., Ross, M., and Dawes, R.M. (1992), "Personal Recall and the Limits of Retrospective Questions in Surveys," in J.M. Tanur (ed.), *Questions About Questions*, New York: Russel Sage, pp. 65–94.

Popper, K.R. (1968), *The Logic of Scientific Discovery*, New York: Harper and Row.

Root-Bernstein, R.S. (1989), *Discovering, Inventing and Solving Problems at the Frontiers of Scientific Knowledge*, Cambridge, MS: Harvard University Press.

Schuman, H. (1966), "The Random Probe: A Technique for Evaluating the Validity of Closed Questions," *American Sociological Review*, 31, pp. 218–222.

Schuman, H., and Kalton, G. (1985), "Survey Methods," in G. Lindzey, and E. Aronson (eds.), *Handbook of Social Psychology*, New York: Random House, Vol. 1, pp. 635–697.

Schwarz, N. (1990), "Assessing Frequency Reports of Mundane Behaviors: Contributions of Cognitive Psychology to Questionnaire Construction," in C. Hendrick, and M.S. Clark (eds.), *Research Methods in Personality and Social Psychology* (*Review of Personality and Social Psychology, Vol. 11*), Beverly Hills, CA: Sage, pp. 98–119.

Schwarz, N., and Hippler, H.J. (1991), "Response Alternatives: The Impact of Their Choice and Ordering," in P. Biemer, R. Groves, L. Lyberg, N. Mathiowetz, and S. Sudman (eds.), *Measurement Errors in Surveys*, Chichester: Wiley, pp. 41–56.

Schwarz, N., and Hippler, H.J. (1995), "Subsequent Questions May Influence Answers to Preceding Questions in Mail Surveys," *Public Opinion Quarterly*, 59, pp. 93–97.

Schwarz, N., Strack, F., Hippler, H.J., and Bishop, G. (1991), "The Impact of Administration Mode on Response Effects in Survey Measurement," *Applied Cognitive Psychology*, 5, pp. 193–212.

Schwarz, N., and Sudman, S. (eds.) (1992), *Context Effects in Social and Psychological Research*, New York: Springer Verlag.

Schwarz, N., and Sudman, S. (1994), *Autobiographical Memory and the Validity of Retrospective Reports*, New York: Springer Verlag.

Schwarz, N., and Sudman, S. (eds.) (1996), *Answering Questions: Methodology for Determining Cognitive and Communicative Processes in Survey Research*, San Francisco: Jossey-Bass.

Strack, F., and Martin, L. (1987). "Thinking, Judging, and Communicating: A Process Account of Context Effects in Attitude Surveys," in H.J. Hippler, N. Schwarz, and S. Sudman (eds.), *Social Information Processing and Survey Methodology*, New York: Springer Verlag, pp. 123–148.

Strack, F., and Schwarz, N. (1992), "Implicit Cooperation: The Case of Standardized Questioning," in G. Semin, and F. Fiedler (eds.), *Social Cognition and Language*, Beverly Hills: Sage, pp. 173–193.

Sudman, S., and Bradburn, N. (1983), *Asking Questions*, San Francisco: Jossey-Bass.

Sudman, S., Bradburn, N., and Schwarz, N. (1996), *Thinking About Answers: The Application of Cognitive Processes to Survey Methodology*, San Francisco, CA: Jossey-Bass.

Tanur, J.M. (ed.) (1992), *Questions About Questions*, New York: Russel Sage.

Tourangeau, R. (1984), "Cognitive Science and Survey Methods: A Cognitive Perspective," in T. Jabine, M. Straf, J. Tanur, and R. Tourangeau (eds.), *Cognitive Aspects of Survey Methodology: Building a Bridge Between Disciplines*, Washington, DC: National Academy Press, pp. 73–100.

Tourangeau, R., and Rasinski, K.A. (1988), "Cognitive Processes Underlying Context Effects in Attitude Measurement," *Psychological Bulletin*, 103, pp. 299–314.

Van der Zouwen, J., Dijkstra, W., and Smit, J.H. (1991), "Studying Respondent-interviewer Interaction: The Relationship Between Interviewing Style, Interviewer Behavior, and Response Behavior," in P. Biemer, R. Groves, L. Lyberg, N. Mathiowetz, and S. Sudman (eds.), *Measurement Errors in Surveys*, Chichester: Wiley, pp. 419–438.

Willis, G., Royston, P., and Bercini, D. (1991), "The Use of Verbal Report Methods in the Development and Testing of Survey Questions," *Applied Cognitive Psychology*, 5, pp. 251–267.

CHAPTER 2

From Theoretical Concept to Survey Question

Joop J. Hox
University of Amsterdam

2.1 INTRODUCTION

Survey methodology has given much attention to the problem of formulating the questions that go into the survey questionnaire. Problems of question wording, questionnaire flow, question context, and choice of response categories have been the focus of much research. Furthermore, experienced survey researchers have written guidelines for writing questions and constructing questionnaires, and cognitive laboratory methods are developing means to test the quality of survey questions.

In comparison, much less effort has been directed at clarifying the problems that occur *before* the first survey question is committed to paper. Before questions can be formulated, researchers must decide upon the concepts that they wish to measure. They have to define what it is that they intend to measure by naming the concept, describing its properties and its scope, and describing how it differs from other related concepts.

Most (if not all) methodologists and philosophers will place the development of concepts for social research in the context of discovery, and not in the context of verification. Consequently, there are no fixed rules that limit the researcher's imagination, and concepts are not rigorously verified or falsified, but they are judged by their fruitfulness for the research process.

The process of concept-formation involves elaborating the concept and defining important sub-domains of its meaning. The next stage is finding empirical indicators for each concept or each subdomain. Empirical indicators

Survey Measurement and Process Quality, Edited by Lyberg, Biemer, Collins, de Leeuw, Dippo, Schwarz, Trewin.
ISBN 0-471-16559-X   © 1997 John Wiley & Sons, Inc.

could be viewed as "not-quite-variables," because at this stage we have not yet written specific questions or proposed a measurement (scale) model. The last stage is the actual question writing. This means that the conceptual content of each indicator must be translated fully and accurately into actual survey questions.

Although this chapter starts with a discussion of some philosophy of science issues, its main focus is the presentation of general strategies and specific techniques that can be used to structure the process of going from theoretical concept to prototype survey question. It is only after this stage that the usual concerns of intended population, question wording, *etcetera* begin to play a role in designing the questionnaire. These problems are not discussed here.

This chapter starts with a discussion of the theoretical substance of scientific concepts. Recent philosophy of science has come to the conclusion that scientific concepts are all laden with theory; this chapter discusses why and the subsequent consequences for conceptualization and operationalization. The next section discusses several approaches that have been proposed for question development, distinguishing between theory driven and data driven approaches. The last section provides a short summary and discussion.

2.2 SCIENTIFIC THEORY AND THE FUZZINESS OF SCIENTIFIC CONCEPTS

Two major aims of scientific research are to enlarge our knowledge of a specific field, and to solve a practical problem. In social science, survey research may be used for both purposes: the information that surveys provide may be used to explore or test theories, but also to shape or decide on policy. In both cases scientific theory plays an important role. Even in the most practical case we are generally not interested in the responses as such, but in their implications at some more abstract level. For instance, we may ask which candidate the respondent is going to vote for. This is a straightforward question, but the research aim is not just producing a frequency distribution, but rather inferring something about the general "voting intention" of the public, for instance, to predict the outcome of an election. In that respect, the question is not straightforward at all; we might attempt to improve our understanding of "voting intention" by asking the respondents how sure they are of their votes, and what other candidates they are also considering. Another example is the conventional question about the respondent's occupation. The objective of that question is usually not to find out what that person does during working hours, but to assign the respondent to a position on a Social-Economic Status (SES) ladder. Again, although the question is straightforward, the concept "SES" is not, and different conceptualizations of "SES" may lead us to different formulations of the occupation question.

The reason for shifting to a more abstract level in designing or interpreting a survey is that scientific concepts are embedded in a more general theory that presents a systematic view of the phenomena we are studying. Instead of tabulating the popularity of candidates for our specific sample of individuals, we try to explain (predict) "voting intention" by linking it to various other concepts that indicate political attitudes. Instead of assigning the value "carpenter" to the variable "occupation," we assign "skilled labor" to the variable "SES."

The terms *concept*, *construct*, and *variable* are often used interchangeably. To understand the differences between these, I refer to a definition given by Kerlinger (1986, p. 9). "A theory is a set of interrelated concepts/constructs, definitions, and propositions that present a systematic view of phenomena by specifying relations among variables, with the purpose of explaining and predicting the phenomena." Thus, a theory is a system of propositions about the relations between concepts and constructs. A variable is a term or symbol to which values are assigned based on empirical observations, according to indisputable rules. The difference between a concept and a construct is small. A concept is an abstraction formed by generalization from similar phenomena or similar attributes. A construct is a concept that is systematically defined to be used in scientific theory (Fiske, 1971; Kerlinger, 1986). For example, "delinquency" is a concept that refers to an individual's tendency to exhibit behaviors that are classified as delinquent. "Deviance" is a construct that refers to an individual's tendency to exhibit behaviors that are classified as deviant. The difference is that "delinquency" can be defined by referring to those acts that our society classifies as delinquent, while "deviance" can only be defined in the framework of some theory (containing other constructs such as "norms" and "social reference group") that defines which behaviors are to be called deviant.

Since constructs and concepts are both theoretical abstractions, I use these terms loosely, with concept referring to abstract terms used in applied research and in early stages of theorizing, and construct referring to a more formal scientific concept used in theoretical research. This terminology emphasizes that the concepts used in social research, even if their name is a commonly used word (e.g., social class, disability, unemployment), are adopted or created by researchers for the purpose of their research. They have a theoretical "surplus meaning" of which the boundaries are not sharply drawn. Theoretical constructs have a rich surplus meaning. If we are working from an applied perspective, we are likely to work with concepts that have little surplus meaning. But there will be *some* surplus meaning, unless we are discussing the measurement of concepts like "sex" or "age" which are very close to their empirical measurement. Consequently, both concepts and constructs must be linked to observed variables by an operational definition that specifies to which variables they are linked and how values are assigned to these variables. This process is often viewed as a translation process; theoretical constructs are translated into observable variables that appear to be a suitable representation of that construct.

2.2.1 Operationism

The scientific/philosophical approach known as *operationism* (Bridgman, 1927) requires that every construct completely coincide with one single observed variable. The operations needed to assign a value to a variable completely define the associated construct. In sociology, Bridgman's ideas were advocated by Lundberg (1939), while in psychology they were influential in the development of behaviorism (Watson, 1924). The essential principle in operationism is that the operational definition *is* the construct, and not a process to observe a construct after it has been defined (Lundberg, 1939). The consequence of the operationist approach is that if we change any aspect of the operational definition we have a new theoretical construct. Thus, if we measure heat with a mercury thermometer and with an alcohol thermometer we are measuring two different things. Similarly, if we change one word in our question, we are measuring a different construct.

The legacy of operationism to social science is a healthy emphasis on explicit definitions and measurement procedures. But the extreme position taken in operationism leads to hopeless problems if we attempt to generalize our results. A related problem in operationism is that we cannot compare two operational definitions to decide which one is better; they are merely different.

2.2.2 Logical Positivism

The various problems and inconsistencies in the operationalist approach have led to the conclusion that it is useful to distinguish between theoretical constructs and observed variables. This is a central notion in the philosophical approach known as *logical positivism*, which originated in the group of philosophers known as the Vienna Circle (cf. Hempel and Oppenheim, 1948). Logical positivism has been highly influential in shaping the ways social scientists think about the relationship between theoretical constructs and observed variables. A lucid exposition of the logical positivist position is given by Carnap (1956). Carnap distinguishes between a *theoretical language* and an *observation language*. The theoretical language contains the assumptions about the theoretical constructs and their interrelations. The observation language contains only concepts that are operationally defined or that can be linked to operationally defined concepts by formal logical or mathematical operations. The observation language should also be completely understandable by all scientists involved; there should be no discussion about the legitimacy of the observations. The connection between the theoretical and the observation language is made by correspondence rules, which are assumptions and inference rules. In Carnap's view the theoretical language is much richer than the observation language. Thus, a scientific theory may contain constructs that are defined solely by their relations with other constructs; in other words, constructs that have no operational definition. The view that the theoretical language is richer than the observation language is summarized by stating that the theoretical language has surplus meaning with respect to the observation language. In the course of research, we may hope to reduce the surplus meaning by extending our knowledge, but there will always be some surplus meaning left.

In sociology, the influence of logical positivism is strong in the work of Northrop (1947), Lazarsfeld (1972), and Blalock (1982). Northrop distinguishes between concepts-by-intuition, which can be immediately understood upon observation, and concepts-by-postulation, which derive their meaning at least in part from theory. The two kinds of concepts are linked by epistemic correlations, which are analogous to Carnap's correspondence rules. Blalock identifies Northrop's concepts-by-intuition as terms in the observational language, and concepts-by-postulation as terms in the theoretical language. Both kinds of terms must be linked by "auxiliary measurement theory," which can be interpreted as the theory that describes the assumptions and inferences that together form Carnap's correspondence rules. In psychology, the influence of logical positivism is reflected in the distinction between theoretical constructs that are embedded in a nomological network, and operational definitions that specify how some theoretical constructs are measured. A key concept in psychometric theory is the *construct validity* of a measure. Construct validity indicates how well a specific measure reflects the theoretical construct it is assumed to measure (Cronbach and Meehl, 1955; Campbell and Fiske, 1959). Just like the epistemic correlation, the construct validity is not a parameter that can be estimated as some specific value; it refers to the general quality of the correspondence rules.

2.2.3 Sophisticated Falsificationism

The views of logical positivism have been strongly attacked by philosophers like Popper, Kuhn, and Feyerabend. Popper is probably best known for introducing the falsification principle; science grows by formulating theoretical conjectures and exposing these theories to the risk of falsification (Popper, 1934). However, Popper (and other philosophers) also criticized logical positivism for its belief that there are such things as direct, unprejudiced observations. As Popper makes clear in many examples (Popper, 1934; see also Chalmers, 1976), observation statements are always made in the language of some theory. Precise observations require explicit and clearly formulated theories. And, if theories are fallible, the observation statements made in the context of those theories are also subject to falsification.

If observations are theory-laden, straightforward falsification is impossible, because both theories and observations may be false. Popper solves this problem by subjecting not only theories but also observation statements to criticism and testing. Observations that have survived critical examination are accepted by the scientific community—for the time being (Popper, 1934). This position has been adopted by most social science methodologists. For instance, Blalock (1968), referring to Northrop, argues that concepts-by-intuition are accepted by consensus among the scientists in the field. Cronbach (1971) lists as one of the possible consequences of negative results that a construct may be abandoned, subdivided, or merged with other constructs. De Groot (1969) explicitly refers decisions about both theoretical constructs and operational definitions to "the scientific forum," that is, to ongoing discussion in the scientific community.

Kuhn (1970) and Lakatos (1978) both deal with the problem of the theory-dependence of observation statements in a slightly different way, by taking into account the way scientific theories evolve. As Chalmers (1976, pp. 63, 71) shows, if Popper's falsification principle had been strictly followed by scientists some of the most powerful theories, including Newton's gravitational theory, would have been rejected in their infancy. Kuhn (1970) maintains that most of the time scientific research is realized in a state of normal science, within a commonly accepted paradigm that contains the commonly accepted assumptions, laws and techniques. Apparent falsifications are regarded as anomalous results, to be resolved by further research and improvements in measurement and experimental techniques. If anomalies proliferate, a scientific revolution takes place, in which the anomaly-ridden paradigm is replaced by a new one. Lakatos (1978) views science as a world of competing research programs. A research program possesses a hard core: the main theoretical assumptions that are not to be rejected. This theoretical core is protected by a belt of auxiliary theory (including unstated assumptions about the accepted research methodology). Faced with a falsification, the theoretical core remains unchanged, and only the auxiliary theory is adjusted. Research programs that consistently lead to interesting new results are said to be *progressive*; programs that fail to do so are said to be *degenerated*. When a research program shows severe signs of degeneration, scientists leave it for a competing program (old theories never die, they just fade away). An example of a research program in Lakatos's sense in survey research is the cognitive approach to studying response behavior. The theoretical core is a set of cognitive theories and accepted information processing processes. The protective belt consists of the various adjustments needed to apply these theories in survey research. A falsification will generally not lead to abandoning the cognitive approach altogether, instead there will be adjustments in the auxiliary theory or research method. As long as the cognitive research program leads to interesting results, researchers will cling to it; when it fails, they will be attracted by another program (if available).

In both Kuhn's and Lakatos's views most research will be conducted in a state that can be described as normal or within a progressive research program. Thus, discussions about theoretical constructs and observations tend to be technical. A critical examination of a theoretical construct will probably focus on its dimensionality (should it be subdivided or merged with another construct) or its nomological network (how it is related to other constructs). New theoretical constructs will fit within a commonly accepted framework, and there are accepted guidelines for translating these into observable measures. Only in a state of theoretical crisis are we in the position described by Feyerabend as "anything goes."

It is interesting to note that operationism and logical positivism regard the fuzziness of theoretical constructs and the theory-dependence of observations as problems that must be eliminated, while Popper, Kuhn and Lakatos regard these problems as the heart of scientific progress. Especially Lakatos, whose views are usually denoted as "sophisticated falsificationism" (Chalmers, 1976), argues strongly that the critical discussion of constructs, their theoretical meaning, and relevant observations of their occurrence, is what brings about scientific progress in the form of new theories or new research questions.

2.3 CONCEPTUALIZATION AND OPERATIONALIZATION

As the previous section illustrates, both philosophers and methodologists have come to the conclusion that there are no pure observations. In the context of measurement in surveys, this means that there are no purely objective survey questions. Instead, there is a continuum that stretches from the theoretical to the observational. Some questions are so close to the observational end that for most purposes we may consider them as perfect measures for the construct. Examples would be questions about age or sex. But even these can be strongly theory-laden. Consider the case of "sex" in a gender study, or "age" in a study about calendar versus biological age, or in life cycle research.

In practice, faced with the problem of elaborating theoretical constructs and constructing questionnaire items, methodologists act as if there were a clear distinction between the theoretical language and the observational language. This leads to the sharp distinction between conceptualization and operationalization. Conceptualization involves concept formation, which establishes the meaning of a construct by elaborating the nomological network and defining important subdomains of its meaning. Operationalization involves the translation of a theoretical construct into observable variables by specifying empirical indicators for the concept and its subdomains. To bridge this gap between theory and measurement, two distinct research strategies are advocated: a theory driven or "top down" strategy, which starts with constructs and works towards observable variables, and a data driven or "bottom up" strategy, which starts with observations and works towards theoretical constructs (cf. Hox and De Jong-Gierveld, 1990).

2.3.1 From Conceptualization To Operationalization: Theory Driven Approaches

The theory driven strategies described in this section are dimension/indicator analysis, semantic analysis, and facet design.

2.3.1.1 *Dimension/Indicator Analysis*
Lazarsfeld (1958, 1972) outlines an approach that starts with a general and often vague concept. This is followed by a process of *concept specification*; the

concept is divided into separate components or dimensions. The subdimensions form a further, more detailed specification of the meaning of the theoretical construct. The subdimensions of a theoretical construct may be derived logically, or they may be based on the empirical results from previous research. Next, a set of empirical indicators is defined for each subdimension. For practical research, a subset of indicators is chosen, and usually summated to a total score for each subdimension. Since in Lazarsfeld's view the indicators for a specific subdimension are highly interchangeable, the precise choice of indicators is not very important.

Lazarsfeld does not explain how we should actually achieve concept specification. Fiske (1971) describes a procedure that is analogous to Lazarsfeld's approach. In Fiske's view, an explicit conceptual specification must lead to a well-defined construct that poses a clear target for operationalization and permits multi-component operationalization. Just as Lazarsfeld, Fiske assumes that we start with a general and probably vague concept. The first step is to identify the theoretical context of this concept. Often, concepts are taken from an existing general theory, which provides the theoretical background. Fiske (1971, p. 97) favors this approach because it builds on what has already been worked out conceptually and on previous empirical research. The researcher's task here is to clarify the theoretical background and the way the concept is extended to fit a specific research problem. Researchers may also formulate their own concept, in which case they must elaborate the theoretical context and demonstrate the fruitfulness of the concept and its advantages over similar concepts.

The second step in Fiske's approach is to delineate the core of the construct, the unique quality to which the construct refers. The theoretical core is the essence of the concept, it must cover all phenomena that the concept refers to. This requirement often leads to a theoretical construct that is much too broad. Such a broad theoretical construct can better be considered as a general label that refers to a conglomerate of subconstructs. In practice, it will usually be preferable to examine the theoretical meaning of the subconstructs, and construct separate measures for each of these. Fiske describes several strategies for defining the core meaning, which can be subsumed under the heading *differentiation*. Thus, it is important to state explicitly what is *not* included in the construct, and the differences with other related constructs. Also, we may attempt to describe the opposite pole of the construct, and decide whether this defines the low end of the scale, or whether it introduces a new quality. For instance, if we want to measure "life satisfaction," is the low end of the scale defined by absence of satisfaction, or by presence of much dissatisfaction?

The third step in Fiske's approach is to construct a measurement instrument. For this, it is helpful to make a number of distinctions before the construction process starts. One is the modalities across which the measure is assumed to generalize: may we expect consistency across self-description, description by others (proxy-measurement), and observation? Guttman's facet design (described in more detail later) can be used to make systematic distinctions within modes, for instance by distinguishing different situations or behavior types. Fiske also recommends explicit decisions about measurement scales and measurement designs.

Evidently Fiske's approach has much in common with Lazarsfeld's. Lazarsfeld's subdimensions resemble Fiske's subconstructs. Lazarsfeld does not make clear why we would want to distinguish subdimensions. Fiske relates this to the research problem. In applied research, we may have a highly specific purpose, for which constructs with limited meaning are adequate. In basic research, we may want to use more general constructs, especially in the early stages of research. Later, we may want to specify subdimensions to achieve higher precision. The choice depends on the research problem and the state of knowledge in a given field, and may also be made on pragmatic grounds.

Although the dimensional/indicator analysis approach appears complicated, it is not difficult to find examples in survey research. An insightful discussion of the specification of several key concepts in social research can be found in Burgess (1986). A good example is the study by Andrews and Withey (1976, pp. 11–14) on subjective well-being. In a general sort of way, we know what well-being is. For some research purposes, we might measure it by asking respondents a single question about their well-being. In general, we would prefer to be more precise. Andrews and Withey proceed by discussing a conceptual model for well-being that distinguishes between different domains of and different criteria for well-being.

Andrews and Withey's domains are based on pilot research about people's concerns, aimed at finding life-domains that are distinguishable and cover most of people's concerns. The result is a list of domains, such as housing, job, family life, and neighborhood. The criteria are the values, standards, or goals by which one judges how one feels about the various domains. Their criteria are further differentiated into absolute and relative criteria, and general and specific. Examples of criteria are standard of living, independence, fun, and safety. Figure 2.1 illustrates a specific conceptualization (cf. Andrews and Withey, 1976, p. 234). The final instrument design also includes considerations about the type of response scales to be used. Given all these distinctions, Andrews and Withey were able to construct a large number of survey questions that tap different aspects of subjective well-being, and to relate these both theoretically and empirically to the theoretical constructs. An example of a question generated by this design is "How do you feel about the beauty and attractiveness of your house?" (Andrews and Withey, 1976, p. 13).

2.3.1.2 Semantic Analysis

Sartori (1984) notes that conceptualization is mediated by language, which makes it important to study the semantic structure of our statements. Verbal theories take the linguistic system through which they are expressed for granted. An important aspect of the linguistic system is the vocabulary. Theories use an existing vocabulary that reflects a global perception of the world. This is important, because what is not named is often not noticed, and the choice of

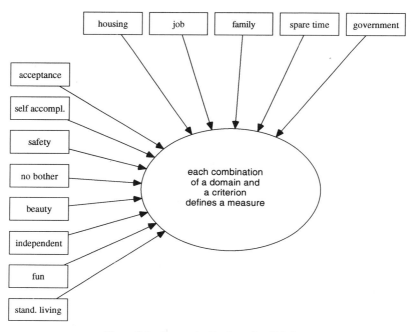

Figure 2.1   Conceptualization of well-being.

the name involves a sometimes far-reaching interpretation. Thus, it is important to examine the semantic network surrounding our theoretical constructs.

The meaning of a theoretical construct can be defined in two general ways. First, we have the semantic *connotation* of the construct, loosely described by Sartori (1984 p. 78) as the associations it has in the minds of the users, or the list of all characteristics included in the construct. Second, we have the semantic *denotation* of the construct, for instance by listing the set of objects to which it refers. Social scientists usually refer to these two definitions as constitutive and operational definitions.

A theoretical construct may be problematic in three ways. First, there may be problems with the connotation. There may be confusion of meanings because the construct is associated with more than one meaning, or two constructs point to the same meaning, and researchers have not made explicit which one(s) they intend. Second, there may be problems with the denotation: it may be vague to which objects or referents the construct applies. And third, there may be terminological problems: we may have chosen a label for the construct that induces us to refer to the wrong characteristics or referents.

The three types of problem must be attacked in the right order. As a first step, we must clarify the constitutive definition of the construct. This is not the same as the dimensional analysis proposed by Lazarsfeld, which is a substantive (theoretical) analysis. It comes closer to the analysis of the core meaning of (sub)constructs proposed by Fiske, which also has a strong semantic component. Sartori (1984, p. 32) describes it as assessing the *defining* (essential) characteristics of a construct, as opposed to the *accompanying* (variable) characteristics. The constitutive definition should be both adequate and parsimonious. It must be adequate in containing enough characteristics to allow a satisfactory delineation of the empirical boundaries and referents (the denotative definition). On the other hand it must be parsimonious in containing no unnecessary accompanying characteristics. One strategy Sartori proposes is to collect characteristics of existing connotative definitions (either from a dictionary or from previous research) and abstract a common core from these. However different various constitutive definitions may seem, the common core of defining properties may be manageably small. Important checkpoints are the occurrence of homonyms and synonyms. Constructs that have more than one meaning should be split, so we have one word for each meaning. Having more than one construct for the same meaning is less damaging, but should be avoided because it encourages ambiguity.

The next step is determining the empirical referents of the construct. This is not identical to providing an operational definition, because we may include empirical referents for which we do not have a measurement operation (yet). The essential part here is the establishment of boundaries and the decision concerning which empirical referents belong to the construct, and which not. Again, the issue is not the measurement operation or classification to be performed in the field. This is solved by specifying an operational definition, with misclassifications to be treated as measurement error. The issue is establishing in principle what properly belongs to the theoretical construct. Difficulties with the empirical boundaries or the empirical referents often point to deficiencies in the constitutive definition of the construct. The problem often is that the core of the construct is not defined with sufficient precision. The obvious solution would be to restrict the meaning of the construct by increasing the number of defining characteristics of the construct. Yet, as Lakatos (1976) shows, *concept stretching* is in many respects a more innovative way to deal with problems concerning the content of a theoretical construct, provided that the consequences of changing the construct are carefully examined.

The third step is to make sure that the verbal label for the construct is understood unequivocally. One consideration is the level of abstractness of the chosen term. Highly abstract terms often have few distinctive characteristics, and refer to a large ensemble of empirical referents. Sartori advises to select terms that do not disturb the semantic relationships with other constructs. A convenient way to evaluate the terminology is the substitution test: if in a constitutive definition, a word can be substituted by another word with a gain in clarity or precision, then the first word is being misused.

An example of semantic analysis is the Riggs (1984) analysis of the construct "development." Riggs starts with a lexical analysis, which shows that "to develop" originally referred to the opposite of "envelop." The modern meaning is "bring out all that is potentially contained in something" (Oxford English

Dictionary). In social science the meaning is more specific: evolution of human systems to more complex, mature, or higher forms, or in political science: transformations of political systems, notably at the macro level. An analysis of earlier definitions of "development" in political science (23 different definitions) leads to the conclusion that in political science the construct "development" is so overloaded with overlapping meanings that it is virtually useless. In a series of definitions "development" is changed into such constructs as "change," "growth," "modernization," and many more. In some instances this improves the original definition by making it less ambiguous.

Semantic analysis does not directly lead to survey questions. However, it does help to disentangle different meanings and to recognize ambiguity in our constructs. Since semantic analysis works with language, it is language dependent. This can be an advantage, for instance in examining the meaning of translated measures in cross-cultural research. For example, a semantic analysis of "happiness" in Dutch leads to an examination of the concept of "fortunate," because the Dutch word *geluk* combines the meanings of the English words "happy" and "lucky." Thus, a Dutch questionnaire about "geluk" could well contain questions about beliefs on being "lucky," while an American questionnaire about "happiness" would be unlikely to contain such questions.

2.3.1.3 Facet Design

A useful device for the systematic analysis of a construct is Guttman's facet design (Guttman, 1954). Facet design defines a universe of observations by classifying them with a scheme of facets with elements subsumed within facets. Facets are different ways of classifying observations, the elements are distinct classes within each facet. The universe of observations is classified using three kinds of criteria: (1) the population facets that classify the population, (2) the content facets that classify the variables, and (3) the common range of response categories for the variables.

For our present goal, we concentrate on the facet structure of the variables. The various content facets can be viewed as a cross-classification, analogous to an analysis of variance design, that specifies the similarities and dissimilarities among questionnaire items. Each *facet* represents a particular conceptual classification scheme that consists of a set of elements that define possible observations. The common *response range* specifies the kind of information required. For instance, attitude questions typically have a response range formulated as: very negative . . . very positive, and value questions typically have a response range formulated as: very unimportant . . . very important.

The content facets must be appropriate for the construct that they define. In selecting the most appropriate facets and elements, the objective is to describe all important aspects of the content domain explicitly and unequivocally. For example, for many constructs it may be useful to distinguish a behavior facet that defines the relevant behaviors, and a situation facet that defines the situations in which the behaviors occur. In facet design, the facet structure is often verbalized by a mapping sentence, which describes the observations in

Table 2.1 Example of Mapping Sentence

"To what extent does person (X) feel that

Source		Reason	
(her own experience)	led her to	(feel healthier	)
(her husband)	believe that	(feel fitter	)
(her doctor)	she would	(be more physically attractive	)
(the media)		(have fewer clothing problems	)
		(suffer less social stigma	)
		(be less anxious in social situations)	
		(feel less depressed	)

Response

if she lost weight, as rated	(not really at all)
	(not very much)
	(to a slight degree)
	(to a fair degree)
	(quite a lot)
	(very much)
	(very much indeed)

where (X) are married women attending slimming groups"

(Source: Gough, 1985, p. 247)

one or more ordinary sentences. An example of a facet design with mapping sentence is Gough's (1985, p. 247) mapping sentence for reasons for attending weight reduction classes, which is given in Table 2.1.

In this facet design the first facet (source) refers to the source of the belief, and the second facet (reason) refers to a specific consequence of losing weight. A facet design as given above can be used to generate questionnaire items. The (X) facet points to a target population of individuals. The source facet has four elements, the reason facet has seven, which defines $4 \times 7 = 28$ questions. For example, combining the first elements of the source and reason facets leads to the survey question "Did your own experience lead you to believe that your health would improve if you lost weight?" (Gough, 1985, p. 257).

Facet design is part of a more general approach called facet theory, which uses the facet structure to generate hypotheses about similarities between items, and for the analysis and interpretation leans heavily on geometric representations (Shye *et al.*, 1994; Borg and Shye, 1995). Facet theory contains no general guidelines to determine the need for introducing specific facets. Instead, it emphasizes that a facet design is a definition, which means that it should not be judged in terms of right or wrong, but whether it leads to productive research. Also, facet designs must be constructed with the research problem firmly in mind. Thus, as a device to specify the meaning of a construct, facet design assumes that we already have a good notion of the empirical domain we want to investigate. For instance, Gough's facet definition given above is derived from the literature and pilot research.

Usually, facet design is employed to specify the denotative meaning of a construct; the design specifies the set of empirical referents, and the actual survey questions are derived from the mapping sentence. However, the distinctions made in a dimension/indicator or semantic analysis can be introduced as distinct facets in a facet design, in which case the facet design also refers to the connotative meaning. In both cases, using a facet design encourages a systematic and exhaustive conceptual analysis. Facet designs can also be used to classify the questions in an existing instrument.

2.3.2 From Operationalization To Conceptualization: Data Driven Approaches

The data driven strategies described in this chapter are content sampling, symbolic interactionism, and concept mapping.

2.3.2.1 Content Sampling

An alternative approach to bridge the gap between theoretical construct and actual survey questions is to assemble a large set of questions which appear relevant to the research problem, collect data, and determine which questions co-vary and thus seem to measure the same construct. This is a thoroughly inductive approach. We assume that we know enough of the subject to be able to assemble questions that share one or more conceptual cores, and use factor analysis or comparable techniques to determine what these common cores are, and which questions measure them. Starting from data, we infer theoretical constructs, which may be altered in subsequent empirical research.

The data driven approach to measurement and conceptualization may seem rather superficial and *ad hoc*. Nevertheless, it can lead to useful standard scales, as exemplified by Schuessler's analysis of well-being scales (Schuessler 1982). Schuessler analyzed over 140 scales used in American sociology between 1920 and 1970, which contained over 9500 questions. He found that many of these scales were *ad hoc*, in the sense that they were used only once or twice. There were many overlapping items. A disturbing finding was that conceptually different scales could have several items in common. Altogether there were about 500 distinct questions. In an effort to construct a set of standard scales to measure well-being, Schuessler selected 237 items. A factor analysis on the responses of a sample of 1522 respondents revealed 17 factors, which in the end led to 12 scales based on 95 items.

As Fiske (1971) points out, if we use the results of a factor analysis to infer constructs and relationships between constructs, including factoring of subsets of questions to search for subfactors, we are working along lines very similar to the dimensional/indicator approach, only the starting point is different. The danger to avoid in this type of data driven approach is *reification* of the factors that are found. That is, factors should not be interpreted as fixed causal entities, but rather as convenient descriptive categories that summarize response consistencies (Anastasi, 1985). From a theoretical viewpoint, the interesting question is why these patterns arise. Verbal labels and interpretations of the factors depend on our explanation for the process that leads to these response contingencies.

In content sampling, it is essential not to miss any important aspects of the research domain, and consequently the usual approach is to assemble a *large* set of items. For instance, Cattell (1973) started the development of a mood scale by taking from the dictionary all words that can be used to describe a mood. A random sample of these words was used to construct questionnaire items, and subsequent research led to a mood scale with several subscales, such as "evaluation" and "arousal." Cattell *et al.* (1970) followed a similar approach in the construction of the Sixteen Personality Factors test (16 PFT). Because one starts with a large set of questions, factor analysis is often used to search for underlying dimensions. The goal of such an analysis is to identify sets of questions that measure the same construct. Both Fiske (1982) and Marradi (1981) have pointed out that if a set of questions measures the same construct, a factor analysis of this set should produce one single factor. So, the recommended analysis approach proceeds in two steps. In a first step, potential sets of questions are identified by exploratory factor analysis or similar means. In the second step, these sets are analyzed separately, to test whether a single factor is indeed sufficient.

In the previous paragraphs, "factor analysis" has been used as a general term for any technique designed to examine the similarities between questionnaire items in terms of dimensionality or scalability, as a first stage in data driven construct formation. As a technique, factor analysis has been criticized severely (e.g., Duncan, 1984) because it does not test dimensionality very strictly, and because it does not actually produce a measurement. Instead, item response models such as the Rasch model are preferred. Still, this does not affect the warnings given above against reification of latent variables and misinterpretation. Searching for sets of items that scale well according to some scaling model such as the Mokken or Rasch model merely identifies observed response consistencies. Theoretical interpretations of the scales depend on our explanation of why these response contingencies arise.

2.3.2.2 Symbolic Interactionism
The social sciences often use concepts that are taken from common language. These concepts often have a manifest function in our social life. As researchers, we may decide to attach a more precise meaning to such a concept to make it more useful in scientific theory (Fiske, 1971; Kerlinger, 1986). Still, if we aim to describe or explain the experiences and behaviors of individuals in society,

we should understand how these concepts are used and understood in social life. Thus, one approach to construct formation is to start with common language concepts as they are used by ordinary people in daily life, and bring these into the scientific discourse.

This approach is congruent with the social research paradigm known as symbolic interactionism. Symbolic interactionism views the social world as a world of interacting individuals, who are constantly negotiating a shared definition of the interaction situation (cf. Blumer, 1969). Symbolic interactionist research attempts to discover individuals' interpretations of social interactions, by developing concepts that describe these interpretations. This leads to a qualitative research strategy that focuses on the processes of interaction and interpretation. The starting points for conceptualization are *sensitizing concepts*; vague notions that do no more than point the research process in some promising directions. Qualitative field research gives these sensitizing concepts a more detailed empirical reference and clears up the connections between the various concepts.

This comes close to Sartori's aim to clarify the denotative and connotative meaning of theoretical constructs (Sartori, 1984). The difference is that Sartori mainly directs his attention to theoretical clarification by semantic analysis, and symbolic interactionists prefer to remain as close as possible to the individuals and their social worlds. This preference is well epitomized by the term "grounded theory" coined by Glaser and Strauss (1967).

Symbolic interactionist research often uses unstructured interviews and naturalistic observation. The data are the recorded material: field notes, interview texts, or observation protocols. Starting from the sensitizing concepts, the development and empirical specification of the concepts proceeds by examining the data and investigating which concepts and basic processes seem to be involved. During the analysis, new data may be collected to amass evidence about the value of new concepts. The output of this investigation is a file of "emerging concepts" that is updated constantly when new insights arise or new data come in. The process is stopped when the concepts appear *saturated*, that is, no new aspects of concepts can be found. The analysis is concluded by an effort to reduce the set of concepts to a smaller set of key concepts, which are then integrated in an overall theory by establishing their relationships.

Symbolic interactionism and other qualitative approaches have not evolved into a unified method of social research. However, they have led to methods that are also useful in developing concepts with a view of further operationalizing these into survey questions. The general procedure is to use open interviews to collect information from respondents, followed by an analysis as outlined above. Methods are available that bring rigor and reliability to the coding process (Miles and Huberman, 1994), and computer programs are available to support the analysis. The use of symbolic interactionist methods to construct survey questions is also discussed by Foddy (1993).

An example of the symbolic interactionist approach is Schaeffer and

Thomson's (1992) description of grounded uncertainty. They start with the observation that respondents in surveys express uncertainty in different ways, such as "don't know," "doesn't apply to me," or "it depends." It appears that such phrases point to two different types of uncertainty: state uncertainty (respondents are uncertain about their true state) and task uncertainty (respondents are uncertain how to communicate their true state). Schaeffer and Thomson use semi-structured telephone interviews with standardized open questions in a study of fertility motivation. Their analysis roughly follows the symbolic interactionist methods outlined above. They arrive at core concepts such as: neutrality, lack of clarity, ambivalence, indecision, and truly mixed expressions. In the end, they recommend to expand the Don't Know category in the closed questionnaire, and to measure the different types of state uncertainty directly.

Another example from a different field is Sternberg's research into the meaning of "intelligence" (Sternberg, 1985). Sternberg asked respondents to list typical "intelligent" and "unintelligent" behaviors. From these lists, 250 behaviors were selected. A factor analysis on a second sample produced three factors: "practical problem solving ability, verbal ability, and social competence" (Sternberg, 1985, p. 60). An interesting observation was that the factor structures of lay people and experts were very similar, which leads to the conclusion that it is possible to pose questions about intelligence to ordinary respondents.

2.3.2.3 Concept Mapping
A disadvantage of the symbolic interactionist approach is the cost in time and effort. Another weakness is its lack of structure, which makes it difficult to separate the respondents' contribution from the researcher's. The basic method, to have respondents formulate their own thoughts on a topic, can be applied in a more direct way. With a view to constructing questionnaire items, the symbolic interactionist research paradigm can be reduced to two leading principles: generation of statements and structuring of statements. The generation stage is explorative, and aims to produce a comprehensive list of statements that ideally should span the entire domain of interest. The structuring stage seeks to reduce the statement list to a smaller list of key concepts subsuming the former.

Focus groups have often been used to find out what concepts respondents use to describe their experiences or reactions. In such focus groups, carefully selected groups of respondents discuss a central topic, and examine it from several different points of view. Typically, focus groups are composed of individuals from the target population. The sampling criterion is not so much *representativeness* as *variance*; a wide variety of individuals is included. Sometimes specific groups are used, such as people who have certain situations or experiences in common. The discussion in the focus group is recorded, and analyzed later to extract statements about the domain of interest.

The statement list can be structured by the researchers, or by collecting more data from the respondents. If the number of statements is small, all pairwise combinations can be presented to respondents who are asked to assign a similarity rating to each pair. The resulting similarity matrix can be subjected to multidimensional scaling or other similarity analyses to determine how the statements cluster into more abstract concepts. Since with k statements, the number of paired comparisons is $(k \times (k - 1))/2$, paired comparisons are not attractive with more than about ten statements. If the number of statements is large, a sorting task is often used to examine the empirical structure of the statements. In a sorting task, the respondents must sort the statements in different clusters of mutually similar statements. The sorting task can be completely unconstrained, in the sense that the respondent simply must put the statements in a number of different clusters, but has complete control over the number of clusters and the number of statements per cluster (usually it is not permitted to put all statements in one cluster, or to produce as many clusters as there are statements). The sorting task can also be constrained, as for example, in natural grouping analysis. Natural grouping analysis starts with all statements in one large group. The respondent must split this group in two, and explain how the two new groups differ. Next, one of the available groups must be split further, and again the respondent must explain how the two new groups differ. This process is repeated until the respondent feels unable to make further splits. By tabulating how often statements are grouped together, it is possible to construct a similarity matrix, which can then be subjected to some kind of similarity analysis.

Various approaches have been proposed to systematize concept mapping, for an overview see Jonassen *et al.* (1993). Trochim (1989) has developed a comprehensive system for concept mapping in focus groups, which involves six steps: (1) developing the focus, (2) statement generation, (3) statement structuring, (4) statement representation, (5) concept map interpretation, and (6) utilization of concept maps. In step 1 "developing the focus" one must specify the primary focus or domain of interest, and select the participants for the focus group. Step 2 "statement generation" is a brainstorming session with the focus group to generate statements that describe all relevant aspects of the domain of interest. Trochim (1989) gives as an example a brainstorming session with as focus ". . . the issues, problems, concerns or needs which the elderly have . . ." (Trochim 1989, p. 4). In step 3 "structuring" the individual participants are instructed to sort the printed statements into different piles "the way that makes sense to you." The participants are also asked to rate the statements as to importance. The individual sorts are combined in a group similarity matrix. Step 4 "statement representation" subjects this similarity matrix to a multi-dimensional scaling, and superimposes a cluster solution on the map to facilitate interpretation. In fact, clusters may be viewed as "emerging concepts." It is also possible to overlay the importance ratings upon the concept map. An example of the concept map from Trochim's (1989) study about the elderly is given in Figure 2.2. Step 5 "concept map interpretation" is again a group activity. The participants discuss possible meanings and acceptable names for each statement cluster. They are also asked to see whether there are plausible groups of clusters

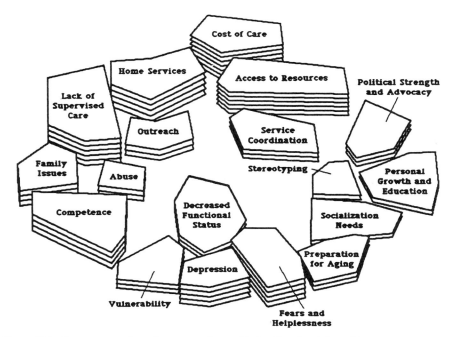

Figure 2.2 Concept mapping of concerns about elderly. (Reprinted from W.M.K. Trochim, An introduction to concept mapping for evaluation and planning, *Evaluation and Program Planning*, 12, pp. 1–16, © (1989), with kind permission from Elsevier Science Ltd., The Boulevard, Langford Lane, Kidlington OX5 1GB, U.K.)

or regions in the cluster space that can be named. Basically, this step attempts to identify relations between concepts in the form of a group-approved concept map. Step 6 "utilization" applies the final concept map to plan evaluations, interventions, *etcetera*. In our case, utilization would mean using the individual statements to translate the concepts into survey questions, and using the concept map to provide a first hypothesis as to the empirical structure of the questionnaire.

2.4 SUMMARY AND DISCUSSION

Theoretical constructs used in social science are often complex, and have an indirect relationship to the empirical observations they are designed to explain. The distance between a theoretical construct and observable phenomena creates the problem that researchers must state how they plan to measure what they are theorizing about. In the context of survey research, the link between theory and data is provided by formulating survey questions that are related to the constructs in the theory at hand.

This chapter presents two global strategies to conceptualization and question construction. The first strategy gives a logical priority to the theory and the constructs embedded in it. The researcher starts with the theoretical constructs, and their attached surplus meanings. To arrive at specific survey questions, it is necessary to clarify the surplus meaning of the construct. Three different approaches are described that guide this process. The first is dimension/indicator analysis. This approach has been advanced by many methodologists, such as Lazarsfeld and Fiske, and is implicit in the writings of many others such as Blalock, Cronbach, and Meehl. Many researchers probably view it as *the* approach. In dimension/indicator analysis we subdivide the central concept into subconcepts and specify empirical indicators for each subconcept. The second approach relies on semantic analysis to specify the various related meanings of a construct, and to find empirical referents for each subconstruct. The third approach is facet design, which lays out all aspects of a construct's content by specifying a facet structure and a common response range.

The second strategy gives logical priority to the empirical observations. Of course, there must be some notion of a central issue, but that can be vague. From the observations, the researcher must extract theoretical constructs. Three different approaches are described that guide this process. The first is content sampling. In this case, there is already a large set of questions that appears to relate to the central issue, or such a set can easily be assembled. Next, factor analysis or related methods are used to condense the large set of questions to a smaller set of theoretical constructs. The two other approaches rely on common language. The first adapts research methods derived from symbolic interactionism to the problem of construct formation. The other uses concept mapping methods, derived from cognitive research.

The merits and demerits of theory driven versus data driven strategies are discussed at length in Hox and De Jong-Gierveld (1990). Note that except for the case when we start with a set of questions to be explored, the various approaches do not directly produce survey questions. In the theory driven strategy, we end with a list of specifications which are (we hope) sufficiently tangible that the translation into survey questions is a comparably routine operation. In the data driven strategy, we end with a list of fairly concrete statements and their relationships to more abstract constructs. Compared to the theory driven approach, we have the advantage that we usually also have raw material that is condensed in these statements. This raw material often contains literal transcripts of respondents' verbalizations, which can often be turned into survey questions. Another advantage of the data driven strategy is that one can start with the respondent's understanding of a particular concept. However, with the data driven approach we run the risk of ending up with theoretical constructs that do not fit well into the theories that we wish to use.

In the end, researchers faced with the job of constructing valid measures may find comfort in Fiske's (1971) position that the actual choice for a specific strategy may not be all important. Constructing valid measures is not a one-time exercise. Rather, it is a continuous process of devising measures, investigating their theoretical and empirical meaning, and devising better measures. For a discussion of the importance of a systematic approach toward constructing measures for survey use see the chapter by Heath and Martin (Chapter 3). At the end of the process, the actual starting point is probably not crucial (cf. Oosterveld and Vorst, 1995). Important is that we must start at some sensible point and let *both* theoretical analysis and empirical tests guide our progress. In doing this, we shuttle back and forth between the theoretical and the empirical ends of a continuum.

The end product of concept formation can be characterized as a *prototype survey question*. The implication is that we still have to address the usual concerns of intended population, question wording, and so on. The issues involved in constructing the questionnaire and testing it in the cognitive laboratory and the field are treated in this section.

ACKNOWLEDGMENTS

I thank Edith de Leeuw, Anthony Heath, Jean Martin, Norbert Schwarz, and Arie van Peet for their comments on earlier versions. I thank the members of the Dutch Research Committee on Conceptualization and Research Design collectively for many stimulating discussions about the topics presented in this chapter.

REFERENCES

Anastasi, A. (1985), "Some Emerging Trends in Psychological Measurement: A Fifty-year Perspective," *Applied Psychological Measurement*, 9, pp. 121–138.

Andrews, F.M., and Withey, S.B. (1976), *Social Indicators of Well-being*, New York: Plenum.

Bridgman, P.W. (1927), *The Logic of Modern Physics*, New York: Macmillan.

Blalock, H.M. Jr. (1968), "The Measurement Problem: A Gap Between the Languages of Theory and Research," in H.M. Blalock, Jr., and A.B. Blalock (eds.), *Methodology in Social Research*, New York: McGraw-Hill, pp. 5–27.

Blalock, H.M. Jr. (1982), *Conceptualization and Measurement in the Social Sciences*, London: Sage.

Blumer, H. (1969), *Symbolic Interactionism. Perspective and Method*, Englewood Cliffs: Prentice-Hall.

Borg, I., and Shye, S. (1995), *Facet Theory: Form and Content*, Newbury Park, CA: Sage.

Burgess, R.G. (1986), *Key Variables in Social Investigation*, London: Routledge.

Campbell, D.T., and Fiske, D.W. (1959), "Convergent and Discriminant Validation by the Multitrait-multimethod Matrix," *Psychological Bulletin*, 56, pp. 81–105.

Carnap, R. (1956), "The Methodological Character of Theoretical Concepts," in H. Feigl, and M. Scriven (eds.), *The Foundations of Science and the Concepts of Psychology and Psychoanalysis*, Minneapolis: University of Minneapolis Press.

Cattell, R.B. (1973), *Personality and Mood by Questionnaire*, San Francisco: Jossey-Bass.

Cattell, R.B., Eber, H.W., and Tatsuoka, M. (1970), *Handbook for the Sixteen Personality Factor Questionnaire*, Champaign, IL: Institute for Personality and Ability Testing.

Chalmers, A.F. (1976), *What Is This Thing Called Science?* Milton Keynes: Open University Press.

Cronbach, L. (1971), "Test Validation," in R.L. Thorndyke (ed.), *Educational Measurement*, Washington, D.C.: American Council on Education, pp. 443–507.

Cronbach, L., and Meehl, P.E. (1955), "Construct Validity in Psychological Tests," *Psychological Bulletin*, 52, pp. 281–302.

De Groot, A.D. (1969), *Methodology. Foundations of Inference and Research in the Behavioral Sciences*, The Hague, NL: Mouton.

Duncan, O.D. (1984), *Notes on Social Measurement. Historical and Critical*, New York: Sage.

Fiske, D.W. (1971), *Measuring the Concepts of Personality*, Chicago: Aldine.

Fiske, D.W. (1982), "Convergent-discriminant validation in measurements and research," in D. Brinberg, and L.H. Kidder (eds.), *Forms of Validity in Research*, San Francisco: Jossey-Bass, pp. 77–92.

Foddy, W. (1993), *Constructing Questions for Interviews and Questionnaires*, New York: Cambridge University Press.

Glaser, B.G., and Strauss, A.L. (1967), *The Discovery of Grounded Theory*, Chicago: Aldine.

Gough, G. (1985), "Reasons for Slimming and Weight Loss," in D. Canter (ed.), *Facet Theory*, New York: Springer, pp. 245–259.

Guttman, L. (1954), "An Outline of Some New Methodology for Social Research," *Public Opinion Quarterly*, 18, pp. 395–404.

Hempel, C.G., and Oppenheim, P. (1948), "The Logic of Explanation," *Philosophy of Science*, 15, pp. 135–175.

Hox, J.J., and De Jong-Gierveld, J.J. (eds.) (1990), *Operationalization and Research Strategy*, Lisse, NL: Swets & Zeitlinger.

Jonassen, D.H., Beissner, K., and Yacci, M. (1993), *Structural Knowledge*, Hillsdale, NJ: Lawrence Erlbaum Associates.

Kerlinger, F.N. (1986), *Foundations of Behavioral Research*, New York: Holt, Rinehart & Winston.

Kuhn, T.S. (1970), *The Structure of Scientific Revolutions*, Chicago: University of Chicago Press.

Lakatos, I. (1976), *Proofs and Refutations*, Cambridge, UK: Cambridge University Press.

Lakatos, I. (1978), *The Methodology of Scientific Research Programmes*, Cambridge, UK: Cambridge University Press.

Lazarsfeld, P.F. (1958), "Evidence and Inference in Social Research," *Daedalus*, 87, pp. 99–130.

Lazarsfeld, P.F. (1972), *Qualitative Analysis. Historical and Critical Essays*, Boston: Allyn & Bacon.

Lundberg, G.A. (1939), *Foundations of Sociology*, New York: Macmillan.

Marradi, A. (1981), "Factor Analysis as an Aid in the Formation and Refinement of empirically useful concepts," in D.J. Jackson, and E.F. Borgatta (eds.), *Factor Analysis and Measurement in Sociological Research*, London: Sage, pp. 11–49.

Miles, M.B., and Huberman, A.M. (1994), *Qualitative Data Analysis: An Expanded Sourcebook*, Newbury Park: Sage.

Northrop, F.S.C. (1947), *The Logic of the Sciences and the Humanities*, New York: World Publishing Company.

Oosterveld, P., and Vorst, H. (1995), "Methods for the Construction of Questionnaires: A Comparative Study," *Proceedings of the International Conference on Survey Measurement and Process Quality*. Alexandria, VA: American Statistical Association, pp. 269–274.

Popper, K.R. (1934), *Logic der Forschung*, Wien: Springer. Reprinted as Popper, K.R. (1968), *The Logic of Scientific Discovery*, New York: Harper & Row.

Riggs, F.W. (1984), "Development," in G. Sartori (ed.), *Social Science Concepts*, Beverley Hills, CA: Sage, pp. 125–203.

Sartori, G. (1984), "Guidelines for Concept Analysis," in G. Sartori (ed.), *Social Science Concepts*, Beverley Hills, CA: Sage, pp. 15–85.

Schaeffer, N.C., and Thomson, E. (1992), "The Discovery of Grounded Uncertainty: Developing Standardized Questions About the Strength of Fertility Motivations," in P.V. Marsden (ed.), *Sociological Methodology* 1992, 22, pp. 37–82.

Schuessler, K.F. (1982), *Measuring Social Life Feelings*, San Francisco: Jossey-Bass.

Shye, S., Elizur, D., and Hoffman, M. (1994), *Introduction to Facet Theory*, Newbury Park, CA: Sage.

Sternberg, R.J. (1985), *Beyond IQ*, Cambridge, MA: Cambridge University Press.

Trochim, W.M.K. (1989), "An Introduction to Concept Mapping for Evaluation and Planning," *Evaluation and Program Planning*, 12, pp. 1–16.

Watson, J.B. (1924), *Behaviorism*, Chicago: University of Chicago Press.

CHAPTER 3

Why Are There so Few Formal Measuring Instruments in Social and Political Research?

Anthony Heath
Nuffield College, Oxford

Jean Martin
Office for National Statistics, U.K.

3.1 INTRODUCTION

Compared with disciplines such as education or psychology there are relatively few measuring instruments which command general assent from survey researchers in sociology and political science. There are a few well known ones such as Inglehart's index of postmaterialist values and Campbell's measures of political efficacy and party identification (which we discuss further below). But more often researchers use their own *ad hoc* measures, providing little description of how they were developed and frequently lacking any formal assessment or validation.

Moreover, even when measures do exist and are widely used, they rarely accord closely to the character that would be expected by psychometrics. The psychometric approach (see, for example, Nunnally, 1978; Zeller and Carmines, 1980; Thorndike, 1982; Spector, 1992) starts with the idea of an underlying theoretical concept or construct which cannot be directly observed and for which a measure is required. Questions (items) are used as imperfect indicators of the underlying concept—imperfect because they are subject to measurement error. In the classical measurement model underlying the psychometric approach

Survey Measurement and Process Quality, Edited by Lyberg, Biemer, Collins, de Leeuw, Dippo, Schwarz, Trewin.
ISBN 0-471-16559-X © 1997 John Wiley & Sons, Inc.

the errors are assumed to be uncorrelated, while the indicators themselves are assumed to be correlated by virtue of being measures of the same underlying concept of interest. (Extensions of the classical model allow for correlated errors.) Thus several items are normally used to measure one underlying concept. Indeed a psychometric measuring instrument or scale may well contain a very large number of items, such as the 60 items of the General Health Questionnaire (Goldberg, 1972) or the 26-item child behavior questionnaire developed by Rutter (1967). In this chapter we refer to instruments constructed according to such psychometric principles as formal measuring instruments.

The psychometric approach not only has an underlying theory of measurement, it also has widely accepted procedures for development and assessment. Development will typically begin with a large number of items which purport to cover the different facets of the theoretical construct to be measured; next, empirical data are collected from a sample of the population for whom the measure is intended. After appropriate analysis of this data, a subset of the original list of items is then selected and becomes the actual multi-indicator measure. This measure will then be formally assessed with regard to its reliability, dimensionality and validity, and it would be usual to have a published report of the measure's performance on these criteria.

In contrast, measures of social and political concepts do not usually go through the same process of development and assessment that is expected in psychometrics, and published reports of their reliability and validity are rare. Moreover, social science instruments tend to be very much shorter than psychometric ones, and even a six-item measure is unusual. Some of the best known instruments in social science are effectively single-item ones. To be sure, a single-item measure (such as that of party identification) which had no formal development or assessment at the time it was devised may subsequently prove successful (Green and Shikler, 1993), but in general there are powerful reasons for preferring a multi-item measure. Blalock for example has argued that

> With a single measure of each variable, one can remain blissfully unaware of the possibility of measurement [error], but in no sense will this make . . . inferences more valid. . . . In the absence of better theory about our measurement procedures, I see no substitute for the use of multiple measures of our most important variables." (Blalock, 1970, p. 111)

The central aim of this chapter is to explore some of the reasons social science practice has made relatively little use of psychometric principles, and to consider how justifiable these reasons are. We address problems in four areas: the nature of the concepts to be measured, the development process, the instrument, and the assessment of validity. We conclude that social and political scientists have in effect been making rather different trade-offs between psychometric principles and other desiderata, and we raise the question of whether these trade-offs should be reconsidered.

The primary focus of this chapter is on instruments to measure social and political attitudes, values and related "subjective" phenomena, for which empirical measures are more or less problematic to obtain. However, many of the principles and problems apply equally well to measures of other types of complex concepts (see, for example, Schwarz, 1990; Schwarz and Sudman, 1994).

3.2   THE NATURE OF THE CONCEPTS TO BE MEASURED

One reason for the relative absence of formal measuring instruments in social and political science, we suggest, is the paucity of clearly specified underlying concepts and the widespread concern of researchers with topical issues which they take to be directly measurable. The psychometric approach assumes that the research begins with a theoretical concept which cannot be directly measured. As Spector has put it, these concepts are often "theoretical abstracts with no known objective reality . . . [they] may be unobservable cognitive states, either individual (e.g., attitudes) or shared (e.g., cultural values). These constructs may exist more in the minds of social scientists than in the minds of their subjects" (Spector, 1992; p. 13). A well known example of an abstract social science concept of this kind is Inglehart's concept of postmaterialism (Inglehart 1971, 1977). The term postmaterialism is a neologism and it is doubtful whether many respondents could accurately identify its meaning. Inglehart himself defines postmaterialists as people for whom "values relating to the need for belonging and to aesthetic and intellectual needs, would be more likely to take top priorities [relative to economic security and to what Maslow terms the safety needs]" (Inglehart, 1971; p. 991-2; note that in his original 1971 paper Inglehart termed these post-bourgeois values). We would not expect the respondents to our surveys to have any direct apprehension themselves of the concept of postmaterialism, or at least not an apprehension which bore much resemblance to that of Inglehart. In Spector's terms, this is a concept which exists more in the minds of the social scientists than in the minds of their subjects.

If we inspect major surveys such as the U.S. National Election Survey or the U.S. General Social Survey, or the British Election Survey and British Social Attitudes Survey, we are likely to find a few instruments measuring underlying theoretical abstractions such as postmaterialism, political efficacy or party identification. However, clearly defined abstract concepts in social science are relatively rare. In many surveys we find a much larger number of attitudinal items towards specific topical issues—issues of which the respondent is assumed to have a direct apprehension. These will often be grouped in "batteries" but we should not confuse such batteries with formal measuring instruments. They are usually simply groupings of items which can be administered in the same format and which are conceived as covering a variety of specific issues in a more general domain rather than as tapping different aspects of an underlying construct. For example, in the British Election Survey the following battery was introduced in 1979 and has been asked (with modifications) ever since

> Now we would like to ask your views on some of the general *changes* that have been taking place in Britain over the last few years. Thinking first about the *welfare benefits* that are available to people today, would you say that these have gone much too far, gone a little too far, are about right, not gone quite far enough, not gone nearly far enough?
>
> Next, using one of the answers on the card, how do you feel about the attempts to ensure equality for women?
>
> Now, using one of the answers on the card again, how do you feel about the right to show nudity and sex in films and magazines?
>
> And how do you feel about people challenging authority?
>
> And how do you feel about recent attempts to ensure equality for coloured people?
>
> How do you feel about the change towards modern methods in teaching children at school nowadays?

While these items could perhaps be treated as a multiple item instrument to measure a single underlying dimension of attitudes towards change, we believe that they were not conceived in this way, nor are there published reports assessing the dimensionality, reliability and validity of the battery as a whole. Instead, these items are assumed to be of individual interest in their own right. Issues such as welfare benefits, equality for women and Blacks were believed to have been ones which potentially may have influenced voters in their choice of political party in 1979, and it is the actual role of these individual issues in which the researchers seem to have been interested (Sarlvik and Crewe, 1983).

To be sure, an item could serve both functions simultaneously. As DeVellis (1991, p. 8) has suggested, measures of voter preferences may be used simply as an atheoretical indicator of how a person will vote at an election or as a way of tapping the theoretical dimension of liberal/conservatism. Individual items of this kind are sometimes used in subsequent analysis to form multiple item scales (as discussed further in the next section), but they do not appear to have been designed originally with this in mind.

While widespread, the use of individual items to measure opinions directly may nonetheless be questioned. In these cases measurement error is effectively ignored and a one to one correspondence between the researcher's and the respondent's interpretation is assumed. However, the vast literature on question wording effects, context effects, interviewer variance and other types of measurement error suggests that these individual items will not be altogether unproblematic. Is it only, as Blalock suggests, "our most important variables" that need multiple-item measures?

3.3 DEVELOPING EMPIRICAL INDICATORS

As we noted at the beginning, the psychometric approach begins with a theoretical concept for which a measure is required. To obtain the measure, a relatively large number of items are designed, from which a smaller subset is eventually selected on the basis of formal criteria. In sociology and political science, on the other hand, it is not uncommon for the researcher to begin with a set of items which have already been included in a survey; the researcher may then use a technique such as factor analysis to determine whether any theoretically interpretable dimensions underlie the available items (see, for example, Witherspoon and Martin, 1993). The "gone too far" battery described above could well be treated in this way, and factor analysis of the items might well pick up more than one dimension. Or the researchers may ransack an existing survey in order to find items which they believe tap a theoretical concept of interest. Marshall and his colleagues, for example, constructed a class consciousness scale in this way from existing items in the international class structure survey (Marshall *et al.*, 1988). Perhaps not surprisingly, such expedients do not always lead to scales with all the desired properties, and the class consciousness scale, for example, has been criticized (Evans, 1992).

In many cases, however, social and political scientists have little choice but to proceed in this way. The conventions of their disciplines emphasize the importance of analyzing large-scale representative samples of the population, but few researchers have direct influence over the content of such surveys, which tend to be controlled by small groups of high-ranking scholars in the field. Younger scholars, therefore, have to be content with re-analysis of the existing items. In contrast, in psychological research national random samples do not have the same dominant position and small-scale research, often on nonrepresentative samples, can obtain high prestige. The General Health Questionnaire, for example, was validated on 200 consecutive attenders at a suburban London general practice (Goldberg and Blackwell, 1970). (We should remember that ideally in the psychometric approach the instrument should be administered to a representative sample of the population in question to establish norms, but this is more common in educational research.)

Even where a development process is feasible, it may not follow the psychometric pattern. Rather than beginning with a large number of items from which a formal instrument may be derived on quantitative criteria, the researcher may begin with qualitative interviews assessing the respondent's own conceptions of the concept at hand. This process will be appropriate where the aim is not so much to measure a concept "which exists more in the minds of the social scientists" but where the aim is precisely to tap respondents' own conceptions. In psychometrics, many concepts will have the former character, but in social sciences with their traditions of *verstehende Soziologie* researchers often also try to tap lay concepts. For example, sociologists have developed instruments to measure both "objective" and "subjective" social class. "Objective" social class is in fact something of a misnomer and refers to sociologists' constructs based on criteria such as market situation (Weber, [1920] 1978) or employment relations (Goldthorpe, 1980), and might be unrecognizable to the sociologists' subjects. In contrast, "subjective" social class refers to the subjects' own conceptions of the class structure and their position in it (Bulmer, 1975).

3.4 THE NATURE OF THE INSTRUMENT

As we noted earlier, even where explicit theoretical concepts exist in social and political science and formal instruments have been developed to measure them, the instruments are notable for their brevity in comparison with psychometric instruments such as the General Health Questionnaire. Campbell's measure of political efficacy illustrates this well. Campbell defined political efficacy as "The feeling that individual political action does have, or can have, an impact upon the political process, i.e., that it is worth while to perform one's civic duties. It is the feeling that political and social change is possible, and that the individual citizen can play a part in bringing about this change" (Campbell *et al.*, 1954; p. 187). The measure of political efficacy then takes the form of a Likert scale and can be regarded as a formal multi-item scale designed to measure a single underlying theoretical concept. It thus conforms in some respects to the psychometric approach, but it is much shorter, containing only the following five items

> I don't think public officials care much what people like me think.
>
> The way people vote is the main thing that decides how things are run in this country.
>
> Voting is the only way that people like me can have any say about how the government runs things.
>
> People like me don't have any say about what the government does.
>
> Sometimes politics and government seem so complicated that a person like me can't really understand what's going on.

To each item a simple binary "agree" or "disagree" response was allowed (although later users of the scale have employed four or five response categories). It should be noted that, although widely used, subsequent research has suggested that the scale may not be a unidimensional measure of personal efficacy and that it may be contaminated by respondents' perceptions of "perceived system responsiveness" (Marsh, 1977), a confusion that was perhaps already present in the original definition of the concept. (We should also note that the scale may suffer serious acquiescence effects; see Schuman and Presser, 1981.)

Similarly, Inglehart's widely used index of postmaterialist values is notable for its brevity. Respondents are simply asked to select their first and second choices from a list of four alternatives:

In politics it is not always possible to obtain everything one might wish. On this card, several different goals are listed. If you had to choose among them, which would be your first choice? Which would be your second choice?

- Maintain order in the nation
- Give people more say in the decisions of government
- Fight rising prices
- Protect freedom of speech

Postmaterialists are then taken to be respondents who give as their two choices "more say in the decisions of government" and "protect freedom of speech," while materialists are those whose two choices are "maintain order in the nation" and "fight rising prices." The remainder constitute a mixed group. The scale itself thus takes only three values. There is a longer and more time consuming version which Inglehart himself recommends because of its greater reliability (Inglehart, 1990; pp. 131–4), but this is rarely used in survey research.

Other measures of classic theoretical concepts are effectively only single-item measures. For example, Klingemann's measure of the left-right dimension takes the following form (Klingemann, 1972)

Many people think of political attitudes as being on the 'left' or the 'right'... When you think of your own political attitudes, where would you put yourself? Please mark the scale in the box.

Respondents are presented with ten points labelled "left" and "right" at either end and asked to choose a point to indicate their positions. The U.S. National Election Study instrument designed to measure positions on the liberal-conservative dimension takes essentially the same form, although with seven points not ten.

Similarly the Michigan measure of party identification contains only two questions

Generally speaking, do you usually think of yourself as a Republican, a Democrat, an Independent, or what?

[If Republican or Democrat] Would you call yourself a strong [PARTY CHOSEN] or not very strong [PARTY CHOSEN]?

[If Independent] Do you think of yourself as closer to the Republican or to the Democratic party?

Responses to the two questions asked of the respondents are often combined to give a single five or seven point scale running from strong Republican to strong Democrat, effectively giving a single item measure (since the individual has only one substantive value on the questions combined).

One of the main reasons for the brevity of these social science measures stems from the uses to which the measures are put and hence the amount of questionnaire space which can be justified. In psychology, measurement theory grew up around the development of tests of individual differences that were designed to produce scores for individuals, not just for groups. Decisions about individuals might be made on the basis of their test scores, for example, about their education, occupation, or treatment. So any deficiency in the measuring instrument or the way in which it was administered might have consequences for an individual. Thus the General Health Questionnaire was designed as a screening device for possible mental illness, as was Rutter's measure of children's behavior. (We should note that it is not uncommon with long scales such as this for shorter subscales to be identified. Rutter's behavior questionnaire, for example, contains neurotic and antisocial behavior subscales. Nevertheless, the overall scale is used to give a score which is taken to indicate the threshold for likely disorder.)

By contrast, measures in social and political science are normally used as research tools to provide measures of association between variables not decisions about individuals. In so far as they provide deficient measures, relationships between variables are distorted or diminished and their explanatory power reduced, but there are no consequences for individual respondents. (Survey research may also be used to provide group measures for planning purposes, and these tend to have greater resources devoted to their development.)

It also follows that research instruments in social science can never be the sole focus of a questionnaire, and will have to compete with the other variables of interest for questionnaire space. It may well be the sheer brevity of Inglehart's short index of postmaterialist values that accounts for its popularity among survey researchers. Much longer scales have been developed in the social sciences such as McClosky and Zaller's 28-item measure of capitalist values and their 44-item measure of democratic values (McClosky and Zaller, 1984), but to our knowledge they have rarely been used by other researchers. Incidentally, the same pressures are at work in psychology, too; even in clinical practice it is common to use a shorter 30-item version of the General Health Questionnaire, and for research purposes an even shorter 12-item version has been constructed.

While these considerations perhaps explain why social scientists will usually be obliged to construct much shorter scales than will psychologists, it is not clear that the drastic economy associated with single item measures such as the left–right scale is justifiable. As we noted earlier, the usual justification for a multiple indicator approach is that, in order to measure a complex theoretical concept which has no immediate meaning to respondents, more than one question needs to be asked. Now it may be that, when we are trying to measure respondents' own subjective conceptions (as Klingemann is attempting with his left–right scale), the conventional psychometric model is no longer so applicable. However, using only one question implies a one to one relationship between the question and the concept of interest. This can rarely be assumed, and certainly only in the case of concepts which are clearly understood by respondents. It is not at all clear that this can be assumed in the case of Klingemann's measure of "left" and "right."

In a methodological study of British respondents in 1986–87, we first administered Klingemann's left–right scale, and found that around 90 percent of respondents were willing to place themselves somewhere on the scale (suggesting that most respondents did in fact attach some meaning to the concepts). We then asked respondents the open-ended questions "What did you understand the word 'left' to mean?" and "What did you understand the word 'right' to mean?" Their answers suggested that only about half shared the political scientist's conception that left and right referred to positions on an ideological dimension. A substantial number took "left" to mean extreme and "right" to mean moderate. This was at a time when attention was largely focused on splits within the Labour party, whose right wing was generally reported by the media to be more moderate; when attention turns to divisions within the Conservative party, it may be that more respondents will believe that "right" means extreme. There were also a substantial number of respondents whose answers did not seem to correspond to any recognizable political science concept (Evans *et al.*, 1996). It would be interesting to know whether the same applied to the U.S. public's interpretation of the terms "liberal" and "conservative."

The case for single-item measures is not entirely persuasive, even where we are dealing with respondents' own conceptions rather than with theoretical abstractions, and it appears that on the standard criteria such as test–retest reliability or construct validity, multi-item measures of, for example, left–right values are more successful than Klingemann's single item measure (Heath *et al.*, 1994).

3.5 VALIDITY

Validity is generally defined in terms of various types of evidence that a measuring instrument is indeed measuring what it purports to measure. The literature on research methods generally distinguishes several different forms of validity—usually face, criterion, predictive and construct validity (although other terms are also used). The psychometric literature tends to emphasize criterion and predictive validity. However, typically in survey research few or at best rather weak attempts are made prior to the development of a measuring instrument to specify what would constitute a test of validity.

Face validity is generally considered the weakest form of validity, yet in the social sciences much emphasis is placed (at least implicitly) on the face validity of individual items and of an instrument as a whole. This is partly because, as explained above, questions are seen as providing information about respondents' opinions on particular issues in their own right, not just as indicators of an underlying construct, and partly because the concepts to be measured are often the respondents' own subjective conceptions rather than concepts which exist only in the minds of the scientists. For psychologists face validity is much less important; what matters is whether individual items "work" in terms of their statistical properties which demonstrate their contribution to the instrument as a whole, and this would appear to be wholly justifiable in the case of unobservable theoretical concepts "which exist more in the minds of social scientists than in the minds of their subjects." In sociology and political science, the fact that researchers are often attempting to measure what might be termed "lay" concepts may make face validity more compelling. For example, if sociologists wish to identify respondents' own conceptions of their place in the class structure, lack of face validity might lead one to doubt whether the instrument could possibly tap the respondents' own conceptions. Even in this case, however, we would regard face validity as a necessary, not a sufficient, condition for an acceptable instrument.

Criterion and predictive validity are given great importance in the field of test development. It is clear that if tests are being used to make decisions about individuals, evidence of their validity in relation to external criteria is crucial. In fields such as medicine or education this often takes the form of demonstrating that the test results agree, say, with the judgement of experts. The General Health Questionnaire, for example, was validated against the Present State Examination, a detailed interview following explicit rules and conducted by a trained psychiatrist.

Both criterion and predictive validity are problematic in the social sciences. Indeed in some cases it is far from clear what the test of criterion validity would be. In the case of unobservable theoretical concepts, it appears to be rare for external criteria—alternative measures of the same concept from an external source—to exist against which a new measure could be validated. In the case of postmaterialism, for example, it is unclear how we are to recognize a postmaterialist other than by responses to the items in the index. This is in contrast to, say, respondents' reports of whether they voted in the last election which are, at least in principle, subject to direct empirical verification, for example by checking self-reports against the official records (Swaddle and Heath, 1989). It may be that for some unobservable concepts there can in principle be no direct measure of criterion validity but only indirect tests of predictive validity via an explicit theory. We would therefore wish to repeat Fisher's dictum "make your theories elaborate" (made in the context of assessing causality and quoted in Cox, 1992; p. 292).

Predictive validity also appears to be problematic in the social sciences, given the shortage of explicit theories about the determinants of behavior. For example, if the measure is of an underlying attitude and attitudes are seen as predispositions to act, the measure might in principle be validated against evidence of relevant behavior. However, other factors than attitudes affect behavior so lack of action congruent with the attitude in question cannot necessarily be viewed as evidence of lack of validity. The "pale green" respondent who lives near a bottle bank may find it easier in practice to recycle bottles than the "dark green" respondent in the countryside who would have to use nonrenewable fossil fuel to drive to a recycling depot, so the failure of a scale of green attitudes to predict recycling might not indicate lack of validity. What is needed is a theory about the relationship between the unobservable concept and observable behavior. Without explicit theory, predictive validation of such concepts is difficult.

Given the perceived difficulties with predictive and criterion validity, researchers in the social sciences who are concerned with the validity of their measures tend to concentrate on construct validity when they conduct an explicit assessment of validity. According to Zeller and Carmines "Construct validation involves three distinct steps. First, the theoretical relationship between the constructs themselves must be specified. Second, the empirical relationship between the measures of the concepts must be examined. Finally the empirical evidence must be interpreted in terms of how it clarifies the construct validity of the particular measure" (Zeller and Carmines, 1980; p. 81).

Social scientists often bypass the first step and proceed directly to examine empirical relationships between the measure and other variables with which it is expected to relate. These are rarely alternative measures of the same construct, and so the relationships between them would not be expected to be strong. For example, a measure of left–right attitudes would be expected to distinguish between members of the Labour and Conservative parties. Or it may take the form of examining the demographic antecedents of the characteristics in question. Postmaterialists, for example, according to Inglehart's theory, would be expected to come from relatively affluent middle class backgrounds.

As we noted earlier, there are relatively few concepts in social and political science which are embedded in explicit theories. If a concept is not part of a well developed theory, then the opportunities for testing construct validity will be reduced. The researcher may offer some general expectations about the variables with which they expect their measure to be correlated. For example in their attempts to validate measures of political knowledge Delli Carpini and Keeter (1993) and Martin *et al.* (1993) suggest that greater political knowledge will be associated with higher levels of education, but there is no formal theory of political knowledge from which a number of more precise tests could be derived. Nor is there any standard criterion for assessing whether a correlation is high enough to establish construct validity. It is difficult to specify in advance the strength of relationships with other variables that would be acceptable as evidence of validity. Whereas there is a rule of thumb that, in measuring reliability, Cronbach's alpha should be at least 0.6 (depending on the number of items in the instrument), there is no comparable rule of thumb with correlations. Indeed, given the well known statistical limitations of measures like the correlation coefficient, there may be cases where a low correlation is acceptable and a high one is not.

We would argue, therefore, that correlations with background variables which are not intended to measure the concept of interest provide only weak evidence of construct validity. Virtually all measures successfully clear the hurdle, and therefore the weeding out of unsatisfactory instruments, and the selection of superior instruments, fails to take place. Without more stringent tests, it is relatively easy for the researcher to defend the use of *ad hoc* measures, and the pressure to use instruments that have been through the full psychometric program of development and assessment is thereby reduced.

We suggest that what is needed is a comparison between alternative measures of the same concept. The kind of procedure we have in mind is illustrated by the comparison between different measures of party identification reported by Abramson and Ostrom (1994). While the Michigan formulation has been the dominant one, there have been alternative ones such as the Gallup formulation. Abramson and Ostrom show that the different specifications have different properties, with the Michigan measure demonstrating more of the properties that the theory had advanced.

3.6   CONCLUSIONS

We have suggested a number of reasons why formal measuring instruments following psychometric principles are rare in sociology and political science:

- Sociology and political science are not disciplines with large numbers of abstract concepts embedded in formal theories; in many cases there are no underlying concepts and individual survey items are taken to be of interest in their own right.
- Sociology and political science are, however, disciplines which attach considerable importance to the analysis of (expensive) large-scale representative surveys; most researchers cannot directly influence the content of these surveys and they therefore construct measures from existing items rather than following the development stages of the psychometric approach.
- Moreover, sociology and political science, unlike psychology or education, are not used to make decisions about individuals but to investigate relationships between variables; hence fewer resources can be justified on any one instrument.
- Finally, social and political sciences offer few opportunities for rigorous validation and hence "anything goes"; most measures, whether constructed on psychometric principles or not, can easily satisfy the rather weak criteria of face and construct validity that are usually all that are expected.

In some of these cases we can see that the trade-offs involved in sociology and political science are rather different from those in education or psychology. First, given a fixed budget, there are resource trade-offs. For example, if we use resources to support preliminary methodological work developing appropriate instruments, we are bound to have fewer resources to devote to the final survey itself. Given the emphasis that social and political scientists place on large-scale representative surveys, which are very expensive to conduct, the natural corollary is that they devote fewer resources to methodological development. It is notable that one major national survey which devotes considerable resources to methodological development—the U.S. National Election Study (US NES)—is singularly well resourced. But it also has its critics who would argue that its sample size is far too small for the proper analysis of a society as diverse as the U.S.A. and that the US NES has to that extent made the wrong trade-off. The dominance of national representative surveys could also be criticized: it is, for example, arguable that the small local studies of electoral behavior which were conducted by the Columbia school (Lazarsfeld *et al.*, 1948; Berelson *et al.*, 1954) made contributions to political science as great as those of the Michigan school based on their national representative surveys.

Second, there is a trade-off between cost/questionnaire space and the length of the scales. Since social scientists are generally concerned to establish relationships between variables rather than to make decisions about individuals, it is perhaps natural that the social scientist wishes to include more, but shorter, scales. Thus the short postmaterialism and party identification measures are still widely used despite numerous criticisms. Inglehart's longer index can be shown to be greatly superior in terms of its test–retest reliability and its construct validity (Heath *et al.*, 1994), but is nevertheless rarely included. However, whether social scientists have made the right trade-offs in preferring the short Inglehart index can be questioned. If we are concerned with relationships betweeen variables, to include in the same analysis a variable measured with high reliability along with a variable such as the short Inglehart index measured with low reliability could seriously bias the results.

Third, in addition to these major trade-offs between psychometric sophistication and other desiderata, there are also trade-offs that have to be made within a psychometric model of scale development. There is a resource trade-off, for example, between development and assessment. Our review has suggested that social science measures are rarely validated against external criteria, cheaper but "softer" methods of obtaining internal assessments of construct validity typically being preferred. Should resources perhaps be diverted from the development phase (fieldwork administering new items) to the assessment phase (fieldwork assessing criterion validity)?

There are also technical trade-offs to be considered, for example between internal reliability and discriminatory power, or between reliability and acquiescence bias. The psychometric test construction literature lists a number of different technical criteria which are used to assess the properties of a measuring instrument. Only one is regularly quoted in the social and political science literature: Cronbach's alpha as a measure of internal consistency (and even this cannot be calculated for single-item instruments). But high values of alpha can be obtained with a scale that contains two dimensions (Cortina, 1993), while an unbalanced scale which suffers from acquiescence bias, such as Campbell's measure of political efficacy, may have much higher internal consistency than a balanced scale.

We would suggest that the decisions which social scientists routinely make about these trade-offs should be scrutinized. It may well be that too much weight is conventionally given to easily quantifiable criteria such as Cronbach's alpha and to procedures that are internal to the survey (e.g., correlations between survey variables to establish construct validity).

It is not our intention in this chapter to be prescriptive, much less to say that any particular research team has made the "wrong" trade-off. Indeed, it is by no means clear that there can be any simple yardstick for judging whether or not the "right" trade-off has been made. There is no simple criterion of intellectual profitability that can be used to judge trade-offs.

Nevertheless, trade-offs are implicitly being made and they should be openly scrutinized rather than taken for granted. As a first step, to increase the transparency of the implicit decisions, we believe that social researchers should provide a lot more information about the properties of their instruments than is currently generally the case. Starting with a clear description of the concept which is to be measured and an account of the criteria for its validation, details of how the concept was operationalized should be given. The information should include evidence of the dimensionality of scales, the contribution of each item to the overall measure and the technical properties of the measure as a whole, e.g., reliability, distribution of scores on the population for which it is intended, stability over time, information relating to measurement bias (or lack of it). Information about validity should include not only correlations with other variables of relevance but, if possible, comparison with other instruments for measuring the same concept and evidence of criterion validity. Not all this information will necessarily be available immediately for a new scale but indication of gaps yet to be filled can be useful.

We would also argue that there should be greater diversity and competition between measures and research strategies so that the consequences of different trade-offs become apparent. Vigorous competition rather than settled orthodoxy is more likely to promote scientific development. We do not therefore advocate the replacement of the current sociological orthodoxy by a psychometric orthodoxy. We would wish to emphasize that even within the psychometric approach there is scope for considerable diversity and competition.

REFERENCES

Abramson, P.R., and Ostrom, C.W. (1994), "Question Wording and Partisanship," *Public Opinion Quarterly*, 58, pp. 21–48.

Bellknap, G., and Campbell, A. (1952), "Political Party Identification and Attitudes Toward Foreign Policy," *Public Opinion Quarterly*, 15, pp. 601–623.

Berelson, B.R., Lazarsfeld, P.F., and McPhee, W.N. (1954), *Voting: A Study of Opinion Formation During a Presidential Campaign*, Chicago: University of Chicago Press.

Blalock, H. (1970), "Estimating Measurement Error Using Multiple Indicators and Several Points in Time," *American Sociological Review*, 35, pp. 101–111.

Bulmer, M. (1975), *Working Class Images of Society*, London: Routledge and Kegan Paul.

Campbell, A., Gurin, G., and Miller, W.E. (1954), *The Voter Decides*, Evanston, Illinois: Row, Peterson.

Cortina, J.M. (1993), "What Is Coefficient Alpha? An Examination of Theory and Applications," *Journal of Applied Psychology*, 78, pp. 98–104.

Cox, D.R. (1992), "Causality: Some Statistical Aspects," *Journal of the Royal Statistical Society*, series A, 155, pp. 291–301.

Delli Carpini, M.X., and Keeter, S. (1993), "Measuring Political Knowledge: Putting First Things First" *American Journal of Political Science*, 37, 4, pp. 1179–1206.

DeVellis, R.F. (1991), *Scale Development: Theory and Applications*, Applied Social Research Methods Series Volume 26, Newbury Park, CA: Sage.

Evans, G.A. (1992), "Is Britain a Class-divided Society? A Re-analysis and Extension of Marshall *et al.*'s Study of Class Consciousness," *Sociology*, 26, pp. 233–258.

Evans, G.A. Heath, A.F., and Lalljee, M. (1996), "Measuring Left–right and Libertarian–authoritarian Values in the British Electorate," *British Journal of Sociology*, 47, 1, pp. 93–112.

Goldberg, D. (1972), *The Detection of Mental Illness by Questionnaire*, Maudsley Monograph 21, Oxford: Oxford University Press.

Goldberg, D., and Blackwell, B. (1970), "Psychiatric Illness in General Practice: A Detailed Study Using a New Method of Case Identification," *British Medical Journal*, 1970, pp. 439–443.

Goldthorpe, J.H. (1980), *Social Mobility and Class Structure in Modern Britain*, Oxford: Clarendon Press.

Green, D.P., and Schickler, E. (1993), "Multiple-measure Assessment of Party Identification," *Public Opinion Quarterly*, 57, pp. 503–535.

Heath A.F., Evans, G.A., and Martin, J. (1994), "The Measurement of Core Beliefs and Values," *British Journal of Political Science*, 24, pp. 115–132.

Inglehart, R. (1971), "The Silent Revolution in Europe: Intergenerational Change in Post-industrial Societies," *American Political Science Review*, 65, pp. 991–1017.

Inglehart, R. (1977), *The Silent Revolution: Changing Values and Political Styles Among Western Publics*, Princeton: University of Princeton Press.

Inglehart, R. (1990), *Culture Shift in Advanced Industrial Society*, Princeton: Princeton University Press.

Klingemann, H.D. (1972), "Testing the Left–right Continuum on a Sample of German Voters," *Comparative Political Studies*, 5, pp. 93–106.

Lazarsfeld, P.F., Berelson, B., and Gaudet, H. (1948), *The People's Choice*, New York: Columbia University Press.

Marsh, A. (1977), *Protest and Political Consciousness*, London: Sage.

Marshall, G., Newby, H., Rose, D., and Vogler, C. (1988), *Social Class in Modern Britain*, London: Hutchinson.

Martin, J., Ashworth, K., Heath, A.F, and Jowell, R. (1993), *Development of a Short Quiz to Measure Political Knowledge*, JUSST Working Paper No. 21, London: SCPR.

McClosky, H., and Zaller, J. (1984), *The American Ethos: Public Attitudes Towards Capitalism and Democracy*, Cambridge, MA: Harvard University Press.

Nunnally, J.C. (1978), *Psychometric Theory*, 2nd edn., New York: McGraw-Hill.

Rutter, M. (1967), "A Children's Behavior Questionnaire for Completion by Teachers; Preliminary Findings," *Journal of Child Psychology and Psychiatry*, 8, pp. 1–11.

Sarlvik, B., and Crewe, I. (1983), *Decade of Dealignment*, Cambridge: Cambridge University Press.

Schuman, H., and Presser, S. (1981), *Questions and Answers in Attitude Surveys: Experiments on Question Form, Wording and Context*, New York: Academic Press.

Schwarz, N. (1990), "Assessing Frequency Reports of Mundane Behaviors: Contributions of Cognitive Psychology to Questionnaire Construction," in C. Hendrick, and M.S. Clark (eds.), *Research Methods in Personality and Social Psychology* (Review of Personality and Social Psychology, Vol 11), Beverly Hills, CA: Sage, pp. 98–119.

Schwarz, N., and Sudman, S. (1994), *Autobiographical Memory and the Validity of Retrospective Reports*, New York: Springer Verlag.

Spector, P.E. (1992), *Summated Rating Scale Construction: An Introduction*, Quantitative Applications in the Social Sciences 82, Newbury Park, CA: Sage.

Swaddle, K., and Heath, A.F. (1989), "Official and Reported Turnout in the British General Election of 1987," *British Journal of Political Science*, 19, pp. 537–570.

Thorndike, R.L. (1982), *Applied Psychometrics*, New York: Houghton Mifflin.

Weber, M. ([1920] 1978), in G. Roth, and C. Wittich (ed.), *Economy and Society: An Outline of Interpretive Sociology*, Berkeley, University of California Press.

Witherspoon, S., and Martin, J. (1993), *Environmental Attitudes and Activism in Britain: the Depth and Breadth of Green World-views*, JUSST Working Paper No 20, London: SCPR.

Zeller, R.A., and Carmines, E.G. (1980), *Measurement in the Social Sciences*, London: Cambridge University Press.

CHAPTER 4

Social Cognition and Responses to Survey Questions Among Culturally Diverse Populations

Timothy Johnson, Diane O'Rourke, Noel Chavez, Seymour Sudman, Richard Warnecke, Loretta Lacey
University of Illinois

John Horm
U.S. National Center for Health Statistics

4.1 INTRODUCTION

Recent years have seen a significant expansion in the variety and number of surveys conducted across national and cultural boundaries. The International Social Survey Program (ISSP) annually collects cross-national survey data from up to 20 countries, with data from the U.S. obtained through the General Social Survey. Fertility surveys sponsored by the United Nations and surveys of cross-national opinions are also common. Many of the methodological pitfalls inherent in uncritically applying survey procedures and instruments designed in one nation or culture to persons in other environments have been well documented (Brislin *et al.*, 1973). What is sometimes less appreciated is that research conducted within what has been referred to as "loose" (Triandis, 1990), or multicultural, societies are subject to many of these same problems. That is, countries like the United States and Canada present researchers with many of the same methodological challenges that were previously reserved for cross-national inquiries. We interpret culture to represent a social group with a shared language and set of norms, values, beliefs, expectations and life experiences.

Survey Measurement and Process Quality, Edited by Lyberg, Biemer, Collins, de Leeuw, Dippo, Schwarz, Trewin.
ISBN 0-471-16559-X © 1997 John Wiley & Sons, Inc.

The past decade has also witnessed an expanding dialogue between survey researchers and cognitive psychologists (Jobe and Mingay, 1991). Increasingly, cognitive methodologies and theoretical models are being employed to investigate the underlying thought processes respondents use when answering survey questions. These approaches have also been found effective in identifying sources of response error often missed using traditional survey methods and to gather information about possible reasons for these errors (Lessler *et al.*, 1989). As such, they suggest a potentially useful framework for assessing cultural differences and commonalities in responding to survey questionnaires (Hines, 1993). The focus of this chapter is an exploration of this issue; our discussion begins with an evaluation of the current state of knowledge regarding potential cultural variations in social cognition that are relevant to survey research, followed by a presentation of original data specifically collected to investigate the relationship between respondent culture and the cognitive processing of survey questions. We will conclude with a discussion of these findings and implications for future research.

4.2   CULTURE AND SOCIAL COGNITION

Although several cognitive models of the survey response process currently exist, it is generally understood that there are four basic tasks in responding to survey questions (Strack and Martin, 1987). These tasks include (a) question interpretation, (b) memory retrieval, (c) judgment formation, and (d) response editing. Cultural norms, values, and experiences may intervene to influence the outcome of each of these cognitive activities.

4.2.1   Question Interpretation

It is likely that cultural perceptions moderate the meanings and interpretations respondents assign to most survey questions. Persons acquire from their social and cultural groups a context for interpreting the environment that may largely define how they comprehend and deal with it (Angel and Thoits, 1987). For example, researchers have found evidence that the concept of pain is culturally learned (Morse and Morse, 1988; Zborowski, 1969; Zola, 1966). Similar findings have been reported regarding cross-cultural differences in other concepts, such as the expression of psychological symptoms (Mirowsky and Ross, 1980). Other examples of cultural differences in perceptions of depressive symptoms are also available (for reviews, see Kleinman and Good, 1985; and Marsella, 1978). It has also been suggested that the differentiation of psychologic and somatic dimensions of depression can only be found among persons in more industrialized cultures (Guarnaccia *et al.*, 1989).

Triandis (1990) has outlined several "cultural syndromes" that may be useful in developing an appreciation of how culture can influence the interpretation of survey items. Among these is the concept of cultural complexity, which is associated with emphasis on time, social role differentiation/specificity, and field dependence/independence. Cultural groups are also known to vary along the dimensions of individualism versus collectivism, emotional control versus emotional expressiveness, masculinity versus femininity, and the acceptability of physical contact. The unique position that each culture holds within this matrix of values is likely to influence respondent comprehension in ways that are difficult to predict. Level of abstraction may also be associated with the degree of cross-cultural variability in question interpretation (Scheuch, 1993).

Much survey research nonetheless continues to ignore the problem of cross-cultural equivalence, making the tacit assumption that concepts and terminology are being equally and equivalently evaluated by members of all respondent groups. This practice, referred to as "category fallacy" (Kleinman, 1977), poses a serious challenge to the comparability of health-related survey data. This issue can be alternately viewed through the distinction in cross-cultural psychology between "etic" and "emic" constructs (Triandis, 1972). Etic constructs are thought to be universal and equally understood across multiple cultural groups. Emic constructs are culture specific, in that they have important meaning within a single cultural group and are either interpreted differently or completely ignored by persons from other cultures. When emic constructs are treated as if they were etic, they are described as "pseudoetic," and they result in a category fallacy. The widespread use of pseudoetic constructs has been an important criticism of survey research for many years (Cicourel, 1964).

Of course, comprehension difficulties may also be attributed to the poor translation of questionnaires. Questions of translation quality and effects will not be reviewed in this chapter. Other aspects of multiple language surveys, however, may also influence respondent comprehension and interpretation and are relevant here (Botha, 1970). Marín et al. (1983), for instance, examined the answers given by Hispanic respondents when completing a self-administered instrument in Spanish or English and found that more complex cognitive structures were used when the questionnaire was completed in the native language. These findings make the corollary point that respondents are better able to comprehend and manipulate information when they are free to do so within a familiar context. Church et al. (1988) also reported findings consistent with the notion that concepts are more refined in one's primary language. Other researchers have reported no differences in questionnaire responses in two different languages (Tyson et al., 1988).

4.2.2 Memory Retrieval

The memory retrieval task involves retrieving either the answer to the question or relevant information that can be used to construct an answer. While some activities are typically recalled as individual events, others are recalled as or through schemas or generalizations. Tulving (1983) has referred to these as "episodic" and "semantic" recall strategies, respectively.

Episodic memories are stored as detailed information about specific and unique events. When searching memory for specific information, respondents also use various types of contextual cues, such as references to particular events, locations, or persons. Each type of cue is thought to access a particular subset of memories, and it has been suggested that the types may vary in their effectiveness (Brewer, 1988). In addition, the use of multiple cues may be most effective for isolating specific memories (Conway, 1990). Retrieval cues, however, only enable respondents to access memories that have been previously encoded (Strube, 1987). Because of the often overwhelming amount of information that individuals must process as they navigate their environment, cultural conditioning and values are likely to influence the subset of information that is identified as important and encoded (Angel and Thoits, 1987). Consequently, cultural factors may influence the types of information that are available for memory cues to help individuals' recall.

In contrast to episodic memories, semantic memories are generic descriptions of a specific class of events and provide no detail about individual episodes. These schemas are likely to be culture specific, representing events as they are perceived and practiced within the respondent's community or larger culture (D'Andrade, 1995). Respondent use of episodic and semantic recall strategies is also believed to vary depending on the frequency and regularity with which they experience various events (Blair and Burton, 1987; Menon, 1994). To the extent that specific events are experienced with varying frequency or regularity across cultures, one would expect to find cross-cultural differences in the formation of schemas and the use of alternative memory retrieval strategies. Little empirical information, however, is currently available.

Substantial work has also been reported about experimental efforts to evoke or discourage the use of alternative recall strategies in order to improve the accuracy of respondent recall (for a review, see Schwarz and Sudman, 1994). Virtually all of the available research, though, has been conducted with white, non-Hispanic populations. One exception is a study conducted by Jobe et al. (1990) in which no cross-racial differences were found in the accuracy of self-reports when respondents were instructed to remember events using forward, backward, or free recall search strategies.

Some evidence from nonexperimental validation studies is available that indicates members of some racial and ethnic minority groups are less likely to accurately recall certain types of information, including health events (Warnecke and Graham, 1976), reports of illicit substance use (Fendrich and Vaughn, 1994), and time use reports (Engle and Lumpkin, 1992). Yet because of limited cognitive research, it remains unclear whether these differences are due to differential motivation to conduct memory searches or to other cultural variations in social cognition, such as differences in question comprehension/interpretation or perceived social desirability.

4.2.3 Judgment Formation

The formation of judgments based on information retrieved from memory is an important aspect of attitude formation and the reporting of behavioral frequencies. In some instances, a precise answer to a survey question can be accessed, and no judgment is required. More often, however, respondents must synthesize the information they have sampled from memory into an appropriate answer. This may involve reconciling and assessing the relative importance of conflicting information and selecting an appropriate frame of reference or standard for forming a judgment. A related process is response formatting, during which selected answers are mapped on to interviewer defined response categories (Anderson, 1981). Some evidence is available regarding cultural variations in norms that may be relevant to each of these processes.

Judgment formation is moderated by respondent access to previously acquired information and attitudes. More frequently used information is thought to be "chronically accessible" in memory, and thus more readily available for use in developing on-the-spot judgments during survey interviews (Dovidio and Fazio, 1992). It is possible that cultural differences in the saliency and use of various concepts will influence the degree to which specific types of information are chronically accessible and thus available to serve as inputs to judgment formation. Currently, little empirical evidence exists with which to assess the cognitive accessibility of information across two or more cultural groups.

Assigning relative weights to the information recovered from memory is an important step in the process of developing opinions. Just as culture is likely to influence the types of information individuals encode in their memory, it is also probable that cultural norms and values dictate to some degree the importance respondents assign to the pieces of information they recover from memory when formulating answers (Douglas and Wildavsky, 1982). For instance, health opinion survey findings suggest that Hispanics in the U.S. are more *fatalistic* regarding cancer than are Anglo respondents (Pérez-Stable *et al.*, 1992). Those findings have been attributed to the culture specific concept of *fatalismo*, or the idea that little can be done to alter one's fate. It is thus reasonable to expect that Hispanic respondents might place greater emphasis on information consistent with a fatalistic orientation when forming opinions relevant to health-related topics. Other research suggests that variations across cultures in probabilistic thinking (e.g., the ability to express thoughts in terms of uncertainty) may also influence response decisions (Wright *et al.*, 1978).

In forming judgments, survey respondents must also frequently select a frame of reference or standard to use in anchoring their answers. As an example, when asked to evaluate themselves as overweight, underweight, or "about right," respondents might choose to use as a reference for their judgments: (a) how much they weighed at some time in the past; (b) how much they weigh in relation to members of a reference group, such as family, friends, or persons of the same age or sex; or (c) the opinions of others (e.g., physicians, family members). Selecting from among these alternative standards may in some cases be culturally mediated. Frames of reference may also evolve as individuals are acculturated into new societies. Hurtado *et al.* (1994) have recently reported that second generation Mexican-Americans compare themselves to a greater variety of people than do more recent immigrants from Mexico.

Empirical research is also available regarding cross-cultural differences in response formatting. Response styles, in particular, have received considerable attention. Findings to date suggest that both African-American and Hispanic survey participants are less likely than Anglo-Americans to qualify their answers (Bachman and O'Malley, 1984; Hui and Triandis, 1989; Marín *et al.*, 1992). Research has also suggested that Asian respondents are less likely to prefer extreme responses (Zax and Takahashi, 1967). Preference for an extreme versus cautious response style has been interpreted as being a consequence of variations in emphasis on sincerity versus modesty in social interactions (Marín and Marín, 1989). Recent findings also suggest that preference for extreme response styles tends to decrease among Hispanics in the U.S. as they become more acculturated (Marín *et al.*, 1992).

A related body of literature is concerned with how respondents map personal judgments on to survey rating scales (Moxey and Sanford, 1992). Central to this research is the use of numeric scales to examine how respondents understand the vague quantifiers often included in survey questions (e.g., "excellent," "fair," "occasionally"). In addition to contextual effects, respondent differences in personal development, attitude and experience have been shown to account for variations in the use of these expressions (Pepper, 1981). No direct evidence of cultural similarity or variability in the use of vague quantifiers, though, is currently available.

4.2.4   Response Editing

Several elements of social cognition may also encourage respondents to "edit" the information they report during a survey. Perceptions of social desirability, concerns about self-presentation, respondent evaluations of the interviewer, and interview language may each influence the quality of the information reported. As with the previous tasks reviewed, each of these editing processes may also be culturally conditioned.

Research has shown that survey respondents edit responses if they feel that certain answers are more socially desirable than others (Bradburn *et al.*, 1979). Socially desirable behaviors, such as exercise and good nutrition, may often be overreported, whereas undesirable behavior, such as smoking and substance abuse, may be underreported. Because personal definitions of social desirability are likely to be culturally mediated, response editing may also vary by topic and cultural tolerance for the behavior (Angel and Thoits, 1987; Marín *et al.*, 1983). Available information supports the notion that definitions of social desirability do vary cross-culturally. Dohrenwend (1966), for instance, found ethnic differences between Puerto Rican respondents and those of African-American, Irish or Jewish origin in the perceived desirability of the symptoms included in an index of mental health. On average, Puerto Ricans rated a series of psychological symptoms as being less undesirable than did members of these other groups. It has been suggested that these differences in perceptions of social desirability may be linked to the prevalence of these symptoms within each cultural group. Other researchers have failed to find differences in the perceived desirability of various psychological traits and behaviors among African-American, Spanish-American and white respondents (Gove and Geerken, 1977).

Ethnic group differences in responses to the Marlowe–Crowne social desirability scale have been reported by Ross and Mirowsky (1984), who found that the likelihood of giving socially desirable answers is associated both with Mexican (in contrast to Anglo-American) culture and minority group status. Relevant to these findings is a common pattern of social interaction in Hispanic cultures referred to as *simpatía*, or the expectation that interpersonal relations will be guided by harmony and the absence of confrontation (Triandis *et al.*, 1984). A similar cultural emphasis on courtesy among Asian populations is also thought to influence responses during some survey interactions (Deutcher, 1973).

Closely related is the concept of respondent acquiescence, which is a form of response editing that may also be related to cultural factors. Acquiescence, or "yea-saying," is defined as the tendency to agree with a statement irrespective of its content (Schuman and Presser, 1981). It has been suggested that this tendency is a strategy for self-presentation that is more commonly used by persons in the types of low status positions often occupied by members of minority groups (Ross and Mirowsky, 1984). Research conducted in the U.S. suggests that Hispanic (Marín *et al.*, 1992) and African-American (Bachman and O'Malley, 1984) respondents may both have a greater tendency to provide acquiescent answers. These findings, though, are not consistent (Gove and Geerken, 1977), and it may be that observed differences in acquiescence across cultural groups are actually indicators of emic question content, rather than group mediated response tendencies. That is, most survey respondents may have a predisposition to acquiesce, or play it safe, when answering questions that are interpreted as vague or confusing within their culturally defined cognitive framework. The level of cross-cultural equivalence within questionnaires, then, may be what determines the degree of differential acquiescence that is expressed. If so, the presence of high amounts of respondent acquiescence may be one measure of a pseudoetic question or instrument.

The racial or ethnic status of the interviewer may also influence the editing process. A substantial literature now exists suggesting that respondents may edit their answers on a selective basis as a consequence of social or cultural distance, defined here as similarity/dissimilarity of ethnic status between survey participant and interviewer (Anderson *et al.*, 1988; Reese *et al.*, 1986; Schaeffer, 1980). These studies demonstrate that respondents also may provide more deferential responses to interviewers of dissimilar ethnicity than to those who are similar. It is unclear from the available evidence, however, whether various cultures are equally likely to respond to "mixed" interviews by editing their responses. In addition, conventional wisdom has long held that race-of-interviewer effects may only be present when survey questions are race relevant (Groves, 1989). Yet few studies are available that examine interviewer effects on non-race-related questions.

The status of women also varies across cultures, and each culture has its own set of normative expectations regarding appropriate and inappropriate communication between the sexes. Consequently, the acceptability of cross-sex survey interactions is also likely to vary across cultural groups. Some traditional societies, for instance, have formal prohibitions against females conversing with males who are not immediate family members (Lerner, 1958). Many codes of interaction, though, are less formal and more difficult to assess. Male respondents from some cultural traditions may be less inclined to comply with an interview request or to take an interview seriously when the interviewer is female. Cultural influences may also be manifested through greater reluctance to discuss or give accurate responses to sensitive or threatening questions. Thus, females with more traditional values would be expected to be less comfortable discussing sexual and health behaviors with male interviewers. Again, little direct evidence is available regarding how cultural differences in sex relations influence survey respondent behaviors.

The language that survey data is collected in may also be related to response editing. Available evidence suggests that bilingual respondents may answer differently depending on the questionnaire's language and the cultural salience of the questions (Bond and Yang, 1982). Differential response patterns across varying language versions of a questionnaire may be due to social desirability, ethnic affirmation (i.e., affirming one's native culture when questioned in a second language), or cross-cultural accommodation. Interview language may also provide additional cues to respondents that influence responses. For instance, a study completed in Kenya found that bilingual respondents reported more modern answers when completing a questionnaire in English, compared to those responding in Gusii (Feldman, 1975).

4.3 METHODOLOGY

The research presented in this section is concerned with evaluating the cognitive processes underlying responses to health survey questions among several culturally diverse groups found in the Chicago metropolitan area. In this study, we employ self-identified racial or ethnic group membership as a proxy indicator of cultural group membership. Among the groups examined were African-Americans, Mexican-Americans, Puerto Ricans and non-Hispanic whites. The two largest Hispanic communities in Chicago were selected in recognition that although there is a core culture common to all Hispanics, there is also considerable heterogeneity, due in part to differences in migration experiences and acculturation (Aday *et al.*, 1980). A total of 423 adults aged 18 through 50 participated in laboratory interviews with structured probes (Jobe and Mingay, 1989) as part of this study. These respondents were stratified such that approximately one-quarter were representatives of each culture (there were 111 African-American, 112 Mexican-American, 92 Puerto Rican, and 108 white non-Hispanic respondents). In recognition that some psychological processes, such as self-presentation and social desirability, may also vary across cultural groups by sex, age, and educational achievement, participants within each of these four groups were further stratified by these three characteristics so their influence could also be investigated.

4.3.1   Instrument Development

A three-step process was used to develop the survey instrument for this study. Initially, a large pool of health questions previously used in national health surveys was selected for consideration. We reviewed these items and approximately 50 questions were included in a pretest instrument. Items were deliberately chosen to produce variation in terms of substantive topics, question types, and response formats. Sets of specific probes were developed next for use with each of these questions. These probes were designed to obtain insights into the underlying thought processes used by participants when answering each survey question and were based upon the experiences of previous researchers (Belson, 1981; Bradburn *et al.*, 1979). Each specific probe was designed to examine one of the four phases of the question response process described above. On average, two to three probes were developed for each health question. The number of probes per question ranged from one to five. The instrument was pretested with 30 participants and subsequently revised prior to main study.

In addition to the information collected during the interviews, participants were also asked to complete three short, self-administered forms: one for demographic information; a second containing 10 items from the short form version of the Marlowe–Crowne social desirability scale (Strahan and Gerbasi, 1972) and an 18-item health locus-of-control scale (Wallston *et al.*, 1978); and a third containing a 12-item acculturation scale (Marín *et al.*, 1987) which was provided only to Mexican-American and Puerto Rican participants. The items in the locus-of-control scale were subsequently used to provide measures of acquiescence and extreme response style.

4.3.2   Recruitment and Interviewing

Respondents were recruited from a variety of sources, including newspaper ads and community organizations. Screened as eligible were persons: (a) not currently employed by a physician, clinic, or hospital; (b) between the ages of 18 and 50; and (c) who self-identified as African-American, Hispanic, or white non-Hispanic. Persons who indicated that they were Hispanic, Latino, or of

Spanish origin were deemed eligible if they also self-identified as Mexican-American (or Mexican or Chicano/a) or Puerto Rican. Persons determined to be eligible were offered a cash incentive to participate.

All interviewing was conducted by a research team that included the study's six investigators and a staff of four research assistants. An important component of our data collection was the fact that this research team was representative of the cultural groups asked to participate in the study. Approximately one-half of each interviewer's work was with members of his or her own ethnic group. All interviews, though, were conducted by same-sex interviewers. All interviews were conducted in English, and averaged approximately one hour in length, ranging from 35 minutes to more than two hours. Participants received a complete description of the study's general goals and the questionnaire's subject matter prior to the interview. Prior to being interviewed, they were also asked to read and sign an informed consent document allowing audiotape recording. Subsequently, all tapes were transcribed, and responses to unstructured probes were reviewed and assigned codes representing the content of each participant's answer. All interviews were conducted between July 1993 and April 1994.

4.4 RESULTS

The results presented in this section are organized by stage of the response process. We highlight both similarities and differences in the cognitive processes employed cross-culturally in responding to survey questions.

4.4.1 Question Interpretation

During each interview, respondents were probed regarding their understanding and interpretation of survey questions using a variety of techniques. One was to simply ask respondents what specific concepts meant to them. For instance, after reporting the amount of stress they had recently experienced, respondents were asked "what does the word 'stress' mean to you?" Interview transcripts were reviewed and responses coded along several dimensions. In this example, five general themes were revealed (see Table 4.1). These included descriptions of specific stress-relevant feelings, health-related problems, economic problems, difficulties with social relationships, and stress-related behaviors (an "other response" category was also coded). Cultural differences were found in the likelihood that each of these themes would be mentioned by respondents. Hispanic respondents, for example, were more likely to cite economic problems when asked about the meaning of "stress" than were persons from other cultural backgrounds. In addition, non-Hispanic whites were less likely to mention difficulties in social relationships when asked to describe stress than were other respondents. These two sets of findings were unaffected by controls for age, sex, and education in logistic regression analyses. Additional analyses found no relationship between respondent income level and descriptions of stressful

Table 4.1   Question Interpretation by Cultural Group

Survey Question: "During the past 2 weeks, would you say you experienced a lot of
                       stress, a moderate amount of stress, relatively little stress, or almost
                       no stress at all?"

Probe: "In this question, what does the word 'stress' mean to you?"

Concept Mentioned	African-American ($N = 110$)	Mexican-American ($N = 112$)	Puerto Rican ($N = 89$)	White non-Hispanic ($N = 118$)
Specific feelings	73.8	73.0	78.9	68.4
Economic problems*	11.8	20.5	22.5	11.1
Specific behaviors	8.2	8.9	11.2	12.0
Health problems	27.3	27.7	28.1	27.8
Social relationships*	16.4	17.9	19.1	6.5
Other problems	23.6	40.2	32.6	32.4

* Chi-square $p < 0.05$.
Note: Responses do not sum to 100% due to multiple answers.

economic problems. Finally, there was no association between acculturation and personal definitions of "stress" among Hispanic respondents.

Differences between Hispanic and non-Hispanic cultures were observed in the responses to other types of probes. For example, respondents were probed regarding the perceived difficulty of concepts used in survey questions. Respondents were asked, "How much control do you think you have over your future health?" followed by a probe that asked, "Do you feel this question about control over future health is one that people would or would not have difficulty understanding?" Greater proportions of respondents of Hispanic background (55.9 percent of Mexican-Americans and 54.2 percent of Puerto Ricans) expressed concern that this concept would pose problems for others than did non-Hispanic respondents (30.2 percent of African Americans and 36.9 percent of non-Hispanic whites). Logistic regression analysis confirmed these differences in perceived conceptual difficulty between Hispanic and non-Hispanic respondents. Use of these projective comprehension probes, however, did not always uncover cross-cultural differences. For instance, although one-quarter of all respondents (25.1 percent) felt that people would have difficulty understanding the Center for Epidemiologic Studies Depression (CES-D) question, "During the past week, how often have you felt you could not get going?" no differences were observed among the four cultural groups being studied.

Another set of analyses examined cultural differences in preferences for various disease labels. In one example, respondents were shown a list of four synonyms for diabetes mellitus and asked to identify which label they would be most likely to use in discussing this condition. Of these, a large majority

(88.3 percent) selected "diabetes." The proportion selecting some other label, such as "sugar diabetes," "sugar," "high sugar," or something else, varied across cultural groups, from a high of 18.3 percent of the studied African-Americans to a low of 6.4 percent of the non-Hispanic whites (12.4 percent of Puerto Rican and 9.7 percent of Mexican-American respondents also selected a label other than "diabetes"). Multivariate analyses confirmed that among the studied participants, African-Americans were more likely than non-Hispanic whites to prefer labels other than "diabetes."

Another technique used to examine potential cultural differences in question interpretation involved asking respondents to use their own words to rephrase a question they had just been asked. One such question asked "On about how many different days did you have one or more drinks during the past 30 days?" A review of interview transcripts revealed that only 30.7 percent of those interviewed were able to accurately describe the question's meaning. No differences in rephrasing accuracy were found across cultural groups.

4.4.2 Memory Retrieval

Information regarding use of alternative recall strategies was collected for a number of questions concerned with behavioral frequencies and other numeric information. Several variations of the question, "How did you remember this number?" were used to probe for this information. For purposes of these analyses, reported retrieval strategies were collapsed into three categories: (1) episodic, or counting, strategies; (2) semantic, or rate/estimation-based, strategies; and (3) guessing. Here, we will focus on a single question: the total number of physician contacts in the past year ("During the last year, how many times did you see or talk to a medical doctor?").

When asked about past year physician contacts, 50.9 percent indicated use of episodic recall, 43.3 percent used semantic recall processes, and 5.8 percent guessed. Additional analysis confirmed previous findings suggesting that choice of retrieval strategy is in large measure dependent on the frequency with which the event is reported. Among those using episodic recall, an average of 2.9 physician contacts were reported. The average was 10.0 contacts among those who indicated use of semantic recall, while 7.1 contacts were reported by those who guessed. No differences were found, however, in the use of these various retrieval strategies by cultural group (see Table 4.2). Episodic recall was reported by approximately one-half of all respondents. The proportions employing semantic retrieval processes ranged from 41.7 to 45.7 percent. In addition, the relationship between reported frequency and type of recall was consistent within each group.

Using a logistic regression model, the effects of cultural group membership on memory retrieval were also examined while controlling for age, sex, and education. In this analysis, the dependent variable was coded to examine effects on use of episodic recall (1 = episodic; 0 = semantic or guessing). Findings again revealed no relationship between cultural group and memory retrieval

Table 4.2   Memory Retrieval by Cultural Group

Survey Question: "During the last year, how many times did you see or talk to a
                        medical doctor?"

Probe: "How did you arrive at this answer?"

	African- American ($N = 101$)	Mexican- American ($N = 103$)	Puerto Rican ($N = 81$)	White non-Hispanic ($N = 96$)
Percent Using Each Strategy				
Episodic	53.5%	49.5%	49.4%	51.0%
Semantic	42.6%	41.7%	45.7%	43.8%
Guessing	4.0%	8.7%	4.9%	5.2%
Mean Physician Contacts				
Episodic	4.0[a]	2.4[a]	2.5[a]	2.6[a]
Semantic	8.7[b]	7.6[b]	12.7[b]	11.5[b]
Guessing	3.5[a, b]	7.6[a, b]	11.5[a, b]	5.4[a, b]
F-Ratio	5.1**	5.5**	7.9***	13.4***

** $p < 0.01$.
*** $p < 0.001$.
Note: Percentages with different superscripts differ significantly ($p < 0.05$) *within* cultural group
on the basis of the least significant difference test. For each cultural group, types of retrieval not
noted by the same superscript indicate a significant difference between them.

processes. The reported number of contacts with physicians, though, remained
inversely associated with the likelihood of employing episodic recall, even after
controlling for all respondent characteristics. Similar patterns were observed
when examining the memory retrieval processes used to answer a number of
other survey questions included in this study.

Respondents who indicated use of an episodic retrieval strategy when asked
about past year contacts with physicians and who reported a minimum of two
contacts were also asked about the order in which they recalled specific
events. Overall, a plurality (44.5 percent) indicated using free recall strategies in
which information about individual events was retrieved in whichever order it
was remembered (e.g., contacts related to more serious health problems first).
Approximately one-third (35.9 percent) indicated use of a backward search
procedure in which the most recent events were first remembered, and one-fifth
(19.5 percent) reported using a forward search process in which events closest
to the start of the recall interval were searched for first and then used as cues
to search for subsequent events. Cultural differences were observed in the use
of these alternative episodic search strategies. Most importantly, 60 percent of
all African-American respondents reported searching memory in a free recall

format, compared with less than 40 percent of the other groups (30.8 percent of Mexican-Americans, 37.5 percent of Puerto Ricans, and 39.4 percent of non-Hispanic whites. Pluralities of Mexican-American (42.3 percent) and non-Hispanic white respondents (45.5 percent) preferred backward search strategies. In contrast, equal proportions of Puerto Rican respondents preferred forward and free recall processes (37.5 percent). A set of logistic regression analyses was employed to examine these differences when controlling for other respondent characteristics and number of physician contacts. Even after controlling for frequency of physician contacts, African-Americans were more likely than Mexican-Americans to prefer free recall strategies. African-American respondents were also less likely than were Puerto Ricans to have used a forward search strategy.

The use of contextual memory cues when retrieving specific elements of information from long term memory was also examined. As an example, after answering the question, "How much did you weigh when you were 16 years old?" respondents were probed as to how they remembered this information. Interview transcripts were reviewed, and specific memory cues mentioned by respondents were systematically coded. For this analysis, these cues were classified as referencing specific locations, events, or social network members. Overall, use of contextual cues was reported by 47.1 percent of all respondents. Use of these cues did not vary by cultural group, as 44.6 percent of Puerto Ricans, 45.4 percent of African-Americans, 46.4 percent of Mexican-Americans, and 51.4 percent of non-Hispanic whites indicated use of one or more contextual cues in answering this question. Event cues were most commonly reported (36.1 percent), followed by location (19.8 percent) and social network (16.8 percent) cues. Proportions using each specific type of memory cue also did not vary by cultural group. Logistic regression models, however, revealed cultural differences in the use of social-network-related memory cues. When controlling for other variables, both African-American and Mexican-American respondents were less likely to make references to specific individuals when recalling their weight at age 16. Additional multivariate analyses confirmed the absence of cultural differences in the use of specific event and location memory cues, as well as the use of contextual cues, in general, after controlling for retention interval (i.e., respondent age minus 16 years) and other demographic variables. Males, however, were found less likely to use social network cues, and more educated respondents were more likely to employ both specific event cues and any type of memory cue.

A linear regression model also assessed the use of multiple memory cues, which were measured as a simple count of the total number of types of cue mentioned. In this analysis, African-Americans were less likely to use contextual memory cues (compared to the contrast group of non-Hispanic whites) when responding to the question about their weight at age 16. Education was also directly associated with the total number of contextual cues reported to have been used. Finally, a similar set of analyses was also conducted to examine the contextual memory cues used to recall other types of autobiographical information, including age when first began smoking, age when first began drinking alcohol, and the most recent episode of marijuana use. In analyses of these reports, no cultural differences in the use of any types of contextual cues were observed.

Finally, we also examined the relationship between level of acculturation among Hispanic respondents and each of the memory retrieval processes examined. In no instance was acculturation to the dominant U.S.-Anglo culture found to be associated with preference for any specific recall or order of search strategy or with the use of any particular type of contextual cue.

4.4.3   Judgment Formation

Cultural variations in the accessibility of information were examined in response to the question, "How concerned are you about getting cancer in the future?" The probe, "Have you ever thought about getting cancer before?" was used to examine group differences in previous consideration of this topic. Non-Hispanic whites and Puerto Ricans were more likely (75.2 percent and 71.1 percent, respectively) than Mexican-Americans (54.4 percent) to have previously thought about getting cancer. Among African-Americans, 65.5 percent had thought about this. These differences remained in multivariate analyses that controlled for age, sex, education, and the respondent's expressed concern about getting cancer. Acculturation was not associated with the accessibility of attitudes regarding cancer risk among Hispanic respondents.

The use of various reference points for developing judgments was examined using probes that asked respondents how they formed answers to specific questions. One such example was a question that asked, "Do you consider yourself now to be overweight, underweight, or about the right weight?" Interview transcripts were reviewed to examine the types of comparisons (or reference points) mentioned by respondents in answering this question. Most commonly used (by 75.4 percent of all respondents) were self-references, in which respondents used themselves as the standard for assessing their weight status. These included references to their current weight, their past weight, or their ideal weight. Less common strategies involved comparisons to social norms for weight (21.5 percent) and comparisons to other specific persons or reference groups, such as family, friends, or others of the same age (17.7 percent). Few cross-cultural differences were found in the use of these alternative strategies in responding to this question.

Cultural differences in response styles were also examined. A measure of extreme response style was developed using the 18-item health locus-of-control scale. Each of these 18 questions had four response options: "strongly agree," "somewhat agree," "somewhat disagree," and "strongly disagree." The extreme response style was represented by a simple count of the number of items respondents answered using either the "strongly agree" or the "strongly disagree" options. All three minority groups were found to be more likely, on average, to select extreme responses when answering these questions. Of the 18

Table 4.3   Response Mapping by Cultural Group

Survey Question: "Would you say your health is excellent, very good, good, fair or poor?"

Probe: "On a scale of 0 to 10, with 0 being the worst possible health and 10 being the best possible health, what number best represents (RESPONSE OPTION) to you?"

	Mean Values			
Response Option	African-American ($N = 109$)	Mexican-American ($N = 114$)	Puerto Rican ($N = 90$)	White non-Hispanic ($N = 108$)
Excellent	9.6	9.5	9.6	9.8
Very good	8.1	8.1	8.0	7.9
Good*	6.4[a, b]	6.5[a]	6.7[a]	6.1[b]
Fair*	4.3[a, b]	4.6[a]	4.4[a]	4.0[b]
Poor***	0.9[a]	1.6[b]	1.2[a, b]	1.4[b]

* *F*-ratio $p < 0.05$.
*** *F*-ratio $p < 0.001$.
Note: Means with different superscripts differ significantly ($p < 0.05$) *across* cultural groups on the basis of the least significant difference test. For each response option, mean values not noted by the same superscript indicate a significant difference between them.

items in the scale, Puerto Rican respondents selected extreme responses an average of 8.7 times, followed by 8.1 extreme responses among Mexican-American respondents and 7.8 among African-Americans. In contrast, non-Hispanic white respondents averaged 6.7 extreme responses. Using analysis of covariance (ANCOVA) to control for respondent age, sex, and education, these cultural differences were found to be significant. In addition, education was inversely associated with preference for extreme responses. Additional multivariate analyses indicated that among respondents of Hispanic origin, degree of acculturation was not related to preference for extreme responses.

Respondent mapping of judgments on to predefined response categories was investigated next. In this exercise, respondents were asked to use an 11-point semantic differential rating scale (0 = worst possible health; 10 = best possible health) to assess the degree of good or ill health represented by each of five precoded responses to the common global health rating question, "Would you say your health is excellent, very good, good, fair, or poor?" Basic findings of this analysis are presented in Table 4.3. Cross-culturally, there was considerable agreement regarding the numeric values assigned to the two most positive health ratings ("excellent" and "very good"). Differences in mapping were observed, though, for more neutral and less positive responses. Compared with

Hispanic respondents (both Mexican-American and Puerto Rican), non-Hispanic whites gave, on average, lower ratings to the responses "good" and "fair." African-Americans assigned lower ratings to the response "poor" than did either Mexican-American or non-Hispanic white respondents. In additional analyses, acculturation was not associated with the mean ratings assigned by Hispanics to four of these five precoded responses. Ratings of the response "good," however, were found to be inversely correlated with degree of acculturation. That is, more acculturated respondents tended to assign lower ratings to the response "good" than did those who were less acculturated.

4.4.4 Response Editing

Projective probes were also employed to examine the likelihood of over- and underreporting behavioral frequencies. Using this technique, respondents were asked whether they believed that people in general might over- or underreport. This indirect approach has been shown to be a better indicator of uneasiness than a direct approach to respondent editing (Bradburn et al., 1979). Two examples will be discussed, one that examines socially desirable and one that examines socially undesirable behaviors. As part of a discussion of respondent vegetable consumption, respondents were asked, "In general, do you feel that people might purposely say they eat more vegetables than they do, say they eat fewer vegetables than they do, or would they try to answer accurately?" A majority (57.8 percent) felt that most people would try to report accurately their consumption of vegetables. Smaller proportions felt that most would overreport (29.3 percent) or underreport (11.1 percent) the amount of vegetables they ate (1.8 percent felt the likelihood was dependent on other factors). Cultural groups did vary in their opinions regarding the likelihood that editing would or would not take place. Just over two-thirds of non-Hispanic white respondents (68.4 percent) expressed the opinion that most people would attempt to answer accurately. Of all Mexican-Americans interviewed, 60.4 percent shared this opinion. Smaller proportions of Puerto Ricans (50.6 percent) and African-Americans (51.0 percent), though, felt that others would try to answer accurately. When examined in a logistic regression model that controlled for sex, age, education, and self-reported vegetable consumption, Puerto Rican, African-American, and Mexican-American respondents all continued to be more likely to feel that response editing would take place when reporting vegetable consumption.

Respondents were also probed regarding perceived over- and underreporting of a less desirable behavior: alcohol consumption. After reporting the number of days during which they drank alcohol in the past month, respondents were asked, "In general, do you feel that people might purposely say they drink more, drink less, or would they try to answer accurately?" Approximately one-third of all respondents (36.4 percent) believed that others would try to provide accurate answers. A majority (54.0 percent) felt that underreporting was more likely, and a small proportion (5.8 percent) believed overreporting was more likely (3.8 percent said reporting behavior would be dependent on other factors). No cross-cultural differences were detected. The proportions feeling that others' responses would be as accurate as possible ranged from 32.9 percent among Puerto Ricans to 41.7 percent among non-Hispanic whites (percentages were 33.9 percent among Mexican-Americans and 36.3 percent among African-Americans). Similarly, the proportions feeling that alcohol consumption would be underreported ranged between 48.5 percent of non-Hispanic whites and 56.9 percent of Mexican-Americans (among African-Americans and Puerto Ricans, the percentages were 55.9 percent and 54.9 percent, respectively). Logistic regression confirmed these findings of no cross-cultural differences. Females and persons who drank more frequently, though, were more likely to believe that others would misreport their alcohol consumption. Among Hispanics in the sample, acculturation was not associated with these perceptions of response editing.

Cultural differences in the provision of socially desirable answers were also examined using the Marlowe-Crowne scale. Of the ten-item version of that scale included in this study, white respondents not of Hispanic origin provided an average of 4.9 socially desirable answers. In comparison, Puerto Rican respondents averaged 5.5, African-American respondents averaged 5.6, and Mexican-American respondents averaged 5.9 socially desirable answers. When examined in a multiple regression model, both Mexican-American and African-American respondents were found to be more likely to give socially desirable answers after controlling for age, sex, and education. Females and older respondents were also more likely to respond in a socially desirable manner. A second multiple regression model was examined to assess the relationship between acculturation and social desirability. Although acculturation was found not to be associated with the Marlowe–Crowne scale, this additional analysis revealed that Mexican-Americans were more likely to provide socially desirable answers than were Puerto Rican respondents.

Respondent acquiescence was next investigated using the 18 items from the health locus-of-control scale. The indicator used was a count of the total number of "strongly agree" or "somewhat agree" responses. (Half of the questions were worded such that agreement indicated an internal and half an external locus-of-control.) Cultural group differences were found in the tendency to provide positive, or acquiescent, answers, regardless of question content. African-American and Mexican-American respondents each gave an average of ten positive responses, followed by an average of 9.4 among Puerto Ricans. Non-Hispanic whites gave an average of 8.8 positive responses. An ANCOVA model confirmed the findings that African-American and Mexican-American respondents gave more acquiescent responses, on average, than did non-Hispanic whites. This analysis also indicated that males and less educated respondents were more likely to yea-say. Among persons of Hispanic origin, degree of acculturation was unrelated to respondent acquiescence.

Potential cross-cultural interviewer effects were also investigated. Projective methods were again used. In the example provided here, respondents were asked

Table 4.4   Response Editing by Cultural Group

Survey Question: "About how many drinks did you usually have in a day on the days that you drank during the past 30 days?"

Probe: "Do you feel this is a question that (RESPONDENT'S CULTURAL GROUP) respondents would be comfortable or uncomfortable talking about with: [a] a (RESPONDENT'S CULTURAL GROUP) survey interviewer; [b] a survey interviewer who is not (RESPONDENT'S CULTURAL GROUP)?"

	African-American (N = 89) %	Mexican-American (N = 95) %	Puerto Rican (N = 78) %	White non-Hispanic (N = 98) %
a. *Interviewer Same Culture***				
Comfortable	88.8	74.7	85.9	92.9
Uncomfortable	11.2	25.3	14.1	7.1
b. *Interviewer Different Culture****				
Comfortable	60.0	60.0	69.4	89.3
Uncomfortable	40.0	40.0	30.6	10.7

** Chi-square $p < 0.01$.
*** Chi-square $p < 0.001$.

a question regarding alcohol consumption during the past 30 days. This question was followed by two projective probes that solicited opinions regarding how comfortable most members of the respondent's cultural group, rather than the respondent *per se*, would feel about being asked the alcohol use question by interviewers from similar and dissimilar backgrounds.

When asked whether persons with their cultural background would or would not be comfortable talking to interviewers with a similar cultural identity about their alcohol consumption, a large majority of all respondents (85.6 percent) felt that most people would be comfortable. As Table 4.4 demonstrates, however, cross-cultural differences were observed. Nine out of ten non-Hispanic white respondents indicated that most people would feel comfortable discussing this topic with a same-culture interviewer. More than 80 percent of African-American and Puerto Rican respondents also felt most persons would be comfortable. However, among Mexican-Americans, a smaller majority felt that way. A logistic regression equation confirmed that compared with non-Hispanic whites, Mexican-American respondents indicated a greater perceived discomfort with discussing alcohol consumption with same-culture interviewers. Females and younger respondents were also more likely to report discomfort in this interview situation. Additional analyses indicated that acculturation was associated with perceived discomfort among Hispanic respondents. That is, those who were less acculturated to the mainstream U.S.-Anglo culture were more likely to express discomfort with same-culture interviews when discussing this topic.

Overall, a smaller majority of respondents (70.1 percent) felt that persons with their cultural background would be comfortable discussing personal alcohol consumption with survey interviewers from a different cultural group. Again, cross-cultural differences were found. While almost 90 percent of all non-Hispanic whites indicated that most would feel comfortable, only 60 percent of African-Americans and Mexican-Americans believed that members of their groups would feel comfortable in this situation. Among Puerto Ricans, approximately 70 percent believed that respondents would feel comfortable. These findings of white minority group differences were confirmed by a logistic regression model that controlled for sex, age, education, and self-reported alcohol consumption. In these analyses, acculturation was not related to perceived discomfort.

4.5   DISCUSSION

This research has indicated the presence of cross-cultural differences in each of the four cognitive tasks examined. Regarding question interpretation, these data suggest that cultural differences should be an important consideration when developing survey questions. In particular, they leave us to conclude that the etic–emic distinction may be more appropriately conceived as a continuum, one that acknowledges the possibility of purely etic or emic concepts while also allowing for varying degrees of parallel meaning across groups. Depending on the concept being discussed, the cultural experience of each respondent will cue the cognitive context within which it is assigned meaning. Some elements of this context, we believe, will be shared across cultural groups, while others will be culture specific. The balance of shared-to-unique contextual elements between cultures will define the degree to which the concept in question is generalizable across cultures.

In addition, our findings suggest that when more than two cultures participate in a survey, the array of culturally mediated contextual elements that underlie question interpretations becomes multidimensional. What is found to be emic and what is shared cross-culturally appear to vary depending on which cultures and which concepts are being investigated. For instance, the interpretations of African-Americans were in some cases most similar to those of Hispanic respondents when interpreting questions. In other examples, their interpretations came closer to approximating those of non-Hispanic whites. Consequently, the challenge of conceptual equivalence becomes even greater in pluralistic societies.

At the same time, considerable effort has been directed by researchers at the problem of cross-cultural equivalence, and a variety of methods have been proposed for addressing it. Among these are the Triandis (1972) "emic + etic" methodology, the cross-cultural semantic differential model (Malpass, 1977), generalizability theory models (Van de Vigver and Poortinga, 1982), the Leung and Bond (1989) factor analytic procedure for identifying strong and weak etic concepts, and decentralizing techniques (Word, 1992). Each of these approaches, of course, has its own advantages and disadvantages. Hence, the application of multiple strategies is most likely to be successful in evaluating and understanding cultural similarities and differences.

As anticipated, no cross-cultural differences in the use of alternative strategies for the retrieval of behavioral frequencies were observed. Rather, respondents across all cultural groups were consistent in their use of semantic processes for the recall of frequently occurring events, and episodic strategies for the retrieval of events that occurred with less frequency. Our analyses did, however, identify cultural differences in other memory retrieval processes. In one example, African-Americans were more likely to prefer free order of recall strategies when conducting an episodic memory search. Most respondents from other cultural groups preferred more linear (either forward or backward) search strategies. There was also some evidence of cultural differences in the use of contextual memory cues to recall autobiographical information, suggesting the possibility of variations in the types of information encoded in long-term memory stores. Consequently, additional research into cultural differences in the memory retrieval task is warranted.

Some cultural differences in judgment formation tasks were also identified in this study group. Most interesting were those related to response formatting. In particular, evidence presented in Section 4.4.3 suggests that respondents from distinct cultural backgrounds may systematically vary in how they apply available questionnaire response options. Future research should be directed at developing models capable of accounting for cross-cultural variability in the use of response choices. As Table 4.3 suggests, vague quantifiers in particular would appear to be susceptible to cross-cultural variability in use.

Our indicators of response editing also provided evidence of cross-cultural differences. Representatives of the cultural groups examined varied significantly in their expression of beliefs and reporting styles thought to be associated with response editing. It is difficult to conclude from these data, however, whether the patterns of findings reflect simple differences in cultural cognition or differences that are a unique, socially conditioned consequence of minority group status in the U.S. Previous work by Ross and Mirowsky (1983) suggests that processes related both to cultural and minority group status contribute to these differences. From this perspective, one could also argue that less evidence of response editing among non-Hispanic whites is a consequence of their privileged position in the U.S. and other Western countries. It is clear that future inquiries into cultural variations in response editing will also need to consider how dominant versus minority group status influences response editing.

Degree of acculturation among Hispanic respondents was found to have few associations with the cognitive processes examined. Because ability to speak

English requires a certain minimum level of acculturation, our sample was truncated along this dimension due to the fact that highly unacculturated respondents were not able to participate. Consequently, the fact that acculturation was unrelated to most of the dependent variables examined should not be taken as evidence that this concept does not influence the cognitive processing of survey information.

While we believe that theoretical models that fully account for cultural differences in the cognitive processing of survey questions are not yet available, some of the cultural syndromes cited earlier may be applied to our findings. For instance, the cultural dimension of collectivism versus individualism can be used to account for some respondent differences in both question interpretation and response editing. As outlined in a recent work by Triandis (1995), non-Hispanic whites are in general more individualist in their orientation, while those from Latino cultures are more collectivist. This framework is consistent with the finding that Mexican-Americans and Puerto Ricans are more likely to emphasize social relationships when conceptualizing "stress" than are non-Hispanic whites (Table 4.1). Likewise, the collectivist–individualist perspective is also useful in interpreting cultural differences in indicators of response editing (Table 4.4). Because individualists deal with strangers or members of out-groups quite easily, it is not surprising to find that non-Hispanic whites are equally comfortable with the idea of discussing a sensitive topic during a survey interview, regardless of whether the interviewer is from the same or another culture. Respondents from more collectivist cultures, in contrast, are less comfortable with the idea of discussing such a topic with interviewers from another culture. Other theoretical models, such as Helson's (1964) adaptation-level theory, may also hold promise as useful frameworks for understanding culturally mediated differences in the processing of survey information.

In reviewing these findings, it is important to acknowledge that they are not the product of a large, representative sample. Rather, they are results from a modest ($N = 423$) number of laboratory interviews that were specifically designed to investigate cultural similarities and differences in the cognitive processing of survey questions. It is also important to recognize that space limitations require that the findings presented are only examples. As such, we have deliberately overemphasized questions that depict cultural differences in order to provide readers with examples of how respondent culture can affect cognitive processes. There are several other examples (not presented here) in which no systematic cultural differences were observed. In addition, each of the cultural groups examined are likely to be less homogeneous than was assumed in these analyses. Subcultural variations in occupational achievement and socioeconomic status, community and religious involvement, ethnic identification and a host of other factors will also influence social cognition and should be investigated.

It should also be recognized that the structured probes used to collect a significant amount of the information presented in this chapter are themselves subject to error. It has, for example, been suggested that respondents may not be able to accurately report all of the mental processes they routinely employ in answering questions, such as those involved in judgment formation (Nisbett and Wilson, 1977). Additionally, it is possible that the very use of these probes may interfere with or influence the cognitive processes routinely employed by respondents when answering survey questions.

The findings in this chapter nonetheless provide important evidence of cross-cultural variation in the cognitive processes underlying responses to survey questions. They reinforce the notion that respondent culture may have a systematic effect on how questions are understood and answers are formulated and reported. An important future goal of the discipline should be the development and empirical assessment of theories to more systematically account for cross-cultural variation in social cognition as applied to survey research. Theories that allow for both culturally mediated and individual variations in the cognitive processing of survey questions will, in all likelihood, be most effective in modeling these processes.

ACKNOWLEDGMENTS

This study was supported by Cooperative Agreement #U83/CCU508663 from the National Center for Health Statistics. The authors gratefully acknowledge the assistance of Gloria Chapa-Resendez, Charles Bright, Jane Burris, Shaunda Bonds, Francisco Perez, and Kendra Uhe in collecting and processing the data presented in this chapter. Thanks are also extended to Marya Ryan and Bernita Rusk for their assistance in manuscript preparation, and Dr. Harry Triandis for his helpful comments on an earlier draft.

REFERENCES

Aday, L.A., Chiu, G.Y., and Anderson, R. (1980), "Methodological Issues in Health Care Surveys of the Spanish Heritage Population," *American Journal of Public Health*, 70, pp. 367–374.

Anderson, N. (1981), *Foundations of Information Integration Theory*, New York: Academic Press.

Anderson, B.A., Silver, B.D., and Abramson, P.R. (1988), "The Effects of the Race of the Interviewer on Race-related Attitudes of Black Respondents in the SRC/CPS National Election Studies," *Public Opinion Quarterly*, 52, pp. 289–324.

Angel, R., and Thoits, P. (1987), "The Impact of Culture on the Cognitive Structure of Illness," *Culture, Medicine and Psychiatry*, 11, pp. 465–494.

Bachman, J.G., and O'Malley, P.M. (1984), "Black-White Differences in Self Esteem: Are They Affected by Response Styles?" *American Journal of Sociology*, 90, pp. 624–639.

Belson, W.A. (1981), *The Design and Understanding of Survey Questions*, Aldershot, UK: Gower.

Blair, E., and Burton, S. (1987), "Cognitive Processes Used by Survey Respondents to Answer Behavioral Frequency Questions," *Journal of Consumer Research*, 14, pp. 280–288.

Bond, M.H., and Yang, K.S. (1982), "Ethnic Affirmation Versus Cross-cultural Accommodation: The Variable Impact of Questionnaire Language on Chinese Bilinguals in Hong Kong," *Journal of Cross-cultural Psychology*, 13, pp. 169–185.

Botha, E. (1970), "The Effect of Language on Values Expressed by Bilinguals," *Journal of Social Psychology*, 80, pp. 143–145.

Bradburn, N.M., Sudman, S., and Associates. (1979), *Improving Interview Method and Questionnaire Design: Response Effects to Threatening Questions in Survey Research*, San Francisco: Jossey-Bass.

Brewer, W.F. (1988), "Memory for Randomly Sampled Autobiographical Events," in U. Neisser, and E. Winograd (eds.), *Remembering Reconsidered: Ecological and Traditional Approaches to the Study of Memory*, Cambridge: Cambridge University Press, pp. 21–90.

Brislin, R.W., Lonner, W.J., and Thorndike, R.M. (1973), *Cross-cultural Research Methods*, New York: Wiley.

Church, A.T., Katigbak, M.S., and Castañeda, I. (1988), "The Effects of Language of Data Collection on Derived Conceptions of Healthy Personality with Filipino Bilinguals," *Journal of Cross-cultural Psychology*, 19, pp. 178–192.

Cicourel, A.V. (1964), *Method and Measurement in Sociology*, Glencoe, IL: Free Press.

Conway, M.A. (1990), *Autobiographical Memory: An Introduction*, Philadelphia: Open University Press.

D'Andrade, R. (1995), *The Development of Cognitive Anthropology*, New York: Cambridge University Press.

Deutcher, I. (1973), "Asking Questions: Linguistic Comparability," in D. Warwick, and S. Osherson (eds.), *Comparative Research Methods*, Englewood Cliffs, NJ: Prentice-Hall, pp. 163–186.

Dohrenwend, B. (1966), "Social Status and Psychological Disorder: An Issue of Substance and an Issue of Method," *American Sociological Review*, 31, pp. 14–34.

Douglas, M., and Wildavsky, A. (1982), *Risk and Culture*, Berkeley: University of California Press.

Dovidio, J.F., and Fazio, R.H. (1992), "New Technologies for the Direct and Indirect Assessment of Attitudes," in J. Tanur (ed.), *Questions About Questions: Inquiries into the Cognitive Bases of Surveys*, New York: Russell Sage Foundation, pp. 204–237.

Engle, P.L., and Lumpkin, J.B. (1992), "How Accurate Are Time-Use Reports? Effects of Cognitive Enhancement and Cultural Differences on Recall Accuracy," *Applied Cognitive Psychology*, 6, pp. 141–159.

Feldman, R.H.L. (1975), "The Effect of Administrator and Language on Traditional-modern Attitudes among Gusii Students in Kenya," *The Journal of Social Psychology*, 96, pp. 141–142.

Fendrich, M., and Vaughn, C.M. (1994), "Diminished Lifetime Substance Use Over Time," *Public Opinion Quarterly*, 58, pp. 96–123.

Gove, W.R., and Geerken, M.R. (1977), "Response Bias in Surveys of Mental Health: An Empirical Investigation," *American Journal of Sociology*, 82, pp. 1289–1317.

Groves, R.M. (1989), *Survey Errors and Survey Costs*, New York: John Wiley & Sons.

Guarnaccia, P.J., Angel, R., and Worobey, J.L. (1989), "The Factor Structure of the CES-D in the Hispanic Health and Nutrition Examination Survey: The Influences of Ethnicity, Gender and Language," *Social Science and Medicine*, 29, pp. 85–94.

Helson, H. (1964), *Adaptation-Level Theory*, New York: Harper and Row.

Hines, A.M. (1993), "Linking Qualitative and Quantitative Methods in Cross-Cultural Survey Research: Techniques from Cognitive Science," *American Journal of Community Psychology*, 21, pp. 729–746.

Hui, C.H., and Triandis, H.C. (1989), "Effects of Culture and Response Format on Extreme Response Style," *Journal of Cross-cultural Psychology*, 20, pp. 296–309.

Hurtado, A., Gurin, P., and Peng, T. (1994), "Social Identities—A Framework for Studying the Adaptations of Immigrants and Ethnics: The Adaptations of Mexicans in the United States," *Social Problems*, 41, pp. 129–151.

Jobe, J.B., and Mingay, D.J. (1989), "Cognitive Research Improves Questionnaires," *American Journal of Public Health*, 79, pp. 1053–1055.

Jobe, J.B., and Mingay, D.J. (1991), "Cognition and Survey Measurement: History and Overview," *Applied Cognitive Psychology*, 5, pp. 175–192.

Jobe, J.B., White, A.A., Kelley, C.L., Mingay, D.J., Sanchez, M.J., and Loftus, E.F. (1990), "Recall Strategies and Memory for Health Care Visits," *Milbank Quarterly*, 68, pp. 171–189.

Kleinman, A. (1977), "Depression, Somatization, and the "New Cross-cultural Psychiatry," *Social Science and Medicine*, 11, pp. 3–10.

Kleinman, A., and Good, B. (1985), *Culture and Depression*, Berkeley: University of California Press.

Lerner, D. (1958), *The Passing of Traditional Society*, New York: Free Press.

Lessler, J., Tourangeau, R., and Salter, W. (1989), "Questionnaire Design in the Cognitive Research Laboratory," *Vital and Health Statistics*, Series 6, No. 1. DHHS Publication No. (PHS) 89–1501, National Center for Health Statistics, Rockville, MD: Public Health Service.

Leung, K., and Bond, M.H. (1989), "On the Empirical Identification of Dimensions for Cross-cultural Comparison," *Journal of Cross-cultural Psychology*, 20, pp. 133–151.

Malpass, R.S. (1977), "Theory and Method in Cross-cultural Psychology," *American Psychology*, pp. 1069–1079.

Marín, G., Gamba, R.J., and Marín, B.V. (1992), "Extreme Response Style and Acquiescence Among Hispanics: The Role of Acculturation and Education," *Journal of Cross-cultural Psychology*, 23, pp. 498–509.

Marín, G., and Marín, B.V. (1989), *Research with Hispanic Populations*, Newbury Park, CA: Sage.

Marín, G., Sabogal, F., Marín, B., Otero-Sabogal, R., and Pérez-Stable, E. (1987), "Development of a Short Acculturation Scale for Hispanics," *Hispanic Journal of Behavioral Sciences*, 9, pp. 183–205.

Marín, G., Triandis, H.C., Betancourt, H., and Kashima, Y. (1983), "Ethnic Affirmation

Versus Social Desirability: Explaining Discrepancies in Bilinguals' Responses to a Questionnaire," *Journal of Cross-cultural Psychology*, 14, pp. 173–186.

Marsella, A.J. (1978), "Thoughts on Cross-cultural Studies on the Epidemiology of Depression," *Culture, Medicine and Psychiatry*, 2, pp. 343–357.

Menon, G. (1994), "Judgments of Behavioral Frequencies: Memory Search and Retrieval Strategies," in N. Schwarz, and S. Sudman (eds.), *Autobiographical Memory and the Validity of Retrospective Reports*, New York: Springer-Verlag, pp. 161–172.

Mirowsky, J., and Ross, C.E. (1980), "Minority Status, Ethnic Culture, and Distress: A Comparison of Blacks, Whites, Mexicans, and Mexican Americans," *American Journal of Sociology*, 86, pp. 479–495.

Morse, J.M., and Morse, R.M. (1988), "Cultural Variation in the Inference of Pain," *Journal of Cross-cultural Psychology*, 19, pp. 232–242.

Moxey, L., and Sanford, A. (1992), "The Communicative Functions of Quantifiers and Their Use in Attitude Research," in N. Schwarz, and S. Sudman (eds.), *Context Effects in Social and Psychological Research*, New York: Springer Verlag, pp. 279–296.

Nisbett, R.E., and Wilson, T.D. (1977), "Telling More Than We Can Know: Verbal Reports on Mental Processes," *Psychological Review*, 84, pp. 231–259.

Pepper, S.C. (1981), "Problems in the Quantification of Frequency Expressions," in D.W. Fiske (ed.), *Problems with Language Imprecision, New Directions for Methodology of Social and Behavioral Science*, 9, pp. 25–41.

Pérez-Stable, E.J., Sabogal, F., Otero-Sabogal, Hiatt, R.A., and McPhee, S.J. (1992), "Misconceptions About Cancer Among Latinos and Anglos," *Journal of the American Medical Association*, 268, pp. 3219–3223.

Reese, S.D., Danielson, W.A., Shoemaker, P.J., Chang, T.K., and Hsu, H.L. (1986), "Ethnicity-of-Interviewer Effects Among Mexican-Americans and Anglos," *Public Opinion Quarterly*, 50, pp. 563–572.

Ross, C.E., and Mirowsky, J. (1983), "The Worst Place and the Best Face," *Social Forces*, 62, pp. 529–536.

Ross, C.E., and Mirowsky, J. (1984), "Socially-Desirable Response and Acquiescence in a Cross-cultural Survey of Mental Health," *Journal of Health and Social Behavior*, 25, pp. 189–197.

Schaeffer, N.C. (1980), "Evaluation of Race-of-Interviewer Effects in a National Survey," *Sociological Methods and Research*, 8, pp. 400–419.

Scheuch, E.K. (1993), "The Cross-Cultural Use of Sample Surveys: Problems of Comparability," *Historical Social Research*, 18, pp. 104–138.

Schuman, H., and Presser, S. (1981), *Questions and Answers in Attitude Surveys: Experiments on Question Form, Wording and Context*, San Diego: Academic Press.

Schwarz, N., and Sudman, S. (eds.) (1994), *Autobiographical Memory and the Validity of Retrospective Reports*, New York: Springer Verlag.

Strack, F., and Martin, L.L. (1987), "Thinking, Judging, and Communicating: A Process Account of Context Effects in Attitude Surveys," in H.-J. Hippler, N. Schwarz, and S. Sudman (eds.), *Social Information Processing and Survey Methodology*, New York: Springer Verlag, pp. 123–148.

Strahan, R., and Gerbasi, K.C. (1972), "Short, Homogeneous Versions of the Marlowe-Crowne Social Desirability Scale," *Journal of Clinical Psychology*, 28, pp. 191–193.

Strube, G. (1987), "Answering Survey Questions: The Role of Memory," in H.-J. Hippler, N. Schwarz, and S. Sudman (eds.), *Social Information Processing and Survey Methodology*, New York: Springer Verlag, pp. 86–101.

Triandis, H.C. (1972), *The Analysis of Subjective Culture*, New York: Wiley-Interscience.

Triandis, H.C. (1990), "Theoretical Concepts That Are Applicable to the Analysis of Ethnocentrism," in R.W. Brislin (ed.), *Applied Cross-cultural Psychology*, Newbury Park, CA: Sage, pp. 34–55.

Triandis, H.C. (1995), *Individualism & Collectivism*, Boulder, CO: Westview Press.

Triandis, H.C., Marín, G., Lisansky, J., and Betancourt, H. (1984), "Simpata as a Cultural Script for Hispanics," *Journal of Personality and Social Psychology*, 47, pp. 1363–1375.

Tulving, E. (1983), *Elements of Episodic Memory*, New York: Oxford University Press.

Tyson, G.A., Doctor, E.A., and Mentis, M. (1988), "A Psycholinguistic Perspective on Bilinguals' Discrepant Questionnaire Responses," *Journal of Cross-cultural Psychology*, 19, pp. 413–426.

Van de Vigver, F., and Poortinga, Y.H. (1982), "Cross-cultural Generalization and Universality," *Journal of Cross-cultural Psychology*, 13, pp. 387–408.

Wallston, K.A., Wallston, B.S., and DeVillis, R. (1978), "Development of the Multidimensional Health Locus of Control (MHLC) Scales," *Health Education Monographs*, 6, pp. 160–170.

Warnecke, R.B., and Graham, S. (1976), "Characteristics of Blacks Obtaining Papanicolaou Smears," *Cancer*, 37, pp. 2015–2025.

Word, C.O. (1992), "Cross-cultural Methods for Survey Research in Black Urban Areas," in A.K. Burlew, W.C. Banks, H.P. McAdoo, and D.A. Azibo (eds.), *African American Psychology: Theory, Research, and Practice*, Newbury Park, CA: Sage, pp. 28–42.

Wright, C.N., Phillips, L.D., Whalley, P.C., Choo, G.T., Ng, K.-O., Tan, I., and Wisudha, A. (1978), "Cultural Differences in Probabilistic Thinking," *Journal of Cross-cultural Psychology*, 9, pp. 285–299.

Zax, M., and Takahashi, S. (1967), "Cultural Influences on Response Style: Comparisons of Japanese and American College Students," *Journal of Social Psychology*, 71, pp. 3–10.

Zborowski, M. (1969), *People in Pain*, San Francisco: Jossey-Bass.

Zola, I.K. (1966), "Culture and Symptoms—An Analysis of Patients' Presenting Complaints," *American Sociological Review*, 31, pp. 615–630.

CHAPTER 5

Reducing Question Order Effects: The Operation of Buffer Items

Michaela Wänke
Universität Heidelberg

Norbert Schwarz
University of Michigan, Ann Arbor

5.1 INTRODUCTION

That preceding questions can influence responses to subsequent questions is well known to survey researchers. In the time since Sayre (1939) reported one of the first systematic investigations of question order effects, survey researchers have documented a wide range of context effects (see Schuman and Presser, 1981; Schwarz and Sudman, 1992; Sudman *et al.*, 1996, chapters 3 to 6; Tourangeau and Rasinski, 1988, for reviews). As this research demonstrates, question order effects are highly varied. Preceding questions may influence subsequent responses in the same direction as the implications of the context question (resulting in assimilation effects) or in the opposite direction (resulting in contrast effects; e.g., Schwarz and Strack, 1991; Tourangeau and Rasinski, 1988). Moreover, they may cause omissions (e.g., Martin *et al.*, 1990) and may change the qualitative nature of response (e.g., Smith, 1992).

To reduce the effect of preceding questions, researchers frequently separate related questions by a number of unrelated items, often referred to as "buffer items." An early discussion of buffer items has been provided by Cantril (1944). He observed that Americans were less likely to endorse that U.S. citizens serve in the French or British army when this question was preceded by a question about serving in the German army, an effect that we address in more detail

Survey Measurement and Process Quality, Edited by Lyberg, Biemer, Collins, de Leeuw, Dippo, Schwarz, Trewin.
ISBN 0-471-16559-X © 1997 John Wiley & Sons, Inc.

below. He further observed that such order effects may be reduced if the questions are separated by a number of buffer items—only to add in the next sentence that buffer items do not always work. To this day, little more is known about the conditions that determine the operation of buffer items. Some studies found no reductions in question order effects as a function of buffer items (e.g., Schuman *et al.*, 1983; Bishop *et al.*, 1984, exp. 2; Bishop, 1987), whereas others found nonsignificant trends (e.g., Tourangeau *et al.*, 1989a). Moreover, even when the introduction of buffer items significantly reduced the emerging context effects compared to no buffer control conditions, the effects were only attenuated but not eliminated (e.g., Bishop *et al.*, 1984, exp. 1; Tourangeau *et al.*, 1989b). Even more disturbing is the observation that buffer items may reverse rather than eliminate context effects, e.g., by turning an otherwise obtained contrast effect into an assimilation effect (e.g., Ottati, *et al.*, 1989).

As these examples illustrate, there is much we need to learn about the operation of buffer items. Because buffer items are intended to reduce the influence of preceding questions on responses given to subsequent ones, any theoretical analysis of their operation must emanate from what we know about the cognitive and communicative processes that underlie the emergence of context effects in the first place. At present, the use of buffer items is typically based on somewhat simplified assumptions about cognitive accessibility. Presumably, preceding questions may bring information to mind that is subsequently used in answering related questions, resulting in a question order effect. Introducing a number of unrelated buffer items increases the temporal distance between the context and the target question, as well as the amount of intervening information that is brought to mind. Both of these factors are expected to decrease the likelihood that the information primed by the context question is still accessible in memory when respondents answer the target question, thus eliminating the order effect. These theoretical assumptions are generally correct and well supported by research in cognitive psychology (see Anderson, 1983; Wyer and Srull, 1989). However, their application to survey measurement is less straightforward than survey practice suggests. On the one hand, the specific nature of order effects not only depends on which information is likely to come to mind, but also depends on how respondents *use* this information. The use of accessible information, however, is itself influenced by the presence or absence of buffer items, resulting in many of the surprises observed in the survey literature. On the other hand, some order effects are not driven by information accessibility *per se* to begin with, rendering them relatively immune to conditions that decrease the accessibility of previously used information.

In this chapter, we review current knowledge about the cognitive and communicative processes that drive the emergence of question order effects and explore the likely effect of buffer items on each of these processes. The organization of our discussion follows respondents' tasks, which include question comprehension, information retrieval and judgment formation, and response formatting and editing. We pay particular attention to the distinction between *information accessibility* and *information use*, which provides a conceptual framework for understanding the conditions under which buffer items attenuate, augment, or reverse the influence of preceding questions. Consistent with the bulk of available studies, most of our examples are drawn from the domain of attitude measurement. At this point, much of what we say has to remain speculative, reflecting a scarcity of tightly controlled experiments. We hope, however, that our conceptual analysis will prove helpful in setting the agenda for future research into the operation of buffer items.

5.2 COGNITIVE SOURCES OF QUESTION ORDER EFFECTS

How do preceding questions influence responses to subsequent questions? A closer examination of the response process helps us to understand how contextual influences come into play. To answer a survey question, respondents solve several tasks (see Schwarz, Chapter 1, for a summary and Strack and Martin, 1987; Tourangeau, 1984, for more detail). First, respondents interpret the question in order to understand what is meant. This comprehension task involves decoding the literal meaning and arriving at a semantic understanding. Understanding the semantic meaning of the question, however, is not sufficient. To provide an informative answer, respondents also have to determine what exactly the interviewer wants to know, i.e., they must infer the pragmatic meaning of the question (e.g., Schwarz, 1994; Strack, 1992). Next, respondents retrieve information that will allow them to answer the question. If the question is an opinion question, they may retrieve a judgment formed at a previous occasion. However, a previously formed judgment that pertains directly to the aspects addressed in the question is rarely accessible and in most cases respondents retrieve information that enables them to form a judgment at the time of the interview. If the question is a behavioral question, an appropriate response is also rarely accessible in memory and respondents again have to retrieve relevant information to arrive at an appropriate estimate (see Sudman *et al.*, 1996, Chapters 7 to 9). Once a judgment is formed, respondents need to report it. Most often, this requires that the judgment is formatted to fit the response alternatives provided by the interviewer. Moreover, some respondents may hesitate to reveal their "private" judgment, and may edit their responses, according to what they perceive as desirable in the specific situation.

Preceding questions may influence respondents' performance at each of these tasks, resulting in question order effects at the comprehension, retrieval, judgment formation, formatting or editing stage. Next, we review the relevant processes at each stage, identifying the likely effect of buffer items. As will become evident, different processes underlie question order effects at different stages and the likely effect of buffer items depends on the specific process involved.

5.2.1 Understanding the Question

At the question comprehension stage, preceding questions may influence respondents' interpretation of the *semantic* as well as *pragmatic* meaning of the question (see Strack, 1992; Strack and Schwarz, 1992, for reviews). These influences can be traced to the activation of semantic knowledge and the operation of conversational norms and may result in assimilation as well as contrast effects on subsequent judgments. We first address how accessible concepts influence the semantic interpretation of ambiguous information and identify conditions under which people try to avoid such influences. We introduce these processes by drawing on research in cognitive social psychology and subsequently discuss their implications for survey interviews. Finally, we turn to how respondents determine the pragmatic meaning of a question.

5.2.1.1 Semantic Meaning

As many studies in cognitive psychology demonstrate, ambiguous information is interpreted in terms of the concept that is most accessible, provided that the concept is applicable (see Higgins, 1989; Wyer and Srull, 1989, for reviews). For example, "crossing the Atlantic in a sailboat" may be perceived as adventurous or as reckless. Higgins *et al.* (1977) observed that subjects were more likely to interpret this behavior as adventurous when the concept "adventurous" rather than "reckless" had been activated by a preceding task. Extending this observation to the survey domain Strack *et al.* (1991, exp. 1) showed how concepts activated by a preceding question can influence the interpretation of a subsequent ambiguous question. They asked respondents to report their opinion on the introduction of a fictitious "educational contribution." This question was either preceded by a question on tuition fees or on financial support for students. As expected, the student subjects in their study inferred that the ambiguous "educational contribution" pertained to "paying money" when preceded by the tuition question, but to "getting money" when preceded by the student support question. As a result, they reported more favorable attitudes in the latter context than in the former.

As this example illustrates, respondents interpret an ambiguous question in terms of the most accessible concept that is applicable to the question at hand. This may be a concept that has been rendered accessible by a preceding question, resulting in an assimilation effect. Considering only the accessibility of the relevant concept in memory, we can expect that intervening buffer items reduce the strength of the context effect, because the accessibility of the relevant concept will decline over time (e.g., Higgins, 1989). The sheer accessibility of information in memory, however, is only part of the story as we shall see repeatedly throughout this chapter. To understand the effect of accessible information, we also need to consider the conditions under which it is, or is not, used.

Importantly, the literature in cognitive psychology suggests that assimilation effects of the type described above are only obtained when subjects are *not* aware that the relevant concept may only come to mind because it was recently used for a different purpose. To understand why this is the case, suppose that you read that "Donald crossed the Atlantic in a sailboat," as did the subjects in Higgins *et al.* (1977). If "reckless" comes to mind, you may assume that this reflects your reaction to the described behavior. But when you are aware that "reckless" may only come to mind because the preceding task made you think of it, you may discount this reaction and may not draw on it in interpreting the behavior. As a result, no assimilation effect would be obtained, reflecting that the highly accessible information is not *used*. Moreover, you may try to correct for the possible influence of the "inappropriate" thought, resulting in an interpretation of the behavior as being less reckless than would otherwise be the case. In fact, several studies observed that such corrections may result in contrast effects, essentially reflecting that people "overcorrect" for influences they consider inappropriate. Different theories have been proposed in this regard, which are beyond the scope of the present chapter (see Martin, 1986; Martin *et al.*, 1990, Higgins, 1989; Schwarz and Bless, 1992a; Strack *et al.*, 1993).

Unfortunately, it is difficult to generalize from this experimental research to the survey setting. Experiments that address that effect of subjects' awareness of the influence usually emphasize that the first task, used to activate the relevant concept, is completely independent from the second task, used to assess the effect of the activated concept. Given the alleged independence of the two tasks, subjects are likely to avoid any undue influence of the first task on the second task, provided that they are aware of such a possible influence. In surveys, however, respondents may perceive a sequence of questions as belonging together. If so, they may see nothing wrong in relying on information that was brought to mind by a related preceding question, even when they are fully aware that this is why it comes to mind. In Strack *et al.*'s (1991) educational contribution study, for example, respondents may have deliberately drawn on the content of preceding questions to determine what is meant by the ambiguous term educational contribution. In fact, such a deliberate use of contextual information is licensed by norms of conversational conduct, which suggests that a sequence of exchanges is meaningfully related to a central theme.

At present, we know little about the conditions under which survey respondents perceive a set of questions as belonging together, thus legitimizing the use of related information in interpreting subsequent questions. We suppose that the default is to consider adjacent questions as belonging together, in which case respondents would be unlikely to consider information that is brought to mind by a preceding question as constituting an undesirable extraneous influence. This default can be called into question by appropriate transitions that introduce subsequent questions as pertaining to a different issue. Likewise, an intervening set of unrelated questions may signal a change in topics. In this case, respondents would presumably try to avoid undue influences of apparently unrelated questions, much as has been observed in the experimental literature.

These considerations suggest that the effect of buffer items depends on respondents' perception of whether a set of questions belongs together or not.

Perceiving questions as related presumably implies that the information brought to mind by one question does not constitute an inappropriate source of information in interpreting a subsequent related question. Hence, respondents should rely on the accessible information, resulting in an assimilation effect. The strength of this assimilation effect should decrease as the number of buffer items increases, reflecting that the intervening buffer items render the information less accessible.

Not so, however, when respondents perceive the questions as unrelated, e.g., because they are separated by buffer items. In this case, respondents who are aware that the accessible information may only come to mind because it was addressed in a preceding question, may not use it. This may not only result in the absence of an assimilation effect, but in the emergence of a contrast effect, as observed in the experimental literature reviewed above. Moreover, respondents' awareness of the possible influence may itself depend on the number of intervening buffer items: The larger the number of buffer items, the less respondents may be aware of the possible influence. If so, we may expect the strongest assimilation effect among apparently unrelated items when the number of buffer items is large enough to reduce respondents' awareness of the possible influence of the preceding question, but not large enough to fully eliminate the information's increased accessibility.

As this discussion indicates, we need to consider the *accessibility* as well as *use* of semantic information to understand the emergence of context effects at the question comprehension stage as well as at the judgment stage, as we see below. The assumption that these effects reflect the sheer accessibility of information in memory holds only when respondents are not aware of a possible influence of a preceding question or do not consider this influence inappropriate. If respondents are aware of a possible influence and consider it inappropriate, they will not use the accessible information in interpreting a subsequent question, resulting in the absence of an assimilation effect. Ironically, buffer items may reduce respondents' awareness of a possible influence, thus rendering assimilation effects more likely. Future research will need to determine which features of a questionnaire influence respondents' perception of the relatedness of questions and the appropriateness or inappropriateness of perceived possible influences. Next, we turn to context effects in respondents' interpretation of the pragmatic meaning of a question.

5.2.1.2 *Pragmatic Meaning*

Understanding the semantic meaning of a question is not enough to answer it. Rather, respondents have to determine which information is of interest to the researcher to provide an informative answer. Much as the semantic interpretation, the pragmatic interpretation is also subject to context effects, which can be traced to the operation of conversational processes. In survey interviews as in daily life, our conversational conduct is governed by the tacit knowledge described in the work of Grice (1975), a philosopher of language (see Clark and Schober, 1992; Schwarz, 1995; Strack and Schwarz, 1992, for reviews and applications to survey research). Central to Grice's analysis of the logic of conversation is the assumption that communicators proceed according to a co-operativeness principle. This principle says that every contribution should be relevant to the aim of the ongoing conversation, and that speakers should not provide information that is irrelevant to the task at hand. Moreover, speakers are required to make their contributions informative, that is, to provide information that the recipient needs, rather than information that the recipient already has, or may take for granted anyway. Conforming to these conversational norms requires a considerable degree of inference to determine which information is "informative" in the specific context given. In the survey interview, this context is, in part, constituted by preceding questions and by the response alternatives presented with a question (which we do not address in the present chapter; see Schwarz and Hippler, 1991, for a review).

The assumption that all contributions are relevant to the goal of the ongoing conversation allows the use of preceding questions in making sense of subsequent ones. As discussed above, we assume that respondents deliberately draw on the information brought to mind by preceding questions as long as the features of the questionnaire or the conversational situation do not suggest that this is inappropriate. Hence, this aspect of conversational norms contributes to the emergence of assimilation effects at the comprehension stage, which reflect assumed similarities of related questions.

However, the request to make one's contribution informative, rather than to reiterate information that has already been given, gives rise to question interpretations that emphasize differences rather than commonalities. For example, Strack *et al.* (1991, exp. 2) asked respondents to report their general happiness and life satisfaction and varied the perceived conversational relatedness of both questions. In one condition, the happiness question was presented as the last question of the first questionnaire, whereas the life satisfaction question was presented as the first question of an ostensibly unrelated second questionnaire, attributed to another researcher. In this case, respondents' reported mean happiness and mean satisfaction were nearly identical and the two measures correlated $r = 0.96$. In another condition, however, both questions were presented at the end of the same questionnaire and were introduced by a joint lead-in that read, "Now, we have two questions about your life." In this case, respondents' reported mean happiness differed from their reported mean satisfaction and the correlation of both measures dropped significantly to $r = 0.75$. A comparison of both conditions suggests that respondents interpreted the happiness and satisfaction questions as having roughly the same meaning when they were presented by different researchers in separate questionnaires. Presenting them with a joint lead-in in the same questionnaire, however, induced respondents to differentiate between happiness and satisfaction, because assigning the same meaning to both questions would have rendered the second one redundant and uninformative.

As this example illustrates, the norm of nonredundancy draws attention to differences between apparently similar questions when respondents try to determine what constitutes an "informative" response. Buffer items are likely to influence this process in two ways. First, they may reduce the perceived conversational belongingness of similar questions, rendering it less likely that respondents recognize the potential redundancy. Presumably, presenting the happiness and satisfaction questions in the same questionnaire would not be sufficient to elicit differential interpretations if both questions were separated by a number of unrelated ones. Second, the specific content of buffer items may influence what is or is not considered "informative," as the next example illustrates.

Schwarz et al. (1991) asked respondents about their marital satisfaction and their general life satisfaction, in different question orders. In one condition, the marital satisfaction question was presented at the bottom of one page of a self-administered questionnaire, followed by the general question at the top of the next page. In this case, both questions correlated $r = 0.67$, reflecting that respondents considered information about their marriages in evaluating their lives in general. In another condition, however, both questions were presented on the same page and introduced by a joint lead-in that read, "We now have two questions about your life. The first pertains to your marital satisfaction and the second to your general life satisfaction." In this case, the correlation dropped to a low and nonsignificant $r = 0.18$. Having just provided information about their marriages, respondents apparently interpreted the general question as if it referred to aspects of their lives that they had not yet reported on. Consistent with this interpretation, a condition in which respondents were explicitly asked how satisfied they were with "other aspects" of their lives, "aside from their relationship," yielded a nearly identical correlation of $r = 0.20$. Thus, assigning both questions to the same conversational context induced respondents to avoid redundancy, resulting in a pragmatic interpretation of the general question as pertaining to other aspects of one's life. That respondents may interpret a more general question to exclude the specific information, when the specific question precedes the general one, may explain such part-whole-contrasts, as have been found in the literature (Schuman and Presser, 1981; Schuman et al., 1981; Kalton et al., 1978; Bishop, et al., 1985).

Suppose, however, that several (rather than only one) specific questions precede the general question. For example, respondents may be asked to report on their job satisfaction, their leisure time satisfaction, and their marital satisfaction before a general life satisfaction question is presented. In this case, they may interpret the general question as a request to integrate the previously reported aspects into an overall judgment, much as if it were worded, "Taking these aspects together, how satisfied are you with your life as a whole?" Interpreting the general question in this way is legitimate from a conversational point of view if several specific questions have been asked. In that case, a general judgment provides "new" information about the relative importance of the respective domains, which are the focus of the conversation. If only one specific question has been asked, however, there is little to sum up, rendering an integrative interpretation illegitimate.

To test these considerations, Schwarz et al. (1991) asked three specific questions, pertaining to their marital satisfaction, leisure time satisfaction, and job satisfaction, before they answered the general question. In this case, introducing all questions by a joint lead-in, thus assigning them explicitly to the same conversational context, did *not* reduce the correlation of marital satisfaction and general life satisfaction. This indicates that respondents adopted a "taking all aspects together" interpretation of the general question if it was preceded by three, rather than one, specific questions. This interpretation is further supported by a similar correlation when the general question was reworded to request an integrative judgment, and a much lower correlation when the reworded question required the consideration of other aspects of one's life (for other examples see Schul and Schiff, 1993; Benton and Daly, 1991).

These findings have important implications for the effect of buffer items on respondents' understanding of the pragmatic meaning of a question. As we have seen, assigning one specific and one general question to the same conversational context induced respondents to interpret the general question as pertaining to aspects *not* covered by the specific question, thus avoiding redundancy. Asking several specific questions prior to the general one, however, changed the pragmatic meaning of the general question from a request for additional information to a request for an integrative *summary* judgment. The latter interpretation, however, is only feasible if the specific questions are related to one another and can plausibly be integrated into one judgment. Had these questions pertained to issues unrelated to the quality of one's life, this interpretation would not have been feasible. Hence, the effect of buffer items on respondents' interpretation of the pragmatic meaning of a question depends on the *specific content* of the buffer items. As we see throughout this chapter, the sheer number of intervening items is much less relevant than their specific content.

The studies reviewed, so far, manipulated respondents' perceptions of the conversational context by introducing related questions with a joint lead-in. Can we assume that these processes operate when both questions simply follow each other, without an explicit lead-in? The answer presumably depends on the similarity of the questions, which determines their perceived relatedness. At least for the case of closely related questions, however, the answer is "yes" as a series of studies by Ottati et al. (1989), shown in Table 5.1, suggests.

In these studies, respondents were asked to report their agreement with general propositions, such as, "People should have the right to express their opinions in public." Each general proposition was preceded by a more specific item pertaining, for example, to the right of free speech for a positively (e.g., Parent Teacher Association) or for a negatively (e.g., Ku Klux Klan) evaluated group. Moreover, each specific item was either separated from the related general item by eight buffer items or directly preceded the general item without buffer items. The results showed a strong effect of item spacing. When the specific and general items were presented without buffer items, respondents reported higher support for the general proposition when the immediately

Table 5.1 Buffer Items May Reverse the Direction of Context Effects: Mean Agreement with General Propositions as a Function of Specific Propositions Pertaining to Favorable or Unfavorable Groups and Buffer Items

	Propositions adjacent	Propositions separated
General proposition preceded by:		
Study 1		
Favorable group	2.19	2.58
Unfavorable group	3.78	2.06
Study 2		
Favorable group	1.96	2.54
Unfavorable group	3.06	1.40
Study 3		
Favorable group	1.86	3.50
Unfavorable group	2.71	1.57

Note: Ratings ranged from -5 (disagree) to $+5$ (agree).

Adapted from Ottati *et al.* (1989), Copyright by American Psychological Association; reprinted by permission.

preceding specific item pertained to a negative rather than positive group. Having just said, for example, that the Ku Klux Klan should be denied freedom of speech, they presumably interpreted the general question as if it were worded, "Aside from the Ku Klux Klan, how about freedom of speech in general?" resulting in higher support for the general proposition. This interpretation was not evoked, however, when the specific and general questions were separated by eight buffer items. In this case, respondents reported less support for the general proposition (e.g., freedom of speech) when the specific item pertained to a negative rather than to a positive group. This reflects that the specific item about the Ku Klux Klan, for example, brought concerns to mind that respondents considered in forming their general judgments, thus reducing support for the general proposition.

As this example illustrates, buffer items may actually *reverse* the direction of order effects. In the present case, this reversal reflects that the number of buffer items was sufficient to reduce the perceived conversational relatedness of the specific and general question and respondents' awareness that their answers may be influenced by what has been brought to mind by the earlier question, as discussed in the preceding section. Had the number of buffer items been larger, it would eventually have been sufficient to reduce the accessibility of the previously activated information, thus eliminating order effects altogether. Without buffer items, however, respondents interpreted the question in line with the norm of nonredundancy, resulting in a contrast effect.

5.2.1.3 *Summary*

In summary, the reviewed research indicates that context effects at the question comprehension stage are not solely a function of information accessibility *per se*. Rather, the specific nature of context effects depends on how respondents use the accessible information in interpreting a question, which, in turn, is a function of conversational norms. As a result, the operation of buffer items is more complex than a simple accessibility model would suggest because buffer items may not only affect the accessibility of applicable information, but also respondents' awareness of the potential influence of preceding questions as well as their perceptions of the conversational relatedness of questions. Understanding this interplay in more detail provides an important agenda for future research.

5.2.2 Forming a Judgment

Once respondents have determined the intended meaning of the question they have to retrieve a previously formed judgment or form a judgment on the spot, based on information accessible at that point in time. Preceding questions may influence either of these possibilities by rendering a previously formed judgment more accessible, by inducing respondents to form a related judgment that they can subsequently draw on, or by rendering information accessible that may be used in forming a new judgment. In the latter case, the information may either bear directly on the target of judgment or may be of a more general nature, pertaining, for example, to general norms. Finally, a respondent may have had the subjective experience that it was easy or difficult to answer certain preceding questions. Such experiences may also affect subsequent judgments under specific conditions. Below, we discuss each of these possibilities in more detail and highlight how buffer items may influence the underlying processes.

5.2.2.1 *General Norms*

A general norm that has received considerable attention in survey research is the norm of even-handedness. This norm holds that "if an advantage (or disadvantage) is given to one party in a dispute, it should be given to the other as well" (Schuman and Ludwig, 1983, p. 112). Consider, for example, the question "Do you think the United States should let Communist newspaper reporters from other countries come in here and send back to their papers the news as they see it?" As Hyman and Sheatsley (1950) observed, agreement with this question was lower when it preceded rather than followed a parallel question that asked whether Communist countries should allow American reporters to report the news as they see it. Given that most of their American respondents endorsed freedom of press for American reporters in Communist countries, this suggests that the question about American reporters evoked the norm of even-handedness, resulting in higher endorsements of freedom of press for Communist reporters as well (see Schuman and Ludwig, 1983 for other examples). This effect was replicated well three decades later (Schuman *et al.*,

1983), with 44.4 percent of an American sample granting freedom of press to Communist reporters when this question was asked first, and 70.1 percent doing so when it was preceded by the American reporter question. More importantly, however, interposing 17 unrelated buffer items attenuated the effect only nonsignificantly, with 66.4 percent of the respondents endorsing freedom of press for Communist reporters.

Unfortunately, it remains unclear what the observed inefficiency of buffer items is due to. On the one hand, one may speculate that the cognitive accessibility of activated general norms has a long decay time, thus rendering 17 buffer items insufficient. Consistent with this possibility, studies that introduced buffer items in questionnaires that assessed other general concepts, such as anomia (Martin, 1980) or anxiety (Hayes, 1964), also failed to obtain a reduction of order effects through buffer items. On the other hand, the wording of the items pertaining to American and to Communist reporters is nearly identical, rendering it almost impossible not to notice that a similar question has been asked before. If so, the second question may have reminded respondents of the first one, thus undermining the effect of buffer items. Hence, we cannot say whether the norm of even-handedness would persist if the second question were phrased in a way less reminiscent of the first. The same holds for the anomia (Martin, 1980) and anxiety (Hayes, 1964) studies, which are also characterized by a high similarity of the relevant items. Future research will need to separate these possibilities by manipulating the surface similarity of the context and target question, bearing in mind that surface similarities have occasionally been found to increase question order effects (e.g., Duncan and Schuman, 1980; Feldman and Lynch, 1988).

Note, however, that the effect of being reminded of the previous question should depend on the process that drives the context effect in the first place. If the context effect is due to a general norm, reminding respondents of this norm should resuscitate the context effect despite the intervening buffer items. But if the context effect is due to the use of previously activated information in interpreting the meaning of the target question, reminding respondents of the context question may result in assimilation or in contrast effects, depending on the conditions discussed in the preceding section. Systematic explorations of these possibilities provide a promising avenue for future research.

5.2.2.2 *Information Bearing on the Target of Judgment*

Many question order effects discussed in the survey literature reflect that preceding questions increase the cognitive accessibility of information that may be used in forming subsequent judgments. The underlying processes have been conceptualized in several models that are consistent with current theorizing in cognitive and social psychology (see Feldman and Lynch, 1988; Feldman, 1992; Schwarz and Bless, 1992a; Schwarz and Strack, 1991; Strack and Martin, 1987; Strack, 1994a; Tourangeau and Rasinski, 1988; Tourangeau, 1992). Our discussion draws on Schwarz and Bless's (1992a; see also Sudman *et al.*, 1996, Chapter 5) inclusion/exclusion model. This model is highly similar to

Tourangeau's (1992) belief-sampling model in its conceptualization of assimilation effects, but has the advantage of offering an explicit conceptualization of contrast effects, which is absent in other models.

The key assumption of the model is that evaluative judgments require a representation of the target (i.e., the object of judgment), as well as a representation of some standard against which the target is evaluated. Both representations are context dependent and include information that is *chronically* accessible as well as information that is only *temporarily* accessible, for example, because it was used to answer a preceding question. How accessible information influences the judgment depends on how it is used, reiterating a theme we already emphasized in the section on question comprehension.

Assimilation Effects

Information that is *included* in the temporary representation that respondents form of the target will result in *assimilation effects*. This reflects that the judgment is based on the information that is included in the representation used. Accordingly, the addition of information with positive implications results in a more positive judgment, whereas the addition of information with negative implications results in a more negative judgment. Not surprisingly, the effect of including a highly accessible piece of information increases with its evaluative extremity. However, the size of assimilation effects is limited by the amount and extremity of other information that is temporarily or chronically accessible and included in the representation formed. Thus, adding a piece of information to a representation that includes only one other piece of information has more effect than adding it to a representation that includes many other pieces of information.

Accordingly, buffer items may influence the emergence and size of assimilation effects in two ways. On the one hand, they may reduce the accessibility of the previously used information X, thus rendering it less likely that X comes to mind when a representation of the target is formed. This is the most commonly held assumption in survey research and it holds most clearly when the buffer items are unrelated to the target item. On the other hand, the buffer items may bring other information (Y and Z) to mind that may be used in forming a representation of the target. This would reduce the effect of X even if X were included in the representation formed of the target, reflecting that this representation would now include a larger set of information (X, Y, and Z). Unfortunately, studies that systematically compare the relative effect of buffer items which do or do not bear on the target are not available.

Examples of the effect of substantively related buffer items include a study by Smith (1981, 1988), who observed that the influence of a previous question was restricted to responses to the first question of a question battery. Conceivably, this first question brought additional diagnostic information to mind that limited the effect of the context question on subsequent judgments (see also Feldman and Lynch, 1988). As another example, consider the Schwarz *et al.* (1991) study on marital satisfaction and life satisfaction, discussed above.

When respondents reported their general life satisfaction prior to their marital satisfaction, the two measures were correlated $r = 0.32$. This presumably reflects that some respondents spontaneously drew on chronically accessible information about their marriage in forming a representation of their life as a whole. When the marital satisfaction question preceded the general life satisfaction question, however, this correlation increased to $r = 0.67$, indicating that the marital satisfaction question rendered information about one's marriage temporarily accessible. As mentioned earlier, however, other respondents had to report their leisure time satisfaction, job satisfaction, and marital satisfaction prior to answering the general question. In this case, the observed increase in correlation between the marital satisfaction and life satisfaction questions was more modest, and the resulting correlation was $r = 46$. This reflects that the questions about one's job and leisure time brought a more varied set of information to mind. Including this information, along with information about one's marriage, in the representation of one's life as a whole attenuated the effect of marital information, as predicted by the model.

In most cases, however, the buffer items used in survey research are unrelated to the target question. According to the inclusion/exclusion model, these buffer items may attenuate assimilation effects to the extent that they reduce the temporary accessibility of information primed by the context question, in line with the usual assumption made by survey researchers. Substantively related buffer items, on the other hand, may further attenuate assimilation effects by rendering additional information accessible that may be included in the mental representation of the target. In addition, related buffer items may reduce the accessibility of prior material due to retroactive interference (Feldman and Lynch, 1988). Hence, different context effects may attenuate one another and this may underlie some of the apparently more successful applications of buffer items.

Contrast Effects

According to the inclusion/exclusion model, the same piece of information that elicits an assimilation effect may also result in a *contrast effect*. This is the case when the information is *excluded* from, rather than included in, the cognitive representation formed of the target.

As a first possibility, suppose that a given piece of information with positive (negative) implications is excluded from the representation of the target category. If so, the representation will contain less positive (negative) information, resulting in less positive (negative) judgments. We call this possibility a *subtraction*-based contrast effect (see Bradburn, 1983; Schuman and Presser, 1981, for related suggestions). The size of subtraction-based contrast effects increases with the amount and extremity of the temporarily accessible information that is excluded from the representation of the target, and decreases with the amount and extremity of the information that remains in the representation of the target. Thus, excluding the same piece of information from a representation that contains many other pieces of information has less effect than excluding it from a representation that contains only a few other pieces. As in the case of assimilation effects, buffer items may influence these processes by decreasing the accessibility of the context information or by rendering other information accessible that may be included in the representation of the target.

As a second possibility, respondents may not only exclude accessible information from the representation formed of the target, but may also use this information in constructing a standard of comparison or scale anchor. If the implications of the temporarily accessible information are more extreme than the implications of the chronically accessible information used in constructing a standard or scale anchor, they result in a more extreme standard, eliciting contrast effects for that reason. The size of these *comparison*-based contrast effects increases with the extremity and amount of temporarily accessible information used in constructing the standard or scale anchor, and decreases with the amount and extremity of chronically accessible information used in making this construction. Thus, adding a given piece of information to the representation of a standard that contains many other pieces of information changes the standard less than adding the same piece to a representation that contains many other pieces. Again, buffer items may influence these processes by decreasing the accessibility of the context information or by increasing the accessibility of other information that may be used in forming a representation of a relevant standard.

Determinants of Inclusion and Exclusion

According to the inclusion/exclusion model (Schwarz and Bless, 1992a), the *default operation* is to include information that comes to mind in the representation of the target. This suggests that we should be more likely to see assimilation rather than contrast effects in survey research, an issue that should be addressed by meta-analyses. In contrast, the exclusion of information needs to be triggered by salient features of the question answering process. In principle, *any* variable that affects the categorization of information is likely to affect the emergence of assimilation and contrast effects, linking the model to cognitive research on categorization processes in general. Schwarz and Bless (1992a) review a host of heterogeneous variables that have been shown to affect context effects in social judgment. These variables can be conceptualized as bearing on three implicit decisions that respondents have to make with regard to the information that comes to mind.

Some information that comes to mind may simply be irrelevant, pertaining to issues that are unrelated to the question asked. Other information may potentially be relevant to the task at hand and respondents have to decide what to do with it. The first decision bears on why this information comes to mind. Information that seems to come to mind for the "wrong reason," e.g., because respondents are aware of the potential influence of a preceding question, is likely to be excluded (e.g., Lombardi *et al.*, 1987; Ottati *et al.*, 1989; Strack *et al.*, 1993). This parallels our previous discussion of the disuse of highly accessible information at the question comprehension stage when respondents are aware of the possible influence of a preceding question. As discussed previously, buffer items may decrease this awareness. For this reason, they may decrease the likelihood of contrast effects and increase the likelihood of assimilation effects, as observed in the Ottati *et al.* (1989) study mentioned above.

The second decision bears on whether the information that comes to mind "belongs to" the target category or not. The content of the context question (e.g., Schwarz and Bless, 1992a), the width of the target category (e.g., Schwarz and Bless, 1992b), the extremity of the information (e.g., Herr, 1986), or its representativeness for the target category (e.g., Strack *et al.*, 1985) are relevant at this stage. None of these variables, however, is likely to be affected by the introduction of buffer items, unless they are sufficient to render the context information inaccessible—in which case respondents do not have to decide how to use it in the first place.

Finally, conversational norms may determine respondents' perceptions of what they are supposed to do with highly accessible information (e.g., Schwarz *et al.*, 1991; Strack *et al.*, 1988), as we have seen in the section on question comprehension. As discussed in that context, respondents try to avoid redundancy and are therefore unlikely to reiterate information that they have already provided. Accordingly, they exclude this information from consideration. As noted earlier, however, the introduction of unrelated buffer items may undermine the perceived conversational relatedness of the questions and may hence undermine the otherwise obtained exclusion of redundant information. Moreover, introducing substantively related buffer items (as in Schwarz *et al.*, 1991) may change the nature of the conversational process, as discussed above.

Whenever *any* of these decisions results in the exclusion of information from the representation formed of the target, it will elicit a contrast effect. Whether this contrast effect is limited to the target or generalizes across related targets depends on whether the excluded information is merely subtracted from the representation of the target or used in constructing a standard or scale anchor. Whenever the information that comes to mind is included in the representation formed of the target, on the other hand, it results in an assimilation effect. Hence, the model predicts the emergence, the direction, the size, and the generalization of context effects in attitude measurement. Moreover, it provides specific predictions regarding the effect of buffer items on the *use* of accessible information, as discussed above. These predictions provide a promising agenda for future experimentation.

5.2.2.3 Recall Experiences

The preceding section focused on the information that is brought to mind and emphasized that the effect of this information depends on how respondents use it. However, our judgments do not only depend on the content we recall. Rather, the implications of this content may be qualified by how easy or difficult it is to bring to mind (see Clore, 1992, for a review). As an example, consider a series of studies reviewed by Bishop (1987). Bishop and his colleagues asked respondents how much they "follow what's going on in government and public affairs." This question was either preceded or followed by a political knowledge question that asked respondents to report what their U.S. representative has done for his or her district, a question that many respondents could not answer. When the knowledge question preceded the political interest question, respondents reported following public affairs less than when it followed the political interest question. Presumably, respondents concluded from their inability to answer the knowledge question that they do not follow public affairs very closely, or else they would know what their representative had done.

In a series of intriguing studies, Bishop (1987) explored the effect of buffer items on this question order effect. In three experiments, the number of unrelated buffer items used was 33, 40, and even 101, taking between 7 and 17 minutes to administer. Despite this unusual length of the buffer sequence, the size of the context effect remained virtually unaffected, in contrast to what most researchers would expect.

How are we to account for this? Presumably, respondents based their judgments about how much they follow politics on the implications of the most accessible relevant information. Given that the buffer items were unrelated, pertaining mostly to community issues and cable TV, the most accessible information relevant to that judgment was the salient experience that one knew little, if anything, about one's representative's record. Hence, Bishop (1987, p. 182) concluded that these effects "will last until the respondent has an experience that changes his or her self-perception, either during the interview or afterwards."

Note that this self-perception explanation presupposes an implicit attribute: respondents have to assume that their lack of knowledge about their representative is due to their own behavior, namely that they do not follow public affairs. Alternatively, however, they might assume that their representative is not doing a good job in keeping them informed. If so, a single buffer item that draws attention to this possibility may accomplish what 101 unrelated items could not. To test this possibility, Schwarz and Schuman (1995) replicated Bishop's (1987) study and introduced a buffer item that asked respondents to evaluate how good a job their representative does in keeping them informed. As in Bishop's studies, the knowledge question reduced self-reported political interest by 18.5 percent. However, the single buffer item pertaining to the representative's public relations work attenuated this effect to about 8.8 percent. Note that this attenuation is not due to any reduction in the cognitive accessibility of respondents' difficulties in answering the knowledge question, as the inefficiency of the previously used 101 buffer items indicates. Rather, the public relations question changed the perceived implications of the experienced difficulty by suggesting that respondents' lack of knowledge was due to their representative's poor public relations work, rather than to their not following public affairs.

Inducing respondents to misattribute their lack of knowledge is only one way to change their self-perception. Martin and Harlow (1992) suggested that the failure to answer a political knowledge question disconfirms one's self-concept. To test this notion, they introduced different buffer questions in an extended replication of Bishop's (1987) studies. One set of questions allowed respondents to confirm their self-concept by expressing their political attitude (cf. Steele, 1988), whereas the other set pertained to nonpolitical issues. As expected, answering two political attitude questions reduced the effect of the political knowledge question, whereas the nonpolitical questions did not. It remains unclear, however, whether this reflects that the political questions allowed respondents to confirm their self-concept or whether answering these questions brought information to mind that indicated that one does, indeed, follow politics at least on some occasions.

In combination, these examples again illustrate that different psychological processes may contribute to the emergence and elimination of question order effects. The differential effect of Bishop's (1987) 101, Martin and Harlow's (1992) two, and Schwarz and Schuman's (1995) single buffer item clearly indicates that the efficiency of buffer items is not a function of their sheer number, but crucially depends on their specific contents.

5.2.3   Formatting

Once respondents have reached a judgment, they face the additional task of formatting it to fit the response alternatives provided by the researcher. Suppose a respondent has arrived at a positive judgment. Is he or she to choose a 7 on a 7-point rating-scale to express this, or a 6 or a 5? Unless the rating scale is calibrated in some way, the numbers are rather meaningless. To calibrate the scale in the context of the ongoing conversation with the researcher, respondents usually anchor the scale with the most extreme stimuli that come to mind, as numerous studies in psychophysics (see Poulton, 1989, for a review) and attitude measurement (see Strack, 1994b, for a review) demonstrate. Thus, a weight of one pound, for example, is rated as heavier when it is presented in the context of lighter weights than when it is presented in the context of heavier weights.

Such "perspective" (Ostrom and Upshaw, 1968) or "range" effects (Parducci, 1982, 1983) reflect that the contextual stimuli serve as a frame of reference and that the scale is anchored with the most extreme stimuli in the set. Note, however, that respondents can only anchor the scale on the extremes of the presented stimulus continuum when they are aware what these extremes are. If only one stimulus is presented at a time, respondents will anchor the scale on extremes recalled from memory, reflecting whatever perspective is accessible to them at the time of judgment, which is partially a function of the content of preceding questions. To enter into the perspective, however, the extreme stimulus must seem relevant to the task at hand, as an example may illustrate. In a classic study in psychophysics, Brown (1953) asked subjects to rate the weight of different objects they were asked to lift. As usual, a given weight was rated as heavier when preceded by a light rather than by a heavy object. This, however, was only the case when the heavy object was perceived as being part of the series of objects to be rated. In contrast, when the subjects were asked to lift a tray of equivalent weight that was in the experimenter's way, their judgments of the target objects remained unaffected. Apparently, the tray was not perceived as part of the stimulus series. Accordingly, it did not enter subjects' representation of the relevant "perspective" and was hence not used to anchor the rating scale.

Again, the content of preceding questions may influence what comes to mind and is used as a scale anchor, as discussed in the preceding section. Buffer items may influence this process by either reducing the cognitive accessibility of previously used information or by undermining the perceived relevance of this information for the new task at hand. In many cases, however, it is difficult to determine whether an observed contrast effect reflects solely a change in rating scale use or a change in the mental representations formed. When the stimuli require little interpretation (as is the case for weights), any emerging contrast effect is likely to reflect solely a change in response language. When the stimuli require more interpretation, on the other hand, changes in the use of rating scales as well as substantive changes in the mental representations formed may contribute to context effects (e.g., Lynch *et al.*, 1991; for an extended discussion see Strack, 1994b).

5.2.4 Editing

Preceding questions, pertaining, for example, to one's religious affiliation or moral values, may bring information to mind that may increase subsequent self-presentation concerns or may influence what respondents consider to be socially desirable answers (DeMaio, 1984). Again, buffer items may decrease these influences if they reduce the cognitive accessibility of the relevant information. At present, however, we are not aware of studies that have addressed the impact of buffer items at the editing stage.

5.3 CONCLUSIONS

Probably the most obvious conclusion to be drawn from the present review is that we know surprisingly little about the operation of buffer items. Although most researchers pay close attention to the spacing of questions in their questionnaires, the decisions made are primarily based on intuition and common sense. Systematic experimentation pertaining to the operation of buffer items is rare. Moreover, many of the experiments reported in the survey literature are difficult to interpret, reflecting that they were not designed to test specific hypotheses about the underlying cognitive and communicative processes. As a result, theoretically informed guidelines for item spacing are difficult to derive from the inconsistent body of evidence. As noted previously, some studies obtained no reductions in question order effects as a function of buffer items (e.g., Schuman *et al.*, 1983; Bishop *et al.*, 1984, exp. 2; Bishop, 1987), whereas others found nonsignificant trends (e.g., Tourangeau *et al.*, 1989a). Moreover, even the most successful applications of buffer items resulted merely in an attenuation of context effects, rather than the hoped for elimination (e.g., Bishop *et al.*, 1984, exp. 1; Tourangeau *et al.*, 1989b). Finally, buffer items reversed rather than eliminated context effects in other experiments, e.g., by turning an otherwise obtained contrast effect into an assimilation effect (e.g., Ottati *et al.*, 1989). This inconsistency is not surprising when we take into account that different context effects are due to different cognitive and communicative processes and are hence unlikely to be attenuated by the same mechanisms. Complicating things further, the buffer items themselves may not only influence the accessibility of previously used information, but may also influence what respondents do with this information when it comes to mind later on.

In general, the accessibility of information decreases as the time since its last use increases (see Wyer and Srull, 1989, for a review). How fast this decay occurs depends on characteristics of the information itself (such as its distinctiveness and vividness) as well as on characteristics of the information that are activated by intervening questions. If two questions about related political issues are separated by 20 questions about consumer products, for example, the previous political question is more likely to come to mind than if the intervening questions had also pertained to political issues, thus rendering the relationship between the context and the target question less distinct. Moreover, any attempt to separate the context and the target question may be futile if both questions are highly similar in content and wording, thus rendering the target question an excellent cue for the retrieval of the context question.

In addition to attenuating context effects by decreasing the accessibility of the context information, buffer items may attenuate the effect of the initial context question by bringing other information to mind that may bear on the target. In this case, the "original" context effect would be attenuated by subsequent context effects, elicited by the buffer items themselves. This possibility renders substantively related buffer items undesirable, even though these buffer items may be most efficient in attenuating the effect of the initial context question.

Complicating things further, buffer items may influence how respondents use the information rendered accessible by the context question, a possibility that has not previously been addressed in the survey literature. Buffer items may do so by influencing respondents' awareness of why some information comes to mind, by influencing the operation of conversational norms, or by influencing the perceived diagnosticity of accessible information. We addressed each of these possibilities in some detail above and summarize them in turn.

We illustrated the effect of buffer items on the perceived diagnosticity of accessible information with Bishop's (1987) studies, which documented that up to 101 unrelated buffer items were insufficient to eliminate the effect of respondents' inability to answer a knowledge question about their representative on their self-reported political interest. In contrast, a single question about the quality of the representative's public relations work reduced the otherwise observed context effect to nonsignificance (Schwarz and Schuman, 1995). This reflects that the single buffer item rendered respondents' lack of knowledge nondiagnostic for the judgment at hand. Hence, buffer items may not only eliminate context effects by decreasing the accessibility of contextual information, but may also do so by changing the perceived implications of the contextual information.

Moreover, the introduction of buffer items may influence the operation of the conversational norm of nonredundancy. Having just answered a question about freedom of speech for the Ku Klux Klan, for example, respondents interpret a subsequent general question about freedom of speech as referring to other speakers. This interpretation is not made when the specific and general questions are separated by several unrelated buffer items, thus disrupting their perceived conversational belongingness. As a result, the Ku Klux Klan question elicited a contrast effect without buffer items, but an assimilation effect with buffer items in Ottati et al.'s (1989) study. Hence, unrelated buffer items may change the interpretation of a general question by reducing the likelihood that the general and a preceding specific question are perceived as belonging to the same conversational context.

Similarly, substantively related buffer items may change question interpretation even under conditions where all questions are explicitly assigned to the same conversational context, e.g., by a joint lead-in. Thus, Schwarz et al. (1991) observed that respondents interpreted a question about general life satisfaction as a request for new information when it was preceded by a single specific question about marital satisfaction, but as a request for an integrative summary judgment when it was preceded by several specific questions, pertaining to satisfaction with different life domains. These differential interpretations elicited a part-whole contrast effect when only one specific question was asked, but a part-whole assimilation effect when several specific questions were asked. This particular change in question interpretation would not be expected when the buffer items are substantively unrelated, thus rendering an "integration" interpretation of the general question inapplicable. As these examples illustrate, the number and content of buffer items may influence the interpretation of the general question in part-whole question sequences, determining respondents' use of accessible information.

Finally, buffer items may have an influence if respondents are aware that some information may only come to mind because it was addressed in a preceding question. In general, individuals exclude information that seems to come to mind for the "wrong" reason from consideration, resulting in contrast effects. If the number of buffer items is sufficient to reduce respondents' awareness of the possible influence but insufficient to eliminate the increased accessibility of the context information, they may reverse an otherwise obtained contrast effect into an assimilation effect. The extent to which awareness of a previous question will result in the deliberate disuse of the information it brought to mind depends on whether the perceived influence is considered undesirable. At present, the conditions that determine this perception are ill understood, as discussed in more detail in Section 5.2.1 on question comprehension.

In summary, buffer items are a considerably more complex feature of questionnaires than the usual focus on accessibility would suggest. Far from only reducing the accessibility of previously used information, buffer items influence what respondents do with the information that comes to mind. As a result, they may attenuate as well as elicit assimilation and contrast effects. We hope that systematic experimental tests of the hypotheses offered in this chapter will eventually contribute to a better understanding of the dynamics of context effects and hence of the operation of buffer items.

REFERENCES

Andersen, J. (1983), *The Architecture of Cognition*, Cambridge: Harvard University Press.

Benton, J., and Daly, J. (1991), "A Question Order Effect," *Public Opinion Quarterly*, 55, pp. 640–642.

Bishop, G. (1987), " Context Effects in Self Perceptions of Interests in Government and Public Affairs," in H.-J. Hippler, N. Schwarz, and S. Sudman (eds.), *Social Information Processing and Survey Methodology*, New York: Springer Verlag, pp. 179–199.

Bishop, G., Oldendick, R., and Tuchfarber, A. (1984), "What Must My Interest in Politics Be If I Just Told You "I Don't Know"?," *Public Opinion Quarterly*, 48, pp. 510–519.

Bishop, G., Oldendick, R., and Tuchfarber, A. (1985), "The Importance of Replicating a Failure to Replicate: Order Effects on Abortion Items," *Public Opinion Quarterly*, 49, pp. 105–114.

Bradburn, N. (1983), "Response Effects," in P.H. Rossi, and J.D. Wright (eds.), *The Handbook of Survey Research*, New York: Academic Press, pp. 289–328.

Brown, D. (1953), "Stimulus Similarity and the Anchoring of Subjective Scales," *American Journal of Psychology*, 66, pp. 199–214.

Cantril, H. (1944), *Gauging Public Opinion*, Princeton: Princeton University Press.

Clark, H., and Schober (1992), "Asking Questions and Influencing Answers," in J. Tanur (ed.), *Questions About Questions*, New York: Russell Sage, pp. 15–48.

DeMaio, T.J. (1984), "Social Desirability and Survey Measurement: A Review," in C.F. Turner, and E. Martin (eds.), *Surveying Subjective Phenomena*, Vol. 2, New York: Russell Sage, pp. 257–281.

Duncan, O., and Schuman, H. (1980), "Effects of Question Wording and Context: An Experiment with Religious Indicators," *Journal of the American Statistical Association*, 75, pp. 269–275.

Feldman, J. (1992), "Constructive Processes as a Source of Context Effects in Survey Research: Explorations in Self-generated Validity," in N. Schwarz, and S. Sudman, (eds.), *Context Effects in Social and Psychological Research*, New York: Springer Verlag, pp. 49–62.

Feldman, J., and Lynch, J. (1988), "Self-generated Validity and Other Effects of Measurement on Belief, Attitude, Intention, and Behavior," *Journal of Applied Psychology*, 73, pp. 421–435.

Grice, H. P. (1975), "Logic and Conversation," in P. Cole, and J.L. Morgan (eds.), *Syntax and Semantics, Vol. 3: Speech Acts*, New York: Academic Press, pp. 41–58.

Hayes, D. (1964), "Item Order and Guttman Scales," *American Journal of Sociology*, 70, pp. 51–58.

Herr, P. (1986), "Consequences of Priming: Judgment and Behavior," *Journal of Personality and Social Psychology*, 51, pp. 1106–1115.

Higgins, E. T. (1989), "Knowledge Accessibility and Activation: Subjectivity and Suffering from Unconscious Sources," in J. S. Uleman, and J. A. Bargh (eds.), *Unintended Thought*, New York: Guilford Press, pp. 75–123.

Higgins, E., Rholes, W., and Jones, C. (1977), "Category Accessibility and Impression Formation," *Journal of Experimental Social Psychology*, 13, pp. 141–154.

Hyman, H. H., and Sheatsley, P. B. (1950), "The Current Status of American Public Opinion," in J. C. Payne (ed.), *The Teaching of Contemporary Affairs*, New York: National Education Association, pp. 11–34.

Kalton, G., Collins, M., and Brook, L. (1978), "Experiments in Wording Opinion Questions," *Journal of the Royal Statistical Society*, 27, pp. 149–161.

Lombardi, W., Higgins, E., and Bargh, J. (1987), "The Role of Consciousness in Priming Effects on Categorization: Assimilation and Contrast as a Function of Awareness of the Priming Task," *Personality and Social Psychology Bulletin*, 13, pp. 411–429.

Lynch, J., Chakravarti, D., and Mitra, A. (1991), "Contrast Effects in Consumer Judgments: Changes in Mental Representations or Anchoring of Rating Scales?," *Journal of Consumer Research*, 18, pp. 284–297.

Martin, E. (1980), "The Effects of Item Contiguity and Probing on Measures of Anomia," *Social Psychology Quarterly*, 43, pp. 116–120.

Martin, L. L. (1986), "Set/Reset: Use and Disuse of Concepts in Impression Formation," *Journal of Personality and Social Psychology*, 51, pp. 493–504.

Martin, E., DeMaio, T., and Campanelli, P. (1990), "Context Effects for Census Measures of Race and Hispanic Origin," *Public Opinion Quarterly*, 54, pp. 551–566.

Martin, L., and Harlow, T. (1992), "Basking and Brooding: The Motivating Effects of Filter Questions in Surveys," in H.-J. Hippler, N. Schwarz, and S. Sudman (eds.), *Social Information Processing and Survey Methodology*, New York: Springer Verlag, pp. 81–96.

Martin, L. L., Seta, J. J., and Crelia, R. A. (1990), "Assimilation and Contrast as a Function of People's Willingness to Expend Effort in Forming an Impression," *Journal of Personality and Social Psychology*, 59, pp. 27–37.

Ostrom, T.M., and Upshaw, H.S. (1968), "Psychological Perspective and Attitude Change," in A.C. Greenwald, T.C. Brock, and T.M. Ostrom, (eds.), *Psychological Foundations of Attitudes*, New York: Academic Press, pp. 217–242.

Ottati, V.C., Riggle, E.J., Wyer, R.S., Schwarz, N., and Kuklinski, J. (1989), "The Cognitive and Affective Bases of Opinion Survey Responses," *Journal of Personality and Social Psychology*, 57, pp. 404–415.

Parducci, A. (1982), "Category Ratings: Still More Context Effects," in B. Wegener (ed.), *Social Attitudes and Psychological Measurement*, Hillsdale, NJ: Erlbaum, pp. 89–105.

Parducci, A. (1983), "Category Ratings and the Relational Character of Judgment," in

H. G. Geissler, H. F. J. M. Bulfart, E. L. H. Leeuwenberg, and V. Sarris (eds.), *Modern Issues in Perception*, Berlin: VEB Deutscher Verlag der Wissenschaften, pp. 262–282.

Poulton, E.C. (1989), *Bias in Quantifying Judgments*, London: Erlbaum.

Sayre, J. (1939), "A Comparison of Three Indices of Attitudes Toward Radio Advertising," *Journal of Applied Psychology*, 23, 23–33.

Schul, Y., and Schiff, M. (1993), "Measuring Satisfaction with Organizations," *Public Opinion Quarterly*, 57, pp. 536–551.

Schuman, H., Kalton G., and Ludwig, J. (1983), "Context and Contiguity in Survey Questionnaires," *Public Opinion Quarterly*, 47, pp. 112–115.

Schuman, H., and Ludwig, J. (1983), "The Norm of Even-Handedness in Surveys as in Life," *American Sociological Review*, 48, pp. 112–120.

Schuman, H., and Presser, S. (1981), *Questions and Answers in Attitude Surveys*, New York: Academic Press.

Schuman, H., Presser, S., and Ludwig, J. (1981), "Context Effects on Survey Responses to Questions About Abortion," *Public Opinion Quarterly*, 45, pp. 216–223.

Schwarz, N. (1994), "Judgment in a Social Context: Biases, Shortcomings, and the Logic of Conversation," in M. Zanna (ed.), *Advances in Experimental Social Psychology*, Vol. 26, San Diego, CA: Academic Press, pp. 123–162.

Schwarz, N. (1995), "What Respondents Learn from Questionnaires: The Survey Interview and the Logic of Conversation," *International Statistical Review*, 63, pp. 153–177.

Schwarz, N., and Bless, H. (1992a), "Constructing Reality and Its Alternatives: Assimilation and Contrast Effects in Social Judgment," in L. Martin, and A. Tesser (eds.), *The Construction of Social Judgment*, Hillsdale, NJ: Erlbaum, pp. 217–245.

Schwarz, N., and Bless, H. (1992b), "Scandals and the Public's Trust in Politicians: Assimilation and Contrast Effects," *Personality and Social Psychology Bulletin*, 18, pp. 574–579.

Schwarz, N., and Hippler, H.-J. (1991). "Response Alternatives: The Impact of Their Choice and Ordering," in P. Biemer, R. Groves, L. Lyberg, N. Mathiowetz, and S. Sudman (eds.), *Measurement Error in Surveys*, Chichester: Wiley, pp. 41–56.

Schwarz, N., and Schuman, H. (1995), "Political Knowledge, Attribution, and Inferred Interest in Politics: The Operation of Buffer Items," University of Michigan: Unpublished manuscript.

Schwarz, N., and Strack, F. (1991), "Context Effects in Attitude Surveys: Applying Cognitive Theory to Social Research," in W. Stroebe, and M. Hewstone (eds.), *European Review of Social Psychology* (Vol. 2), Chichester: Wiley, pp. 31–50.

Schwarz, N., Strack, F., and Mai, H.-P. (1991), "Assimilation and Contrast Effects in Part-Whole Question Sequences: A Conversational Logic Analysis," *Public Opinion Quarterly*, 55, pp. 3–23.

Schwarz, N., and Sudman, S. (1992), *Context Effects in Social and Psychological Research*, New York: Springer-Verlag.

Smith, T. (1981), "Can We Have Confidence in Confidence? Revisited," in D.F. Johnston (ed.), *Measurement of Subjective Phenomena*, U.S. Bureau of the Census, Special

Demographic Analyses, CDS-80-3, Washington, DC: U.S. Government Printing Office, pp. 119–189.

Smith, T. (1988), "*Ballot Position: An Analysis of Context Effects Related to Rotation Design*," GSS Methodological Re. No. 55, Chicago: NORC.

Smith, T. W. (1992), "Thoughts on the Nature of Context Effects," in N. Schwarz, and S. Sudman (eds.), *Context Effects in Social and Psychological Research*, New York: Springer Verlag, pp. 163–186.

Steele, C. (1988), "The Psychology of Self-Affirmation: Sustaining the Integrity of the Self," in L. Berkowitz (ed.), *Advances in Experimental Social Psychology*, Vol. 21, San Diego: Academic Press, pp. 261–302.

Strack, F. (1992), "Order Effects in Survey Research: Activative and Informative Functions of Preceding Questions," in N. Schwarz, and S. Sudman (eds.), *Context Effects in Social and Psychological Research*, New York: Springer Verlag, pp. 23–34.

Strack, F. (1994a), *Zur Psychologie der standardisierten Befragung* [*The Psychology of Standardized Questioning*], Heidelberg: Springer Verlag.

Strack, F. (1994b), "Response Processes in Social Judgment," in R. Wyer, and T. Srull (eds.), *Handbook of Social Cognition*, Vol. 2, Hillsdale, NJ: Erlbaum, pp. 287–322.

Strack, F., and Martin, L. (1987), "Thinking, Judging, and Communicating: A Process Account of Context Effects in Attitude Surveys," in H.-J. Hippler, N. Schwarz, and S. Sudman (eds.), *Social Information Processing and Survey Methodology*, New York: Springer Verlag, pp. 123–148.

Strack, F., Martin, L.L., and Schwarz, N. (1988), "Priming and Communication: The Social Determinants of Information Use in Judgments of Life-Satisfaction," *European Journal of Social Psychology*, 18, pp. 429–442.

Strack, F., and Schwarz, N. (1992), "Implicit Cooperation: The Case of Standardized Questioning," in G. Semin, and F. Fiedler (eds.), *Social Cognition and Language*, Beverly Hills: Sage, pp. 173–192.

Strack, F., Schwarz, N., Bless, H., Kübler, A., and Wänke, M. (1993), "Awareness of the Influence as a Determinant of Assimilation versus Contrast," *European Journal of Social Psychology*, 23, pp. 53–62.

Strack, F., Schwarz, N., and Gschneidinger, E. (1985),"Happiness and Reminiscing: The Role of Time Perspective, Mood and Mode of Thinking," *Journal of Personality and Social Psychology*, 49, pp. 1460–1469.

Strack, F., Schwarz, N., and Wänke, M. (1991), "Semantic and Pragmatic Aspects of Context Effects in Social and Psychological Research," *Social Cognition*, 9, pp. 111–125.

Sudman, S., Bradburn, N., and Schwarz, N. (1996), *Thinking About Answers: The Application of Cognitive Processes to Survey Methodology*, San Francisco: Jossey-Bass.

Tourangeau, R. (1984), "Cognitive Science and Survey Methods: A Cognitive Perspective", in T. Jabine, M. Straf, J. Tanur, and R. Tourangeau (eds.), *Cognitive Aspects of Survey Methodology: Building a Bridge Between Disciplines*, Washington, DC: National Academy Press, pp. 73–100.

Tourangeau, R. (1992), "Attitudes as Memory Structures: Belief Sampling and Context

Effects," in N. Schwarz, and S. Sudman (eds.), *Context Effects in Social and Psychological Research*, New York: Springer Verlag, pp. 35–47.

Tourangeau, R., and Rasinski, K. (1988), "Cognitive Processes Underlying Context Effects in Attitude Measurement," *Psychological Bulletin*, 103, pp. 299–314.

Tourangeau, R., Rasinski, K., Bradburn, N., and D'Andrade, R. (1989a), "Carryover Effects in Attitude Surveys," *Public Opinion Quarterly*, 53, pp. 495–524.

Tourangeau, R., Rasinski, K., Bradburn, N., and D'Andrade, R. (1989b), "Belief Accessibility and Context Effects in Attitude Measurement," *Journal of Experimental Social Psychology*, 25, pp. 401–421.

Wyer, R., and Srull, T. (1989), *Memory and Cognition in Its Social Context*, Hillsdale, NJ: Erlbaum.

CHAPTER 6

Designing Rating Scales for Effective Measurement in Surveys

Jon A. Krosnick
The Ohio State University

Leandre R. Fabrigar
Queen's University at Kingston

6.1 INTRODUCTION

Rating scales are omnipresent in contemporary surveys measuring subjective phenomena such as attitudes and beliefs.[1] Beginning with the pioneering work of Thurstone and Chave (1929), Likert (1932), and their contemporaries, a long line of research has documented the utility of rating scales for this purpose. As such, rating scales play central roles in empirical studies by psychologists, sociologists, political scientists, economists, and many other social scientists. Yet remarkably, research methods and questionnaire design textbooks rarely provide specific, practical advice about how best to construct rating scales for particular purposes.

Such advice seems potentially useful because designing a rating scale entails making many different decisions. One must establish how long a scale will be and whether it will include a midpoint. One must decide whether to include verbal labels on all points or to label some with numbers only. If one chooses to use verbal labels, decisions must be made about precisely which words to

---

[1] Although the term "scale" is often used to refer to a battery of items measuring a single construct, we use the term "rating scale" in this chapter to refer to the response options used in a single question employing a rating format.

Survey Measurement and Process Quality, Edited by Lyberg, Biemer, Collins, de Leeuw, Dippo, Schwarz, Trewin.
ISBN 0-471-16559-X © 1997 John Wiley & Sons, Inc.

use. If one uses numeric labels, they must be chosen as well, and a no-opinion response option can be either included or omitted. For lack of an alternative approach, most questionnaire designers make these decisions by relying on their intuition, so different investigators sometimes end up employing very different procedures.

Most survey methodologists are unaware, however, that there is a very large empirical literature, scattered across eight decades and various different social sciences, that offers scientific bases for making these design decisions. At present, there is much work ongoing that will synthesize this literature, as well as research on many related topics of interest to questionnaire designers (e.g., see Krosnick and Fabrigar, forthcoming). In this chapter, we preview some of what is being uncovered in that literature. Specifically, we focus on three principal issues of interest in rating scale design: number of scale points, verbal versus numeric labeling, and inclusion of no-opinion options. We begin with a brief discussion of the criteria we use to assess the quality of a rating scale's design: reliability and validity. We then review some of the literature on each of our three principal topics.

6.2 EVALUATING DATA QUALITY

Research methods textbooks in the social sciences routinely point to two fundamental aspects of the quality of a measure: reliability and validity. Because there are many excellent and extensive published discussions of these issues elsewhere, there is no need to define or refine these criteria here. However, it does seem worthwhile to note the inherent strengths and weaknesses in methods used to assess item reliability and validity in surveys in order to help us make sense of the evidence we consider below that may at times appear to be contradictory.

6.2.2 Reliability

In conducting surveys to measure attitudes, we presume that attitudes are latent constructs that reside in the minds of individuals. No single question can tap such a construct perfectly, because answers to any question will be a function of a variety of extraneous factors other than the attitude itself. But the construct nonetheless resides in the individual's mind, and the question provides a reading of it, subject to the influence of the other forces. A question's reliability can be gauged by assessing the consistency of its readings, either by the consistency of results obtained when the same person is asked the same question on multiple occasions (what we will call *longitudinal reliability*), or by the consistency of results obtained from asking the person a series of different questions intended to tap the same attitude on one occasion (called *cross-sectional reliability*).

Each method has drawbacks. A lack of correspondence in answers to an item over time can be observed for either of two reasons: the measure may be unreliable, or the attitude itself may have changed during the intervening time interval. The shorter the time interval, the less likely attitude change is to occur, but the more likely it is that the person will remember his or her answer to the question during its first administration and simply repeat it, thus artifactually inflating apparent reliability.

A series of questions intended to measure the same attitude on a single occasion may be perceived by a respondent as just that, and he or she may strive to provide answers that appear to be consistent with one another across the items. Or respondents may doubt that a researcher would intentionally ask a series of questions about exactly the same issue, so the respondents may attempt to infer fine distinctions between the questions and thus exaggerate differences between them. Given these drawbacks of each method, we should not expect every study to yield perfectly pure results. Rather, we should look for trends in findings obtained by a variety of different methods.

6.2.2 Validity

The validity of a measure refers to the accuracy with which it taps the construct of interest. Reliability and validity are related to one another causally, in that a measure with no reliability at all cannot be valid. However, a measure that is highly reliable may nonetheless be quite inaccurate. Its consistency may come instead from a fundamental insensitivity or from tapping a construct other than the one of interest.

The validity of survey measures can be assessed in a number of ways. For instance, what we will call *correlational validity* is the degree to which a given measure can predict other variables to which it should be related. Presumably, a stronger relation would indicate greater validity. So, for example, one attitude measure would be considered to be more valid than another if the former predicts attitude-relevant behavior better. Discriminant validity is the degree to which a measurement approach can differentiate between constructs that are presumed to be distinct from one another. Therefore, the validity of a rating scale is presumably greater if it is better able to detect differences in perceptions of different objects. Of course, the utility of these approaches to validity assessments depends upon the researcher's ability to correctly identify criteria that should be predicted by a measure and to identify objects that should be evaluated differently from one another. Consequently, it is again most useful to look for trends in results across large sets of studies that are heterogeneous in terms of design.

6.3 NUMBER OF SCALE POINTS

One important decision that must be made when constructing a rating scale is how many scale points to include, and a number of theoretical issues enter into this determination. It is important to distinguish between bipolar scales (i.e., scales reflecting two opposing alternatives with a clear conceptual midpoint) and unipolar scales (i.e., scales reflecting varying levels of some construct with no conceptual midpoint and with a zero point at one end). Attitudes can be thought of as bipolar constructs, because they range from extremely positive to extremely negative, with neutral as a specific midpoint, representing neither positive nor negative. The amount of importance a person attaches to a particular attitude he or she holds is an example of a unipolar construct: it ranges from zero importance to some maximum level, and there is no precise midpoint.

There are various reasons to believe that more scale points will generally be more effective than fewer. This is because people's perceptions of their attitudes presumably range along a continuum of extremely positive to extremely negative. In order to translate a point on that continuum on to a categorical response scale, the set of points must presumably represent the entire continuum. A scale with only three options (e.g., favor, oppose, and neither) does not allow people to say that they favor something slightly. Thus, it is ambiguous as to whether a person with a slight leaning should select "favor" (which might imply stronger positivity than is the case) or "neither." A more refined scale with more points will presumably permit such moderate individuals to express their stands precisely and comfortably. Furthermore, the more scale points there are, the more a person can distinguish his or her attitude toward one object from his or her attitude toward another object. Similarly, more scale points permit a researcher to make more subtle distinctions among individuals' attitudes toward the same object. Thus, longer scales have the potential to convey more useful information.

On the other hand, using too many scale points may reduce the clarity of meaning of the response options. For scales with only a few options (e.g., the 3-point scale with favor, oppose, and neither), the meaning of these options is quite clear. But when the number of options becomes quite large (e.g., a 101-point thermometer scale), the meaning of any particular point is less precise. This may lead to less consistency within and between individuals concerning the meanings they attach to particular response options. In addition, including too many response options may make it more difficult for respondents to decide where they fall on a scale and thereby encourage respondents to shortcut the effort they expend via satisficing (Krosnick, 1991).

Thus, the optimal scale length would seem to be a moderate one. But just exactly how long should it be? With bipolar constructs, one might imagine that the scale should be set up to represent cognitive categories likely to capture the differentiations people make naturally. So with regard to attitudes, for example, people might be inclined to think of their liking of an object as being either slight, moderate, or substantial. If we think of three such points to the left of a midpoint on an attitude scale and three such points to the right, a 7-point scale appears optimal. It is also possible that people might only naturally distinguish between slight and substantial leaning to one side or the other, thus implying that a 5-point scale may be optimal.

In the case of unipolar scales, using too few scale points probably compromises information gathered, and too long of a scale probably compromises clarity of meaning. But it is hard to anticipate what the optimal length will be. It seems likely that people can readily conceive of zero, a slight amount, a moderate amount, and a great deal along any unipolar continuum. And perhaps they are comfortable making slightly finer distinctions, so the optimal scale length might be expected to fall between 4–7 points.

6.3.1 Reliability

Much of the empirical research exploring the effect of scale point number on measurement quality has investigated its influence on reliability. This research has used a variety of approaches, including secondary analyses of existing data and direct experimental comparisons.

Although some studies have failed to find a relation between the number of points on bipolar scales and cross-sectional reliability (Bendig 1954a), most studies indicate that such reliability is greatest for scales with approximately 7 points. For example, Masters (1974) experimentally compared 2-, 3-, 4-, 5-, 6-, and 7-point scales and found that reliability increased up to 4 points and leveled off thereafter. Birkett (1986) experimentally compared 2-, 6-, and 14-point scales and found that 6-point scales had the highest reliability. Similarly, Komorita and Graham (1965) found 6-point scales to be more reliable than 2-point scales. In one study that examined longitudinal reliability, 7- to 9-point scales appeared to be more reliable than shorter ones (Alwin and Krosnick, 1991).

Investigations of cross-sectional reliability in unipolar scales suggest that the optimum number of scale points is between 5 and 7 points. Although some studies failed to find systematic relations between scale point number and reliability (Bendig, 1953; Matell and Jacoby, 1971; Peterson, 1985), other studies found that cross-sectional reliability is greater for 4-point than 2-point scales (Watson, 1988), 5-point than 7- or 11-point scales (McKelvie, 1978), and 5- to 7-point than 3- or 9-point scales. In an investigation of longitudinal reliability (Jacoby and Matell, 1971; Matell and Jacoby, 1971), 7- and 8-point scales had greater reliability than scales with 3 to 6 points or scales with 9 to 19 points.

6.3.2 Validity

A substantial amount of research has also examined the effect of scale point number on the validity of measurement. Much of this research has used computer simulations to examine how transforming data from continuous representations of relations to representations with discrete scale points distorts known patterns of data. With a few exceptions (Martin, 1973, 1978), these simulations suggest that distortion in data decreases as the number of scale points increases, but that this improvement is relatively modest beyond 5 to 7 points (Green and Rao, 1970; Lehmann and Hulbert, 1972; Ramsay, 1973).

Studies of correlational validity are generally consistent with the above-mentioned reliability patterns, even though some studies have produced contradictory evidence (e.g., Smith and Peterson, 1985). For instance, Matell and Jacoby (1971) found greater correlational validity for 7- and 8-point unipolar scales than for shorter or longer scales. Similarly, Rosenstone *et al.* (1986) found that 5-point bipolar scales predicted conceptually related variables better than 3-point bipolar scales.

Another set of studies relevant to assessing the effect of scale point number on validity has examined the susceptibility of rating scales to context effects (Wedell and Parducci, 1988; Wedell *et al.*, 1990). As expected, the effect of context on ratings of different stimuli along various dimensions of judgment (e.g., happiness of faces, size of geometric objects, happiness of life events) varies as a function of scale point number. Although results depend on the nature of the judgment being made, context effects generally weaken as the number of scale points increases, but these reductions are relatively modest beyond 7 scale points.

Another criterion for evaluating scales is the amount of information they yield about differences between attitudes within and across respondents. Needless to say, scales that are too short cannot reveal much about the distinctions a person makes among a large set of objects. Consistent with this notion, a number of studies showed that longer scales conveyed more useful information up to 7–9 points (Bendig, 1954b; Bendig and Hughes, 1953; Garner, 1960), and information transfer appears to decrease for scales of 12 points or longer (McCrae, 1970a, 1970b). Interestingly, the proportion of scale points used stays approximately constant across scales of between 4 and 19 points, but the additional number of points used on scales longer than 7 points seems not to convey additional substantive information (Matell and Jacoby, 1972).

6.3.3   Magnitude Scaling

A rather sizable literature has evolved proposing a radically different approach to rating scale construction under the label of "magnitude scaling" (e.g., Lodge, 1981). This approach involves the use of scales of infinite length, thus not restricting respondents in the extent to which they can differentiate among objects. For example, people could be shown a line four inches long and could be told that this represents the amount they like oranges. Then, they could be asked to draw another line indicating how much they like apples, with shorter lines indicating less liking and longer lines indicating more liking. In addition to the visual mode, this sort of scaling can be done by having people adjust the pitches of tones they hear or squeeze dynamometers to report comparative judgments in terms of the amount of pressure they apply.

As appealing as this method is in principle, however, it is difficult to administer practically, especially in typical survey settings. Also, magnitude scaling only reveals the ratios among stimuli in terms of their placement on an evaluative dimension, not the absolute levels of people's placement of them. That is, you can learn that a person likes apples twice as much as oranges, but you cannot tell whether apples or oranges fall on the positive or the negative side of his or her attitude continuum. Furthermore, a number of studies that compared the reliability and validity of magnitude estimation data with that obtained from traditional, categorical rating scales showed the latter to be superior (e.g., Kaplan *et al.*, 1979; Miethe, 1985). Thus, at the moment, it appears that the traditional approach is preferable.

6.3.4 Midpoints

Another important design issue is whether to include a middle alternative or not. On bipolar dimensions, a middle alternative can take a number of forms, including "neither" on a favor/oppose or agree/disagree or like/dislike scale, or a "status quo" endorsement on a scale representing change, ranging from a large increase to a large decrease (e.g., in U.S. defense spending). On a unipolar scale, a midpoint presumably represents a moderate position, though its meaning is not as precise as is the case for the bipolar dimensions.

Use of a middle alternative on a bipolar dimension can be justified if one believes that some individuals truly have neutral positions and that forcing them to respond in one direction or the other will add to measurement error. However, many people might lean slightly in one direction or the other but might select an offered midpoint because it provides an easy choice that requires little effort and is easy to justify. One scenario by which this might occur was proposed by Krosnick (1991).

According to Krosnick (1991), the cognitive tasks required of survey respondents are typically quite burdensome if people attempt to be fully diligent (a behavior he called survey *optimizing*). Consequently, individuals may sometimes look for ways to avoid expending this effort while maintaining the appearance of answering responsibly (a response behavior he called survey *satisficing*). According to Krosnick (1991), a respondent who wishes to satisfice will look for a cue in the question suggesting how to do so. If no such cue is obvious, then a respondent may choose to optimize instead.

Among various other cues, Krosnick (1991) pointed to scale midpoints. Specifically, he argued that when such points represent the status quo (e.g., "keep government spending on defense at present levels"), respondents will find them easy to defend if pressed by an interviewer. Consequently, offering a midpoint may discourage people from taking sides and may encourage them to satisfice, whereas if no midpoint is offered, respondents might optimize instead. Thus, offering a midpoint may forego collection of useful data.

Empirical research in this area has documented a variety of effects of offering scale midpoints. First, people seldom spontaneously offer midpoint responses when they are not legitimated, but substantial proportions select them when included explicitly in questions (Ayidiya and McClendon, 1990; Bishop, 1987; Kalton *et al.*, 1980; Schuman and Presser, 1981). Attraction to midpoints when offered is sometimes greatest among respondents who are lower in cognitive skills (Narayan and Krosnick, 1996) and lower in the personal importance of the topic involved (Krosnick and Schuman, 1988; Schuman and Presser, 1981, pp. 173–175; though see Stember and Hyman, 1949–1950). But although one might imagine that such people flip mental coins to select responses when no middle alternative is offered, the distributions of opinions offered by these individuals often depart significantly from a flat shape (Ayidiya and McClendon, 1990; Bishop, 1987; Kalton et al., 1980; Schuman and Presser, 1981). Thus, their responses do not appear to be haphazard.

Some additional studies have examined the effect of midpoint presence on reliability and validity, but with contradictory results. For example, Alwin and Krosnick's (1991) and Andrews's (1984) secondary analyses suggest that scales with midpoints were less reliable than those without. Experimental studies, however, have generally failed to find any consistent pattern of increases or decreases of reliability according to the presence or absence of midpoints (Masters, 1974; Jacoby and Matell, 1971; Matell and Jacoby, 1971). Although Kalton et al. (1980) found that including midpoints had no influence on correlational validity (as gauged by associations between attitudes and demographics), Schuman and Presser (1981) found that associations among attitudes were strengthened when middle alternatives were included, and Stember and Hyman (1949–1950) found that the effect of interviewer bias decreased with the inclusion of a midpoint. Thus, including a midpoint may at times increase data quality and at other times not.

6.3.5   Ease of Administration

Another criterion with which to assess the effectiveness of various scale lengths is the ease and success with which they can be administered to respondents. Matell and Jacoby (1972) found that it takes longer to administer scales of longer lengths, but sizable increases in time only occur after scales reach 13 points or longer. Of the studies that examined rates at which people properly completed all rating scales in a set, one found better completion rates for 4-point scales than 2-point scales (Ghiselli, 1939), but the other found no difference between 3- and 7-point scales (Smith and Peterson, 1985).

6.3.6   Conclusion

Taken as a whole, this research suggests that the optimal length of a rating scale is 5 to 7 points, because scales of this length appear to be more reliable and valid than shorter and longer scales. With respect to midpoints, it is less clear whether researchers should include midpoints. Evidence on validity is mixed, and the theory of satisficing suggests that including midpoints may decrease measurement quality. Therefore, we look forward to further studies on this matter to yield a clear resolution. In the meantime, however, it seems sensible to include midpoints when they are conceptually demanded (e.g., in a question about whether defense spending should be increased, decreased, or kept the same), and steps should be taken elsewhere in the questionnaire to minimize the likelihood of satisficing (see Krosnick, 1991).

6.4 LABELING SCALE POINTS

Another important decision in the construction of scales is whether to label all scale points with words or to label some scale points with numbers only. Obviously, in order for any rating scale to have meaning, it is necessary to at least label the endpoints of a scale. But researchers can decide not to label other scale points, and many routinely make this choice.

Using scales with only the endpoints labeled verbally presumably has two major advantages. First, numeric values may be more precise than verbal labels, which may suffer from the inherent ambiguity of language. Second, numeric scale values are presumably easier for respondents to hold in memory (e.g., during telephone interviews) than more complex verbal labels. Thus, responding to scales with only the endpoints verbally labeled may be less cognitively demanding than scales that are fully labeled.

However, there are also reasons to expect that verbally labeling all scale points might improve data quality. Because people rarely express complex conceptual meaning in everyday conversation via numbers, verbal labels might be a more natural (and therefore easier) method for respondents to express themselves. Additionally, numbered scale points have no inherent meaning, other than to suggest equal divisions between concepts established by verbal labels. Thus, including labels on all scale points could help to clarify the meaning of scale points and thereby increase the ease with which people can make reports and the precision of them.

6.4.1 Reliability

A number of empirical studies have compared fully labeled and partially labeled scales in terms of reliability. Studies of cross-sectional reliability found it to be unaffected by verbal labels (Finn, 1972; Madden and Bourdin, 1964) or to be lower for scales with more verbal labels (Andrews, 1984). However, studies of longitudinal reliability have consistently found it to be increased by verbal labels (Alwin and Krosnick, 1991; Krosnick and Berent, 1993; Zaller, 1988). Furthermore, these improvements in reliability were most pronounced among people with low to moderate education, just the individuals who can presumably benefit most from greater clarity (Krosnick and Berent, 1993). And using verbal labels to make the endpoints of a rating scale seem farther apart also increases reliability (Bendig, 1955). Thus, verbal labels seem to be an asset in this regard.

6.4.2 Validity

Studies of verbal labels also indicate that they increase the validity of obtained data. Krosnick and Berent (1993) and Dickinson and Zellinger (1980) found stronger correlations between attitudes and other variables when the former were measured using more verbal labels, especially among respondents with relatively little formal education. In their secondary analyses, Andrews (1984) and Alwin and Krosnick (1991) found greater true score variance in fully labeled items than in partially labeled items, indicating greater validity of the former. Raters of the same objects tend to agree more when the rating scales employed have more verbal labels (Barrett *et al.*, 1958; Bendig, 1953; Peters and McCormick, 1966). Furthermore, more differentiation between different rating dimensions and different objects is apparent when more verbal labels are employed (Barrett *et al.*, 1958; Bendig and Hughes, 1953; Bernardin *et al.*, 1976). And finally, ratings are less susceptible to context effects when more verbal labels are employed (Wedell *et al.*, 1990). Thus, much evidence suggests that fully labeling scales improves measurement validity.

6.4.4   Respondent Satisfaction

A number of studies that have assessed respondent satisfaction found that most people preferred to use rating scales with more verbal labels (Dickinson and Zellinger, 1980; Wallsten *et al.*, 1993; Zaller, 1988) and believed such scales to be more valid measurement instruments (Dickinson and Zellinger, 1980).

6.4.4   Selecting Verbal Labels

Although fully labeled scales appear to be more reliable and valid than partially labeled scales, the benefits of verbal labeling are obviously contingent on selecting appropriate labels. If verbal labels are to be useful, they must have reasonably precise meanings for respondents. It is also important that the labels one chooses reflect relatively equal intervals along a continuum, particularly if an analyst is to capture all variance in the latent construct and plans to treat the results as an interval-level variable in statistical analysis.

A number of studies have investigated the meanings that people attach to verbal labels reflecting liking, amount, frequency, and likelihood—four dimensions routinely asked about in surveys. These studies have generally asked respondents to assign a numerical value (e.g., on a scale from 0 to 100 or via magnitude scaling) to various verbal quantifiers (e.g., excellent, good, fair). Many such labels seem to have meanings that are very stable across different samples and consistent over periods of many years and across different methods of rating (see Table 6.1 for an illustration of some results). Longitudinal reliability for these scale values has also been found to be quite high, with test–retest correlations frequently greater than .90. Thus, it is possible to find verbal labels with sufficiently precise meanings to construct useful scales.

Some researchers have looked at this literature and emphasized the imprecise nature of verbal labels suggested by research findings (e.g., Pepper, 1981). And it is certainly true that the meanings of some terms can change significantly depending upon various factors. For example, the absolute meaning of a frequency quantifier can change depending on the type of event being described

Table 6.1   Scale Values for Verbal Labels Assessing Liking

Verbal label	Myers and Warner (1968)	Wildt and Mazis (1978)	French-Lazovik and Gibson (1984)	Stone and Schkade (1991)
Excellent	93	91	91	99
Very good	78	82	80	
Good	67	70	58	73
Fair	43	49		48
Poor	21	17	16	23
Very poor	12	5	11	1

(e.g., earthquakes happening once a year might be described as "often," whereas going to the movies once a year might be described as "rarely"). Furthermore, people interpret the meanings of labels in the context of the other labels offered in a scale, presuming that they are intended to be equally spaced from one another. And different social groups sometimes interpret labels in systematically different ways (see, e.g., Schaeffer, 1991). However, we do not view this evidence with quite as much concern as do some other observers; although there is random and systematic variation in the meanings of verbal labels to respondents, many labels appear to have sufficiently universal meanings to be very useful for improving attitude measurement.

6.4.5   Selecting Numeric Labels

Even if a scale is fully verbally labeled, one may nonetheless be tempted to include numbers as well, partly to facilitate coding afterwards. Interestingly, on self-administered questionnaires, these numbers might have effects, just as numbers may affect ratings on partially labeled scales. This is because respondents sometimes use the specific numeric values to infer the meanings of scale points or verbal labels. For example, when respondents were asked how successful their lives have been, using an 11-point scale ranging from 0 to 10 with the endpoints labeled "not at all successful" and "extremely successful," approximately 34 percent selected a value ranging from 0 to 5 (see Schwarz and Hippler, 1991; Schwarz et al., 1991). In contrast, when the same endpoints were used for an 11-point scale with numeric values ranging from -5 to $+5$, only 13 percent selected one of the logically equivalent values ranging from -5 to 0. This difference apparently occurred because respondents inferred that the label "not at all successful" meant the absence of success when paired with the number "0" but meant the presence of failure when paired with the number "-5."

Consequently, if numbers are to be used on rating scales, they should be selected carefully to reinforce the intended meaning of the scale points.

6.4.6   Conclusion

It seems that fully labeled scales are more reliable and valid than partially labeled scales. Based on this, we recommend that researchers verbally label all scale points when it is practical to do so and avoid the use of numbers alone. However, researchers should be certain that they select labels that have relatively precise meanings for respondents and that reflect equal intervals along the continuum of interest. This should be done by using labels that have been previously scaled and are known to possess good psychometric properties. When such labels are not available, researchers should conduct pretesting to identify appropriate labels. Because numeric values can alter the meaning of labels, researchers should probably avoid using them altogether and simply present verbal response options alone.

6.5   NO-OPINION FILTERS

When we ask attitude questions in surveys, we usually presume that respondents' answers reflect opinions that they previously had stored in memory. And if a person had not stored a pre-existing opinion about the precise attitude object of interest, the question itself presumably prompts him or her to draw on relevant stored information in order to concoct a reasonable, albeit new, evaluation of the object. Consequently, whether based upon a pre-existing attitude or a newly formulated one, responses presumably reflect the individual's orientation, favorable or unfavorable, toward the object.

6.5.1   The Non-Attitude Hypothesis

Converse (1964) proposed a very different possibility. He argued that respondents sometimes simply answer attitude questions randomly in surveys. His principal evidence was correlations between reports of the same attitudes made by a panel of respondents who were interviewed first in 1956, again in 1958, and finally in 1960. These correlations were remarkably weak: across eight policy issues, tau-betas ranged from approximately .28 (for federal housing) to approximately .47 (for school desegregation) and averaged .37. Because these correlations did not become notably weaker as the time interval between measurements increased from two years to four years, he concluded that no true attitude change had occurred. Furthermore, the distributions of opinions that the respondents expressed were remarkably consistent over the four-year period, which reinforced this same conclusion.

What, then, was responsible for the weak over-time correlations between these attitude reports if not attitude change? For some issues, Converse (1964) asserted, large portions of Americans had no pre-existing opinions but felt pressure to report attitudes in the surveys, even when they had absolutely no relevant information with which to formulate meaningful judgments. The mere presence of an interviewer and his or her persistence at asking questions presumably indicated to these respondents that they were expected to have opinions and that the researcher needed to know what those opinions were. Because respondents preferred not to look foolish by having to admit ignorance on numerous topics, Converse argued, respondents concocted answers even when they had no knowledge at all on which to base them.

6.5.2 Other Supportive Evidence

In fact, a great deal of other research, done both before and after Converse's (1964) chapter was published, is consistent with his conclusions. For example, one set of studies found comparably low levels of over-time consistency in attitude reports (e.g., Jennings and Markus, 1984; Jennings and Niemi, 1978). And latent class analyses of over-time data indicated that an average of between 50 and 70 percent of respondents had nonattitudes on any given public policy issue (Brody, 1986; Taylor, 1983).

Other research supporting the nonattitudes conclusion showed that attitude reports usually predict reports of other attitudes very weakly (for a review, see Kinder and Sears, 1985) and often predict relevant behavior rather poorly (Wicker, 1969). Furthermore, the results of studies showing strong attitude–behavior consistency may actually be artifactual consequences of correlated measurement error (Budd, 1987; Budd and Spencer, 1986). Still other supportive findings were reported by Gill (1947), Hartley (1946, pp. 26-31), Ehrlich and Rinehart (1965), Schuman and Presser (1981), Bishop et al. (1986), and others showing that people often report attitudes toward fictitious objects or objects so obscure that people are extremely unlikely to have heard of them.

Converse's (1964) interpretation of his initial evidence has received a great deal of criticism by observers who have claimed that other readings are equally plausible. Similarly, the evidence we just described as consistent with the nonattitudes hypothesis can also be interpreted in ways that are quite different (see Krosnick and Fabrigar, forthcoming). Thus, this literature does not provide definitive support for the nonattitude notion. However, its consistency with that hypothesis has led researchers to explore means by which nonattitude reports might be prevented.

6.5.3 Types of No-Opinion Filters

In fact, the only method that has been considered empirically to any significant degree is no-opinion filtering. The notion here is that respondents may report nonattitudes partly because the form of survey questions encourages them to do so. As Schuman and Presser (1981, p. 299) pointed out, respondents generally "play by the rules of the game," meaning that they choose among the response alternatives offered by a closed-ended question rather than offering reasonable answers outside the offered set. If a question does not explicitly include a "don't know" (DK) or "no-opinion" (NO) option, that might imply to respondents that they are expected to have opinions and therefore encourage them to report nonattitudes. Thus, when such a response option is explicitly legitimated, significantly larger proportions of respondents would be expected to admit having no opinion.

Many studies have reported evidence consistent with this expectation (e.g., Ehrlich, 1964; Bishop et al., 1980a, 1980b, 1982, 1983; 1986; Schuman and Presser, 1979, 1981; Schuman and Scott, 1989). Furthermore, filters can be phrased in various different ways, and these different approaches yield understandably different results. Quasi-filters involve simply including an explicit DK or NO response option at the end of a question asking respondents about their attitudes. So, for example, a question might ask: "Do you favor or oppose legalized abortion, or don't you have an opinion on this?" In contrast, full filters involve a question that precedes the attitude measure, inquiring about whether the respondent has an opinion on the issue. Only if he or she answers affirmatively is the attitude question then asked. Full filters attract significantly more DK responses (Bishop et al., 1983; Schuman and Presser, 1981).

Full filters can be phrased in ways that vary in terms of their apparent encouragement and legitimation of DK responses, and this variation seems to be consequential as well (Bishop et al., 1983). For example, one could ask "Do you have an opinion on this or not?" or "Have you been interested enough in this to favor one side over the other?" or "Have you thought much about this issue?" or "Have you already heard or read enough about it to have an opinion?" By focusing on a person's interest in, thoughts about, or exposure to information on a topic, the latter three filters presumably make it easier for respondents to admit that they have not considered the topic previously and therefore have no opinion on it. Consistent with this presumption, Bishop et al. (1983) found that filters can be ordered in terms of their strength: the blunt filter gathered slightly fewer DK responses than the "interested" filter, which gathered fewer DKs than the "thought" and "heard or read" filters.

6.5.4 The Meaning of No-Opinion Responses

This evidence is encouraging about the ability of no-opinion filters to prevent nonattitude reporting. However, this conclusion presumes that DK responses mean that respondents truly have no opinion and would guess purely randomly or would provide equally meaningless answers via some other method if forced to offer an opinion. But is this necessarily the case?

In fact, people may provide DK responses for many reasons other than having no attitudes at all (e.g., see Dunnette et al., 1956; Edwards and Ostrom, 1971; Klopfer and Madden, 1980; Krosnick, 1991; Smith, 1984). First, DK responses may reflect ambivalent attitudes, which may be especially difficult to report when no scale midpoint is offered to respondents. Second, DK responses could occur when a person has only neutral thoughts or feelings about an object but no neutral response option is offered (e.g., in an agree/disagree question). Third, selection of a DK alternative may reflect that the respondent does not understand the question being asked. Fourth, respondents may know approximately where they fall on an attitude scale (e.g., around 6 or 7 on a 1–7 scale), but because of ambiguity in the meaning of the scale points, they may be unsure of exactly which to choose. Finally, DK responses may reflect satisficing: even though respondents may have preformed opinions or information with which to concoct a reasonable opinion on the spot, lack of motivation or ability to do so may lead these individuals to choose the DK option in order to avoid doing the mental work involved (Krosnick, 1991).

Many studies suggest that don't know or no-opinion responses can have all of these different meanings (Coombs and Coombs, 1976; Dunnette *et al.*, 1956; Feick, 1989; Smith, 1984). Furthermore, some studies suggest that ambivalence is more often the cause of DK responses than nonattitudes (e.g., Ehrlich, 1964). Thus, if respondents who say "no-opinion" were forced to answer substantively instead, their responses might well not be primarily random noise.

6.5.5 Do No-Opinion Filters Improve Data Quality?

Given this backdrop, an interesting question here is whether offering a no-opinion option reduces the amount of random variation in the attitude reports obtained. Of course, nonattitude reports could be based upon systematic response biases, which would not necessarily increase random variation. But if Converse (1964) was correct that some nonattitude reports are purely random, then this randomness should weaken relations between variables. A variety of methods has been used to address this issue, and all of this work has produced either mixed evidence or clear disconfirmation of the random nonattitudes perspective.

For example, Gilljam and Granberg (1993) asked survey respondents three different questions tapping attitudes toward building nuclear power plants. The first of these questions included a no-opinion filter, and 15 percent of respondents selected it. The other two questions, asked later in the interview, did not include no-opinion filters, and only 3 and 4 percent of respondents, respectively, volunteered "don't know" responses to them. Thus, the majority of respondents who initially said "don't know" offered opinions on the later two questions. At issue, then, is whether these later responses reflected real opinions or were nonattitudes.

To address this issue, Gilljam and Granberg (1993) examined two indicators: the strength of the correlation between these two latter attitude reports, and their ability to predict people's votes on an actual nuclear power referendum in a subsequent election. The correlation between answers to the latter two items was .41 ($p < .001$) among individuals who said "don't know" to the first item, as compared to a correlation of .82 ($p < .001$) among individuals who answered the first item substantively. Similarly, answers to the second two items correctly predicted an average of 76 percent of subsequent votes by people who initially said "don't know," as compared to a 94 percent accuracy rate among individuals who answered the first item substantively.

Thus, the nonattitudes perspective received mixed support, in that the filter apparently separated out people whose opinions were, on average, of less quality than others' opinions but were not purely random.

Two other nonexperimental studies taking a different investigative approach produced similarly mixed evidence. Andrews (1984) and Alwin and Krosnick (1991) meta-analyzed the correlates of the amount of random measurement error in numerous survey items, some of which included no-opinion filters and others of which did not. Andrews (1984) found that the amount of random error was significantly less when a no-opinion filter was included than when it was not, but Alwin and Krosnick (1991) found just the opposite.

One set of studies producing a clear challenge to the NO filter notion gauged changes in associations between target items and other variables. If offering a NO option leads some respondents to select it and thereby prevents them from providing poor quality data, then associations between variables should become stronger. Bishop *et al.* (1979) did a secondary analysis of existing surveys that had asked similar questions of different national samples in either filtered or unfiltered forms. These authors found a trend toward strengthened associations for filtered rather than unfiltered forms, but their comparisons involved confounding of time, survey administration form, question context, and more. In Schuman and Presser's (1981) 19 experiments on no-opinion filtering, no meaningful shifts in correlations occurred in the vast majority of cases. Two cases did turn up significant association shifts, but neither was consistent with Converse's expectations in terms of the directions of the shifts. Furthermore, Schuman and Presser (1981) found no cases in which relations between attitudes and respondents' education, interest in politics, age, or gender were altered by filtering. Thus, this evidence is clearly inconsistent with the notion that no-opinion filters improve the quality of data obtained.

A final pair of studies explored the effect of experimental variations in the presence or absence of NO filters on reliability. It would seem necessary for reliability to be increased by filters in order for the nonattitudes perspective on filters to be validated. However, using cross-sectional data, McClendon and Alwin (1993) found no greater reliability when NO filters were included in questions than when they were not. And Krosnick and Berent (1990) found no significant change in the over-time consistency of attitude reports depending upon whether no-opinion filters were present or absent in questions.

6.5.6 An Alternative Perspective

Taken together, these studies call into question the notion that NO filters successfully remove nonattitudes from the data obtained by a survey question. As surprising as these results are from Converse's (1964) perspective, they are quite understandable in light of the notion of survey satisficing (Krosnick, 1991). Offering a no-opinion option may not only fail to improve the quality of data obtained but may also forego collection of useful data. Specifically, such a response option might constitute a cue encouraging respondents who are otherwise disposed to satisfice to do so by saying "don't know." If, instead, the no-opinion option was not offered and no other cue was apparent, these respondents might choose not to satisfice and would optimize instead. Thus, useful data could be collected from these individuals, and offering the no-opinion option would preclude obtaining these data.

If this view is correct, then attraction to a NO option when offered should be greatest under the conditions thought to foster satisficing: low respondent ability and motivation to optimize, and high task difficulty. Consistent with this reasoning, attraction is greatest among respondents with the lowest levels of cognitive skills (Bishop *et al.*, 1980a; Schuman and Presser, 1981; Schuman and Scott, 1989) and among people who consider an issue to be less personally important (Bishop *et al.*, 1980a; Schuman and Presser, 1981, pp. 142–143). However, these patterns can be viewed as consistent with Converse's (1964) nonattitudes perspective, because these individuals are precisely those who would seem most likely to be uninformed on an issue and to report nonattitudes. Therefore, additional evidence is needed before the meanings of these associations can become clear.

Although not optimal for testing the satisficing notion, some other studies have provided supportive evidence simply by correlating the frequency of no-opinion responses to a single question form with various aspects of task difficulty. Klare (1950) and Converse (1976) showed that no-opinion rates increased as the complexity of the language in a question increased. Converse (1976) also found higher no-opinion rates when questions contained long explanations, when questions required respondents to predict the future rather than simply describing the past or present, for dichotomous questions than for politimous questions (presumably because the former pose more difficulty in terms of describing moderate or qualified opinions), and for questions regarding foreign affairs than for those addressing domestic affairs (presumably because of the greater remoteness of the former and lower knowledge levels likely to be associated with their topics). Furthermore, although Ferber (1966) found no relation between no-opinion rates and position of a question in the questionnaire (in a study that did not manipulate question order experimentally), Culpepper *et al.* (1992) did find high NO rates later in a questionnaire when question order was experimentally manipulated.

Yet another line of empirical inquiry that shares the spirit of the satisficing view was done by Hippler and Schwarz (1989). These investigators proposed that strongly worded no-opinion filters might suggest to respondents that a great deal of knowledge is required to answer an attitude question and thereby discourage respondents who wish to optimize from answering substantively. Hippler and Schwarz (1989) demonstrated that respondents did indeed infer from the presence and strength of a no-opinion filter that follow-up questioning would be more extensive, would require more knowledge, and would be more difficult. Respondents who said they had no opinion when this option was offered were sometimes also able to report opinions on the same issue later in the questionnaire when the no-opinion option was omitted. Furthermore, these latter responses appeared to be systematic, rather than the result of mental coin flipping.

Still another line of research that calls into question the validity of no-opinion responses is work that has used the proportion of people saying they had no opinion on an issue as an index of the strength of the public's will on the issue. These studies have examined the correspondence between public preferences on a policy issue and government action on that issue, presuming that when the public feels more strongly, government will act more in accord with its wishes. In fact, this expectation has been confirmed using direct questions asking people how strongly they feel about an issue or how important it is to them (e.g., Conover *et al.*, 1982; Schuman and Presser, 1981). But rates of no-opinion responses are unrelated to correspondence between public opinion and government action (Brooks, 1990). This, too, calls into question the notion that no-opinion responses genuinely reflect lack of opinions.

6.5.7 Conclusion

Overall, then, this evidence does not provide a solid basis for recommending that no-opinion filters be included in attitude questions in order to avoid collecting nonattitudes. Rather, it seems that some real attitudes are missed by the inclusion of such filters. Therefore, we view this evidence as supporting the recommendation that such filters be omitted when possible. Of course, if people are to be asked about an obscure issue, on which many are likely to have no information at all, offering a filter may be appropriate and beneficial. But asking people about such obscure matters may not be of much interest to researchers, anyhow.

It is important to note, however, that we believe the essential notion standing behind Converse's (1964) nonattitudes hypothesis is correct: Many people who report attitudes in surveys do not have deeply rooted preferences that shape their thinking and behavior. But we do not think that no-opinion filters are the best ways to identify those individuals. Rather, such individuals are probably best identified by measuring the strength of people's attitudes directly. This can be done in many different ways (see, e.g., Krosnick and Abelson, 1992; Petty and Krosnick, 1995), and the choice of method can be consequential both practically and theoretically. For example, people can be asked how certain they are of their opinions or how important the issues are to them personally or how knowledgeable they are on the topic. We encourage readers to examine the attitude strength literature and to use techniques developed therein to address the nonattitudes problem.

6.6 EPILOGUE

The final word has certainly not been spoken on the design decisions we have considered in this chapter. The number of studies exploring each issue is not great enough to justify final conclusions, and there is occasional disagreement between studies, suggesting that some interacting variables may determine the conditions under which some approaches are better than others. Therefore, there is much work left to be done exploring these issues in future research. We hope that scholars doing questionnaire research will take the opportunity of every data collection effort to incorporate methodological experiments whenever possible. The more we know about differences in measurement approaches, the more effective we all can be in the future in designing our measurement tools.

REFERENCES

Alwin, D. F., and Krosnick, J. A. (1991), "The Reliability of Survey Attitude Measurement: The Influence of Question and Respondent Attributes," *Sociological Methods and Research*, 20, pp. 139–181.

Andrews, F. M. (1984), "Construct Validity and Error Components of Survey Measures: A Structural Modeling Approach," *Public Opinion Quarterly*, 48, pp. 409–442.

Ayidiya, S. A., and McClendon, M. J. (1990), "Response Effects in Mail Surveys," *Public Opinion Quarterly*, 54, pp. 229–247.

Barrett, R. S., Taylor, E. K., Parker, J. W., and Martens, L. (1958), "Rating Scale Content: I. Scale Information and Supervisory Ratings," *Personnel Psychology*, 11, pp. 333–346.

Bendig, A. W. (1953), "The Reliability of Self-ratings as a Function of Amount of Verbal Anchoring and of the Number of Categories on the Scale," *Journal of Applied Psychology*, 37, pp. 38–41.

Bendig, A. W. (1954a), "Reliability and the Number of Rating Scale Categories," *Journal of Applied Psychology*, 38, pp. 38–40.

Bendig, A. W. (1954b), "Transmitted Information and the Length of Rating Scales," *Journal of Experimental Psychology*, 37, pp. 303–308.

Bendig, A. W. (1955), "Rater Reliability and the Heterogeneity of the Scale Anchors," *Journal of Applied Psychology*, 39, pp. 37–39.

Bendig, A. W., and Hughes, J. B. (1953), "Effect of Amount of Verbal Anchoring and Number of Rating-scale Categories Upon Transmitted Information," *Journal of Experimental Psychology*, 46, pp. 87–90.

Bernardin, H. J., LaShells, M. B., Smith, P. C., and Alvares, K. M. (1976), "Behavioral Expectation Scales: Effects of Developmental Procedures and Formats," *Journal of Applied Psychology*, 61, pp. 75–79.

Birkett, N. J. (1986), "Selecting the Number of Response Categories for a Likert-type Scale," *Proceedings of the American Statistical Association*, pp. 488–492.

Bishop, G. F. (1987), "Experiments with the Middle Response Alternative in Survey Questions," *Public Opinion Quarterly*, 51, pp. 220–232.

Bishop, G. F., Oldendick, R. W., Tuchfarber, A. J., and Bennett, S. E. (1979), "Effects of Opinion Filtering and Opinion Floating: Evidence from a Secondary Analysis," *Political Methodology*, 6, pp. 293–309.

Bishop, G. F., Oldendick, R. W., and Tuchfarber, A. J. (1980a), "Experiments in Filtering Political Opinions," *Political Behavior*, 2, pp. 339–369.

Bishop, G. F., Oldendick, R. W., and Tuchfarber, A. J. (1983), "Effects of Filter Questions in Public Opinion Surveys," *Public Opinion Quarterly*, 47, pp. 528–546.

Bishop, G. F., Oldendick, R. W., Tuchfarber, A. J., and Bennett, S. E. (1980b), "Pseudo-opinions on Public Affairs," *Public Opinion Quarterly*, 44, pp. 198–209.

Bishop, G. F., Tuchfarber, A. J., and Oldendick, R. W. (1986), "Opinions on Fictitious Issues: The Pressure to Answer Survey Questions," *Public Opinion Quarterly*, 50, pp. 240–250.

Brody, C. J. (1986), "Things Are Rarely Black and White: Admitting Gray into the Converse Model of Attitude Stability," *American Journal of Sociology*, 92, pp. 657–677.

Brooks, J. E. (1990), "The Opinion-policy Nexus in Germany," *Public Opinion Quarterly*, 54, pp. 508–529.

Budd, R. J. (1987), "Response Bias and the Theory of Reasoned Action," *Social Cognition*, 5, pp. 95–107.

Budd, R. J., and Spencer, C. P. (1986), "Lay Theories of Behavioural Intention: A Source of Response Bias in the Theory of Reasoned Action?" *British Journal of Social Psychology*, 25, pp. 109–117.

Conover, P. J., Gray, V., and Coombs, S. (1982), "Single-issue Voting: Elite-mass Linkages," *Political Behavior*, 4, pp. 309–331.

Converse, J. M. (1976), "Predicting No Opinion in the Polls," *Public Opinion Quarterly*, 40, pp. 515–530.

Converse, P. E. (1964), "The Nature of Belief Systems in Mass Publics," in D. E. Apter (ed.), *Ideology and Discontent*, New York: Free Press, pp. 206–261.

Coombs, C. H., and Coombs, L. C. (1976), "'Don't Know': Item Ambiguity or Respondent Uncertainty?," *Public Opinion Quarterly*, 40, pp. 497–514.

Culpepper, I. J., Smith, W. R., and Krosnick, J. A. (1992), "The Impact of Question Order on Satisficing in Surveys," paper presented at the Midwestern Psychological Association Annual Meeting, Chicago, Illinois.

Dickinson, T. L., and Zellinger, P. M. (1980), "A Comparison of the Behaviorally Anchored Rating and Mixed Standard Scale Formats," *Journal of Applied Psychology*, 65, pp. 147–154.

Dunnette, M. D., Uphoff, W. H., and Aylward, M. (1956), "The Effect of Lack of Information on the Undecided Response in Attitude Surveys," *Journal of Applied Psychology*, 40, pp. 150–153.

Edwards, J. D., and Ostrom, T. M. (1971), "Cognitive Structure of Neutral Attitudes," *Journal of Experimental Social Psychology*, 7, pp. 36–47.

Ehrlich, H. L. (1964), "Instrument Error and the Study of Prejudice," *Social Forces*, 43, 197–206.

Ehrlich, H. L., and Rinehart, J. W. (1965), "A Brief Report on the Methodology of Stereotype Research," *Social Forces*, 43, pp. 564–575.

Feick, L. F. (1989), "Latent Class Analysis of Survey Questions that Include Don't Know Responses," *Public Opinion Quarterly*, 53, pp. 525–547.

Ferber, R. (1966), "Item Nonresponse in a Consumer Survey," *Public Opinion Quarterly*, 30, pp. 399–415.

Finn, R. H. (1972), "Effects of Some Variations in Rating Scale Characteristics on the Means and Reliabilities of Ratings," *Educational and Psychological Measurement*, 32, pp. 255–265.

French-Lazovik, G., and Gibson, C. L. (1984), "Effects of Verbally Labeled Anchor Points on the Distributional Parameters of Rating Measures," *Applied Psychological Measurement*, 8, pp. 49–57.

Garner, W. R. (1960), "Rating Scales, Discriminability, and Information Transmission," *Psychological Review*, 67, pp. 343–352.

Ghiselli, E. E. (1939), "All or None Versus Graded Response Questionnaires," *Journal of Applied Psychology*, 23, pp. 405–413.

Gill, S. (1947), "How Do You Stand Sin?," *Tide* (March 14), pp. 72.

Gilljam, M., and Granberg, D. (1993), "Should We Take Don't Know for an Answer?" *Public Opinion Quarterly*, 57, pp. 348–357.

Green, P. E., and Rao, V. R. (1970), "Rating Scales and Information Recovery—How Many Scales and Response Categories to Use?" *Journal of Marketing*, 34, pp. 33–39.

Hartley, E. L. (1946), *Problems in Prejudice*, New York: Kings' Crown Press.

Hippler, H.-J., and Schwarz, N. (1989), "'No-Opinion' Filters: A Cognitive Perspective," *International Journal of Public Opinion Research*, 1, pp. 77–87.

Jacoby, J., and Matell, M. S. (1971), "Three-point Likert Scales Are Good Enough," *Journal of Marketing Research*, 8, pp. 495–500.

Jennings, M. K., and Markus, G. B. (1984), "Partisan Orientations Over the Long Haul: Results from the Three-wave Political Socialization Panel Study," *American Political Science Review*, 78, pp. 1000–1018.

Jennings, M. K., and Niemi, R. G. (1978), "The Persistence of Political Orientations: An Overtime Analysis of Two Generations," *British Journal of Political Science*, 8, pp. 333–363.

Kalton, G., Roberts, J., and Holt, D. (1980), "The Effects of Offering a Middle Response Option with Opinion Questions," *The Statistician*, 29, pp. 65–79.

Kaplan, R. M., Bush, J. W., and Berry, C. C. (1979), "Category Rating Versus Magnitude Estimation for Measuring Levels of Well-Being," *Medical Care*, 17, pp. 501–525.

Kinder, D. R., and Sears, D. O. (1985), "Public Opinion and Political Action," in G. Lindzey, and E. Aronson (eds.), *The Handbook of Social Psychology*, Vol. 2, pp. 659–741.

Klare, G. R. (1950), "Understandability and Indefinite Answers to Public Opinion Questions," *International Journal of Opinion and Attitude Research*, 4, pp. 91–96.

Klopfer, F. J., and Madden, T. M. (1980), "The Middlemost Choice on Attitude Items: Ambivalence, Neutrality, or Uncertainty," *Personality and Social Psychology Bulletin*, 6, pp. 97–101.

Komorita, S. S., and Graham, W. K. (1965), "Number of Scale Points and the Reliability of Scales," *Educational and Psychological Measurement*, 25, pp. 987–995.

Krosnick, J. A. (1991), "Response Strategies for Coping with the Cognitive Demands of Attitude Measures in Surveys," *Applied Cognitive Psychology*, 5, pp. 213–236.

Krosnick, J. A., and Abelson, R. P. (1992), "The Case for Measuring Attitude Strength in Surveys," in J. M. Tanur (ed.), *Questions About Questions: Inquiries into the Cognitive Bases of Surveys*, New York: Russell Sage Foundation, pp. 177–203.

Krosnick, J. A., and Berent, M. K. (1990), "The Impact of Verbal Labeling of Response Alternatives and Branching on Attitude Measurement Reliability in Surveys," paper presented at the American Association for Public Opinion Research Annual Meeting, Lancaster, Pennsylvania.

Krosnick, J. A., and Berent, M. K. (1993), "Comparisons of Party Identification and Policy Preferences: The Impact of Survey Question Format," *American Journal of Political Science*, 37, pp. 941–964.

Krosnick, J. A., and Fabrigar, L. R. (forthcoming), *Designing Good Questionnaires Effectively*, New York: Oxford University Press.

Krosnick, J. A., and Schuman, H. (1988), "Attitude Intensity, Importance, and Certainty and Susceptibility to Response Effects," *Journal of Personality and Social Psychology*, 54, pp. 940–952.

Lehmann, D. R., and Hulbert, J. (1972), "Are Three-point Scales Always Good Enough?" *Journal of Marketing Research*, 9, pp. 444–446.

Likert, R. (1932), *A Technique for the Measurement of Attitudes*, New York: Columbia University Press.

Lodge, M. (1981), *Magnitude Scaling: Quantitative Measurement of Opinions*, Beverly Hills, CA: Sage Publications.

Madden, J. M., and Bourdon, R. D. (1964), "Effects of Variations in Scale Format on Judgment," *Journal of Applied Psychology*, 48, pp. 147–151.

Martin, W. S. (1973), "The Effects of Scaling on the Correlation Coefficient: A Test of Validity," *Journal of Marketing Research*, 10, pp. 316–318.

Martin, W. S. (1978), "Effects of Scaling on the Correlation Coefficient: Additional Considerations," *Journal of Marketing Research*, 15, pp. 304–308.

Masters, J. R. (1974), "The Relationship Between Number of Response Categories and Reliability of Likert-type Questionnaires," *Journal of Educational Measurement*, 11, pp. 49–53.

Matell, M. S., and Jacoby, J. (1971), "Is There an Optimal Number of Alternatives for Likert Scale Items? Study I: Reliability and Validity," *Educational and Psychological Measurement*, 31, pp. 657–674.

Matell, M. S., and Jacoby, J. (1972), "Is There an Optimal Number of Alternatives for Likert-scale Items? Effects of Testing Time and Scale Properties," *Journal of Applied Psychology*, 56, pp. 506–509.

McClendon, M. J., and Alwin, D. F. (1993), "No-Opinion Filters and Attitude Measurement Reliability," *Sociological Methods and Research*, 21, pp. 438–464.

McCrae, A. W. (1970a), "Information Measures in Perceptual Experiments: Some Pitfalls Encountered by Kintz, Parker, and Boynton," *Perceptual Psychophysics*, 7, pp. 221–222.

McCrae, A. W. (1970b), "Channel Capacity in Absolute Judgment Tasks: An Artifact of Information Bias?," *Psychological Bulletin*, 73, pp. 112–121.

McKelvie, S. J. (1978), "Graphic Rating Scales—How Many Categories?," *British Journal of Psychology*, 69, pp. 185–202.

Miethe, T. D. (1985), "The Validity and Reliability of Value Measurements," *Journal of Psychology*, 119, pp. 441–453.

Myers, J. H., and Warner, W. G. (1968), "Semantic Properties of Selected Evaluation Adjectives," *Journal of Marketing Research*, 5, pp. 409–412.

Narayan, S., and Krosnick, J. A. (1996), "Education Moderates Some Response Effects in Attitude Measurement," Public Opinion Quarterly, 60, pp. 58–88.

Pepper, S. (1981), "Problems in Quantification of Frequency Expressions," in D. Fiske (ed.), *New Directions for Methodology of Social and Behavioral Science: Problems with Language Imprecision*, San Francisco: Jossey-Bass, pp. 25–41.

Peters, D. L., and McCormick, E. J. (1966), "Comparative Reliability of Numerically Anchored Versus Job-task Anchored Rating Scales," *Journal of Applied Psychology*, 50, pp. 92–96.

Peterson, B. L. (1985), *Confidence: Categories and Confusion*, (Report No. 50), Ann Arbor, MI: General Social Survey Project.

Petty, R. E., and Krosnick, J. A. (1995), *Attitude Strength: Antecedents and Consequences*, Hillsdale, NJ: Erlbaum Associates.

Ramsay, J. O. (1973), "The Effect of Number of Categories in Rating Scales on Precision of Estimation of Scale Values," *Psychometrika*, 38, pp. 513–532.

Rosenstone, S. J., Hansen, J. M., and Kinder, D. R. (1986), "Measuring Change in Personal Economic Well-being," *Public Opinion Quarterly*, 50, pp. 176–192.

Schaeffer, N. C. (1991), "Hardly Ever or Constantly? Group Comparisons Using Vague Quantifiers," *Public Opinion Quarterly*, 55, pp. 395–423.

Schuman, H., and Presser, S. (1979), "The Open and Closed Question," *American Sociological Review*, 44, pp. 692–712.

Schuman, H., and Presser, S. (1981), *Questions and Answers in Attitude Surveys*, New York: Academic Press.

Schuman, H., and Scott, J. (1989), "Response Effects Over Time: Two Experiments," *Sociological Methods and Research*, 17, pp. 398–408.

Schwarz, N., and Hippler, H. (1991), "Response Alternatives: The Impact of Their Choice and Presentation Order," in P. Biemer, R. Groves, L. Lyberg, N. Mathiowetz, and S. Sudman (eds.), *Measurement Errors in Surveys*, New York: Wiley and Sons.

Schwarz, N., Knauper, B., Hippler, H., Noelle-Neumann, E., and Clark, L. (1991), "Rating Scales: Numeric Values May Change the Meaning of Scale Labels," *Public Opinion Quarterly*, 55, pp. 570–582.

Smith, T. W. (1984), "Nonattitudes: A Review and Evaluation," in C. F. Turner, and E. Martin (eds.), *Surveying Subjective Phenomena* (vol. 2), New York: Russell Sage.

Smith, T. W., and Peterson, B. L. (1985), "The Impact of Number of Response Categories on Inter-item Associations: Experimental and Simulated Results," paper presented at the American Sociological Association Meeting, Washington, DC.

Stember, H., and Hyman, H. (1949–1950), "How Interviewer Effects Operate Through Question Form," *International Journal of Opinion and Attitude*, 3, pp. 493–512.

Stone, D. N., and Schkade, D. A. (1991), "Numeric and Linguistic Information Representation in Multiattribute Choice," *Organizational Behavior and Human Decision Processes*, 49, pp. 42–59.

Taylor, M. C. (1983), "The Black-and-White Model of Attitude Stability: A Latent Class Examination of Opinion and Nonopinion in the American Public," *American Journal of Sociology*, 89, pp. 373–401.

Thurstone, L. L., and Chave, E. J. (1929), *The Measurement of Attitudes*, Chicago: University of Chicago Press.

Wallsten, T. S., Budescu, D. V., Zwick, R., and Kemp, S. (1993), "Preferences and Reasons for Communicating Probabilistic Information in Verbal or Numeric Terms," *Bulletin of the Psychonomic Society*, 31, pp. 135–138.

Watson, D. (1988), "The Vicissitudes of Mood Measurement: Effects of Varying Descriptors, Time Frames, and Response Formats on Measures of Positive and Negative Affect," *Journal of Personality and Social Psychology*, 55, pp. 128–141.

Wedell, D. H., and Parducci, A. (1988), "The Category Effect in Social Judgment: Experimental Ratings of Happiness," *Journal of Personality and Social Psychology*, 55, pp. 341–356.

Wedell, D. H., Parducci, A., and Lane, M. (1990), "Reducing the Dependence of Clinical Judgment on the Immediate Context: Effects of Number of Categories and Type of Anchors," *Journal of Personality and Social Psychology*, 58, pp. 319–329.

Wicker, A. W. (1969), "Attitudes Versus Actions: The Relationship of Verbal and Overt Behavioral Responses to Attitude Objects," *Journal of Social Issues*, 25, pp. 41–78.

Wildt, A. R., and Mazis, M. B. (1978), "Determinants of Scale Response: Label Versus Position," *Journal of Marketing Research*, 15, pp. 261–267.

Zaller, J. (1988), "Vague Questions Get Vague Answers: An Experimental Attempt to Reduce Response Instability," unpublished manuscript, University of California at Los Angeles.

CHAPTER 7

Towards a Theory of Self-Administered Questionnaire Design

Cleo R. Jenkins
U.S. Bureau of the Census

Don A. Dillman
Washington State University

7.1 INTRODUCTION

Our understanding of self-administered questionnaire design clearly remains in its infancy. Although recommendations for design have been offered (e.g., U.S. General Accounting Office, 1993; Dillman, 1978), few systematic efforts have been made to derive principles for designing self-administered questionnaires from relevant psychological or sociological theories.

One notable exception is a paper by Wright and Barnard (1975). They reviewed the behavioral research, particularly on language and comprehension, and presented ten rules for designing forms. In a later paper, Wright and Barnard (1978) write that the problems of completing self-administered questionnaires fall into two classes: problems with the language used and problems arising from the way information is arranged spatially. This statement suggests that the spatial arrangement of information is not "language." However, it is more precise to label both as graphic language and to further subdivide them into "verbal" versus "non-verbal" language. One reason for suggesting that the term "graphic non-verbal language" be used is that it is fully encompassing. Not only does it intimate that respondents extract meanings and cues from the

Survey Measurement and Process Quality, Edited by Lyberg, Biemer, Collins, de Leeuw, Dippo, Schwarz, Trewin.
ISBN 0-471-16559-X   © 1997 John Wiley & Sons, Inc.

spatial arrangement of information, but it includes other important visual phenomena, such as color and brightness.

Our understanding of verbal language as used in surveys has blossomed in the last few decades (e.g., Jobe *et al.*, 1993; Martin, 1993; Schwarz and Hippler, 1991; Jobe *et al.*, 1990; Lessler, 1989; Converse and Presser, 1986; Belson, 1981). Although much of this work involves interviewer-administered questionnaires (that is, verbal language presented aurally), some of it is applicable to mail questionnaire design (that is, verbal language presented visually). Schwarz *et al.* (1991) summarize the major distinctions between these two channels of presentation and conclude that most question wording effects are likely to be relatively independent of the channel of presentation. On the other hand, Schwarz *et al.* conclude that the channel of presentation is likely to moderate question order, context, and memory effects. The researchers suggest that the following differences in the two presentation channels are responsible for these effects: (1) varying amounts of control over the temporal order in which information is presented (which they term "sequential versus simultaneous presentation"), (2) time pressure, (3) non-verbal interviewer–respondent interaction, (4) interviewer explanations, (5) perception of interviewer characteristics, (6) perceived confidentiality, (7) external distractions, and (8) the differential self-selection of respondents.

Although much research has been directed at understanding the verbal aspects of surveys, in comparison only a limited body of research addresses the non-verbal issues specific to self-administered questionnaires (e.g., Dillman *et al.*, 1993; DeMaio and Bates, 1992; DeMaio *et al.*, 1987; Rothwell, 1985). Jenkins and Dillman (1993) stated 20 design principles, all of which emphasize the visual presentation of information. Although the principles themselves are more comprehensive than those by Wright and Barnard, neither paper attempts to consolidate and organize the information into a working model of self-administered questionnaire design. More importantly, none of the research incorporates the concepts and theories of relevant visual disciplines, such as pattern recognition and the Gestalt Grouping Laws.

Our purpose in this chapter is to contribute to the development of a theory of questionnaire design for mail, self-administered questionnaires. We define and discuss relationships among concepts important to such a theory. These concepts are used to propose a working model along with several principles for guiding the questionnaire design process, thus extending our previous work. Finally, we present results from an initial test of U.S. census questionnaires that adhered to these design principles.

7.2 RESPONDING TO SELF-ADMINISTERED QUESTIONNAIRES: A CONCEPTUALIZATION

In Tourangeau's model (1984), as well as other models of the survey interview process, the first step is specified as comprehending the question. Depending on the model, different steps follow, but generally they are retrieval of the relevant facts, judgment, and finally, the response.

Although comprehending the question is the first step in an interviewer-administered survey, the task is different in a self-administered survey. In a self-administered survey, respondents must first perceive the information before they can comprehend it. Once respondents perceive the information, they must comprehend the layout (the visual aspect) of the information as well as the wording (the verbal aspect). Furthermore, respondents must comprehend much more than just the wording of the survey questions and response categories. In a self-administered survey, respondents are often given introductory material and instructions. Also, they must comprehend directions that are meant to guide them through the questionnaire.

In an interviewer-administered questionnaire, the interviewer plays a critical role in the perceptual process. In contrast, the entire onus of perception is on the respondent in a self-administered format. Although we have learned that errors arise as a result of this process (Jenkins and Ciochetto, 1993; Jenkins *et al.*, 1992), we have not developed procedures for controlling these errors. Clearly we need a keen understanding of the perceptual process to exert control over it.

When respondents are asked to complete a self-administered questionnaire, they are asked to perform a task that from their perspective may be different from the task we wish them to perform. From the respondent's perspective, the task may be similar to asking them to view a picture; they are free to start anywhere and to make their own decisions as to which parts of the picture to examine and in what order.

However, from our perspective, this viewing method is detrimental, for it gives us very little control over the perceptual process. From our perspective, it would be best if respondents started at a specified place, read prescribed words (in order to comprehend the question or stimulus) in the order in which we intend, provide answers to each stimulus, and move sequentially through the questionnaire. In general, we do not want respondents to mark answers without having fully read and understood the questions and accompanying instructions, nor do we want them to pick and choose which questions get answered and in what order.

In order to control the above process, it seems important that we have some understanding of graphic language, for this is the channel through which communication takes place. In the following discussion, we begin by categorizing questionnaires in terms of the graphic language they use. After this, we go on to discuss cognition, especially as it applies to visual perception, and motivation, for these are the actual processes we wish to influence.

7.2.1 Graphic Language

Graphically speaking, questionnaires are complex. In a tutorial, Twyman (1979) provides a classification matrix for graphic language. Along the horizontal axis, graphic language is divided into seven methods of configuration, or ways in which information can be configured on a page: (1) pure linear, (2) linear interrupted, (3) list, (4) linear branching, (5) matrix, (6) non-linear directed viewing, and (7) non-linear most options open. The vertical axis is made up of four modes of symbolization, that is, ways in which the information may be symbolized: (1) verbal/numerical, (2) pictorial and verbal/numerical, (3) pictorial, and (4) schematic. This schema gives rise to a matrix comprising 28 graphic language categories. Text, for instance, uses a linear-interrupted method of configuration and a verbal/numerical mode of symbolization, whereas a picture would fall into the pictorial, non-linear, most-options-open method of configuration and the pictorial symbolization category.

Questionnaires do not perfectly conform to any one category. Often they use more than one configuration, such as linear interrupted (for the questions themselves), list (for the answer options) and matrix (for answer tables). To further complicate matters, questionnaires tend to use several modes of symbolization—verbal (for the questions and answers) and schematic (for the answer boxes and symbolic directions). Unfortunately, most of the graphic language literature is applicable to only a few of the graphic language categories: to a large extent, text and pictures (e.g., Waller, 1985 and 1987; Hartley, 1985; Spencer, 1968; Tinker, 1965; Buswell, 1935) and to a much lesser extent, graphs (e.g., Tufte, 1990 and 1983).

7.2.2 Cognition and Visual Perception

Cognition refers to all the processes by which sensory input is transformed, reduced, elaborated, stored, recovered, and used. *Perception* involves using previous knowledge to gather and interpret the stimuli registered by the senses (Matlin, 1994; Glass and Holyoak, 1986; Neisser, 1967). *Pattern recognition* is a particular perceptual process that involves identifying a complex arrangement of sensory stimuli. Obviously, to make sense of the information presented on a questionnaire, respondents must be able to see patterns.

Pattern recognition is accomplished through two complementary subprocesses: *bottom-up* and *top-down processing*. There are a number of theories about how bottom-up processing occurs, but essentially, they all focus on the arrival of stimulus. "Bottom-up" is a metaphorical phrase meant to convey the notion of processing physical stimulus through our senses alone. In contrast, top-down processing emphasizes the role of context and expectations in identifying a pattern. In this case, our conceptual knowledge about how the world is organized helps us to identify patterns.

If in Figure 7.1, you see a "c" on the left and a "d" on the right, you have just engaged in pattern recognition and most likely you used bottom-up processing. That is, the physical stimulus in this picture is clear enough that you did not need to invoke higher top-down processes to correctly recognize the letters here. However, the task is not as clear in Figure 7.2. For a purely bottom-up perspective, place your hand over the first and third columns of letters in this figure. What do you see? Let go, and you now have an example

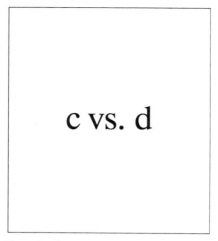

Figure 7.1 Example of pattern recognition using bottom-up processing.

Figure 7.2 Example of pattern recognition using top-down processing.

of where by using top-down processing, that is, by placing well-founded expectations on the middle letter, you can quickly decipher the phrase, "The cat sat."

As respondents begin to fill out a questionnaire, they apply both bottom-up and top-down processing, only no doubt the top-down processing at this point comes from other more common experiences, such as reading or looking at objects. Most respondents are not experienced at filling out questionnaires, and it seems quite clear that we should keep this in mind when we design questionnaires.

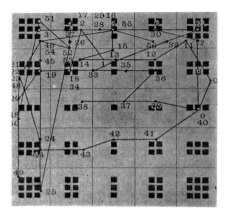

Figure 7.3 Illustration of preferred eye-movement positions when all information on a page is of equal visual interest. Reprinted, by permission, from Brandt, H.F., *The Psychology of Seeing*, p. 31. Copyright © 1945 by The Philosophical Library, New York, NY.

One would expect, however, that the act of filling out the questionnaire would influence the top-down processing so that as respondents move along, they will begin to associate particular visual information with particular requests. This argues for a consistent application of visual information.

The world extends 360 degrees around us. However, our field of vision spans about 210 degrees, and is sharp only within 2 degrees (Kahneman, 1973). When people are presented with simultaneous tasks, like the need to perceive a 210-degree field, they must focus their *attention* on one task at a time. In stationary visual perception, this corresponds to choosing a place to look first, then moving to a second place, and so on. This approach requires a detailed analysis of a field, which we carry out so effortlessly and automatically that we are not even aware of being engaged in this step-wise process. In contrast, *preattentive processing* involves the automatic registration of features at a global or holistic level (Neisser, 1967). It occurs when individuals survey an entire visual field, instantaneously recognizing the features enough to grossly make sense of the scene.

Both preattentive processing and attention are necessary for visual perception and both suggest that we must pay attention to how visual information is presented at both the macro- and micro-level on a questionnaire. Respondents should be able to glance at a questionnaire and as a result of preattentive processing be able to quickly understand where to start, and generally where they are expected to go. And then the path should continue to be unambiguous as respondents attend to the details of the questionnaire.

Brandt (1945) discovered that humans have preferred positions in *eye movements* (see Figure 7.3). He constructed a card with squares that were symmetrically located about a locus. The results of his study reveal that if a field is divided into four quadrants of equal visual interest, subjects' eyes will naturally fall in the upper left-hand quadrant, closer to the center of the page

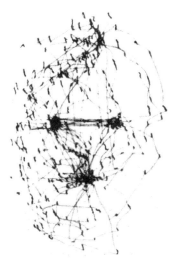

Figure 7.4 Illustration of an eye-movement trail when all information is not of equal visual interest. Reprinted, by permission, from Yarbus, A.L., *Eye Movements and Vision*, p. 180. Copyright © 1967 by Plenum Publishing Corporation, New York, NY.

than the extreme left-hand corner. The least preferred space was the lower right-hand quadrant. Brandt also discovered that successive movement of the eyes following the initial fixation is toward the left and upward. He suggests that the left-side preference is most likely due to our habits of reading a printed page from left to right and to the fact that most people are left-brain dominant. As questionnaire designers, we are in a position to take advantage of these tendencies.

Most of what we look at, however, is not of equal visual interest. In this case, we know that people will focus on areas which are physically informative, like high contrast areas, or on areas of ecological significance (Kahneman, 1973). Figure 7.4, for instance, depicts the typical eye movement trail of someone looking at a face in which the eyes and the mouth have attracted a great deal of attention.

How we look at pictures is important, but it is not all we need to know to design questionnaires. Since we predominantly use linear-interrupted text to communicate, we also need to understand how we read. Lima (1993) and Rayner (1993) note that contrary to what we might think, we do not read text smoothly. Rather, our eyes fixate about every eight to nine letter spaces, corresponding to the 2-degree visual angle mentioned earlier. This is known as the *foveal* region. Generally, our eyes are fixated about 90 percent of the time. The remainder of the time, our eyes move very rapidly between fixations in what are known as *saccades*.

There is evidence to suggest that the region to the right of the region on which our eyes are fixated, the *parafoveal* region, plays an important role in

Figure 7.5   A picture based primarily upon the visual element of shape.

our ability to recognize words during reading. Previewing the initial letters of the next word helps us to recognize that word more quickly (Lima, 1993). We will explain this in greater detail later, but it suffices to say that in effect we deprive readers of the information to the right of a fixated word when conceptually connected information is presented at separate locations on a questionnaire.

The eyes and their movements are the mechanism by which we take in our surroundings. What the eyes take in, however, are *visual elements.* Visual elements are first described in terms of *brightness and color*, then *shape*, and finally *location.* When information from these levels is encoded, perceptual representation occurs (Glass and Holyoak, 1986). Figure 7.5, for instance, depicts a relatively simple line drawing of a house that is based primarily upon the visual elements of shape and location. In comparison, the building blocks in Figure 7.6 rely on an additional visual element: contrast. The problem with these building blocks is that they impede our ability to make out this picture clearly. Their perimeters contribute a high level of spatial frequency, that is, too many lines. If, however, you squint at this picture, which will help to diminish the perimeters, Abraham Lincoln should come into view.

Bringing together the visual elements leads to an important visual outcome— that is, we are able to distinguish between *figure* and *ground*. The recognition of a figure against a background depends on contrast, especially contrast between brightness (or color) and shapes (see Figure 7.7).

According to Gestalt psychologists, a number of perceptual principles guide our understanding and interpretation of figures (Wallschlaeger and Busic-Snyder, 1992; Castner and Eastman, 1984; Zusne, 1970). Look at Figure 7.8 for 10 seconds. Now cover it and try to remember what you saw. According to the

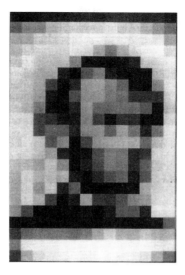

Figure 7.6   A picture based primarily upon the visual element of brightness.

Figure 7.7   Illustration of figure and ground. (Adapted from Rubin, 1958.)

Law of Pragnanz you should remember the first column better than the others. The *Law of Pragnanz* states that figures with simplicity, regularity, and symmetry, such as those in the first column, will be more easily perceived and remembered than the irregularly shaped polygons in the remaining columns.

The *Law of Proximity* states that when similar figures are located in close proximity to each other, we tend to see them as belonging to the same group. Do the dashes in Figure 7.9 appear to be horizontal or vertical lines? What if

Figure 7.8   Illustration of the Law of Pragnanz.

Figure 7.9   Illustration of the Law of Proximity.

you turn the figure sideways? Figure 7.10 is a good example of the grouping laws in operation. The first example depicts two lines of different shapes. That is because according to the *Law of Similarity*, we tend to see similar shapes as belonging together. Finally, if you see a white triangle in Figure 7.11, you are acting according to the *Law of Closure*.

These laws provide further evidence for paying close attention to how we display information on a page, for respondents will extract meaning from how the information is shaped, shaded, and grouped. The real challenge for us in all this is to learn how to visually communicate our intentions.

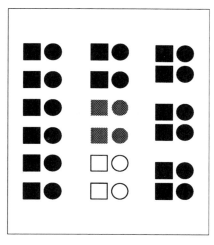

Figure 7.10   Illustration of the Gestalt Grouping Laws in operation.

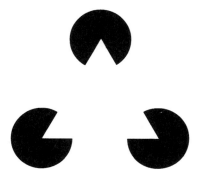

Figure 7.11   Illustration of the Law of Closure. Reprinted, by permission, a figure originally published in Kanizsa, G., *Organization in Vision*, p. 74. Copyright © Praeger, an imprint of Greenwood Publishing Group, Inc., Westport, CT., and adapted by Crick, F., *The Astonishing Hypothesis*, p. 28, Charles Scribner's Sons, Macmillan Publishing Company (presently Simon and Schuster), New York, NY.

7.2.3   Motivation

Neisser (1967) points out that although cognitive psychology is concerned with all human activity rather than some fraction of it, the concern is from a particular point of view. He asserts that other viewpoints are equally legitimate and necessary, such as dynamic psychology: "Instead of asking how a man's actions and experiences result from what he saw, remembered, or believed, the dynamic psychologist asks how they follow from the subject's goals, needs, or instincts . . ."

The motivational aspect of responding to questionnaires concerns whether respondents desire to read the questions and to formulate and express appropriate answers. In a larger sense, motivational considerations influence whether respondents even begin the process of filling out the questionnaire, and whether it gets returned to the researcher. Consequently, we need to understand what motivates respondents to answer surveys, and how questionnaires can be designed to positively influence this process.

Dillman (1978) has argued from a social exchange perspective (also see Goyder, 1988) that people are more likely to complete a mail questionnaire if they expect that the costs to them of completing it are less than the expected rewards to themselves or groups with which they identify. This perspective leads to recommendations to reduce perceived costs by making the questionnaire appear quick and easy to complete and to avoid information that may embarrass the questionnaire recipient (e.g., a question that is hard to understand) or subordinate them to the survey sponsor. Among the recommendations for increasing rewards are including explanations of the study's usefulness to the respondent or groups with which the respondent is likely to identify, including questions that are likely to be salient or of interest to the recipient, and laying out the questionnaire in a format that is easy and encourages the respondent to get a sense of progress from moving with ease through its pages. Visual aspects of questionnaire construction are accorded substantial significance under this perspective for improving response to mail surveys.

Cialdini (1984) has argued more generally that people decide whether to perform a requested task on the basis of the inherent attractiveness of that task *and* other social or psychological influences, including

- reciprocation: the tendency to favor requests from those who have previously given something to you,
- commitment and consistency: the tendency to behave in a similar way in situations that resemble one another,
- social proof: the tendency to behave in ways similar to those like us,
- liking: the tendency to comply with requests from attractive others,
- authority: the tendency to comply with requests given by those in positions of power, and
- scarcity: the tendency to attach greater value to rare opportunities.

Groves *et al.* (1992) provide examples of how each of these can be used to encourage survey participation. Although most of the examples refer to interviewer behavior or the implementation process, some can be applied to questionnaire design. For example, the fact that people tend to comply with requests from attractive others suggests that respondents may be more likely to answer an attractive questionnaire than an unattractive one.

Finally, the literature on opinion change (Petty and Capioppo, 1986) suggests that when a topic is of high personal relevance, subjects will change their opinions based on an in-depth review of a message. However, when the topic is not important to subjects, they will rely on a heuristic review, such as the credibility of the source. This literature suggests that if a questionnaire is not really important to respondents, then we probably are not going to persuade them to complete it by presenting them with an in-depth, highly logical, persuasive discussion of why they should complete it. Instead, we should rely on other means.

7.3 PRINCIPLES FOR DESIGNING SELF-ADMINISTERED QUESTIONNAIRES

Using concepts from the preceding section on cognition and visual perception as well as motivation, it is our ultimate goal to develop a series of theoretically based principles that can be empirically tested, as part of the effort to develop an overall theory of good questionnaire design. In this section, we provide examples of such principles and illustrate their application to selected self-administered questionnaires, the task to which we now turn.

To accomplish this goal, it is useful to introduce a heuristic device which we have found beneficial for guiding our design efforts, that is, the distinction between *navigational guides* and *information organization*. In essence, decisions about navigational guides are aimed at motivating and guiding the respondent through a questionnaire in a particular way in the absence of an interviewer who would otherwise perform that task. Decisions about information organization concern what a respondent would hear if he/she were being interviewed.

7.3.1 Designing Navigational Guides

To some degree, navigational features of self-administered questionnaires have been discussed previously. For instance, in 1978, Dillman provided practical advice on how to construct a questionnaire so as to guide respondents through it. More recently, DeMaio and Bates (1992) carried out an experiment during the 1990 U.S. census, known as the Alternative Questionnaire Experiment, which was largely about how to structure a questionnaire to ensure its completion.

The visual perception literature provides conceptual underpinnings for such design efforts. It suggests that respondents should be able to glance at a questionnaire and as a result of preattentive processing be able to quickly locate the starting point and the path they are expected to take. Additionally, the path should continue to be unambiguous as respondents attend to the details of the questionnaire. The following principles are meant to accomplish these goals.

Principle 1. Use the visual elements of brightness, color, shape, and location in a consistent manner to define the desired navigational path for respondents to follow when answering the questionnaire.

The questionnaire in Figure 7.12 used for a 1992 pretest of the National Survey of College Graduates (Shettle *et al.*, 1993) successfully manipulates the visual

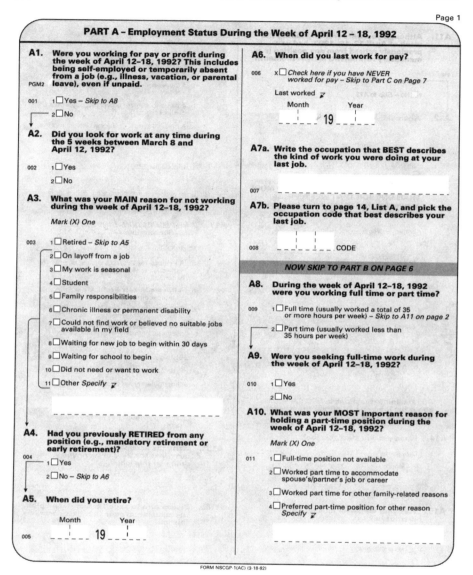

Figure 7.12 Example of good use of navigational guides from the 1992 pilot study for the National Survey of College Graduates.

presentation of information. Because they have different brightnesses and shapes, the questions, answer categories, and other information on the page unambiguously appear as *figures* in the foreground, whereas the formless shaded area (printed on the original in light blue, 10 percent of fixed color) appears to extend behind the figures as *background*.

An important attribute of this questionnaire is that the question numbers are prominently displayed. They stand out because they are located on the left-hand side of the reading columns, a highly visible area, and because they are set apart somewhat from the questions.

Another important attribute is that the beginning of the questionnaire is clearly marked "A1." This questionnaire does not place information before the "A1" that is likely to confuse respondents about where to begin.

Also, this questionnaire format uses contrast to its advantage. To begin with, both the question numbers and the questions are in bold type to enhance their salience and define the navigational path for respondents to follow. And then there is the highly effective contrast between the shaded background and the other information on the page, especially the white answer spaces.

Not only that, but the white answer boxes are equally sized simple shapes that are vertically aligned, all of which is in keeping with the Gestalt grouping laws (the Laws of Similarity and Proximity). As a result, the elements of shape, brightness, and location give rise to a well-defined regular pattern on the questionnaire that is immediately evident at the pre-attentive level and that can be used to guide respondents through the form at the attentive level.

Based on both the Gestalt Grouping Laws and the graphic language literature, one can make an argument for listing closed-ended answer categories vertically and scaling categories horizontally. As demonstrated in Figure 7.12, grouping closed-ended categories vertically—a form of typographical cuing— should give respondents the correct impression that the categories are distinct entities. On the other hand, grouping them horizontally might give the false impression that they are continuous, much the same way a sentence is a continuous thought that runs across the page horizontally. By the same token, the grouping of scaling answer categories vertically may give respondents the wrong impression that the categories are unassociated, when in fact they are continuous.

Recent evidence supporting this possible underlying construct comes from an experiment by Gaskell *et al.* (1994). They demonstrate that the choice of response alternatives can influence responses to questions about the frequency of vaguely defined target events. In addition, they show that the size of the observed shifts depends upon the presence of priming questions and the orientation of the response scales. They conclude that either a horizontal orientation of the response scale or priming questions (or both) may increase the effect of the response alternatives. They speculate that the horizontal presentation of the scale may make the question more distinctive. Perhaps, as suggested above, the horizontal orientation of the scale provides different, and in fact more accurate, information to the respondent—that the response options actually lie along a continuum.

Generally, questionnaires are printed in one color, that is, black print on white paper. The questionnaire presented in Figure 7.12 has two colors: black print on a light blue field printed on white paper. A number of researchers have speculated that certain colored paper (green, yellow, pink) would appeal to respondents more than others. For the most part, this has not proven true (Phipps *et al.*, 1991; Crittenden *et al.*, 1985). One exception is a metanalysis by Fox *et al.* (1988) in which green rather than white questionnaires were found to have significantly increased mail response rates. Until more evidence is forthcoming, however, we tentatively think that visual elements (such as a uniformly colored background) which do not serve to guide respondents through the form will have little effect on respondents.

Another good feature of the Survey of College Graduates questionnaire is that the key and source codes needed for processing are a deeper shade (100%) of the blue background color. Theoretically, the subtle contrast between the codes and the background should help make these codes less visible to the respondent, while still being easily read by processing personnel. In addition, because the source codes are located on the far left of the page, outside the foveal and parafoveal view of the respondent, they are even more likely to be ignored. However, this may not be the ideal spot for the source codes. According to the eye-movement literature, information is most likely to be ignored if it is placed in the lower right-hand corner of the viewing space. In the case of A1, this would be to the far right of the "No" response and just left of the dividing line in the middle of the page.

Another minor visual imperfection in this questionnaire is the use of dotted lines beneath the write-in answer spaces. Since the Law of Closure states that respondents will connect the dotted lines anyway, why not just make them solid from the start? However, one reason they can be made solid without adding visual clutter is because this questionnaire uses lines sparingly. Remember, too much detail impedes our ability to interpret visual data. Lines are used mainly to delineate the colored areas from the white areas; they are *not* used to delineate one question from another, which leads to unnecessary visual clutter. Other than the line running down the middle, the space is open with a strong and consistent separation between figure (what we want respondents to see) and background (what we do not want them to see).

The questionnaire in Figure 7.13, used in the 1993 U.S. Census of Agriculture, violates many of the principles that Figure 7.12 exemplifies. It conflictingly invites a respondent's eye to different areas of the questionnaire for different reasons. Rather than working in unison to achieve a desired outcome, information competes for the respondent's eye. At the very least, there appear to be three different competing areas:

- The upper-left hand quadrant, which according to the eye-movement literature, is the eye's preferred position for looking at information when all of the information on the page is of equal interest.
- The high contrast areas, which according to the eye-movement literature attract our attention. The boldest information on this questionnaire is: the "Census Use Only" boxes, which are of the least relevance to respondents, the reverse contrast "Section 1" heading, and the answer box on the right-hand side of the page associated with number 4.

DUE BY FEBRUARY 1, 1993   OMB No. 0607-0722: Approval Expires 09/30/94

FORM **92-A0202** (10-1-91)	U.S. DEPARTMENT OF COMMERCE BUREAU OF THE CENSUS

NOTICE – Response to this inquiry is required by law (title 13, U.S. Code). By the same law, YOUR REPORT TO THE CENSUS BUREAU IS CONFIDENTIAL. It may be seen only by sworn Census employees and may be used only for statistical purposes. Your report CANNOT be used for purposes of taxation, investigation, or regulation. The law also provides that copies retained in your files are immune from legal process.

UNITED STATES CENSUS OF AGRICULTURE

AG CENSUS USA

In correspondence pertaining to this report, please refer to your Census File Number (CFN)

92-A0202

INFORMATION COPY

- **Use reasonable estimates** – If records are not available.

- **Time extension** – Send request to address on bottom of page 11.

- **Correspondence** – Include 12-character Census File Number (CFN) found on address label.

- **Duplicate forms** – Enter extra CFN(s) below and return with your completed report.

A   –

A   –

Please correct errors in name, address, and ZIP Code. ENTER street and number if not shown.

CENSUS USE ONLY	035	036	037	038	039	040	041	042

SECTION 1   S1   **ACREAGE IN 1992** – Report land owned, rented, or used by you, your spouse, or by the partnership, corporation, or organization for which you are reporting. *Include ALL LAND, REGARDLESS OF LOCATION OR USE – cropland, pastureland, rangeland, woodland, idle land, house lots, etc.*

If the acres you operated in 1992 changed during the year, refer to the INFORMATION SHEET, section 1.

		None	Number of acres
1. All land owned		☐	043
2. All land rented or leased FROM OTHERS, including land worked by you on shares, used rent free, in exchange for services, payment of taxes, etc. *Include leased Federal, State, and railroad land. (DO NOT include land used on a per-head basis under a grazing permit.) Also complete item 5 below.*		☐	044
3. All land rented or leased TO OTHERS, including land worked on shares by others and land subleased. *Also complete item 6 below.*		☐	045

4. Acres in "THIS PLACE" – ADD acres owned (item 1) and acres rented FROM OTHERS (item 2), then SUBTRACT acres rented TO OTHERS (item 3), and enter the result in this space.

For this census report, these are the acres in "THIS PLACE." If the entry is zero, please refer to the INFORMATION SHEET, section 1.

046

	None	Number of landlords
5. If you rented land FROM OTHERS (item 2), how many of landlords did you have?	☐	065

	None	Number of acres
6. If you rented land TO OTHERS (item 3), how many of the acres rented or leased TO OTHERS did you own?.	☐	053

7. Did you have any grazing permits on a per-head basis?   054   1 ☐ YES – *Mark (X) all boxes which apply* . . . .

2 ☐ NO – *Go to item 8*

3 ☐ Forest Service
4 ☐ Taylor Grazing, Sec. 3 (BLM)
5 ☐ Indian Land
6 ☐ Other – *Specify*

8. LOCATION OF AGRICULTURAL ACTIVITY FOR "THIS PLACE"

		County name	State	Number of acres
a. In what county was the largest value of your agricultural products raised or produced?	Principal county →			056
b. If you also had agricultural operations in any other county(ies), enter the county name(s), etc.	Other counties			057
				058
				059

PENALTY FOR FAILURE TO REPORT   *CONTINUE ON PAGE 2* ➡

Figure 7.13 Example of poor use of navigational guides and information organization from the 1993 U.S. Census of Agriculture.

- A third competing area is the question 1 beneath the Section 1 heading. Cognitive research with both the Service Based Enumeration Questionnaire (Gerber and Wellens, 1994) and the Public School Questionnaire (Jenkins *et al.*, 1992) demonstrates that the first question often attracts respondents' eyes.

If respondents decide to begin with question 1, they are likely to miss many important instructions, including those in the upper left-hand quadrant about estimates and time extensions, the confidentiality statement in the upper right-hand corner, the instruction to "report land owned, rented, or used by you, your spouse, or by the partnership, corporation, or organization for which you are reporting," and the instruction about what to do if the acreage the respondent operated changed during the year.

We need to avoid giving respondents the erroneous impression that information is not important. The best way to do this is to make certain that all of the visual cues are "saying" the same thing, like "Start here" or "Read this." Otherwise respondents are reasonably confused about which messages to follow and in which order the messages should be followed.

We know from psychological experiments with subjects in which brain activity is monitored through the use of PET scans that confusion results in greater brain activity, presumably because subjects need to engage in greater cognitive processing to solve a problem. This suggests that we should be interested in more than just outcomes. It is in our best interest to be sensitive to and concerned with minimizing processing times as well.

Principle 2. When established format conventions are changed in the midst of a questionnaire, prominent visual guides should be used to redirect respondents.

Ideally, the visual elements should be used consistently, as emphasized in Principle 1, but occasionally, it may be necessary to violate this principle. In those cases, special care needs to be taken to redirect respondents' attention. For example, the questionnaire shown in Figure 7.14 from a 1994 survey of new Washington residents (Dillman *et al.*, 1995) uses a common question format. This format involves prominently identifying questions with "Q-x" designations, writing the questions in dark ink, and then listing categories below the questions, vertically. This format, however, would consume a great amount of space if used for questions that repeat themselves for each of many items, as in question 8. In addition, this page has the added complexity of including certain questions (3, 4, and 7) that do not apply to some respondents.

In general, top-down processing lends efficiency to our ability to interpret visual stimuli. It enables us to more quickly perceive our visual world with less effort by exploiting expectations formed by past experience. However, this system can break down when the visual world deviates from our expectations, for example when a change is made in the way visual elements are used.

Figure 7.14 illustrates this problem. In questions 1 through 7 respondents are supposed to answer to the left of the answer categories (stubs) usually, by circling a number. In question 8, a change was made to circling a word (rather than a number) placed to the right of stubs that are listed similar to answer choices in previous questions, for example, "A. For me to accept a new job." When used in a pilot study, this questionnaire resulted in several people either

Q1. **In what month and year did you move to (or back to) Washington State?**

_____ MONTH and _____ YEAR

(If you moved here before 1990, you do not need to complete the rest of the
questionnaire. However, please mail it back so we can remove your name from
the mailing list.)

Q2. **Are you a <u>first time</u> Washington resident?** *(Please circle number of your answer.)*

1   YES
2   NO⟶ **Q3.**   **(If No) What year did you leave here to live someplace else?**

_____ YEAR

Q4.   **All together, how many years have you lived in Washington?**

_____ YEARS IN WASHINGTON

Q5. **What is the most important reason why you recently moved here?**

_____

_____

Q6. **Did you move here alone or with others?** *(Please circle number of answer.)*

1   ALONE
2   WITH OTHER(S) ⟶ **Q7.**   **Who did you move here with?** *(Circle*
number of all who moved here with you.)

1   Spouse
2   Partner, with whom you live in a
marriage-like relationship
3   Child or children _____ NUMBER
4   Other *(Please specify)*

_____

Q8. **Were any of the following job-related considerations involved in your decision to move**
to Washington? *(Please circle yes or no for each item.)*

A.   For me to <u>accept</u> a new job . . . . YES   NO
B.   For someone else in my household to <u>accept</u> a new job . . . . YES   NO
C.   For me to <u>look for</u> new work . . . . YES   NO
D.   For someone else in my household to <u>look for</u> new work . . . . YES   NO
E.   A <u>transfer</u> for me by current employer . . . . YES   NO
F.   A <u>transfer</u> for someone else in my household by current employer . . . YES   NO

Figure 7.14 Example of poor use of navigational guides to change answering formats within a
questionnaire from the 1994 Washington State University pilot study of new state residents.

circling the letters (or starting to answer and then erasing) to the left of the
stubs, rather than making the desired switch to circling actual answers on the
right.

One way to correct this problem is to place numbers to the left of the stubs
and ask respondents to "circle the answers which apply." Although this solution
might seem simple, there is evidence that doing so encourages a primacy effect,
that is, biases towards the selection of answers from early in the list. Asking
respondents to respond either "yes" or "no" to each stub may avoid this bias.

Q1. **In what month and year did you move to (or back to) Washington State?**

_____ MONTH and _____ YEAR

(If you moved before 1990, you do not need to complete the rest of the questionnaire. However, please mail it back so we can remove your name from the mailing list.)

Q2. **Are you a <u>first time</u> Washington resident?** _(Please circle number of your answer.)_

1   Yes
2   No ⟶

> **Q3.** **(If No) What year did you leave here to live someplace else?**
>
> _____ YEAR
>
> **Q4.** **All together, how many years have you lived in Washington?**
>
> _____ NUMBER OF YEARS

Q5. **Did you move here alone or with others?** _(Please circle number of answer.)_

1   Alone
2   With Other(s) ⟶

> **Q6.** **Who did you move here with?** _(Circle number of all who moved here with you.)_
>
> 1   Spouse or partner
> 2   Child or children   _____ NUMBER
> 3   Other _(Please specify)_
>
> _____

Q7. **What was the most important reason why you moved here?**

_____

_____

Q8. **Did any of the following job-related considerations influence your decision to <u>move to Washington</u>?** _(Please circle YES or NO for each item.)_

	YOURSELF		YOUR SPOUSE OR PARTNER (if you have one)	
To accept a new job	YES	NO	YES	NO
To look for new work/job	YES	NO	YES	NO
To start/take over a business	YES	NO	YES	NO
A military transfer	YES	NO	YES	NO
A transfer by current employer (except military)	YES	NO	YES	NO

Figure 7.15 Example of good use of navigational guides to change answering formats in the revised questionnaire of the 1994 Washington State University study of new state residents.

Figure 7.15 demonstrates another solution. In this revision of question 8, attention is directed away from the left-hand side of the answer categories by removing the letters (which a few respondents treated like numbers) and placing an area of dense information on the right to attract the respondent's attention. That is done by the dark "hat" over the answer choices and by shading behind the answer choices. This page was further revised by using lines to visually demarcate "screened" questions from those to be answered by everyone, and aligning the demarcated spaces vertically.

7.3.2 Achieving Good Information Organization

Many authors have written extensively about principles for achieving desirable information organization in survey questions, many of which apply equally to both self-administered and interview surveys (Sudman and Bradburn, 1982; Dillman, 1978; Wright and Barnard, 1975). Examples of such principles are: ask questions in the affirmative using short sentences, avoid the use of double negatives, and be sure that all answer choices are mutually exclusive. However, space constraints in self-administered questionnaires may result in additional information organization problems. These concerns are addressed in the following principles.

Principle 3. Place directions where they are to be used and where they can be seen.

The following information is representative of what sometimes appears on a separate page at the beginning of a questionnaire.

> To get comparable data, we will be asking you to refer to the week of April 15, 1993, when answering most questions.
>
> Unless otherwise directed, mark one box for each question.
>
> When answering questions that require marking a box, please use an "x."
>
> To answer questions which require an occupational code see pages 12 and 13 of the questionnaire.
>
> When finished please return this questionnaire to the address shown above.

Such information is often provided at the start of questionnaires in a well-intentioned effort to avoid repetition and "simplify" the questionnaire, but the result is exactly the opposite. The information is given at a point where none of it can as yet be used, and on many different topics. Cognitive research with the Public School Questionnaire shows that in general respondents either never read the beginning information or they read it and forget it by the time they need it (Jenkins *et al.*, 1992). Memory studies have shown that information tends to be remembered in "chunks." Each item of information is likely to be remembered best, and subsequently used, if it appears as part of the question where it applies. For example, the reference date might be introduced in this way at the first question where it is to be used.

> "Were you working for pay (or profit) during the week of April 15, 1993?"

Then the date can be repeated at points where it is needed. Similarly each of the other directions can be imparted at the point where they are first used.

Principle 4. Present information in a manner that does not require respondents to connect information from separate locations in order to comprehend it.

The 1992 U.S. Census of Agriculture included this question, followed by a listing of 12 crops.

"Were any of the following CROPS harvested from "THIS PLACE" in 1992?"

As shown in Figure 7.16, the column headings to the right specify "acres harvested," "quantity harvested," and "acres irrigated." One problem with this format is that the question stem does not correspond with what respondents are being asked to provide. The question requires a "yes" or "no" response, whereas the answer categories suggest that respondents are really being asked to report numerical data. Only by making the effort to perceive and then integrate each separate part can the meaning of the question be deciphered.

Based on what we know about reading and memory, presenting *conceptually connected but physically disconnected* information is problematic for the following reasons. First, it requires a greater effort on the part of respondents to perceive the information that is out of both the respondent's foveal and parafoveal view. Confirmation of this comes from cognitive research with both the Public School questionnaire and the Census of Construction Industries questionnaire in which respondents were likely to overlook information that was not presented in a natural reading format. More importantly, this was likely to lead to the misreporting of information (Jenkins *et al.*, 1992; DeMaio and Jenkins, 1991).

Second, a separate presentation of conceptually related information leads to an increase in processing times because respondents do not have the opportunity to preview the information in the parafoveal view.

Finally, a separate presentation requires respondents to store separate pieces of information in short-term memory long enough to integrate them. Theoretically, one would expect this to burden short-term memory more than if the pieces are already consolidated. Therefore, one way to simplify the respondent's task is to ask a comprehensive question, which both visually and logically consolidates the information for the respondent. For instance, the agricultural question would become:

For each of the following crops, how many acres and bushels (or other measures of yield) were harvested from "THIS PLACE" in 1992? In addition, please indicate how many of the acres for each crop were irrigated.

INSTRUCTIONS – Please report your crops in the appropriate section. Use section 7 to report ONLY those CROPS NOT listed in sections 2 through 6 and section 8. (DO NOT include crops grown on land rented to others.)

SECTION 2 — Were any of the following CROPS harvested from "THIS PLACE" in 1992?

		None	Acres harvested	Quantity harvested		Acres irrigated
1.	Corn (field) for grain or seed (Report quantity on a dry shelled-weight basis.)	☐	067	068	Bu.	069
2.	Corn (field) for silage or green chop	☐	070	071	Tons, green	072
3.	Soybeans for beans	☐	088	089	Bu.	090
4.	Beans, dry edible	☐	554	555	Cwt.	556
5.	Wheat for grain	☐	545	546	Bu.	547
6.	Oats for grain	☐	076	077	Bu.	078
7.	Barley for grain	☐	079	080	Bu.	081
8.	Rye for grain	☐	686	687	Bu.	688
9.	Sorghum for grain or seed (including milo)	☐	082	083 {----- OR ----- Bu. 1 Cwt.	084
10.	Sorghum for silage or green chop (DO NOT include sorghum-sudan crosses.)	☐	085	086	Tons, green	087
11.	Popcorn	☐	662	663	Lbs., shelled	664
12.	Tobacco – all types	☐	094	/10	095	Lbs.	096	/10
13.	Potatoes, Irish	☐	097	/10	098	Cwt.	099	/10

SECTION 3 — Was any DRY HAY, GRASS SILAGE, HAYLAGE, or GREEN CHOP cut or harvested from "THIS PLACE" in 1992? Include sorghum-sudan crosses and hay cut from pastures.

S3   1 ☐ YES – Complete this section   2 ☐ NO – Go to section 4

If cuttings were made for both dry hay and grass silage, haylage, or green chop from the same fields, report the acreage in the appropriate items under DRY HAY and also under GRASS SILAGE, HAYLAGE, and GREEN CHOP.

	Acres harvested	Quantity harvested		Acres irrigated
1. DRY HAY (If two or more cuttings of dry hay were made from the same acres, report acres only once, but report total tons from all cuttings.)				
a. Alfalfa and alfalfa mixtures for hay or dehydrating	103	104	Tons, dry	105
b. Small grain hay – oats, wheat, barley, rye, etc.	106	107	Tons, dry	108
c. Other tame dry hay – clover, lespedeza, timothy, bromegrass, Sudangrass, millet, etc.	109	110	Tons, dry	111
d. Wild hay	112	113	Tons, dry	114
2. GRASS SILAGE, HAYLAGE, and GREEN CHOP (If two or more cuttings were made from the same acres, report acres only once, but report total tons from all cuttings.)	115	116	Tons, green	117

SECTION 4 — Were any NURSERY and GREENHOUSE CROPS, MUSHROOMS, sod, bulbs, flowers, flower seeds, vegetable seeds and plants, or vegetables under glass or other protection GROWN FOR SALE on "THIS PLACE" in 1992?

S4   1 ☐ YES – Complete this section   2 ☐ NO – Go to section 5

	None	Area irrigated		
		Square feet	Acres	Tenths
1. Nursery and greenhouse crops irrigated in 1992	☐	477	478	/10

2. From the list below, enter the crop name and code for each crop grown.	Crop name	Code	Square feet under glass or other protection in 1992	Acres in the open in 1992		Sales in 1992	
				Whole acres	Tenths	Dollars	Cents
				1	/10	2 $	00
If more space is needed, use a separate sheet of paper.				1	/10	2 $	00
				1	/10	2 $	00

Crop name	Code	Crop name	Code	Crop name	Code	Crop name	Code
Bedding plants (include vegetable plants.)	479	Cut flowers and cut florist greens	485	Foliage plants	707	Vegetable and flower seeds	500
Bulbs, corms, rhizomes, or tubers – dry	482	Nursery crops – ornamentals, fruit and nut trees, and vines	488	Potted flowering plants	710	Greenhouse vegetables	503
				Mushrooms	494	Other – Specify	506
				Sod harvested	497		

FORM 92-A0202 (10-1-91)

Page 2

Figure 7.16 Example of poor information organization from the 1993 U.S. Census of Agriculture.

Principle 5. Ask people to answer only one question at a time.

The above question, however, also illustrates the problem of asking more than one question at a time, which so often occurs on mail questionnaires in an attempt to save space. Because acres and yield are likely to be interconnected in the respondent's mind, and acres irrigated connects to total acres, the respondent may be able to provide the desired information, but not without effort. Such a question might be restructured into a temporal sequence, as follows:

> For each of the following crops, please write down how many acres of each was harvested from "THIS PLACE" in 1992.
>
> Next, indicate the quantity of each crop that was harvested.
>
> Finally, please write down how many of the acres, if any, were irrigated.

An even more complicated multiple-part question taken from another questionnaire where it was used to save space is the following:

> "How many of your employees work full-time with health insurance benefits, how many are full-time without health insurance benefits, and what is the average length of time each type of employee has worked for this firm?"

The respondent is being asked to think of four questions at once. This question can be divided into its constituent parts somewhat more easily than the previous one:

1. How many of your full-time employees receive health insurance benefits and how many do not?
2. Of those who receive health insurance, what is the average length of time they have worked for this firm?
3. Of those who do not receive health insurance, what is the average length of time they have worked for this firm?

7.3.3   A Combined Application of Information Organization and Navigational Guide Principles

Principles of the nature discussed here have been experimentally tested on survey questionnaires to some degree. Insight into the results that might be achieved from adhering to these principles can be inferred from some experiments conducted by the U.S. Bureau of the Census.

The Decennial Census Questionnaire shown in Figure 7.17 was changed from the matrix format shown there to one in which respondents were given an individual space in which to respond, as shown in Figure 7.18. The matrix questionnaire clearly violates our model of organizing information and navigating respondents through a form properly. In order to comprehend the questions and answers, respondents need to connect row stubs, column headings, and their intersecting information, all of which are located at separate places on the questionnaire.

Furthermore, it is left up to respondents to determine in what order to answer

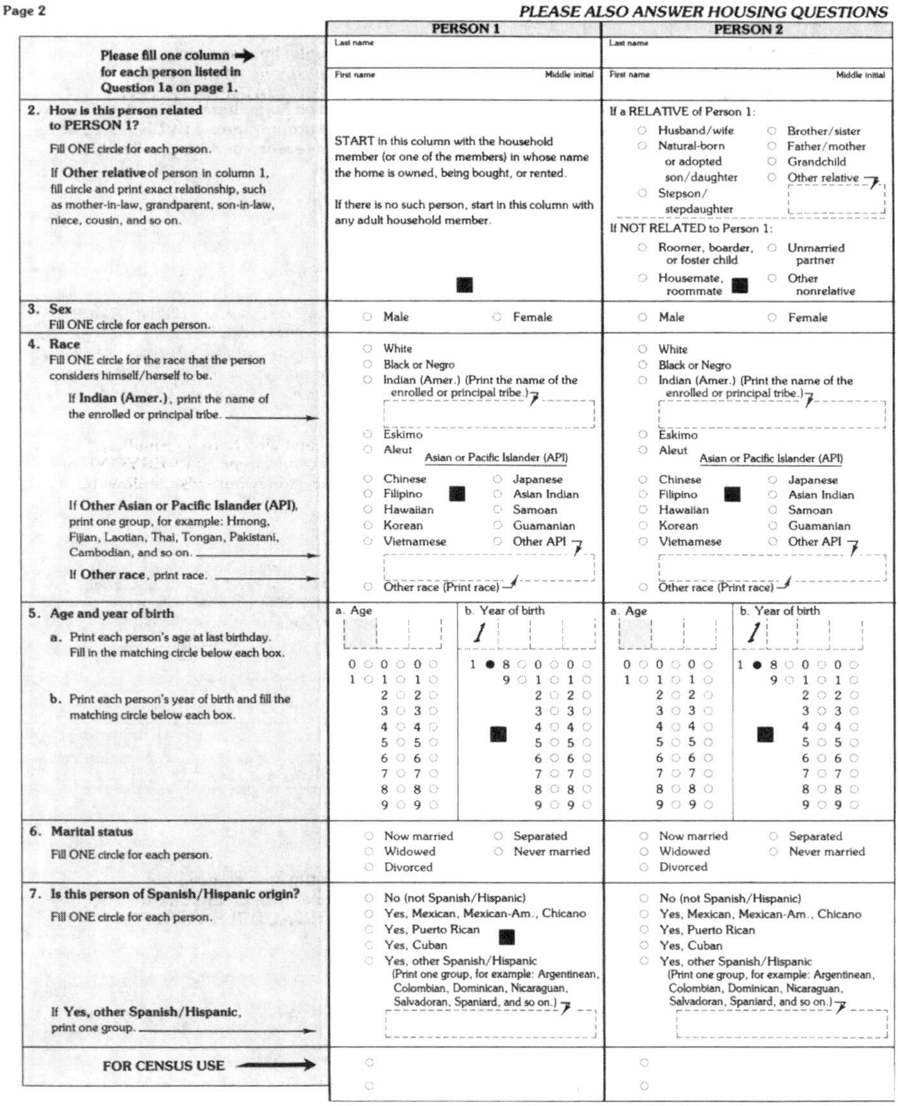
Page 2

PLEASE ALSO ANSWER HOUSING QUESTIONS

	PERSON 1	PERSON 2
Please fill one column → for each person listed in Question 1a on page 1.	Last name First name / Middle initial	Last name First name / Middle initial
2. How is this person related to PERSON 1? Fill ONE circle for each person. If Other relative of person in column 1, fill circle and print exact relationship, such as mother-in-law, grandparent, son-in-law, niece, cousin, and so on.	START in this column with the household member (or one of the members) in whose name the home is owned, being bought, or rented. If there is no such person, start in this column with any adult household member.	If a RELATIVE of Person 1: ○ Husband/wife ○ Natural-born or adopted son/daughter ○ Stepson/stepdaughter ○ Brother/sister ○ Father/mother ○ Grandchild ○ Other relative If NOT RELATED to Person 1: ○ Roomer, boarder, or foster child ○ Housemate, roommate ○ Unmarried partner ○ Other nonrelative
3. Sex Fill ONE circle for each person.	○ Male ○ Female	○ Male ○ Female
4. Race Fill ONE circle for the race that the person considers himself/herself to be. If Indian (Amer.), print the name of the enrolled or principal tribe. → If Other Asian or Pacific Islander (API), print one group, for example: Hmong, Fijian, Laotian, Thai, Tongan, Pakistani, Cambodian, and so on. → If Other race, print race. →	○ White ○ Black or Negro ○ Indian (Amer.) (Print the name of the enrolled or principal tribe.) ○ Eskimo ○ Aleut Asian or Pacific Islander (API): ○ Chinese ○ Filipino ○ Hawaiian ○ Korean ○ Vietnamese ○ Japanese ○ Asian Indian ○ Samoan ○ Guamanian ○ Other API ○ Other race (Print race)	○ White ○ Black or Negro ○ Indian (Amer.) (Print the name of the enrolled or principal tribe.) ○ Eskimo ○ Aleut Asian or Pacific Islander (API): ○ Chinese ○ Filipino ○ Hawaiian ○ Korean ○ Vietnamese ○ Japanese ○ Asian Indian ○ Samoan ○ Guamanian ○ Other API ○ Other race (Print race)
5. Age and year of birth a. Print each person's age at last birthday. Fill in the matching circle below each box. b. Print each person's year of birth and fill the matching circle below each box.	a. Age: 0–1 / 0–9 / 0–9 circles b. Year of birth: 1 ● / 8–9 / 0–9 / 0–9 circles	a. Age: 0–1 / 0–9 / 0–9 circles b. Year of birth: 1 ● / 8–9 / 0–9 / 0–9 circles
6. Marital status Fill ONE circle for each person.	○ Now married ○ Widowed ○ Divorced ○ Separated ○ Never married	○ Now married ○ Widowed ○ Divorced ○ Separated ○ Never married
7. Is this person of Spanish/Hispanic origin? Fill ONE circle for each person. If Yes, other Spanish/Hispanic, print one group. →	○ No (not Spanish/Hispanic) ○ Yes, Mexican, Mexican-Am., Chicano ○ Yes, Puerto Rican ○ Yes, Cuban ○ Yes, other Spanish/Hispanic (Print one group, for example: Argentinean, Colombian, Dominican, Nicaraguan, Salvadoran, Spaniard, and so on.)	○ No (not Spanish/Hispanic) ○ Yes, Mexican, Mexican-Am., Chicano ○ Yes, Puerto Rican ○ Yes, Cuban ○ Yes, other Spanish/Hispanic (Print one group, for example: Argentinean, Colombian, Dominican, Nicaraguan, Salvadoran, Spaniard, and so on.)
FOR CENSUS USE →	○ ○	○ ○

Figure 7.17 Example of poor information organization from the 1994 U.S. Decennial Census using a matrix format.

questions, a situation which does not usually occur in interviews. Respondents can answer all of the questions about one person at a time, in which case they will need to work down the columns. Or they can answer the same question for each person, in which case, they will need to work across the rows. In any

Page 3

PERSON 1.

NAME (Print) Last First Initial

1. What is this person's sex?

☐ Female
☐ Male

2. What is this person's marital status?

☐ Now married
☐ Widowed
☐ Divorced
☐ Separated
☐ Never married

3. What is this person's date of birth?

Month Day Year

(It is important to answer both Questions 4 and 5)

4. Is this person of Spanish/Hispanic origin?

☐ No, not Spanish/Hispanic
☐ Yes, Mexican, Mexican-Am., Chicano
☐ Yes, Puerto Rican
☐ Yes, Cuban
☐ Yes, other Spanish/Hispanic
(Print group) →

5. What is this person's race?

☐ White
☐ Black or Negro
☐ American Indian *(Print name of enrolled or principal tribe)* →
☐ Eskimo
☐ Aleut
☐ Chinese
☐ Japanese
☐ Filipino
☐ Asian Indian
☐ Hawaiian
☐ Samoan
☐ Korean
☐ Guamanian
☐ Vietnamese
☐ Other Asian or Pacific Islander
(Print race) →
☐ Some other race
(Print race) →

PERSON 2.

NAME (Print) Last First Initial

1. What is this person's sex?

☐ Female
☐ Male

2. What is this person's marital status?

☐ Now married
☐ Widowed
☐ Divorced
☐ Separated
☐ Never married

3. What is this person's date of birth?

Month Day Year

4. How is this person related to Person 1?

☐ Husband/wife
☐ Natural-born or adopted son/daughter
☐ Stepson/stepdaughter
☐ Brother/sister
☐ Father/mother
☐ Grandchild
☐ Other relative *(Print exact relationship)*
☐ Roomer, boarder, or foster child
☐ Housemate, roommate
☐ Unmarried partner
☐ Other nonrelative

(It is important to answer both Questions 5 and 6)

5. Is this person of Spanish/Hispanic origin?

☐ No, not Spanish/Hispanic
☐ Yes, Mexican, Mexican-Am., Chicano
☐ Yes, Puerto Rican
☐ Yes, Cuban
☐ Yes, other Spanish/Hispanic
(Print group) →

6. What is this person's race?

☐ White
☐ Black or Negro
☐ American Indian *(Print name of enrolled or principal tribe)* →
☐ Eskimo
☐ Aleut
☐ Chinese
☐ Japanese
☐ Filipino
☐ Asian Indian
☐ Hawaiian
☐ Samoan
☐ Korean
☐ Guamanian
☐ Vietnamese
☐ Other Asian or Pacific Islander
(Print race) →
☐ Some other race
(Print race) →

FORM DF-1A (10-14-93)

(If only 1 or 2 persons should be counted here, please turn to page 6, Step 4.)

Figure 7.18 Example of improved information organization and better use of navigational guides for a U.S. Decennial Census questionnaire based on an individual space format. White answer spaces on a light blue background field printed on white paper were used.

event, they are given a choice, but provided little guidance for making the choice.

Not only that, but the black squares, which were placed on the questionnaire as optical scanning guides but are not important to respondents, are the most dominant feature. Then to make matters worse, color (contrast) is not used effectively. As can be seen in Figure 7.17, the white answer spaces are not easily distinguished from the white background. In this case, respondents must rely on only two visual clues, the shape and location of the answer circles, to identify the answer spaces. In the Survey of College Graduates, respondents were given three elements—shape, location, and contrast—to guide them. Remember, the recognition of a figure against a background depends on contrast, especially contrast between brightness (or color) and shapes. Remember as well that too much visual detail impedes our ability to interpret visual data. Unfortunately, many lines are used to form the rows and columns of the matrix questionnaire shown in Figure 7.17.

With the individual space format, shown in Figure 7.18, respondents no longer need to connect information from separate locations on the questionnaire. Now they need only answer one question about one person at a time, and they need not deviate from moving down the page or to the top of the next column in search of the next question and response categories. In other words, this format effectively uses the visual element of location to guide the respondent through the form.

This format also uses contrast effectively. To begin with, the words "Person 1, Person 2," etc. are the dominant information points. Also, the white answer spaces are easily distinguishable from the blue background and the blue person spaces are easily distinguished from one another. Finally, the individual space format is a much more open format than the matrix format. The individual space revisions required extending the questionnaire from the one page ($10\frac{1}{2} \times 28''$) matrix used in the 1990 census to an eight-page booklet of $8\frac{1}{2} \times 11''$ pages. When tested experimentally, the booklet format achieved a completion rate of 66.8 percent versus only 63.4 for the matrix form, a statistically significant improvement of 3.4 percent (Dillman *et al.*, 1993).

These individual-space procedures were also used to revise a much longer census form, which in the matrix format consisted of 20 pages. In the revised 28-page format, a response improvement of 4.1 percentage points from 51.8 percent to 55.9 percent was achieved (Dillman *et al.*, 1994).

This experiment also included a treatment that provided improved navigational guides, but stayed within the matrix format of the 1990 census form. Shown in Figure 7.19, this redesign made better use of the visual element of contrast than the original matrix. This matrix used light blue background fields, and provided white answer spaces. Also, a 100 percent blue color with reverse printing was used to identify the person columns, making this the most dominant navigational guide on the page. These limited changes, and retention of the matrix format allowed staying within the original length of 20 pages. This treatment achieved an intermediate completion rate of 54.4 percent, 2.6 percentage points higher than the 1990 form, an amount that suggests improvement, but is still within sampling error.

PERSON 1 **PERSON 2**

1. **Please fill one column for each person listed in Question 1 on page 3.** Begin with the household member (or one of the members) in whose name the home is owned, being bought, or rented.

Last name

First name — Middle initial

Last name

First name — Middle initial

2. **What is this person's sex?**

☐ Male ☐ Female

☐ Male ☐ Female

3. **What is this person's marital status?**
Mark (X) ONE box for each person.

☐ Now married ☐ Separated
☐ Widowed ☐ Never married
☐ Divorced

☐ Now married ☐ Separated
☐ Widowed ☐ Never married
☐ Divorced

4. **How is this person related to person 1?**
Mark (X) ONE box for each person.

If Other relative of person in column 1, mark (X) and print exact relationship, such as mother-in-law, grandparent, son-in-law, niece, cousin, and so on.

This information is not needed for Person 1.

If a RELATIVE of Person 1:
☐ Husband/wife
☐ Natural-born or adopted son/daughter
☐ Stepson/stepdaughter
☐ Brother/sister
☐ Father/mother
☐ Grandchild
☐ Other relative ⟋

If NOT RELATED to Person 1:
☐ Roomer, boarder, or foster child
☐ Housemate, roommate
☐ Unmarried partner
☐ Other nonrelative

5. **What is this person's age and year of birth?**
a. Print each person's age at last birthday.
b. Print each person's year of birth.

Age   Year
a. [ ] b. [1][ ]

Age   Year
a. [ ] b. [1][ ]

6. **Is this person of Spanish/Hispanic origin?**
Mark (X) the "No" box if **not** Spanish/Hispanic.

If **Yes, other Spanish/Hispanic**, print one group.

☐ No (not Spanish/Hispanic)
☐ Yes, Mexican, Mexican-Am., Chicano
☐ Yes, Puerto Rican
☐ Yes, Cuban
☐ Yes, other Spanish/Hispanic (Print one group, for example: Argentinean, Colombian, Dominican, Nicaraguan, Salvadoran, Spaniard, and so on.) ⟋

☐ No (not Spanish/Hispanic)
☐ Yes, Mexican, Mexican-Am., Chicano
☐ Yes, Puerto Rican
☐ Yes, Cuban
☐ Yes, other Spanish/Hispanic (Print one group, for example: Argentinean, Colombian, Dominican, Nicaraguan, Salvadoran, Spaniard, and so on.) ⟋

7. **What is this person's race?**
Mark (X) ONE box for the race that the person considers himself/herself to be.
If **Indian (Amer.)**, print the name of the enrolled or principal tribe.

If **Other Asian or Pacific Islander (API)**, print one group, for example: Hmong, Fijian, Laotian, Thai, Tongan, Pakistani, Cambodian, and so on.

☐ White
☐ Black or Negro
☐ Indian (Amer.) (Print the name of the enrolled or principal tribe) ⟋

☐ Eskimo
☐ Aleut
        Asian or Pacific Islander (API)
☐ Chinese ☐ Japanese
☐ Filipino ☐ Asian Indian
☐ Hawaiian ☐ Samoan
☐ Korean ☐ Guamanian
☐ Vietnamese ☐ Other API ⟋

☐ Other race (Print race) ⟋

☐ White
☐ Black or Negro
☐ Indian (Amer.) (Print the name of the enrolled or principal tribe) ⟋

☐ Eskimo
☐ Aleut
        Asian or Pacific Islander (API)
☐ Chinese ☐ Japanese
☐ Filipino ☐ Asian Indian
☐ Hawaiian ☐ Samoan
☐ Korean ☐ Guamanian
☐ Vietnamese ☐ Other API ⟋

☐ Other race (Print race) ⟋

Page 4

Figure 7.19 Another revision of the U.S. Decennial Census questionnaire which utilized changes in navigational guides, but retained the matrix approach. White answer spaces on a light blue background field printed on white paper were used.

7.4 CONCLUSION

We view this chapter as a beginning. Ponder for a moment the large body of research targeting whether or not a "don't know" category should be included as one of the response categories on a questionnaire. Now consider the dearth of information for the multitude of decisions that face self-administered questionnaire designers, for example, should we include an instruction here or not, and if so, what should it look like? For the most part, designers have had to rely on convention and "common sense" because issues relating to the non-verbal aspects of questionnaire design have generally been neglected.

We are convinced, however, that it is not only possible but necessary to develop a set of scientifically derived principles for guiding this process. To better define where errors occur in this process and what we need to do to avoid these errors, we have reviewed the cognitive literature, especially as it applies to visual perception, and the motivation literature. Combining the concepts we uncovered, such as pattern recognition and the Gestalt Grouping Laws, with the existing empirical research on self-administered questionnaires has enabled us to propose a working model along with principles for design.

The model that has emerged is composed of two major decision-making components: the first component is aimed at encouraging respondents to follow a prescribed path through the questionnaire, which we call designing navigational guides. The second component refers to the choice of words for formulating questions and answers and the prescribed sequence in which respondents are expected to process them. This we call achieving good information organization. Finally, we presented initial results from a test of a U.S. census questionnaire that seem to support the model.

We need to continue the development of a set of scientifically derived and experimentally proven principles to guide the construction of self-administered questionnaires. Much remains to be considered and tested. We hope other survey methodologists will expand upon this discussion, and will begin to test these and other principles to determine their potential influence on response rates, processing times, and the accuracy of self-administered responses.

ACKNOWLEDGMENTS

Opinions expressed in this chapter are those of the authors and should not be construed as representing those of either the U.S. Bureau of the Census or Washington State University. At the time this chapter was written Don Dillman was Senior Survey Methodologist, Office of the Director, at the U.S. Bureau of the Census. We would like to thank Elizabeth Martin and Jon Krosnick for helpful comments on a previous draft.

REFERENCES

Brandt, H.F. (1945), *The Psychology of Seeing*, New York: The Philosophical Library.
Belson, W. (1981), *The Design and Understanding of Survey Questions*, Gower.

Buswell, G.T. (1935), *How People Look at Pictures*, Illinois: The University of Chicago Press.

Castner, H.W., and Eastman, J.R. (1984), "Eye-Movement Parameters and Perceived Map Complexity," *The American Cartographer*, Vol. 11, No.2, pp. 107–117.

Cialdini, R.B. (1984), *Influence: The New Psychology of Modern Persuasion*, New York: Quill.

Converse, J.M., and Presser, S. (1986), *Survey Questions: Handcrafting the Standardized Questionnaire*, Newbury Park: Sage Publications, Inc.

Crittenden, W.F., Crittenden, V.L., and Hawes, J.M. (1985), "Examining the Effects of Questionnaire Color and Print Font on Mail Survey Response Rates," *Akron Business and Economic Review*, 16, pp. 51–56.

DeMaio, T.J., and Bates, N. (1992), "Redesigning the Census Long Form: Results from the 1990 Alternative Questionnaire Experiment," *Proceedings of the Section on Survey Research Methods, American Statistical Association.* pp. 784–789.

DeMaio, T., and Jenkins, C. (1991), "Questionnaire Research in the Census of Construction Industries," *Proceedings of the Section on Survey Research Methods, American Statistical Association*, pp. 496–501.

DeMaio, T.J., Martin, E.A., and Sigman, E.P. (1987), "Improving the Design of the Decennial Census Questionnaire," *Proceedings of the Section of Survey Research Methods, American Statistical Association.* pp. 256–261.

Dillman, D.A. (1978), *Mail and Telephone Surveys: The Total Design Method*, New York: Wiley-Interscience.

Dillman, D.A., Dillman, J.J., Baxter, R., Petrie, R., Miller, K., and Carley, C. (1995), "The Influence of Prenotice vs. Follow-up Letters on Response Rates to Mail Surveys Under Varied Conditions of Salience," paper presented at the Annual Meetings of the American Association for Public Opinion Research.

Dillman, D.A., Reynolds, R.W., and Rockwood, T.H. (1991), "Focus Group Tests of Two Simplified Decennial Census Questionnaire Design Forms," *Technical Report* 91–39, Pullman: Washington State University, The Social and Economic Sciences Research Center.

Dillman, D.A., Sinclair, M.D., and Clark, J.R. (1993), "Effects of Questionnaire Length, Respondent-Friendly Design, and a Difficult Question on Response Rates for Occupant-Addressed Census Mail Surveys," *Public Opinion Quarterly*, 57, pp. 289–304.

Dillman, D.A., Treat, J.B., and Clark, J.R. (1994), "The Influence of 13 Design Factors on Response Rates to Census Surveys," *Proceedings of the Annual Research Conference*, U.S. Bureau of the Census, Washington, D.C.

Felker, D.B. (1980), *Document Design: A Review of the Relevant Research*, Washington, DC: American Institutes for Research.

Gaskell, G.D., O'Muircheartaigh, C., and Wright, D.B. (1994), "Survey Questions About Vaguely Defined Events: The Effects of Response Alternatives," *Public Opinion Quarterly*, 58, pp. 241–254.

Gerber, E.R., and Wellens, T.R. (1994), "Cognitive Evaluation of the Service Based Enumeration (SBE) Questionnaire: A Tale of Two Cities (and One Mobile Food Van)," Internal U.S. Bureau of the Census Report, Center for Survey Methods Research, September 15.

Glass, A.L., and Holyoak, K.J. (1986), *Cognition*, New York: Random House.

Goyder, J. (1988), *The Silent Minority*, Oxford, U.K.: Oxford University Press.

Groves, R.M., Cialdini, R.B., and Couper, M.P. (1992), "Understanding the Decision to Participate in a Survey," *Public Opinion Quarterly*, 56, pp. 475–495.

Hartley, J. (1985), *Designing Instructional Text*, New York: Nichols.

Jenkins, C., and Ciochetto, S. (1993), "Results of Cognitive Research on the Multiplicity Question from the 1991 Schools and Staffing Survey Student Records Questionnaire," Report submitted to the National Center for Education Statistics, U.S. Bureau of the Census Memorandum, February 10.

Jenkins, C., Ciochetto, S., and Davis, W. (1992), "Results of Cognitive Research on the Public School 1991–92 Field Test Questionnaire for the Schools and Staffing Survey," Internal U.S. Bureau of the Census Memorandum, June 15.

Jenkins, C.R., and Dillman, D.A. (1993), "Combining Cognitive and Motivational Research Perspectives for the Design of Respondent-Friendly Self-Administered Questionnaires," revision of a paper presented at the Annual Meetings of the American Association for Public Opinion Research.

Jobe, J.B., Tourangeau, R., and Smith, A.F. (1993), "Contributions of Survey Research to the Understanding of Surveys," *Applied Cognitive Psychology*, 7, pp. 567–584.

Jobe, J.B., White, A.A., Kelley, C.L., Mingay, D.J., Sanchez, M.J., and Loftus, E.F. (1990), "Recall Strategies and Memory for Health Care Visits," *Milbank Memorial Fund Quarterly/Health and Society*, 68, pp. 171–189.

Kahneman, D. (1973), *Attention and Effort*, New Jersey: Prentice-Hall Inc.

Lessler, J.T. (1989), "Reduction of Memory Errors in Survey Research: A Research Agenda," *Bulletin of the International Statistical Institute*, Paris, France, pp. 303–322.

Lima, S.D. (1993), "Word-Initial Letter Sequences and Reading," *American Psychological Society*, 2, pp. 139–142.

Martin, E. (1993), "Response Errors in Survey Measurements of Facts," *Bulletin of the International Statistical Institute*, Firenze, Italy, pp. 447–464.

Matlin, M. (1994), *Cognition*, Fort Worth: Harcourt Brace Publishers.

Neisser, U. (1967), *Cognitive Psychology*, New York: Meredith Corporation.

Petty, R.E., and Cacioppo, J.T. (1986), *Communication and Persuasion: Central and Peripheral Routes to Attitude Change*, New York: Springer-Verlag.

Phipps, P.A., Robertson, K.W., and Keel, K.G. (1991), "Does Questionnaire Color Affect Survey Response Rates?", *Proceedings of the Section of Survey Research Methods, American Statistical Association*, pp. 484–488.

Rayner, K. (1993) "Eye Movements in Reading: Recent Developments," *American Psychological Society*, 2, pp. 81–85.

Rothwell, N. (1985), "Laboratory and Field Response Research Studies for the 1980 Census of Population in the United States," *Journal of Official Statistics*, 1, pp. 137–148.

Rubin, E. (1958), "Figure and Ground," in D.C. Beardslee, and M. Wertheimer (eds.), *Readings in Perception*, New York: D. Van Nostrand Company, Inc.

Schwarz, N., and Hippler, H. (1991), "Response Alternatives: The Impact of Their Choice and Presentation Order," in P.P. Biemer, R.M. Groves, L.E. Lyberg, N.A.

Mathiowetz, and S. Sudman (eds.), *Measurement Errors in Surveys*, New York: John Wiley and Sons, pp. 41–71.

Schwarz, N., Strack, F., Hippler, H.-J., and Bishop, G. (1991), "The Impact of Administration Mode on Response Effects in Survey Measurement," in J. Jobe, and E. Loftus (eds.), *Cognitive Aspects of Survey Methodology*, Special issue of *Applied Cognitive Psychology*, 5, pp. 193–212.

Shettle, C., Mooney, G., and Giesbrect, L. (1993), "Evaluation of Using Incentives to Increase Response Rates in a Government Survey," paper presented at the Annual Meetings of the American Association for Public Opinion Research.

Spencer, H. (1968), *The Visible Word*, New York: Hasting House.

Sudman, S., and Bradburn, N.M. (1982), *Asking Questions: A Practical Guide to Questionnaire Design*, San Francisco: Jossey-Bass.

Tinker, M.A. (1965), *Bases for Effective Reading*, Minneapolis: University of Minnesota Press.

Tourangeau, R. (1984), "Cognitive Science and Survey Methods" in T.B. Jabine, M.L. Straf, J.M. Tanur, and R. Tourangeau (eds.), *Cognitive Aspects of Survey Methodology: Building a Bridge Between Disciplines*, Washington, DC: National Academy Press, pp. 73–100.

Tufte, E.R. (1990), *Envisioning Information*, Chesire, Connecticut: Graphics Press.

Tufte, E.R. (1983), *The Visual Display of Quantitative Information*, Cheshire, Connecticut: Graphics Press.

Twyman, M. (1979), "A Schema for the Study of Graphic Language (Tutorial Paper)," *Processing of Visible Language*, 1, pp. 117–150.

U.S. General Accounting Office (1993), *Developing and Using Questionnaires*, Document No. 10.1.7, Washington, D.C.: U.S. General Accounting Office/Program Evaluation and Methodology Division, pp. 189–197.

Waller, R. (1985), *Designing Usable Texts*, Orlando: Academic Press, Inc.

Waller, R. (1987), "The Typographic Contribution to Language: Towards a Model of Typographic Genres and Their Underlying Structure," thesis submitted for the degree of Doctor of Philosophy, Department of Typography and Graphic Communication, University of Reading.

Wallschlaeger, C., and Busic-Snyder, C. (1992), *Basic Visual Concepts and Principles for Artists, Architects, and Designers*, Dubuque, IA: Wm.C. Brown Publishers.

Wright, P., and Barnard, P. (1975), "Just Fill in This Form—A Review for Designers," *Applied Ergonomics*, 6, pp. 213–220.

Wright, P., and Barnard, P. (1978), "Asking Multiple Questions About Several Items: The Design of Matrix Structures on Application Forms," *Applied Ergonomics*, 9, pp. 7–14.

Zusne, L. (1970), *Visual Perception of Form*, New York: Academic Press.

SECTION B

Data Collection

CHAPTER 8

Data Collection Methods and Survey Quality: An Overview

"You are under no obligation to finish, but you are not allowed to quit"
Rabbi Tafron, Saying of the Fathers, Chapter 2

Edith de Leeuw
Vrije Universiteit, Amsterdam

Martin Collins
City University Business School, London

8.1 INTRODUCTION

Administrative censuses date back to Roman times. Surveys as we now know them have a much shorter history, of around 100 years. In 1886, Charles Booth undertook his study of the *Labour and Life of the People of London*. But Booth interviewed informants, School Attendance Officers, on the living conditions of school children and their families. Around 1900, Rowntree started work on a survey of the city of York, entitled *Poverty: A Study of Town Life*. Rowntree went further than Booth by collecting interview data *directly* from the families themselves; in doing this he was one of the first to conduct a face to face interview survey (cf. Moser and Kalton, 1979). The first recorded mail survey in fact predated this, being attributed to Karl Marx who, in 1880, mailed out questionnaires to French workers.

Modern survey methods have their roots in the 1930s, when three major aspects of the survey process were developed and refined. At the U.S. Bureau

Survey Measurement and Process Quality, Edited by Lyberg, Biemer, Collins, de Leeuw, Dippo, Schwarz, Trewin.
ISBN 0-471-16559-X © 1997 John Wiley & Sons, Inc.

of the Census sampling techniques were tried out and sampling schemes developed. Commercial polling firms and market research institutes, such as Gallup and Roper, were instrumental in the development of data collection techniques. And at universities methods for the analysis of survey data were developed by pioneers such as Stouffer and Lazersfeld (cf. Babbie, 1973).

An important step in the early 1960s was the introduction of telephone interviewing. Around 1970 the first computer assisted telephone interviewing (CATI) systems were developed in the U.S.A., principally in market research agencies (cf. Nicholls and Groves, 1986). Telephone survey methodology came of age in 1987 with an international conference and a subsequent monograph published in 1988 (Groves et al., 1988). Meanwhile, Dillman (1978) did much to increase the respectability of mail surveys and contributed greatly to the enhancement of data quality in such surveys.

More recently, the focus has shifted from developing data collection techniques towards concern with measurement and measurement error. Groves (1989) makes a strong contribution synthesizing psychometrics, statistics and social sciences in order to produce a comprehensive review of survey errors and possible causes. He emphasizes error estimation and the incorporation of survey error and survey costs. In 1990 an international conference was dedicated to measurement errors in surveys; the resulting monograph was published in 1991. In that monograph Lyberg and Kasprzyk (1991) presented an overview on data collection methods and measurement error.

It is a long road from Caesar Augustus and Karl Marx to Lars Lyberg and Daniel Kasprzyk, and during the journey survey researchers have learned to use new instruments and improve old ones. But a researcher's lot is not an easy one and once again major changes in technology and society force us to adapt. New technologies and advances in computer technology have made various new forms of computer assisted data collection possible. Besides CATI, computer assisted personal interviewing (CAPI) is now used on a large scale and new forms of computer assisted self-interviewing (CASI) are within reach of survey practitioners. These new technologies have also influenced the control of interviewer errors and the monitoring of the interview process. At the same time, major changes in society are reflected in more surveys of special populations and the necessity to adapt standard survey methods to the needs of special groups (e.g., the elderly, children, ethnic groups).

In addition to such changes in survey type and data collection methods, we have also seen a shift in the way researchers look at survey error. As in studies of other scientific phenomena, there are distinct stages. The first stage is description and classification as found in the seminal work of Deming (1944). The second stage is a search for causes and the forming of theories; an excellent example is the aforementioned book by Groves (1989). The last stage is coping, controlling, and improving. We are now moving into this third stage and interest has shifted to managing survey errors: the study and practice of survey quality and quality control.

The section on data collection of this monograph contains contributions on the effects of new data collection techniques on survey data quality (Nicholls *et al.*, Chapter 9), speech recognition applications for survey research (Blyth, Chapter 10), evaluation methods for interviewer performance in CAPI surveys (Couper *et al.*, Chapter 11), interaction coding of interviewer and respondent behavior in a validity study (Dykema *et al.*, Chapter 12), interview mode and sensitive questions (Jobe *et al.*, Chapter 13), and children as respondents (Scott, Chapter 14). In this introduction we will provide a discussion of quality in social surveys and data collection methods. We build on the work of Lyberg and Kasprzyk (1991) but with a change of emphasis as is reflected in the change of monograph title: from measurement error in 1991 to survey quality in 1996. In Section 8.2, we restrict our discussion of the concept of quality to social surveys. We then move on to discuss the effects of the two major changes mentioned above: the effect of computer technology on survey data collection (Section 8.3) and the way in which changes in society emphasize the need for surveys of special populations and topics (Section 8.4). One topic that has remained important over the years is that of interviewing, especially interviewer evaluation and training and the interaction between interviewer and respondent (Section 8.5). We end (in Section 8.6) with a research agenda for the future.

8.2 QUALITY IN SOCIAL SURVEYS

There can be no dispute that we should aim for high quality surveys. But what exactly defines high quality? Can there be any absolute standards? And where does quality of data collection fit into the total survey model?

Such questions are not new. In his review of errors, Deming (1944) already stressed that absolute accuracy is a mythical concept and that it is more profitable to speak of tolerance bands or limits of likely error. He further pointed out that allowable limits must vary from case to case, depending on the resources available and the precision needed for a particular use of the data. Here we have an early reference to the needs of a survey designer to take into account both available resources and survey purpose.

Deming also provided an early warning of the complexity of the task facing the survey designer. In addition to the relatively well understood effects of sampling variability, we have to take into account a number of other factors that affect the ultimate usefulness of a survey. Deming listed thirteen factors, including sampling error and bias, interviewer effects, method of data collection, nonresponse, questionnaire imperfections, processing errors and errors of interpretation.

Further focus was added by Hansen *et al.* (1953, 1961), whose model of survey errors centered on the concept of the *individual true value* of a given variable for each population element. This model differentiates between variable error and systematic bias and offers a concept of *total error*.

Kish (1965) refers to the total error as the root mean squared error and proposes it as a replacement for the simpler but incomplete concept of the standard error. He also sets out a classification of sources of bias in survey data, differentiating between errors of nonobservation (e.g., nonresponse) and observation (e.g., in data collection and processing).

Discussion of survey errors has taken place in a variety of different contexts— sampling statistics, psychology, economics—often using different terminology and frameworks. Groves (1989) attempts to reconcile these different perspectives and reduce the communication problems among disciplines through a generalized classification scheme developed from Kish's taxonomy. At the core is the concept of mean squared error: the sum of error variance (squared variable errors) and of squared bias. Variable (random) errors are those that take on different values over replications on the same sample, while bias (nonrandom errors) is made up of the components of error viewed to be constant over replications. Both types of error can be divided into errors of nonobservation (coverage, nonresponse, sampling) and errors of observation. The latter, also referred to as response or measurement errors, can arise from the respondent, the interviewer, the instrument, and the mode of data collection. Processing and coding errors are not included in Groves's system. Groves also reviews designs to estimate error and summarizes literature from different disciplines concerning possible causes of error.

An important facet of the treatment by Groves is his willingness to incorporate cost considerations into the discussion of survey quality. Recognizing that not everyone approves of this inclusion, he offers the following justification: "The explicit inclusion of cost concerns may seem antithetical to discussions of quality of survey data. In the crudest sense it may even evoke thoughts of attempts to 'cut corners'. This book argues that this reaction confuses 'cost efficiency' with 'low cost'. In contrast the position taken here is that only by formally assessing the costs of alternative methods (jointly measuring quality) can the 'best' method be identified" (Groves, 1989, p. 6). In this treatment, Groves is recognizing not only that cost is a constraint on the research designer but also that cost efficiency may be a component of "quality" in the design.

Cost is only one of a number of issues that may contribute to survey design considerations. Other important factors may be timeliness or continuity of information. To this we can add a range of other considerations in the search for an optimal data collection method given a research question and given certain restrictions. This often depends on the focus of the customer (cf. Dippo, Chapter 20). The basic research question defines the population under study and the types of questions that should be asked. Questions of survey ethics and privacy regulations may restrict the design. Important practical restrictions are time and funds available and other resources (skilled labor, administrative resources, experience, computer hardware and software). Within the limits of these restrictions difficult decisions have to be made concerning, for example, the acceptable level of nonresponse or the acceptable level of measurement error.

It is easy to see that such considerations must constrain survey design. For example, Alwin (1977, p. 132), in reviewing survey errors recognizes that "decisions which affect the ultimate quality of the survey data are often governed by practical considerations rather than theoretical and/or methodological ones." We would urge a different viewpoint: that efficient use of any resource should be regarded as a "quality" of a survey approach itself. We believe in the survey researcher's version of Occam's razor, that matching the survey process to the resources available should be seen not just as a constraint on the designer but rather as an essential quality to be sought in the design.

In setting this framework for assessing data collection quality, we have to remind ourselves, fifty years on, of Deming's note that the quality of the data should reflect the quality demanded by the intended uses of the data. Also we have to recognize that data collection quality is only one aspect of total survey design, competing for attention and resources with all other aspects (see O'Muircheartaigh, Introduction and Dippo, Chapter 20 in this volume).

8.3 MEANS OF DATA COLLECTION IN SURVEYS AND COMPUTERIZATION

8.3.1 Technological Changes: Their Influence on Data Collection

Whether computer assisted data collection methods should be used for survey data collection is no longer an issue for debate. Most professional research organizations are now employing at least one form of computer assisted data collection. The use of these new technologies has led to a completely new vocabulary full of abbreviations. The following taxonomy will introduce the reader to the abbreviations and terminology that are commonly used. A full overview of the influence of new technologies on data quality will be given by Nicholls et al. (Chapter 9).

It is important to make a distinction between the method or means of data collection (e.g., interview, self-administered questionnaire) and the technology used to capture the data. For instance, in a face to face interview an interviewer can use paper and pencil, a tape recorder, or a computer. In their overview of data collection methods, Lyberg and Kasprzyk (1991) review six means of paper and pencil data collection in surveys: face to face interviewing, telephone interviewing, self-administered questionnaires and diary surveys, the use of administrative records, and direct observation. Computerized equivalents exist for all methods; some like computer assisted telephone interviewing are widespread and relatively long established, others like electronic data exchange are still new.

Computer assisted methods for survey research are often summarized under the global terms *CADAC* (Computer Assisted DAta Collection), *CASIC* (Computer Assisted Survey Information Collection), and *CAI* (Computer Assisted Interviewing); in this context the traditional paper and pencil methods

Table 8.1   Taxonomy of Computer Assisted Interviewing Methods

General name: CADAC (Computer Assisted Data Collection), CASIC (Computer Assisted Survey Information Collection, CAI (Computer Assisted Interviewing)

Specific method	Computer assisted form
Face to face interview	CAPI (Computer Assisted Personal Interviewing)
Telephone interview	CATI (Computer Assisted Telephone Interviewing)
Written questionnaire including	CASI (Computer Assisted Self-administered Interviewing) Synonyms CSAQ (Computerized Self-Administered Questionnaire) and PDE (Prepared Data Entry)
mail survey	DBM (Disk by Mail) and EMS (Electronic Mail Survey)
panel research and diaries	CAPAR (Computer Assisted Panel Research), Teleinterview, Electronic diaries
various self-administered (no interviewer)	TDE (Touchtone Data Entry), VR (Voice Recognition), ASR (Automatic Speech Recognition)
Direct observation	Various techniques: people meters and TV appreciation panels, barcode scanning, computer assisted video coding
Administrative records	Electronic data exchange, data mining

are often denoted by *PAPI* (Paper And Pencil Interviewing). Table 8.1 presents a systematic overview of the various computer assisted interviewing techniques.

The computer assisted forms of telephone interviewing (CATI) and face to face interviewing (CAPI) are well known and hardly need an introduction; computer assisted forms of self-administered questionnaires are less widespread. Nicholls, *et al.* (Chapter 9) provide a short introduction on this topic. Blyth (Chapter 10) gives a full description of the development of Automatic Speech Recognition.

Electronic equivalents for data collection using administrative records or direct observation are not so widely recognized as the above methods. Still especially for direct observation, electronic methods are developing rapidly. In media research "people meters" are used to record viewing behavior, in consumer research automated barcode scanning at the check-out or in-home provides data about buying behavior, and several statistical agencies collect data to calculate monthly consumer price indices using trained observers with handheld computers. In the behavioral sciences sophisticated tools have been developed for observing human and animal behavior. For instance, event recorders enable observers to register separate events in the field with an accurate duration record, and integrated computer assisted coding and video systems have been developed for detailed behavior coding and subsequent data analysis (cf. Noldus *et al.*, 1989; Noldus, 1991).

In the near future administrative records will be exchanged electronically. In establishment research by statistical agencies, electronic data interchange is a potential swift and reliable way of collecting administrative data. Social science data archives already have a long tradition of transporting data by electronic means. In the old days data were transported by tape; now researchers can search libraries and catalogs through Gopher or a Web browser—menu-driven packages that allow Internet users to look for information worldwide. In addition, they can use FTP (File Transfer Protocol) to access remote computers and retrieve data files from these computers (cf. Crispen, 1995). For instance, the Steinmetz Archive in Amsterdam offers these services through the internet. NSD, the Norwegian data archive, acts as a central electronic access or "top menu" for all European data archives. Through its World Wide Web server, which is actually a linked system of menus and documents, one can reach Gopher and Web servers of all European data archives and most data archives outside Europe.

8.3.2 Mode Effects and Mode Comparisons

Choosing a particular administration mode is just one part of the survey process, but it can have important consequences for the total survey design. Schwarz (Chapter 1) gives a comprehensive review of the implications of particular choices for questionnaire construction. Special attention is given to the influence of the channel of communication (audio and visual) and the temporal order (sequential vs random access) in which the material is presented.

Hox and de Leeuw (1994) studied the nonresponse associated with mail, telephone and face to face surveys. Their meta-analysis covered 45 studies that explicitly compare the response obtained using a mail, telephone, or face to face survey. On average, face to face interviews produce the highest completion rate (70%), telephone interviews the next highest (67%) and mail surveys the lowest (61%). These differences are in accordance with the review of Goyder (1987).

The influence of mode of data collection on data quality has been extensively studied for traditional paper and pencil face to face interviews, telephone interviews, and self-administered (mail) questionnaires. De Leeuw (1992) presents a meta-analysis of known experimental comparisons of these data collection methods. Comparing mail surveys with both telephone and face to face interviews, de Leeuw found that it is somewhat harder to have people answer questions in mail surveys: both the overall nonresponse and the item nonresponse are higher in mail surveys. However, when the questions are answered, the resulting data tend to be of better quality. In particular, mail surveys performed better with more sensitive questions (e.g., more reporting of drinking behavior, less item nonresponse on income questions). When face to face and telephone surveys were compared only small effects were discovered. Face to face interviews result in data with slightly less item nonresponse. No differences were found concerning response validity (record checks) and social desirability. In general, similar conclusions can be drawn from *well-conducted* face to face and telephone interview surveys (de Leeuw, 1992). This conclusion is in accordance with Groves's statement that the most consistent finding in studies comparing face to face and telephone surveys is the lack of differences (Groves, 1989, p. 551). De Leeuw's meta-analysis of mode effects did not extend to examination of the potential influence of the technology of interviewing (e.g., paper and pencil vs computer assisted interviewing). Jobe *et al.* (Chapter 13) report a well-controlled study on the influence of both interview method and technology on answers to sensitive questions.

Influenced by the work of Schwarz and his colleagues (Schwarz *et al.*, 1991) an interest developed in the influence of data collection mode on context effects and response order effects. Until now, only a limited number of comparisons have been performed, focusing on differences between mail and telephone surveys. These studies suggest that context effects are order dependent in interview surveys, due to the strict sequential presentation of questions. However, in mail surveys or with self-administered questionnaires context effects are not order dependent. In fact, in some studies of self-administered questionnaires subsequent questions have influenced the response to a preceding question (cf. Schwarz, Chapter 1). There are some indications that recency effects (i.e., higher endorsement of response categories last in a list) are found in interview surveys that depend only on audio communication channels (e.g., telephone surveys, face to face surveys that do not apply showcards), while primacy effects (i.e., higher endorsement of items first in a list) are more common in modes that make use of visual presentation of stimuli (cf. Schwarz and Hippler, 1991).

Direct mode comparisons of different computer aided data collection methods (CADAC) are very rare. Most of the literature concerns comparisons between a paper and pencil and a computer assisted form of the *same* data collection method. In these cases the influence of different technologies are compared, controlled for data collection method (cf. Nicholls *et al.*, Chapter 9). In the near future the new computer technologies will be prevalent and computer assisted modes can be compared while controlling for the technology used (i.e., all are computer assisted). Meanwhile, we can only extrapolate the main findings on comparisons of paper and pencil survey methods to the computer aided forms of data collection methods.

For respondents in a telephone interview nothing changes when a research institute switches from paper and pencil telephone surveys to CATI. For the interviewers the task becomes less complex, because administrative duties have been taken over by the computer. As a result, the differences, if any, point toward a slight advantage for CATI, for instance because there are fewer routing errors. In CAPI the computer is visible to the respondent, who might react to its presence. However, very few adverse reactions and no reduction in response rates have been reported. It seems safe to assume that the main findings concerning mode differences between telephone and face to face surveys are also valid for the computer aided versions of these survey techniques. This means that with well-trained interviewers and the *same* well-constructed structured questionnaire, both CAPI and CATI will perform well and differences in data quality will be extremely small. Of course, it should be noted that CAPI has a greater potential than CATI, just as paper and pencil face to face interviews have a greater potential than paper and pencil telephone interviews (e.g., using visual stimuli and special techniques such as card sorting).

There are several forms of computer aided self-administered questionnaires (CASAQ, see Section 8.3.1). All these variations have in common that the question is read from a screen and the answer is entered into the computer by the *respondent*. Just as in paper and pencil self-administered questionnaires the respondents answer the questions in a private setting, which reduces a tendency to present themselves in a favorable light. There is some evidence that CASAQ produces fewer socially desirable answers than CAPI when sensitive questions are asked (Beckenbach, 1995). Furthermore, in a CASAQ session the respondent and not the interviewer paces the questions. However, the respondent is not the only locus of control. The computer program controls the order of the questions and the number of questions visible at a given time. The respondent is generally not allowed to go back and forth at will as can be done in a paper and pencil questionnaire. In this sense a CASAQ session more resembles an interview than a self-administered questionnaire. Therefore, differences in context effects caused by the order or sequencing of the questions will cease to exist, effects caused by differences in channel capacity (audio vs visual presentation) will endure. And CASAQ will resemble CATI and CAPI regarding well-known context sequencing effects.

8.4 SURVEYING SPECIAL POPULATIONS

8.4.1 Changes in Society and the Need for Specialized Surveys

The introduction of the computer changed the technology of data collection. Now, major changes in demography and society force survey researchers to redesign their surveys in other respects. Specialized surveys are becoming more important for the survey industry, either as surveys into special or sensitive topics (e.g., AIDS, sexual behavior, fertility) or as surveys of special populations (e.g., children, the elderly, ethnic groups). General methodological literature on specialized surveys is sparse, but for certain subpopulations more knowledge is available than for others. A prime example is surveys of the elderly. The baby-boomers who dominated the 1960s and 1970s are becoming the elderly of tomorrow—in the developed countries the 65+

age group is expected to grow from 12% in 1990 to 19% in 2025[1]. Because the elderly form a distinct, relatively homogeneous and growing group, researchers have invested in the development of survey methods for the elderly. For a summary of survey methods for the elderly see Herzog and Rodgers (1992).

When we look at immigrants and culturally and ethnically diverse groups a different picture emerges: different countries have different culturally diverse groups and even within one country several distinct groups, each with its own culture can be distinguished. Knowledge on survey methods for these special groups does exist but it is scattered and situation and group specific. In the next paragraphs we will try to integrate this diverse knowledge and offer a general framework for special surveys.

8.4.2 Designing a Specialized Survey

To successfully design and implement a special survey it is extremely important to analyze the research goal. What do we want from this special survey? What makes the basic research question special? Why is the group under study special? It is not enough merely to turn automatically to a standard face to face interview and leave the problems to the interviewer: special populations need specially adapted surveys. The analysis of issues should not focus on just one aspect of the survey, for instance the writing of the questionnaire or the selection and training of interviewers, but should embrace the *total* survey design. Key stages in the design process are thus: establishing familiarity with the special population concerned; adopting general guidelines for a high quality survey; listing the phases in the survey; analyzing each phase in terms of the possible effect of the special population; adjusting each phase accordingly; and discussing all aspects of the total survey design with knowledgeable colleagues and with key informants.

If this process is followed, all stages of the survey design will have been handcrafted for the purpose. This will mean extra work compared to a run-of-the-mill standard survey, but less than is often feared. In the remainder of this section we discuss separate phases in a special survey design with the emphasis on data collection. The illustrations and examples of surveys of the elderly are taken from Herzog and Rodgers (1992) and Broese van Groenou *et al.* (1995), of surveys of children from Scott (Chapter 14), and of surveys of ethnic minorities (specifically North African groups in the Netherlands) from Meloen and Veenman (1990) and Schothorst (1993).

---

[1] Source: Demographic Yearbook 1991, United Nations. Developed countries are Northern America, Australia, New Zealand, Japan, all regions in Europe, and in the former Union of Soviet Socialist Republics. For the world as a whole the percentage of those aged 65 and over was 6.2 in 1990 and is expected to rise to 9.7 in 2025.

Phase 1: Define the Target Population and Analyze What Makes This Group Special

Scott (Chapter 14) emphasizes that appropriate methodologies for interviewing children must take into account their age. Different age groups differ not only in cognitive development but also in emotional development and in the acquisition of the social skills necessary for the cognitive and communication tasks that underlie the question–answer process (cf. Schwarz, Chapter 1). The younger the child, the less likely she/he is to know what a survey is, or how a "good" respondent behaves. A survey of preschool children needs an entirely different approach from a survey of primary school children, or of early adolescents.

Similarly, in surveying the elderly it is important to define the target population. Older adults are more likely to experience cognitive or health problems that impair their responses or restrain them from responding at all. These problems are more common among the older old (over 80). In surveying ethnic groups other problems arise, depending on the definition of the target population. For instance, when country of birth is the criterion variable, language problems and ignorance about the survey process will often be encountered. When parents' country of birth is the key to the definition and second generation respondents are interviewed, these problems will be less prevalent.

Phase 2: Mode Choice

Often special *adaptations* of the face to face interview or of self-administered forms are used, especially with sensitive topics or in special situations. Telephone interviews are often used for screening a large sample in order to find the target sample. The choice of mode is never simple and, when surveying special groups, the standard modes must be adapted. Such adaptation is the key to a successful special survey. For instance, interviewers may have to be specially trained in the customs of special groups or to handle anticipated problems. Some examples of adaptations of self-administered forms are, in a survey of the deaf, group administration of a written questionnaire with a sign language interpreter to offer assistance; in surveys of children, questionnaires in combination with a "Walkman" on which the questions are read aloud; in a survey of sexual contacts, the use of intricate security measures such as handing the respondent a questionnaire and stamped envelope and offering to walk together to the nearest mailbox.

Phase 3: Sampling

In general surveys, household samples are perhaps the most commonly used. But in studying special populations this form is rarely used: one example is Scott's survey of the children in a sampled household. Elsewhere, we find a wide range of varying and often creative sampling techniques, many of them nonprobability methods. Examples range from random sample screening phase, through the use of existing lists sampling frames, to network sampling and capture–recapture techniques. Sudman *et al.* (1988) discuss sampling rare populations in general; Kalsbeek (1989) gives a review of sampling techniques for AIDS studies; Czaja *et al.* (1986) discuss network sampling to find cancer patients; and Kalton and Anderson (1989) review sampling techniques useful when studying the elderly.

Phase 4: Questionnaire
The questionnaire must, of course, follow the basic precepts, some of which may be especially critical, for instance the need to maintain a clear structure and to use simple language in surveys of the elderly or a group that may experience language problems. But further adaptation may well be needed when conventional survey wisdom is not appropriate to a specific group. For instance, standard rating scales may not be appropriate: studies of North African migrants in the Netherlands report that a five-point scale is answered with yes/no answers and recommend the use of two-point or at most three-point rating scales. For elderly respondents, seven- or nine-point scales may be too demanding and unfolding procedures may be preferred. Another simple illustration is that standard showcards may be of no help unless supported by the interviewer reading out the content: for instance, among first generation migrants from North Africa illiteracy is high; elderly respondents may have reading difficulties.

Pretests of question wording and question order are extremely important. Even well-tested standard survey questions may not work well with certain special groups. For example, the elderly are less willing than younger groups to compare themselves with others ("Compared to others your age are you. . . ."); strict religious groups are often not willing to answer questions about the future ("This is not for me to say, God/Allah will decide").

It is advisable to identify the key concepts in a questionnaire and have a shortened version covering this core available when studying the elderly or handicapped. If health problems prevent them from answering the long version, the short version can be presented to either the respondent or a proxy. When studying non-native speakers, a translated version of the questionnaire is advisable, rather than relying on the translation skills of interviewers or interpreters. Computer interviewing can greatly ease this task (cf. Bond, 1991).

Phase 5: Training and Selecting Interviewers
When studying ethnic minorities it is advisable to use interviewers who are not only bilingual but also bicultural. Matching the sex of interviewer and respondent is often necessary. If it is not possible to recruit a large specialized interviewer panel, one should at least train a small number of bicultural and bilingual interviewers who can be assigned to special cases. No special type or age of interviewer seems most effective for studying the elderly. The main conclusion is that interviewers should look very reliable, as elderly people can be more concerned with victimization.

Every interviewer should of course receive a sound basic interviewer training (e.g., Institute for Social Research, 1976, 1983; McCrossan, 1991). Additional training is advisable to acquaint interviewers with problems often encountered with special groups of respondents, for instance how to overcome hearing or visual problems with the elderly, to avoid digression by the elderly, to avoid influencing a respondent's answers whenever extra interviewer assistance is needed, or to cope with the presence of others (husbands, children, nurses). When surveying a culturally different group, a short course on differences in culture may help interviewers who are not already bicultural: conversational rules may differ; a longer introduction may be needed; it may be difficult or even impossible to interview women and children without a representative of the head of the family present; particular days or times of day may be inappropriate for an interview. Finally, some background knowledge of survey methods and the theory behind general interview rules will help interviewers in a nonstandard situation and guide them to behave optimally when forced to deviate.

Because of the difficulties interviewers may experience during their work with special populations more intensive supervision is necessary. This means not only more contacts but also a more coaching attitude from supervisors. Supervisors should partially be mentors for the interviewers and teach them to cope with distress and extra interview burden. This helps minimize fieldwork problems and interviewer turnover and will increase fieldwork efficiency.

In sum: surveying special populations successfully requires a handcrafted high-quality survey. Wide discussion and analysis of what makes a group and a survey special, together with adequate pretesting, allows peculiar difficulties and solutions to emerge. And, perhaps most important of all, publication of successful adaptations and shared experience are needed.

8.5 INTERVIEWER EVALUATION AND TRAINING

8.5.1 Interviewer Errors

Changes in technology and society may have changed data collection but one topic maintains its high profile, that of interviewer quality, evaluation and training. In a survey interview, be it by telephone or face to face, by computer or pencil and paper, there are two important actors: the respondent and the interviewer. Of these two, the interviewer is most directly under the control of the survey researcher and researchers have for a long time tried to measure and minimize interviewer error. The basic strategies used are the selection of interviewers, training, monitoring of performance and the design of robust questionnaires and survey procedures that reduce the risk of error. Even the best interviewer will make errors when a question is incomprehensible and no standard procedures are laid out by the researcher. For an excellent review of ways to reduce interviewer-related error see Fowler (1991).

In this monograph several authors discuss aspects of questionnaire design to reduce interviewer error. Krosnick and Fabrigar (Chapter 6) review the literature on rating scales, while Jenkins and Dillman (Chapter 7) give an overview of questionnaire design. Nicholls *et al.* (Chapter 9) discuss how computer assisted data collection techniques can help to prevent interviewer errors; Couper *et al.* (Chapter 11) show how technology gives us a new and useful tool to monitor interviewer performance.

The prerequisites for efficient error reduction strategies are the availability of detection methods and knowledge of interviewer characteristics and performance which influence interviewer error. In the next sections we concentrate on methods to statistically detect interviewer errors, on methods for monitoring compliance with interviewer training, and on interaction coding as a tool to detect and describe interviewer errors.

8.5.2   Statistical Detection and Modelling of Interviewer Error

In general, variation in answers due to interviewers cannot be distinguished from variation due to respondents. To investigate the independent effect of the interviewer, it is necessary to control and interpenetrate interviewer assignments. In its simplest form we may simply assign respondents at random to interviewers; more commonly we will use a "nested" design with respondents in one or more primary sampling units being allocated randomly to a few interviewers (for examples, see Collins, 1980). The most commonly used index for estimating interviewer error is the intraclass correlation. In essence, a random effect ANOVA model is used to estimate the interviewer variance component, and the ratio of between interviewer variance to the total variance of a measure is estimated. The intraclass correlation has a minimum of 0 and a maximum of 1. A value of 0 means that answers to a particular question are unrelated to the interviewer, a value of 1 that they are completely determined by the interviewer. For a more detailed discussion see Groves (1989, chapters 7 and 8).

In well conducted face to face interviews the intraclass correlation is generally small, averaging around 0.02, but is larger for a proportion of survey variables (cf. Collins, 1980). In centrally controlled telephone surveys, lower values averaging below 0.01 are reported (cf. Hox *et al.*, 1991). However, the total effect of the interviewers on the overall variance of a survey statistic is a function of both the intraclass correlation and the interviewers' workload. Even small interviewer effects can have an important effect on the quality of survey data if the average workload is large (for an illustration see Groves *et al.*, 1988, p. 364).

Studies of interviewer variance only tell how much variance is attributable to the interviewers, they do not tell which interviewer characteristics are responsible. Traditionally there were two statistical approaches to this. In the first approach the effect of explanatory interviewer variables is investigated by splitting the interviewer sample (e.g., male vs female, trained vs untrained) and estimating the intraclass correlation for each subsample. The second approach is to disaggregate interviewer variables (e.g., sex, age, training) to the respondent level and perform a statistical analysis to assess the effect of the interviewer variables, ignoring the statistical problems created by the fact that respondents are nested within interviewers.

Recently, these methods have been superseded by hierarchical linear or multilevel models, which take into account the hierarchical structure of the data, with respondents nested within interviewers, and make statistically correct inferences about interviewer and respondent effects. Multilevel models are especially attractive if practical field restrictions make it impossible to randomly assign respondents to interviewers, because they allow the analyst to control statistically instead of experimentally by modeling the interviewer effects conditional on the respondent effects. Hox (1994) gives a lucid tutorial on how to use multilevel models when analyzing interviewer effects. For a practical application see Hox *et al.* (1991).

In most surveys interviewer demographic characteristics have been found to be unrelated to data quality. However, interviewer demographic characteristics do have effects when they are related to the survey topic or the questions asked. The only consistent findings on interviewer characteristics are for interviewer race, especially when race-connected issues are concerned. There are also some reports of an effect of interviewer's sex when gender-related topics are studied. Mixed results have been found concerning interviewer age and experience. Training interviewers does seem to have a positive effect on interviewer skills, although systematic research on this topic is scarce. Task-oriented supervision after training is needed to make sure that interviewers remain motivated to use the skills learned. Both direct monitoring in a centralized setting or the use of tape recordings fulfill this goal. Couper *et al.* (Chapter 11) describe how computerized data collection techniques offer new ways to monitor and evaluate interviewer behavior. For a general overview of interviewer-related error see Groves (1989, chapter 8) and Fowler (1991).

8.5.3   The Interview Process and Interaction Coding

To successfully control and reduce interviewer error one should not only be able to statistically detect and model interviewer errors but should also have information on the mechanisms whereby these errors arise. For instance, studies of interaction coding can reveal situations that interviewers are not prepared for. These studies can also show us what features of the interview, be it the question and the questionnaire or the interaction between respondent and interviewer, undermine standardization of the interview (Schaeffer, 1995). Knowledge about what actually happens in the interview helps researchers to develop theories of the interview process and guides them in improving interviewer training.

Interest in the interview process, and especially in the effects of interviewer behavior, dates back to the work of Hyman *et al.* (1954) and still commands a great deal of interest. In this field, interaction coding, classifying the verbal behavior of interviewer and respondent, is a common approach. The approach, originally developed in social psychology, was introduced into survey research by Cannell and others (e.g., Cannell *et al.*, 1968). Since then, it has been applied in several ways: to investigate the appropriateness of an interviewer's contribution to the interview and interviewers' adherence to training instructions (e.g., Cannell *et al.*, 1975), to examine the consequences of different interviewing styles for the quality of the data (e.g., Van der Zouwen *et al.*, 1991), and to identify questions causing problems within a pretest interview (e.g., Morton-Williams and Sykes, 1984; Oksenberg *et al.*, 1991). A more detailed review of the approach, especially as applied to question testing, will be found in Fowler and Cannell (1996).

Early uses of the technique tended to look at behavior in a fragmented way. Following the 1990 Tucson conference, Sykes and Collins (1992) sought to extend the field by analyzing sequences of behavior, comparing them with simple models of the interview as a stimulus-response process and defining forms of "non-straightforward" sequences departing from such a model. This form of analysis, then laborious and cumbersome, has been recently developed and improved by Van der Zouwen and Dijkstra (1995) and by Dijkstra and Draisma (1995).

In this work the assumptions seem to be that major departures from a simple model of the interview process are indicative of problems with interviewer performance, question design, or both. Some would argue that this is not necessarily the case, that an element of "negotiation" between interviewer and respondent may contribute to improved data quality. Dykema *et al.* (Chapter 12) provide a rare example of a test of this hypothesis against validated data, suggesting that—under some circumstances—nonstandard interviewer behavior may indeed be a positive contribution. There is a great need for more such work, extending into other measurement areas and integrating the concept of sequence analysis.

8.6 A RESEARCH AGENDA

In this chapter we have sought to build on the work of Lyberg and Kasprzyk and the 1990 conference in Tucson. The conference on Survey Measurement and Process Quality in Bristol, 1995 allowed us to consider and absorb research that has taken place since the 1990 Tucson conference, to take stock and to leave an agenda for further research. This is not easy—we would not have forecast confidently the changes that have taken place in the last five years and can be little more confident when looking into the future. We can only point to some topics that seem priorities to us now.

Further Development of CAI
Computer assisted interviewing (CAI) has been rapidly adopted in the field; some would say too rapidly. Early technical complaints about CAI and

especially about CAPI (e.g., slow response times, problems with entry of long verbal responses, weight of the "portable" computer) have been remedied by developments in hardware and software. Much research energy is put into the development of highly sophisticated software that maximizes the machine's capabilities. But human factors seem neglected. Reading from a screen and typing answers demands different perceptual and motor skills than going through a paper questionnaire. Question texts are harder to read on a monitor and interviewers also report that it is harder to grasp the overall structure of the questionnaire. To improve further the quality of CAI data, such human factors warrant more attention, both in empirical research and in software development to maximize the system's support of the interviewer. A promising start has been made by Edwards et al. (1995) who discuss CAPI screen designs that anticipate interviewer problems.

CAI Effects

As reported above and by Nicholls et al. (Chapter 9) the relatively few direct studies of the effects on data quality of introducing computer assistance into a given mode of data collection are uniformly reassuring, with almost no evidence of CAI effects on data. Discussion with users of data series that have undergone the change seem to reveal a somewhat different picture, with complaints about discontinuities in time series and effects on particular questions. This apparent contradiction might reflect the fact that most comparative studies have looked at relatively straightforward data collection exercises or, more likely, that they look across many and highly varied question types. Then, isolated problems may be missed. On the other hand, isolated problems may present great, even insurmountable difficulties to some users. A broad parallel can be seen in earlier comparisons of face to face and telephone interviewing. Sykes and Collins (1988), for example, report few mode effects in a range of general comparative studies but also report many effects in one special study, of a sensitive topic (alcohol) and dominated by particular question types (open questions and rating scales). In terms of our research agenda, the implication is that there is a need for more study of the usually isolated instances of CAI effects: searching for them and seeking understanding of how they arise and how they might be controlled.

Interviewer Skills

Whenever a new technology is used in interviewing, there is always concern with the influence it has on the interview situation and whether good rapport between interviewer and respondent can be maintained. Also, concern with the handling of answers to open questions is expressed. In response to these concerns it is argued that well-trained and experienced interviewers do not experience any difficulties. Woijcik et al. (1992) sketch a training program for CAPI but methods of training interviewers for the special needs of computer assisted interviewing need further development. The development and sharing of training programs for CAPI interviewers is thus a point for this agenda.

Interviewer skills necessary for the use of new technologies—both CAI and telephone technology—will merit more attention.

Interviewer Behavior
As we have mentioned, investigation of the interaction between interviewers and respondents has remained a topic of interest to researchers. Some new questions in this area are raised by Dykema *et al.* (Chapter 12), suggesting that some departures from straightforward sequences of question and response behavior may serve to improve measurement quality. The acknowledged limitations of their results point clearly to the need to extend research linking behavior coding with measured data quality.

Specialized Surveys
Specialized surveys are becoming more important for the survey industry. Knowledge is now emerging on how to implement special surveys, but this knowledge is sparse and scattered. Researchers should share this knowledge. Even the mere documentation of problems encountered has value. We hope that in the near future more theoretical and empirical research on this subject will be undertaken.

Sensitive Questions
We must expect to see continuing increases in public concern with privacy and in demands for surveys of sensitive topics. Since the quality image of many survey organizations depends on their ability to collect sensitive data, this topic should remain on the research agenda. In particular, we may look for more research into the use of self-administered computer assisted interviewing and other methods that enhance feelings of privacy.

Effects on Industry Structure
The increased emphasis on computer assisted data collection and the large investment this requires may give advantage to larger supplying agencies and have major effects on industry structure. Smaller agencies may have to merge or to subcontract data collection to larger agencies. Such changes would not necessarily affect data quality but do seem to justify some attention by researchers in the field of data collection and data quality.

Agency Effects on Data Quality
Increased pressure towards competition for regular or continuing surveys can lead to changes in the supplying agencies. Variations between agencies in terms, for example, of fieldwork strategies, contact strategies for nonrespondents, interviewer training, and supervision or editing and coding can affect data quality. This effect would be worthy of investigation. Hierarchical or multilevel analysis with agency as one level and with respondents nested within agencies could be a promising tool in the future estimation and correction of these agency effects.

This agenda may appear at times to stray outside the data collection focus of this chapter, into instrument design and data processing. This is inevitable, in that many issues in data collection emerge at the interfaces with other stages of the survey process. It also reflects our view that quality in data collection can only be considered in the context of total survey design, resource constraints and planned usage.

ACKNOWLEDGMENTS

We thank Abby Israëls, Liz McCrossan, Jeannette Schoorl, Yolanda Schothorst, and Ron Cjaza for their support in finding relevant material for this overview. We also thank Joop Hox and Norbert Schwarz for their insightful comments and constructive suggestions on earlier drafts of this chapter. While the chapter has greatly benefited from their help, the final product and especially its errors remain the full responsibility of the authors.

REFERENCES

Alwin, D.A. (1977), "Making Errors in Surveys; An Overview," *Sociological Methods and Research*, 6, pp. 131–149.

Babbie, E.R. (1973), *Survey Research Method* (chapter 3), Belmont: Wadsworth.

Baker, R.P. (1992), "New Technologies in Survey Research: Computer Assisted Personal Interviewing," *Social Science Computer Review*, 10, pp. 145–157.

Beckenbach, A. (1995), "Computer-assisted Questioning: The New Survey Methods in the Perception of the Respondents," *Bulletin de Methodologie Sociologique (BMS)*, 48, pp. 82–100.

Bond, J.R.P. (1991), "Increasing the Value of Computer Interviewing," *Proceedings of the 1991 ESOMAR Congress*, Trends in Data Collection, pp. 737–753.

Broese van Groenou, M.I., Van Tilburg, T.G., de Leeuw, E.D., and Liefbroer, A.C. (1995), "Data Collection," in P.M. Knipscheer, J. de Jong Gierveld, T.G. van Tilburg, and P.A. Dykstra (eds.), *Living Arrangements and Social Networks of Elderly Adults*, Amsterdam: VU-publishers, pp. 185–197.

Cannell, C.F., Fowler, F.J., and Marquis, K.H. (1968), "The Influence of Interviewer and Respondent Psychological and Behavioral Variables on the Reporting in Household Interviews," *Vital and Health Statistics*, Series 2, no. 26.

Cannell, C.F., Lawson, S.A., and Hausser, D.L. (1975), *A Technique for Evaluating Interviewer Performance*, Michigan: Institute for Social Research, University of Michigan.

Collins, M. (1980), "Interviewer Variability: A Review of the Problem," *Journal of the Market Research Society*, 22, pp. 77–95.

Crispen, P.D. (1995), "Roadmap for the Information Superhighway Internet Training Workshop (Map 11–19)." Only available in electronic form through internet. Send e-mail letter to LISTSERV@UA1VM.UA.EDU with the command GET MAP PACKAGE F=MAIL.

Czaja, R., Snowden, C., and Casady, R. (1986), "Reporting Bias and Sampling Error in a Survey of a Rare Population Using Multiplicity Rules," *Journal of the American Statistical Association*, 81, pp. 411–419.

De Leeuw, E.D. (1992), *Data Quality in Mail, Telephone, and Face to Face Surveys*, Amsterdam: T.T.-publikaties.

Deming, W.E. (1944), "On Errors in Surveys," *American Sociological Review*, 9, pp. 359–369.

Dillman, D.A. (1978), *Mail and Telephone Surveys: The Total Design Method*, New York: Wiley.

Dijkstra, W., and Draisma, S. (1995), "A New Method for Studying Verbal Interactions in Survey Interviews," *Proceedings of the International Conference on Survey Measurement and Process Quality, Bristol, U.K., Contributed Papers*, pp. 296–301.

Edwards, B., Sperry, S., and Schaeffer, N.C. (1995), "CAPI design for improving data quality," *Proceedings of the International Conference on Survey Measurement and Process Quality, Bristol, U.K., Contributed Papers*, pp. 168–173.

Fowler, F.J. (1991), "Reducing Interviewer-related Error Through Interviewer Training, Supervision, and Other Means," in P.P. Biemer, R.M. Groves, L.E. Lyberg, N.A. Mathiowetz, and S. Sudman (eds.), *Measurement Errors in Surveys*, New York: Wiley, pp. 259–278.

Fowler, F.J., and Cannell, C.F. (1996), "Using Behavioral Coding to Identify Cognitive Problems with Survey Questions," in N. Schwarz, and S. Sudman (eds.), *Answering Questions: Methodology for Determining Cognitive and Communicative Processes in Survey Research*, San Francisco: Jossey-Bass, pp. 15–36.

Goyder, J. (1987), *The Silent Minority*, Cambridge: Blackwell.

Groves, R.M. (1989), *Survey Errors and Survey Costs*, New York: Wiley.

Groves, R.M., Biemer, P.P., Lyberg, L.E., Massey, J.T., Nicholls, W.L. II, and Waksberg, J. (eds.) (1988), *Telephone Survey Methodology*, New York: Wiley.

Hansen, M.H., Hurwitz, W.N., and Madow, W.G. (1953), *Sample Survey Methods*, Vol. II, New York: Wiley.

Hansen, M.H., Hurwitz, W.N., and Bershad, M.A. (1961), "Measurement Error in Censuses and Surveys," *Bulletin of the International Statistical Institute*, 38, pp. 359–374.

Herzog, A.R., and Rodgers, W.L. (1992), "The Use of Survey Methods in Research on Older Americans," in R.B. Wallace, and R.F. Woolson (eds.), *The Epidemiologic Study of the Elderly*, Oxford: Oxford University Press, pp. 60–89.

Hox, J.J. (1994), "Hierarchical Regression Models for Interviewer and Respondent Effects," *Sociological Methods and Research*, 22, pp. 300–318.

Hox, J.J., and de Leeuw, E.D. (1994), "A Comparison of Nonresponse in Mail, Telephone, and Face to Face Surveys," *Quality and Quantity*, 28, pp. 329–344.

Hox, J.J., de Leeuw, E.D., Kreft, I.G.G. (1991), "The Effect of Interviewer and Respondent Characteristics on the Quality of Survey Data: A Multilevel Model," in P.P. Biemer, R.M. Groves, L.E. Lyberg, N.A. Mathiowetz, and S. Sudman (eds.), *Measurement Errors in Surveys*, New York: Wiley, pp. 439–459.

Hyman, H., Feldman, J., and Stember, C. (1954), *Interviewing in Social Research*, Chicago: Chicago University Press.

Institute for Social Research (ISR) (1976), *Interviewer's Manual, revised edition,* Ann Arbor: ISR, University of Michigan.

Institute for Social Research (ISR) (1983), *General Interviewing Techniques; A Self-instructed Workbook for Telephone and Personal Interviewer Training,* Ann Arbor: ISR, University of Michigan.

Kalsbeek, W.D. (1989), "Samples for Studies Related to AIDS", in F.J. Fowler (ed.), *Health Survey Research Methods,* Department of Health and Human Services, publication no. (PHS) 89–3447, pp. 199–203.

Kalton, G., and Anderson, D.W. (1989), "Sampling Rare Populations," in M.P. Lawton, and A.R. Herzog (eds.), *Special Research Methods for Gerontology,* Amityville: Baywood, pp. 7–30.

Kish, L. (1965), *Survey Sampling,* New York: Wiley.

Lyberg, L., and Kasprzyk, D. (1991), "Data Collection Methods and Measurement Error: An Overview," in P.P. Biemer, R.M. Groves, L.E. Lyberg, N.A. Mathiowetz, and S. Sudman (eds.), *Measurement Errors in Surveys,* New York: Wiley, pp. 237–257.

Meloen, J.D., and Veenman, J. (1990), *Het Is Maar De Vraag,* Lelystad: Koninklijke Vermande (in Dutch).

Morton-Williams, J., and Sykes, W. (1984), "The Use of Interaction Coding and Follow-up Interviews to Investigate Comprehension of Survey Questions," *Journal of the Market Research Society,* 26, pp. 109–127.

Moser, C.A., and Kalton, G. (1979), *Survey Methods in Social Investigation* (Chapter 1), Aldershot: Gower.

McCrossan, L. (1991), *A Handbook for Interviewers,* Office of Population Censuses and Surveys, London: HMSO.

Nicholls II, W.L., and Groves, R.M. (1986), "The Status of Computer Assisted Telephone Interviewing: Part I," *Journal of Official Statistics,* 2, pp. 93–115.

Noldus, L.P.J.J. (1991), "The Observer; A Software System for Collection and Analysis of Observational Data," *Behavior Research Methods, Instruments and Computers,* 23, pp. 415–429.

Noldus, L.P.J.J., Van de Loo, E.H.L.M., and Timmers, P.H.A. (1989), "Computers in Behavioral Research," *Nature,* 341, pp. 767–768.

Oksenberg, L., Cannell, C., and Kalton, G. (1991), "New Strategies of Pretesting Survey Questions," *Journal of Official Statistics,* 7, pp. 349–365.

Schaeffer, N.C. (1995), "A Decade of Questions," *Journal of Official Statistics,* 11, pp. 79–92.

Schothorst, J. (1993), "Lessen in en uit allochtonen onderzoek," *Comma,* 10, pp. 20–23 (in Dutch).

Schwarz, N., and Hippler, H.-J. (1991), "Response Alternatives: The Impact of Their Choice and Presentation Order," in P.P. Biemer, R.M. Groves, L.E. Lyberg, N.A. Mathiowetz, and S. Sudman (eds.), *Measurement Errors in Surveys,* New York: Wiley, pp. 41–56.

Schwarz, N., Strack, F., Hippler, H.-J., and Bishop, G. (1991), "The Impact of Administration Mode on Response Effects in Survey Measurement," *Applied Cognitive Psychology,* 5, pp. 193–212.

Sudman, S., Sirken, M., and Cowan, C. (1988), "Sampling Rare and Elusive Populations," *Science*, 240, pp. 991–996.

Sykes, W., and Collins, M. (1988), "Effects of Mode of Interview: Experiments in the U.K.," in R.M. Groves, P.P. Biemer, L.E. Lyberg, J.T. Massey, W.L. Nicholls II, and J. Waksberg (eds.), *Telephone Survey Methodology*, New York: Wiley, pp. 301–320.

Sykes, W., and Collins, M. (1992), "Anatomy of the Survey Interview," *Journal of Official Statistics*, 8, pp. 277–291.

Van der Zouwen, J., and Dijkstra, W. (1995), "Non-trivial Question–answer Sequences in Interviews: Types, Determinants, and Effects on Data Quality," *Proceedings of the International Conference on Survey Measurement and Process Quality, Bristol, U.K., Contributed Papers*, pp. 296–301.

Van der Zouwen, J., Dijkstra, W., and Smit, J.H. (1991), "Studying Respondent–interviewer Interaction: The Relationship Between Interviewing Style, Interviewer Behavior, and Response Behavior," in P.P. Biemer, R.M. Groves, L.E. Lyberg, N.A. Mathiowetz, and S. Sudman (eds.), *Measurement Errors in Surveys*, New York: Wiley, pp. 419–438.

Weeks, M.F. (1992), "Computer-assisted Survey Information Collection: A Review of CASIC Methods and Their Implications for Survey Operations," *Journal of Official Statistics*, 24, pp. 445–466.

Woijcik, M.S., Bard, S., and Hunt, E. (1992), *Training Field Interviewers to Use Computers: A Successful CAPI Training Program*, Information Technology in Survey Research Discussion Paper 8, Chicago: NORC.

CHAPTER 9

The Effect of New Data Collection Technologies on Survey Data Quality

William L. Nicholls II
U.S. Bureau of the Census

Reginald P. Baker
National Opinion Research Center, University of Chicago

Jean Martin
Office for National Statistics, U.K.

This chapter examines the effects of new data collection technologies on survey data quality. It describes the most significant new technologies currently in use for census and survey data collection and summarizes research studies on the effects of collection technology on survey coverage, nonresponse, and measurement error. It concludes by offering general observations about the direction and magnitude of technology effects and by proposing new objectives for their study.

9.1 OVERVIEW OF NEW DATA COLLECTION TECHNOLOGIES

The focus of this chapter is on technologies of survey data collection. Most of these technologies perform data collection, capture, and editing in a single process. They replace the paper and pencil methods formerly used in face to face interviews, telephone interviews, and self-administered questionnaires

Survey Measurement and Process Quality, Edited by Lyberg, Biemer, Collins, de Leeuw, Dippo, Schwarz, Trewin.
ISBN 0-471-16559-X © 1997 John Wiley & Sons, Inc.

as well as the procedures to capture and edit data from paper forms. This chapter does not consider technologies used only in post collection capture and processing, such as optical character recognition and batch editing.

The best known family of new collection technologies is *computer assisted interviewing*, or *CAI*, which replaces paper and pencil (P&P) forms with a small computer or computer terminal. The survey questions are displayed on the computer screen, and the interviewer (or respondent) enters the answers directly through the computer keyboard. CAI systems differ in their capabilities, but most control the sequence of questions based on prior entries, modify question wording to reflect information already received, and check responses for consistency and completeness. Demographic or household surveys typically present the questions and check their entries one at a time in a standard sequence. Economic or establishment surveys typically employ multiple-item screens, perform editing at the end of screens, and permit freer movement across items and screens to accommodate the order in which establishments maintain their records.

In the most common forms of CAI, an interviewer operates the computer, reading questions from the computer screen and recording the respondent's spoken answers. *Computer assisted telephone interviewing (CATI)* was developed in the early 1970s to support telephone interviewing from centralized telephone facilities. Current CATI systems not only perform CAI interviewing but handle call scheduling and case management as well (Nicholls and Groves, 1986). *Computer assisted personal interviewing (CAPI)* emerged in the late 1980s when lightweight, laptop microcomputers made face to face CAI interviews feasible in respondents' homes (Baker, 1990). Field interviewers equipped with laptop microcomputers also have been employed in *decentralized CATI* from interviewers' homes (Bergman *et al.*, 1994).

Self-administered forms of CAI also emerged during the 1980s. In *computer assisted self-interviewing (CASI)* the respondent reads the survey questions from the computer screen and enters his/her own answers on the keyboard. An interviewer may bring the computer to the respondent's home or the respondent may be invited to a site equipped with a computer. A field worker is present to assist at the start, but the respondent operates the computer on his/her own (Duffy and Waterton, 1984; O'Reilly *et al.*, 1994; Mitchell, 1993). Scherpenzeel (1995) has proposed the acronym *CASIIP*, with the added *IP* for "interviewer present," for this form of CAI, but this chapter will use the simpler acronym CASI.

A related form of self-administered CAI without an interviewer present is frequently referred to as a *computerized self-administered questionnaire (CSAQ)*. The data collection organization asks the respondent to answer on his/her own personal computer (PC) or computer account. When the software is mailed on a floppy disk, this method is called *disk by mail (DBM)*. Alternately, the CSAQ software may be transmitted by local area network, modem to modem telecommunications, electronic mail, or it may be remotely accessed from electronic bulletin boards or World Wide Web sites (Kiesler and Sproull, 1986; Downes-LeGuin and Hoo, 1994; Sedivi and Rowe, 1993). When the data collection organization also provides the respondent with a home computer, this method has been called *the teleinterview* (*TI*) (Saris, 1990).

Two new data collection technologies require neither a human interviewer nor respondent access to a computer screen and keyboard. The first is *touchtone data entry* (*TDE*) which uses an ordinary touchtone telephone as a data collection instrument. In common applications, respondents call a toll free number at the data collection organization. The system reads voice digitized survey questions to the respondent who enters his/her answers using the telephone keypad (Clayton and Harrell, 1989). The digitized voice may repeat each answer for respondent confirmation or correction. In principle, TDE can employ complex question routing, tailored wording, and edit checks similar to that of CAI. In practice, TDE has primarily been used in simple government panel surveys of establishments, such as businesses, making little use of these capabilities. TDE also has been tested for surveys of the general public (Frankovic *et al.*, 1994; McKay and Robinson, 1994).

Voice recognition (*VR*) is similar to TDE but accepts spoken rather than keyed responses. A computer reads voice digitized survey questions to the respondent, analyzes the spoken answer, and records its recognized meaning. Recognition failure may prompt a repeat of the question or a related probe. Depending on the sophistication of the system, respondents may be asked to answer with a simple Yes or No, single digits, continuous numbers, spoken words, and spelled names. Small vocabulary VR (Yes, No, and digits) may serve as a backup or replacement for TDE (Clayton and Winter, 1992). Large vocabulary VR may become a replacement for reverse CATI interviewing, where respondents call the survey agency (Blyth and Piper, 1994). Like TDE, VR is used primarily for call-in, panel surveys of establishments. It also is under investigation for call-in interviews of consumer panels (Chapter 10) and as one of several means of answering the year 2000 U.S. census (Appel and Cole, 1994).

Since CATI, CAPI, CASI, CSAQ, TDE, and VR are the most commonly used new collection technologies, their data quality effects have been most frequently studied. Very little is known about the data quality effects of other emerging technologies, such as *electronic data interchange*, or *EDI* (Ambler and Mesenbourg, 1992). Additional emerging or envisioned technologies have the potential to revolutionize survey data collection if they become broadly disseminated. They include such new telephone options as screen phones and video phones, interactive television, and personal digital assistants (Baker, 1995; Ogden Government Services, 1993). Since little or no systematic evidence is available on their data quality effects, they are not considered further in this chapter.

9.2 SURVEY DATA QUALITY AND TECHNOLOGY

Survey managers may choose a survey's data collection technology for a variety of reasons. Cost efficiency will be most important to some, timely release of estimates to others, and survey data quality to still others. Previous summaries of the effect of new data collection technologies on survey costs, timeliness, and data quality have been prepared by Nicholls and Groves (1986), Groves and Nicholls (1986), Saris (1990), Weeks (1992), de Leeuw (1994), Baker *et al.* (1995), and Martin and Manners (1995).

This chapter focuses exclusively on *survey data quality* defined in its traditional sense *as an absence of survey error*. Cost efficiency, timely reporting, and data quality defined in terms of customer satisfaction are often best understood within the operating budgets, time constraints, and priorities of specific surveys. Partly for this reason, more generalizable empirical evidence exists on the impact of new technologies on survey errors than on their effects on costs and timeliness. Even when costs and timeliness are primary in choosing a collection technology, an understanding of the technology's effects on survey errors may suggest ways to make maximum use of its error reducing features while guarding against those errors to which it is most vulnerable.

Our perspective originates with what Bradburn and Sudman (1989) have called the "scientific basis of surveys." A survey following this model selects a probability sample from a carefully defined population, uses a standardized instrument (or questionnaire) to ensure that the same information is obtained for each relevant sample member, collects information from a very high proportion of the selected sample, and employs sampling theory to draw inferences about the population from the sample data. The same model applies whether the data are collected by interviewing or by respondent self-reporting. It also applies whether the collection mode is paper and pencil, a microcomputer, or a telephone touchtone keypad. The use of innovative collection technologies is not a substitute for proper survey procedures.

The survey model generally recognizes two classes of survey error: sampling error and nonsampling error. Sampling error is primarily a function of sample design and sample size. Nonsampling error has more complex origins. Biemer (1988) and Groves (1989) have classified nonsampling errors into three major categories: coverage error, nonresponse error, and measurement error. These errors arise from a variety of sources including the survey collection method, the survey instrument, the interviewer, and the respondent. When all major forms of nonsampling survey error have been controlled and minimized, the survey data can be described as having high quality.

Research into the effects of survey collection methods on survey data quality has focused primarily on comparisons of three alternative paper and pencil data collection modes: the face to face interview, the telephone interview, and the self-administered (usually mail) questionnaire (Sudman and Bradburn, 1974; Lyberg and Kasprzyk, 1991; de Leeuw, 1993). The literature on mode effects has shown, for example: (1) that telephone surveys are more vulnerable to coverage error than face to face interview surveys where telephone subscribership is less than universal; (2) that mailed questionnaire surveys generally attain lower unit and item response rates than telephone or face to face interview surveys; (3) that when self-administered questionnaires are answered, they generally obtain fuller reports of sensitive behaviors than either telephone or face to face interviews.

The term, "mode effects," is sometimes used to describe differences in coverage, nonresponse, or measurement error between computer assisted and P&P collection methods. This can be confusing. A data collection method involves both a collection mode (face to face, telephone, or self-administration) and a collection technology (paper and pencil or computer assisted). Changing either may affect data quality. A change from mailed questionnaires to CATI modifies both the mode and the technology. A very careful research design is necessary to distinguish their separate effects.

To assess the effects of collection technologies on survey data quality, survey treatments must be compared which differ in collection technology but which have the same collection mode, questionnaire, and other essential survey conditions. These experimental conditions can be closely approximated only in controlled studies undertaken to assess technology effects. Weaker evidence is sometimes available from: studies that concurrently modify two or more design elements, such as collection technology and questionnaire, at the same time under controlled conditions; before and after studies conducted when a survey's technology is changed; and operational experiences when current respondents are moved from one collection method to another (e.g., CAPI to CASI) in midinterview. Information presented in this chapter is based on these types of evidence.

9.3 COVERAGE ERROR

Coverage error occurs when some persons are omitted from the list or frame used to identify members of the study population (Groves, 1989). The use of CAI rather than P&P collection methods poses no special coverage problems for telephone or face to face interview surveys. Exactly the same sampling frame and field sampling methods can be employed for both. This also is true of interviewer introduced CASI. By contrast, CSAQ, TDE, and VR limit population coverage when they permit participation only by respondents who have specialized equipment. Those without that equipment may be omitted from the survey.

Both TDE and VR require a telephone, and for TDE this must be a touchtone telephone. In countries such as Sweden, where telephones are nearly universal and almost all are touchtone, this is not a problem. But where telephone subscribership or touchtone dialing is not prevalent (Trewin and Lee, 1988), TDE may be used only selectively, if at all. For example, in 1993 about 20 percent of U.S. households had rotary or pulse telephones and 5 percent had no telephone at all. Rotary telephones were concentrated in older neighborhoods, among the elderly, and selected geographic regions (Frankovic *et al.*, 1994). Even among U.S. businesses, about 10 percent do not have a touchtone phone (Clayton and Werking, 1994). Replacement of TDE with small vocabulary VR eliminates the need for touchtone handsets but may face new coverage limitations since affordable, efficient VR software is not readily available in all languages.

CSAQ coverage error is potentially more severe. PC ownership has been estimated at 26 percent of U.S. households in 1994 (U.S. Bureau of the Census, 1995), 24 percent of British households in 1993 (Office of Population Censuses and Surveys, 1995), and typically smaller percentages in other countries. PC ownership also is concentrated among upper-income, urban professionals. A direct solution to these major coverage biases was devised by the Sociometric Research Foundation (Saris and de Pijper, 1986) by placing a small computer and modem in each respondent's home. The respondent's television set serves as the computer monitor and the modem transmits blocks of survey questions and respondent's answers through the respondent's telephone. The requirements for an appropriate TV, telephone, and high respondent motivation may limit coverage but perhaps no more than for similar consumer panels using P&P methods.

The case for CSAQ establishment surveys is stronger, since PC ownership and use is much greater for establishments than households. Personal computers are almost universal in large U.S. businesses. Even among businesses with less than 100 employees, two-thirds have PCs (Ogden Government Services, 1993). Similar levels of PC business use are anticipated in other industrialized nations. Even among large establishments, however, a readiness for CSAQ surveys is not assured. Since PCs differ in capacities, drives, modems, and operating systems, CSAQ respondents generally must be prescreened to identify those able to access and run a survey's CSAQ software. Those unable or unwilling to reply by CSAQ must be reached by some other collection method or the bias of their omission accepted.

9.4 NONRESPONSE ERROR

Nonresponse error results from failure to collect information from units in the designated sample. *Unit nonresponse* is the failure to obtain any information at all from a designated sample unit. *Item nonresponse* is the loss of specific data items for participating units.

9.4.1 Unit Nonresponse

Unit nonresponse rates often are partitioned into noncontact rates and noncooperation (refusal) rates. There is little evidence that the new collection technologies affect noncontact rates, although in the past some researchers thought they might. For example, Groves and Nicholls (1986) and Weeks (1988) argued that CATI's automated call scheduling should reduce noncontact rates relative to those of P&P telephone surveys, but this possibility has never been confirmed in a controlled study. Early CATI and CAPI tests also appeared to increase noncontact rates when operational problems slowed the field work, giving interviewers less time to make needed calls. When the operational problems were solved, the effect disappeared (Catlin and Ingram, 1988; Bernard, 1990; Kaushal and Laniel, 1993).

The respondent's willingness to participate in surveys using new technologies is a more important issue. Although some survey managers initially expressed concerns that CATI or CAPI respondents would object to having their answers entered directly into a computer, research studies proved them wrong. CATI respondents typically were found unaware of, or indifferent to, the use of a computer during telephone interviews, while most CAPI respondents were found to prefer CAPI to P&P methods or to be indifferent to the technology used (Groves and Nicholls, 1986; Wojcik and Baker, 1992; van Bastelaer et al., 1988). Comparisons of CATI and CAPI refusal rates with those of equivalent P&P control groups typically have found no significant differences between them (Catlin and Ingram, 1988; Baker et al., 1995; Bergman et al., 1994; de Leeuw, 1994; Manners, 1991).

Investigations of respondent acceptance of interviewer introduced CASI are less consistent, perhaps because the setting and basis of comparison vary from study to study. In an early study, Duffy and Waterton (1984) reported a higher refusal rate for household CASI than for a P&P household interview. More recent studies in laboratory and classroom settings find that respondents prefer CASI to P&P self-administered forms (O'Reilly et al., 1994; Mitchell, 1993). Controlled response rate comparisons of CASI and P&P questionnaires in field settings are lacking. When CASI is introduced into the middle of a CAPI interview, Lessler and O'Reilly (1994) report that most women of childbearing age accept it with few, if any, problems. Among older respondents, Couper and Rowe (1995) found that many CASI interviews actually were completed by (or with direct assistance of) the interviewer.

Since CSAQ can only be used with those specialized populations who have PCs (or other appropriate computing access), the few comparisons of CSAQ with P&P questionnaire response rates are difficult to generalize. In a study of university students and staff, all with active computer accounts, Kiesler and Sproull (1986) report a lower response rate for CSAQ than for a comparable P&P questionnaire. Operationally, U.S. governmental agencies have successfully moved establishment respondents (in the petroleum industry and in state governments) from P&P forms to CSAQ without major nonresponse losses after a prescreening to identify those with the appropriate computing resources (Kindel, 1994; U.S. Energy Information Administration, 1989; Sedivi and Rowe, 1993). In a review of the disk by mail CSAQ literature, Downes-LeGuin and Hoo (1994) argue that computer knowledgeable professionals often prefer CSAQ to mail questionnaires because they seem more interesting, easier to complete, and less time consuming. For selected professional populations (e.g., LAN managers) with access to DOS-based computers, Downes-LeGuin and Hoo report CSAQ response rates of 40–60 percent compared with 15–40 percent for comparable mailed questionnaires. They conclude, however, that characteristics of the study population, questionnaire length, and follow-up procedures are more important to response rates than use of CSAQ.

Respondent acceptance of TDE and small vocabulary VR, as pioneered in the U.S. Bureau of Labor Statistics' (BLS) Current Employment Statistics (CES) Survey, also must be interpreted in terms of the overall survey design. This monthly continuing panel of employers was originally conducted exclusively by mail with response rates of about 50 percent at first closing. Samples of respondents were moved to CATI to improve response rates and timely reporting. After six months, respondents in the CATI samples were moved to TDE with CATI follow-up where necessary. TDE was found acceptable to almost all CATI respondents with touchtone phones, the high 80 percent CATI response rates were maintained, while costs and sample attrition were reduced (Clayton and Harrell, 1989; Rosen *et al.*, 1993). TDE did not increase response rates over mail questionnaires by itself. Rather it efficiently sustained high response rates initially obtained with CATI. When BLS next offered small vocabulary VR as an option for the TDE respondents, they generally preferred VR to TDE for its ease of use and perceived shorter administration time (Clayton and Winter, 1992). BLS has now developed alternative strategies to move virtually all CES mail respondents to TDE and VR, but each strategy requires careful respondent preparation and staged nonresponse follow-up (Clayton and Werking, 1994). Similar TDE procedures have been developed and tested at the U.S. Census Bureau with comparable success (Bond *et al.*, 1993).

TDE and VR are now recognized in the U.S. as collection technologies well suited to the reporting of small sets of numeric data by continuing establishment panels. Although respondent acceptance of these same technologies by household respondents seems more problematic, they have not been tested in controlled field comparisons with alternative collection methods.

9.4.2 Item Nonresponse

Item nonresponse arises when: (1) a respondent fails to provide a substantive answer to a survey item or omits the item in a self-administered form; or (2) an interviewer fails to ask an applicable question of a respondent or to record the answer given.

All of the new collection technologies are designed to minimize item nonresponse errors by computer controlled routing through the instrument and checks of each item's entry before presenting the next. Even when alternative forms of movement are possible, such as backing to prior items or skipping over a section and returning to it later, the inherent logical sequence of items can be enforced. Such self-administered technologies as CASI, CSAQ, TDE, and VR may have the further advantage of prompting an explicit "refused" or "don't know" rather than leaving all missing data blank as in P&P forms. These features do not necessarily encourage respondents to reply to questions they do not want to answer or feel unprepared to answer.

One of the most consistent conclusions of the CAI literature is that CAI can eliminate virtually all respondent and interviewer omissions of applicable items but typically provides little or no reduction in rates of explicit refusals, don't knows, and no opinions. This result applies to CATI, CAPI, CASI, and CSAQ and to both factual and opinion questions (Baker *et al.*, 1995; Catlin and Ingram, 1988; de Leeuw, 1994; Groves and Nicholls, 1986; Manners *et al.*, 1993; Martin *et al.*, 1993; O'Reilly *et al.*, 1994; Weeks, 1992). The magnitude of the reduction in item nonresponse depends on the nature of the survey, the effective use of CAI routing and editing, and the complexity of the survey instrument. Nonresponse is reduced most where skip pattern errors are the predominant form of P&P item nonresponse (Baker *et al.*, 1995; Catlin and Ingram, 1988; Manners *et al.*, 1993). Similar reductions in item nonresponse have not been demonstrated for TDE and VR, perhaps because of their relatively simple questionnaires and infrequent use of edit checks during data collection.

9.5 MEASUREMENT ERROR

Groves (1989) defines measurement error as "inaccuracies in responses recorded on the survey instruments." These inaccuracies may be attributable to the interviewer, to the respondent, or to the questionnaire. A collection technology may affect measurement error either by operating through these traditional sources or by weaknesses of the technology itself.

9.5.1 General Approaches to Measurement Error

Three approaches have traditionally been used to assess levels of measurement error associated with alternative data collection methods (Biemer, 1988; Groves, 1989; Lyberg and Kasprzyk, 1991.)

Validation
The first is to compare data collected by each method with data from other sources, for instance, administrative records. The use of record checks in comparisons of data collection technologies has been rare and not especially informative when attempted. In a study of college students, Mitchell (1993) found no significant differences between CASI and P&P questionnaires in the accuracy of reported college grade point averages, scholastic aptitude test scores, and number of failed courses when university records served as the validation instruments. Grondin and Michaud (1994) matched reports of various income sources from successive P&P and CAI telephone interviews

against Canadian Internal Revenue Service records and found greater agreement for the CATI reports. However, concurrent questionnaire changes introduced with CATI confound interpretation of this result.

Split Sample Treatments

When validation criteria are not available, distributions from different survey treatments may still be compared to identify significant differences between them. The nature of the differences may suggest which survey method was least subject to bias and error. About 15 studies have examined CAI and P&P survey distributions obtained under otherwise comparable conditions (Baker *et al.*, 1995; Bergman *et al.*, 1994; Bernard 1990; Catlin and Ingram, 1988; Groves and Mathiowetz, 1984; Groves and Nicholls, 1986; Heller, 1993; Kaushal and Laniel, 1993; Manners, 1990; Manners and King, 1995; Martin *et al.*, 1993; Mitchell, 1993; Tortora, 1985; van Bastelaer *et al.*, 1988). With only a few exceptions, differences between CAI and P&P distributions are rare, small, and typically no more common than would be expected by chance. This conclusion applies to comparisons between CAPI, CATI, CASI, CSAQ, and each of their P&P equivalents. It also applies to demographic and establishment surveys, to personal and household background characteristics, to measures of labor force participation and unemployment, to reports of personal and household income, to political opinions and social attitudes, to job satisfaction measures, and to psychological test scores.

This conclusion must be qualified in three ways. First, it does not necessarily apply to sensitive items where social desirability reporting may be involved or to situations where questionnaire changes are introduced with a conversion to CAI. These exceptions are explored in sections that follow. Second, since many CAI vs P&P comparisons have been based on samples of only a few hundred cases each, they could not be expected to detect small differences between their distributions. Third, comparisons of CAI and P&P data have generally focused on *t*-tests of central tendency and broad differences measured by chi-square analyses. Few studies have compared the dispersion of CAI and P&P distributions or examined subtle measures of opinion acquiescence and extremity. One exception is Martin *et al.* (1993) who found more frequent use of extreme values in attitude scales in CAPI than in P&P face to face interviews.

Comparisons of TDE with mailed questionnaire distributions have been less common but yield similar results. A sample of respondents to the U.S. Manufacturers' Shipments, Inventories, and Orders (M3) Survey was asked to provide their monthly reports to the U.S. Census Bureau both by mail and by TDE for three months. Bond *et al.* (1993) report no meaningful differences between their distributions. Phipps and Tupek (1991) found essentially no differences between TDE and mail questionnaire reports to the BLS Current Employment Survey after longitudinal edit checks eliminated infrequent gross errors from the TDE replies.

Response Variance, Interviewer Variance, and Response Stability

A third way to assess the relative vulnerability of two collection methods to measurement error is to compare their response variances, interviewer variances, or response stability. This approach begins with the model of a survey response as consisting of a true value, a response error deviation from that true value, and a bias term. Response errors may be assessed by reinterviewing a survey respondent shortly after the initial interview using the same instrument, reference period, and survey conditions. Differences between the initial interview and reinterview values are regarded as estimates of response errors. Interviewer contributions to error are estimated by the between-interviewer contributions to variance in surveys where the assignment of respondents to interviewers has been randomized. Panel response stability implies control of measurement errors in panel surveys, where a respondent's answers are subject both to response errors and to interviewer errors at each panel wave.

Relatively few studies have compared response variances, interviewer variances, or panel response stability across collection technologies. Bergman *et al.* (1994) compared CAI and P&P decentralized telephone interviews in the Swedish Labour Force Survey and found no difference between their response variances on labor force items. In the U.S. Current Population Survey, CAI response variances of labor force variables were found generally equal to or smaller than P&P response variances of the same variables (Rogers, 1991; Weller, 1994). Other differences between the CAI and P&P treatments and the before and after design of the second study, however, limit unambiguous interpretation of the CPS results. Groves and Magilavy (1980) reported smaller interviewer variances for CATI than P&P telephone interviews for the same survey questionnaire, although the differences were not statistically significant. Martin *et al.* (1993) compared the response stability of attitude items for three sequences of consecutive face to face interviews (P&P both times, P&P followed by CAPI, and CAPI followed by P&P) and found that response stability was significantly greater when one of the waves used CAPI. Martin and her colleagues conclude that CAPI is the slightly more reliable technology.

Although none of the response variance, interviewer variance, and response stability studies provides more than suggestive evidence that CAI improves data quality, taken together they imply that CATI and CAPI data are at least equal in quality with P&P data and perhaps marginally superior. As use of these new technologies grows, especially in large government surveys, opportunities to replicate and expand these analyses may prove the best means of assessing the overall consequences of CAI for reduction of measurement error.

9.5.2 Interviewer Errors

Since a change from P&P to CAI has its most obvious effect on the activities of the interviewer, the interviewers' acceptance of this new technology, their ability to use it effectively, and its consequences for traditional methods of field quality control have been common subjects of speculation and research.

Interviewer Acceptance

When CATI was first introduced in the 1970s, researchers feared that telephone interviewers would find it difficult, constraining, or detracting from their professional skills (Coulter, 1985; Groves and Mathiowetz, 1984). Early studies provided conflicting reports of interviewer acceptance; but as CATI systems increased in power and efficiency and interviewers gained familiarity with this new technology, the great majority of telephone interviewers came to prefer CATI to P&P methods. Many believed that CATI enhanced the professionalism of their positions while reducing its clerical burden (Nicholls and Groves, 1986; Weeks, 1992).

When CAPI was proposed for face to face interviewing in the 1980s, the issue of interviewer acceptance was raised again. In many countries, field interviewers were women over 40 years of age with relatively little prior computer experience. The weight and poor screen visibility of early portable computers raised concerns that CAPI might aggravate or produce health problems (Blom *et al.*, 1989; Couper and Burt, 1993). Some survey managers feared that CAPI might require replacement of their experienced field staffs with new hires better suited to computer interviewing. Pilot tests of CAPI in the mid to late 1980s were not always reassuring. Some reported operational success and interviewer enthusiasm; others noted frequent interviewer complaints about hardware and software. Early models of portable microcomputers were heavy and slow, some early CAPI software was primitive and vulnerable to entry errors, and methods of transmitting completed data back to headquarters by disk or modem were not fully reliable (Baker, 1990; Bernard, 1990; Brown and Fitti, 1988).

Broad interviewer acceptance of CAPI was achieved by a combination of lighter and more powerful hardware, enhanced software, improved data management and communications, and carefully developed interviewer training. With thorough preparation and special attention to trainees requiring the most help, survey organizations found that virtually all experienced interviewers could be trained in CAPI methods. Once trained, most interviewers preferred CAPI to P&P interviewing (Blackshaw *et al.*, 1990; Couper and Burt, 1993; Edwards *et al.*, 1993; Wojcik and Baker, 1992).

Interviewer Entry

A principal advantage of computer assisted interviewing is that it permits range, consistency, and editing of data for completeness as they are being collected. When an entry is out of range or inconsistent with prior data, the interviewer can be prompted to clarify it with the respondent immediately. By contrast, in P&P interviewing the interviewer must rely upon his/her own alertness and memory to ensure that the respondent's answers are probed for apparent errors and correctly entered on the paper form. Errors which escape the interviewer's attention must be caught in post interview field or office editing, key entry checks, or batch computer edits. And once those errors are detected, additional steps are required to resolve them, such as respondent recalls, manual or computer imputations, or conversion of failed values to missing data codes. As Manners (1990) has observed, CAI enhances quality by "data editing taking place where it is most likely to be successful—in the interview."

The effectiveness of CAI on-line editing has been assessed by submitting comparable CAI and P&P data sets to the same batch computer edits and comparing their rates of edit failure. In an early comparison of CATI and P&P telephone interviews in an agricultural survey, Tortora (1985) reported that batch edits identified 76 percent fewer items requiring corrections for the CATI data than for comparable P&P data. Manners *et al.* (1993) reported 75 percent fewer items requiring correction in a controlled comparison of CAPI with P&P face to face interviewing for a complex financial survey. The effectiveness of CAI on-line editing also has been documented by Bernard (1990) for a labor force survey and by Sigman *et al.* (1993) for an establishment survey. Since on-line editing is often an optional feature of CAI, such gains will occur only when on-line edits are built into the interview.

Although CAI is generally recognized as reducing item omissions and data inconsistencies, some writers have expressed concerns that it may be uniquely vulnerable to interviewer recording errors. Groves and Nicholls (1986) have suggested that entering a response on a computer may be more error prone than checking a box on paper form. Lyberg and Kasprzyk (1991) have noted that since interviewers generally are not skilled typists, they might truncate verbatim responses or antagonize respondents with their slow entry of lengthy text. The available evidence does not support these views. Interviewers now may be better prepared for CAI than previously assumed. In a sample of U.S. Census Bureau P&P interviewers, Couper and Burt (1993) found that about half began CAPI training with moderate or extensive computer experience and almost three-fourths described themselves as fast or slow touch typists. Others were experienced hunt-and-peck typists.

Systematic studies of interviewer entries of CAI closed items are reassuring. Based on double monitored CATI interviews (Kennedy *et al.*, 1990) and tape recorded CAPI interviews (Dielman and Couper, 1995), interviewer errors of 0.6 and 0.1 percent of entered items have been reported. An earlier study of P&P mock interviews suggests that P&P entry errors are at least as common (Rustemeyer, 1977) and in combination with skip pattern errors may well exceed CAI entry errors by a substantial margin.

The quality of CATI and CAPI entries for open questions has been examined in five studies conducted in four countries. None reported meaningful or consistent differences between CAI and P&P entries of text answers based on such criteria as length in words, number of different responses in multiresponse open questions, and the ability to code entries made (Bergman *et al.*, 1994; Bernard, 1990; Catlin and Ingram, 1988; Gebler, 1994; Kennedy *et al.*, 1990). Both occupation and industry descriptions and broad opinion items have been examined in these studies.

Supervisory Quality Control

Field supervisors of P&P interviewers have traditionally used a variety of methods to control the quality of interviewers' work. These methods include: (a) keeping records of response rates and clerical error rates by interviewer; (b) clerical review of returned interview forms for completeness, consistency, and evidence of falsification; (c) observing interviewer performance in respondents' homes; and (d) monitoring centralized telephone interviews through the telephone system.

With CATI, record keeping can be automated to provide daily or real-time updates of interviewer performance. Audio-visual monitoring, available with many CATI systems, permits the supervisor to see what is on the interviewer's CATI screen while listening to the interview telephone conversation. This aids the supervisor in identifying interviewing problems and focusing corrective actions where most needed (Couper *et al.*, 1992). Systematic monitoring also provides excellent protection from such severe interviewer failures as falsification of interviews or omitting questions and guessing their answers. Although audio-visual monitoring is frequently cited as one of CATI's strongest tools to enhance data quality (Groves and Nicholls, 1986; Weeks, 1992; and de Leeuw, 1994), its effectiveness has never been demonstrated in a controlled study comparing the data quality of monitored interviews with comparable interviews not monitored.

The introduction of CAPI in face to face interviews is more disruptive to traditional supervisory practices. Clerical review of paper forms is neither possible nor necessary, and supervisory field observation of CAPI interviewers is more difficult because supervisors often cannot keep pace with experienced CAPI interviewers while simultaneously making notes about their performance (Bernard, 1990; Blackshaw *et al.*, 1990). As a result, supervisors may feel that they have less control over CAPI than P&P interviewers. At the same time, automatic date and time stamping of CAPI interviews, daily transmission of completed interviews to headquarters, and analysis of CAPI keystroke files, as described in Chapter 11, are providing new quality control methods for face to face interviewing. The effectiveness of the new methods of supervisory quality control remains to be demonstrated in controlled studies, but that is equally true of other quality control methods. Since the value of supervisory quality control is often regarded as self-evident or too difficult to measure separately, its effects on data quality may be identifiable only in before and after designs when supervisory practices change.

9.5.3 Respondent Errors

This section considers how the new technologies affect self-administered data entry and respondent reporting of sensitive information.

Respondent Entry

Only a handful of studies have compared the entry errors of respondents using the new technologies with their entry errors using P&P forms. O'Reilly *et al.* (1994) report fewer respondent errors in CASI than in P&P self-administered forms. Similar results have been reported for CASI by Kiesler and Sproull (1986) and by Mitchell (1993). In establishment panel surveys, Bond *et al.* (1993) found fewer respondent recording errors for TDE than for P&P mailed forms. Phipps and Tupek (1991) report initial TDE entry error rates of 1.2 to 2.5 percent per item in the BLS CES. These rates were reduced to zero after longitudinal editing for readily identifiable TDE entry mistakes, such as numbers with too many digits.

In household TDE surveys more problems have been encountered. McKay and Robinson (1994) observed that 5 of 35 recruited subjects were unable to complete a TDE version of the U.S. Current Population Survey on the first try in a laboratory setting, and 2 of 23 made errors repeating their TDE report from home. In a national sample recruited to report public reactions to a television broadcast by TDE, Frankovic *et al.* (1994) report that a substantial proportion of respondents (especially among the elderly) had to be transferred to a live operator, apparently because they found TDE frustrating or difficult.

In voice recognition surveys, mistakes made by the respondent are probably less important than voice recognition failures by the system, which are considered in a later section.

Reporting Socially Sensitive Behavior

Survey methods differ in their ability to elicit sensitive information from respondents on such topics as sexual behavior, drug and alcohol use, participation in illegal activities, and income. While recent studies find no consistent difference between face to face and telephone interviews in this respect, there is strong evidence that self-administered forms are better at eliciting sensitive information than interviewers (Chapter 13 and de Leeuw, 1993). The question in this chapter is whether CAI methods encourage the reporting of sensitive behaviors more than the P&P form of the same mode.

In telephone surveys, the use of CATI rather than P&P forms is unlikely to have any effect on the reporting of sensitive behavior because respondents typically are unaware that a computer is being used. In face to face surveys, the CAPI laptop is fully visible, and two studies have shown that respondents often view CAPI as providing more confidentiality than P&P interviews (Baker *et al.*, 1995; van Bastelaer *et al.*, 1988.) Whether that translates into more accurate reporting with CAPI is uncertain.

One study—the National Longitudinal Survey of Youth Cohort (NLS/Y)—found fuller reporting of some sensitive topics with CAPI than in comparable, P&P face to face interviews (Baker *et al.*, 1995). In the 11th wave of the NLS/Y, significantly more alcohol-related problems were reported in CAPI than in P&P interviews. The same items were not repeated on the 12th NLS/Y wave, but among newly added items significantly higher levels of contraceptive use were reported by men in CAPI than in P&P interviews. At the same time, no significant differences were found between CAPI and P&P reports of women's current and former pregnancies and health-related behavior during pregnancy. Differences between CAPI and P&P reports of respondent income and liquid assets also approached statistical significance. These results have not yet been replicated by other studies. If CAPI encourages reporting on sensitive topics, the effects seem to be small and inconsistent.

Many researchers currently believe that CASI, rather than CAPI, is the most promising technology to elicit responses on sensitive topics. Like other self-administered collection methods, CASI should reduce privacy concerns and minimize social desirability reporting. Kiesler and Sproull (1986) hypothesized that "the reduced social context ... in the electronic survey will make the research setting impersonal and anonymous, and ... respondents will become self-centered, and relatively unconcerned with social norms and with the impression they give others." In a controlled study of university students and staff, Kiesler and Sproull found significantly higher levels of candid reports on sensitive topics in an electronic questionnaire than in comparable P&P forms. Subsequent studies summarized by Mitchell (1993) and her own investigation comparing CASI and P&P self-administered questionnaires with samples of college students were unsuccessful in replicating the Kiesler and Sproull findings.

In a laboratory study, O'Reilly *et al.* (1994) compared three technologies in the reporting of sensitive topics: a P&P questionnaire, CASI with questions displayed on the computer screen, and audio-CASI with questions read to the respondent by a digitized voice through a headset. A Greco-Latin square experimental design rotated each respondent through all three technologies and three sensitive topics: sexual behavior, drug and alcohol use, and personal income. Significant differences (at the 0.10 level) were found between P&P and both CASI treatments on selected sexual and drug use items but not on income. The two forms of CASI (standard and audio) were about equally effective in obtaining reports of sensitive behaviors, but most respondents preferred audio-CASI as a reporting method.

In Chapter 13, Jobe and his colleagues describe a study that examined the effects of three survey design factors on women's reporting of abortions and sexual behavior. The three factors were: survey mode (self-administered or interview); technology (CAI or P&P); and interview site (respondent's home or a neutral site, such as a library or fast food restaurant). Based on analyses of a variety of sensitive topics, these investigators conclude that survey mode (self-administration rather than interview) was the most important design factor to encourage reporting of sensitive behavior. Their analyses suggested that technology (CAI vs P&P) at best produced small, subtle effects in reporting sensitive behavior that varied with circumstances. For example, CAI significantly increased the reported number of sex partners when women were interviewed in their own homes but significantly decreased reported sex partners when interviewed in neutral sites, perhaps by calling public attention to the interview and thereby reducing privacy. Since the sample was insufficient to support separate analyses of CAPI and CASI, the conclusions for CAI in this study refer to a mixture of both.

The most effective combination of survey design features to encourage reporting of sensitive behavior is unresolved at this writing and under continuing investigation. In the interim, several U.S. research organizations have converged on audio-CASI as the current method of choice for data collection on sensitive topics since it combines both self-administration and CAI in a form which also may be appropriate for persons with limited reading skills (Lessler and O'Reilly, 1995).

9.5.4 Instrument Effects

CAI instruments offer a variety of standard and optional features to assist the interviewer or respondent in completing the interview. Automatic routing avoids the complex skip instructions of P&P forms which required looking back at earlier entries in the interview to see which item to ask next. Tailored question wording in CAI may remind respondents of information they already reported, narrow questions to the respondent's specific situation, or personalize items about different household members. Data from prior interviews may be incorporated in survey questions, prompts, and edit checks to permit dependent interviewing. In TDE and VR such features are infrequently used although they are available in principle. In CAI, these design features are frequently used although research evidence of their effectiveness to improve data quality has been lacking.

Several recent research studies have provided such evidence. Baker *et al.* (1995) compared CAPI and P&P distributions of 445 items from the 12th round of the NLS/Y, covering topics of marital history and status, education, military background, labor force status, job training, immigration, and income. About 13.5 percent of the differences were significant at the 0.05 level. The most frequent significant differences occurred at items where the P&P interviewer was required to check previous answers or to refer to a household roster or calendar to know what question to ask next or how to ask it. These steps are time consuming and error prone. By making them unnecessary, CAI apparently is performing its job of guarding the interviewer from these distractions.

Similar evidence is provided by Bergman *et al.* (1994) from pilot tests for the Swedish Labour Force Survey. The P&P interview required interviewers to calculate the difference between hours worked in the reference week and in a normal week to determine which question to ask next. In CATI the calculations and routing were automatic. The greater clustering around 40 hours Bergman and his colleagues found in the P&P interviews suggests that some P&P interviewers favored values that avoided difficult hand calculations.

Automated routing, tailored questions, and on-line editing can be based not only on previous entries in the current interview but on information preloaded in the current CAI data record from prior interviews, administrative records, or other sources. In household surveys, use of prior wave data is generally called "dependent interviewing." In establishment surveys, it is usually described as "use of historical data." Dependent interviewing has been used in P&P surveys, and Hill (1994) observed that it appears "to improve the empirical validity of noisy measures in panel surveys." But CAI facilitates more timely capture and transfer of prior wave data and its controlled disclosure in the current wave. In P&P interviews, as Pafford and Coulter (1987) have observed, the interviewer's use of historical data cannot be controlled and may invite response bias and interviewer variability. With CAI the prior data can be withheld until the current value is reported and then disclosed if needed in an edit, prompt, reminder, or query.

Dippo *et al.* (1992) have described uses of dependent CAI interviewing in a redesign of the U.S. Current Population Survey (CPS) which have contributed to major data quality improvements. The original P&P CPS instrument asked occupation and industry of employment independently each month. Small variations in respondent descriptions of the same occupations and industries often resulted in large month to month changes between occupation and industry code categories. CAI pilot tests using controlled disclosure of prior wave data reduced month to month changes for three-digit occupation codes from 39 to 7 percent, a level much closer to independent estimates of true change. Similar results occurred for industry codes. Grondin and Michaud (1994) describe similarly effective use of dependent interviewing in developing CAI methods for the Canadian Survey of Labour and Income Dynamics (SLID).

Dependent interviewing is only one of several powerful interviewing options made available through CAI. In market research, randomization of response category order or question order is used to control well-known order effects in opinion and consumer preference items. Table look-up or pick-list routines can access on-line databases to prompt and clarify responses about the type of automobile owned, the specific names of drugs used, or the prices paid for specific items in local fast food restaurants (Groves and Nicholls, 1986; Weeks, 1992). These features make surveys possible which conventional P&P methods cannot attempt. The most important contributions of CAI to survey data quality may occur by extending the range of data which can be collected with high quality.

While survey designers may employ new CAI features to expedite interviewing and enhance data quality, they must guard against unintended changes when adapting a P&P questionnaire to CAI. The extensive literature on questionnaire design has repeatedly shown that relatively minor changes in question wording and sequence, the placement and order of response categories, and questionnaire formatting may have substantial effects on response distributions (Schuman and Presser, 1981; Schwarz and Hippler, 1991). The transition to CAI does not free the questionnaire designer of these effects. Bergman *et al.* (1994) describe an example in decentralized telephone interviewing where identically worded items resulted in significantly different CAI and P&P

distributions because the P&P response options were grouped in three columns, while the CAI response options were arranged in a single column. When the CATI format was revised to match the P&P layout, the difference in response distributions vanished. Baker and his colleagues also traced several significant differences between CAI and P&P distributions in the NLS/Y to relatively minor differences in question wording or formatting. If consistency of estimates is an important goal in converting a P&P survey to CAI, the CAI questionnaire should parallel the P&P questionnaire as closely as possible.

Differences between CAI and P&P response distributions may also arise from unintended changes in field processing steps, such as the treatment of "other" categories used to record responses outside provided fixed alternatives. In traditional P&P methods, the interviewer or office clerks typically review the "others" to delete irrelevant answers and to reclassify meaningful responses, where possible, to new or existing categories. CATI and CAPI interview data often are sent directly to the next stage of processing without clerical editing. This differing treatment of "others" has produced significant differences between CAI and P&P labor force distributions reported both by Cahoon and Russell (1991) and by Baker *et al.* (1995). These examples represent incomplete transitions from P&P to CAI. CAI systems can be designed to replicate all P&P survey steps, including the editing of "other" responses; or first-stage CAI "other" responses can be routed to second or third tiers of fixed alternatives.

9.5.5 Technology Errors

Data collection technology can itself be a source of measurement error and of nonresponse. Some technologies are incapable of collecting some forms of data a survey may require, and any collection technology may produce errors when that technology malfunctions.

Technological Constraints
A CATI or CAPI interview generally can ask and record any type of survey item possible in a P&P interview. These include check-answer, multiple choice, numeric, and open-ended items, as well as extended pick-lists and randomized items not possible with P&P. But P&P methods generally remain the best option for lengthy, discursive replies required in subjective interviews and exploratory studies. Automatic routing, filling, and editing also permit more complex self-administered data collection with CASI and CSAQ than has been possible with P&P forms; but the number and type of CASI and CSAQ items easily accommodated may depend on respondent keyboard skills and their comfort level with computers.

TDE places more severe limits on the survey items accommodated. Touch-tone telephone keypads permit single-key entry only of numbers and a few special symbols. This greatly limits their use for alpha text, such as names, addresses, and descriptions of occupations. TDE applications also appear limited to 5- to 10-minute sessions even with motivated respondents in

establishment surveys. Until quite recently, general purpose, speaker independent voice recognition systems could accommodate only digits, Yes, and No.
More powerful systems, some operating only in university laboratories, are now
capable of recognizing hundreds of possible responses at each prompt, continuous numbers, spelled names, and commands to repeat a question. The
maximum length of a VR interview acceptable to respondents has not been
studied.

Hardware and Software Failures
All data collection technologies are vulnerable to hardware failures, programming mistakes, and set-up errors which may result in significant data losses.
Unless appropriate steps are taken, interviewing assignments, callback records,
and completed interview data may be lost in hardware or software crashes or
in transmission failures between central headquarters and dispersed CAPI or
CATI interviewers. Questionnaire set-up mistakes may result in items over-
writing each other's answers or in omission of crucial questions for applicable
respondents. Insufficient computing power, unanticipated downtime, or inadequate battery life in portable microcomputers may slow interviewing progress
and threaten response rates.

 Virtually all major data collection organizations have experienced one or
more of these problems as they began feasibility and operational testing of
CATI and CAPI (Bergman *et al.*, 1994; Bernard, 1990; Brown and Fitti, 1988;
Catlin and Ingram, 1988; Grondin and Michaud, 1994; Presser, 1983). However,
in each cited instance, the survey organization found ways to prevent or
minimize the effect of technical problems and proceeded to stable production.
Over the years, computer hardware, general purpose CAI systems, and
communication protocols have improved. Experienced survey organizations
have learned to employ redundant hardware at critical points, to use sophisticated backup and recovery routines, and to require systematic testing of CAI
instruments with improved diagnostic and debugging tools. At the U.S. Census
Bureau, losses of CAI cases through technical failures are now estimated as less
frequent than losses of paper forms misplaced by P&P interviewers or in the
mail.

Voice Recognition Error Rates
Voice recognition is clearly susceptible to measurement error when it fails to
recognize a respondent's spoken answer. Clayton and Winter (1992) report an
error rate of 0.6 percent for U.S. small vocabulary VR systems (digits, Yes, and
No) which is on a par with BLS experiences with other collection and capture
methods. Appel and Cole (1994) report a recognition error rate of about 1
percent in their feasibility testing of medium vocabulary VR with volunteer
respondents. Blyth and Piper (1994) cite error rates of 2 to 5 percent. These
variations in VR error rates are not surprising since these studies differ greatly
in the complexity of the spoken responses they permit and in the recognition
methods they employ. Appel and Cole have focused their work on specific

answer sets anticipated for each separate census-type question. Blyth and Piper are attempting to build a generalized speech recognizer based on syllables which is usable for a wide variety of surveys. Since both these latter estimates were obtained in feasibility tests, it is unclear how predictive they will be of recognition rates in production survey use.

9.6 SUMMARY AND CONCLUSIONS

This chapter has reviewed the research literature to assess the effects of new data collection technologies on survey data quality. Somewhat different conclusions are appropriate for CATI, CAPI, and CASI than for CSAQ, TDE, and VR.

9.6.1 CATI, CAPI, and CASI

CATI, CAPI, and CASI were developed in part to improve data quality. Even when cost savings, time savings, or processing convenience was the primary objective, enhanced data quality typically was a promised by-product, virtually guaranteed by computer controlled routing and on-line editing. Those promised data quality gains have now been demonstrated by major reductions in item nonresponse and post interview edit failures, at least when CAI's capabilities are used effectively. Equally important, these new technologies have not detracted from survey data quality in any significant or consistent way. None of the dire predictions made about CATI and CAPI when they were first introduced proved true after they passed from feasibility testing to production. Based on the published evidence, virtually no respondents object to their use, few if any interviewers feel their jobs are degraded, entry errors do not increase, the quality of text answers is not reduced, and losses of case and item data do not exceed those of P&P methods. Individual surveys may encounter such problems in unusual applications or if appropriate precautions are not taken, but the weight of available evidence to date suggests these are not general problems of CAI methods.

Some early advocates of CATI and CAPI anticipated additional data quality enhancements by easing the interviewer's clerical tasks, providing smoother delivery of computer-tailored questions, and by other subtle changes in the interview process. Where such subtle effects have been identified, they have been small. Further data quality gains were expected in CATI from automated call scheduling and audio-visual monitoring, but if these benefits were realized they remain unsubstantiated by controlled studies. Although moving a survey from P&P to CAI generally improves its data quality, those improvements have not been as large, broad, or well documented as the early proponents of CAI methods anticipated. The enthusiasm of early advocates may have prompted unrealistic expectations of the effect CAI could have on basic sources of survey bias and error. Moreover, unintended changes in questionnaire design or field procedures in moving from P&P to CAI may produce small changes in survey estimates. In brief, computer assisted methods do not provide a panacea (or even a general palliative) for survey noncoverage, nonresponse, and measurement error. Nor do they free the survey manager and survey designer from a need to understand the broader survey literature on questionnaire design, field work procedures, and process quality.

At the same time, CAI methods provide a platform for new approaches to data collection that could yield important future gains in the control and reduction of survey error. The data quality consequences of controlled dependent interviewing, on-line databases, and audio-CASI remain to be fully assessed; and these may only be the first wave of new methods CAI will add to the methodological toolbox. Object-oriented questionnaire design, applications of artificial intelligence, and multimedia data collection may be worthy of careful examination; and they may be followed by currently unimagined computer-based approaches to data collection.

If this perspective is correct, then it may be time to call a halt to studies comparing CAI and P&P methods, especially when the CAI survey is merely a computerized version of P&P procedures. Such studies can tell us little more than we already know. They are becoming irrelevant as P&P surveys move to CAI at an accelerating pace and increasingly employ procedures possible only with CAI. A new research emphasis is needed to evaluate the effectiveness of the newest CAI tools and to compare alternative CAI survey designs with each other. As CAI becomes the new standard for survey data collection, the principles and knowledge base of survey methodology should be refocused to inform choices of collection mode, questionnaire design, and field work procedures within this new environment.

9.6.2 CSAQ, TDE, and VR

The newer technologies of CSAQ, TDE, and VR present a more complicated case. When a research organization provides sampled respondents with both CSAQ software and hardware, as in the teleinterview, the benefits of CSAQ probably resemble those of CAPI or CASI. When respondents are asked to provide their own equipment, the survey designer must begin by addressing potentially serious coverage problems, either by restricting applications to specialized study populations known to possess that equipment or by relying on multimethod data collection with CATI or P&P options as an alternative.

Successful implementations of CSAQ, TDE, and small vocabulary VR data collection can be found in monthly establishment panels reporting small sets of numeric data items to government agencies, usually with mailed questionnaires as a backup and CATI or other prompting methods for nonresponse. The primary objective in developing CSAQ, TDE, and VR for these applications was not to *enhance* data quality but to reduce costs and to speed collection and processing without *reducing* data quality. In this, their developers report success by maintaining high response rates, minimizing panel attrition, and minimizing problems of entry error.

The use of disk by mail CSAQ to conduct surveys of specialized professional populations known to have computers seems to be a more qualified success. It is replacing low response rate paper questionnaire surveys with not quite so low response rate disk by mail CSAQ surveys, while gaining the benefits of respondent data entry and CAI's computer-controlled branching and on-line editing. Extension of CSAQ, TDE, and VR to surveys of the general public currently seems more problematic and clearly in need of further innovative development and careful research.

REFERENCES

Ambler, C., and Mesenbourg, T. M. (1992), "EDI-Reporting Standard of the Future," *Proceedings of the Annual Research Conference*, U.S. Bureau of the Census, pp. 289–297.

Appel, M.V., and Cole, R. (1994), "Spoken Language Recognition for the Year 2000 Census Questionnaire," paper presented at the American Association for Public Opinion Research Annual Conference, Danvers, MA.

Baker, R.P. (1990), "What We Know About CAPI: Its Advantages and Disadvantages," paper presented at the American Association for Public Opinion Research Annual Conference, Lancaster, PA.

Baker, R.P. (1995), *The Role of Current and New Technologies in Polling and Public Opinion Research*, a report of the National Opinion Research Center to the U.S. Congress Office of Technology Assessment.

Baker, R.P., Bradburn, N.M., and Johnson, R.A. (1995), "Computer-Assisted Personal Interviewing: An Experimental Evaluation of Data Quality and Cost," *Journal of Official Statistics*, pp. 413–431.

Bergman, L.R., Kristiansson, K.-E., Olofsson, A., and Safstrom, M. (1994), "Decentralized CATI Versus Paper and Pencil Interviewing: Effects on the Results in the Swedish Labour Force Surveys," *Journal of Official Statistics*, 10, 2, pp. 181–185.

Bernard, C. (1990), *Survey Data Collection Using Laptop Computers*, INSEE Report 01/C520, Institut National de la Statistique et des Etudes Economiques, Paris.

Biemer, P.P. (1988), "Measuring Data Quality," in R.M. Groves, P.P. Biemer, L.E. Lyberg, J.T. Massey, W.L. Nicholls II, and J. Waksberg (eds.), *Telephone Survey Methodology*, New York: Wiley, pp. 273–282.

Blackshaw, N., Trembath, D., and Birnie, A. (1990), "Developing Computer Assisted Interviewing on the Labour Force Survey: A Field Branch Perspective," *Survey Methodology Bulletin*, Social Survey Division, U.K. Office of Population Censuses and Surveys, 27, pp. 6–13.

Blom, E., Blom, R., Carlson, H., Henriksson, E., and Marstad, P. (1989), *Computer Assisted Data Collection in the Labour Force Surveys: Report of Technical Tests*, a report of the CADAC Project, Statistics Sweden.

Blyth, W., and Piper, H. (1994), "Developing Speech Recognition Applications for

Market Research," paper given at the 47th ESOMAR Marketing Research Congress, Davos, Switzerland.

Bond, D., Cable, G., Andrews, S., and Hoy, E. (1993), "Initial Results for Touchtone Data Entry Usage in the Manufacturers' Shipments, Inventories, and Orders Survey," *Proceedings of the Annual Research Conference*, U.S. Bureau of the Census, pp. 525–545.

Bradburn, N.M., and Sudman, S. (1989), *Polls and Surveys: Understanding What They Tell*, San Francisco: Jossey-Bass, pp. 1–11.

Brown, J., and Fitti, J.E. (1988), *Report of the 1987 Automated National Health Interview Survey Feasibility Study: An Investigation of Computer Assisted Personal Interviewing*, a report jointly prepared by the U.S. National Center for Health Statistics Division of Health Interview Statistics and the U.S. Bureau of the Census.

Cahoon, L., and Russell, G. (1991), "Effects of CATI on Estimates from Demographic Surveys Conducted by the Bureau of the Census," paper presented at the U.S. Bureau of the Census Joint Advisory Committee Meeting, October 31–November 1, Alexandria, VA.

Catlin, G., and Ingram, S. (1988), "The Effects of CATI on Costs and Data Quality: A Comparison of CATI and Paper Methods in Centralized Interviewing," in R.M. Groves, P.P. Biemer, L.E. Lyberg, J.T. Massey, W.L. Nicholls II, and J. Waksberg (eds.), *Telephone Survey Methodology*, New York: Wiley, pp. 437-450.

Clayton, R.L., and Harrell Jr, L.J. (1989), "Developing a Cost Model for Alternative Data Collection Methods: Mail, CATI, and TDE," *Proceedings of the Section on Survey Research Methods*, American Statistical Association, pp. 264–269.

Clayton, R.L., and Werking, G. (1994), "Integrating CASIC into the Current Employment Survey," *Proceedings of the Annual Research Conference*, U.S. Bureau of the Census, pp. 738–749.

Clayton, R.L., and Winter, D.L.S. (1992), "Speech Data Entry: Results of a Test of Voice Recognition for Survey Data Collection," *Journal of Official Statistics*, 8, 3, pp. 377–388.

Coulter, R. (1985), *A Comparison of CATI and Non-CATI on a Nebraska Hog Survey*, SRS Staff Report 85, Washington, DC: U.S. Department of Agriculture, Statistical Reporting Service.

Couper, M.P., and Burt, G. (1993), "Interviewer Reactions to Computer-Assisted Personal Interviewing (CAPI)," *Proceedings of the Annual Research Conference*, U.S. Bureau of the Census, pp. 429–454.

Couper, M.P., and Rowe, B. (1995), "Evaluation of a Computer-Assisted Self-Interviewing (CASI) Component in a CAPI Survey," paper presented at the American Association for Public Opinion Research Annual Conference, Fort Lauderdale, FL.

Couper, M.P., Holland, L., and Groves, R.M. (1992), "Developing Systematic Procedures for Monitoring in a Centralized Telephone Facility," *Journal of Official Statistics*, 8, 1, pp. 62–78.

De Leeuw, E.D. (1993), "Mode Effects in Survey Research: A Comparison of Mail, Telephone, and Face to Face Surveys," *BMS*, 41, pp. 3–15.

De Leeuw, E.D. (1994), "Computer Assisted Data Collection Data Quality and Costs:

A Taxonomy and Annotated Bibliography," *Methods and Statistics Series*, Publication 55, University of Amsterdam, Department of Education.

Dielman, L., and Couper, M.P. (1995), "Data Quality in a CAPI Survey: Keying Errors," *Journal of Official Statistics*, 11, 2, pp. 141–146.

Dippo, C., Polivka, A., Creighton, K., Kostanish, D., and Rothgeb, J. (1992), "Redesigning a Questionnaire for Computer-Assisted Data Collection: The Current Population Survey Experience," paper presented at the Field Technology Conference, St. Petersburg, FL.

Downes-LeGuin, T., and Hoo, B.S. (1994), "Disk-by-Mail Data Collection for Professional Populations," paper presented at the American Association for Public Opinion Research Annual Conference, Danvers, MA.

Duffy, J.C., and Waterton, J.J. (1984), "Under-Reporting of Alcohol Consumption in Sample Surveys: The Effect of Computer Interviewing in Fieldwork," *British Journal of Addiction*, 79, pp. 303–308.

Edwards, B., Bitner, D., Edwards, W.S., and Sperry, S. (1993), "CAPI Effects on Interviewers: A Report from Two Major Surveys," *Proceedings of the Annual Research Conference*, U.S. Bureau of the Census, pp. 411–428.

Frankovic, K., Ramnath, B., and Arnedt, C.M., (1994), "Interactive Polling and American's Comfort Level with Technology," paper presented at the American Association for Public Opinion Research Annual Conference, Toronto.

Gebler, N. (1994), "Comparing Results from Open-Ended Questions from CAPI and PAPI," paper presented at the 1994 International Field Technologies Conference, Boston, and private correspondence from the author.

Grondin, C., and Michaud, S. (1994), "Data Quality of Income Data Using Computer Assisted Interview: The Experience of the Canadian Survey of Labour and Income Dynamics," *Proceedings of the Section on Survey Research Methods*, American Statistical Association, pp. 839–844.

Groves, R.M. (1989), *Survey Errors and Survey Costs*, New York: Wiley.

Groves, R.M., and Magilavy, L.J. (1980), "Estimates of Interviewer Variance in Telephone Surveys," *Proceedings of the Section on Survey Research Methods*, American Statistical Association, pp. 622–627.

Groves, R.M., and Mathiowetz, N.A. (1984), "Computer Assisted Telephone Interviewing: Effects on Interviewers and Respondents," *Public Opinion Quarterly*, 48, 1, pp. 356–369.

Groves, R.M., and Nicholls II, W.L. (1986), "The Status of Computer-Assisted Telephone Interviewing: Part II–Data Quality Issues," *Journal of Official Statistics*, 2, 2, pp. 117–134.

Heller, J.-L. (1993), "The Use of CAPI and BLAISE in the French Labour Force Survey," *Proceedings of the Second International BLAISE Users Conference*, London.

Hill, D.H. (1994), "The Relative Empirical Validity of Dependent and Independent Data Collection in a Panel Survey," *Journal of Official Statistics*, 10, 4, pp. 359–380.

Kaushal, R., and Laniel, N. (1993), "Computer Assisted Interviewing Data Quality Test," *Proceedings of the Annual Research Conference*, U.S. Bureau of the Census, pp. 513–524.

Kennedy, J.M., Lengacher, J.E., and Demerath, L. (1990), "Interviewer Entry Error in CATI Interviews," paper presented at the International Conference on Measurement Errors in Surveys, Tucson, AZ.

Kiesler, S., and Sproull, L.S. (1986), "Response Effect in the Electronic Survey," *Public Opinion Quarterly*, 50, pp. 402–441.

Kindel, C.B. (1994), "Electronic Data Collection at the National Center for Education Statistics: Successes in the Collection of Library Data," *Journal of Official Statistics*, 10, 1, pp. 93–102.

Lessler, J.T., and O'Reilly, J.M. (1995), "Literacy Limitations and Solution for Self-Administered Questionnaires to Enhance Privacy," *Statistical Policy Working Paper 23*, Council of Professional Associations in Federal Statistics, Bethesda, MD, pp. 453–469.

Lyberg, L.E., and Kasprzyk, D. (1991), "Data Collection Methods and Measurement Error: An Overview," in P.P. Biemer, R.M. Groves, L.E. Lyberg, N.A. Mathiowetz, and S. Sudman (eds.), *Measurement Errors in Surveys*, New York: Wiley, pp. 237–257.

Manners, T. (1990), "The Development of Computer Assisted Interviewing (CAI) for Household Surveys: The Case of the British Labour Force Survey," *Survey Methodology Bulletin*, Social Survey Division, U.K. Office of Population Censuses and Surveys, 27, pp. 1–5.

Manners, T. (1991), "The Development and Implementation of Computer Assisted Interviewing (CAI) for the British Labour Force Survey," *Proceedings of the Section on Social Statistics*, American Statistical Association, pp. 312–316.

Manners, T., Cheesbrough, S., and Diamond, A. (1993), "Integrated Field and Office Editing in Blaise: OPCS's Experience of Complex Financial Surveys," in OPCS, *Essays on Blaise*, Office of Population Censuses and Surveys, London.

Manners, T., and King, J. (1995), "A Controlled Trial of CAPI and PAPI Modes of Data Collection and Editing: U.K. Family Expenditure Survey," paper presented at the International Conference on Survey Measurement and Process Quality, Bristol, U.K.

Martin, J., and Manners, T. (1995), "Computer Assisted Personal Interviewing in Survey Research," in R. Lee (ed.), *Information Technology for the Social Scientist*, London: UCL Press Ltd., pp. 52–71.

Martin, J., O'Muircheartaigh, C., and Curtis, J. (1993), "The Use of CAPI for Attitude Surveys: An Experimental Comparison with Traditional Methods," *Journal of Official Statistics*, 9, 3, pp. 641–661.

McKay, R.B., and Robinson, E.L. (1994), "Touchtone Data Entry for CPS Sample Expansion," *Proceedings of the Section on Survey Research Methods*, American Statistical Association, pp. 509–511.

Mitchell, D.L. (1993), "A Multivariate Analysis of the Effects of Gender and Computer vs. Paper/Pencil Modes of Administration on Survey Results," unpublished doctoral dissertation of the College of Administration and Business, Louisiana Tech University.

Nicholls II, W.L., and Groves, R.M. (1986), "The Status of Computer-Assisted Telephone Interviewing: Part I—Introduction and Impact on Cost and Timeliness of Survey Data," *Journal of Official Statistics*, 2, 2, pp. 93–115.

Ogden Government Services (1993), *U.S. Bureau of the Census Technology Assessment of Data Collection Technologies for the Year 2000: Final Technology Assessment Report*, a report prepared for the U.S. Bureau of the Census Year 2000 Staff.

Office of Population Censuses and Surveys (1995), *General Household Survey 1993*, London: Her Majesty's Stationary Office.

O'Reilly, J.M., Hubbard, M.L., Lessler, J.T., Biemer, P.P., and Turner, C.F. (1994), "Audio and Video Computer Assisted Self Interviewing: Preliminary Test of New Technologies for Data Collection," *Journal of Official Statistics*, 10, 2, pp. 197–214.

Pafford, B.V., and Coulter, D. (1987), "The Use of Historical Data in a Current Interview Situation: Response Error Analysis and Applications to Computer-Assisted Telephone Interviewing," *Proceedings of the Third Annual Research Conference*, U.S. Bureau of the Census, pp. 281–298.

Phipps, P.A., and Tupek, A.R. (1991), "Assessing Measurement Errors in a Touchtone Recognition Survey," *Survey Methodology*, 17, pp. 15–26.

Presser, S. (1983), "Discussion of papers by House and Morton and by Morton and House," *Proceedings of the Section on Survey Research Methods*, American Statistical Association, pp. 142–143.

Rogers, P. (1991), "1989 and 1990 Current Population Survey CPS/CATI Reinterview Response Variance Report," a memorandum of the U.S. Census Bureau, Demographic Statistical Methods Division.

Rosen, R.J., Clayton, R.L., and Wolf, L. L. (1993), "Long Term Retention of Sample Members Under Automated Self-Respose Data Collection," *Proceedings of the Section on Survey Research Methods*, American Statistical Association, pp. 748–752.

Rustemeyer, A. (1977), "Measuring Interviewer Performance in Mock Interviews," *Proceedings of the Social Statistics Section*, American Statistical Association, pp. 341–346.

Saris, W.E. (1990), *Computer-Assisted Interviewing*, Newbury Park: Sage Publications.

Saris, W.E., and de Pijper, W.M. (1986), "Computer Assisted Interviewing Using Home Computers," *European Research*, 14, pp. 144–152.

Scherpenzeel, A.C. (1995), *A Question of Quality: Evaluating Survey Questions in Multi-Trait-Multimethod Studies*, Leidschendam, Royal PTT, The Netherlands.

Schuman, H., and Presser, S. (1981), *Questions and Answers in Attitude Surveys: Experiments on Question Form, Wording, and Context*, New York: Academic Press.

Schwarz, N., and Hippler, H.-J. (1991), "Response Alternatives: The Impact of Their Choice and Presentation Order," in P. P. Biemer, R. M. Groves, L. E. Lyberg, N. A. Mathiowetz, and S. Sudman, (eds.), *Measurement Errors in Surveys*, New York: Wiley, pp. 41–56.

Sedivi, B., and Rowe, E. (1993), "Computerized Self-Administered Questionnaires by Mail or Modem: Initial Technical Assessment," a report to the CASIC Committee on Technology Testing, U.S. Bureau of the Census.

Sigman, R., Dorinsk, S., and Tinari, R. (1993), "Comparison of Telephone Interviewing Methods in the Quarterly Apparel Survey," ESMD Report Series ESMD-9302, U.S. Bureau of the Census.

Sudman, S., and Bradburn, N. M. (1974), *Response Effects in Surveys*, Chicago: Aldine.

Tortora, R.D., (1985), "CATI in an Agricultural Statistical Agency," *Journal of Official Statistics*, 1, 3, pp. 301–314.

Trewin, D., and Lee, G. (1988), "International Comparisons of Telephone Coverage," in R.M. Groves, P.P. Biemer, L.E. Lyberg, J.T. Massey, W.L. Nicholls II, and J. Waksberg (eds.), *Telephone Survey Methodology*, New York: Wiley, pp. 9–24.

U.S. Bureau of the Census (1995), *Current Population Survey–November 1994, Computer Ownership/Usage Supplement*, Table 1.

U.S. Energy Information Administration (1989), "PEDRO—Respondent User Guide to the Petroleum Electronic Data Reporting Option," Version 3.0, U.S. Department of Energy.

Van Bastelaer, R.A., Kerssemakers, F., and Sikkel, D. (1988), "A Test of the Netherlands' Continuous Labour Force Survey with Hand-Held Computers: Contributions to Questionnaire Design," *Journal of Official Statistics*, 4, pp. 141–154.

Weeks, M.F. (1988), "Call Scheduling with CATI: Current Capabilities and Methods," in R.M. Groves, P.P. Biemer, L.E. Lyberg, J.T. Massey, W.L. Nicholls II, and J. Waksberg (eds.), *Telephone Survey Methodology*, New York: Wiley, pp. 403–420.

Weeks, M.F. (1992), "Computer-Assisted Survey Methods Information Collection: A Review of CASIC Methods and Their Implications for Survey Operations," *Journal of Official Statistics*, 8, 4, pp. 445–465.

Weller, G.D., (1994), "The 1994 Current Population Survey Response Variance Report," a memorandum of the U.S. Census Bureau, Demographic Statistical Methods Division.

Wojcik, M.S., and Baker, R.P. (1992), "Interviewer and Respondent Acceptance of CAPI," *Proceedings of the Annual Research Conference*, U.S. Bureau of the Census, pp. 619–621.

CHAPTER 10

Developing a Speech Recognition Application for Survey Research

Bill Blyth
Taylor Nelson AGB plc, London

10.1 INTRODUCTION

Automatic Speech Recognition (ASR) is a new, rapidly developing and, for survey research, largely untried technology. The purpose of this chapter is to discuss the potential types and scales of application of this technology, based on the development and trial of an interactive ASR interview prototype.

Firstly, the major ASR variants are described for the lay reader in so far as they differ in their potential for survey research. Secondly, consideration is given to the questions that ASR use raises for survey design and methodology. Thirdly, some of the collaborative program of research between Marconi Speech & Information Systems, The Hirst Research Centre and Taylor Nelson AGB is described with regard to questionnaire design. Finally, conclusions are drawn as to the future development and application of ASR to survey research.

10.1.1 How ASR Works

ASR can be carried out in a number of ways. The common approaches are detailed below.

10.1.1.1 The Basic Method
Automatic Speech Recognition is what its name says—the automatic recognition by a machine—some form of data processing intelligence—of human speech, i.e., words. The physiology of human speech generation is extremely

Survey Measurement and Process Quality, Edited by Lyberg, Biemer, Collins, de Leeuw, Dippo, Schwarz, Trewin.
ISBN 0-471-16559-X © 1997 John Wiley & Sons, Inc.

complex. Indeed it is so complex that no two individuals speak exactly the same, although at times different individuals can sound the same to human listeners. However, the basis of communication is that speakers of a common language can understand each other if they are from common or similar accent regions.

The acoustic waveforms that make up the sounds of human speech can be represented through time using a spectrogram. The phonemes into which the word is broken down are the smallest units of speech whose absence or substitution might make a difference of meaning to the word. We shall return to phonemes below. ASR works by recognizing the *pattern* that the spectrogram of a distinct word makes. First the computer is *trained* to recognize patterns of different words by repeated iteration. Subsequent test words are compared with a databank of trained models and the pattern that makes the closest fit is identified. The accuracy of the recognition is the percentage of times that such matching is correct.

A number of different pattern matching techniques are in use. The recognition approach used by the software we have employed was based on Hidden Markov Models.

10.1.2 ASR Variants

The concern of this chapter is not with the details of the mathematics of different ASR modeling techniques. Rather we are concerned with the application of ASR in practice. However, some aspects of the underlying concepts affect the applications that are possible. Some key factors are listed below.

Speaker Dependent or Independent
Because of the way humans are physically built, the basic way a specific individual speaks does not vary a great deal. Thus for an individual the patterns made by each word are very consistent. To *train* a speech recognizer for one specific speaker requires only three or four repetitions of each word (*word tokens*) we wish to use. This is called *speaker dependent* modeling.

However, the variability between two speakers is considerable. Men, women and children all have different ranges. Regional or class accents also vary considerably. Thus the sample of *tokens* of each word that is required to produce reliable models is much greater, and to maximize accuracy different models may be needed for regions or sexes. Such models are called *speaker independent*. They require considerably more effort and cost.

Speech Medium
A speaker who is physically present at the recognizer gives a complete speech spectrogram. However, if the speaker is not physically present—and in most *speaker independent* applications this will be the case—the speech will have been transmitted by some medium. The medium used for the transmission will affect the spectrogram in some way—typically the telephone will truncate the upper

and lower parts of the frequency band. These parts of the frequency band are important for distinguishing some phonemes and their truncation increases the difficulty of recognition substantially.

Background Noise

Speech recognizers are not programmed to recognize only the human voice. Any sound that is present during speech offering will be included in the pattern. Thus silence—the gap between words—is another pattern which needs to be modeled and recognized. Consideration must be given to ensure low background noise.

Whole Word or Sub-Word Modeling

Initially speech scientists approached recognition modeling by analyzing the spectrographic patterns of whole words. Whilst this approach is direct and suited to high-volume, unchanging applications of speech recognition, it lacks flexibility. Every time a model is required for a *new* word data have to be collected to enable the model to be built. In the case of speaker dependent modeling this task is not too difficult. However, in the case of speaker independent modeling the data collection exercise is relatively large, takes time and increases costs.

To overcome this difficulty some researchers working in the ASR area have developed sub-word modeling. This approach models smaller units than words. A number of different approaches are possible and in use. The key point of this approach is that rather than train for individual words one can build up a sample database of different sub-word units which is comprehensive. In this fashion once the models have been built for each of the sub-word units models can be constructed for any whole word model that one desires.

Our collaborators at GEC Marconi have been developing the theory underlying sub-word modeling over a number of years.

Vocabulary Size

The matching of speech against a bank of word models involves a substantial number of calculations. The larger the possible recognition vocabulary, the greater the number of calculations. As vocabulary increases, either the time taken to do the necessary computing increases or the size of the available processor must be increased. This has cost implications.

It is possible to have different vocabularies at different stages of the word matching processing if the necessary software support has been put in place, and this will reduce the recognition matching load.

Early examples of speaker independent ASR were able to use only small vocabularies. What is meant by vocabulary is the number of potential words that can be recognized at any one point in a dialogue. The greater the number of words that might be offered at any point in a dialogue, the greater the number of necessary computations. Unless the models are simplified or the processors' power increased, this manifests itself in lengthy pauses whilst the data are digested. Such breaks in a dialogue artificially increase the time required and inhibit user rapport.

Improvements in software and hardware have largely eliminated the restrictions on vocabulary size. In our testing we used a list of over two hundred alternatives as possible answers to one question without difficulty. This is exceptional and was designed specifically to test the ability of our system to handle large vocabularies. Most survey questions have fewer than ten pre-coded answers and thus vocabulary size is not a great issue. The ability of the software to filter out extraneous words or noises is more important and is the focus of much software development.

10.2 ASR VARIANTS AND SURVEY RESEARCH

ASR is a rapidly evolving technology. As with all such evolutionary technologies when you start generally determines where you start. However, the immediate public face of ASR—speaker dependent, physically present applications—is not the route that survey research will probably take. These early applications— PC command systems, hands free word processing, will in all likelihood give a misleading impression of the tractability of the technology for survey applications.

Survey applications will certainly require speaker independent models. Similarly they will be distanced applications—telephone or the like.

When we were introduced to ASR technology in 1990 we were unable to find any published references to work in this area that had been carried out by survey researchers. Thus in designing our project we first decided upon principles and let the development work emanate from those principals. Our experience in designing effective interviewing methods weighed in heavily. In the interim, work on ASR has developed simultaneously on a number of fronts: survey research, customer support, sales and ordering systems. Practitioners are starting to publish material and it is apparent that similar principles of application design are emerging across different areas and languages.

However, at the time of writing (other than our own work), the only published references we have found to the use of ASR for survey research still are by the U.S. Bureau of the Census (Appel and Cole, 1994; Nicholls and Appel, 1994; Jenkins and Appel, 1995). In Europe we have been told of some limited applications of the *grunt–no grunt* variety of speech recognition. However, this lowest level of modeling is unsuitable for viable survey application.

Nicholls and Appel (1994) distinguish between small (digits or yes/no), medium (up to 100 words) and large (100–1000 words) vocabulary ASR applications. Our own work went straight to large vocabulary applications because our technology partner had such systems already available.

Clearly ASR could be used across a wide range of survey applications: as an alternative to touch-tone data entry (TDE) when collecting basic numeric data; for use by interviewers telephoning results in from the field; for large scale administrative data collection; as an alternative to self-completion paper diaries in panel research.

If the technology can be made good enough to mimic a human interviewer then the range of possible survey research applications is extremely wide. Our concern has been how that should be achieved.

10.2.1 Key Issues

Successful survey research application of ASR requires modern telecommunications networks and low-cost and reliable technology. Robust, speaker independent, large vocabulary modeling algorithms will be needed. Wide use will demand tractability, respondent acceptance and speedy and straightforward implementation tools. Finally, applications must be demonstrably replicable.

In our development the technology requirements were all assumed. The availability of suitable telecommunications networks varies internationally but is not a problem for most industrialized countries. The cost and reliability of the technology are improving continuously. Of greater concern is the existence of an agreed set of standards enabling complete inter-operability between products from different suppliers. However, there are encouraging signs that this issue is being addressed and will be eliminated somewhat quicker than it has been in other areas of technology.

Part of our work was to test the modeling algorithms but here we assumed their adequacy and designed an application based on that assumption. Our priority has been to understand the factors that will affect the acceptability of ASR by respondents and from this identify what is required to meet the last two needs: straightforward implementation and replicable applications.

10.3   MAN–MACHINE DIALOGUE ISSUES

At the heart of the question regarding how ASR might or should be used in survey research is the issue as to whether it is regarded as an interviewing technique or rather as some form of data entry for self-completion questionnaires. A number of subsidiary issues interact with this, including the facilities of the ASR system being used, its sophistication, the familiarity of the respondents with ASR, the readiness of the researcher to discard interviews or introduce human editing and, of course, the complexity of the survey. However, it is our belief that, irrespective of these subsidiary issues, it is the interviewer versus data entry perspective which is the most important and which will determine both future usage and development.

For the particular trial development reported here, our approach has been to use ASR as an interactive device: to attempt to make the respondents feel as though they are talking to another real person! In doing this the human factor considerations are extremely complex. At their root is the fundamental question of how humans communicate. The study of this area is developing rapidly and involves input from many different scientific disciplines. Few of these are currently familiar to survey researchers and they can be hard to understand. However, when looked at in a different light, what is occurring is the systematic analysis and description of the most basic and innate skills of survey researchers—phrasing, asking, and listening. A general overview of the human factors issues in ASR, together with a large bibliography, is given by Karis and Dobroth (1991) and in the report of the European Union funded EAGLES project (1994).

To consider the application of ASR to survey research, it is easier to discuss the issues by going back a stage or two. Communication between two individuals who are physically present depends on a whole host of factors in addition to speech: facial expression, posture, hand movement, interaction and so forth. Familiarity introduces learning and often succinctness into this process. Between strangers, conversation provides a greater structure and a greater emphasis on words rather than other physical manifestations. However, gender, appearance, accent and posture will all affect the style and content of communication. In the interview context the transactional and value aspects become more important and it is their explicit or implicit exploitation in survey design, interviewer selection, training and questionnaire design that produce *good* survey research.

With telephone interviewing the visual cues are removed but telephone interviewing develops its own range of techniques to both recognize respondent *types* and facilitate the interview administration. A telephone interviewer will instinctively classify respondent sex, broad age band, any regional accent, an indication of intelligence or education based on vocabulary, grammar and syntax, and other personal information given as part of the interview itself or an aside or embellishment to an answer. The tone of the respondent will give some indication of their general personality characteristics as well as some indication of feelings about being interviewed.

However, in everyday life one rarely *listens* to a conversation. One is either part of it—reacting instinctively—or a listener, overhearing the conversation of others. Even in the latter situation one rarely listens to the essentials of the conversation itself but rather to the content of the conversation and what it tells us about the lives and experiences of the participants. A consequence of this absence of listening to conversation is that we have no way to describe it in any transitive fashion. We have no way of portraying a good or a bad conversation.

A conversation, even one which is as relatively formalized as a telephone interview, is an extremely complex transaction. Speed, tone, intonation, and accent all play a part in any telephone conversation. In ASR applications what is done instinctively by one speaker responding to another has to be replaced and formalized into a set of instructions for a computer to follow. Thus when two people are talking, one to the other, it may very well be acceptable for one to interrupt the other. Typically that is the way a dialogue would be programmed, even when it is inappropriate. How do we resolve this problem?

We believe that successful interactive man–machine dialogue—as opposed to command or instructional systems—will only occur if we can design dialogues that incorporate the key parts of everyday interpersonal speech. The problem is directly analogous to music. A musical score tells us which *notes are being played*, by which instrument, when, how fast, how loud, and with what tone. A musical score also tells us *what is not playing*—how long the silences are. Most people cannot read music, but they do know when something is out of tune. There is general accord about what is a good or catchy tune. Those who are trained to read music can, if they are fortunate, hear the music in their minds and, by a mix of experience, training and rules, know whether what is portrayed is in harmony or not.

10.4 SETTING UP AN ASR SURVEY

The bulk of the work involved in setting up an ASR survey is in the dialogue design and implementation. Figure 10.1 shows sequentially the steps involved in the process. For our project development this involved a very close liaison with both the speech recognition scientists and the engineers providing the prompts, recognition device, and telecommunications platform. This was a necessity, given the state of the art; in the medium term, the need for such close liaison and involvement will disappear.

10.4.1 Dialogue Design

Our objective is to conduct telephone interviews without the interviewer. ASR replaces the ear but not the brain; the dialogue software has also to carry out the tasks of the brain. Our approach to dialogue design has been instinctive and has drawn on our long experience of designing questionnaires, interviewing, and questionnaire analysis. Since we completed the work reported here we have become aware of the beginning of more formalized dialogue design protocols which have their roots in speech science (Fraser, 1994).

The interviewer not only carries out the observed acts, but is continuously carrying out the hidden acts (see Figure 10.2). Similarly our ASR technology and supporting software must do both. The list in detail is considerable.

The starting point for an ASR survey dialogue is a normal questionnaire design task: deciding which questions to ask; wording and order; specifying pre-codes and routing; testing logic flow forwards and backwards. Then, new tasks arise.

From the *observable*, for example, we have to consider the preamble. We have to choose the type of voice to be used—synthetic, male or female; if human, the age, regional accent or other classifiers. We have to decide: how quickly and how loudly do we ask; how much time do we allow for response; how and

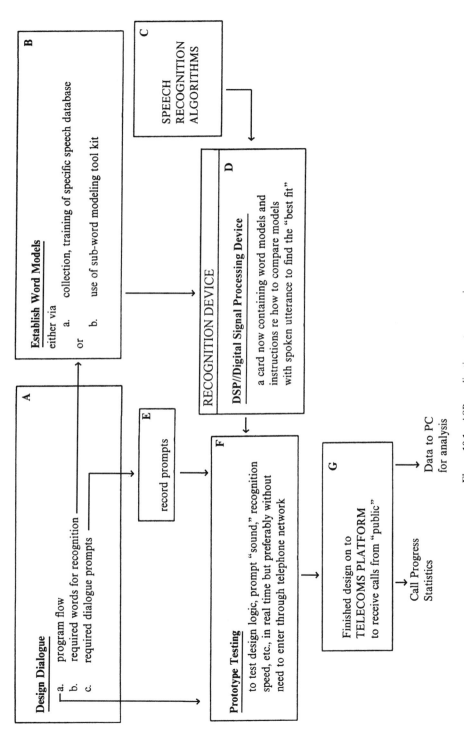

B

Establish Word Models
either via
 a.   collection, training of specific speech database
 or
 b.   use of sub-word modeling tool kit

C

SPEECH
RECOGNITION
ALGORITHMS

A

Design Dialogue
 a.   program flow
 b.   required words for recognition
 c.   required dialogue prompts

D

RECOGNITION DEVICE

DSP//Digital Signal Processing Device

 a card now containing word models and
 instructions re how to compare models
 with spoken utterance to find the "best fit"

E

record prompts

F

Prototype Testing

to test design logic, prompt "sound," recognition
speed, etc, in real time but preferably without
need to enter through telephone network

G

Finished design on to
TELECOMS PLATFORM
to receive calls from "public"

→ Call Progress
 Statistics

→ Data to PC
 for analysis

Figure 10.1   ASR application set-up overview.

256

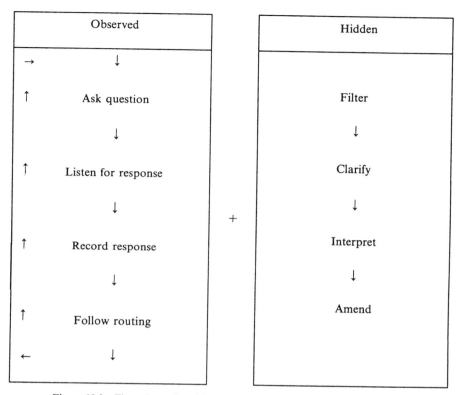

Figure 10.2   Flow chart of problems that can occur in ASR dialogue design.

when do we prompt? From the *hidden* actions, we must determine recognition criteria and ways of confirming recognition, e.g., *"did you say Devon?"* We need ways to cope with problems: reacting to no response, retrieving cases of non-recognition, providing escape routes for respondents.

This list is by no means comprehensive but it is illustrative of the issues that arise in ASR dialogue development. The principles of piloting and feedback apply equally in ASR survey research. The challenge is the identification of the best approach. Possible methods are described in the conclusions section.

10.4.2   Question and Prompt Recording

Question and prompt recording is an extremely important area for survey applications. For some applications e.g., directory enquiries or electronic banking, users may either have no alternative or, because they are highly committed to the other benefits the system provides—time saving, convenience, cost reduction—they are indifferent as to whether the system sounds human and interested or robotic and monotone. Until technology diffusion has made

ASR applications as commonplace as remote TV controls or answering machines, to attract and engross a sample selected to be representative of a population, the experience has to be made as pleasurable as possible. Questions should be asked with the correct inflexion and the voice should be warm, friendly and unforbidding: we have found that a woman's voice is better. Since few people have all the qualities required for good speaking, it may be desirable to use an actor or someone with speech training. No repeated prompt should sound identical to a previous prompt; it should be rephrased or additional words or guidance should be added. Testing reactions to the prompt *sound* and *tone* is crucial. A preamble should include some instruments for system control—e.g., repeat, stop, and go back.

10.4.3 Prototype Testing

Prototype testing is more than logic testing. It is a test flight which starts off doing the simple things and then goes all the way through all the difficulties that can occur in real life. Some of these difficulties will occur naturally during the testing. The opportunity for others may not arise and they need to be artificially introduced, e.g., high levels of background noise; purposefully awkward respondents; telephones left off receivers after call termination. Exhaustive prototyping is crucial in an ASR application. Errors or incomplete programming may be impossible to retrieve once the main research program has started and may be undetected unless every call is being monitored. Since in actual applications this would be self-defeating, the testing, combined with the progress statistics and quality control procedures have to suffice.

10.4.4 Progress Statistics and Quality Control

Care needs to be taken in specifying the scale and nature of the progress statistics and quality control. Listening in to a random sub-sample of calls is useful but, in addition, one wants data about numbers of attempted and completed calls, flow statistics through the system, and recognition levels at key questions. Unfortunately, until a body of case study experience has evolved, the derivation of guidelines and benchmarks for this will have to be subjective.

10.5 PILOT STUDY DESIGN AND EXECUTION

For our test amongst members of the public we wanted a survey designed to test areas of specific interest to us and our partners. We wanted to test complex dialogue with large vocabularies developed using sub-word modeling. We wanted to test over a broad cross section of the population who had no previous experience of ASR. We wanted a topic that most adults could talk about; which was unintrusive and interesting. Eventually we chose the subject of holidays.

10.5.1 Sample Design and Contact

We were not concerned to test the population's overall attitude to survey research. Our interest is in ASR compared to other *successful* data collection techniques. Because of this and because of the need to have a controlled demographic balance we decided to draw our sample from adult women who had been interviewed for another survey by telephone in the previous 12 months. The sample was restricted to women aged 16–64, from three social grades: B, C1, C2, living in London and the South East of England. The purpose of the sample refinement was to reduce the variability in accents, so requiring only one set of word models, and to maximize the probability of holiday taking and thus eligibility for participation. Respondents were contacted by telephone, the project was described, and respondents were asked whether they would be willing to participate. If they agreed, they were sent a briefing document. No incentive was employed. The briefing document (see Appendix) is extremely succinct. Lessons from earlier development testing suggested the need to minimize the high-technology nature of the exercise. That this was eventually successfully achieved was indicated by a number of respondents saying "*thank you*" and "*goodbye*" at the end of the interview.

10.5.2 Dialogue Design

Prior to the adult pilot study we had carried out a small scale pilot study using interviewers as respondents. The results of this work have been described in earlier papers by Blyth and Piper (1994a, 1994b). From this work we developed a number of principles and software models which we could apply in our adult pilot study. Amongst these were that prompts should be short to be understood and acted upon. Respondents should be provided with a number of basic control words enabling them to control the system where necessary: "repeat," "help," and "stop." A sense of progression through the call was needed: letting respondents get a feeling of being absorbed into a never-ending loop induces anger, despair, or mental claustrophobia. This can be avoided by eliminating unvarying repetition or by adding some small piece of additional information as you proceed.

As an example, in prompting for confirmation of a recognized interviewer serial number, the original message would be: "*Is that , 1 2 3 4 5 ?*" If the respondent says "*Repeat*" the next system prompt would be: "*Is your interviewer number ,12345 ?*" Questions must be designed such that the range of valid responses contains as few confusable words as possible. It is not possible to design questions whose valid responses are always unambiguously recognizable. The level of confusability varies, not surprisingly, depending on the words which are used. For any set of words a *confusion matrix* can be derived from the sub-word models. This gives the probability with which one word can be recognized as another. The semantic differential ranking used in our pre-pilot employed five words with very low intra-confusion potential: *Terrible, Poor, Fair, Good, Excellent.*

It is possible for the dialogue designer to evaluate the risk of confusion against the importance of the response. Response can be confirmed by simply saying: "*I heard , 1 2 3 4 5. Is that correct?*" But to be repeatedly asked if you said what you said can be extremely irritating. Such response confirmation breaks the flow of the questionnaire as well as increasing length and cost. We have therefore reserved confirmation for key digit strings and countries. Where substitution error has occurred strategies for correction can vary. In the example of the five digit serial number above, if the respondent says recognition is incorrect, our strategy has not been to ask for repetition of the string, but rather ask, for each individual number sequentially, if it is correct, and if it is incorrect, substitute the next highest probable match from the confusion matrix.

Our objective in handling the strategy in this manner has been to avoid implying that the respondent is in some way wrong or faulty. Rather the computer *system* has made a simple mistake. Building up respondent confidence is crucial for what we regard as the key commercial applications of the technique.

10.5.3 The Pilot Questions

The pilot questions were straightforward. They can be summarized and paraphrased as:

1. Have you had a holiday so far this year?	Yes	→	3
	No	→	2
2. Are you planning a holiday in the remainder of this year?	Yes	→	3
	No		Close
3. Where did/will you go?		→	4
4. Have you been before?		→	5
5. Was it/will it be a package?		→	6
6. How will you go?		→	7
7. Length of holiday			Close

The full flow chart for this questionnaire runs to over 10 pages. The questionnaire itself takes under two minutes to answer. An extract from the flow chart is shown in Figure 10.3.

All the questions, prompts and reiterations were recorded in the appropriate future tense. Trips abroad were filtered sequentially; Europe, U.S.A., and rest of world. This reflects the incidence of holiday taking and also reduces the potential for recognition confusion. In addition to full country names we also included the names of popular destinations, e.g., Majorca, to minimize repetition. However, one cannot always have perfect foresight. In the method of travel question we omitted "aircraft" from the list of eligible recognizable responses. Clearly, over time, experience builds up a lexicon of probable valid utterances. In this particular pilot study, in order to let the flow go as smoothly as possible we used minimum levels of confirmation. Thus, for example, on the "which country visited" question answers were divided into those with high probability to visit and low probability. Confirmation as shown in the flow diagram was sought only for those of low probability. We have a particularly good recording of a journalist trying the system in a forthright manner getting the response— "*did you say Uzbekistan?*"—yes!

10.5.4 Pilot Results

Some 150 interviews were conducted in July and August 1994. Tapes were made of all calls whether successful or only attempted. Overall the quality of the results was surprisingly high and much better than we had anticipated. Recognition itself was not a problem; countries such as Thailand or Ireland were successfully distinguished and identified. Over 60 percent of the interviews were usable without editing or transcription.

Less than 5 percent of respondents had significant problems. Listening showed that these were less a manifestation of technology-phobia but more a misapprehension that ASR requires one to speak as if to an aurally and intellectually disadvantaged child: too slowly and too loudly.

The problems that did occur were in the mechanistic application of the dialogue design, for example, when respondents started talking before the question was finished, or alternatively pausing longer than *allowed* before answering. One instance was caused by the anticipated answer set vocabulary being incomplete—*aircraft* as a transport mode had been omitted.

Whilst the manner in which the pilot sample was drawn prevents complete generalization, we concluded that ASR will be viable for survey research data collection and that ASR application will not be demographically restricted.

10.6 THE FUTURE

Doing and hearing is believing. The potential for ASR in survey research and elsewhere seems great, perhaps sometimes frightening in a Huxleyesque view of the future. It is necessary to damp down our enthusiasm and reflect objectively. What do we actually know?

We know that ASR is a viable technology whose cost is decreasing; that standardization and inter-operability are close; that ASR is multi-lingual. It provides the opportunity for consistently applied *interactive* interviewing and a single site could cover the world. But application is an art form, not a science; researchers need new skills and resources to use it. ASR has relatively high capital cost and low running cost (compared with CAPI's high capital and high running costs).

There remain significant barriers to development. We need open architecture telecommunications platforms, standard chip design for alternative word models and comprehensive and representative word databases. These are ASR technology issues. Beyond these, for everyday ASR survey research applications

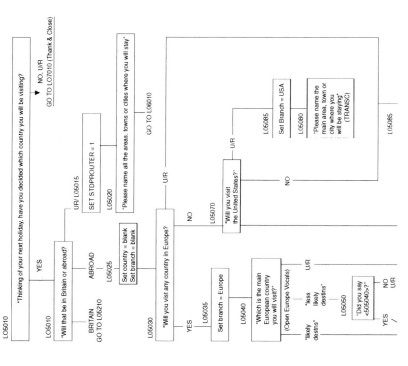

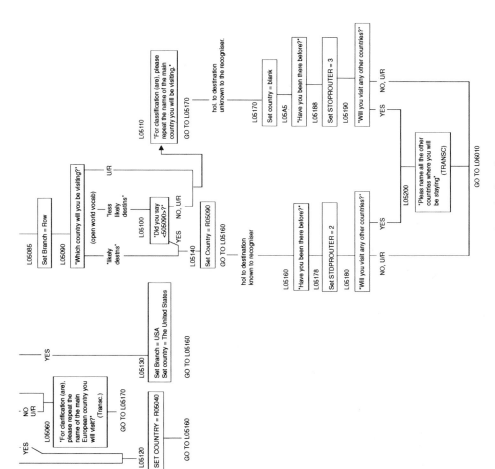

Figure 10.3 Extract from questionnaire flow chart.

263

we need flexible, low cost, and *expert* dialogue design tools. The commercial world has short lead times for getting surveys into the field. This requires dialogue design to become akin to CATI questionnaire design, i.e., a relatively simple task that does not require advanced linguistic skills.

ACKNOWLEDGMENTS

I would like to thank Reg Baker, Mick Couper, Jean Martin and Bill Nicholls for their comments and suggestions. I also wish to thank Heather Piper for her effort and determination in turning a concept into a reality and the efforts of our collaborators at GEC Marconi.

REFERENCES

Appel, M.A., and Cole, R. (1994), "Spoken Language Recognition of the Year 2000 Census Questionnaire," paper presented at the Annual Conference of the American Association for Public Opinion Research.

Blyth, W.G., and Piper, H. (1994a), "Speech Recognition—A New Dimension in Survey Research," *Journal of The Market Research Society*, 36, 3, pp. 183–204.

Blyth, W.G., and Piper, H. (1994b), "Developing Speech Recognition Applications for Market Research," *Proceedings of the Annual Conference of ESOMAR*.

EAGLES: Expert Advisory Group on Language Engineering Standards (1994), Interim report, EU Framework 3.

Fraser, N.M. (1994), "Spoken Language Understanding: The Key to Designing the Perfect Dialogue for Order Processing," *Proceedings of Voice*, Cologne.

Jenkins, C.R., and Appel, M.A. (1995), "Respondents' Attitudes Towards a U.S. Census Voice-Recognition Questionnaire," paper presented at the International Field Directors & Technologies Conference, Deerfield Beach, Florida.

Karis, D., and Dobroth, K.M. (1991), "Automating Services with Speech Recognition over the Public Switched Telephone Network: Human Factor Considerations," *IEEE Journal on Selected Areas in Communication*, 9, 4.

Nicholls II, W.L., and Appel, M.A. (1994), "New CASIC Technologies at the U.S. Bureau of the Census," *Proceedings of the Section on Survey Research Methods, American Statistical Association*, pp. 757–762.

APPENDIX

TAYLOR NELSON	**VOICELINE** - Instruction Card

Here's what to do:

* **CALL FREEPHONE 0800 563**
  and answer a few short questions on this year's holidays

* **Your REFERENCE NUMBER is**

  <u> 4 </u> . <u> 8 </u> . <u> 5 </u> . <u> 5 </u>

Helpful Hints:

* Just speak naturally, but with **SHORT ANSWERS**, not sentences - say "YES" or "SPAIN" rather than 'yes we did' or 'We went to Spain this year'!

* Please let the question finish before you begin to answer - there should be plenty of time

* If you need to, try saying

  REPEAT
  HELP   or
  STOP -   this if you've made a mistake: it will take you back a few questions to let you try again.

Figure A.10.1 Respondent instructions.

VOICELINE - HOLIDAYS 1994

British Regions Map

	Code	Region
Scotland		
	No. 1	Highlands and Islands
	No. 2	Grampian
	No. 3	Tayside
	No. 4	Central
	No. 5	Fife
	No. 6	Strathclyde
	No. 7	Lothian
	No. 8	Dumfries and Galloway
	No. 9	Borders
England		
	No. 10	Cumbria
	No. 11	Northumbria
	No. 12	North West
	No. 13	Yorkshire
	No. 14	Heart of England
	No. 15	East Midlands
	No. 16	East Anglia
	No. 17	London
	No. 18	The West Country
	No. 19	Southern
	No. 20	South East
Wales		
	No. 21	North Wales
	No. 22	Mid Wales
	No. 23	South Wales

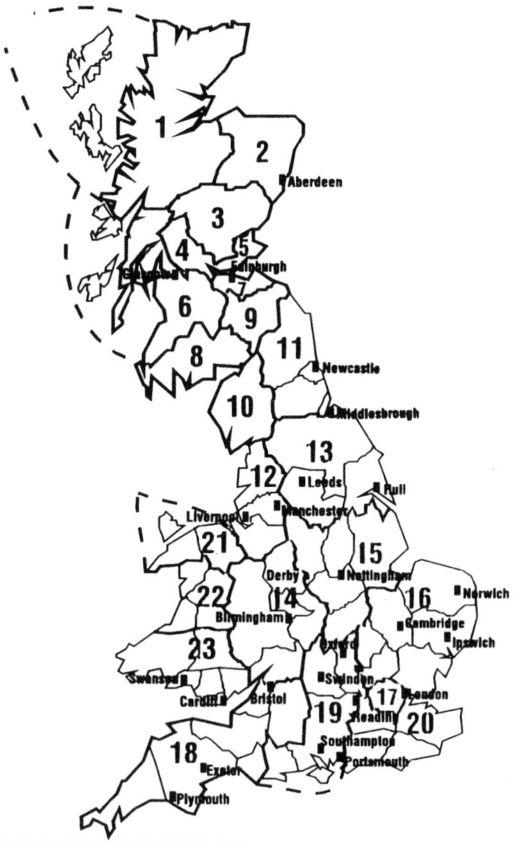

Figure A.10.2 Respondent instructions.

CHAPTER 11

Evaluating Interviewer Use of CAPI Technology

Mick P. Couper, Sue Ellen Hansen, and Sally A. Sadosky
Survey Research Center, University of Michigan

In light of the shift from paper and pencil interviewing (PAPI) to computer assisted personal interviewing (CAPI), new ways need to be found to evaluate interviewer performance. We focus here on the effects the CAPI instrument may have on interviewer performance. Other important features of an interviewer's work such as the ability to gain cooperation, production efficiency and interviewing skills will not be addressed. Indeed, it seems clear that the introduction of CAPI has not diminished the need for these critical interviewing skills (Couper and Burt, 1994). Nonetheless, the focus here is on the additional skills that may be required of interviewers in interacting with a CAPI system.

Traditional measures of interviewer performance on paper and pencil surveys are no longer sufficient in a CAPI environment. Not only are some of the tools no longer valid, but the nature of the task has changed, requiring new measures of performance. For example, in many PAPI surveys supervisors review completed questionnaires for legibility, completeness of responses, answers within range, correctness of skips, and so on. This is unnecessary using CAPI, as the automated instrument ensures completeness, follows the correct skip patterns and performs range and other error checks (see Nicholls *et al.*, Chapter 9). Because many of these functions are now automated, however, does not mean that the interviewer does not make errors in using the CAPI instrument. What it may mean is that many of the remaining errors have become harder to detect with conventional tools.

One set of traditional tools for evaluating interviewer performance that

Survey Measurement and Process Quality, Edited by Lyberg, Biemer, Collins, de Leeuw, Dippo, Schwarz, Trewin.
ISBN 0-471-16559-X   © 1997 John Wiley & Sons, Inc.

works for both PAPI and computer assisted surveys is behavior coding of taped interviews in face to face surveys (Cannell *et al.*, 1975) and monitoring of telephone survey interviews (Couper *et al.*, 1992). These methods yield relatively rich information on the interaction between the two humans engaged in the interview (the interviewer and the respondent), but provide little if any insight into the interaction between the human (primarily the interviewer) and the computer. Ideally, a CAPI system (both the hardware and software) should be an unobtrusive presence in the interview, a tool to facilitate the smooth and successful completion of the interview. To the extent that interacting with the computer causes difficulty for the interviewer, it may intrude in the interaction between interviewer and respondent. In the same way that difficulties in the interviewer–respondent interaction may harm the data obtained (see, for example, Suchman and Jordan, 1990), so too may the interaction between interviewer and computer in a CAPI survey.

How can the interaction between interviewer and computer effectively be studied? Laboratory procedures are commonly used in software evaluation to observe the interaction between user and computer, as they are for the interviewer–respondent interaction. However, these are largely artificial environments in which the user or subject is in a heightened state of awareness, a situation which threatens the generalizability of the findings to field settings. Observing natural interactions in a field setting is not only expensive, but the uncontrolled environment also makes comparisons across interviewers difficult. Scripted mock interviews offer some advantage over both these approaches by providing a relatively natural setting and a consistent set of stimuli to all interviewers (Rustemeyer, 1977). Furthermore, they share the advantages of laboratory approaches in allowing for embedded tests in the instrument to evaluate interviewer performance on specific aspects of CAPI use.

Computer assisted interviewing provides an additional source of data in that a record can be obtained of all keys pressed by the interviewer as he or she moves through the instrument. Although routinely captured by many CAPI systems, these keystroke files (also called audit trails or trace files) are primarily used for testing and debugging purposes (see House and Nicholls, 1988), rather than for evaluating interviewer behavior or performance. One weakness of using keystroke files for this purpose is that they tend to produce large volumes of free-format data that are difficult to reduce to manageable and meaningful levels. It is thus important to develop methods to efficiently manage these data to aid our understanding of interviewer–computer interaction and the types of errors that are likely to occur.

Although CAPI surveys require different methods of evaluating interviewer performance than PAPI surveys, they also provide in part the tools to facilitate such an evaluation. We discuss here the use of keystroke files to explore interviewer interaction with the CAPI system. First, we combine the use of mock interviews with an analysis of keystroke files. The use of a consistent set of questions and answers in the mock interview facilitates the detailed analysis of keystroke files. Second, we examine keystroke files from production interviews to explore interviewer variation in the use of various CAPI functions in a less controlled field setting.

Little is known about the types and frequency of errors interviewers make when using a computerized survey instrument. Identifying the frequency, type and source of such errors can be used to inform efforts to reduce the errors made in using CAPI, whether through training or the design of more effective CAPI instruments (see Couper, 1995). The focus is thus on an evaluation of the utility of these tools for measuring and evaluating interviewer performance in computer assisted interviewing.

11.1 ERRORS IN HUMAN–COMPUTER INTERACTION

Why should we be concerned about interviewer errors in a CAPI survey? There is some evidence that computer assisted interviewing improves data quality over paper and pencil data collection (see Nicholls et al., Chapter 9), and indeed this is why many have adopted the new technology. CAPI instruments are powerful tools in automatically routing the interviewer through the instrument, in performing range and consistency checks, and in ensuring that all applicable items are completed. Interviewers are clearly able to comply with the requirements of computer assisted interviewing, but there is little information on the extent to which they are doing so efficiently or in the ways intended by the survey designer.

Studies of interviewer keying errors in CATI (Kennedy et al., 1990) and CAPI (Dielman and Couper, 1995) reveal that, for fixed response or closed-ended questions, keying errors that are undetected during the interview are extremely rare. However, studies of computer use in other settings (e.g., spreadsheet use, word processing, programming) reveal that errors occur far more frequently than might be expected (see Shneiderman, 1992). One reason for this apparent contradiction may be that in computer assisted interviewing a narrow definition of errors is used, referring only to the outcome of the task (i.e., the final data set) rather than the process. Another reason may be the low complexity of the computer assisted interviewing task relative to other uses of computers. However, even in relatively closed systems like CAPI where the interviewer has little control over the process and a relatively limited range of actions to perform, errors can have an effect on data quality, efficiency and user satisfaction. To the extent that an interviewer is thwarted in the attainment of a particular goal (e.g., changing a previous answer), the result may range from increased frustration to abandonment of the original goal (e.g., leaving the answer unchanged).

Given this, it is important to investigate the interaction between interviewer and computer in CAPI in order to explore the kinds of difficulties interviewers have and what they do to overcome them. Focusing only on the data recorded in the completed interview may underestimate the problems interviewers face, whether through shortcomings in the instrument itself, interface problems between the interviewer and the system, insufficient training, or inadequacies on the part of the interviewer.

We thus employ a broad view of error, by which we mean any (temporary) non-attainment of a goal. We focus on all types of difficulties interviewers experience as they interact with the computer during a survey interview, regardless of whether or not the end result is a correct answer. This includes initial errors that are detected and corrected, as well as other suboptimal paths to goal attainment or inefficient use of the CAPI functions available to the interviewer.

A number of error classification schemes have been proposed for evaluating human computer interaction (e.g., Reason, 1990; Rasmussen *et al.*, 1987). Many of these follow Norman's (1983) distinction between mistakes and slips. A slip occurs when a correct intention is executed wrongly, while a mistake is an incorrect intention where the action conforms to the intention. For example, if a person deletes a file accidentally by pressing the wrong keys, it is a slip. If a person intends to delete a file, but later realizes it is needed, a mistake has been made (see Frese and Altmann, 1989). Further elaborations of this scheme have been proposed. For example, Zapf *et al.* (1992) distinguish between three levels of action regulation: the intellectual level (at which thought errors, memory errors and judgment errors are made), the level of flexible action patterns (at which errors of habit, omission or recognition are made) and the sensorimotor level (which includes typographical errors). Mistakes typically occur at the highest level (intellectual). Zapf *et al.* (1992) note that in systems of relatively low complexity (such as computer assisted interviewing), errors at the lower levels (e.g., sensorimotor errors) are more likely to occur. It is often difficult to distinguish between the various classes of error, as to do so requires knowing the intentions of the user. Such intentions are difficult to divine from behavioral data. Nonetheless, by understanding the levels at which various types of error occur, appropriate error reduction strategies can be developed.

We examine the keystroke files from a CAPI survey within the general framework of Norman's (1983) classification in order to learn what types of error are made, or what difficulties are encountered by interviewers as they use the CAPI instrument. Although we refer throughout to interviewer errors, this does not imply that all errors are necessarily *caused* by interviewers. Errors can be attributable to the interviewer (e.g., through poor training, inattention or lack of motivation), the computer (e.g., hardware or software failures or design flaws) or the interface (e.g., a mismatch between interviewers' needs and computer requirements).

11.2 DESIGN AND DATA COLLECTION

This research was conducted as part of the first wave of the Study on Asset and Health Dynamics of the Oldest Old (AHEAD), a national survey of adults aged 70 and older conducted by the Survey Research Center (SRC) at the University of Michigan. The sample consists of approximately 9,500 households and 12,000 individuals. Both telephone and face to face interviewing were used by field interviewers, using laptop computers to collect the survey data. We refer to both interviewing modes as CAPI to emphasize the dispersed nature of the data collection. The AHEAD instrument was programmed using SurveyCraft, developed by Microtab Systems of Australia.

A total of 137 interviewers were trained for the AHEAD study in fall 1993. Shortly after training, and before the start of production interviewing, each interviewer completed an identical scripted mock interview with his/her field supervisor. Interviewers were instructed to treat these interviews as actual or "live" interviews and the supervisors were trained to play the role of the respondent to ensure that an equivalent stimulus was presented to all interviewers. This was done over the telephone, with the interviewers using the CAPI instrument on the laptop computer. In addition, each mock interview was tape recorded by the supervisor to ensure that instructions had been followed. The scripted mock interviews were designed to include a number of specific tests of both interviewing skills (e.g., probing) and CAPI functions (e.g., entering a "don't know" or refusal, changing an answer on a previous screen).

Out of the 137 interviewers trained for the AHEAD study, all completed the mock interview after training. However, complete usable keystroke files were received for only 125 interviewers; keystroke files for six mock interviews were not transmitted correctly, and an additional six were received but incomplete.

A similar loss of cases was experienced for production keystroke files. These were obtained on a flow basis from the field as part of each interviewer's regular transmission. Of the 8,223 completed AHEAD interviews, keystroke files were obtained for only 7,222 (or 88 percent) of the interviews. Of these, 754 did not start at the beginning of the interview (i.e., the interview was resumed after an earlier suspension). These cases were dropped, leaving 6,468 production keystroke files for analysis. This is an average of 47.6 cases per interviewer (standard error = 2.03).

We should note that this was the first test of saving and transmitting keystroke files in a production survey using the SurveyCraft software, and the first production implementation of the CAPI Sample Management System (CSMS) developed at SRC. Subsequent enhancements have been made to both SurveyCraft and CSMS to facilitate the collection, review and analysis of keystroke files in future studies.

Given these losses, we caution about inference to the full set of AHEAD interviews. Analyses to explore differences between AHEAD interviews with and without corresponding keystroke files reveal no differences by interviewer characteristics. The missing cases do not appear to cluster by interviewer survey or computer experience. There is a slight tendency for the missing keystroke files to come from interviews completed later in the survey period, and from those obtained from reluctant respondents. If anything, this suggests that the level of errors may be slightly underestimated from the available cases, but that comparisons of keystroke behavior by interviewer characteristics should be little affected by the losses.

11.3   ANALYSES

The keystroke files for the mock interviews were examined in two ways. First, given that a number of tests were embedded in the instrument, a coding scheme was developed to summarize interviewer performance on each of the tests. Second, summary counts of a variety of function key presses and key combinations used by each interviewer were obtained. These counts were then compared against a "template" or model keystroke file representing the most efficient route through the instrument. In the case of equally acceptable alternative approaches, the method emphasized in interviewer training was adopted for the creation of the template file.

Given the large volume of production keystroke files received and the wide diversity of situations encountered by interviewers in the field, the analysis of these cases began with aggregate counts of various keystroke behaviors. Identification of outliers then led to more detailed examination of a smaller set of production keystroke files.

These two approaches to keystroke file analysis complement each other. The first involves a detailed review and coding of each keystroke file, in similar fashion to behavior coding. This yields rich data on what happened at various points in the interview. We restrict the use of this method to the series of tests explicitly included in the mock interview and to production cases identified as containing particular problems. The second approach is less time-consuming and labor intensive, but yields data that are less rich in detail. This approach aggregates selected keystroke behaviors across interviews or interviewers. The aggregate keystroke counts can reveal how many times a particular key was pressed by an interviewer, but does not inform us whether the use of that key was appropriate at that point or even successful (i.e., achieved the desired goal). Only by a detailed examination of the surrounding keystrokes can we infer the intention of the interviewer. However, the aggregate counts can help identify interviews or interviewers that may require closer attention using more detailed procedures. Each method of analysis thus focuses on different types of keystroke behavior.

The function key and other key combinations for special functions for the AHEAD instrument are presented in Table 11.1. Other functions are available in the SurveyCraft system (such as jump back options and various editing functions); however, interviewers were not trained on these for the AHEAD study.

Table 11.1 Function Keys Used in AHEAD Instrument

Key	Function
[F1]	Invokes question-specific help (QxQ). [Esc] exits the help screen
[F2]	Invokes the comment or note field. [Enter][Enter] returns to the question
[F3]	Invokes pop-up menu in household roster screens
[F4]	NO FUNCTION
[F5]	Suspends the interview and saves the data
[F6]	NO FUNCTION
[F7]	Advances to next unanswered question after backup [F9]
[F8]	NO FUNCTION
[F9]	Backs up one question
[F10]	Restores a previously entered response (used when backing up)
[Alt D]	Don't know
[Alt R]	Refused
[Tab],[Shift Tab]	Movement between items in the roster screen
[Up],[Down],[Left],[Right]	Movement in text entry, comment fields and roster screens
[Backspace]	Destructive backspace
[Esc]	Cancels an entry or exits from help screen

11.4 EVALUATION OF MOCK INTERVIEW KEYSTROKE FILES

A detailed examination of the tests embedded in the mock interview is discussed elsewhere (Couper *et al.*, 1994). We present a brief summary here to illustrate the utility of mock interviews combined with keystroke files for the evaluation of interviewer performance.

Generally the keystroke files from the mock interview tests reveal that interviewers have little difficulty using the most common CAPI functions, even shortly after training. However, there appears to be some variation in the use of specific CAPI functions in particular instances called for in the mock interview. For example, three of the tests were designed to have interviewers use the [F1][Esc] key combination to consult question specific help. While few of the interviewers used the help screen on all three tests, only 7 percent did not use such help at all during the mock interview.

Similarly, the successful use of the [F9] key to back up and change a previous response ranged from 82 to 95 percent across three tests in the mock interview. However, 20 of the 125 interviewers who completed all three tests failed to make a correction at least once.

The mock interview keystroke files were also used to produce aggregate counts of various keystroke sequences for each interviewer for comparison against the model mock interview. This comparison reveals that certain functions (e.g., [F1] for help, [F9] to back up, and [F7] to return after backup) were used less often than expected. On the other hand, 51 percent of interviewers used the [F2] key to enter a comment more often than was required during the mock interview, while 43 percent used [F2] fewer times than required. Similarly, [F3] to pop up a list of codes was used more frequently than required by 60 percent of interviewers, and [Alt D] for a "don't know" response was used more frequently by 74 percent of interviewers.

The mock interview keystroke files can also be compared to the model keystroke file to determine how "efficient" interviewers are in using the CAPI functions. On average, interviewers used 1,563 keystrokes to complete the mock interview, fewer keys than in the model keystroke file (1,603). However, it should be noted that the latter includes detailed comments (including notes on interviewer behavior such as repeating the question, probing, etc.), as well as lengthy responses to the two open-ended questions (industry and occupation). In fact, almost a third (530/1,603) of the total keystrokes in this file are textual responses in comments or open-ended questions. Another indicator of possible inefficient keystroke use is that interviewers on average used cursor keys 45.2 times, while their use was required in only a single instance in the template file. Thus, while it appears that interviewers are able to use the critical CAPI functions shortly after training, they do so with varying frequency and efficiency. Furthermore, they may enter textual comments (marginal notes) less often than expected, but appear to take time making corrections to such entries.

11.5   EVALUATION OF PRODUCTION KEYSTROKE FILES

Very little is known about the extent to which interviewers use the various CAPI functions that are provided. Using interviewer debriefing questionnaires, Baker (1992) found that interviewers reported using the help screen less than once in every 100 interviews (0.6 percent), while in 22.4 percent of the interviews they corrected or changed a response. Other than that study, we have no information on the frequency with which various CAPI functions are used. Such information would be useful for a number of reasons. For example, we suspect that a great deal of effort is put into developing help screens and training interviewers how to use them. However, if the use of such screens is so rare, we may reconsider the investment. Similarly, we can identify the most commonly used interviewer activities, and appropriately focus our training. Unfortunately these measures do not provide such information as (a) whether the interviewer used the function successfully or not, (b) whether the function met their needs (e.g., whether the help provided was useful), (c) whether infrequent function use reflects lack of need or interviewer inability to use a function, or (d) what other functions interviewers would have liked or needed but were not

Table 11.2   Use of Various CAPI Functions in Production Interviews

Keystroke	Percent of interviews with at least one use	Percent of interviewers with at least one use	Average over all interviews	Average over first five interviews	Average over last five interviews
[F1] Help	20.6	87.5	0.43	0.73	0.28
[F2] Comment	92.0	100.0	10.57	11.91	9.42
[F3] Popup menu	66.6	100.0	7.70	7.91	7.13
[F7] Return after backup	16.2	84.6	0.35	0.52	0.28
[F9] Back up	93.8	100.0	7.47	8.05	7.30
[F10] Restore current response	30.2	94.1	1.35	1.65	1.13
[Bs] Backspace	85.9	97.8	15.53	17.40	13.31
[Esc] Cancel	84.7	97.8	5.98	6.15	6.08
Erroneous function keys	6.3	86.0	0.12	0.16	0.13
Cursor keys	—	—	47.67	55.96	41.51
All function keys	—	—	28.40	31.47	26.00
Comment text	—	—	258.00	284.70	232.20
Total keystrokes	—	—	1322.35	1369.45	1245.29

provided. Nonetheless, measures of CAPI function usage provide at least initial information to guide further research and development on optimal instrument design for CAPI, and to evaluate the effective use of these functions by interviewers.

Table 11.2 presents summary measures of various keystroke behaviors in AHEAD production interviews. The percentages in the first column suggest that the help screen is invoked far more frequently (in 21 percent of all interviews) than the Baker (1992) data suggest. The same appears true for the backup function (to correct a previous response). Furthermore, these two functions are used on average less frequently in production interviews than in the mock interviews. This is consistent with the general decline in the use of these and other function keys noted from the last two columns in Table 11.2.

The use of [F2] to enter a comment or note is a frequent occurrence, used on 92 percent of all interviews. Similarly, the vast majority of cases involve backing up to change responses, although the use of [F7] to efficiently jump forward again is much less frequent. From the second column in Table 11.2, it can be seen that most interviewers are able to use the functions at least once. For example, only 12.5 percent of interviewers never invoked online help.

The remaining columns in Table 11.2 present average uses of various keystrokes for all production interviews, and for the first and last five interviews completed by each interviewer, respectively. It is expected that the use of some functions may increase over time as interviewers become more comfortable with the laptop computer; however, other functions (such as the use of online help), and erroneous or inefficient use may decline over the course of the study. The total number of keystrokes declines from the first to last five interviews, suggesting more efficient use of the interviewing software as the study progresses. This is supported by the decrease in use of cursor keys and the backspace key (also used for correction). Functions that should be used throughout the survey period (such as [F2] for comments, [F9] for backing up to change a previous response, or [F3] for the pop-up menu in the roster) show less decline over time, which is encouraging. We should be cautious about attributing cause to any effects detected. For example, decline in the use of online help may be because (a) interviewers no longer need to use the information provided, (b) they found it not to be useful, or (c) over time they forgot how to use the function.

The infrequent occurrence of erroneous function keys is noteworthy. These are function keys ([F4], [F6], [F8]) that are not activated for the AHEAD instrument, as well as [F5] when not used in context (to suspend the interview). Such keys appear in only 6.3 percent of all interviews, and on average once in every eight interviews (1/0.12). However, 86 percent of all interviewers pressed one of these invalid function keys at least once. Thus, despite the apparent rarity of these events, few interviewers are immune from committing such errors. These are worth further investigation (see Section 11.7).

Again from Table 11.2, the average amount of text entry for all marginal comments or notes (average number of characters per interview) declines from first to last interviews. Whether this suggests that interviewers are becoming more efficient in their use of comments, or whether they are increasingly taking shortcuts is not known. It is interesting to note that 19.5 percent of all keystrokes in AHEAD interviews are text entries in comment fields, while 3.6 percent of all key presses are cursor keys, and 2.1 percent are function keys. This provides an idea of the relative frequency with which various activities occur.

While these aggregate counts may be useful for allocating design and training resources to increase the use of desired functions and decrease the occurrence of inefficient behaviors and the erroneous use of function keys, they reveal little of possible variation in interviewer performance in interacting with the CAPI instrument. We explore this issue further in the next section.

11.6 VARIATION IN INTERVIEWER KEYSTROKE BEHAVIOR

In order to be useful as tools for evaluating interviewer performance, we should expect the keystroke files to reveal systematic differences by interviewer attributes related to CAPI performance (see Couper and Burt, 1994). We believe the largest differences should occur in the early stages of interviewing, and thus focus our analysis on the first five interviews conducted by each interviewer.

We would expect that the number of comments entered would be positively

Table 11.3   Average Keystrokes in First Five Cases by Survey Experience

| | Interviewer survey experience | | |
Keystroke	Less than one year	One to five years	More than five years
[F1] Help	1.19[a]	0.80	0.48[b]
[F2] Comments	14.06	12.65	10.31
Cursor keys	61.39	57.16	52.58
All function keys	33.37	34.56[a]	27.66[b]
Erroneous function keys	0.11	0.24	0.11
Comment text	351.9[a]	299.5	242.9[b]
Total keystrokes	1428.2	1408.2	1307.7
(Number of interviewers)	(23)	(56)	(57)

Note: Means with different superscripts across rows are significantly different at $p < .05$.

related to survey experience. Novice interviewers may focus their attention on the question answering process, and not recognize the need for marginal notes. On the other hand, experienced interviewers may be better judges of which comments are necessary, thus eliminating extraneous comments. Table 11.3 presents selected keystroke behaviors by three categories of survey experience. These data suggest that the use of marginal comments declines slightly with survey experience, as does the use of online help. More experienced interviewers also appear to do less editing (as evidenced by the decline in cursor key use). Although these general trends hold for all three variables, given the relatively small numbers of interviewers, they do not always reach statistical significance ($p < .05$). The amount of text entered in comments declines significantly with survey experience. It also appears that those with extensive survey experience make less use of function keys than either those with little or moderate experience (although only statistically significant for the latter). These trends may explain the decline in total keystrokes with increasing survey experience, suggesting that experienced interviewers may be more efficient users of the CAPI instrument.

As far as computer experience and typing skills are concerned, hypotheses can go in either direction. On the one hand, it is expected that experienced computer or keyboard users would make fewer keying errors (i.e., they would press fewer erroneous function keys, and make less use of cursor keys for corrections). On the other hand, this group may be less intimidated by the computer, and hence be more inclined to use various function and cursor keys to make corrections.

Table 11.4   Average Keystrokes in First Five Cases by Computer Experience

Keystroke	Interviewer computer experience		
	None	Moderate	Extensive
[F1] Help	0.36[a]	1.03[b]	0.76[b]
Cursor keys	42.39[a]	49.38	0.76[b]
Backspace	13.47[a]	15.86	20.26[b]
All function keys	28.29	33.98	31.73
Erroneous function keys	0.094	0.18	0.19
Total keystrokes	1308.2	1360.6	1405.8
(Number of interviewers)	(34)	(36)	(66)

Note: Means with different superscripts across rows are significantly different at $p < .05$.

The average keystroke counts by computer experience presented in Table 11.4 tend to support the latter hypothesis. Interviewers with extensive computer experience (i.e., those who use computers on a regular basis) make greater use of correction keys (cursor keys and backspace) than those with less experience. This suggests that more experienced computer users may feel more comfortable moving around the instrument and making changes. Total keystrokes also increase with computer experience.

The trends for self-assessed typing skills (not shown) are in a similar direction to those for computer experience. The use of cursor keys and the backspace key increases with typing skill, suggesting greater editing activity among more experienced typists. This may suggest that they are less careful about entering responses, sacrificing initial accuracy for speed. However, self-described "slow touch typists" use fewer function keys, enter fewer comments and less text in comments, and have fewer overall keystrokes on average than the other two groups. This may partly be explained by the constraints of a small laptop keyboard on fast touch typing, suggesting that reasonable familiarity with the keyboard may be optimal.

No clear trends were detected for erroneous function key usage by various interviewer attributes. One explanation may lie in the relative rarity of these events. Another is that only a portion of the erroneous function key presses is captured using aggregate counts. There may be numerous instances where a valid function key (e.g., [F1]) is pressed in the wrong context (e.g., interviewer trying to press [F2]). Some evidence of this can be found in the difference between the total number of [F1] key presses, and those used in combination with [Esc], the appropriate use of online help. Whereas from Table 11.2 it is seen that [F1][Esc] to invoke help was used an average of 0.73 times in the first five interviews, the [F1] key itself appeared an average of 1.02 times. This suggests that about a quarter of the [F1] uses may not have successfully invoked help. Similarly, about one in twelve uses of the [F2] key did not result in a comment being entered.

Nonetheless, these comparisons show that differences in the use of the CAPI instrument by interviewer attributes are modest at best. Furthermore, they tend to decline over time. When averages over the last five cases or over all cases are compared on these attributes, the significant differences found in Tables 11.3 and 11.4 tend to disappear. Thus, the errors and inefficiencies discussed earlier do not seem to be clustered among interviewers with the specific characteristics examined.

11.7 EXPLORATION OF ERRONEOUS FUNCTION KEY USE

As noted earlier, the aggregate counts of keystroke behaviors across production interviews only reveal broad trends in the use of the CAPI system. However, these counts can help identify individual interviews for detailed review. This was done for the 408 interviews on which one or more erroneous function keys were used. Each of these files was individually reviewed to explore the context within which the invalid key was pressed, and to try to determine what the interviewer was attempting to do at the time.

We should caution that the type of error (slip or mistake) is inferred from surrounding keystroke behavior, and is thus susceptible to misinterpretation of intention. We err on the side of classifying an error as a mistake unless it is clear that a slip occurred. In other words, an error is classified as a slip only if an adjacent numeric or function key was the intended target. Mistakes occur when the interviewer is not sure what the correct function key is for the action they wish to carry out, while slips assume that the interviewer may have the correct intention, but presses the wrong key. For some of the errors it is unclear what the interviewer's intended goal was, as they may have given up the intended action.

The 408 interviews with one or more erroneous function keys yield a total of 489 separate instances of erroneous usage. A summary of these is presented in Table 11.5. Consistent with Zapf *et al.* (1992), we find that more errors are occurring at the lower levels of functioning. However, there is considerable variation across the four function keys. For example, many of the errors using [F8] are due to its close proximity to the frequently used [F9] key for backing up. Similar errors for [F5] are rare, given that it is bracketed by two other invalid function keys ([F4] and [F6]).

In the laptop computer used for AHEAD interviewing, the function keys are directly adjacent to the numeric keys, and slightly offset horizontally. This means that the [F6] key appears above and to the right of the [6] key. This may explain the preponderance of errors in which function keys are pressed instead of adjacent numeric keys. Given the close proximity of the keys, and the size of keys on the laptop computer, in many instances interviewers appear to be pressing two keys virtually simultaneously, while on other occasions the

Table 11.5   Classification of Erroneous Function Key Usage

Type of error	Percentage of errors				
	[F4]	[F5]	[F6]	[F8]	All
Slips:					
Function key instead of adjacent key	35.2	—	1.9	66.4	34.2
Function key instead of adjacent numeric key	15.2	38.3	73.1	11.2	23.3
TOTAL	50.4	38.3	75.0	77.6	57.5
Mistakes:					
Function key instead of nonadjacent function key	5.6	3.3	—	—	3.5
Correcting ID	1.1	31.7	7.7	0.9	5.5
Intention unknown	43.0	26.7	17.3	21.5	33.5
TOTAL	49.6	61.7	25.0	22.4	42.5
(Number of errors)	(270)	(60)	(52)	(107)	(489)

function keys appear to be pressed instead of the corresponding numeric key. Where the function key pressed in error has no consequence (e.g., [F6]), no effect is felt (other than the entry not being accepted). However, where the erroneously pressed key has some other action, the consequences may be of greater concern.

The classification of errors is especially difficult for those involving the [F4] key. A large proportion of the [F4] uses with unknown intention (65 of the 116 instances) occurs in conjunction with a [1] or [5] for a "yes" or "no" response respectively, but with no other unusual keystroke behavior, and cannot be attributed to problems of adjacency. The only explanation we have is that the instruction "[F4]—Search" appeared on each screen, between the question and entry fields. Although this had no function in the AHEAD instrument, its presence on almost every screen may have led interviewers to attempt to use it. In contrast, most of the slips involving [F4] appear to be in the roster screens where it was pressed instead of [F3] to invoke a pop-up menu.

A separate account is kept of mistakes made during the entry of the 13-digit sample ID at the start of the interview. Given the ease with which errors could be made entering this string of numbers, it seems that interviewers had difficulty correcting the entry of the ID. It appears that interviewers are commonly using [F5] to cancel the entry of an ID and start over, possibly because of confusion between the suspend ([F5]) and cancel ([Esc]) functions.

We should note that this analysis does not focus on the inappropriate use of valid function keys. In other words, the same errors could be occurring with the [F1] or [F2] keys as with [F4] or [F8]. Some evidence for this can be found in the difference between total [F2] counts and the number of comments actually produced (see Section 11.6). An examination of all keystrokes for all interviews reveals 1,523 instances of [F2] presses that were not productive of comments. These errors may be more deleterious, as they lead to actions not intended by the interviewers, whereas in most cases the four function keys discussed here would simply not be acted on by the CAPI system.

11.8 DISCUSSION AND CONCLUSIONS

This chapter presents a preliminary evaluation of keystroke files as tools for assessing interviewer performance using CAPI. In that respect, it seems clear from both the mock and production interviews that interviewers are able to use the CAPI functions on which they are trained. However, a more detailed examination of both the mock and production keystroke files reveals a variety of errors (both slips and mistakes) and inefficient behaviors committed by almost all interviewers in using the CAPI system. While the effect on data quality may not be large (in most cases interviewers are able to recover from the error or work their way around the difficulty), these errors may reduce interviewer efficiency, increase interviewer frustration, and have a negative effect on the interviewer–respondent interaction. Most of these difficulties would go undetected using current tools to study interviewer behavior in a CAPI survey.

It appears that keystroke files may reveal as much (or more) about inadequacies of the instrument or system as they do differences in interviewer performance or capabilities in using the system. We believe that keystroke files are useful tools for both purposes. We discuss first the issue of error reduction as revealed by the analysis of keystroke file data, then return to the issue of interviewer evaluation.

What can we do with the information obtained from the keystroke files? We believe there are three general strategies to error reduction in CAPI that can be informed by the keystroke data: error prevention, error detection and error recovery. Some of these are directed at slips, while others may target the reduction of mistakes.

A variety of actions (including software and hardware design and training) can be taken to *prevent* the occurrence of many of the errors detected. It seems clear that many of the mistakes in function key use could be reduced by providing key templates (either on-screen or affixed to the laptop keyboard). Although most interviewers know how to use the various functions, there are times when they may forget or become confused about certain CAPI functions, particularly when confronted with rare or difficult interviewing situations.

Reliance on recognition rather than recall of functions may help reduce the cognitive load associated with using the CAPI instrument (see Couper, 1995).

A number of the slips that were detected could be attributed to the design of laptop computer keyboards. These results suggest that the close proximity of function and numeric keys is a problem, especially given the preponderance of numeric entry in CAPI interviews. Other things being equal, the separation of these two rows of keys on the laptop computer may reduce erroneous function key use. While this issue may be less tractable for CAPI, it suggests that such slips are less likely to occur in CATI using full-size computer keyboards.

Good software design could also prevent certain errors. The mapping of functions to keys should be done with careful consideration of the types of error that interviewers commit. For example, the spacing of keys (e.g., using [1] and [5] rather than [1] and [2] for "yes" and "no" respectively) could reduce the number of slips in entering survey responses. Similarly, in AHEAD the [F5] key to suspend a case is surrounded by function keys that are not regularly used, possibly reducing the number of unintended suspensions. Even so, following Norman's (1993) assertion that the more deleterious an action, the more difficult it must be to perform, the suspend function has been changed to [Shift-F5], reducing the likelihood of slips leading to suspended interviews.

Analysis of keystroke files can also help reduce errors by informing interviewer training in subsequent studies. Areas in which interviewers experienced some difficulty in the AHEAD instrument included movement and editing in the roster screens, moving backwards and forwards through the instrument to change answers, and correcting entry of the case ID. Interviewers could be given more practice in these operations during training for other surveys.

The second approach to error reduction lies in the timely *detection* of errors. This is already a part of many CAPI instruments, using range and consistency checks, warning interviewers of invalid entries, disabling certain function keys or other key combinations (e.g., [Ctrl-Alt-Delete] to reboot the computer), and so on.

Error *recovery* proceeds from the assumption that despite the best efforts of all concerned, errors will inevitably occur. Frese and Altmann (1989) argue for a focus on error management, and suggest that trainees should be taught to deal with errors, and not merely to avoid them. In addition to the types of errors made, errors can be distinguished by their potential consequences. Some errors (particularly inefficiencies) have few critical effects on data quality. Others, however (even if they are sensorimotor errors), can have potentially deleterious consequences (e.g., accidental deletion of files). This suggests different strategies for error recovery. The judicious use of confirmation prompts for critical actions, facilities to undo unintended actions, and training in correcting errors or previous entries are all approaches to facilitate error management in CAPI surveys.

What about the many instances of interviewer inefficiency that were detected, particularly in the editing of textual entries? It appears that much of this editing is not only time-consuming but also largely unnecessary. One approach may be to tell interviewers to ignore typographical errors in text entry, concentrating rather on readable and codeable entries. This was in fact done during training for this study, but it appears that interviewers insist on editing text responses. This may be a residual effect of paper and pencil interviewing, where legibility and completeness of text entries were key elements in interviewer evaluation. Given that most organizations do not permit interviewers to get back into the instrument after the interview is completed, it may be inevitable that they try to get it right the first time. An alternative approach is to accept that such editing will occur, and give the interviewers tools to facilitate this task. For example, SurveyCraft has a variety of editing functions, but interviewers are not informed about them or trained to use them. As they become more comfortable using CAPI, interviewers may be given increased access to such tools.

Finally, we return to the issue of keystroke files as tools for the evaluation of interviewer performance. Despite the successful implementation of CAPI on a number of studies, and interviewers' positive reactions to the use of a laptop computer for interviewing, there is much we still do not know about how (well) interviewers use the systems and instruments we provide for them. For instance, we may spend a lot of resources and effort to provide interviewers with online context specific help, but these data suggest such facilities are not used very frequently. Analysis of keystroke files permits us to determine the extent to which interviewers use the various CAPI functions that are provided, and the success with which they do so.

Although their limitation should be acknowledged, keystroke files should be seen as one tool among many available to survey managers to evaluate the performance of their interviewers. They should be used in conjunction with other measures to assess interviewer competence and skill in carrying out CAPI surveys. While some methods of evaluation focus on the outcome of the interviewing process, keystroke files (along with other methods such as behavior coding) focus on the process itself.

There are a number of ways in which this information can be used for interviewer evaluation. First, as we have noted, keystroke files can be used in conjunction with scripted mock interviews as one tool to judge the readiness of interviewers for production interviewing. Second, supervisors can (and already do) use the "playback" function of audit trails or keystroke files to review the progress of individual interviews. Third, the keystroke files can be used to produce aggregate data to permit comparisons of interviewer behavior across a large number of production cases, using statistical process control principles and procedures. Finally, this process can be used to detect outlying cases or inappropriate keystroke sequences that warrant further investigation and appropriate intervention.

We have demonstrated that the utility of keystroke files goes beyond the evaluation of interviewer performance. Keystroke files also provide data that can inform the design of CATI and CAPI instruments to facilitate the task of the interviewers, and to understand the process of human–computer interaction in computer assisted interviewing. Not only are they useful for a variety of purposes but these data can be collected at low marginal cost (relative to data collection costs). As our understanding of these data increases, and as analytic tools for summarizing and analyzing keystroke files are developed, their utility may be further enhanced.

The automation of survey data collection has produced additional challenges for the evaluation of interviewer performance, but at the same time it also provides us with the tools to facilitate such evaluation. Keystroke files are one source of data that can be routinely collected in an automated environment at little additional cost. Other examples include time stamp information and detailed call record data. These can all facilitate our understanding of how interviewers interact with the computer instruments they use, in order to reduce errors and improve the efficiency of computer assisted interviewing.

ACKNOWLEDGMENTS

This chapter is based on research supported in part by the National Agricultural Statistics Service of the U.S. Department of Agriculture through a cooperative agreement with the University of Michigan; and through internal Center support to the Survey Methodology Program. We thank the AHEAD staff (Willard Rodgers, P.I.) and Field Section for their assistance in collecting the keystroke files and administering the mock interviews; Qian Yang for her assistance in the preparation and analysis of the keystroke files; and Bill Connett for his overall support and advice on this project. We also thank Edith de Leeuw for her insightful comments on early drafts. The usual disclaimers of responsibility apply.

REFERENCES

Baker, R.P. (1992), "New Technology in Survey Research: Computer-Assisted Personal Interviewing (CAPI)," *Social Science Computer Review*, 10, pp. 145–157.

Cannell, C.F., Lawson, S.A., and Hausser, D.L. (1975), *A Technique for Evaluating Interviewer Performance*, Ann Arbor: Institute for Social Research.

Couper, M.P. (1995), "What Can CAI Learn from HCI?," *Seminar on New Directions in Statistical Methodology*, Washington: Federal Committee on Statistical Methodology (Statistical Policy Working Paper No. 23), pp. 363–377.

Couper, M.P., and Burt, G. (1994), "Interviewer Attitudes Toward Computer-Assisted Personal Interviewing (CAPI)," *Social Science Computer Review*, 12, pp. 38–54.

Couper, M.P., Holland, L., and Groves, R.M. (1992), "Developing Systematic Procedures for Monitoring in a Centralized Telephone Facility," *Journal of Official Statistics*, 8, pp. 63–76.

Couper, M.P., Sadosky, S.A., and Hansen, S.E. (1994), "Measuring Interviewer Behavior

Using CAPI," *Proceedings of the Survey Research Methods Section, American Statistical Association*, pp. 845–850.

Dielman, L., and Couper, M.P. (1995), "Data Quality in a CAPI Survey: Keying Errors," *Journal of Official Statistics*, 11, pp. 141–146.

Frese, M., and Altmann, A. (1989), "The Treatment of Errors in Learning and Training," in L. Bainbridge, and S.A. Ruiz Quintanilla (eds.), *Developing Skills with Information Technology*, Chichester: Wiley, pp. 65–86.

House, C.C., and Nicholls II, W.L. (1988), "Questionnaire Design for CATI: Design Objectives and Methods," in R.M. Groves, P.P. Biemer, L.E. Lyberg, J.T. Massey, W.L. Nicholls II, and J. Waksberg (eds.), *Telephone Survey Methodology*, New York: Wiley, pp. 421–436.

Kennedy, J.M., Lengacher, J.E., and Demerath, L. (1990), "Interviewer Entry Error in CATI Interviews," paper presented at the International Conference on Measurement Errors in Surveys, Tucson, AZ.

Norman, D.A. (1983), "Design Principles for Human–Computer Interfaces," *Proceedings of CHI '83: Human Factors in Computing Systems*, New York: ACM, pp. 1–10.

Norman, D.A. (1993), "Design Rules Based on Analyses of Human Error," *Communications of the ACM*, 26, pp. 254–258.

Rasmussen, J., Duncan, K., and Leplat, J. (eds.) (1987), *New Technology and Human Error*, Chichester: Wiley.

Reason, J. (1990), *Human Error*, Cambridge: Cambridge University Press.

Rustemeyer, A. (1977), "Measuring Interviewer Performance in Mock Interviews," *Proceedings of the Social Statistics Section, American Statistical Association*, pp. 341–346.

Shneiderman, B. (1992), *Designing the User Interface: Strategies for Effective Human-Computer Interaction*, (2nd. edn.), Reading, MA: Addison-Wesley.

Suchman, L., and Jordan, B. (1990), "Interactional Troubles in Face-to-Face Survey Interviews," *Journal of the American Statistical Association*, 85, pp. 232–241.

Zapf, D., Brodbeck, F.C., Frese, M., Peters, H., and Prümper, J. (1992), "Errors in Working with Office Computers: A First Validation of a Taxonomy for Observed Errors in a Field Setting," *International Journal of Human–Computer Interaction*, 4, pp. 311–339.

CHAPTER 12

The Effect of Interviewer and Respondent Behavior on Data Quality: Analysis of Interaction Coding in a Validation Study

Jennifer Dykema
University of Wisconsin, Madison

James M. Lepkowski
Institute for Social Research, Ann Arbor

Steven Blixt
MBNA-America, Wilmington

12.1 BACKGROUND AND OBJECTIVES

Recent research regarding the accuracy with which interviewers administered questions in major government surveys presents sobering news for advocates of standardized interviewing. Reporting on systematic coding of tape recorded interviews from a national survey of medical expenditures and a second survey of consumer expenditures, Oksenberg *et al.*, (1996) found that face to face interviewers changed the wording of over one-half of the questions they administered. This occurred despite the fact that interviews were conducted by

Survey Measurement and Process Quality, Edited by Lyberg, Biemer, Collins, de Leeuw, Dippo, Schwarz, Trewin.
ISBN 0-471-16559-X © 1997 John Wiley & Sons, Inc.

professionally trained interviewing staffs from three reputable survey organizations. Numerous other studies have found smaller, yet still substantial incidents of changes in the wording of questions in a variety of surveys (Bradburn et al., 1979; Cannell et al., 1968; Cannell and Oksenberg, 1988; Fowler and Mangione, 1986; Marquis, 1971). Several of these studies have uncovered other notable violations of standardized interviewing, including directive interviewer probing and inappropriate feedback, as well as questions with high levels of respondent behaviors such as interruptions, requests for clarification, and inadequate answers.

Considered in the context of research that shows even small changes in the wording of a question may be related to substantial differences in the distribution of responses (Schuman and Presser, 1981), and that nonstandard probing and feedback increase error variation (Cannell et al., 1981; Collins, 1980; Mangione et al., 1992), these results raise concerns over data quality both in the studies being evaluated and in surveys in general. To critics of standardized interviewing, the results support the notion that the method itself is unworkable, an unrealistic and ineffective way of gathering meaningful information (Mishler, 1986; Suchman and Jordan, 1990). For survey designers the issues are immediate and practical; how does one respond: Hire better interviewers? Improve training and supervision? Revamp questions? Adopt new interviewing procedures? Commitment to any of these options involves significant cost with no assurance that the remedy will improve the quality of the data. In order to make proper methodological decisions, researchers must know more about the effect of nonstandardized behaviors on the accuracy of responses.

What effect do misread questions have on accuracy? Are requests for clarification or interruptions by respondents simply manifestations of response difficulty, or should they be considered attempts by motivated respondents to report accurately? Does adherence to principles of standardized interviewing enhance, detract from, or make no difference to the accuracy of reports of factual data? These issues are fundamental to the way interviewers are trained, questionnaires written, and survey interactions analyzed. Nevertheless, virtually no empirical evidence exists linking interviewer or respondent behavior to accuracy in reports from surveys (Groves, 1989).

In this chapter we use systematic coding of the behaviors of interviewers and respondents, complemented by a medical record check, to examine the implications of these behaviors for the accuracy of responses in a standardized health interview. The objective of the research is to examine empirical data from a survey about events and behaviors in order to assess the importance of standardization, the consequences of its breakdown, and the behavior of respondents. Findings from this inquiry can improve our understanding of the results of interaction coding for pretests and assist in the design of optimal interviewing procedures. A brief review of the rationale for the standardized interview and the role of interaction in the interview process provides a framework for our inquiry.

12.1.1 Standardized Interviewing

The standardized interview, for decades the preeminent model for survey research data collection, is a hybrid of science and conversation. Characterized as "conversation with a purpose" by proponents (Bingham and Moore, 1924, cited by Cannell and Kahn, 1968), assailed by critics as "a fragile, technical object that is no longer viable in the real world of interaction" (Suchman and Jordan, 1990, p. 241), the standardized interview remains a complex subject, periodically debated yet still not well understood.

Standardized interviewing has as its goal controlling and reducing the interviewer's contribution to measurement error (Groves, 1989; Schaeffer, 1991). Believing that the generalizability of results depends on tight control over the entire interview process, adherents write questions designed to convey a shared, fixed meaning. They insist that interviewers administer these items exactly as worded and in a prescribed order. Thus an ideal interview proceeds as a smoothly flowing series of stimuli and responses. When situations arise that interrupt this flow, interviewers are expected to respond in specified ways to bring matters back under control (Brenner, 1982, 1985; Sykes and Collins, 1992).

While standardization has been credited with reducing the variable component of measurement error (Beatty, 1995; Cannell *et al.*, 1981), recent critics have found much to fault in its procedures (Mishler, 1986; Suchman and Jordan, 1990). Among the complaints are contentions by critics that standardized interviewing induces error by constraining effective communication. Mishler (1986, p. 7) argues that "the nature of interviewing as a form of discourse between speakers has been hidden from view by a dense screen of technical procedures." Viewing tapes of standardized interviews, Suchman and Jordan (1990) criticize the standardized interview for not only failing to take advantage of the conventions and resources of conversation, but for attempting to control and suppress them. They argue that limitations on the interaction between interviewers and respondents compromise rather than promote accuracy and urge a more collaborative approach in which interviewers are granted both freedom and responsibility to help negotiate the meaning of questions and to detect and repair misunderstandings.

In addition, some practitioners, in response to the finding that deviations from the exact reading of questions are not infrequent, suggest that the traditional standardized interview model may be inappropriate in some situations (Oksenberg *et al.*, 1992). This may be especially true for those kinds of question that ask about events and behaviors or other kinds of factual information. Both critics as well as these practitioners have hearkened back to the words of Paul Lazarsfeld (1935, p. 4) who recommended that interviewers use discretion in phrasing questions because, "it seems to us much more important that the question be fixed in its meaning, than in the wording." While acknowledging the inherent difficulty of achieving true standardization, other survey research investigators defend its efficacy. Several point out that critics tend to overlook the forces that led to standardization in the first place (Tourangeau, 1990).

Discussing how talk in interviews compares to talk in other situations, Schaeffer (1991, p. 371) notes that "the artificiality of the standardized interview does not necessarily invalidate responses" and, further, that "people express important information about themselves in situations odder than a survey interview."

12.1.2 Interviewer and Respondent Roles

Interviewers are trained in service of an ideal (Sykes and Collins, 1992). They typically cannot change the wording of the questions, must remain nondirective and are to avoid any social interaction outside the realm of behavior that is directly related to the task of interviewing (Brenner, 1982; Fowler and Mangione, 1990; Sykes and Collins, 1992). At the same time, they are also expected to manage the response process in order to maintain a socially effective interaction with the respondent. On a basic level, this means the interviewer must persuade the respondent to take part in the interview and then keep the respondent engaged until it is completed. The interviewer may use verbal or, in face to face interviews, nonverbal methods of indicating attention and interest in order to motivate the respondent to report adequately. Balancing measurement responsibilities with the social and motivational requirements of the interview can compromise performance (Fowler and Mangione, 1990).

Part of what makes standardized interviewing difficult is that the definition of the interaction as a *survey interview* is compelling only to the interviewer (Brenner, 1982). Respondents have no formal training for their role. Their understanding of and compliance with interview "rules" is simply assumed. Given the conversational setting of the interview, respondents can break the question and answer sequence in any number of ways, including by requesting clarification, giving inadequate answers, or digressing on unrelated topics.

12.1.3 Interaction Coding

The principles of standardization presume an ideal framework in which interviewers and respondents interact in a specific way. Deviations from this ideal interaction are believed to lead to poorer quality data. During a series of studies conducted during the 1960s and 1970s, researchers led by Cannell and colleagues at the University of Michigan began applying various observational techniques to the survey interview in order to capture these deviations. Adapted from interaction process analysis used to investigate small group processes and structure (Bales, 1950), this methodology, known as interaction or behavior coding, relies on direct systematic observation of the question and answer process.

With the development of interaction coding, researchers began shedding light on the "black box" of the interaction between interviewers and respondents (Wilcox, 1963). The most important discovery was that survey interviews frequently did not proceed as researchers might have expected based on less direct means of observation. Questions were often not administered as specified, nor were other techniques used according to instructions (Brenner, 1982; Morton-Williams, 1979). Research revealed that interviewers tended to probe directively when the initial question did not elicit a codeable response (Brenner, 1982; Collins, 1980; Mangione et al., 1992). For example, in an investigation of the effects of reinforcing feedback, Cannell et al., (1981) found that positive feedback was more likely to be administered by interviewers in response to undesired than desired respondent behavior. Respondents were also found to frequently display behaviors such as asking for clarification of particular terms or providing answers that did not meet the objectives of the question. The failure of interviewers to conform to instructions or of respondents to perform to ideals was seen as evidence that a question or question sequence failed (Cannell et al., 1992; Morton-Williams and Sykes, 1984; Oksenberg et al., 1991).

Studies by Fowler (1992) and Mangione et al., (1992) provide evidence that deviations from ideal role behavior are related to characteristics of the questions, and that by studying these deviations for a particular question, we can draw inferences about the quality of data that question is likely to generate. For example, in the first study, questions identified as having key terms that were poorly defined based on high levels of requests for clarification and inadequate answers were redesigned to communicate their objectives more clearly. When the survey was repeated and responses from the initial and new versions of the questions were compared, it was found that requests for clarification and inadequate answers dropped substantially. In the second study, researchers analyzed the behavior of interviewers and found a direct relationship between questions that resulted in inadequate answers and error by the interviewers, such as directive probing and failure to probe. Further, questions that were difficult, asked for opinions or open-ended responses were more likely to lead to incorrect behavior by the interviewers, especially more probing. The rationale underlying studies like these that use interaction coding is that the occurrence of nonstandardized behaviors indicates problems that may affect the quality of the data. Although researchers have examined interaction and interviewer variability, there is a notable lack of research examining the association between the behavior of interviewers and respondents and subsequent accuracy.

12.1.4 Interaction Coding and the Accuracy of Responses

The effects of interviewer and respondent behavior on accuracy may be in any one of three directions. First, there may be no relationship between behaviors measured by interaction coding and response accuracy. It is possible that when an interviewer changes the wording of a question it does not significantly affect a respondent's final answer. Similarly, respondent behaviors such as interruptions, requests for clarification, and inadequate answers may reflect individual response styles, or represent harmless artifacts of question structures or response tasks. This finding would have implications for question design, the structure of the interviewer role, and interviewer training, as well as for the analysis of pretest and behavior coding results (see Chapter 1, this volume). If there is no relationship between behaviors and accuracy, then measuring behavior in the interview is clearly not very useful in predicting which questions will produce the least accurate data.

On the other hand, interviewer and respondent behaviors that interaction coding systems identify as potentially problematic may indeed be related to the accuracy of responses. If this is correct, then questions that are associated with a high rate of nonstandardized behaviors should be those that are most likely to be answered inaccurately.

A third possibility, however, is that interviewer and respondent behaviors considered "problem indicators" are associated with more accurate answers. Respondents who ask for clarification, or who explain the specifics of their situation rather than select a response option, may be harder working, more attentive, and more accurate than respondents who do not exhibit these behaviors. If this premise is true, codes like requests for clarification are perhaps better indicators of characteristics of respondents such as their motivation to answer accurately than they are of problems with particular questions such as unclear concepts or awkward phrasing. Question sequences in which an exact reading of the question is followed by an immediately codeable answer may be those that lack salience for respondents or are easy to answer without consideration. Thus, we might expect to find better reporting at questions with high levels of certain respondent behaviors.

12.2 DESCRIPTION OF THE HEALTH FIELD STUDY

12.2.1 Study Specifications

The Health Field Study (HFS) is a validation survey concerning health and health care utilization (see Krause and Lepkowski, 1994, for a full description of the HSF). Conducted in 1993 by the Survey Research Center at the University of Michigan for the National Center for Health Statistics, the primary objective was to evaluate questions currently in use or under consideration for the redesign of the National Health Interview Survey (NHIS). While many of the items selected for the HFS are replications or adaptations of existing NHIS questions, the questions in our analysis were written specifically for the HFS. The study also incorporated an experiment on interviewing in which several cognitive and motivational devices were employed. Forty-five interviewers were trained to administer the HFS. Half of these interviewers had no previous interviewing experience. Interviewers were randomly assigned to the experimental conditions. Figure 12.1 shows the exact wording of the questions used in this analysis. Special instructions read to respondents who received the experimental version of the questionnaire are shown in brackets.

A total of 2,006 respondents were interviewed in their homes on a variety of health-related topics. Respondents were persons ages 14 and older drawn

Hospover	The next questions ask about any hospital stays you may have had recently. [Since the information these questions ask for may be difficult to remember, please use any of your own records or calendars to help you answer the questions more accurately.] As you can see on this calendar, we are interested in the time from (CURRENT MONTH) 1ˢᵗ, 1992 until yesterday. Please think back carefully over this time in answering the next questions.
	Since (CURRENT MONTH) 1ˢᵗ, 1992 have you been a patient in a hospital <u>overnight</u>? (YES/NO)
6movisit	I see that the total number of visits you have had to medical doctors or assistants during the past 6 months is (TOTAL). Is this number correct? (YES/NO)
12movisit	I see that the total number of visits you have had to medical doctors or assistants during the past 12 months is (TOTAL). Is this number correct? (YES/NO)
4weeks	Including visits to hospital emergency rooms, urgent care centers, doctor's offices or HMO's, and to other places that you already told me about, how many of these visits were (was this visit) during the <u>past 4 weeks</u> shown on this calendar? (NUMBER OF TIMES)
2weeks	(Was this visit/How many of these visits were) during the <u>past two weeks</u> shown on this calendar? (NUMBER OF TIMES)
Reason	What was the reason for this visit? (Can you tell me more about that?) (OPEN QUESTION)
Genproc	I'm going to ask you a series of questions about different procedures you may have had done during <u>your last visit</u> to a medical doctor or assistant. This includes x-rays, lab tests, surgical procedures, and prescriptions. For each of these areas, I'll ask you whether or not it happened, and whether you paid any of your own money to cover the costs. First, during your last visit to a medical doctor or assistant, did you have an x-ray, CAT scan, MRI, or NMR? (YES/NO)
Labtest	During your last visit to a medical doctor or assistant, did you have any lab tests done that required blood, urine, or other body fluids? (YES/NO)
Surgproc	During your last visit to a medical doctor or assistant, did you have any surgical procedures? (YES/NO)
Othrtest	I've asked you a number of questions about x-rays, lab tests, or surgical procedures that you may have had done at your last visit. This is an important area for our research. Can you think of <u>any other</u> tests or procedures you had done at your last visit to a medical doctor or assistant that you have not already had a chance to tell me about? (YES/NO)

Figure 12.1 Exact wording of the questions used in the analysis.

from members of a Health Maintenance Organization (HMO) in a large metropolitan area. The sampling frame included individuals who had at least one health care utilization during the past 16 months. Youths in ages 14 to 17, African-Americans, and persons 65 years and older were oversampled.

Nearly all interviews were audiotaped, except those for which the respondent did not give permission or there was a failure in the recording process. Respondents were also asked if they would allow their medical records to be accessed. Of the 1,834 tape recordings available, a total of 455 interviews were selected at random within strata defined by the respondent's age, gender, and race, and the interviewer's assignment to experimental conditions.

Validation information was provided by the HMO for hospital stays, visits to the doctor, lab tests, chronic conditions, injuries, and health insurance coverage. We then compared each respondent's report with his or her medical record to determine whether or not the survey report and the medical record agreed. Studies that examine the correspondence between self-reports and information in medical records make the assumption that the medical records are complete and that the information has been measured without error. In practice, however, this condition is difficult to achieve because of coverage errors associated with health care obtained outside the participating system, variation in record keeping practice, lost or misplaced records and other reasons. Although the HFS was designed to anticipate general problems associated with any record check study as well as issues specific to this study, we recognize that we are not completely justified in treating the medical record data as true values as we do in this analysis.

12.2.2 Interaction Coding System

For this study we selected interaction codes that had been tested over a number of investigations (Cannell *et al.*, 1968; Cannell and Robison, 1971; Mathiowetz and Cannell, 1980; Morton-Williams, 1979). We reduced our coding scheme to those behaviors that could be coded reliably and coders were given detailed instructions for how to apply the codes (see Blixt *et al.*, 1994). Figure 12.2 presents the codes we used. Under question-asking codes, a *substantive change* is one that appears to change the meaning of the question, results in the omission of consecutive words that are not prepositions or articles, or involves paraphrasing. *Skips Q* indicates a question that the interviewer was supposed

---

Interviewer Behavior - Question-Asking (Assign one code only)

Substantive change	Makes a substantive change in reading question as written
Skips Q	Skips applicable question
Verifies Q	Verifies, states or suggests an answer
Reads wrong Q	Reads question that was not supposed to be read

Respondent Behavior (Code all that apply)

Interrupt	Interrupts question with an answer
Uncertain	Expresses uncertainty about question, requests clarification
Qualified	Qualifies answer
Uncodeable	Response does not meet question objectives, uncodeable
Don't know	Offers a don't know response
Refusal	Refuses to answer

Figure 12.2 HFS question-level interviewer–respondent interaction coding scheme.

to read but did not. *Verifies Q* means that rather than phrasing the question *as* a question, the interviewer states or suggests what he or she assumes to be the answer. Although *skips Q* can be the result of a simple mistake, both this behavior and *verifies Q* tend to occur when the respondent has provided information at a previous item that the interviewer believes either directly or indirectly answers the current question. The final question-asking code, *reads wrong Q*, refers to the relatively rare instances when the interviewer reads a question that according to the interview schedule should not have been read.

With regard to the respondent behavior codes, the code *interrupts* was assigned if the respondent interrupted with an answer during the initial reading of the question. A code for *uncertain* was assigned at requests for a repeat of the question or clarification of a term. A *qualified* answer meets the objectives of the question (e.g., for a fixed choice question, the interviewer can check a box), but the respondent accompanies the answer with a qualifier such as "probably" or "about." An *uncodeable* answer is one that does not completely answer the question (e.g., the response categories are "much better" and "somewhat better" and the respondent says "better"). We analyze the effect of *don't know* and *refusal* codes on accuracy when a final answer was eventually provided by the respondent. In addition to assigning codes, coders wrote brief notes documenting the nature of *substantive changes* and recorded the respondent's verbatim answer when assigning codes for *uncertain* and *uncodeable* responses. The notes are useful in diagnosing the source and the seriousness of the problem.

Six persons with extensive interviewing experience and backgrounds in behavior coding or interviewer monitoring were trained to do the coding. During training and throughout production, coders independently coded sections of the questionnaire in order to assess intercoder reliability and to monitor performance. Sixty-one interviews that included a total of 3,324 administrations of the questions were analyzed and the Kappa statistic was computed. The Kappa statistic is a measure of interrater agreement and provides the ratio of the difference between the observed and expected levels of agreement to the proportion of agreement that is unexplained: $\kappa = (P_{obs} - P_{exp})/(1 - P_{exp})$ (see Fleiss, 1981, for a more thorough discussion of the Kappa statistic). The reliability scores for all of the question-asking and respondent behavior codes fell into the range of good to excellent agreement (Fleiss, 1981). The values of the question-asking codes were as follows: *substantive change*, $\kappa = .79$; *skips Q*, $\kappa = .95$; *verifies Q*, $\kappa = .68$; and there was perfect agreement for the code for *reads wrong Q*. With regard to the respondent behavior codes: *interrupts*, $\kappa = .80$; *uncertain*, $\kappa = .92$; *qualified*, $\kappa = .82$; *uncodeable*, $\kappa = .90$; *don't know*, $\kappa = .87$; and *refusal*, $\kappa = .89$.

Once the record matching and behavior coding operations were completed, these two data sets were merged. This analysis examines the association between the interviewer and respondent behavior codes and the accuracy of survey reports.

12.3 ANALYSIS

12.3.1 Description of the Dependent Variables

Ten items for which validation information was available were selected from the HFS for this analysis (see Figure 12.1). Five were yes/no questions asking whether respondents had been hospitalized overnight during the past year or had general procedures, lab tests, surgical procedures, or other tests performed during their last visit to a medical doctor or assistant. Four questions asked about the number of health visits the respondent made during specific reference periods such as the past six months, twelve months, four weeks, and two weeks. The remaining item was an open question asking about the reason for the most recent health care visit.

Dichotomous dependent variables were created for each of these questions indicating whether the medical record and survey report concurred. For the five yes/no items a value of 1 was assigned for the dependent variable whenever there was *disagreement* between the survey report and the medical record. For example, a 1 was assigned if the medical record failed to confirm a hospitalization reported by the respondent or contained an unreported hospitalization. Respondents' answers to the questions asking about various procedures were characterized as inaccurate if the medical record either failed to reveal a procedure mentioned during the interview or contained a procedure that was not reported. For questions about the number of contacts with a health provider, respondents were randomly assigned to a series of questions that used either a six- or twelve-month reference period. This series asked about visits to emergency rooms, urgent care centers, doctor's offices and any other facilities during the past six or twelve months. Following these questions the interviewer tallied the total number of visits, verified the accuracy of the total, and made any necessary corrections. Accuracy of reports for the number of doctors' visits in the preceding six- or twelve-month period of time was computed by subtracting the number of visits listed in the medical record from the survey report. Any value other than 0 was coded as inaccurate. Respondents who reported any visits in the past six or twelve months were asked how many of those visits were in the past four and two weeks; we constructed indicators of inaccuracy for these four-week and two-week reports similar to those described above. Finally, the dependent variable for the reason for the most recent health care visit was constructed by comparing the coded survey report to the set of ICD-9 diagnostic codes contained in the medical record. If none of the reasons mentioned by the respondent matched *any* of the ICD-9 diagnostic codes in the record, the report was judged inaccurate (see Jay *et al.*, 1994 for a detailed description of the process involved in matching HFS survey data to the medical records for the reason for the last health care visit).

Panel A in Table 12.1 presents the proportion of inaccurate reports for each of the dependent variables. Questions regarding health care visits made during the past six or twelve months show the highest proportion of inaccurate

Table 12.1 Descriptive Statistics for Dependent Variables and Interviewer–Respondent Interaction Codes

	Dependent Variables									
	Hospover	6movisit	12movisit	4weeks	2weeks	Reason	Genproc	Labtest	Surgproc	Othrtest
Panel A										
Proportion inaccurate	.115	.717	.845	.247	.157	.642	.180	.335	.040	.196
Panel B										
Substantive change	.199	.120	.091	.210	.294	.086	.111	.045	.033	.108
Skips Q/Verifies Q/Reads wrong Q	.004	.034	.010	.029	.067	.016	.003	.003	.003	.010
Interrupt	.055	.043	.030	.041	.046	.009	.028	.045	.013	.033
Uncertain	.051	.090	.071	.105	.062	.038	.111	.020	.043	.075
Qualified	.007	.056	.091	.046	.036	.009	.010	.013	.005	.003
Uncodeable	.027	.030	.081	.054	.052	.062	.033	.038	.015	.048
Don't know	.035	.034	.036	.044	.036	.011	.060	.050	.015	.050
Refusal	.018	.017	.025	.020	.021	.029	.005	.005	.000	.003
Panel C										
Any R codes	.146	.236	.274	.256	.206	.139	.219	.161	.076	.196
Top quartile of Q-asking errors	.428	.496	.437	.447	.471	.437	.416	.405	.400	.392
Bottom three quartiles of Q-asking errors	.000	.075	.075	.072	.082	.090	.087	.085	.085	.083
Top quartile of R behaviors	.469	.451	.468	.439	.450	.455	.434	.425	.421	.416
Bottom three quartiles of R behaviors	.000	.170	.152	.170	.147	.156	.181	.176	.173	.169
Sample size[a]	453	243	206	453	453	455	322	310	322	311

[a] The number of cases in the analysis varies from this total due to missing data.

297

responses, with the longer twelve-month period being the least accurate. The reason for last visit was also reported inaccurately more than half of the time, according to our coding criteria. Questions about health care visits made more recently, within the past four or two weeks, showed fewer disagreements with the medical records than the longer six- or twelve-month reference periods. Reports from the series of items concerning procedures done at the last visit and any hospitalizations generally agreed with the medical record, with 64 to 96 percent of respondents reporting accurately at these questions. However, as indicated earlier, other factors may influence these estimates. For example, based on responses to a series of questions assessing care from a provider, just under one-fifth of the respondents reported visiting a provider outside of the HMO which would not be captured in the medical record.

12.3.2 Model Construction and Results

Interaction coding results were compiled for each of the ten items (see Panel B in Table 12.1 for the proportion of cases assigned various question-asking codes or a specific respondent behavior code for each of the dependent variables). Overall, the frequency of interviewer reading changes and respondent behavior codes in these questions is relatively low, perhaps at least partially the result of extensive pretesting that took place prior to the field period. For example, the questions with the highest levels of reading errors—hospitalizations and visits during the past four weeks or two weeks—were misread less than a third of the time.

Once the dependent variables were created and behavior coding data tabulated, three multivariate models were constructed. In each model the accuracy of response was regressed on various respondent and interviewer behaviors. Each model also includes controls for variables that are potentially confounding, including characteristics of the respondent such as age, education, gender and race, and characteristics of the interviewer such as age, education, gender, race and the number of years of experience as an interviewer. Where relevant we also use controls for various method factors including the experimental manipulations. Because our dependent variables are binary we perform logistic regressions and present the odds of obtaining an inaccurate response for the ten questions of interest. We indicate statistical significance at three levels, .10, .05 and .01. Several variables were dropped from the model either due to collinearity or because the coefficient could not be estimated due to an inadequate number of degrees of freedom.[1]

Model 1: Individual Interviewer and Respondent Interaction Codes at the Current Question
To construct the first model, we adopted the most common approach found in the literature and created question by question predictors based on the

---

[1] Logistic regression analyses were carried out using STATA, release 3.1. A description of the problem of "one-way causation by a dummy variable" is provided on pages 469–471 in the Stata Reference Manual: Research 3.1, 6th edition, College Station, TX, Stata Corporation, 1993.

various interaction codes. These variables were given values of 1 or 0 depending on whether a particular code was or was not assigned during the interaction between the interviewer and respondent. Model 1 simultaneously includes measures for whether the interviewer made a *substantive change* in reading the question, whether the respondent *interrupted*, expressed *uncertainty*, *qualified* an answer, gave an *uncodeable* response or said *don't know* or *refused* to answer the question prior to providing a final response. Table 12.2 presents the results from logistic regressions of inaccuracy in reports for each of the dependent variables on the individual interaction codes and various control variables. The cell entries show the exponentiation of the coefficients or the odds of obtaining an inaccurate response relative to an accurate response when a particular code is assigned. For example, we can interpret the value of 1.12 for a *substantive change* in question-asking for the question about hospitalizations in the following manner: Relative to interviewers who do not make *substantive changes* reading this question, those interviewers who make *substantive changes* at this question are 12 percent more likely to obtain an inaccurate response.

From an examination of the independent variables in Model 1, it appears that substantive deviations from the exact reading of the question have virtually no effect on accuracy. The only question for which departure from the exact reading of the question reaches statistical significance is the number of visits to health care providers in the past six months. In this case, changes are negatively associated with error such that interviewers who make *substantive changes* when reading the question are 62 percent *less* likely to obtain an inaccurate answer. With regard to the respondent behavior codes, *interrupting* during the reading of the question about specific lab tests is significantly related to providing an inaccurate answer. We also find that respondents who are unsure or struggle over their answers are more likely to provide answers that are inaccurate. Specifically, *qualified* responses and answers in which the respondent says *don't know* at some point in the interaction predict inaccuracy for four of the five yes/no questions concerning health utilization. *Qualified* answers are also related to inaccuracy for the question about visits to doctors or assistants during the past four weeks. *Uncodeable* answers appear to be related to inaccurate reports about hospitalization and surgical procedures, but these findings are only of borderline significance. Finally, expressions of *uncertainty* are not significantly associated with accuracy for any of the ten questions presented in the table.

Model 2: Summary Level Interviewer–Respondent Interaction Codes at the Current Question
Rather than focusing on each code assigned at the question, Model 2 includes analyses for two at the current question summary indicators. This is an especially useful analysis for examining the effect of the respondent codes. Unlike the question-asking codes, more than one respondent behavior code can be assigned at a particular question if it is warranted (see Panel C in Table 12.1 for the proportion of respondents who were assigned a respondent behavior

Table 12.2 Odds Ratios (exp^{b}) from Logistic Regressions of Inaccuracy in Reports about Health Care Utilization on Individual Interviewer–Respondent (IR) Interaction Codes, Model 1

					Dependent Variables					
IR Codes	Hospover	6movisit	12movisit	4weeks	2weeks	Reason	Genproc	Labtest	Surgproc	Othrtest
Substantive change	1.12	0.38**	1.44	0.91	0.83	0.92	0.44	0.33	nc	0.80
Interrupt	0.48	0.40	0.36	0.96	0.70	0.46	0.64	4.39**	nc	0.47
Uncertain	1.62	1.51	3.58	1.68	0.74	0.55	0.99	nc	2.39	1.00
Qualified	21.52**	3.08	nc	4.79***	2.25	2.91	1.43	6.31	nc	nc
Uncodeable	4.45*	3.23	nc	1.97	0.50	0.70	2.65	0.42	9.02*	2.11
Don't know	0.93	1.95	0.35	1.63	0.73	0.90	6.20***	5.67***	2.85	3.16**
Refusal	0.47	0.72	nc	1.77	3.94	1.59	nc	nc	nc	a

* $p < .10$; ** $p < .05$; *** $p < .01$.

Model includes controls for respondent and interviewer characteristics and experimental design variables.

[a] The independent variable was dropped from the model due to collinearity.

[nc] Not calculable due to an inadequate number of degrees of freedom.

code for each of the dependent variables). The first variable we include in the model, "*any Q-asking errors*," includes any of the possible errors interviewers can make in presenting the question. This dummy variable is coded 1 if the interviewer made *any* errors reading the question exactly as worded, including skipping the question, verifying, stating or suggesting an answer, or reading the wrong question. The second variable "*any R-codes*" is also a dummy variable and indicates whether *any* of the codes for the various respondent behaviors were assigned. Panel A in Table 12.3 shows the results from logistic regressions of each of the dependent variables regressed on the two summary indicators and the other control variables. As in the previous table we find that the relationship between the question-asking codes and the accuracy of the survey report only reaches statistical significance for the question about visits during the past six months. Thus, none of the errors made by interviewers appears to be systematically related to accuracy in our analysis. However, in eight out of the ten tests shown in the table, respondent codes are positively associated with inaccurate responses. The relationship is significant for the questions regarding the number of visits during the past twelve months and four weeks and for the questions about general procedures, lab tests, and surgical procedures. Respondents who exhibited any behaviors during the asking of these questions were between two to five times more likely to report inaccurately.

Model 3: Cumulative Measures of Interviewer–Respondent Interaction Codes
In the third and final model we examine the cumulative effect that the interaction between the interviewer and respondent has on the accuracy of reporting at the current question. We do this by creating a *cumulative* measure for the question-asking and respondent behavior codes. For question-asking errors this cumulative measure is a proportion that represents the total number of questions the interviewer administered in error divided by the total number of questions asked up to, but not including the current question. Similarly, the cumulative code for any respondent behavior is the total number of questions in which the respondent exhibited any kind of behavior divided by the total number of questions asked up to the current question. For example, the question asking about hospitalizations appears as the fourth question in the self-report portion of the interview schedule. If the respondent interrupted the interviewer during the reading of the two preceding questions, the value of the cumulative code for respondent behavior would be .67 because there were a total of three preceding questions, two of which were interrupted.

We then truncate the full distribution of the cumulative scores and create two dummy variables. The first, "*top quartile of Q-asking errors*" is coded 1 if the interviewer is in the top 25 percent of the cumulative distribution of question-asking errors.[2] A value of 1 for this variable signifies that the

---

[2] Because hospitalizations appear as the fourth question in the instrument the distribution for the cumulative measure is truncated. For interviewer behaviors we contrast the top 30 percent versus the bottom 60 percent; for respondent behaviors we contrast the top 50 percent versus the bottom 50 percent.

Table 12.3 Odds Ratios (exp^b) from Logistic Regressions of Inaccuracy in Reports about Health Care Utilization on Summary Level Measures of Interviewer–Respondent Interaction Codes, Model 2 and Model 3

| | Dependent Variables | | | | | | | | | |
	Hospover	6movisit	12movisit	4weeks	2weeks	Reason	Genproc	Labtest	Surgproc	Othrtest
Panel A: Model 2										
Any Q-asking errors	0.95	0.33**	1.60	0.85	0.69	0.99	0.37	0.42	nc	0.90
Any R-codes	1.58	1.27	3.93**	2.39***	0.97	0.79	2.49**	1.85*	4.49**	1.70
Panel B: Model 3										
Top quartile of Q-asking errors	0.42**	0.81	0.44	0.79	0.40**	0.71	0.84	1.43	0.38	0.85
Top quartile of R behaviors	2.55***	1.07	2.55*	1.96***	1.99**	1.41	1.20	0.97	0.93	2.48**

* $p < .10$; ** $p < .05$; *** $p < .01$.
Models include controls for respondent and interviewer characteristics and experimental design variables.
^{nc} Not calculable due to an inadequate number of degrees of freedom.

interviewer is among the worst interviewers in terms of deviating from the prescribed standardized interviewer role which dictates that questions be read exactly as worded. The second variable, *"top quartile of R-behaviors"* is coded 1 if the respondent is in the top 25 percent of the distribution for exhibiting various respondent behaviors. A value of 1 for this variable signifies that among all respondents this respondent appears to have the most problems with the questions as evidenced by interruptions, expressions of uncertainty, etc. (Panel C in Table 12.1 provides the mean of the cumulative measure for dummy variables for the top quartile and the lower three quartiles of interviewer and respondent behaviors.)

Panel B in Table 12.3 shows the odds ratios from the logistic regression model of inaccuracy in reports about health care utilization on the indicators representing interviewers who misread the questions most frequently and respondents who most frequently exhibit coded behaviors. These models also include controls for the various characteristics of the respondents and interviewers described above. Cumulative errors in question-asking up to the current question for both hospitalizations and two-week visits are significantly related to accuracy such that interviewers who habitually make changes when reading the preceding questions obtain better data at the current question.

Respondents who consistently perform the behaviors presented in Table 12.2 are much more likely to provide an answer that is inaccurate for questions on hospitalizations, four-week visits, two-week visits, and other tests. For the interviewer behaviors, the pattern that we see early in the interview appears to level off for later questions. The question about two-week visits appears as the 18th item in the self-report section of the HFS interview; the question assessing the reason for the last visit is the 46th item in the interview. After the question about two-week visits we do not find a consistent or significant relationship between interviewers' errors in reading questions and the accuracy of respondents' reports. In contrast to the findings regarding interviewer behavior, the cumulative effect of frequently exhibiting various respondent behaviors does not appear to plateau. Respondents who most often indicate problems with the items are over two times more likely to provide an inaccurate response when reporting about other tests, the 60th question in the interview, than respondents who are not in the top quartile.

12.4　CONCLUSIONS

Question-Asking
In general, in our study there appears to be no consistent relationship between departure from the exact reading of questions and the accuracy of responses. This finding is compatible with research that has found that errors in question-asking are not significantly related to other components of measurement error (see Groves and Magilavy, 1986; Mangione *et al.*, 1992). *Substantive changes* in question-asking are only significantly related to accuracy for one of

our ten items, the question asking whether the respondent's estimate of the number of visits to medical doctors or assistants in the past six months is correct. Surprisingly, this association is in the opposite direction from what advocates of standardization would predict: substantive changes in wording at this question are related to *more* accurate responses. Examining the remaining items, there are no consistent generalizations that can be drawn about the impact of errors made in reading questions as the direction of the relationships varies. Adding other types of interviewer behaviors, such as reading the wrong question, verifying answers, and skipping questions, as we did in Model 2 also appears to have no generalizable influence on accuracy. Again the six-month question is the only item that shows a significant relationship between any errors in administering these questions and accuracy. Finally, we examined what the cumulative effect of errors made in administering questions has on the accuracy of responses at a particular question. We found statistically significant relationships for hospitalizations and visits in the past two weeks such that interviewers who frequently depart from administering the questions exactly as worded are approximately 60 percent less likely to obtain an inaccurate response.

Our interpretation of these findings is that some interviewers alter questions in such a way as to more clearly communicate the objectives of the question. Analysis of notes made by coders in assigning the code for a *substantive change* shows that these changes frequently involve omitting parts of a question that the interviewer perhaps feels are inapplicable to the respondent's situation, or are specifications of reference periods or other criteria that the respondent already understands. Other common changes involve rewording, restating, or elaborating on terms and instructions that the interviewer may believe are unclear in the context of the current interaction. Taken with our results showing that "errors" in administering questions had either no significant association or a positive association with accuracy of reporting at these ten questions, analysis of the content of changes suggests that interviewers try and some become adept at tailoring questions to a specific respondent's situation, and that they sometimes modify items which they anticipate may pose problems for the respondent. However, we are cautious in applying this finding more generally as the ability to tailor questions may vary with interviewing experience and familiarity with the instrument.

Respondent Behaviors

In contrast to our findings regarding the behavior of interviewers, some behaviors by respondents appear to be more consistently related to inaccuracy. For example, respondents who provide a *qualified* or *don't know* answer indicate that they have doubts about whether their final answer is correct. Our results suggest these doubts are frequently well-founded. Where the results are significant, respondents who indicate doubt are anywhere from three to twenty-one times more likely to provide an inaccurate response than respondents who do not express such doubts. While these respondent behaviors are not the actual cause of inaccuracy, they are an indication of problems in answering the question.

Interruptions are significantly related to inaccuracy for the question about lab tests. In this question respondents are first asked, "did you have any lab tests," which is followed by a set of examples, "that required blood, urine, or other fluids?" Respondents who *interrupt* are four times more likely to provide an answer that is incorrect because they probably do not consider these examples when answering this item. At every other question respondents who *interrupt* are more likely to provide accurate responses, although these differences are not statistically significant for any of the questions. These are probably respondents for whom the questions are either highly salient or else inapplicable.

Perhaps more interesting than the respondent codes that predict inaccuracy is one code that does not. The code for respondents' expressions of *uncertainty* about the meaning of the question shows no predictive power in this analysis even though behavior coding practitioners consider these expressions to be highly indicative of a problematic question (Fowler, 1992). One explanation for this may be that by expressing *uncertainty* these respondents actively seek the information they require in order to answer adequately. Similarly *uncodeable* answers appear to be only marginally related to accuracy in these data. From an analysis of notes provided by coders we believe there is a clear reason for this. Like responses that indicate *uncertainty*, *uncodeable* answers are usually negotiated in a series of exchanges between the respondent and the interviewer and result in a final codeable answer.

Based on the findings of the summary indicators for *any* respondent codes in Model 2, we believe that while these individual behaviors may occur infrequently they should be taken seriously. For five of the ten items in the analysis, the summary code for *any* respondent behaviors is significantly related to inaccurate reports, an indication of its value in identifying problematic questions. This finding may reflect the problem of the "classroom question" in which questions by only one or two students indicate that the class at large, or in this case respondents in general, do not understand. In our study the questions about twelve-month and six-month visits and the reason for the last visit had the highest proportion of inaccurate responses. However, in the analysis presented in Table 12.2 which examines behaviors for the current question only, we find that respondent behavior codes only identify the question about six-month visits as problematic. None of the interviewer or respondent codes is significantly related to inaccuracy for the questions about twelve-month visits and the reason for the last visit. On the other hand, the summary measure for whether any respondent behavior has occurred at the question did identify these items. This demonstrates the importance of including a summary measure to identify questions that are potentially problematic.

Among respondents who consistently display problematic behaviors as indicated by their placement in the top quartile of the behavior coding scores, we find that for half of the questions, they have significantly larger errors than the respondents who fall below this level. This seems to imply that when the stimulus and response chain is frequently broken by respondents the quality of the data suffers.

Implications for Current Practice

These findings suggest both strengths and weaknesses of standardized interviewing and interaction coding and point toward possible improvements in these methods. Regarding the principles of standardized interviewing, our evidence does not support the assumption that all questions must be read exactly as written in order to obtain accurate information. On the contrary, our findings suggest that at some questions, interviewers' adaptations lead to improved reporting by respondents. However, a number of qualifications are in order. First, we again caution that interviewers in our study made relatively few errors when administering these questions. Second, it is entirely possible that characteristics of these questions—their topic and structure—may allow them to work equally well whether they are read exactly as written or tailored for a particular respondent. Substantive changes made to questions on different topics or poorly constructed items may have different outcomes. Third, our questions ask for relatively objective kinds of information. For questions that seek subjective information such as attitudinal items, minor changes in wording may be more likely to alter the meaning of the question. For example, while research comparing the levels of interviewer variability in administering factual and attitudinal questions provides mixed results, some studies show factual questions are less susceptible to interviewer effects (Collins, 1980; Collins and Butcher, 1983; O'Muircheartaigh, 1976). Finally, and perhaps most important, while it appears that *some* interviewers are able to adapt questions to obtain more accurate information, deviations by interviewers increases interviewer variability and therefore, the variable component of measurement error.

Regarding the respondent behavior codes, our findings provide support for the use and relative importance of each code in this study and for conducting studies of interaction coding to measure data quality in general. Codes for initial *don't know* responses and final *qualified* answers emerge from our analyses as the strongest predictors of inaccuracy. While this might seem obvious, the frequency with which these highly predictive behaviors occur at any particular item is usually unavailable to the survey data analyst. Researchers who rely on the frequency of final don't knows or a review of qualifiers recorded by interviewers in the questionnaire booklet cannot get a full picture of the extent of the problem. By contrast, responses that are classified as indicating *uncertainty* or are *uncodeable*, two codes that have been heralded by practitioners of interaction coding for their ability to detect otherwise unidentified question problems, proved much less predictive of response accuracy. Finally, the code for *interruptions* appears useful in identifying questions that have a structure that might disturb the stimulus and response flow, even if the interruption does not contribute to inaccuracy in responding to that current item.

Part of the reason we do not find any relationship between accuracy and certain predictors probably lies in our method of analysis. Our approach in analyzing the accuracy of reports on various health topics was to simplify the process by dichotomizing the outcomes and testing whether any basic associations were present. Other approaches could be used to analyze these data and while we have not done so here we think it is important for future researchers to explore other models (see, for example, Blixt and Dykema (1995), Sykes and Collins (1992) and Van der Zouwen and Dijkstra (1995) for alternative methods to analyze interactional sequences).

Certainly more research is needed. Questions from other studies for which survey and validation data are available must be analyzed in order to discern consistent interactional patterns. Are the patterns we observe here limited to our data or are they generalizable to other surveys that assess the frequencies of certain behaviors? Additional research is also necessary to explore the efficacy of various interactional coding schemes, including those that assess other kinds of interviewer behaviors such as probing and administering feedback.

Taken as a whole our research suggests that the comparison of interview behaviors with validation data should improve the interpretation of pretest and interaction coding results, and provide more empirical evidence on which to base recommendations regarding instrument design and interviewer training.

ACKNOWLEDGMENTS

Part of this research was supported by a Cooperative Agreement with The National Center for Health Statistics through the Association of Schools of Public Health (#S054-10/14). The ideas presented here do not necessarily represent the views of these agencies. The authors gratefully acknowledge the help of Nora Cate Schaeffer and Mick Couper who provided comments on earlier drafts.

REFERENCES

Bales, R.F. (1950), *Interaction: A Method for the Study of Small Groups*, Reading, MA:, Addison-Wesley.

Beatty, P. (1995), "Understanding the Standardized/Non-Standardized Interviewing Controversy," Journal of Official Statistics, Vol. 11, pp. 147–160.

Bingham, W., and Moore, B. (1924), *How to Interview*, New York: Harper and Row.

Blixt, S., and Dykema, J. (1995), "Before the Pretest: Question Development Strategies," *Proceedings of the Section on Survey Research Methods*, American Statistical Association, pp. 2:1142–1147.

Blixt, S., Lepkowski, J.M., Belli, R.F., Cannell, C., Giamalva, L., and Buckmaster, J. (1994), "Behavior Coding Results for the Health Field Study," Research Report, Survey Research Center, The University of Michigan.

Bradburn, N.M., Sudman, S., and Associates. (1979), *Improving Interviewing Methods and Questionnaire Design: Response Effects to Threatening Questions in Survey Research*, San Francisco: Jossey-Bass.

Brenner, M. (1982), "Response Effects of Role-restricted Characteristics of the Interviewer," in W. Dijkstra, and J. van der Zouwen (eds.), *Response Behavior in the Survey Interview*, pp. 131–165, London: Academic Press.

Brenner, M. (1985), "Survey Interviewing," in M. Brenner, J. Brown, and D. Canter (eds.), *The Research Interview: Uses and Approaches*, pp. 9–36, London: Academic Press.

Cannell, C.F., Blixt, S., and Oksenberg, L. (1992), "Characteristics of Questions that Cause Problems for Interviewers or Respondents," paper presented at the American Association for Public Opinion Research, St. Petersburg, FL.

Cannell, C.F., Fowler, F.J., and Marquis, K.H. (1968), "The Influence of Interviewer and Respondent Psychological and Behavioral Variables on the Reporting in Household Interviews," *Vital and Health Statistics*, Series 2, No. 26, Washington, DC: U.S. Government Printing Office.

Cannell, C.F., and Kahn, R. (1968), "Interviewing," in G. Lindzey, and E. Aronson (eds.), *The Handbook of Social Psychology*, Vol 2, pp. 526–595, Reading, MA: Addison-Wesley.

Cannell, C.F., Miller, P.V., and Oksenberg, L.F. (1981), "Research on Interviewing Techniques," in S. Leinhardt (ed.), *Sociological Methodology*, pp. 389–437, San Francisco: Jossey-Bass.

Cannell, C.F., and Oksenberg, L. (1988), "Observation of Behavior in Telephone Interviews," in R. Groves, P. Biemer, L. Lyberg, J. Massey, W. Nicholls II, and J. Waksberg (eds.), *Telephone Survey Methodology*, pp. 475–495, New York: Wiley.

Cannell, C.F., and Robison, S. (1971), "Analysis on Individual Questions," in L. Lansing, S. Withey, and A. Wolfe, (eds.), *Working Papers on Survey Research in Poverty Areas*, Chapter 11, Ann Arbor, MI: Institute for Social Research, The University of Michigan.

Collins, M. (1980), "Interviewer Variability: A Review of the Problem," *Journal of the Market Research Society*, Vol. 22, pp. 77–95.

Collins, M., and Butcher, B. (1983), "Interviewer and Clustering Effects in an Attitude Survey," *Journal of the Market Research Society*, Vol. 25, pp. 39–58.

Fleiss, J.L. (1981), *Statistical Methods for Rates and Proportions*, 2nd edition, New York: Wiley.

Fowler, F.J. (1992), "How Unclear Terms Affect Survey Data," *Public Opinion Quarterly*, Vol. 56, pp. 218–231.

Fowler, F.J., and Mangione, T.W. (1986), "Reducing Interviewer Effects on Health Survey Data," Washington, DC: National Center for Health Services Research.

Fowler, F.J., and Mangione, T.W. (1990), *Standardized Survey Interviewing*, Newbury Park, CA: Sage Publications.

Groves, R.M. (1989), *Survey Errors and Survey Costs*, New York: John Wiley.

Groves, R.M., and Magilavy, L.J. (1986), "Measuring and Explaining Interviewer Effects in Centralized Telephone Surveys," *Public Opinion Quarterly*, Vol. 50, pp. 251–266.

Krause, N.M., and Lepkowski, J.M. (1994), "Year III of the Cooperative Agreement: Initial Studies of Issues Related to the Redesign of the 1995 National Health Interview Survey," Ann Arbor, MI: Institute for Social Research, The University of Michigan.

Jay, G.M., Belli, R.F., and Lepkowski, J.M. (1994), "Quality of Last Doctor Visit Reports: A Comparison of Medical Records and Survey Data," *Proceedings of the Section on Survey Research Methods*, American Statistical Association, pp. 362–367.

Lazarsfeld, P.F. (1935), "The Art of Asking Why: Three Principles Underlying the Formulation of Questionnaires," *National Marketing Review*, Vol. 1, pp. 1–7.

Mangione, T.W., Fowler, F.J., and Louis, T.A. (1992), "Question Characteristics and Interviewer Effects," *Journal of Official Statistics*, Vol. 8, pp. 293–307.

Marquis, K.H. (1971), "Purpose and Procedure of the Tape Recording Analysis," in J. Lansing, S. Withey, and A. Wolfe (eds.), *Working Papers on Survey Research in Poverty Areas*, Ann Arbor, MI: Institute for Social Research, The University of Michigan.

Mathiowetz, N.A., and Cannell, C.F. (1980), "Coding Interviewer Behavior as a Method of Evaluating Performance," *Proceedings of the Section on Survey Research Methods*, American Statistical Association, pp. 525–528.

Mishler, E. (1986), *Research Interviewing*, Cambridge, MA: Harvard.

Morton-Williams, J. (1979), "The Use of 'Verbal Interaction Coding' for Evaluating a Questionnaire," *Quality and Quantity*, Vol. 13, pp. 59–75.

Morton-Williams, J., and Sykes, W. (1984), "The Use of Interaction Coding and Follow-up Interviews to Investigate Comprehension of Survey Questions," *Journal of the Market Research Society*, Vol. 26, pp. 109–127.

Oksenberg, L., Beebe, T., Blixt, S., and Cannell, C. (1992), "Research on the Design and Conduct of the National Medical Expenditure Survey Interviews," Ann Arbor, MI: Institute for Social Research, The University of Michigan.

Oksenberg, L., Cannell, C., and Blixt, S. (1996), "Analysis of Interviewer and Respondent Behavior in the Household Survey," AHCPR Publisher National Medical Expenditure Survey, Methods 7, No. 96–16, Rockville, MD: Public Health Service.

Oksenberg, L., Cannell, C., and Kalton, G. (1991), "New Strategies for Pretesting Survey Questions," *Journal of Official Statistics*, Vol. 7, pp. 349–365.

O'Muircheartaigh, C.A. (1976), "Response Errors in an Attitudinal Sample Survey," *Quality and Quantity*, Vol. 10, pp. 97–115.

Schaeffer, N.C. (1991), "Conversation with a Purpose—or Conversation? Interaction in the Standardized Interview," in P. Biemer, R. Groves, L. Lyberg, N. Mathiowetz, and S. Sudman (eds.), *Measurement Errors in Surveys*, pp. 367–391, New York: Wiley.

Schuman, H., and Presser, S. (1981), *Questions and Answers in Attitude Surveys: Experiments on Question Form, Wording, and Context*, New York: Academic Press.

Suchman, L., and Jordan, B. (1990), "Interactional Troubles in Face-to-Face Survey Interviews," *Journal of the American Statistical Association*, Vol. 85, pp. 232–241.

Sykes, W., and Collins, M. (1992), "Anatomy of the Survey Interview," *Journal of Official Statistics*, Vol. 8, pp. 277–291.

Tourangeau, R. (1990), "Comment to Suchman and Jordan, 1990," *Journal of the American Statistical Association*, Vol. 85, pp. 250–251.

Van der Zouwen, J., and Dijkstra, W. (1995), "Trivial and Non-Trivial Question–Answer Sequences: Types, Determinants and Effects of Data Quality" paper presented at the International Conference on Survey Methods and Process Quality, Bristol, U.K.

Wilcox, K.R. (1963), "Comparison of Three Methods for the Collection of Morbidity Data by Household Survey," University of Michigan, School of Public Health, Department of Epidemiology and the Tecumseh Community Health Study, Ann Arbor, MI, pp. 1–69.

CHAPTER 13

Effects of Interview Mode on Sensitive Questions in a Fertility Survey

Jared B. Jobe and William F. Pratt
U.S. National Center for Health Statistics

Roger Tourangeau, Alison K. Baldwin, and Kenneth A. Rasinski
National Opinion Research Center, Chicago

13.1 INTRODUCTION

Since 1973, the National Survey of Family Growth (NSFG) has been conducted for the U.S. National Center for Health Statistics (NCHS). The NSFG obtains information from a sample of American women on a number of reproductive health issues, some of which can be considered to be sensitive from the perspective of the respondent (for a discussion of the methods of the NSFG, see Judkins *et al.*, 1989; for a discussion of sensitive topics, see Renzetti and Lee, 1993). Lee and Renzetti (1993) define a sensitive topic as "one that potentially poses for those involved a substantial threat, the emergence of which renders problematic for the researcher and/or the researched the collection, holding, and/or dissemination of the research data." Sensitive data from the NSFG concern use of contraceptives, infertility, age at first intercourse, number of sexual partners, frequency of intercourse, treatment for sexually transmitted diseases (STDs), and abortions. In addition to the sensitive questions currently on the NSFG, new questions will be added about sexual and drug-related behaviors in response to the widespread concern about the behavioral aspects of HIV infection. Although respondents are willing to answer these sensitive

Survey Measurement and Process Quality, Edited by Lyberg, Biemer, Collins, de Leeuw, Dippo, Schwarz, Trewin.
ISBN 0-471-16559-X © 1997 John Wiley & Sons, Inc.

questions (item nonresponse is generally less than 1 percent on the NSFG), the accuracy of the answers is largely untested and is open to question. Although survey methodologists have long been concerned about the accuracy of self-reports regarding sensitive behaviors, little is known about the level of accuracy of most survey data on sensitive topics.

Women report substantially fewer sexual partners than men (e.g., Gurman, 1989; Smith, 1992). Summarizing work relevant to the AIDS epidemic, Catania *et al.* (1990, p. 339) stated that "Most sex research is based on self-reported sexual behavior of unknown validity." Respondents report less than half the abortions reported by providers (Jones and Forrest, 1992; Tanfer, 1990). Regarding drug use, a secondary analysis by Mensch and Kandel (1988) found systematic differences among the results of three national surveys. Thus, the conclusion of Catania *et al.*—that self-reports are of unknown validity—could be applied to research on drug use, abortions, and other sensitive topics. These data indicate a need for methodological research into the factors underlying response validity on sensitive topics.

Survey methodologists have generally hypothesized that a major source of error in reports of sensitive behaviors is deliberate misreporting (e.g., Bradburn *et al.*, 1979). Accordingly, most research designed to improve answers to sensitive questions has investigated methods of reducing the perceived threat of the questions to respondents.

One method of reducing perceived threat is to provide an anonymous data collection situation. Self-administered questionnaires generally result in more reports of sensitive behaviors than face to face interviews; telephone interviews lie somewhere in between (for a review, see Bradburn, 1983). More recently, computer assisted self-administered questionnaires have obtained results consistent with those observed with conventional self-administration (Waterton and Duffy, 1984).

Responses to sensitive questions are also affected by the question format and question wording (Bradburn, 1983). Open question formats and longer questions result in more reports of sensitive behaviors (Bradburn *et al.*, 1979). This result was consistent with the results of Marquis and Cannell (1971). Moreover, computer assisted personal interviewing may increase reports of some sensitive behaviors (Bradburn *et al.*, 1991).

There were two major purposes of this project. The first was to investigate the feasibility of obtaining detailed information about HIV/AIDS risk behavior within the context of a fertility survey and the second was to improve abortion estimates from household surveys such as the NSFG. The data collected in this project are also expected to contribute to our knowledge about methods to improve the quality of sensitive data collected in health surveys, and also to the general literature on survey design.

This project was a collaboration between NCHS and the National Opinion Research Center (NORC), and involved two phases. Phase I explored the cognitive processes that led to misreporting of sensitive behaviors. It included focus groups, cognitive interviews, and a small-scale pilot study of a computer assisted self-administered interview. Phase II consisted of a large-scale field experiment carried out under realistic survey conditions in urban Chicago, with oversampling of young and minority women (for a more complete description of all phases of the project, see Pratt *et al.*, in press). Independent variables were the interviewer staff, question order, mode of data collection, method of administration, and site of interview. Dependent variables included rates of self-reports of abortions, drug use, sexual partners, sexually transmitted diseases, and condom use. This chapter addresses only Phase II.

13.2 METHOD

13.2.1 Sample

The sample was selected from two sources. Most of the respondents came from an area probability sample designed to represent women between 15 and 35 years of age living in Chicago. The remainder of the sample was selected from the rosters of two cooperating health clinics; these women were known to have had abortions at those facilities. Across both sources, a total of 1,942 women were selected for the study.

Probability Sample
A multistage area probability sample was selected and screened to identify women in the eligible age range, 15 to 35 years of age. A total of 6,325 occupied dwelling units were selected for screening. Screening interviews were completed at 4,659 of these units, for a response rate of 73.7 percent; interviewers requested information on sex, age, race, and Hispanic background about each person residing at the dwelling unit. The screening interviews averaged about 20 minutes to complete, and yielded information on 10,998 persons; of these, 3,141 were within the age range eligible for the experiment.

Within the area probability sample, the selection of persons for the experiment required several steps. In the first step, each household with an eligible person was placed in one of six strata, based on sex, age, and minority group membership. (A small comparison sample of men in the same age range was also selected; the results for the males are described by Tourangeau *et al.*, 1996.) Once each household had been assigned to a single stratum, a systematic sample of households was selected; the use of a systematic procedure assured that the sample for each stratum appropriately represented all of the area segments. Within households that contained more than one member of the stratum to which the household had been assigned (e.g., two eligible teenage girls), a single person was selected at random for the experiment. Altogether, 1,210 women were selected for the experiment from the area probability sample.

Clinic Sample

The second sample was selected from among 1,088 women, ages 15 through 40, who were selected from the rosters of two cooperating health clinics and were known to have had an abortion. (The total number of women was actually fewer than 1,088; some names were later found to be duplicates.) We obtained each woman's name, address, telephone number, race, ethnicity, age, and the clinic where the abortion had been performed. The selection process for the clinic cases was considerably simpler than that for the area probability cases. Each woman on the clinic lists was classified by stratum, and then a random sample was selected from each stratum. Initially, 544 women were selected for the experiment; subsequently, an additional 188 women were selected, bringing the total sample of clinic cases to 732.

13.2.2 Experimental Design

We manipulated five variables in this experiment, in a completely crossed design. Two of the variables, staff and version, were attempts to enhance the medical/health context of the interview; we hypothesized that respondents would be more willing to discuss sensitive topics in a survey if the respondents considered the questions to involve medical/health issues.

We varied the staff, comparing nurses with regular NORC field interviewers. We hypothesized that nurses would elicit more reports of sensitive behaviors than regular field interviewers. Only female interviewers were used. The versions of the questionnaire varied the order in which two sets of abortion questions were asked: pregnancy history questions first or medical procedures questions first. In the former case, questions about abortion were first asked as part of a series of questions about the respondent's pregnancy history; in the latter case, questions about abortion were first asked as part of a series of questions about medical procedures affecting reproduction. We hypothesized that more abortions would be reported by respondents receiving the medical procedures questions first. In addition, we hypothesized that asking medical procedures questions and pregnancy history questions would result in a higher composite (combined) abortion estimate, that is, abortions reported in response to either one set of questions or the other.

We varied the mode of data collection, comparing paper and pencil to computer interviews, and the method of administration, comparing interviewer-administered to self-administered interviews. Crossing the mode of data collection and method of administration resulted in four groups: interviewer-administered paper and pencil questionnaire (PAPI); interviewer-administered computer assisted personal interview (CAPI); paper and pencil self-administered questionnaire (SAQ); and computer assisted self-administered questionnaire (CASI). We hypothesized that respondents in the SAQ and CASI conditions would report higher levels of sensitive behaviors.

We varied the site of data collection, conducting interviews either in the respondent's home or at a neutral site. Respondents interviewed at neutral sites were paid an incentive of $40. We hypothesized that respondents would be more open at neutral-site interviews, where other members of the household could not overhear their answers. (It should be noted that using this design, the interview site and incentive were confounded.) For all five variables, we were particularly interested in main effects, as opposed to higher order interactions.

13.2.3 Instruments

Two versions of the questionnaire were used, as described above. Both questionnaires began with a calendar on which respondents were to list three or four important personal events. Demographic questions were next. These were followed by the medical procedures and pregnancy history questions in counterbalanced order; both of these series of questions included items on abortion. The pregnancy history questions were those usually used on the NSFG, and asked the respondent to list all her pregnancies in order and to report certain data about each pregnancy, including its outcome (i.e., live birth, stillbirth, ectopic pregnancy, miscarriage, or abortion). The medical procedures questions were developed for this experiment and asked whether the respondent had any of a number of medical procedures affecting reproductive health; six of the procedures were methods for inducing an abortion: dilation and curettage (D and C) to end a pregnancy; dilation and evacuation (D and E) or suction curettage to end a pregnancy; injection of saline or prostaglandin to end a pregnancy; hysterectomy to end a pregnancy; hysterectomy during a pregnancy; and abortion, type unknown.

Both versions of the questionnaire also contained an extensive series of questions about the respondent's sexual partners. Respondents were asked when and with whom they first had sexual intercourse, and whether it was voluntary. They were asked to report the number of sexual partners they had had during the previous year, during the previous five years, and during their lifetimes. They were also asked the duration of their relationships with their first four partners and with their most recent partner.

Both versions of the questionnaire contained questions designed to ascertain whether respondents had had a sexually transmitted disease. In the section of questions on medical conditions, respondents were asked whether they had had chlamydia, gonorrhea, genital warts, genital herpes, or syphilis. In the section of questions on medical procedures, respondents were asked whether they had received antibiotic treatment for sexually transmitted diseases or genital infections. The questionnaires also included items asking the respondents how often they had used condoms in the last year. Respondents were also asked how many times in the last 30 days they had sexual intercourse, and on how many of those occasions a condom was used.

Both versions of the questionnaire contained questions about illicit drug use. The initial drug question asked whether the respondents had ever used any illegal drug, and follow-up questions asked about their use of marijuana, amphetamines, barbiturates, tranquilizers, psychedelics, cocaine, crack, and heroin. Another series of questions, for users of injectable drugs, asked how they cleaned their needles and related drug paraphernalia, and how often they shared them with other users.

13.2.4 Procedure

To ensure that the interviewers were unaware of the source from which each woman had been selected, the principal investigator selected the subsample of individuals for participation in the experiment. Each person was then randomly assigned to a treatment cell, the two samples were merged, and the case list for each of the 27 regular NORC interviewers and the 13 nurse interviewers was constructed. In addition to this double-blind procedure and the normal NCHS and NORC confidentiality provisions, we offered one further safeguard to the clinic cases—all respondents were asked to sign a permission form after the interview had been completed, giving the investigators access to their medical records. All data obtained from women in the clinic sample who refused to sign the form were purged from the data file prior to analyses. These procedures were developed after negotiations with the Institutional Review Boards for NCHS and one of the clinics.

13.3 RESULTS

In this section we begin the description of the results of the project with the response rates, and we then describe the results of the levels of reporting for the abortions, drug use, sexual partners, sexually transmitted diseases, and condom use. The major finding was that self-administration resulted in reporting larger numbers of sexual partners in the past year, past five years, and lifetime, more sexually transmitted diseases, and more condom use in the past 30 days (see Table 13.1).

13.3.1 Response Rates

The initial sample consisted of a total of 1,942 women. After adjusting for losses due to duplicate names, errors in screening, and respondents' moving out of the Chicago area, 1,914 women were eligible for the experiment. Complete and valid interviews were obtained for 1,059 female respondents, 705 from the probability sample, and 354 from the health clinics. The overall response rate was thus 55.3 percent. Because failure to contact members of the sample accounted for more than three-fourths of the nonrespondents, we analyzed the proportion of cases who were contacted and the proportion who completed an interview once they were contacted, as well as the overall rates of participation.

Two of the five experimental variables had significant bivariate relations with overall rates of participation in the study—the type of interviewer to which the case was assigned ($\chi_1^2 = 24.9$; $p < .001$) and the site of the interview ($\chi_1^2 = 17.8$;

Table 13.1   Mean Sexual Partners and Rates of Sexually Transmitted Diseases Assessed by Method of Administration

	Method of administration		
	Self-administered	Administered by interviewer	Ratio
Number of sexual partners			
Past year	1.71	1.44	1.19
Past five years	3.87	2.82	1.37
Lifetime	6.51	5.43	1.20
Condom use			
Past 30 days	46.7%	35.3%	1.32
Past year	23.8%	17.9%	1.33
Sexually transmitted diseases	22.0%	17.0%	1.29

Note: The entries in the top panel are the mean numbers of sexual partners reported; those in the middle panel are the average percent of the time condoms were used in the last 30 days and the percent of respondents saying they always used condoms during the past year; the entries in the bottom panel are the proportion of respondents reporting a sexually transmitted disease. All statistics are for women only.

$p < .001$). A higher percentage of assigned cases was completed by the regular field interviewers (60.8%) than was completed by the nurses (49.5%). It should be noted, however, that 27 regular interviewers were available compared to 13 nurse interviewers; the total number of cases assigned to the two groups was identical. The site of the interview was also related to the response rates; 60.1 percent of the cases assigned to be interviewed outside their homes completed an interview versus 50.5 percent of those assigned to an at-home interview. It seems likely that this difference reflects, at least in part, the $40 reimbursement offered to those selected to be interviewed outside their homes.

Difficulties in finding and contacting members of the sample accounted for the bulk of the nonresponse; of the 857 nonrespondents, 666 were never contacted during the field period, or were still pending at the end of it. Nurses (58.9 percent) were less successful than field interviewers (71.3 percent) in contacting their assigned cases ($p < .001$), although again it should be noted that this may be a function of the number of available interviewers. The rate of contact was also significantly related to the source of the case. Only 57.3 percent of the clinic cases were located during the field period versus 70.0 percent of the women selected from the area probability sample ($\chi_1^2 = 32.3$; $p < .001$). Once contacted, the two groups of women had nearly equal cooperation rates—85.3 percent of the clinic cases and 84.4 percent of the area probability cases who were contacted completed an interview.

The overall cooperation rate, given that the case was contacted, was 84.7

percent. Only the site variable had a significant bivariate relation to the cooperation rate. For the outside-the-home cases, the cooperation rate was 88.8 percent; for the at-home cases, the rate was only 80.4 percent ($\chi_1^2 = 17.1$; $p < .001$). (Logit models that include only main effect terms for the experimental variables give an adequate fit to both the contact and cooperation rate data.)

Of the 354 respondents from the clinic sample, the records of 54 were purged; 48 refused to sign a post-interview consent form for medical records verification, and 6 had missing consent forms. Therefore, 1,005 cases, 705 from the probability sample and 300 from the clinic sample remained for inclusion in the analyses. Both samples were combined for the analyses reported here, except for analyses of the abortion results.

13.3.2 Levels of Reporting

13.3.2.1 Abortion Reporting
Table 13.2 displays the percent of women from each sample who reported an abortion in response to each series of abortion questions. Table 13.2 also contains a composite variable, which is the percent of respondents who reported an abortion in response to either set of questions.

Response to Pregnancy History Question
We fit fully saturated logit models to the data on abortion reporting, repeating the analysis for all of the women and just for those women selected from the clinics. Within the sample as a whole, none of the experimental factors or any of their interactions had a significant effect on the proportion of women who reported an abortion. Among the clinic cases, only a significant, though uninterpretable, four-way interaction emerged (involving the site, staff, questionnaire version, and mode of administration variables).

Responses to the Medical Procedures Questions
We conducted logit analyses of abortion reporting in response to the medical procedures items. The analysis of the whole sample revealed no main effects for the experimental variables. No effects emerged when the analysis was restricted to women in the clinic sample.

Composite Estimates
Although there were no effects of the experimental variables, the composite estimates of abortion reporting resulted in higher estimates than either set of questions alone (Table 13.2). In the clinic sample the highest estimate occurred when the medical procedures questions were first, whereas in the area probability sample the highest estimate occurred when the pregnancy history questions were first. Thus, only in the clinic sample was there some limited support to our hypothesis that placing the medical procedures items first reinforces the legitimacy and emotional neutrality of asking about abortion experience.

Table 13.2   Percent of Women Reporting at Least One Abortion, Question Content by Question Sequence Within Samples

	Clinical sample question content			
	Medical procedures	Pregnancy history	Composite	n
Question sequence				
Total sample	68.7	72.5	76.4	284
Medical procedures 1st	71.4	75.9	80.5	133
Pregnancy history 1st	66.2	69.5	72.8	151

	Area probability sample question content			
	Medical procedures	Pregnancy history	Composite	n
Question sequence				
Total sample	22.5	20.4	23.9	681
Medical procedures 1st	19.1	17.6	20.5	346
Pregnancy history 1st	26.0	23.3	27.5	335

13.3.2.2   Rates of Reported Drug Use

Overall, 42.4 percent of the sample reported having used illegal drugs. The most popular illegal drug was marijuana/hashish: 96.2 percent of those who reported using drugs reported marijuana use. The second most commonly used drug was cocaine: 31 percent of those who reported using drugs reported using cocaine; this is 13 percent of the total sample. Few respondents reported using heroin, only 4.5 percent of users, and 1.9 percent of all respondents. Most of those who did use heroin answered "no" when asked directly if they had injected illegal drugs, so the follow-up questions about intravenous drug use practices had too few answers to analyze.

Neither logistic regression models nor chi-square tests revealed any significant main effects of the experimental variables on rates of reporting drug use. However, both types of tests showed that the interaction of staff and method of administration significantly affected reported drug use. Whether a nurse or field interviewer was assigned to the respondent made no difference in reporting rates when the questions were self-administered, but when interviewers asked the questions, the regular field interviewers elicited higher rates of reported drug use than did the nurses (44.6 percent to 35.9 percent; $\chi_1^2 = 3.79$, $p = .05$).

13.3.2.3   Number of Reported Sexual Partners

Because the data on the number of reported sexual partners for the last year, last five years, and lifetime were strongly skewed, we added .5 to all values and used a logarithmic transformation prior to performing the analyses of variance. For ease of interpretation, we have used the untransformed data when reporting means for the various groups.

Respondents who completed self-administered questionnaires reported higher numbers of sexual partners than respondents who received interviewer-administered questionnaires. These differences were significant at the one-year, five-year, and lifetime intervals. For the previous year, respondents who completed self-administered questionnaires reported a mean of 1.71 sexual partners, whereas respondents who received interviewer-administered questionnaires reported a mean of 1.44 sexual partners ($F(1,949) = 6.50$, $p = .01$). For the five-year period, respondents who completed self-administered questionnaires reported a mean of 3.87 sexual partners, whereas respondents who received interviewer-administered questionnaires reported a mean of 2.82 sexual partners ($F(1,951) = 4.33$, $p = .04$). For the lifetime period, respondents who completed self-administered questionnaires reported a mean of 6.51 sexual partners, whereas respondents who received interviewer-administered questionnaires reported a mean of 5.43 sexual partners ($F(1,922) = 6.45$, $p = .01$). No other main effects were significant.

During home interviews, more sexual partners were reported by respondents interviewed using computer assisted questionnaires than by respondents interviewed using paper and pencil questionnaires, whereas during neutral-site interviews, more sexual partners were reported by respondents interviewed using pencil-and-paper questionnaires. This finding was confirmed by a significant interaction of the mode of data collection by site for both the past year and the lifetime sexual partners dependent variables; for the five-year question the effect was in the same direction but not statistically significant. Table 13.3 displays the relevant means.

For the previous year, respondents interviewed at home reported fewer sexual partners on paper and pencil questionnaires than on the computer assisted ones (1.36 vs 1.84), whereas respondents interviewed outside the home reported more partners on the paper and pencil than on the computer assisted questionnaires (1.68 vs 1.43; $F(1,949) = 4.10$, $p = .04$). Similarly, on the lifetime

Table 13.3 Average Number of Sexual Partners Reported by Mode and Site

	At home		Outside the home	
	Paper	Computer	Paper	Computer
One year	1.36	1.84	1.68	1.43
Five years	2.81	4.51	3.33	2.74
Lifetime	5.06	7.48	6.26	5.08

Note: Numbers are based on raw counts.

partners question, respondents interviewed at home reported fewer partners on the paper and pencil questionnaires than on the computer assisted questionnaires (5.06 vs 7.48), whereas respondents interviewed at neutral sites showed the opposite pattern, reporting more partners on the paper and pencil than on the computer assisted questionnaires (6.26 vs 5.08; $F(1,922) = 4.39$, $p = .04$). Overall levels of reporting are consistently higher using computer assisted questionnaires, although not significantly so. It is possible that bringing computers into the respondents' homes fostered a sense of the importance or objectivity of the survey, thereby promoting fuller reporting of sexual partners. Outside their homes, especially in public places, the presence of the computer may have made respondents feel conspicuous, inhibiting them somewhat.

13.2.2.4 *Rates of Reported Sexually Transmitted Diseases*
More respondents who answered self-administered questionnaires reported a sexually transmitted disease than those who answered questions administered by an interviewer (22.0 percent to 17.0 percent, respectively). The effect of method of administration was significant ($\chi_1^2 = 3.98$, $p = .05$). This result is consistent with the results for reporting the numbers of sexual partners, where higher numbers of partners are reported using self-administered questionnaires. However, these figures do not include people with values of "Don't know" on the sexually transmitted disease variable. If they are included, the effect of method of administration is only marginally significant ($\chi_2^2 = 4.71$, $p = .09$). It should be noted that an answer of "Don't know" may be a nondisclosure answer. No other main effects or interactions were significant.

13.2.2.5 *Rates of Reported Condom Use*
We computed the ratio between the two items concerning condom use and sexual intercourse in the last 30 days to construct a variable representing the percentage of condom use in the previous month. We then performed a six-way analysis of variance (the five experimental variables and race as a covariate) to examine the proportion of times that a condom was used during intercourse in the past 30 days. Respondents who reported that they had never had sexual intercourse or had not had sexual intercourse in the last 30 days were excluded from the analysis, so the number of cases analyzed was 641 (approximately 64 percent of the female cases).

Significantly more condom use was reported with self-administered questionnaires (47 percent) than with interviewer-administered questions (35 percent). The main effect for the method of administration variable was significant ($F(1,580) = 9.33$, $p < .005$). This result is consistent with the results for reporting of sexually transmitted diseases and number of sexual partners, where higher reporting was found using self-administered questionnaires. No interactions were significant.

Data from the question on condom use in the past year were categorical: respondents were asked whether in the past year they had used condoms every time, most of the time, sometimes, rarely, or never. Those who had not had

Table 13.4   Reported Condom Use in the Past Year

	Every time	Most times	Sometimes	Rarely	Never	n
Mode of data collection						
Paper and pencil	17.9%	20.4%	13.8%	10.5%	37.4%	457
Computer	23.8%	17.5%	14.3%	13.7%	30.7%	446
Method of administration						
Self-administered	24.3%	18.4%	14.6%	11.1%	31.6%	452
Interviewer-administered	17.3%	19.5%	13.5%	13.1%	36.6%	451
Total	20.8%	18.9%	14.1%	12.1%	34.1%	903

intercourse in the previous year were excluded, resulting in 903 cases remaining for analysis.

We observed main effects for two experimental variables—whether the interview was computer assisted and whether the questions were self-administered or administered by an interviewer. Table 13.4 presents a summary of these results. In general, the respondents reported more condom use in the computer assisted groups: 23.8 percent of the respondents interviewed with computers and 17.9 percent of the respondents interviewed with paper and pencil questionnaires reported they used condoms every time; by contrast, 37.4 percent of the respondents interviewed with paper and pencil questionnaires and 30.7 percent of the respondents interviewed with computers reported that they never used condoms. The difference was significant ($\chi_4^2 = 9.56$, $p = .05$).

The effect of mode of administration was only marginally significant ($\chi_4^2 = 8.10$, $p = .08$). Respondents who answered self-administered questionnaires reported slightly higher rates of condom use than those who answered questions administered by an interviewer, just as they did in the analysis of condom use in the previous 30 days. Of the respondents in the self-administered groups, 24.3 percent reported using condoms every time, compared to 17.3 percent of respondents in the interviewer-administered condition; the "never" responses were 31.6 percent in the self-administered conditions and 36.6 percent in the interviewer-administered conditions.

The results for previous month and previous year, taken together, are somewhat surprising. If condom use is a socially desirable behavior, we would expect to see higher rates of condom use when an interviewer administers the questions, as respondents seek to appear more socially responsible to the interviewer. However, for the respondents in this study, using condoms appears to be somewhat embarrassing—like admitting that one has had multiple sexual partners or a sexually transmitted disease. It is also worth noting that out of all the possible answers to this question, answering "never" reveals the least specific information about one's sexual life and thus protects the respondent's privacy to the greatest degree. This might make a "never" answer more attractive to respondents who receive interviewer-administered questionnaires.

13.4 DISCUSSION

In this experimental project our goals were to investigate the feasibility of obtaining detailed information about HIV/AIDS risk behavior and improve abortion estimates. The experiment was designed to test the effectiveness of new field procedures and compare their effectiveness with customary field procedures. The new field procedures were expected to improve respondent participation, candor, and reporting levels by enhancing the privacy of the interview or the medical context of the survey or both. In this section we review the major findings from the experiment. We begin with a summary of the response rates. We then summarize the effects of self-administration, the effects of computerization and interview site, and finally the effect of the type of interviewer. Throughout the discussion, we describe areas where further work is needed.

13.4.1 Response Rates

The overall cooperation rates, given that the respondents were contacted, were generally high, both for the clinic and area probability samples; cooperation rates were higher for neutral-site interviews. The cooperation rates compared favorably to the NSFG. Moreover, the item nonresponse rate for the sensitive questions was very low; new sensitive items had nonresponse rates as low as the old sensitive items. These data indicate that this survey had a generally high acceptance rate by the respondents, and suggest that it is very feasible to conduct in-depth surveys of sensitive topics relating to HIV/AIDS high risk behaviors and abortions.

13.4.2 Effects of Self-Administration

The variable with the most consistent effect on the level of reporting was method of administration, especially in reporting sexual behaviors. Respondents who completed self-administered questionnaires reported more sexual partners, more sexually transmitted diseases, and greater use of condoms than those who responded to questions read by an interviewer. These findings are summarized in Table 13.1, which displays the ratio between the levels of reporting under the self-administered and interviewer-administered conditions. As Table 13.1 shows, the levels of reporting are substantially higher—from 19 to 37 percent higher—when the questions are self-administered. The results from the focus groups and cognitive interviews in the early stages of the project suggest that this difference may reflect reduced respondent concern about disclosures to other household members (who might otherwise overhear the interview), reduced concern about the reactions of the interviewer, or both (Pratt *et al.*, in press).

These findings on the effect of self-administration are quite consistent with the results of earlier comparisons of self-administered questionnaires with face to face interviews carried out by field interviewers. The largest studies comparing the two methods of administering survey questions are the ones by Schober et al. (1992) and by Turner et al. (1992). In both of these studies, self-administration resulted in greater reporting of drug use; however, the effect of self-administration was largest for answers to questions about recent drug use and was minimal when the questions concerned lifetime drug use. Unfortunately, our experiment included only items on lifetime drug use; the absence of effects for self-administration on drug reporting in this study is thus consistent with the findings from these previous studies. Recently, Boekeloo et al. (1994) reported that patients at a sexually transmitted disease clinic reported more high-risk sex behaviors to 2 of 16 questions for audio CASI compared to a written self-administered questionnaire. Both were superior to interviewer-administered questionnaires. They also found fewer missing responses with the audio CASI.

To our surprise, we failed to find a significant effect on abortion reporting with self-administration. London and Williams (1990) found that a self-administered questionnaire increased abortion reporting for some groups of NSFG respondents. Moreover, Mosher and Duffer (1994) in the 1994 NSFG Pretest, found that for respondents receiving in-home interviews, 25 percent of the respondents using audio CASI reported abortions compared to 14 percent of the respondents who received an interviewer-administered questionnaire. Note that the 14 percent reporting level in their control group is lower than any reporting levels in this experiment. Our levels of reporting ranged from 17.6 to 26.0 percent. Based on the concerns of respondents in our earlier focus groups (Pratt et al., in press), we were particularly careful to write our advance letter to emphasize the legitimacy of our survey; this may have resulted in our higher baseline levels of reporting abortions. Thus, our results may have been affected partially by a ceiling effect.

Another possible interpretation of the results is that self-administration allowed respondents to pace the interview, allowing more time to recall their behaviors and events. Additional recall time may have had no effect on major events such as abortions, but a larger effect on behaviors such as sexual partners and condom use. Some early studies of mode effects indicate that computer assistance slows the pace of the interview (e.g., Waterton and Duffy, 1984).

Although we found no effects of self-administration, the abortion results were, however, consistent with one of our hypotheses. We found that asking about abortions both in the medical procedures context and in the pregnancy history context resulted in higher composite estimates than those obtained from either set of questions taken alone. In the Mosher and Duffer (1994) study, the self-administered abortion questions occurred after abortion questions had been asked by the interviewer. This might indicate that asking the question twice increases the level of reporting. Further research is needed to determine whether this is the reason composite estimates are so consistently higher than either alone.

Within our clinic sample, 75 percent reported at least one abortion in their lifetime. This result contrasts with that of Jones and Forrest (1992), who found that respondents reported less than half of the abortions reported by providers. Several factors could account for this disparity. Jones and Forrest were analyzing national figures; by contrast, our sample was restricted to women from a single large city, where both public policy and private attitudes may be more conducive to candor in survey responses. Another difference is that Jones and Forrest examined abortion reports for the past year; our figures are for the respondent's lifetime. In addition, our sample had a higher rate of nonresponse than the NSFG sample that Jones and Forrest examined; moreover, about 14 percent of the women who completed interviews in the clinic sample are excluded from our analysis because they refused to sign the permission form. Thus, the very women most likely to deny having had an abortion may have opted out of our sample.

13.4.3 Effect of Computerization and Interview Site

Neither computerization nor interview site by itself had any consistent effects on levels of reporting among the respondents. Past investigations of computer assisted interviewing have tended to emphasize its effects on item nonresponse, timeliness, and cost rather than on the answers that are obtained (e.g., Baker, 1990). Only a few studies have reported effects of computer assisted data collection on levels of reporting. The experiment comparing CAPI with conventional paper and pencil data collection on the National Longitudinal Study of Labor Market Behavior/Youth Cohort found that more respondents reported using birth control with CAPI than with paper and pencil interviewing (Baker and Bradburn, 1991; Bradburn et al., 1991). Several other studies have shown effects on reporting for computer assisted self-administration, but in these studies it is impossible to disentangle the effects of computerization from those of self-administration (e.g., Waterton and Duffy, 1984). We suspect that computerization by itself has little effect on the answers respondents give, a conclusion consistent with much of the previous literature on computer assisted telephone interviewing (Groves and Mathiowetz, 1987; Bradburn et al., 1991). Nevertheless, computerization may have effects on the respondent and on the interviewer and the effectiveness of administering the questionnaire. For example, computerization may improve the accuracy with which questions are read and it controls the selection of what question to read, reducing errors in skip patterns.

The overall effects of site are not clear because of the $40 incentive paid to respondents interviewed at neutral sites. Mosher and Duffer (1994) found that incentive, site, and method of administration affected levels of responding to abortion questions; however, their design was not completely crossed, so that the relative effects of each variable cannot be determined. Further research is needed to determine the relative effects of interview site and incentive payments for answering sensitive questions.

13.4.4   Effect of Interviewer

The type of interviewer showed very weak effects on all dependent variables. However, for the questions about illicit drug use, the type of interviewer did have an effect on reporting, interacting with the method of administration. When the respondent completed the questionnaire herself, the type of interviewer made no difference, but when the interviewer administered the questionnaire, more drug use was reported to field interviewers than to nurse interviewers. This interaction is again partially contrary to our hypothesis. We thought that respondents would be more forthcoming about sensitive behaviors with a nurse interviewer than with a field interviewer. In retrospect, respondents may be reticent to be forthcoming about their drug use with a medical person because drug use is known to have harmful effects on the body. Respondents may be particularly embarrassed to tell a medical person that they have abused their bodies. Further research might investigate this interaction to determine whether respondents would be less willing to admit to medical interviewers that they have engaged in other unhealthy behaviors, such as lack of exercise, and eating a high-fat diet.

The procedures used in this study indicate that having nurses as interviewers may not have been particularly salient to respondents. Nurse interviewers introduced themselves as nurses, but they did not wear uniforms, carry any medical paraphernalia, or perform nursing tasks. Two alternative methods of investigating nurses as interviewers are possible. First, nurses could be inter-viewers in an experiment in which respondents come to a medical facility where they are interviewed in an examination room. Second, if nurses are to function efficiently as field interviewers, they must be better trained as interviewers and have better props to emphasize the fact that they are nurses, for example, they should wear their uniforms and perhaps carry a bag. They could also begin the interview by performing tasks that would be associated with nursing such as taking blood pressure.

13.4.5   Summary

Our project indicates that sensitive questions can be asked in population-based surveys such as the NSFG. We obtained good cooperation rates and obtained higher levels of reporting to several sensitive questions about sexual behaviors using self-administered questionnaires and obtained higher composite abortion estimates using medical procedures questions in addition to pregnancy history questions to ask about abortions.

ACKNOWLEDGMENTS

The research reported here was supported by Contract 200-91-7099 from NCHS to NORC. Funds for this project were provided by the Office of the Associate

Director HIV/AIDS, Centers for Disease Control and Prevention, the Center for Population Control, National Institute of Child Health and Human Development, and NCHS.

The authors gratefully acknowledge the late Kathy London, Tom W. Smith, Christine Bachrach, and Norman Bradburn for their valuable contributions during the design of this study; Lisa Lee, Susan Heine, and Cheryl Gilbert played indispensable roles in the collection of the data presented here; Gwen Merker and Ellen Schwartzbach trained the interviewers and oversaw the collection of data in the experiment; our field managers, Ezella Pickett, Debra Garrison, and Sue Martinson directly supervised the interviewers; Ron Dorsey and Geoff Walker were responsible for the development of the data collection systems; Steven Pedlow assisted with the data analysis; Joan Law served as project manager; and Martin Collins, Jacqueline Scott, and Monroe G. Sirken gave valuable suggestions on earlier versions of this manuscript.

REFERENCES

Baker, R.P. (1990), "What We Know About CAPI: Its Advantages and Disadvantages," paper presented at the Annual Meeting of the American Association of Public Opinion Research, Lancaster, PA.

Baker, R.P., and Bradburn, N.M. (1991), "CAPI: Impacts on Data Quality and Survey Costs," *Proceedings of the Public Health Conference on Records and Statistics*, DHHS Publication No. PHS 92-1214, Washington: Public Health Service, pp. 459–464.

Boekeloo, B.O., Schiavo, L., Rabin, D.L., Conlon, R.T., Jordan, C.S., and Mundt, D.J. (1994), "Self-Reports of HIV Risk Factors by Patients at a Sexually Transmitted Disease Clinic: Audio vs. Written Questionnaires," *American Journal of Public Health*, 84, pp. 754–760.

Bradburn, N.M. (1983), "Response Effects," in P. Rossi, J. Wright, and A. Anderson (eds.), *Handbook of Survey Research*, New York: Academic Press, pp. 289–328.

Bradburn, N., Frankel, M., Hunt, E., Ingels, J., Schoua-Glusberg, A., Wojcik, M., and Pergamit, M. (1991), "A Comparison of Computer-Assisted Personal Interviews with Personal Interviews in the National Longitudinal Survey of Labor Market Behavior—Youth Cohort," *Proceedings of the U.S. Bureau of the Census Annual Research Conference*, Washington, DC: U.S. Bureau of the Census, pp. 389–397.

Bradburn, N., Sudman, S., and Associates (1979), *Improving Interview Method and Questionnaire Design: Response Effects to Threatening Questions in Survey Research*, San Francisco: Jossey-Bass.

Catania, J., Gibson, D., Chitwood, D., and Coates, T. (1990), "Methodological Problems in AIDS Behavioral Research: Influences on Measurement Error and Participation Bias in Studies of Sexual Behavior," *Psychological Bulletin*, 108, pp. 339–362.

Groves, R. M., and Mathiowetz, N. (1987), "A Comparison of CATI and non-CATI Questionnaires," in O. Thornberry (ed.), "An Experimental Comparison of Telephone and Health Interview Surveys," *Vital and Health Statistics*, Series 2, No. 106, DHHS Publication No. 87-1380, Washington, DC: U.S. Government Printing Office, pp. 33–39.

Gurman, S. (1989), "Six of One . . . ," *Nature*, 342, p. 12.

Jones, E., and Forrest, J. (1992), "Underreporting of Abortions in Surveys of U.S. Women: 1976 to 1988," *Demography*, 29, pp. 113–126.

Judkins, D.R., Mosher, W.D., and Botman, S. (1989), "National Survey of Family Growth: Design, Estimation, and Inference," *Vital and Health Statistics*, Series 2, No. 109, DHHS Publication No. 89-1383, Washington, DC: U.S. Government Printing Office.

Lee, R.E., and Renzetti, C.M. (1993), "The Problems of Researching Sensitive Topics: An Overview and Introduction," in Renzetti, C.M., and Lee, R.E. (eds.), *Researching Sensitive Topics*, Newbury Park, CA: Sage, pp. 3–13.

London, K. A., and Williams, L. B. (1990), "The Use of a Self-Administered Questionnaire for Improving Abortion Reporting in the 1988 National Survey of Family Growth," paper presented at the Annual Meeting of the Population Association of America, Toronto, Ontario, Canada.

Marquis, K., and Cannell, C. (1971), "Effect of Some Experimental Techniques on Reporting in the Health Interview," *Vital and Health Statistics*, National Center for Health Statistics, Series 2, No. 41, DHEW Publication No. 1000, Washington, DC: U.S. Government Printing Office.

Mensch, B., and Kandel, D. (1988), "Underreporting of Substance Use in a National Longitudinal Youth Cohort: Individual and Interviewer Effects," *Public Opinion Quarterly*, 52, pp. 100–124.

Mosher, W.D., and Duffer, A.P., Jr. (1994), "Experiments in Survey Data Collection: The National Survey of Family Growth Pretest," paper presented at the Annual Meeting of the Population Association of America, Miami.

Pratt, W.F., Tourangeau, R., Jobe, J.B., Rasinski, K.A., London, K.A., Baldwin, A.K., and Smith, T.W. (in press), "Asking Sensitive Questions in a Health Survey," *Vital and Health Statistics*, Washington, DC: U.S. Government Printing Office.

Renzetti, C.M., and Lee, R.E. (eds.) (1993), *Researching Sensitive Topics*, Newbury Park, CA: Sage.

Schober, S., Caces, M.F., Pergamit, M., and Branden, L. (1992), "Effects of Mode of Administration on Reporting of Drug Use in the National Longitudinal Survey," in C.F. Turner, J.T. Lessler, and J. C. Gfroerer (eds.), *Survey Measurement of Drug Use: Methodological Studies*, Rockville, MD: National Institute on Drug Abuse, pp. 267–276.

Smith, T. (1992), "Discrepancies Between Men and Women in Reporting Number of Sexual Partners: A Summary from Four Countries," *Social Biology*, 39, pp. 203–211.

Tanfer, K. (1990), "Contraception and Abortion in the NSFG: A Critical Review," *National Survey of Family Growth: Mission for the 1990s: Conference Proceedings*.

Tourangeau, R., Rasinski, K., Jobe, J.B., Smith, T.W., and Pratt, W.F. (1996), "Sources of Error in a Survey on Sexual Behavior," Manuscript submitted for publication.

Turner, C., Lessler, J., and Devore, J. (1992), "Effects of Mode of Administration and

Wording on Reporting of Drug Use," in C. Turner, J. Lessler, and J. Gfroerer, (eds.), *Survey Measurement of Drug Use: Methodological Studies*, Rockville, MD: National Institute on Drug Abuse, pp. 177–220.

Waterton, J., and Duffy, J. (1984), "A Comparison of Computer Interviewing Techniques and Traditional Methods for the Collection of Self-Report Alcohol Consumption Data in a Field Survey," *International Statistical Review*, 52, pp. 173–182.

CHAPTER 14

Children as Respondents: Methods for Improving Data Quality

Jacqueline Scott
University of Cambridge

14.1 INTRODUCTION

The sentiment that children should be seen and not heard could not be more inappropriate for the current era in which there is a growing demand for research that involves interviewing children. Children are no longer viewed as nonpolitical or economically irrelevant beings and the construction of childhood that focused on the intellectual limitations and social and sexual ignorance of children is now recognized as being overstated, if not simply untrue. Perhaps the change is due to market forces, children are now indisputably a major economic force in society. It is not just that they are consumers in their own right, but the importance of "pester power" in families is now well-recognized by market research. Yet it is not just monetary interests which are putting the spotlight on children. Society is becoming more concerned in general with children's issues and there is an increasing interest in children's rights. The construction of childhood that views children as "incomplete adults" is coming under attack and there is a new demand for research that focuses on children as actors in their own right.

The French historian Philip Ariès (1962) suggested that modern Western childhood is unique in the way that it quarantines children from the world of adults, so that childhood is associated with play and education rather than work and economic responsibility. The quarantine of childhood is reflected in the exclusion of children from statistics and other social accounts (Qvortrup,

Survey Measurement and Process Quality, Edited by Lyberg, Biemer, Collins, de Leeuw, Dippo, Schwarz, Trewin.
ISBN 0-471-16559-X © 1997 John Wiley & Sons, Inc.

1990) and there exists very little material that directly addresses the experience of childhood at the societal level. In surveys of the general population, children have been usually regarded as "out of scope" and samples typically only include respondents over the age of 16 or 18. Interviews with children have been seen as the special preserve of child development specialists, child psychiatrists or educational specialists, and the like, rather than a legitimate domain for general purpose surveys.

14.2 SURVEY RESEARCH AND THE EXCLUSION OF CHILDREN

Survey practice has tended to follow the "quarantine" approach with children being, at best, the subject of proxy information and, at worst, invisible. Moreover, much of the research that does take children into account is concerned with the effect of children on adult lives, rather than focusing on children as social actors in their own right. Panel studies of households, for example, are conducted as if children are auxiliary members, whose presence contributes to measures of household size, density, the labor market participation of mothers, household income, and the like. Research that is interested in children *per se* is relatively rare.

Even child development studies have often been conducted on the assumption that a child should be studied but not heard, with information gathered from the responsible parent, nearly always the mother. Yet, as any parent will attest, children do have voices, they express opinions, they observe and judge, and they exert a crucial influence on the way families and households function. Moreover, there is often a very large gulf between parental observations about their child and the child's own perceptions. Furthermore, parents may be the last people to know about things of consequence in their children's lives, especially once their children reach adolescence. Of course, for some topics the responsible parents are better placed to provide information about children than the children themselves (e.g., health diagnosis), but for questions tapping the child's own viewpoint, proxy information is clearly inadequate.

In this chapter I argue that the best people to provide information on the child's perspective, actions, and attitudes are children themselves. Children, as many studies have shown, are good respondents if questioned about events that are meaningful to their lives. One basic point I wish to make is that the quality of data we collect about society will be enhanced if we tap information that can be provided by children as respondents. Moreover, if children are viewed as actors rather than as incomplete adults, then this has a direct bearing on the types of questions that are relevant to ask. For example, standard definitions of work that are confined to adult activity fail to include the child's own work contributions (Morrow, 1994). Yet, children's ability to perform household chores and care for younger siblings can be crucial to the household economy, when both parents are at work (Solberg, 1990). Similarly, time budget studies often ignore the fact that children have their own time (for an exception, see Timmer *et al.*, 1985). Even studies on the costs of children have tended to view children as items on the parent's budget, rather than as economic actors who exercise considerable clout in family expenditure on food and consumer durables. Including children as respondents can, therefore, improve data quality on a whole range of social and economic issues by providing a more accurate and complete account of social life.

Yet interviewing children does pose some particular practical and methodological problems and the current knowledge about children as respondents is very fragmented. There are many problems to be solved when the respondents are children, including problems of language use, literacy and different stages of cognitive development. There is also a heightened concern about data quality, with some skepticism about whether an adult interviewer can obtain reliable and valid accounts from children, especially in areas where the information may be sensitive and subject to adult sanctions and control. In addition, issues of confidentiality and ethics become especially important when interviewing minors. Yet, as this chapter shows, there are solutions to such problems that deserve consideration, given the potential benefits of collecting data directly from children themselves.

In this chapter, I discuss some of the accumulated knowledge regarding the techniques for interviewing children face to face in large scale surveys and the strategies for optimizing the measures used and the quality of the resulting data. The chapter draws together diverse practical knowledge from both qualitative and quantitative research on appropriate methods of interviewing children. This practical knowledge comes from a very wide range of disciplines, including psychology, anthropology, education, criminology, and sociology. In addition, in order to illustrate the constraints and practical challenges of including children in an ongoing large-scale general purpose survey, I describe, as a case study, the development and implementation of a young person's questionnaire for children aged 11–15 in the British Household Panel Study. First, however, I consider why, in our supposedly child-oriented society, children are so often ignored by large-scale, general population, survey research.

The exclusion of children in surveys has at least four distinct causes. First is the inertia of practice. Most studies, even when their subject matter requires information about children, only interview adult respondents. Second, children may be omitted because of the tendency to accredit adults with greater knowledge, experience, and power (Backett and Alexander, 1991). Third, interviewing children is viewed as too problematic to be worth the possible pay off. Interviewing minors poses both practical and ethical issues which researchers might wish to avoid. The fourth reason is ignorance or perhaps a half-truth. Children are commonly believed to lack the communication, cognitive and social skills that are the prerequisite of good respondents. Recent research on children's cognitive and social development, however, suggests that children make better respondents than previously believed.

14.3 CHILDREN'S COGNITIVE AND SOCIAL DEVELOPMENT

One of the most important, detailed, and controversial theories of intellectual development is that associated with the Swiss psychologist Jean Piaget. Piaget's theory of cognitive development is particularly useful to survey researchers, because he was concerned with building a general theory of cognitive development rather than focusing on cognitive styles and individual differences (Phillips, 1981). According to Piagetian theory, children's thinking develops through a fixed sequence of stages evolving from less to more logical thought. This theory has been subject to much challenge and amendment in the light of subsequent research. It is now generally recognized that, even at a specific age, children's abilities vary. Neo-Piagetian theories and recent empirical research have concentrated on this variation in children's abilities and have provided a far richer and more complex view of how children think (Steward *et al.*, 1993). Nevertheless, children's cognitive capacity clearly does increase with age and the rudimentary levels of cognitive development remain relevant for understanding the question and answer process and for highlighting the ways in which children may differ from adult respondents.

As survey research is so heavily dependent on language, the only developmental periods which are pertinent to this chapter are the preoperational, concrete and formal operational periods. The preoperational period (aged 2–6) is characterized by the symbolic use of language, but has at least two inherent limitations that affect the child's capacity to answer questions concerning the social world. First, the child is egocentric in his or her world view and has a simplistic representation that is centered on him- or herself. Second, and related to egocentricism, is the notion of centering, whereby the child's attention is focused so exclusively on one aspect of a situation that he or she is unable to process information about any other aspect. These limitations, if true, pose very serious constraints on the types of questions that can be asked of children under seven.

There has been much criticism of Piaget for underestimating the reasoning ability of very young children (e.g., Donaldson, 1978). The assumption that children under the age of seven are very bad at communicating because they are highly egocentric has been successfully challenged. Experimental research has clearly demonstrated that even preschool children are able to appreciate someone else's point of view, can make social judgements, and even identify false intentions and beliefs (Astington *et al.*, 1988). However, even if children of the pre-school age are not as limited in their ability to reason as once thought, there are still overwhelming limitations of comprehension and verbal memory, that make interviewing children of this age highly problematic.

Dramatic changes in the characteristic of thought occur between the preoperational and concrete operational period, which extends from about age seven to about age eleven. At this stage, children are not only able to take on the view of others, but they are also capable of logical thinking and deductive reasoning, even if their thought processes are still tied to the concrete operations of their immediate world. For children under eleven, visual stimuli can be especially useful in the questioning process, because pictures make the issue far more concrete than verbal representation alone. Aids to memory can also be used to good effect, as children tend to forget even a relatively limited set of response options. Children's performance on memory tasks improves markedly with age and, by eleven, children's ability to remember is not so different from adults (although the information content of memory is much more limited).

Once children reach the formal operational stage (11 onwards), they begin to realize that two people may interpret the "same" facts very differently and that beliefs are inherently subjective. Whereas younger children tend to believe that if two people differ in their interpretation of some event, then at least one, or possibly both, must be mistaken. In contrast, most adolescents, like adults, realize that diversity of opinion is intrinsic to the process of knowledge.

The relativity of opinion is taken for granted in most attitudinal surveys. Standard forms of question wordings such as "Some people believe ... while others think ..." could be confusing for preadolescents. Moreover, it would be difficult, if not impossible, to explain to children under the age of eleven that the survey researcher is interested in exploring the diversity of beliefs or behavior in the population in general. Thus while, with relatively little adaptation, surveys designed for adults can be used with adolescents, different methods of interviewing are more appropriate for younger children.

14.4 DIFFERENT METHODS FOR DIFFERENT AGE GROUPS

Research methods that involve children as respondents have to take account of the wide range of cognitive and social development that depends primarily on age, but also on the sex, socio-economic background, and ethnicity of the child. Standard questionnaire techniques are clearly inappropriate with preschool children, whose self-image is faint and who have difficulty in distinguishing between what is said and what is meant (Robinson, 1986). But what methods can be used instead? One alternative is the clinical method that Piaget himself used, whereby the investigator adapts the questions in the light of the child's own answers. Another method that is commonly advocated as a means of obtaining information from preschool-aged children is the use of projective techniques, such as drawings. It is important to note that artwork usually supplements rather than replaces verbal communication. It is not only that the pictures are usually stimulated by some verbal instruction but also, unless the child provides a verbal commentary, the adult's interpretation of what the drawing represents might be hopelessly flawed.

Once children have reached the age of seven, it is possible to use both individual and group semistructured interviews with children. The classic study *Lore and Language of Schoolchildren* involved interviews with more than 5,000 children (from seven to early teens) and revealed a distinctive childcentered culture of customs and beliefs (Opie and Opie, 1959). One problem is that while preteen children can and do tell us about themselves, they have also mastered the art of impression management and, like adults, will tend to edit their answers (Fine and Sandstrum, 1988). Thus the Opies found that if they asked about superstitions, children said (as they are expected to say) that all superstitions are silly. But probing the child's own perspective revealed a world of half-beliefs and superstitious practices that invest children with some degree of control over the unpredictability of everyday experiences.

Thus, once children have reached the age where they make good respondents, they are also adept at controlling what they reveal. This is aptly illustrated by a study of children aged seven onwards to investigate the strategies used to persuade parents and other adults to buy them things (Middleton *et al.*, 1994). Using group discussions in school, the researchers found that children reported using begging, repetition, direct action, bribery, part-payment, negotiation, threats, and actions, each with varying degrees of success. The range of techniques reported by seven and eight year olds was already large and not much was added to the persuasion repertoire after the age of eleven. The authors note that, if anything, the younger children were less reticent in discussion than teenagers. By adolescence, young people are wary of revealing their secrets to an adult.

By adolescence (aged eleven onwards) it becomes possible to use a standardized questionnaire instrument, although problems of literacy, confidentiality and context have to be taken into account. Often the instuments are very similar to the ones used with adults and, with adult help, standardized instruments can be successfully used with even younger children. In order to identify problems with comprehension and ambiguities in question wording, to detect flippancy and boredom, and to discover discrepancies between the children's understanding and the researcher's intent, pretesting the survey instrument is crucial. A variety of pretest methods can be useful, including cognitive techniques such as asking the child to "think aloud," coding of nonverbal behaviors, and even video analysis of the interview interactions. Certainly, most questionnaires developed for adults or older children, will need some adaptation before they are suitable to use with younger children.

It is also sometimes necessary to adapt standard interview practice. For example, when respondents are children, it may be appropriate to provide interviewers with more leeway than is normal. Children tend to ask for more guidance than adults, especially when they are unsure what a question means. In such circumstances, it is preferable for interviewers to paraphrase the question than give the standard response "whatever it means to you." Standard interview practice might also have to be modified to protect children's privacy and confidentiality, especially in settings where children are subject to what might be called "adult imperialism" (as in schools). Unfortunately, confidentiality issues can also become real ethical dilemmas if children reveal behaviors that put them at risk (Stanley and Sieber, 1992).

Table 14.1 presents, by way of illustration, summary information for seven major social surveys that collect information from children using structured questionnaires in Britain and the United States. The different surveys have

Table 14.1 Seven Major Surveys Interviewing Children in Britain and the United States

Study name and survey type	Country and year	Sample	Method of data collection	Context
British Household Panel Study	Britain 1994– annual	Children aged 11–15 in panel households $N \approx 900$	Walkman tape self-administered	Home interview
British Social Attitudes	Britain 1994–	Children aged 12–17	Face to face	Home interview
National Child Development Study Longitudinal Cohort	Britain 1965, 1969, 1974, 1981, 1991	Children born in one week of 1958 $N \approx 16,000$	Face to face	Home interview
Twenty-07 Longitudinal Cohort	West Scotland 1987/8	Youngest cohort aged 15 $N \approx 1,000$	Face to face and self-completion	Home interview
Monitoring the Future Cross-section	U.S.A. 1975– annual	High-school seniors $N \approx 16,000$	Self-completion	School-based
National Survey of Children Longitudinal	U.S.A. 1976, 1981	Childrn aged 7–11 $N \approx 2200$	Face to face	Home interview
National Longitudinal Survey of Youth	U.S.A. 1979– annual	Youth aged 14–21 $N \approx 12,600$	Face to face	Home interview

337

different strengths and weaknesses, depending on the context and method used, as we discuss below.

14.5 THE IMPORTANCE OF CONTEXT IN INTERVIEWING CHILDREN

Children's social worlds span many different settings but home and school are two of the most important. Context is especially important in interviewing children because the expression of the child's personality, in terms of behavior and attitudinal preferences, is often highly context dependent. The same child could be boisterous and outspoken at home, but shy and reserved at school. Thus *where* the interviews are carried out is quite likely to influence the *way* children respond. In addition, the interviewer setting is important because the social meaning children will attach to concepts such as work or honesty may differ depending on whether children are at home or at school. The mode of interview is also very important in terms of data quality. Whether the interview is face to face, telephone, or self-completion may enhance or reduce the likelihood of different response biases such as social desirability or response contamination. (These and other response biases are discussed more fully in the next section.)

Interviewing children in schools is, on the whole, more cost effective than interviewing children in the home. One problem of classroom surveys is that they usually rely on self-completion schedules, which can encounter difficulties with literacy and motivation. Motivation is often less of a problem with younger children who may even approach a survey questionnaire as if it were a test. This test-taking mentality, although likely to enhance what is perceived to be the "correct" response, may be beneficial in making children pay greater attention to the questions. A main drawback of school-based interviewing is that children of all ages are likely to be influenced by the proximity of class mates. Even if answers are supposedly confidential, children are likely to quiz one another on their responses and may be tempted to give answers that win favor with the peer group.

Interviewing children in the home is more time-consuming and therefore more costly. It also runs the risk of children's answers being influenced by the presence of parents or siblings. Even if the interviewer is instructed to interview the child in private if possible, complete privacy is often impractical or elusive in the home. Home interviews, however, are usually carried out in person, which allows: (1) more routing complexity; (2) use of visual aids and show cards; and (3) explicit interviewer prompts for further information, when needed. All three are particularly important when interviewing young adolescents. Routing is needed to ensure that children at different stages of social development are asked appropriate questions. Visual aids are useful when there are vocabulary problems and limited attention spans. Interviewer prompts are essential when inadequate answers are given, because of lack of communication skills.

New interviewing techniques using Computer Assisted Personal Interviewing (CAPI) methods add further enhancements that could be used to good effect with younger respondents. Not only do they make complex routing relatively effortless for the interviewer, but they also provide the opportunity to incorporate videos and other visual and audio stimuli that reduce the need to rely so heavily on verbal questions and answers.

Telephone interviewing can be a far more economical alternative and, at least in the United States, has proved effective with children aged eleven and older (Reich and Earls, 1990). The success of interviewing young people by telephone is not surprising given the amount of time teenagers spend on the telephone confiding in friends. But a major draw-back with the telephone interview is the possible lack of privacy. This is particularly crucial when interviewing children and may limit the usefulness of telephone interviews as a means of collecting sensitive information.

One relatively novel method of collecting sensitive information from children is the diary method. Diaries are also good for collecting information that is too detailed for retrospective reports to be reliable. In principle, the method should be useful if the format can be made sufficiently simple and internal checks for accuracy can be devised. The Office for National Statistics (ONS) in Britain has recently been investigating the feasibility of children keeping expenditure diaries with a view to introducing children's diaries into the adult Family Expenditure Survey (Jarvis, 1994). The research is in its preliminary stages, but youth diaries are seen as a way of significantly improving the quality of family expenditure estimates.

14.6 ARE CHILDREN GOOD RESPONDENTS?

An old proverb says it is only children and fools who tell the truth. In contrast, Belloc's cautionary tales tell us of Matilda who told such "dreadful lies it made us gasp and stretch one's eyes." In this section, three questions will be considered. First, are children any more or less reliable than adult respondents? Second, how can we evaluate the quality of responses? Third, how can we improve data quality when children are respondents?

There is a seeming reluctance to take children's responses at face value, perhaps because children's opinions are seen as especially pliable and susceptible to suggestion. This is an area that has come under the glare of public scrutiny in recent times, as there is mounting concern about the reliability of children's testimonies in cases concerning child abuse and the like (Fincham et al., 1994; Ceci and Bruck, 1994). There is little reason to discredit children as respondents, however, because in highly traumatic circumstances children, like adults, have been known to lie or display memory distortion. Moreover, modern psychological and medical evidence suggests that children are more reliable as witnesses than previously thought, and reliability can be increased by skillful interviewing (Spencer and Flin, 1990). The interviewing advice is very familiar to survey researchers: give the child unambiguous and comprehensible instructions at the start of the interview; avoid leading questions; explicitly permit "don't know" responses to avoid best guesses; and interview the child on home ground, if possible.

There is growing evidence to suggest that the best source of information about issues pertinent to children is the children themselves. While parents and teachers can provide useful insights into child behavior, the direct interviewing of children provides a far more complete account of the child's life. For example, in the health domain, young children often report depressed or anxious symptoms of which their parents appear to be unaware. School age children report far more fears than their mother's accounts reveal (Tizard, 1986). Older children may be involved with alcohol or drugs without their parents' knowledge. Yet, when it comes to younger children's own behavioral problems, parents can be more forthcoming than the children (Reich and Earls, 1990). For many areas of research, therefore, it is better to gather information from multiple sources, as any one account may be biased (Tein *et al.*, 1994).

14.7 IMPROVING DATA QUALITY

The quality of data that results from interviewing children will depend on a number of different factors. First and most basic is the appropriateness of the research topic and measures used. In designing suitable measures for young children, researchers have to, at minimum, ensure that the questions really do measure the desired concept; that the questions are unambiguous, and that children interpret the questions in the way the researcher intended. Research concerning the question and answer process with children as respondents is extremely sparse. However, the research clearly suggests that the clarity of questions influences the quality of the data, especially for younger children and that complex questions are problematic regardless of the child's age (De Leeuw and Otter, 1995).

We also know that both younger children and adolescents alike tend to respond to adult questioning, whether or not they know the answer, or have an attitude on the issue at hand (Parker, 1984; Weber *et al.*, 1994). This can make for low stability on issues where the children's knowledge is limited or their attitudes are noncrystallized (Vaillancourt, 1973). Thus in order to achieve meaningful data, questions have to be pertinent and relevant to the children's own experience or knowledge. However, when this condition is met it is clear that even quite young children can make good respondents, as the following two studies illustrate.

In the United States, a specially adapted self-report instrument measuring antisocial behavior was administered to boys as young as seven (Loeber and Farrington, 1989). In order to obtain independent measures of behavior, information was also collected from the primary parent or caretaker. When questioning the boys, interviewers first checked whether each item was clearly understood, by probing for examples. Only those items which the child could interpret were included in the subsequent inventory. Information was collected on whether the child had ever engaged in each kind of antisocial behavior, and if so whether he had done so in the past six months. Bounded recall methods were used to establish the six-month period, using Christmas, school terms and events from personal life.

Not surprisingly, there were marked differences between seven-year olds and ten-year olds in understanding the meaning of questions, with skipping school, for example, only understood by 75 percent of American first-graders (aged 6–7) and almost 100 percent of fourth-graders (aged 10–11). The most prevalent antisocial behavior was hitting siblings (a concept well understood by all). The boys' estimates were fairly consistent with parental reports, but for most behaviors there was higher correlation between parental and child reports at aged seven than at aged ten. One possible interpretation is that older children are less reliable. This, however, would be at odds with nearly all other empirical evidence, and a far more likely interpretation is that parents know less about the behavior of older children. Thus the different correlations reflect different states of parental knowledge.

In an Australian study of children in families, even primary-school-aged children (aged 8–9) were able to give articulate and informative responses to questions about objective family circumstances as long as the questions were about the here and now, or very recent past (Amato and Ochiltree, 1987). On the other hand, this study provides clear evidence that questions that are outside the child's own experience, such as what parents do at work, are problematic. The quality of data for this sort of question improves markedly with older children (aged 15–16), who give answers that are more in line with the parental response.

Asking questions that are meaningful to the child's own experience is not, however, sufficient to guarantee data quality. A second factor that is fundamental to improving data quality concerns the child's willingness and ability to answer the questions and articulate his or her subjective experience. This in part depends on the appropriateness, number and order of the response alternatives. In a rare split ballot experiment with school children (aged 10 and over), evidence was found of primacy effects in items where a response set required a choice among five or more options (Hershey and Hill, 1976). A similar problem occurred when using multi-item picture stimuli. Pictures are often considered useful because not only are they non-verbal but they also hold the limited attention of younger children. However, pictures do not ease the basic decision-making process and, when interviewing children, responses are likely to be less prone to measurement error if the choices are kept simple.

Children's responses will also be subject to the standard biases that have been relatively well researched in the question and answer process for adults— things like context effects, social desirability, acquiescence bias, and the like.

However, it is important not to simply assume that findings applicable to adults will generalize to children. For example, it has been claimed that children may be less susceptible than adults to social desirability bias. However, the validity of this claim depends on the definition of social desirability. Social desirability is often defined in adultcentric terms. Moreover, children's own ideas about social desirability are heavily context dependent. In one situation, children might be tempted to downplay their reports of deviant behavior and cigarette or drug use but in another be prone to exaggeration.

Moreover, things like context effects and acquiescence bias may well take a different form at different stages of the life course depending on the subject matter at issue. For example, the norm of reciprocity which exerts a powerful influence on adults to answer contiguous questions in an even-handed way (Schuman and Presser, 1981) may not have the same moral imperative for younger children. On the other hand, it has been claimed that young children are particularly suggestible and interviewers approbation or disapproval can have a marked effect. Nevertheless, experimental work found little evidence of acquiescence bias among older school-aged children (Hershey and Hill, 1976). This is an area where clearly more research is needed but, until we have more evidence, it seems good practice to include internal consistency checks, where possible, when interviewing children.

A third set of issues concerns the children's motivation to give careful and truthful answers. In this regard, the interviewer and the rapport between interviewer and child are crucial. A good relationship can encourage more forthcoming responses, especially when children are convinced that their responses are truly confidential. However, the interviewer and the relation between interviewer and child can also be a source of error. For instance, interviewers who are intimidating or impatient may inhibit children in a way that has damaging consequences for data quality. None of these factors is distinctive to children—they all apply to adult respondents as well—but achieving optimal conditions might require somewhat different solutions for children.

In the following section, I discuss the development of the Young Person's Survey in the British Household Panel Study (BHPS), to illustrate one way in which a general purpose survey can be adapted to incorporate the different approach that is needed with children.

14.8   BHPS YOUNG PEOPLE'S SURVEY

The British Household Panel Study was launched in 1991, with the remit of monitoring microsocial change in Britain through the 1990s and beyond. The survey is carried out by National Opinion Polls on behalf of the University of Essex. The BHPS involves an annual survey of each member of a nationally representative sample of at least 5,000 households, making a total of approximately 10,000 individuals. For the first time, in 1994 (for the fourth wave), children aged 11–15 who are members of the panel households, are being interviewed in person. Almost 800 children are eligible and they are spread across some 600 households. These children will then automatically become adult sample members when they reach the age of 16 (Scott *et al.*, 1995).

Interviewing children in their homes, as we have seen, is often ruled out as being too time-consuming and expensive. For the BHPS this posed no problem as we were already interviewing the adult members of the households. Interviewing children at home, however, does pose special problems because it is not always possible to carry out the interview in private. Our main concern was to find a way of guaranteeing that the children's answers would be private and confidential. Not surprisingly, children may well find it difficult to be frank and honest if their parents or siblings are present. But how can children be interviewed in such a way as to guarantee complete confidentiality, when their answers might well be overheard? Standard face to face interviews are not sufficiently private and self-completion methods were ruled out as unsuitable given some children's limited reading skills.

We turned, therefore, to a new and relatively untested interview method that involved prerecording the young person's questionnaire on a personal Walkman. Listening to the questions on tape has two significant advantages. First, unlike in a standard interview, no one else can hear the questions. Moreover because the answer booklet only contains response categories, the method ensures complete privacy. Second, unlike the standard self-administered questionnaire, the method overcomes the problem of literacy, which can be particularly acute when the respondents are children.

The Walkman interview method was first developed and tested in the United States, for the Youth Risk Behavior Surveillance System Questionnaire for the National Center for Health Statistics (Camburn *et al.*, 1991). Early use of the Walkman technique demonstrated that it is possible to ask for sensitive information from children, while ensuring that the answers would be private and confidential. This feature is particularly important to the BHPS because the young person's component of the survey relates to health attitudes and behaviors within the family, including such highly sensitive material as drug use and mental health. It seems likely, however, that for almost any topic it is important to ensure privacy from other family members. Children from eleven onwards are particularly sensitive to privacy issues in the home, as is manifest by the common "private keep out" signs on bedroom doors.

Having decided on the appropriate mode of interview, it was still necessary to test the application of the Walkman method in the British context. There were many issues to be resolved. What was the preferred voice type and the optimal speed of reading, for the prerecorded interview? What question formats were most appropriate for the method and could open-ended questions be used? Our pretest phase had two parts: first we used qualitative group interviews with children to give feedback on the use of Walkmans and second we piloted the structured interview in the home setting.

14.9   USING FOCUS GROUPS TO DEVELOP THE SURVEY INSTRUMENT

The children's component of the BHPS has been commissioned by the Health Education Authority to focus specifically on the changing nature of health-related attitudes and behaviors of young people and their families. We had very little previous research to draw on for background information in formulating our questions. Given the lack of knowledge about this area, it made sense to use more qualitative methods to help inform the development of the structured survey instruments. Using focus groups was one method of eliciting information from children themselves that was undoubtedly preferable to guesswork (Morgan, 1993).

Turning to the literature on focus groups, we found surprisingly little guidance about conducting focus groups of children. Most of the material to date has been produced by market researchers. Conventional wisdom regarding "best practice" for interviewing children in a group setting suggests that: (1) children should be interviewed in restricted age groups as otherwise older children will dominate; (2) boys and girls should be interviewed separately as they have such different communication styles; (3) groups should be small, with no more than eight children at maximum. Thus, in order to ensure group identity and cohesion we separated the groups by gender, age (11–13 years and 13–15 years), and socio-economic category. The groups were conducted in three different parts of England: the South, the Midlands and the North. The recruiting and conduct of the focus groups were carried out by the qualitative division of National Opinion Polls, who used a combination of doorstep screening and snow-balling to fulfill the recruitment criteria.

Group sessions were conducted in the interviewer's home. The children's focus groups lasted approximately two hours, with a break for a fast food snack at half time. The group leader followed a detailed discussion guide, which included topics such as freedom and rules in the house, family communication and sources of advice, health beliefs and practices, anxiety and depression, and future aspirations and expectations. For some of the topics visual materials were used to stimulate discussion (for example, a picture of Munch's painting "The Scream" was used to probe feelings of anguish). In addition, some semi-structured questions were included that we hoped might prove suitable measures for this age group. For example, we showed a card with a range of smiley faces to see whether young people could identify their state of happiness with respect to different aspects of life. We feared that the smiley faces might be insulting to children of this age range, who are so sensitive to being treated as "kids." However, to our surprise, the scale worked extremely well, prompting some very sophisticated discussion of mood states and changes. Even children as young as eleven have remarkable insight into impression management and self-presentation in everyday life and drew attention to the fact that answers might be different depending on whether the question wanted to know how you feel inside, how you are trying to appear to other people, or how other people perceive you to be (Scott et al., 1995).

A further use of the focus groups was to inform the development of the Walkman method of interviewing. At the end of the focus groups, Walkmans were handed out to each participant together with a short self-completion booklet in which they could record their answers. This short test interview was designed to provide feedback on the voice type, the speed of question delivery, and the clearest design for the answer booklet. It was also important to test out the appropriate format of questions and, in particular, whether children could handle open as well as closed questions in the taped interviews. Group discussion then gave more general feedback on the Walkman method. The only technical problems were caused by adult ineptitude with the machines. Fortunately young people are very familiar with Walkman sets and are able to manage well without adult intervention.

Feedback from these groups indicated that young people were very sensitive to the quality of the voice rather than having a preference for a particular sex, age, or accent. There was, however, less consensus about the optimal speed of question delivery and preferred interview pace. In order to give adequate time for the children to answer, we initially repeated both questions and answers on the prerecorded questionnaires, but this caused considerable irritation to some children. Therefore, we subsequently recorded the question and response only once, but repeated the main thrust of the question. Children do vary considerably in the amount of time they need to respond. A taped interview, however, offers some advantages over the personal interview, in this respect. Children can "pause" and "rewind" the tape, whereas they might be embarrassed or reluctant to ask an interviewer to repeat the question or to allow them more time to think. Simple questions with numeric responses proved particularly suitable for these taped interviews, which rely on verbal memory. However, children as young as eleven are able to deal with open-ended questions, although the mean length of answer tends to increase with age, and younger children tend not to elaborate answers, unless specifically probed.

The focus groups showed beyond doubt that, given the right encouragement, young people are more than willing to say what they think. Thus the groups were both useful in their own right, providing a wealth of qualitative materials to analyze and they were also extremely useful in developing the quantitative research instruments and method. Nevertheless focus groups are clearly no substitute for a structured pretest that replicates as closely as possible the real household interview setting.

14.10 INTERVIEWING YOUNG PEOPLE AND PARENTS AT HOME

The structured household pilot survey was crucial in determining the appropriate length of interview for children of this age group. The questionnaire had been prerecorded with 83 questions in all, of which three were open-ended. The interview on average lasted 30 minutes. Surprisingly, interviewers reported no problems with this length of interview, even for our youngest respondents. For pretest purposes we had asked interviewers to be present when the children were doing their Walkman interview, so they could observe any difficulties. Interviewers were simultaneously conducting the parental interviews. This proved disastrous, as the children's presence inhibited the parent's responses. For our main survey, therefore, children are encouraged to do the Walkman interview in a different room, where possible.

Certain questions had to be modified as a result of our pretest experience. For example, we had followed standard advice and depersonalized possible threatening questions by asking children to respond in terms of "people of my age." Some children, however, are very literal in their interpretation and tried to guess the age of the script reader (or interviewer). Children are used to an adultcentric world and survey questions have to be extremely explicit that it is the child's own views that are wanted. We also found that we had not taken sufficient account of the complexity of family life today and questions that refer to mother or father are often inherently ambiguous when children are living with step-parents or parent substitutes. Vocabulary has not yet caught up with the reality of family situations and it is best to allow respondents to use their own terminology for relationships.

One concern that is usual in contacting children at home is the need to deal with parents as gatekeepers. In our survey, as parents are themselves participants, this rarely causes any problem. Moreover, on the whole the children are very keen to participate and may, if anything, help household response rates by increasing the enthusiasm of other household members. The ethics that apply to interviewing children need, if anything, to be more stringent than with adults. Children are relatively powerless in society and, despite the attention given to children's rights, they have relatively little recourse to official channels of complaint. It is therefore very important that researchers are particularly conscious of their ethical responsibilities when interviewing children. In particular, special attention needs to be paid to explaining the research purpose in a comprehensible fashion and obtaining informed consent from the children themselves.

The BHPS children's questionnaire is preceded with the same statement of confidentiality and voluntary participation as we use with our adult respondents. Following our practice with adult respondents, young people are offered an incentive (a money voucher) to take part in the survey. It has been questioned whether "bribing" children with money is ethical. Our position is that the incentive is acknowledging that we attach the same worth to children as to our adult respondents. Unfortunately, monetary considerations hindered our ideological commitment and children are paid at a lower rate than adult participants. Thus, even as respondents, children experience life's inequities.

14.11 SUMMARY AND CONCLUSIONS

Interviewing children is no fad. Improved data about children are essential in a society where children's role as consumers and citizens is being taken increasingly seriously in the economy, in law, and in social policy. In this chapter we have taken it as axiomatic that it is only by interviewing children directly that we can understand children's social worlds. Moreover, we have argued that general population surveys that omit accounts of children provide biased estimates of many important social variables. However, interviewing children does pose distinctive methodological problems that could impinge on the quality of the data.

We suggested that while structured questions are not appropriate for younger children because of cognitive and language limitations, by preadolescence, children are quite capable of providing meaningful and insightful information. However, children are also well-versed in techniques of impression management and issues of privacy and confidentiality are especially important if children are to give honest and complete answers. The setting of interviews is also important, in this respect, and it is crucial to find ways of avoiding response contamination both in the home and at school.

We stressed the difficulties involved in ensuring that survey instruments take account of the wide range of developmental stages of childhood. Pretests are especially important as children's understanding of a question can be quite different from that which the researcher intended. It is also important that questions are not posed from an adultcentric perspective. One useful way of investigating the children's own understanding of an issue is to use qualitative, in-depth group discussions, prior to developing the structured questionnaire. Another way is to use cognitive pretest methods, such as think-aloud techniques, to probe how the question is understood and why a particular answer is given.

We described the development of the Young People's Survey of the British Household Panel Study to illustrate some of the challenges and possible solutions in interviewing children in a large-scale general household survey. Children's focus groups proved a very successful way of identifying appropriate questions for obtaining sensitive information. The groups also provided useful feedback on technical aspects of using a prerecorded interview and self-completion answer booklet instead of the more standard face to face or self-completion methods. Taped interviews have two main advantages for use with children: they overcome literacy problems and they ensure privacy and confidentiality when interviewing in the home. The common wisdom that asking questions is a useful way of getting information as long as the respondent is able and willing to answer applies just as much, or perhaps even more strongly, when the respondent is a minor. Consternation about the reliability of children as witnesses in areas such as child abuse might stimulate further research in this area. However, the danger is that such research would highlight very specific problems, rather than providing a more general understanding of the question and answer process with children. Methodological research on conditions that enhance the ability of children to be good respondents is extremely sparse. We hope that it will expand rapidly as the practice of interviewing children in large-scale surveys becomes more widespread. Asking children is likely to be one of the best ways of learning how to improve the quality of our data. It is important, however, not to overemphasize data quality issues with respect to children. Data quality is *always* an issue, regardless of the age of the respondent. Interviewing children is essential given that the resulting data clearly provide a far more complete picture of many aspects of the social world than can be achieved by interviewing adults alone.

ACKNOWLEDGMENTS

I wish to acknowledge the support of the Economic and Social Research Council and the Health Education Authority. I am also very grateful to Edith de Leeuw, Jean Martin, Jared Jobe, Virginia Morrow, and Duane Alwin for helpful comments on an earlier draft, and to Terry Tostevin for help with library research.

REFERENCES

Amato, P., and Ochiltree, G. (1987), "Interviewing Children About Their Families: A Note on Data Quality," *Journal of Marriage and Family*, 49, pp. 669–675.

Ariès, P. (1962), *Centuries of Childhood*, Harmondsworth: Penguin.

Astington, J.W., Harris, P., and Olsen, D. (1988), *Developing Theories of Mind*, Cambridge: Cambridge University Press.

Backett, K., and Alexander, H. (1991), "Talking to Young Children About Health: Methods and Findings," *Health Education Journal*, 50, 1, pp. 34–38.

Camburn, D., Cynamon, M., and Harel, Y. (1991), "The Use of Audio Tapes and Written Questionnaires to Ask Sensitive Questions During Household Interviews," paper presented at National Field Directors' Conference, San Diego, CA.

Cedi, S.J., and Bruck, M. (1994), "How Reliable Are Children's Statements? . . . It Depends," *Family Relations*, 43, 3, pp. 255–257.

De Leeuw, E.D., and Otter, M.E. (1995), "The Reliability of Children's Responses to Questionnaire Items: Question Effects in Children's Questionnaire Data" in J.J. Hox *et al.* (eds.), *Advances in Family Research*, Amsterdam: Thesis.

Donaldson, M. (1978), *Children's Minds*, London: Fontana.

Fincham, F., Beach, S., Moore, T., and Diener, C. (1984), "The Professional Response to Child Sexual Abuse: Whose Interests Are Served?" *Family Relations*, 43, 3, pp. 244–254.

Find, G., and Sandstrum, K. (1988), *Knowing Children: Participant Observation with Minors*, Sage Qualitative Research Methods Series, No. 15.

Hershey, M., and Hill, D. (1976), "Positional Response Set in Pre-Adult Socialization Surveys," *Social Science Quarterly*, 56, pp. 707–714.

Jarvis, L. (1994), "The Feasibility of Children Keeping FES Expenditure Diaries: A Qualitative Study," *Survey Methodological Bulletin*, 35, 1, pp. 1–2.

Loeber, R., and Farrington, D. (1989), "Development of a New Measure of Self-Reported Antisocial Behavior for Young Children: Prevalence and Reliability," in M. Klein (ed.), *Cross-National Research in Self-Reported Crime and Delinquency*, Dordrect: Kluwer Academic Press.

Middleton, S., Ashworth, K., and Walker, R., (eds.) (1994), *Family Fortunes: Pressures on Parents and Children in the 1990s*, London: CPAG.

Morgan, D. (1993), "Using Qualitative Methods in the Development of Surveys," *Social Psychology, Newsletter*, 19, 1, pp. 1–2.

Morrow, V. (1994), "Responsible Children? Aspects of Children's Work and Employment Outside School in Contemporary UK," in B. Mayall (ed.), *Children's Childhoods Observed and Experienced*, London: The Falmer Press, pp. 128–143.

Opie, I., and Opie, P. (1959), *The Lore and Language of Schoolchildren*, Oxford: Oxford University Press.

Parker, W. (1984), "Interviewing Children: Problems and Promise," *Journal of Negro Education*, 53, 1, pp. 18–28.

Phillips, J. (1981), *Piaget's Theory: A Primer*, San Francisco: W.H. Freeman and Co.

Qvortrup, J., (1990), "A Voice for Children in Statistical and Social Accounting: A Plea for Children's Right to be Heard" in A. James, and A. Prout (eds.), *Constructing and Reconstructing Childhood*, London: The Falmer Press.

Reich, W., and Earls, F. (1990), "Interviewing Adolescents by Telephone: Is It a Useful Methodological Strategy?" *Comprehensive Psychiatry*, 31, 3, pp. 211–215.

Robinson, W.P. (1986), "Children's Understanding of the Distinction Between Messages and Meaning: Emergence and Implications" in M. Richards, and P. Light (eds.), *Children of Social Worlds*, Cambridge: Polity Press.

Schuman, H., and Presser, S. (1981), *Questions and Answers in Attitude Surveys*, New York: Academic Press.

Scott, J., Brynin M., and Smith, R. (1995), "Interviewing Children in the British Household Panel Study," in J.J. Hox *et al.* (eds.), *Advances in Family Research*, Amsterdam: Thesis.

Solberg, A. (1990), "Negotiating Childhood: Changing Constructions of Age for Norwegian Children," in A. James, and A. Prout (eds.), *Constructing and Reconstructing Childhood*, London: The Falmer Press.

Spencer, J., and Flin, R. (1990), *The Evidence of Children*, London: Blackstone Press Limited.

Stanley, B., and Sieber, J. (eds.) (1992), *Social Research on Children and Adolescents: Ethical Issues*, Newbury Park: Sage.

Steward, M., Bussey, K., Goodman, G., and Saywitz, K. (1993). "Implications of Developmental Research for Interviewing Children," *Child Abuse and Neglect*, 17, pp. 25–37.

Tein, J.-Y., Roosa, M., and Michaels, M. (1994), "Agreement Between Parent and Child Reports on Parental Behaviours," *Journal of Marriage and the Family*, 56, 2, pp. 341–355.

Timmer, S., Eccles, J., and O'Brien, K. (1985), "How Children Use Time" in T. Juster, and F. Stafford (eds.), *Time, Goods and Well-Being*, Ann Arbor: Institute of Social Research, pp. 353–382.

Tizard, B. (1986), "The Impact of the Nuclear Threat on Children's Development" in M. Richards, and P. Light (eds.), *Children of Social Worlds*, Cambridge: Polity Press.

Vaillancourt, P. (1973), "Stability of Children's Survey Responses," *Public Opinion Quarterly*, 37, pp. 373–387.

Weber, L., Miracle, A., and Skehan, T. (1994), "Interviewing Early Adolescents: Some Methodological Considerations," *Human Organizations*, 53, 1, pp. 42–47.

SECTION C

Post-Survey Processing and Operations

CHAPTER 15

Some Aspects of Post-Survey Processing

Lars Lyberg
Statistics Sweden

Daniel Kasprzyk
U.S. National Center for Education Statistics

15.1 INTRODUCTION

The design, development, and implementation of a sample survey comprise myriad aspects and components. While all components contribute directly to the end products of the survey, some, such as sampling and question-naire design, have been more prominent in the research literature than others, which appear to be less statistical and at times less research oriented. One group of components, a set of operations and processes, takes place after the survey, sample, and questionnaire are designed and after the data are collected. These operations, which we will call the post-survey data processing operations, consist of a number of processes whereby the collected data are processed and prepared for use and interpretation by the user. Processing operations can vary between surveys and systems but most of the time the following are included: editing, coding, data entry, imputation, weighting, estimation, and analysis. Some of these operations may affect total survey error considerably.

The relationship between post-survey processing and other survey operations is not always obvious. For instance, the increased use of computers in survey research allows some processing operations at the time of data collection. For this reason, some survey statisticians, when classifying survey processes, include

Survey Measurement and Process Quality, Edited by Lyberg, Biemer, Collins, de Leeuw, Dippo, Schwarz, Trewin.
ISBN 0-471-16559-X © 1997 John Wiley & Sons, Inc.

data collection or parts of it in data processing. In this monograph, quality issues related to data collection as well as estimation and analysis are treated in their own sections. The post-survey processing operations treated in this chapter include editing, coding and data entry. The perspective on these has changed during the last decade or so. These operations, once very labor intensive and error prone, are increasingly becoming automated, resulting in a consistently applied set of edit and coding specifications. Three lines of current development will gradually change the view and implementation of these post-survey data processing activities.

First, automation of operations like editing and coding has decreased the need for large and temporary workforces conducting manual operations for large social and economic surveys. These manual operations are usually implemented following some training and study of specifications as described in manuals. Thus, automation of manual processes has the potential to improve data quality since computers implement edit and coding rules in a more consistent and uniform manner than manual workers.

Second, computer assisted data collection has created opportunities to merge several processing operations and perform them at the time of data collection, thus improving productivity and data quality. Typically, in a Computer Assisted Survey Information Collection (CASIC) environment, some editing and coding functions can be performed at the time of the data collection, thus saving time and costs and increasing quality through immediate access to the respondent. The traditional approach to data collection has different units editing, coding, and keying with each unit following its own set of specifications. Many modern statistical organizations have replaced the traditional approach with other systems where these operations have been merged and modified. See Bethlehem (Chapter 16), Keller (1995) and Shanks (1989). Current thinking emphasizes the need for optimizing a processing system rather than the individual components, like editing and coding. This thinking represents a considerable change in perspective by data collection organizations with the resulting surveys requiring a physical and mental integration of data collection, management and delivery of data. The discussion concerns the integration of post-survey processing components that increasingly use several modes of collection.

Third, traditional quality assurance of operations like coding and keying relies heavily on acceptance sampling schemes guaranteeing a prespecified average outgoing quality. Such schemes, originally developed by Dodge and Romig (1944), do not emphasize employee feedback. Using these methods as the major quality function can also be criticized from other perspectives, namely that they can be costly, inefficient and they diminish employee responsibility. Ideas stemming from Deming (1986) and the total quality management movement identify approaches related to implementing a standardized process at the beginning rather than inspecting, working in teams trying to identify root causes of problems rather than having quality departments checking on employees, and using methods aiming for continuous improvement rather than keeping quality at a stable level. We will touch upon these developments in the remainder of this chapter.

This chapter then reviews the post-survey processing activities of editing, coding, and data entry, contrasting the traditional methods of operation with current thinking. It will describe current methods, problems associated with the specific activity, and finally the integration of these activities.

15.2 EDITING

There does not appear to be a standard definition of editing, but a working group of U.S. federal statisticians (Federal Committee on Statistical Methodology, 1990) defined it in the following way:

> Procedure(s) designed and used for detecting erroneous and/or questionable survey data (survey response data or identification type data) with the goal of correcting (manually and/or via electronic means) as much erroneous data (not necessarily all of the questioned data) as possible, usually prior to data imputation and summary procedures.

This definition encompasses a number of stages in the process of implementing a survey and can occur at various times during the survey cycle. The definition is consistent with the goals of editing as given by Granquist (1984): (1) to provide information about data quality; (2) to provide information about future survey process improvements; and (3) to simply "clean up" the data. However, the definition does explicitly emphasize the third goal, the "cleaning up" aspect.

Broadly speaking, editing is seen as a tool to identify erroneous or suspect data; it can encompass many different activities. As pointed out by Pierzchala (1990), editing can be a validating procedure or a statistical procedure. It can be done at the record level or at some level of aggregation. It can also take place at different stages of the survey, from the time of data collection through the summarization of data. Finally, it can be done manually or by means of computer programs.

When viewed as a validating procedure, editing is a set of methods aimed at identifying inconsistencies and suspicious values and then, if deemed necessary, correcting the value. This type of editing takes place on the individual record or individual questionnaire. When viewed as a statistical procedure, editing is based on a statistical analysis of the respondent data; thus, this type of editing may be characterized as between-record checks aimed at detecting outliers in univariate or multivariate distributions, reviews of aggregate data based on comparisons with historical data or customizing edit limits for individual sample units, based on historical data. In either case, the edit rules or the checks specified to identify erroneous data ought to be developed with subject matter experts who know the data, concepts, and relationships within the data.

The validating procedure is the traditional micro-editing approach which typically means one or more of the following procedures were followed: (1) clerks review a sample of questionnaires to check on the quality of the data collected by the field representatives to provide feedback on omissions, errors, and misunderstood instructions; (2) clerks review selected key items for legibility and consistency on all questionnaires; (3) a computer program is run in batch with prepared checks to identify suspicious or inconsistent data and invalid or missing entries. A reconciliation procedure involving respondent contacts or other information gathering is usually necessary to reach a decision whether data should be corrected or not. Macro-editing checks are used on aggregates of data, so that suspicious subsets of records can be identified. The purpose is to trace systematic inconsistencies and errors to individual records.

15.2.1   Examples of Editing Methods

Editing can be performed manually or automatically and at various stages of the survey process. It is common to have some manual editing as part of the data preparation before data are captured electronically. Manual edits often consist of a quality review of a sample of an interviewer's questionnaires or mail questionnaires followed by legibility checks, missing data checks for key items, and checks on certain reporting standards, such as having a number to the right of the decimal point and rounding to whole numbers when appropriate.

Automated editing is performed after data have been entered or in connection with data collection. These edit specifications are always survey-specific and involve procedures for completeness checks, range of variables checks, and skip instruction checks. In batch editing, records are checked for inconsistencies, impossibilities, and suspicious cases. This procedure usually results in a file of error messages that serves as the basis for potential corrections. In interactive editing, the results of the editing are shown on a display terminal and the editor decides to correct or override the error message. Modern CASIC technology allows some editing to be performed at the time of the interview while the respondent is still available for reconciliation purposes. Other typical checks within this type of micro-editing include acceptable range of variable values, that prespecified statements of type "if, then" hold and that, for instance, the sum of individual items equals a reported total.

On a multirecord level, statistical editing and macro-editing can be used. Statistical editing uses distributions to identify outliers as well as inliers (data falling within prespecified ranges and are suspiciously consistent). Macro-editing is performed on aggregates of records. If inconsistencies and other suspicious cases are found, they are traced to the individual records comprising the aggregate. The technique is similar to ones used in sampling rare items. Granquist (1994) and Houston and Bruce (1993) provide overviews of macro-multiediting methods including:

1. the aggregate method, where a number of key aggregates are reviewed and checked; suspicious aggregates are flagged and individual records comprising the aggregate are reviewed;

2. the Hidiroglou-Berthelot method, where acceptance bounds are derived from a statistical analysis and automatically calculated from the reported data; and

3. the top-down method, where values of the edits which are functions of the weighted keyed-in values are sorted from the top to the bottom of the list; manual review is started from the top or bottom of the list and continues until there is no effect on estimates.

Modifications of these methods include the box-plot method and gred (interactive graphical editing). The latter uses graphics and ideas from exploratory data analysis for problems such as outlier detection (see e.g., Bienias *et al.*, 1995). The general idea is to review and edit the worst data points and stop when editing makes no appreciable difference in the estimates.

Thus, the first aspect of editing identifies potential errors. The second part is the correction or adjustment phase. In some instances adjustments can be made after consulting the respondent, but in other cases the item may need to be blanked and an imputation done, i.e., suspicious, obviously erroneous, and missing values are replaced by more reasonable values. The imputation literature is quite extensive. See, for example, Little and Rubin (1987), Kalton and Kasprzyk (1982, 1986), Madow *et al.* (1983), Pierzchala (1990), and Ferguson (1994). Because of the number of observations and the extent of missing data, these procedures are usually automated. They might rely on deduction (the value is determined from other record values or past reports), models (functions of values determine the missing value), donors (values from a donor record are used for the imputation), and the simulation of human actions (expert data editing system).

As for the last type of procedure, a relatively new paradigm has recently come into play. This paradigm, Artificial Neural Networks, uses the human brain as a starting point. In this approach, a task like editing can be solved by connecting neurons in networks. Given a model of neurons it is possible to represent network knowledge by assigning different weights to different neuron connections. In practice, such an editing model would have to be a learning one, for instance by observing examples of pairs of records before and after expert editing. These ideas are discussed by Roddick (1993) and Nordbotten (1995).

15.2.2 Some Problems Associated with Editing

Editing can be a very costly and time-consuming task. In some surveys, where editing consumes 20%–40% of the entire survey budget (Federal Committee on Statistical Methodology, 1990; Granquist and Kovar (Chapter 18)). There are also additional costs to the user community if extensive editing delays survey processing and the release of the data. A well-known example is the World Fertility Survey, where extensive editing delayed the publication of results by one year. Thus, it is very important, when designing editing procedures, to focus on the total survey error. If budget shares for editing are between 20% and 40% and with fixed costs for the conduct of the data collection, it is easy to see there are limited fiscal resources left for addressing other error sources.

This concern is more problematic in light of studies showing that many editing tasks do not contribute to improved data quality or better estimates. These examples, typically found in the domain of economic surveys, are described in Bethlehem (1987), Boucher (1991), Linacre and Trewin (1989), and Corby (1984). Other evaluation studies show that the Pareto principle is at work; i.e., few errors are responsible for the majority of actual value changes as a result of editing (see, for instance, Greenberg and Petkunas, 1986). These results suggest that in economic surveys the editing function ought to be more productive. Data on hit-rates, i.e., the proportion of error messages that result in value changes, indicate that hit-rates as low as 20%–30% are not uncommon in micro-editing (Linacre and Trewin, 1989; Lindström, 1991). In recent surveys of school districts, schools, principals, and teachers the percent having one or more responses changed between the preliminary file and the pre-edit file (designed to identify inconsistencies, invalid entries, and critical missing data) ranged from 1.6% for public school principals to 34% for Bureau of Indian Affairs schools (Gruber et al, 1994).

Concerns have been raised that a substantial amount of overediting takes place in many surveys and survey organizations. Editing is indeed essential but it can be reduced and implemented more efficiently. As pointed out by Granquist and Kovar (Chapter 18), the overuse of editing might be due partly to its relative simplicity compared to other efforts to control survey quality. These concerns have largely been articulated in the context of business and establishment surveys. Studies of the type mentioned above have seldom appeared in the demographic survey literature. Thus, we cannot say with any certainty that overediting is a significant and costly problem in demographic surveys, even though we recognize that a number of aspects of the implementation of a demographic survey explicitly focus on various types of edits. We should note, however, that it is always true that survey researchers ought to focus more specifically on total survey error, shifting resources from editing to other survey functions if warranted. The integration of the editing function into other survey tasks will be investigated further in the future. For instance, in computer assisted data collection, interviewing, coding, data entry, and editing can be brought together. Macro-editing techniques also make it possible to reduce the extensiveness of preliminary edits and shift much of the editing to the analysis stage.

15.3 CODING

Coding is a process whereby raw survey data, usually in the form of responses to open-ended questions, are classified and transformed into a form that can be used in the final stages of the survey, like estimation, tabulation, and analysis. In most surveys the coding operation is highly dependent on the skills, knowledge, and judgment of individuals trained to code survey responses. The coding operation can be extensive, error-prone, costly, and time-consuming. Examples of variables that are coded are occupation, industry, education and socio-economic status.

The typical coding process has two aspects: first, the researcher must specify categories to be used in coding; second, unstructured verbal responses are coded into the prespecified categories where each category has a code number. The key to these code numbers is called the nomenclature or dictionary. The key is usually supplemented by a set of coding instructions which relate the verbal descriptions to the code numbers. To the extent coding is a clerical operation, the process can be very complex and difficult to control. Nowadays, most survey organizations are either doing research on automated coding or are using some type of automated or computer assisted coding.

15.3.1 The Nomenclature

The nomenclature or code list is a list of numbered categories intended to represent the full array of possible responses to a questionnaire item. Nomenclatures are actually classification systems whose purpose is to group objects or responses that are similar in some sense. Such systems already exist in other sciences, like zoology and astronomy for classification of species and stars. Statistical nomenclatures have been developed to improve the collection, tabulation, and presentation of statistical data; these statistical nomenclatures also improve the analysis and comparison of societal phenomena by means of statistical data. Most nomenclatures are built hierarchically, where the code list consists of several levels and each level represents an increasing amount of detail.

Nomenclatures for key societal concepts, such as industry and occupation, are powerful coordination tools, because of their capacity to promote uniformity and comparability of data series. As society changes, however, these must be revised: for instance, occupations begin and end; new industries form while other industries die; production processes and concepts change over time; statistical systems are integrated, etc. When building nomenclatures, several criteria regarding harmonization, continuity and homogeneity must be taken into account. Building and revising nomenclatures is a tedious, difficult task with the final product always a compromise between theoretical and practical aspects, as a result of discussions and negotiations between different interest groups.

A major problem with this process has been the lack of communication between statistical specialists, subject matter people, and coders, those who apply the nomenclature. Some errors in coding occur because the theoretical aspect of the nomenclature does not coincide with reality. For instance, a nomenclature for occupations might distinguish between telephone operators working in telephone companies and telephone operators working in other businesses. It might be very difficult, however, to ask survey questions that allow respondents and survey researchers to clarify this distinction. Error studies show that coding errors are not uniformly distributed across categories. Here, too, the Pareto principle is at work. Some associations between responses and categories are more clear than others, resulting in skewed distributions of errors per category or variability per category. See Lyberg (1981) and Hansen *et al.* (1964) for descriptions of such studies.

Therefore, categories are sometimes combined to avoid errors due to the fact that it is difficult to distinguish between category descriptions. More collaboration between the theoretical and the practical side is needed. One effort in this direction is discussed by Conrad (Chapter 17) where expert systems as tools for improving consistency in classification processes are described.

15.3.2 Different Kinds of Coding

Three kinds of coding systems exist, ranging from (1) strictly manual coding to (2) varying degrees of manual coding mixed with computer coding, i.e., computer assisted coding systems (CAC), and (3) automated coding. Manual coding can be implemented by trained coders, interviewers, or respondents. The more complex the variable to be coded the greater is the need for specialists trained specifically to perform this task. Manual coding systems, as the name implies, rely strictly on coders with paper, pencil, and manuals to complete the process. In CAC, a coder assigns codes while working interactively with a computer. Computer assisted coding resembles manual coding except for the fact that the coder has access to a number of facilities offered by the computer, like help screens, decision tables, auxiliary information, etc.

In automated coding, a computer program assigns codes to the extent possible and residual cases are handled by CAC or manual coding. Today, many statistical organizations use all three types of coding to varying degrees. A computer program for automated coding can handle a relatively large portion of survey responses. It is not uncommon that this portion is as large as 70%–80%. See Lyberg and Dean (1992) for an overview. The program, if well designed, produces sets of codes with small error rates and at low cost.

The basic features of any automated coding system are: (1) constructing a computer-stored dictionary that replaces the nomenclature used in manual coding; (2) entering survey responses into the computer by more or less free format keying; (3) matching responses to the dictionary and assigning codes; (4) transferring uncoded cases to manual coding or CAC; and (5) evaluating the coding quality resulting in, for instance, dictionary updates. Automated coding systems are in place at many agencies around the world in a variety of applications. The U.S. Census Bureau uses an expert system called the Automated Industry and Occupation Coding System (AIOCS). It tries to simulate manual coding by identifying informative words and less informative words, synonyms, misspellings, and abbreviations. When matching is exact, code assignment is straightforward. When matching is inexact, code numbers are assigned using probabilistic weights. Descriptions of the system are found in Chen *et al.* (1993) and Gillman *et al.* (1993). See Lyberg and Dean (1992) for a review of error rates.

Statistics Sweden uses a simple matching algorithm in its applications. Dictionary entries are based on a mixture of empirical response patterns and subjectively chosen descriptions. The subjectively chosen descriptions come directly from descriptions given in nomenclatures and manuals. Numerous studies have shown respondents tend not to use such descriptions but rather other descriptions which are variants of the ones found in official lists. Therefore, it is extremely important that the computer-stored dictionary covers the empirical pattern generated by the respondents themselves. Statistics Sweden has done that by deliberately including variants due to respondents' use of slang, common misspellings, etc. The "empirical" part of the dictionary is much more efficient than the part that consists of the official list of descriptions. The system is used in combination with CAC and results have been very successful (Lyberg and Dean, 1992). For example, the 1978 Swedish Household Expenditure Survey had an overall coding rate of 65% and a coding error rate of 1%. The 1985 Household Expenditure Survey experienced an automated coding rate of 82% and saw cost savings above 5%. Similarly, the coding of occupation in the 1980 and 1985 Swedish censuses experienced about a 72% automated coding rate and roughly 10% lower costs when compared to manual coding.

Statistics Canada has developed a system called Automated Coding by Text Recognition (ACTR). This is a generalized system that can perform both automated coding and CAC in two languages, English and French. It uses word standardization techniques to match input text files to a reference file of phrases, the output resulting in a code for the text. It requires the user to provide a reference file of phrases or text and their associated codes. System descriptions are found in Wenzowski (1988) and Ciok (1993). Miller (1994) provides a brief history of ACTR development and applications. Version 1, the mainframe version, has been successfully applied in the 1991 Canadian Census of Population, the 1991 Canadian Census of Agriculture, a transportation survey, and a travel survey. All applications have observed cost-savings, and improved quality and consistency of the coding (Miller, 1994).

The Australian Bureau of Statistics has developed the Australian Standard Classification of Occupations (ASCO) CAC System. This system is very efficient and probably superior to a fully automated system since it has many features that allow relatively untrained coders to perform in a way consistent with manual coding experts. The system has a user-friendly interface, providing potential matches and on-line help screens, shortened data-entry descriptions, and fast searching and matching procedures. The 1986 Australian census application cost approximately 1.6 million Australian dollars compared to a 3 million dollar cost for the manual approach. Consistency with expert manual coding was above 95% (Lyberg and Dean, 1992). A more complete description of the system is found in Embury (1988 a, b).

The French office, INSEE, has developed two systems: COLIBRI which is a CAC system and QUID which is a fully automated system (Lorigny, 1988). COLIBRI searches a register of establishments based on name and address; the likely candidates are ranked and displayed on the screen for the coder. Economic activity codes can be assigned in about 70% of the cases. The overall error rate for economic activity is approximately 4%. QUID is a CAC system based on a large dictionary of empirical responses coded by experts. In a QUID application of coding socio-economic status, about 90% of the records were coded with an error rate of between 5% and 10%. Other agencies using CAC include Statistics New Zealand, Statistics Netherlands, and the U.K. Office for National Statistics.

15.3.3 Some Problems with Coding

Despite development efforts associated with automated coding and computer assisted coding (CAC) and despite some good practical applications, problems still exist. First, manual coding will always be needed; residual coding of text and phrases that the automated coding cannot code will require manual effort; for the best quality data, manual coders ought to be expert coders, classifying responses accurately, consistently, and uniformly; however, as coding becomes more automated an experienced group of coders—of sufficient size for typical applications—will be increasingly difficult to maintain in an organization. Second, current automated and CAC systems are still under development and testing; the technology, while offering great hope for uniform and consistent coding and potential cost savings, has not been uniformly adopted.

As for manual coding, quality issues must be emphasized even more in the future. Traditional coding control is based on verification schemes of a dependent or independent nature. Although it has been shown that dependent verification, where a control coder inspects an assigned code and decides whether it is correct or not, is very inefficient it is still used by many organizations. The alternative, independent verification, where two or more coders assign codes independently and where the majority code is assigned, is much more efficient, but used less often. As in the editing case, this might occur because of the relative simplicity associated with administering dependent verification compared to independent verification (see Lyberg, 1981).

Coding verification, whether dependent or independent, is usually performed by means of traditional acceptance sampling, where inspection schemes that guarantee a prespecified average outgoing quality are administered. Such schemes do not emphasize feedback to coders, and they have been criticized by Deming (1986) and others. These criticisms are given in Biemer and Caspar

(1994) who discuss efforts at the Research Triangle Institute (RTI) to improve the quality of manual coding operations while at the same time reducing cost. The criticisms are based on the following:

1. Inspection is expensive; it is better to apply total quality management principles and do it correctly the first time rather than rely on inspection schemes to determine whether the job was done correctly.

2. It is impossible to use inspection to bring quality into a process; quality must be part of the process from the start. Sampling schemes can only maintain a certain stability regarding outgoing quality; they cannot improve it.

3. Inspection is not perfect. Errors of the first and second kind generate a need for adjusting the sampling schemes (Minton, 1972).

4. Inspection performed by an inspection group takes responsibility away from the coders.

5. Empirical studies show coders are responsible for a relatively small portion of coding problems. The system is responsible for the large portion; thus, trying to adjust the behavior of individual coders is not necessarily a useful or practical strategy.

5. Coder feedback is not helpful in an environment where statistical control is performed by an inspection department. Feedback, if it comes at all, comes too late.

Biemer and Caspar (1994) suggest a different paradigm for improving coding, one based on a cyclic process of comparing coding performance with preferred performance, sorting nonconforming observations by type and applying a Pareto analysis, working with coding teams to identify the reasons for the nonconforming observations, and then, finally, implementing suggested measures for improvement. In 1991, the industry error rate and the occupation error rate at RTI were 17% and 21% respectively. Introduction of the new process has dramatically reduced these rates to 4% and 5%, respectively, while at the same time cost has decreased and productivity has increased.

There is need for more emphasis on quality issues in automated coding as well. Most research on automated coding has concentrated on achieving high coding rates. This allocation of resources is understandable. If acceptably high coding rates cannot be achieved, there is no need for quality efforts. Of course, if we look at cases where codes have been assigned based on exact matches with dictionaries, quality is almost perfect by definition. Problems occur with codings based on inexact matches.

One issue that needs to be studied more closely is the concept of true code. The basic assumption is that every element belongs to one and only one category. Sometimes it is stated that an element can be assigned to more than one category depending on interpretation. This is not the case. Coding is a one-to-one mapping which does not give room for more than one category per element. However, in practice, there are situations where it is very difficult to perform this mapping due to vague element descriptions. A good classification scheme should be able to resolve such problems by introducing special categories to which vague cases are assigned. The reason why the coding of elements is sometimes thought to be part of a one-to-many mapping process is that the nomenclature is defective, i.e., it does not include categories for vague descriptions. In practice, true code often has to be defined operationally, for instance as the majority code among experts.

Other issues that need to be observed more closely include the differences in quality between cases coded by exact match, inexact match, CAC, and pure manual; the effects of coding errors on data analysis; quality and cost issues related to dictionary size and dictionary contents; and better articulated decision rules to help program areas determine their "best" strategy to implement, that is, what is the combination of approaches that yields the best data at lowest cost.

15.4 DATA CAPTURE

Data capture is the phase of the survey where information recorded on a questionnaire is converted to a format that can be interpreted by a computer. Data may be captured by keying responses contained on paper and pencil questionnaires, by mark character recognition, optical character recognition and even touch-tone and voice recognition entry. Recently, new forms of data capture technology have evolved, including facsimile transmission, electronic data interchange (EDI), and e-mail transmission through the Internet.

Data entry keying is a tedious and labor intensive task; many organizations try to avoid this activity, relying instead on more modern technologies like computer assisted telephone interviewing (CATI), computer assisted personal interviewing (CAPI), and optical character recognition (OCR). Keying is not a very error-prone operation. For instance, the error level in the Fourth Followup Survey of the National Longitudinal Study, based on a quality control sample of questionnaires, was about 1.6% (Henderson and Allen, 1981). Similarly, error levels in the 1988 U.S. Survey of Income and Program Participation were in the neighborhood of 0.1% (Jabine et al., 1990). These rates relate to errors in individual variables.

In a study designed to determine the quality of keying of long-form questionnaires in the 1990 U.S. Census, keyers committed keystroke mistakes (or omissions) in 0.62% of the fields in the initial stages of production keying (U.S. Bureau of the Census, 1993). Computer technology has initiated new concerns about the number of errors made by interviewers, presumably unskilled in data keying, who conduct computer assisted personal interviews. Dielman and Couper (1995), however, report a keying error rate of 0.095% in a study focused on keying errors in a computer assisted personal interview environment, thereby providing some evidence that keying errors are not a significant problem. In a Swedish study where error rates were observed in terms of percent of records, the average error rate per record was 1.2% (see Lyberg *et al.*, 1977).

The U.S. Census Bureau pioneered the use of FOSDIC (Film Optical Sensing Device for Input to Computers), a process which uses a machine capable of reading "information" from a microfilmed copy of an appropriately designed questionnaire and transferring this information to tape for processing. FOSDIC uses a beam of light to identify whether circles associated with a response to a question have been marked (Brooks and Bailar, 1978). The procedure has an average error rate of about 0.02% (Brooks and Bailar, 1978). With a similar technology, the U.S. National Center for Education Statistics (NCES) optically scanned the questionnaires from the High School and Beyond Study (Jones *et al.*, 1986). These processes have been very effective when used for multiple choice questions; however, as workloads and labor costs increase, a need exists for capturing large volumes of handwritten responses accurately and inexpensively.

Optical character recognition (OCR) has been suggested as a solution to the data capture problem when the data are handwritten responses. The U.S. Census Bureau has sponsored several conferences whose aims were to determine the state of the art in the optical character recognition industry and to stimulate new research. The general findings of the conferences indicated that determining the feasibility of OCR is dependent on the application. The requirements of some applications require considerably less sophistication than others. A conference summary unequivocally states that the accuracy of optical character recognition systems for handwriting has improved dramatically over the last few years, and that "machine performance in reading words and phrases may now be good enough to decrease the cost and time needed to carry out a Census without decreasing the accuracy of the results" (Geist *et al.*, 1994). The U.S. Census Bureau plans to study the cost, accuracy, and speed of OCR technology as part of the 1995 Census Test (U.S. Bureau of the Census, 1994).

OCR errors are of two types: substitution and rejection. Rejects must be corrected and are, therefore, expensive to handle but add no error if corrected properly. Substitutes introduce errors, of course, but usually only minor ones. In a study in connection with the Swedish 1970 Census where handwritten codes were scanned, an estimated 0.14% of the digits were substituted. This is a seemingly very good result but as noted above feasibility is dependent on application. In this case, the process of writing digits was a highly standardized one. The digits were written by coders who had received special training in writing OCR digits. Obviously, this is a situation far removed from one where, say, the general population is asked to write down data on income, etc. for OCR processing (for details, see Lyberg *et al.*, 1977).

Recently, Statistics Sweden and some other agencies have started experimenting with scanning entire forms that are photographed in a scanner. The information is interpreted by a computer. No data entry in the usual sense is necessary and some editing can be done during the scanning. Documents can be stored and retrieved during editing and correction. This research, reported in Blom (1994), is still in its infancy. The step from in-house produced OCR digits to information provided in free-format respondent handwriting is a big one and reject rates range from 7%–35% on position level in initial studies performed at Statistics Sweden.

Even though error rates associated with traditional data capture methods are not large, statistical agencies continue to look for ways to reduce cost and increase quality in this process. If the application allows, new technologies attempt to reduce or eliminate the data entry aspect of the survey. For example, in touchtone data entry the respondent uses a touchtone telephone to call the data collector's computer system, and the respondent reports numeric responses to questions by pressing touchtone telephone buttons. Similarly, voice recognition systems are speaker-independent and they recognize certain verbal responses of the respondent. Facsimile transmissions that translate incoming FAX responses through intelligent character recognition systems also avoid the data entry process (Werking and Clayton, 1995). Finally, collecting data electronically by the U.S. National Center for Education Statistics (Kindel, 1994) and the U.S. Census Bureau (Ambler *et al.*, 1995) illustrates how statistical agencies are responding to the need to lower costs of data collection.

15.5 INTEGRATION ACTIVITIES

As pointed out by Bethlehem (Chapter 16), Shanks (1989), Keller (1995), Pierzchala (1990), Baker (1994), and Weeks (1992) there is a need for integrating data processing operations between themselves and with other survey tasks. Traditionally data processing is carried out in a centralized fashion with separate processes for each operation. It is done sequentially with many different groups of people involved, much transferring of data between the groups, and a substantial number of specifications.

The advent of new technology makes it easier to carry out data processing in a decentralized fashion and, as discussed above, technology can reduce the need for monotonous manual work. On the other hand, using new technology in survey processing will change the way organizations work. Interactive processing will become more prevalent as batch-processing becomes less prevalent. Processing cycles will be reduced in size. Each step in the processing will have its own software, so it will be very important that relationships between these softwares be almost seamless. This can be accomplished in various ways. Statistics Netherlands has a control center (Chapter 16) which constitutes a user-friendly shell around the software required for survey data processing. The control center is capable of putting data and metadata in any format required and users do not have to specify data formats and conversions between data formats. Similar work on developing generalized survey processing software is being conducted at other agencies as well, for instance Statistics Canada (Turner, 1994) and Statistics Sweden (Blom, 1994).

15.6 ENDNOTE

The traditional data processing operations of editing, coding, and data capture will benefit from rapid developments in information technology. Various forms of computer assisted survey information collection make it possible to implement procedures differently and to integrate several operations. Problems still exist with our lack of understanding of basic survey processes and error structures, particularly with inefficient editing procedures and coding errors. Before we can use the full potential of these new technologies the basic operations ought to be understood and improved substantially.

REFERENCES

Ambler, C., Hyman S., and Mesenbourg T. (1995), "Electronic Data Interchange," in B. Cox, D. Binder, B.N. Chinnappa, A. Christianson, M. Colledge, and P. Kott (eds.), *Business Survey Methods*, Chapter 19, pp. 339–354, New York: Wiley.

Baker, R. (1994), "Managing Information Technology in Survey Organizations," *Proceedings of the Annual Research Conference*, U.S. Bureau of the Census, Washington, DC, pp. 637–646.

Bethlehem, J. (1987), "The Data Editing Research Project of the Netherlands Central Bureau of Statistics," Staff report, Netherlands Central Bureau of Statistics.

Biemer, P., and Caspar, R. (1994), "Continuous Quality Improvement for Survey Operations: Some General Principles and Applications," *Journal of Official Statistics*, 10, pp. 307–326.

Bienias, J.L., Lassman, D.M., Scheleur, S.A., and Hogan, H. (1995), "Improving Outlier Detection in Two Establishment Surveys," *Statistical Policy Working Paper* 23, Office of Management and Budget, pp. 137–151.

Blom, E. (1994), "Building Integrated Systems of CASIC Technologies at Statistics Sweden," *Proceedings of the Annual Research Conference*, U.S. Bureau of the Census, Washington, DC, pp. 623–636.

Boucher, L. (1991), "Micro-Editing for the Annual Survey of Manufactures: What is the Value Added?" *Proceedings of the Annual Research Conference*, U.S. Bureau of the Census, Washington, DC, pp. 765–781.

Brooks, C., and Bailar B. (1978), "An Error Profile: Employment as Measured by the Current Population Survey," *Statistical Policy Working Paper* 3, Office of Management and Budget, Washington, DC.

Chen, B.-C., Creecy, R., and Appel, M. (1993), "On Error Control of Automated Industry and Occupation Coding," *Journal of Official Statistics*, 9, pp. 729–745.

Ciok, R. (1993), "The Results of Automated Coding in the 1991 Canadian Census of Population," *Proceedings of the Annual Research Conference*, U.S. Bureau of the Census, Washington, DC, pp. 747–765.

Corby, C. (1984), "Content Evaluation of the 1977 Economic Censuses," *Statistical Research Division Report*, Series No CENSUS/RR-84-29, U.S. Bureau of the Census, Washington, DC.

Deming, E. (1986), *Out of the Crisis*, MIT Press.

Dielman, L., and Couper, M. (1995), "Data Quality in a CAPI Survey: Keying Errors," *Journal of Official Statistics*, 11, pp. 141–146.

Dodge, H.F., and Romig, H.G. (1944), *Sampling Inspection Tables*, New York: Wiley.

Embury, B. (1988a), "The ASCO Computer Assisted coding System," paper presented at the Social Research Conference, University of Queensland, Brisbane.

Embury, B. (1988b), "The Methodology of Occupation Coding in the 1991 Census," unpublished manuscript, Australian Bureau of Statistics.

Federal Committee on Statistical Methodology (1990), "Data Editing in Federal Statistical Agencies," *Statistical Policy Working Paper* 18, Office of Management and Budget, Washington, DC.

Ferguson, D. (1994), "An Introduction to the Data Editing Process," paper presented at the Conference of European Statisticians, Work Session on Statistical Data Editing, Cork, Ireland, October.

Geist, J., Wilkinson, R., Janet, S., Grother, P., Hammond, B., Larsen, N., Klear, R., Matsko, M., Burges, C., Creecy, R., Hull, J., Vogl, T., and Wilson, C. (1994), "The Second Census Optical Character Recognition Systems Conference," Technology Administration, National Institute of Standards and Technology, U.S. Department of Commerce, Gaithersburg, MD, May.

Gillman, D., Appel, M., and Jablin, C. (1993), "Certification of Decennial Automated I&O Coder for Current Surveys," *Proceedings of the Annual Research Conference*, U.S. Bureau of the Census, Washington, DC, pp. 766–782.

Granquist, L. (1984), "On the Role of Editing," *Statistical Review*, pp. 105–118.

Granquist, L. (1994), "Macro-Editing—A Review of Methods for Rationalizing the Editing of Survey Data," Conference of European Statisticians, *Statistical Standards and Studies*, No. 44, United Nations.

Greenberg, B., and Petkunas, T. (1986), "An Evaluation of Edit and Imputation Procedures Used in the 1982 Economic Census in Business Division," *1982 Economic Censuses and Census of Government Evaluation Studies*, Washington, DC: U.S. Bureau of the Census, pp. 85–98.

Gruber, K., Rohr, C., and Fondelier, S. (1994), "1990–91 Schools and Staffing Survey: Data File User's Manual," vol. 1, NCES 93-144-I, Office of Educational Research and Improvement, U.S. Department of Education, Washington, DC.

Hansen, M., Hurwitz, W., and Pritzker, L. (1964), "The Estimation and Interpretation of Gross Differences and the Simple Response Variance," in C. R. Rao (ed.), *Contributions to Statistics*, presented to Professor P.C. Mahalanobis on the occasion of his 70th birthday, pp. 111–136, Pergamon Press.

Henderson, L., and Allen, D. (1981), "NLS Data Entry Quality Control: The Fourth Followup Survey," National Center for Education Statistics, Office of Educational Research and Improvement, Washington, DC.

Houston, G., and Bruce, A.G. (1993), "Gred: Interactive Graphical Editing for Business Surveys," *Journal of Official Statistics*, 9, pp. 81–90.

Jabine, T., King, K., and Petroni, R. (1990), *SIPP Quality Profile*, U.S. Department of Commerce, Bureau of the Census, Washington, DC.

Jones, C., Sebring, P., Crawford, I., Spencer, B., Butz, M., and MacArthur, H. (1986), "High School and Beyond: 1980 Senior Cohort, Second Followup (1984)," Data

File User's Manual, CS 85-216, Center for Statistics, Office of Educational Research and Improvement, Washington, DC.

Kalton, G., and Kasprzyk, D. (1982), "Imputing for Missing Survey Responses," *Proceedings of the Survey Research Methods Section*, American Statistical Association, pp. 22–31.

Kalton, G., and Kasprzyk, D. (1986), "The Treatment of Missing Survey Data," *Survey Methodology*, 12, pp. 1–16.

Keller, W. J. (1995), "Changes in Statistical Technology," *Journal of Official Statistics*, Vol. 11, No. 1, pp. 115–127.

Kindel, C.B. (1994), "Electronic Data Collection at the National Center for Education Statistics: Successes in the Collection of Library Data," *Journal of Official Statistics*, Vol. 10, No. 1, pp. 93–102.

Linacre, S.J., and Trewin, D.J. (1989), "Evaluation of Errors and Appropriate Resource Allocation in Economic Collections," *Proceedings of the Annual Research Conference*, U.S. Bureau of the Census, Washington, DC, pp. 197–209.

Lindström, K. (1991), "A Macro-Editing Application Developed for PC-SAS," *Statistical Journal of the United Nations Economic Commission for Europe*, pp. 155–165.

Little, R.J.A., and Rubin, D.B. (1987), *Statistical Analysis with Missing Data*, New York: Wiley.

Lorigny, J. (1988), "QUID, A General Automatic Coding Method," *Survey Methodology*, 14, pp. 289–298.

Lyberg, L. (1981), "Control of the Coding Operation in Statistical Investigations—Some Contributions," *Urval*, No. 13, Statistics Sweden.

Lyberg, L., and Dean, P. (1992), "Automated Coding of Survey Responses: An International Review," *R&D Report*, 1992:2, Statistics Sweden.

Lyberg, L., Felme, S., and Olsson, L. (1977), *Kvalitetsskydd av data*, Liber (in Swedish).

Madow, W.G., Olkin, I., and Rubin, D.B. (eds.) (1983), *Incomplete Data in Sample Surveys*, Vol. 1, 2 and 3, New York: Academic Press.

Miller, D. (1994), "Automated Coding at Statistics Canada," *Proceedings of the International Conference on Establishment Surveys*, American Statistical Association, pp. 931–934.

Minton, G. (1972), "Verification Error in Single Sampling Inspection Plans for Processing Survey Data," *Journal of the American Statistical Association*, 67, pp. 46–54.

Nordbotten, S. (1995), "Editing Statistical Records by Neural Networks," *Journal of Official Statistics*, Vol. 11, No. 4, pp. 393–414.

Pierzchala, M. (1990), "A Review of the State of the Art in Automated Data Editing and Imputation," *Journal of Official Statistics*, Vol. 6, No. 4, pp. 355–378.

Roddick, L.H. (1993), "Data Editing Using Neural Networks," Memo, Statistics Canada.

Shanks, J.M. (1989), "Information Technology and Survey Research: Where Do We Go From Here?" *Journal of Official Statistics*, Vol. 5, No. 1, pp. 3–21.

Turner, M.J. (1994), "General Survey Processing Software: An Architectural Model," *Proceedings of the Annual Research Conference*, U.S. Bureau of the Census, Washington, DC, pp. 596–606.

U.S. Bureau of the Census (1993), "Memorandum for Thomas C. Walsh from John H. Thompson, Subject: 1990 Decennial Census-Long Form (Sample Write-In) Keying Quality Assurance Evaluation," M. Roberts author, October 18.

U.S. Bureau of the Census (1994), "Memorandum for Distribution from Susan Miskura and John Thompson, Subject: Evaluation Requirements Document for the 1995 Census Test Research Objective: Imaging for Data Capture (DCS 2000)," September 7.

Weeks, M.F. (1992), "Computer-Assisted Survey Information Collection: A Review of CASIC Methods and Their Implications for Survey Operations," *Journal of Official Statistics*, Vol. 8, No. 4, pp. 445–466.

Wenzowski, M. (1988), "ACTR—A Generalized Automated Coding System," *Survey Methodology*, 14, pp. 299–317.

Werking, G., and Clayton, R. (1995), "Automated Telephone Methods for Business Surveys," in B. Cox, D. Binder, B.N. Chinnappa, A. Christianson, M. Colledge, and P. Kott (eds.), *Business Survey Methods*, Chapter 18, pp. 317–338, New York: Wiley.

CHAPTER 16

Integrated Control Systems for Survey Processing

Jelke Bethlehem
Statistics Netherlands

16.1 INTRODUCTION

It is the task of national statistical offices to collect data from persons, households, farms, and establishments and to transform these data into useful statistical information. The growing demand for statistics calls for an organized and efficient approach to the entire process of data collection, data analysis, and publication. At the same time, high quality standards in terms of accuracy, timeliness, and relevance must be maintained.

The traditional survey process suffers from a number of drawbacks which affect the quality of the statistics produced. This chapter discusses the role of information technology in improving the quality of statistical information. Nowadays, there is ample software for almost every step in the survey process. However, software packages requiring their own data formats and control languages hinder an efficient process flow. One solution could be designing the ideal package that does it all. However, such a package does not exist, and most likely will never exist. This chapter proposes a different approach: the *integrated control system*.

An integrated control system is an intelligent and user friendly shell which encompasses all software required for processing survey data. The control system deals with the data and metadata (the description of the data) in a standard format. It is capable of translating the data and metadata into any format required by the individual programs included in the system. The users are relieved of the burden of having to specify several data formats, nor do they

Survey Measurement and Process Quality, Edited by Lyberg, Biemer, Collins, de Leeuw, Dippo, Schwarz, Trewin.
ISBN 0-471-16559-X © 1997 John Wiley & Sons, Inc.

need to worry about conversions between data formats. Despite the many ways that the data are actually formatted, to the user it appears as if there is only one way of specifying the data; the rest is taken care of by the control system.

An integrated control system contains software modules for various data processing tasks. Essential modules are programs for computer assisted data collection, data entry, and data editing. These modules are required to create a data set. Additional modules may be included, such as programs for tabulation, imputation, weighting, and analysis. The control system is flexible and open, allowing its users to include or exclude software modules. In that way, the control system can be modified for use in a specific environment. By having the proper modules in the control system, it can be used by all survey processing departments of the statistical office, thus promoting standardization and integration.

A system capable of processing a variety of surveys must include a facility to specify the various surveys. Here, the metadata description plays a crucial role. The metadata describe the data to be collected, processed, and published. Once a proper metadata description of a survey is available in machine readable form, the control system has the knowledge it needs to transform the data and metadata to any format required by the various software modules.

As of yet, there are no statistical offices that have implemented fully integrated control systems. However, several offices have developments moving in that direction. This chapter discusses the concept of the integrated control system and its effects on the quality of survey results. Section 16.2 is devoted to survey data quality and its many dimensions. Using a generic description of the traditional survey process, it is shown where and how new developments in information technology can help improve the process. Section 16.3 describes some approaches to improving quality in business processes. Because of the similarities between business processes and survey processes, these approaches can be applied for improving quality in surveys. Section 16.4 introduces the concepts of concentration, standardization, and integration. Section 16.5 describes how these concepts are, to a greater or lesser extent, implemented by several major statistical offices in the world. Section 16.6 discusses the possible effects on the organization of implementing a new approach. Section 16.7 attempts to establish the possible effects on the quality of statistical data. Section 16.8 concludes with some possible future challenges.

16.2 QUALITY IN SURVEYS

Surveys are carried out to collect information about a specific population. Results are presented in the form of statistics, i.e., estimates of unknown population characteristics. The statistics will rarely correspond exactly to the unknown values of the population characteristics. Every survey operation is affected by errors, and the magnitude of these errors affects the accuracy of the results. This is the first aspect of quality that is discussed.

Generally, two broad categories of errors are distinguished: sampling errors and nonsampling errors. Sampling errors are due to the fact that only a sample is observed and not the entire population. Every replication of a survey experiment would result in a different sample, and therefore in different estimates. The survey researcher controls the sampling error through a proper sampling design and a sufficiently large sample, which in turn lead to unbiased estimates with small standard errors. Of course, the available budget may impose restrictions on the size of the sample.

Nonsampling errors comprise all survey errors originating from sources other than the use of a sampling mechanism. Nonsampling errors are due either to obtaining the wrong answers to the survey questions (observation errors), or failing to get the answers (nonobservation errors). Examples of observation errors are measurement errors, processing errors, and overcoverage errors. Examples of nonobservation errors are nonresponse and undercoverage. Nonsampling errors are difficult to control. The best way to deal with nonsampling errors is preventive measures at the design stage. One can think of a well considered selection of the sampling frame, careful design of the questionnaire, adequate training of interviewers, and thorough pilot surveys. Nevertheless, many causes of nonsampling errors are inherent to the survey process, and as such are virtually impossible to avoid.

Another aspect of quality in survey research is timeliness. To make well informed decisions, policy makers require up-to-date statistical information. This means that the time interval between survey design and publication of the results must be as short as possible. Traditionally, the production of statistical information is a complex and time-consuming process. Different departments and computer systems may be involved in the process. The process is also sequential. This means that the next step cannot be carried out before the previous step is completely finished. Particularly for large surveys, it may take months, if not years, before the results are ready for publication. The longer it takes to publish the results, the less useful they are.

A final aspect of quality to be mentioned here is the cost of the survey. The survey designer should always try to minimize the costs involved in obtaining accurate and timely statistics. Money can only be spent once, and when it is spent, it should help improve quality.

For greater insight into the effects of various survey processing activities on the quality of the results, Statistics Netherlands carried out a Data Editing Research Project in 1984, see Bethlehem (1987). A careful evaluation was conducted of data editing activities in a number of different surveys: large and small surveys, and social and economic surveys. Although considerable differences were observed between surveys, some general characteristics could be identified. The traditional data editing process is summarized in Figure 16.1. At Statistics Netherlands, this process has now been modernized, but similar processes still exist at several statistical offices in the world.

After collection of the forms, subject matter specialists manually checked the forms for completeness. Skipped questions were answered, if necessary and

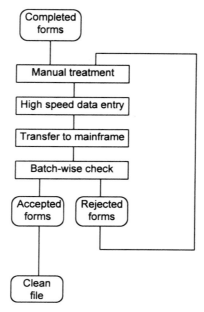

Figure 16.1   The traditional approach to data editing.

possible, and obvious errors were corrected on the forms. Sometimes, forms were manually copied to new forms to prepare for the subsequent step of fast ("heads down") data entry. Next, the forms were transferred to the data entry department. Data typists entered the data to the computer at high speed without checking errors. The computer was a dedicated system for data entry. After data entry, the files were transferred to the mainframe computer system. A tailor-made error detection program was run on the mainframe, and detected errors were printed on a list. These lists with errors were sent to the subject matter department. Specialists investigated the error messages, consulted corresponding forms, and attempted to replace apparently incorrect answers by more acceptable answers. Lists with modifications were sent to the data entry department, and data typists entered the corrections on the data entry computer. The file with corrections was transferred to the mainframe computer. Corrected records and already present correct records were merged. This cycle of batch-wise error detection and manual correction was repeated until the number of rejected forms was considered to be sufficiently small. After that, the resulting "clean" data set with accepted forms was ready for tabulation and further analysis. Detailed investigation of this process for the selected surveys led to a number of conclusions. These conclusions are summarized below:

- Various people in different departments were involved. Many people dealt with the information: respondents filled in forms, subject matter specialists checked forms and corrected errors, data typists entered data in the computer, and programmers from the computer department constructed editing programs. Transfer of material from one person/department to another could be a source of error, misunderstanding, and delay.

- Different computer systems were involved. Most data entry was carried out on dedicated minicomputer systems and data editing programs ran on a mainframe computer. Furthermore, there was a variety of desktop PCs (running under MS-DOS) and other systems. Transfer of files from one system to another caused delay, and incorrect or incomplete specification and documentation could produce errors.

- Not all activities were aimed at quality improvement. A lot of time was spent just preparing forms for data entry, and not on correcting errors. Subject matter specialists had to clean up forms to avoid problems during data entry. The most striking example was assignment of a code for "unknown" to unanswered questions.

- The process was cyclic: from one department to another, and from one computer system to another. The cycle of data entry, automated checking and manual correction was in many cases repeated three times or more. Due to these cycles, data processing was very time-consuming and expensive.

- Nearly every step required its own specification of the data. Although essentially the same, the "language" of the specification could be completely different for every department or computer system involved. The questionnaire itself was a first specification. The next one was for the data entry program. The automated checking program required another specification of the data. For tabulation and analysis, e.g., using the statistical package SPSS, again another specification was needed. All specifications resulted in a description of variables, valid answers, routing and possibly valid relationships to be checked.

If these conclusions are translated to terms of quality, it becomes clear that the problems in the data editing phase have a substantial effect on quality: the process is time-consuming, it is costly, and error correction is attempted at too late a stage, and therefore is not effective.

16.3 IMPROVING THE SURVEY PROCESS

There is a lot of similarity between the statistical production process and business processes in general. Therefore, approaches to improving business processes can also be applied to the survey process. Two such approaches are discussed here. One is the theory of Deming on improving quality in industrial production, and the other is the more recent Business Process Redesign (BPR) approach.

Quality control has always been an important issue in industrial production. The theory of Deming (1986) about improving quality and productivity in industry is well known. Many of his famous 14 points for management also apply to the production of statistical information. One of these points states that one should cease dependence on mass inspection. Inspection of the final product to improve quality is too late, ineffective and costly. Quality must be built in at the design stage. These statements particularly apply to data collection and data editing. By trying to detect and correct errors in questionnaire forms well after they have occurred, one fails to locate the source of the error and, consequently, these errors cannot be eliminated.

Some of the problems mentioned by Deming can be solved by introducing computer assisted interviewing techniques. The rapid advent of the microcomcomputer in the last decade made it possible to use microcomputers for computer assisted interviewing. The paper questionnaire is replaced by a computer program containing the questions to be asked. The computer takes control of the interviewing process. It performs two important activities.

1. *Route control.* The computer program determines the next question to be asked and displays that question on the screen. Such a decision may depend on the answers to previous questions. Hence it relieves the interviewer of the task of having to take care of the correct route through the questionnaire. As a result, it is no longer possible to make routing errors, i.e., to ask irrelevant questions or fail to ask relevant questions.

2. *Error checking.* The computer program checks the answers to the questions. Range checks are carried out immediately after entry, and consistency checks are made after entry of all relevant answers. If an error is detected, the program gives a warning, and one or more of the answers involved can be modified. The program will not proceed to the next question until all detected errors have been corrected.

Computer assisted interviewing started in the 1970s with CATI (Computer Assisted Telephone Interviewing), see Nicholls and Groves (1986) and Groves and Nicholls (1986) for an overview. The emergence of small, portable computers in the 1980s made CAPI (Computer Assisted Personal Interviewing) possible. An account of early experiments is given by Bemelmans-Spork and Sikkel (1985), and Van Bastelaer *et al.* (1987). More recent is the use of CASI (Computer Assisted Self-Interviewing). This type of interviewing is particularly important for business surveys. Wieringa and Fokké (1992) and Stol (1993) describe CASI applications.

Computer assisted interviewing has proved very successful in social and demographic surveys. There are three major advantages compared to traditional interviewing.

- By applying this type of interviewing, it becomes possible to detect data entry errors and inconsistencies during the interview. Since both the interviewer and the respondent are available when these problems are detected, it is the best moment to correct them. In this way, computer assisted interviewing produces more accurate data.

- Computer assisted interviewing integrates three steps in the survey process: data collection, data entry, and data editing. Since interviewers use computers to record the answers to the questions, they take care of data entry during the interview. If all necessary checks are also carried out during the interview, many errors can be detected and corrected, thus making a separate data editing step superfluous in many surveys. After all interviewers' files have been combined into one data file, the information is "clean," and therefore ready for further processing. Thus, computer assisted interviewing reduces the length and the cost of the survey process.
- Computer assisted interviewing makes the work of the interviewers easier. Since the computer is in charge of determining the proper route through the questionnaire, the interviewers can concentrate on asking the questions.

Computer assisted interviewing has increased the efficiency of the survey process and the quality of the results. The effect of CATI surveys is described by, for example, Tortora (1985), Groves and Nicholls (1986), and Catlin et al. (1988). For CAPI surveys, see Van Bastelaer et al. (1987) and Sebestik et al. (1988). Weeks (1992) presents an overview of research on CAPI and CATI techniques.

Application of computer assisted interviewing in economic surveys is not as widespread as in social surveys. The respondents in business surveys usually do not have the requested information readily available. They need some time to consult administrative records, computer files, and possibly other persons and departments. In such cases, paper forms may be more appropriate.

Even if computer assisted interviewing is not possible in a business survey, gains in efficiency can be obtained by combining data entry and data editing. This technique is referred to as CADI (Computer Assisted Data Input), or CADE (Computer Assisted Data Entry). CADI provides an interactive and intelligent environment for combined data entry and data editing of paper forms by subject matter specialists. Two strategies can be applied in processing the data. In the first strategy, the subject matter employees work through a pile of forms with a microcomputer, processing the forms one by one. They enter the data, and after completion of the form, they activate the check option to test for all kinds of errors. Detected errors are reported and explained on the screen. They can correct errors by consulting the form, or by contacting the supplier of the information. After elimination of all errors, a clean record is written to file. If they do not succeed in producing a clean record, they can write the record to a separate file of problem records. Specialists can deal with these problematic cases later on, also with a CADI system. This approach is efficient for surveys with relatively small and complex questionnaires.

In the second strategy, data typists use the CADI system to enter data beforehand without much error checking. After completion, the CADI system checks all records in a batch run, and flags the incorrect ones. Then subject matter specialists take over. They handle the incorrect records one by one, and try to correct the detected errors. This approach works well for surveys with many small and simple questionnaires.

Another approach to improving quality in business processes is Business Process Redesign (BPR). The concepts of Business Process Redesign (or re-engineering) are well described in the book by Hammer and Champy (1994). They claim that many of today's companies operate on principles formulated more than two centuries ago by Adam Smith in his book *The Wealth of Nations*. These principles worked in an expanding market for mass products and services, but fail as survival techniques in today's quickly changing markets and competitive climate. The traditional approach focuses on splitting business processes into simple tasks. The advantage of such a task-oriented approach is that the relatively simple tasks do not require advanced skills or knowledge. However, simple tasks demand complex processes to knit them together. For two hundred years, companies have accepted the inconvenience, inefficiencies and costs associated with these complex processes in order to be able to take advantage of the benefits of simplified tasks.

Typically, a task-oriented process involves different people in different departments. The processes are sequential in the sense that one department cannot start working on a task until another department has finished its task. Moreover, some business processes have iterative components. The process goes through a cycle several times before it can be completed. Such sequential and iterative processes are time-consuming and error prone.

People in a task-oriented organization are not process-oriented. They focus on tasks, jobs, structures, but not on processes. Because of this process fragmentation, no one oversees the process as a whole, and no one seems to be responsible for it. People participating in the business process are inward-oriented. The department is their world, and the boss is the most important person to satisfy. They do not seem to be aware of the existence of customers.

Preaching quality is not sufficient to improve business processes. Even if each separate task is carried out efficiently, the process as a whole need not be efficient. Application of state-of-the-art information technology will not help if it is only used to automate current tasks. Other things are needed to accomplish substantial improvements. Business Process Redesign means the fundamental rethinking and radical redesign of business processes to achieve dramatic improvements in critical contemporary measures of performance, such as cost, quality, service, and speed.

BPR means getting rid of the task-oriented approach and organizing the work around the process. It means breaking with old traditions and it means breaking traditional rules. BPR avoids process fragmentation by concentrating the work where it makes sense. Consequently, the traditional crossing of departmental boundaries will not slow down the process any more, and also the sequential and iterative nature of traditional business processes can be substantially reduced.

BPR promotes the abandonment of specialization by letting the various activities be carried out by generalists instead of specialists. Several different jobs can be combined into one job. Steps in the process can be performed in a natural order. This makes it possible to remove or reduce extensive check and control steps that do not add to the value of the product. By removing the sequential nature, it becomes possible to carry out the work in a parallel fashion, thus speeding up the process.

BPR stresses the importance of the creative use of information technology. However, new developments in information technology should not be used just to automate the tasks in the traditional processes. This will only lead to marginal performance improvements. Far more substantial improvements can be obtained by using computer technology in a fundamentally new way.

There is a striking degree of similarity between the descriptions of the traditional survey process and that of the traditional business process. Both processes share the same problems: process fragmentation, focusing on tasks instead of processes, involvement of several departments, sequentially and cycling. Statistics Netherlands followed a BPR-like approach in redesigning its statistical production processes. This resulted in a new way of processing surveys. In fact, this was a case of BPR *"avant la lettre."*

16.4 TOWARDS A NEW SURVEY PROCESS

Both the conclusions of the Data Editing Research Project and the principles of Business Process Redesign stress the importance of abandoning the task-oriented survey process and replacing it by a radically different approach that looks at the process as a whole. Such a change requires a proper organizational structure, work environment, and tools. This section describes the changes that took place at Statistics Netherlands and other national statistical offices. All these organizations aim at satisfying one or more of the following conditions:

1. *Concentration.* All data processing activities with respect to a specific survey should be concentrated as much as possible in one department. Preferably this should be the subject matter department.

2. *Standardization of hardware.* All data processing activities should be carried out as much as possible on the same type of computer platform.

3. *Standardization of software.* All data processing activities must be carried out with standard software instead of tailor-made software.

4. *Integration:* All software required for data processing must be part of an integrated system, using a machine readable metadata definition containing all required information about the survey. This metadata definition must be used by all systems and departments as the main source of information about the survey.

By satisfying these conditions, some of the disadvantages of the traditional approach are eliminated. Since all work is carried out by a survey team from one department with standard software and hardware, time-consuming transfer of data between departments and computer platforms is avoided. The survey team integrates various activities, thereby removing the cycles from the process.

A control system integrates the various software packages required for the survey process. Nowadays, the computer is used in almost every step of the survey process. There is ample software for each type of activity: data capture programs for interviewing with laptop computers, database programs for storing and managing the data, specialized tabulation programs, powerful packages for complex analysis, and presentation programs producing high quality graphics. Most programs focus on one specific task. One package can do data entry, but another one may be needed for statistical analysis. There are packages that try to cover several, if not all, steps in the process. In practice this does not work very well. A package may be very good for one task, but a different task might require a different package.

Having to work with several different computer packages in one survey is not efficient. Each package requires its own data format and data description. That could mean repeated conversion of data files and set-up files. This costs time, consumes computer resources, and is a potential source of errors. It would be ideal to have a perfect package that could manage the whole survey process. This is an illusion. Handling statistical data has so many diverse aspects that it is impossible to cover everything in one package. Moreover, new ideas evolve in statistics as well as in information technology in rapid succession. Therefore, these ideas can never be implemented at once in one system. The ultimate survey processing package will most probably never become reality.

An integrated control system offers a different approach to solving the problem. The concept is based on the idea that survey processing is essentially a set of activities for the manipulation of data and metadata. The data are the answers to the questions on the questionnaire form, and in the course of the process they are transformed into useful statistical information. The control system integrates all survey software in a user friendly shell. The survey staff use this shell to define the survey, to collect data, and to process data. For traditional data collection, the system must be able to generate a paper questionnaire form, and corresponding data entry and data editing programs. For computer assisted interviewing programs, the system must be able to produce programs for CAPI, CATI, or CASI.

An integrated control system should not be a fixed entity, but a flexible system. Users should be able to include their own components. In-house software development is an expensive and time-consuming activity that must be kept to a minimum. Whenever useful standard software is available on the market, it should be used. The control system must offer the facility of including such software in the shell.

The integrated control system should be in charge of communication between its various components. It must be able to give each component information about the survey data to be processed. Here, there is a vital role for the metadata.

The concept of metadata is not well defined in the literature. Following Berg *et al.* (1992), a general, but rather vague, definition could be "data about data." For use in an integrated control system, metadata is defined in a more practical way, as a formal statement of the information required for collecting, processing and publishing survey data. Such a metadata definition should include the following:

- *Definition of survey variables.* Each variable must have an identifying name, a domain of valid values, a code list (for qualitative variables), a question text required to obtain values (possibly in different languages), and other texts to document the variable.
- *Data model.* The data model describes the relationships between variables in terms of groups, hierarchies, and replication. For example, groups can be nested in other groups, and also groups may be replicated a number of times.
- *Route instructions.* Route instructions define the order in which, and the conditions under which, the questions (variables) are asked. Such instructions see to it that only relevant questions are asked, and that irrelevant questions are skipped.
- *Relationships.* Whenever relationships impose restrictions on the values of variables, these restrictions must be specified to carry out consistency checks on the collected data. Usually, inconsistencies are caused by errors in the data. Detection and correction of such errors may improve data quality.
- *Computations.* Not every survey variable is the direct result of a response to a question. Sometimes, new variables are derived from other variables by means of arithmetical expressions. Furthermore, it must be possible to replace a missing or wrong answer by a "synthetic" answer. Such an answer may be the result of a computation.
- *Links to other files.* For some surveys it is important to make links to data files of other or previous surveys. That makes it possible to import information from other surveys. This facility can also be used to compare data from panel surveys.

Note that this metadata definition could be extended to include more information about the survey, like the definition of the sampling design, and relationships between the survey variables and variables in other surveys.

The starting point of processing a survey with an integrated control system is the definition of a data model. The data model is the backbone of the metadata definition. What do data models look like? Experience shows that many survey researchers see the world as hierarchical: there are households, households consist of members, members have health problems, etc. If this is the natural way in which people think about surveys, a survey design instrument must be able to handle hierarchical data models in a straightforward and easy way. It should promote a modular approach to data model design in which hierarchies are implemented by means of nesting simple data structures.

The main problem with large and complex data structures is to keep track of the overall structure. It is impossible to keep in mind several thousands of variables simultaneously. What is needed is a way of grouping them, allowing concentration on one variable, or on a limited group of variables at a time, and to see at a glance their relevant structural relationships. To solve this problem, the submodel concept is introduced. A submodel is a construct for grouping metadata in a modular way. A submodel can be seen from the inside as a group of closely related variables that are independent of all other variables. From the outside a submodel can be seen as a unit of information that can be nested in other submodels, or repeated a number of times.

Since a data model is hierarchical, its structure can be displayed as a tree. A simple example is given to illustrate this approach. Readers familiar with the usual file managers running under MS-DOS (Dosshell) or Windows (File manager) will recognize the resemblance between the way data models are visualized and the way in which file managers visualize directory structures.

The data model example describes a household survey. It is a hierarchical model with two levels. The highest level represents the household with only three variables: the address (street and town), and the size of the household. Households consist of persons, all having a number of characteristics. These characteristics are recorded in a set of variables: name, sex, marital status, age, number of children (for women), employed (yes/no), and for employed people a job description. A possible data model for this survey is given in Figure 16.2.

The data tree on the left of Figure 16.2 describes the data model at the highest level. This gives a good view of the general structure. There are three fields (*Street*, *Town*, and *HHSize*), and one submodel (*PERSON*). To distinguish basic fields from submodels, names of submodels are written in capitals. The plus sign (+) indicates a submodel the basic fields of which are not displayed. If a submodel is unfolded, i.e., all fields in the submodel are made visible, the data tree on the right-hand side with all the details is obtained. The data model

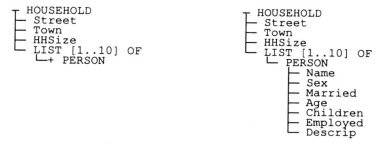

Figure 16.2   A hierarchical data model.

contains the construct *LIST* [*1..10*] *OF* to indicate that the underlying structure can be repeated at most ten times.

Submodels are a powerful instrument for designing complex data models in a modular way. Use of submodels has a number of advantages. The structure of the model is made much clearer. It also simplifies its development and maintenance. Another advantage is re-usability. A well designed and well tested submodel can easily be included in data models of other surveys. That reduces development time and enhances intersurvey co-ordination.

Data models like the one in Figure 16.2 do not contain all the information required for survey processing. A metadata specification not only has to reflect the structure of the data model, but it should also provide detailed information about the variables themselves, rules that apply for processing the variables, and all other information that is required to go through the survey process. The metadata specification language of the integrated control system must take care of all this.

16.5 DEVELOPMENTS AT SOME NATIONAL STATISTICAL OFFICES

Most national statistical offices aim at improving their survey processes. In this section, some of the developments at a number of major statistical offices are described. The general trend is that standard software is used wherever possible, and transfer of data and data descriptions between software is made as simple as possible. Several authors point out how the efficiency of the statistical process can be increased, and its costs reduced, by integrating the various steps. For example, see Baker (1987), Brakenhoff *et al.* (1987), Outrata and Chinnappa (1989), and Bethlehem (1992).

In 1990, the U.S. Bureau of the Census concluded that it was behind in applying new computer technologies in its survey operations. It had no general automation policy. Highly decentralized software development resulted in specialized systems for each survey, thus causing a waste of resources. The response of the Census Bureau was the CASIC concept. CASIC stands for Computer Assisted Survey Information Collection. By maximizing the use of automation and telecommunication, the CASIC effort aims at integrating the various steps in the survey process. Present activities of the Census Bureau concentrate on the data collection stage. Not only computer assisted interviewing techniques (CAPI, CATI, and CASI) are considered, but also other promising technologies like touch-tone data entry (TDE), and voice recognition entry (VRE). Integrating subsequent steps in the survey process (coding, imputation, weighting, and tabulation) has not been taken up yet. This is considered to be a long-term goal. Integration and standardization are expected to improve cost effectiveness, enhance data quality, and reduce data processing time. An overview of the CASIC activities of the Census Bureau is given by Mattchett *et al.* (1994).

Statistics Canada has been developing standard survey software for more than 20 years. The main reason for this strategy is cost. Software development is considered expensive. This also applies to standard software. However, the approach of developing standard software is less costly than repeated custom development. Cost reduction does not apply to software development alone. It also affects survey system design and training. The approach of Statistics Canada has always been to provide tools that automate selected steps in the survey process. Examples are GSAM (for sampling), DC2 (for data capture), GEIS (for editing and imputation), ACTR (for automated coding), GES (for estimation), Oracle (a relational database system), and SAS (for statistical analysis). The standard packages can work together, but no attempt has been made to develop an integrated system. Each package has its own specification language. Statistics Canada did not implement a policy with respect to hardware standardization. Instead it emphasizes portability to multiple platforms to provide platform independence. The future architecture will be one of a combination of commercial packages and in-house developed packages, where integration is taken care of by standard graphical user interfaces, standard protocols, and a common scripting language. For more details, see, e.g., Outrata and Chinnappa (1989) and Turner (1994).

The experience of the Office for National Statistics (ONS) in England is described by Manners and Diamond (1994) and Martin (1993). The ONS realized in an early stage the importance of moving from paper questionnaires to electronic ones. Development work on CAPI started in 1987. Although the initial costs of the change were high (due to, e.g., investments in laptop computers and training), considerable savings were realized in a relatively short period of time. The ONS concentrates on integrating the various activities in the data collection phase. Standard software (Blaise) is used to integrate coding into the interview. Since data entry, data editing and coding are no longer necessary in the office, costs are reduced. Standardizing and integrating the data collection process make it easier for the ONS to compete with other survey organizations. There are fewer development costs, and development time is shorter. The wheel is not reinvented for each survey. Data quality is higher. No attempts are reported to integrate the data collection activities with other parts of the survey process.

Developments in Sweden are described by Blom (1994). Statistics Sweden faced a government decision of decentralizing the statistical system. A number of different authorities now have funds, and they can commission any agency to carry out a survey. This forces Statistics Sweden to be more competitive on the statistics market. Also, budget constraints force Statistics Sweden to carry out research on more cost effective data collection and data processing methods. Computer assisted interviewing techniques (CAPI and CATI) are particularly emphasized. Work on other data collection techniques includes CASI, TDE (touch-tone data entry), scanning of paper forms, and fax reporting. These techniques have not yet been integrated.

Introduction of CASIC technologies in the Australian Bureau of Statistics

(ABS) was somewhat hastened by the removal of "heads down" data entry activities due to restrictions placed on the amount of time that can be spent on intensive keyboard work. Using CADI techniques, data entry has become one task in a multiskilled environment. A standard environment has been developed for CADI work. The IPS system has a client/server architecture with a Windows based front end, and an Oracle database as back end. Recently, Blaise was selected as the standard package for CAPI and CATI. OCR and OMR techniques have been used in some surveys. A trial with TDE was not successful. There is no integration of the various data collection techniques and software for other steps in the survey process. An overview is given by Tozer and Jaensch (1994).

The policy of Statistics Netherlands has always been one of standardization and integration of all steps in the survey process. There is standard software for each activity, and integration is realized by the Blaise Control Centre. This is an implementation of the concept of the integrated control system. It is based on the Blaise III system for computer assisted survey processing. The system uses machine readable metadata that can automatically be transformed into many different formats, depending on the software that needs it. Blaise III supports computer assisted interviewing, and traditional paper form interviewing. It also allows for multi-mode surveys. The Blaise Control Centre contains a number of additional modules, like Manipula for data manipulation, Bascula for adjustment weighting, and Abacus for tabulation. For data analysis, the commercial packages SPSS and Stata are used. The Control Centre dispatches the formatting of both data and metadata. More information about the approach of Statistics Netherlands can be found in Bethlehem and Keller (1992), Bethlehem (1992), and Schuerhoff (1993).

16.6 EFFECTS ON THE ORGANIZATION

A radical redesign of the survey process has consequences for the structure of the organization and the employees working in it. This section discusses some of the possible effects of the implementation of a policy of concentration, standardization, and integration.

Concentration of all survey activities to one department (to the extent possible) implies a new division of labor. Data processing activities that were previously carried out by the computer department will be moved to a subject matter department. Consequently, the computer department will change from an EDP (electronic data processing) department into an IT (information technology) department responsible for providing and maintaining the proper infrastructure (hardware and software) for workers elsewhere in the organization.

Concentration means that statisticians and subject matter specialists who are relatively inexperienced in computing are given the responsibility for and control over survey processing systems. They must have a proper, user friendly work environment to carry out this work successfully. The integrated control system plays a vital role in this.

Concentration does not mean that computer experts have to be included in the survey teams. It means that statisticians should cease to be specialists. They should become generalists with broad knowledge of and experience in all aspects of the survey process, including computing. The consequence of this approach is that a lot of effort has to be put into training activities. As a result, the workers in the subject matter departments will experience greater variety in their work. They will become more process-oriented, and cease to be so task-oriented. Because they exert more control over the organization of their work, they will be able to produce higher quality results.

Of course, it should be realized that the success of the new approach ultimately depends on its acceptability by the workers in the subject matter departments. There is an important task for the general management of the statistical office to support such a change, and to convince the workers of the advantages, not only for their own work, but also for the organization as a whole.

Standardization means both standardization of hardware and software. With respect to the hardware platform, the obvious choice is microcomputers. Survey processing is carried out in a de-centralized way by relatively inexperienced computer users, and this requires user friendliness at a relatively low price. Moreover, there is an abundance of useful software available.

Introduction of microcomputers in an organization should not be done without careful thought. Although the use of microcomputers seems to open new ways towards efficient information processing, it also may be a source of new problems. If every organizational unit is free to select and purchase its own type of computer, the automation infrastructure may easily get out of control, and turn into chaos. This calls for a strong policy on standardization.

Attention should be paid to the way in which the microcomputers are used in the organization. Distribution of a lot of stand alone microcomputers may seem a simple solution, but it is not the proper one. Here are some potential problems.

- It is very easy to copy (confidential) data files from local hard disks to floppy disks, which creates a data security problem.

- Activities like making backups and archiving are often neglected by the users in the subject matter departments. This may result in loss of valuable data.

- Communication between departments (e.g., sharing data files) is only possible by exchanging floppy disks. This is inefficient.

- Distribution of new releases of software packages, including their documentation, is often cumbersome in a large organization with a lot of stand alone microcomputers.

The solution of many, if not all, of the above-mentioned problems, is to connect the microcomputers in local area networks (LANs). This makes it possible for the IT department to archive and back-up in a centralized way. Also, version control and updating software can more easily be realized in such an environment. Distribution and installation of new software releases on a LAN is straightforward, since, with one command, a new version can be uploaded to all fileservers.

Efficient use of the new automation infrastructure requires rules with respect to the use of software by the subject matter departments. Giving these departments complete freedom to choose software to include in the integrated control system would result in an unworkable situation. Although such a system could solve some of the problems, many inefficient and ineffective situations would remain. Different software would be used for the same tasks, or different versions of the same software, resulting in incompatible data files. Furthermore, users from different departments would not speak the same "language" any more, and the need for training and support would increase dramatically. And of course purchasing many different packages is a lot more expensive than a site license for one package.

It is clear that efficient production of high quality statistics with an integrated control system calls for a policy on standardization of automation tools. The main advantages of standardization are summarized in the following points:

- more return on investments in training;
- more widespread knowledge and exchange of experience;
- consequently, fewer resources spent on support and maintenance;
- better exchangeability of hardware, software and data;
- fewer development costs with respect to survey systems;
- fewer problems in selecting new hardware/software;
- cost savings through discounts on large quantities of hardware/software.

The right choice of standards is of great importance and consequence. The choice should take into account the possibility that in the future the application software tools can be transported to another infrastructure, without the obligation of large-scale conversions. One consequence of a standardization policy might be a reduced flexibility. Every change can generate a great deal of extra work, and a wrong choice can have severe consequences. So, it is wise not to change the software standards too often, and also new versions of existing standards should not be installed too often.

The concept of integration is generally accepted as a means to improve the efficiency and the quality of the survey process. An integrated control system is the ultimate implementation of this concept. The advantage for the users is that they need to speak only one "language" to define and process the survey, and that is the metadata definition language. The metadata definition is created in the design phase of the survey, and stored in machine readable format. Every component of the control system retrieves its information about the survey from this metadata definition. If necessary, the control system can translate the metadata into any format that is required in the survey process. The control system also hides all technical aspects with respect to data storage formats from the user. The users only have to specify how they want to store the data, e.g., in a relational database system, and the control system takes care of all technical details.

Introducing concepts like concentration, standardization, and integration means redesigning the work processes. This affects both the organization itself and the people within it. Therefore, it is important that the general management of the statistical office supports the changes. The management must not only authorize and motivate the new approach to survey processing, but also persuade the workers to accept the radical changes the redesign will bring.

The concepts of concentration, standardization, and integration are very attractive. Still, one should be aware of practical problems that may hinder the implementation of these concepts in the organization. One problem is that people generally are reluctant to change their ways of working. They are used to the current systems and procedures, and they see no reason for change. Another potential source of problems is the change from tailor-made systems to standard packages. Tailor-made systems are always better tuned for the application at hand. Therefore, change to a standard package will seem a step back. Of course, this view disregards the development and maintenance costs of tailor-made systems, nor does it appreciate the organization-wide advantages of the use of standard systems.

16.7 EFFECT ON QUALITY

There are no examples (yet) of completely implemented integrated control systems, although some statistical agencies are in various stages of realizing at least part of the idea. Therefore, it is difficult to give concrete figures on the effect of integrated control systems on the quality of surveys.

The available information on survey costs seems to indicate that integrated control systems can reduce the total survey costs. The Subcommittee on Data Editing in Federal Statistical Agencies (1990) reports that editing activities consume a substantial part of the costs of traditional survey operations. In many surveys editing generates at least 20% of the total costs. Particularly in economic surveys, the data editing costs can be 40% or more. Blom (1994) makes a similar statement about the situation in Sweden, where 40% of the survey costs are devoted to data acquisition, and 20% to editing and coding.

A major move towards cost reduction could be the introduction of computer assisted interviewing techniques. Clayton and Werking (1994) characterize this change as a transition from labor intensive activities to capital intensive technology. Since labor costs continue to rise, while nonlabor costs decline, this change has a positive effect on the reduction of survey costs. Some authors, e.g., Weeks (1992), and Nicholls and Groves (1986) indicate that initial development costs of CAPI and CATI surveys are not lower than those of traditional surveys. This is mainly caused by high investments in hardware, software, and training. However, Martin (1993) notes that the return period of the initial investments was only two years at ONS. Large cost savings were accomplished by elimination of data entry, coding, and batch-wise editing. Nicholls and Groves (1986) argue that computer assisted interviewing can be particularly cost effective for continued and expanding surveys. Weeks (1992) mentions cost reductions in re-use of survey instruments, and also in a large and complex survey. All this indicates that an environment allowing for easy development, maintenance, and re-use of survey instruments contributes to a substantial reduction of the total survey costs. The integrated control system is an attempt to offer such an environment.

Statistics Netherlands has already come quite far in the implementation of an integrated control system. In the last decade, the ideas of concentration, standardization, and integration have been realized. The change from mainframes to microcomputers, and the policy of hardware and software standardization made it possible to reduce the survey costs substantially. For example, the costs of the automation infrastructure have been reduced by 20% from 1987 to 1990. At present, the costs are even lower. The new environment has made it possible to maintain the same production level at a time in which the organization had to face substantial budget cuts and personnel reductions. For details, see Bethlehem (1992).

As for data accuracy, it is more difficult to get concrete evidence on the advantages of an integrated control system. The advantages of computer assisted interviewing techniques have become quite clear over the last decade. Weeks (1992) gives an overview of the literature. Studies report significant reductions of critical errors, routing errors, and less "don't know" and item nonresponse. Some studies on traditional surveys mention that 90% of the errors are due to failure to record the answers of the respondents, and such errors are not possible in computer assisted interviewing.

The extra increase in data accuracy due to the use of an integrated control system comes from two sources.

- Since conversion of data and metadata from one step in the survey process to the next is automatically taken care of using a unique metadata definition, consistency between steps is guaranteed. The error prone manual preparation of set-ups and control files for the various software components is avoided.
- An integrated control system promotes re-use of survey instruments. Particularly, the modularity of the survey definition language makes it possible to develop, test, and maintain separate survey definition building blocks that can be used in several surveys.

A final aspect of data quality is the timeliness of survey results. Most authors agree that concentration, standardization and integration improve the time-liness of survey results. In some cases, reductions of 50% of the total processing time have been realized. See e.g., Bethlehem (1992), Blom (1994), Clayton and Working (1994) , and Tozer and Jaensch (1994).

Use of an integrated control system leads to even greater improvements in timeliness. Development of new survey instruments is quicker when standard survey definition modules are used. Furthermore, less time is needed for training interviewers when they work in a standardized environment. The same goes for all other people involved in the survey process. Finally, the smooth transition from one processing step to another reduces total processing time.

Bethlehem (1992) also reports notable benefits for the workers in the subject matter departments. The change increased the scope and variety of their work. They appreciated the increased autonomy and responsibility in organizing their work. It led to reduced specialization, the acquisition of a wider range of skills, and the possibility of adapting and devising systems for their own purposes. These advantages generally outweighed the need, in some cases, to undertake lengthy periods of training, and even overcame a longstanding reluctance on the part of some to work with computers.

16 8   THE FUTURE

Integrated control systems can help statistical offices to streamline their production processes. Through the common data language, and the possibilities of transferring data and metadata from one program to the other, the efficiency of the survey process is increased. This will improve the timeliness and the quality of the produced statistical information.

Integrated control systems are not only able to process traditional surveys in a better way, they also make complex surveys easier to conduct. Therefore, such systems hold promise for even more and better statistical information in the future.

Integrated control systems will improve the work of statistical offices, but they are by no means a panacea. Still, control systems may contribute valuable solutions. Two more or less related problems are mentioned briefly, and some indication is given of what control systems can do to address such problems.

The first problem is that of dissemination of statistical information. Recent developments in data communication led to increased demand for statistical information in electronic form. Media like the electronic highway (the Internet) are highly suitable for providing access to information in this way. Such a new electronic dissemination approach requires the implementation of a co-ordinated output database containing all accessible data and metadata. The metadata component of the integrated control system could be extended to incorporate features that are needed in the output side of the system. For example, aspects that need to be taken care of are aggregation activities and disclosure protection.

The second problem is that of integration of statistics. Most statistical offices publish statistical information that comes from many different surveys. Inconsistencies in survey definitions make it very hard to combine results from different surveys. Particularly, the easy electronic access to statistical information makes this problem more prominent. An integrated control system could play a role in offering a standardized environment for survey definitions. Particularly, features like a modular survey definition language can help in this respect. The modules make it possible to develop standard sets of questions, that can, after careful testing, be included in some kind of library of approved questionnaire modules. By making such standard sets available to all questionnaire designers, it will be very attractive for them to use these sets instead of developing a whole new set of questions.

REFERENCES

Baker, R.P. (1987), "Information Systems in Survey Research," *Proceedings of the Annual Research Conference*, U.S. Bureau of the Census, Washington, DC, pp. 166-177.

Bemelmans-Spork, E.J., and Sikkel, D. (1985), "Data Collection with Hand-held Computers," *Bulletin of the International Statistical Institute*, Book III, topic 18.3, pp. 1–16.

Berg, G.M. van der, Feber, E. de, and Graaf, P. de (1992), "Analysing Statistical Data Processing," in *New Technologies and Techniques for Statistics, Proceedings of the Conference*, Luxembourg: Eurostat, pp. 102–111.

Bethlehem, J.G. (1987), "The Data Editing Research Project of the Netherlands Central Bureau of Statistics," *Proceedings of the Annual Research Conference*, U.S. Bureau of the Census, Washington, DC, pp. 194–203.

Bethlehem, J.G. (1992), "A New Approach to Statistical Information Processing," *CBS-report*, Voorburg: Statistics Netherlands.

Bethlehem, J.G., and Keller, W.J. (1992), "The Blaise System for Integrated Survey Processing," *Survey Methodology*, 17, pp. 43–56.

Blom, E. (1994), "Building Integrated Systems of CASIC Technologies at Statistics Sweden," *Proceedings of the Annual Research Conference and CASIC Technologies Interchange*, U.S. Bureau of the Census, Washington, DC, pp. 623–634.

Brakenhoff, W.J., Remerswaal, P.W.M., and Sikkel, D. (1987), "Integration of Computer Assisted Survey Research," *CBS select* 4, *Automation in Survey Processing*, Voorburg, Statistics Netherlands, pp. 13–26.

Catlin, G., Ingram, S., and Hunter, L. (1988), "The Effects of CATI on Data Quality: A Comparison of CATI and Paper Methods," *Proceedings of the Annual Research Conference*, U.S. Bureau of the Census, Washington, DC, pp. 291–299.

Clayton, R., and Werking, G. (1994), "Integrating CASIC into the Current Employment Statistics Survey," *Proceedings of Annual Research Conference and CASIC Technologies Interchange*, U.S. Bureau of the Census, Washington, DC, pp. 738–750.

Deming, W.E. (1986), *Out of the Crisis*, Cambridge: Cambridge University Press.

Groves, R.M., and Nicholls II, W.L. (1986), "The Status of Computer Assisted Telephone Interviewing: Part II—Data Quality Issues," *Journal of Official Statistics*, 2, pp. 117–134.

Hammer, M., and Champy, J. (1993), *Reengineering the Corporation, A Manifesto for Business Revolution*, London: Nicholas Brealey Publishing.

Martin, J. (1993), "PAPI to CAPI: the OPCS Experience," *Essays on Blaise, Proceedings of the Second International Blaise Users Conferences*, London: OPCS, pp. 96–117.

Manners, T., and Diamond, A. (1994), "Integrated Field and Office Editing in Blaise: OPCS's Experience of Complex Financial Surveys," *Proceedings of the Annual Research Conference and CASIC Technologies Interchange*, U.S. Bureau of the Census, Washington, DC, pp. 732–737.

Mattchett, S.D., Creighton, K.P., and Landman, C.R. (1994), "Building Integrated Systems of CASIC Technologies at the Bureau of the Census," *Proceedings of the Annual Research Conference and CASIC Technologies Interchange*, U.S. Bureau of the Census, Washington, DC, pp. 573–595.

Nicholls II, W.L., and Groves, R.M. (1986), "The Status of Computer Assisted Telephone Interviewing: Part I—Introduction and Impact on Cost and Timeliness of Survey Data," *Journal of Official Statistics*, 2, pp. 93–115.

Outrata, E., and Chinnappa, N. (1989), "General Survey Functions Design at Statistics Canada," *Bulletin of the International Statistical Institute*, International Statistical Institute, Book II, pp. 219–238.

Schuerhoff, M.H. (1993), "Blaise as a Statistical Control Centre," *Bulletin of the International Statistical Institute*, International Statistical Institute, Book 2, pp. 273–282.

Sebestik, J., Zelon, H., DeWitt, D., O'Reilly, J.M., and McGowan, K. (1988), "Initial Experiences with CAPI," *Proceedings of the Annual Research Conference*, U.S. Bureau of the Census, Washington, DC, pp. 357–365.

Stol, H.R. (1993), "An Architecture for EDI in Business Surveys based on the Use of Blaise," *Essays on Blaise* 1993, *Proceedings of the Second International Blaise Users Meeting*, London: OPCS, pp. 143–153.

Tortora, R.D. (1985), "CATI in an Agricultural Statistical Agency," *Journal of Official Statistics*, 1, pp. 301–314.

Tozer, C., and Jaensch, B. (1994), "Use of OCR Technology in the Capture of Business Survey Data," *Proceedings of the Annual Research Conference and CASIC Technologies Interchange*, U.S. Bureau of the Census, Washington, DC, pp. 717–731.

Turner, M.J. (1994), "General Survey Processing Software: an Architectural Model," *Proceedings of the Annual Research Conference and CASIC Technologies Interchange*, U.S. Bureau of the Census, Washington, DC, pp. 596–606.

Van Bastelaer, A.M.L., Kerssemakers, F.A.M., and Sikkel, D. (1987), "A Test of the Continuous Labour Force Survey with Hand-held Computers, Interviewer Behaviour and Data Quality," *CBS select 4, Automation in Survey Processing*, Voorburg: Statistics Netherlands, pp. 37–54.

Weeks, M.F. (1992), "Computer-Assisted Survey Information Collection: A Review of CASIC Methods and Their Implications for Survey Operations," *Journal of Official Statistics*, 8, pp. 445–465.

Wieringa, J., and Fokké, G. (1992), "The Housing Census: A Blaise Application in an External Context," *Essays on Blaise, Proceedings of the First International Blaise Users Meeting*, Voorburg: Statistics Netherlands, pp. 207–215.

CHAPTER 17

Using Expert Systems to Model and Improve Survey Classification Processes

Frederick Conrad
U.S. Bureau of Labor Statistics

17.1 INTRODUCTION

Classification is ubiquitous throughout the survey process. Most analyses require that respondents' experiences, opinions, or knowledge be assigned to categories in order to be treated as data. Sometimes this is done by the respondents themselves. For example, when counting their expenditures for a particular product category—perhaps "meals in restaurants"—respondents must first determine whether individual purchases involve members of that product category—should "going out for dessert" be figured in the response? On other occasions, data are classified by the interviewer. For example, the interviewer might need to indicate whether the respondent lives in an "apartment building" or "private home." On still other occasions, classification is carried out by classification specialists, sometimes called coders, who assign open-ended statements, for example, about work or educational background, to categories in a formal classification scheme, such as an occupational or field of study classification system.

By assigning open-ended responses to a category, the researcher is essentially signaling that they share certain critical features. Similar responses can thus be aggregated and tallied, even if the responses differ from one another in some of their features. In principle, a response can be uniquely associated with a single category, though, in practice, the ambiguous mappings between responses

Survey Measurement and Process Quality, Edited by Lyberg, Biemer, Collins, de Leeuw, Dippo, Schwarz, Trewin.
ISBN 0-471-16559-X   © 1997 John Wiley & Sons, Inc.

and categories is one of the factors which makes the classification process challenging. Reliability between coders is not perfect, and the agreement rates can vary from one category to the next (Cantor and Esposito, 1992).

When open-ended responses are relatively short—just a few words in length—they are typically classified by coding clerks, some time after the data have been collected (see, for example, Campanelli *et al.*, Chapter 19). The clerks do this either manually or with computer support, such as on-line coding manuals or software that suggests possible categories (codes) from which the clerk selects one. In some survey operations, the computer actually assigns codes to responses, though in even the best of these systems, some responses still need to be coded clerically.

Computer support for such classification tasks has been justified on the grounds that, relative to manual coding, it makes the process more reliable, makes additional resources available to the coder, and does so without increasing costs (e.g., Andersson and Lyberg, 1983). Whether the coding is automated, manual, or some combination of the two, a simple, open-ended response is usually classified by looking it up in a dictionary where it is associated with a small number of codes—possibly a single code.

Some surveys involve more complex classification which is based on large numbers of variables, possibly elicited through unscripted dialogue with a respondent. Classification of this sort is performed by professionals who have acquired subject area expertise. They investigate hunches (classification hypotheses) by looking for relevant evidence much as a detective would. Because many factors are potentially relevant, they focus on just those factors that, experience tells them, promise to help in making the decision.

The kinds of data required for such decisions may vary between categories. For example, the difference between a "material handler" and a "fork lift operator" may depend on the equipment they operate but the relevant criterion for distinguishing an "optometrist" from an "ophthalmologist" is more likely to be their education and certification, though equipment could be pertinent as well. This chapter is concerned with this kind of classification expertise and a type of software that can support such tasks, expert systems.

We have developed two prototype expert systems that model complex classification performed at the U.S. Bureau of Labor Statistics. These are used throughout the chapter to illustrate general concepts about expert system technology in the survey process. The next section reviews recent research and development in automated coding. By and large this work has been applied to the classification of short, open-ended responses and so contrasts our work with expert systems. Section 17.3 presents some of the basic concepts of expert system technology and discusses why expert systems are appropriate for relatively complex classification. Then, Section 17.4 describes our prototype expert systems. The section is intended to characterize the kinds of tasks for which this technology is appropriate. In Section 17.5, we present representative results from an evaluation study and discuss some of the practical considerations of developing this kind of software in survey organizations.

17.2   AUTOMATED AND COMPUTER-ASSISTED CODING

Since the 1970s, computers have been used to code simple, open-ended responses to survey questions. These responses usually consist of a few words each, for example, "I work in a cannery—at the loading dock." The response is classified on the basis of known statistical relationships between its content words and categories in the classification scheme. The task does not require knowledge of the subject area. This contrasts with more complex classification where subject matter expertise is required and can be provided through expert systems. To provide a backdrop for our discussion of expert systems, the current section reviews some of the major developments in software for coding simple responses. The general idea involves matching the response string to an entry (or entries) in a dictionary in which each entry is associated with a category or numerical code. The innovation in this research has occurred in developing algorithms for constructing and searching dictionaries of this sort.

This approach has been applied primarily to industrial and occupational coding (e.g., Andersson and Lyberg, 1983; Appel and Hellerman, 1983; Appel and Scopp, 1987; Van Bastelaar et al., 1987; Campanelli et al., Chapter 19), but has also been used to classify descriptions of products (Andersson and Lyberg, 1983), field of study (Bobbitt and Carroll, 1993; Ciok, 1993; Pratt and Burkheimer, 1994; Pratt and Mays, 1989), library book loans (Andersson and Lyberg, 1983) and demographic characteristics such as ethnic origin and language (Ciok, 1993). This effort has been carried out throughout the world, primarily in government statistical agencies (for reviews see Bethke and Pratt, 1989, Gillman and Appel, 1994, and Lyberg and Dean, 1990).

The various efforts to automate the coding process can be divided into "autocoding" (or "batch") processing—applications that entirely automate the coding process—and "computer-assisted" (or "interactive") coding systems—software that provides human coders with pertinent information which the coders then incorporate into their classification decision making. Some coding systems can be run in both batch and interactive modes (e.g., ACTR, the system developed by Statistics Canada). Sometimes the two are used in conjunction: a large data set is submitted to batch processing in a first pass; those cases that cannot be coded by automated means are classified interactively in a second pass. Unlike manual coding of these data, which usually occurs after the data have been collected, automated coding (in all of the above forms) can take place during the interview in both CATI (Computer Assisted Telephone Interviewing) and CAPI (Computer Assisted Personal Interviewing) environments (Pratt and Burkheimer, 1994).

Dictionaries, which associate possible responses with codes, can be derived from coding manuals or previously coded data sets. Dictionaries sometimes uniquely map content words (or phrases consisting of such words) to codes, though on other occasions a word or phrase is associated with multiple codes. Dictionaries are sometimes constructed manually (e.g., Pratt and Mays, 1989), but for large classification systems the process is usually automated.

The U.S. Census Bureau's Automated Industry and Occupation Coding System or AIOCS (Appel and Hellerman, 1983; Appel and Scopp, 1987) relies on a computer generated dictionary (O'Reagan, 1972). The words from all of the phrases in the coding manual are converted to a standard form (basically the root word stripped of affixes) and a dictionary is assembled from all of these standardized words. The dictionary includes weights for each word determined by the number of different codes with which it is associated. Links are created between the words in the dictionary and the phrases in the coding manual that contain those words. The links and the phrases themselves form the knowledge on which the classification is based. In general, no subject matter knowledge is involved.

The actual coding process (as opposed to dictionary construction) involves matching response words and phrases to a dictionary entry. This is done through either dictionary algorithms or weighted algorithms (Andersson and Lyberg, 1983). A dictionary algorithm requires a response string (a word or phrase in standard form) to match a single dictionary entry (sometimes independent of word order). While the match need not always be exact (e.g., Pratt and Mays, 1989), it must be close enough to an entry so that the match is unambiguous. If such a match is found, the response string is assigned the associated code. This is an effective strategy when responses are simple; as the response strings become longer, possibly spanning the answers to several questions, they are more likely to match multiple entries.

Weighted algorithms make it possible to code such ambiguous matches with acceptable accuracy. In AIOCS, the words in the dictionary are weighted according to how well they predict a particular code: a word that is associated with only a single code receives a large weight while the weight is smaller for a word associated with many codes. The algorithm selects candidate phrases from the dictionary based on a score which is computed from the amount, percent, and weight of matching words (Appel and Scopp, 1987). If it identifies a single phrase then the code associated with that phrase is assigned to the response string. If the algorithm selects multiple phrases, then the one with the highest score is chosen, assuming it exceeds the score of the closest competitor by a specified ratio, and the response is classified accordingly. It is possible that the algorithm will not identify any phrases with an acceptable score, in which case the response is coded manually.

Certain systems include a type of knowledge that enables them to circumvent the algorithmic matching process. AIOCS performs a "logical analysis" in which rules map particular keywords directly to codes. If such keywords are present in the response field, they trigger the appropriate rule and the code is assigned without the usual lexical search. The number of keywords is kept small to ensure that the approach is efficient. If the set of keywords were large, then complex matching issues would compromise its value. The approach has proven beneficial: by one recent evaluation, AIOCS assigned codes using logical analysis at a much lower error rate than was found overall (Gillman and Appel, 1994).

A coding system for classifying field of study that uses a comparable approach is described by Pratt and Mays (1989). The abbreviated names of academic departments offering particular courses were mapped to the field of study categories by rules that were general across institutions ("CS" $\Rightarrow$ computer science) unless an institution used unconventional abbreviations in which case rules were developed that were specific to an institution (at institution 12345 "CS" $\Rightarrow$ classical studies). These researchers were able to tune the system's effectiveness by manually re-ordering the rules until they found an optimal sequence.

The logical analysis in AIOCS and mapping of abbreviated department names in the Pratt and Mays (1989) system are both examples of a "rule-based" approach, in that knowledge is encapsulated in a set of rules which are invoked under specific circumstances. They both exploit relationships between responses and codes that experts know to be the case, but which are not the result of executing an algorithm. In addition, they improve performance by avoiding computationally intensive matching procedures. This type of decision making is a version of the reasoning that drives expert systems, discussed in the remaining sections.

A range of performance has been reported for the numerous automated and computer assisted coding systems (Bethke and Pratt, 1989). The two measures that are jointly used to evaluate a system are (1) coding or production rate and (2) agreement rate (or its inverse, error rate). The first of these refers to the proportion of cases that are coded (regardless of accuracy) and the second is an accuracy measure, usually derived by comparing computer generated codes to "truth codes," that is, the judgments of coding experts, for some test set of responses.

Without updating the dictionary, one can increase the coding rate by relaxing certain matching requirements, although this comes at the expense of lower agreement rates. It is hard to compare performance of different coding systems because the criteria for acceptable levels depend on the particular survey. In general, however, rates are considered acceptable if they are comparable to or better than the results of manual coding, and the amortized cost of developing and maintaining the systems is lower than personnel costs for the same task.

One new technology that has performed well in pilot coding studies is neural networks (see, for example, Bechtel and Abrahamson, 1991 and Rummelhart and McClelland, 1986, for general descriptions of the technology). Neural network technology is a type of classification software that "learns" the associations between a set of inputs and a set of outputs. In survey coding tasks, the inputs are the words in an open-ended response and the outputs comprise a code. Each association between response words and a code is represented as a network, where every unit (or node) is either an input, output or mediating ("hidden") unit. Several layers of units may intervene between the inputs and outputs. The units are neuron-like objects which influence the activation of the other units to which they are linked—usually in the next layer of the network. The exact effect of one unit on another is governed by weighting the link between them.

The network is "trained" on a set of example cases where both input and output are known (i.e., the responses have been manually coded by experienced personnel). A learning algorithm repeatedly adjusts the weights until the network produces an output pattern suitably close to the known output for each example case. (This is similar, in spirit, to a regression analysis in which the residuals are minimized by a least squares method.) Once the network has been trained, it can code cases that have not previously been encountered. An automated coding system based on neural network technology (Raud and Fallig, 1993) was tested with Census Bureau industry and occupation data and produced about the same overall coding rate as AIOCS (48%) though its agreement rate was slightly lower (87% as opposed to about 90% for AIOCS) (Gillman and Appel, 1994).

Another technology that has performed well for simple coding tasks is memory-based reasoning (MBR) (Stanfil and Waltz, 1986). MBR computes the similarity of each unclassified case to every case in a database of classified cases. Similarity between any two cases is usually implemented as a measure of feature overlap—usually some variant of the "nearest neighbor" metric—where each case consists of multiple features. In a survey coding application, the features might include the content of open-ended responses, demographic variables like age and gender, as well as combinations of all those features. Although the similarity metric itself is computationally simple, the number of such comparisons can become astronomical for large databases. Because of this, MBR is intended to run on parallel computers where each case is a separate (physical) processing unit, making the number of comparisons far less relevant to performance time than it is on serial hardware.

MBR technology was tested on Census Bureau industry and occupation data (Creecy et al., 1992). The MBR approach was more productive than AIOCS (61% coding rate with an overall agreement rate of about 88%) and was constructed in a fraction of the person months required to build AIOCS. However, the associated costs have led to some skepticism about its practicality (Gillman and Appel, 1994). MBR, like neural network techniques, requires a training set for any new domain; the clerical expense involved in certifying that the codes are accurate can be substantial. In addition, MBR must run on a massively parallel computer; one was borrowed for the test by Creecy et al., (1992), but currently, these each cost over a million dollars.

17.3 CONCEPTS BEHIND EXPERT SYSTEMS

The automation of simple coding tasks has been well served by the technological approaches described in the previous section. For the most part, the applications described earlier take advantage of the relationship between the words used by respondents and the way those words are usually coded. There is another category of survey classification tasks which relies on larger amounts of information, some of which may not be explicitly stated by respondents. In light of this extra complexity, it is generally not possible to create a dictionary of mappings between responses and codes.

In these more complex cases, the classification specialist may infer unstated information based on knowledge of the subject area. Often the process is driven by a working hypothesis about the likely classification; the specialist decides which data to investigate based on their relevance to the hypothesis. This approach to classification has many of the hallmarks of expert problem solving (e.g., Glaser and Chi, 1988). As it turns out, it is a model of expert problem solving that serves as the foundation for expert systems (e.g., Kahn, 1988), suggesting that there is a good fit between complex classification and expert system technology.

An expert system is a piece of software that solves problems much as a human expert would. For example, several classic expert systems were developed to diagnose medical problems based on the knowledge and diagnostic strategies of physicians (e.g., Buchanan and Shortliffe, 1984). Expert systems are the oldest and most stable technology to emerge from research in artificial intelligence (AI). In one survey of AI practitioners (Hayes-Roth and Jacobstein, 1994) expert systems were deployed in 27 broad industrial categories, with about 30% of the applications in finance and manufacturing. There are several well known success stories in which expert systems have increased productivity, improved quality and timeliness of products, and saved money well beyond their development costs (see, for example, Leonard-Barton, 1987).

The general kinds of problems that have been addressed by this technology include diagnosis, classification, scheduling, configuration, monitoring, and design (Kelly, 1991). While several expert systems have been developed for various statistical applications (see Gale et al, 1993 for a review), the technology has seen relatively little use in survey research. The one application of which we are aware, other than our own, is in the area of questionnaire design (Halfpenny et al, 1992; Taylor et al., 1992).

What distinguishes these problems from those to which more conventional techniques have been applied is that (1) there may not be an optimal solution—or at least it may be hard to evaluate optimality—but there may be many acceptable solutions, and (2) solutions require intimate familiarity with the problem domain rather than abstract formulae. This situation-specific knowledge is often referred to as heuristics.

17.3.1 Heuristic Knowledge

Heuristics are rules of thumb that lead to acceptable solutions, although the exact rationale for their success may be unknown to the problem solver. They just work. They are not guaranteed to produce the right answer but usually do, and do so in relatively few steps. For example, a seasoned taxi driver might know that it is wise to avoid a particular tunnel at certain hours of the day and certainly knows this without having constructed a quantitative model of traffic flow.

Heuristics are often compared to algorithms, which are precise computational recipes, guaranteed to produce the expected outcome. If one bakes a cake by following a recipe to the letter, one expects to produce the cake promised by the recipe. Algorithms are derived from "first principles," that is, from what is known about the fundamental properties of the problem. Heuristics, in contrast, are based on experience—they have been shown to work on many previous occasions, though their success cannot necessarily be explained on theoretical grounds. Algorithms tend to be general purpose. They are tailored to a particular situation when specific values are passed to them. Heuristics tend to be narrower in scope than algorithms, defined in terms of particular conditions.

Consider the traffic example above. A heuristic might consist of a simple set of rules such as "If you are in a hurry to reach the airport and the current time is between three and six o'clock PM, do not use the tunnel to reach the expressway. Instead, cross the river on a bridge and take Carson Street to the West End where there is an entrance to the expressway." To reach the same conclusion algorithmically, one would need to consider a vast number of parameters (typical traffic volume, tunnel capacity, likelihood of breakdowns, presence of road construction, etc.) and determine how these factors interact with one another in general. The algorithm could then be applied to specific values, such as times of the day, in order to produce a recommendation on when to use the tunnel. In this problem, familiarity with particular city streets leads to a far simpler solution than would an algorithm.

Heuristics are especially helpful for solving a class of problems whose solution time seems to grow exponentially as a function of size. These are known as NP-complete problems (which stands for "non-deterministic polynomial-time-complete" problems; see Poundstone, 1988, Chapter 9 for an accessible discussion). One goal of AI, and in particular, expert systems, has been to slow the growth of solution time for NP-complete problems. The key has proven to be informal knowledge about the domain, that is, heuristics (Nilson, 1980). Again, the solutions may not be optimal but they are serviceable. Many of the methods developed for solving NP-complete problems have proven useful for other applications, even if these problems are less complex. Heuristic methods simplify a problem by focusing on just those variables that, experience dictates, are relevant to the solution.

Because heuristics are informal and learned through experience, they are often not documented. One benefit of building expert systems is that the process creates a permanent record of this knowledge—even if the software is never used. Of course, running the expert system ensures that the knowledge is applied systematically and reliably.

17.3.2 Components of an Expert System

The knowledge in expert systems is generally represented as IF-THEN rules: IF certain conditions are present, THEN take a particular action or reach a particular conclusion. In this approach, small packets of knowledge are invoked when particular circumstances of the problem occur. The collection of rules required to solve problems in the particular content area is known as the knowledge base or rule base, hence the technique for developing expert systems is called "rule-based" programming.

Because rules are specific to a particular problem area, a knowledge base for automotive trouble shooting would share virtually nothing with a knowledge base for occupational classification. However, expert systems in these two different areas could have in common the reasoning mechanism that determines which rules to apply and the particular sequence in which to apply them. This mechanism is known as the inference engine. When the inference engine applies or "fires" a rule, that changes the current state of the problem—presumably to a state closer to a solution. The inference engine then searches for other rules in the knowledge base that might be applied under the new circumstances and continues this cycle until the problem is solved, or until it determines the problem cannot be solved by the current knowledge base.

Inference engines come in two varieties, backward chaining and forward chaining. Different types of problems are better suited for one or the other of these. A backward chaining inference engine begins with a hypothesis about the solution to the problem and the reasoning process seeks evidence to support the hypothesis. Backward chaining is appropriate for problems where there is a small set of possible hypotheses, such as diagnosis or classification tasks where the categories correspond to hypotheses. In contrast, forward chaining inference engines construct solutions in the absence of hypotheses. Strictly on the basis of the data describing the problem state, rule actions manipulate and create data that eventually comprise a solution. Appropriate problems for forward chaining are scheduling or configuration tasks where there are many possible solutions, none of which is known in advance.

In addition to the knowledge base and inference engine, an expert system requires data on which to operate and some mechanism to store the data elements. Data describe the problem state and determine which rules are eligible to be fired. This is because the condition elements of a rule (the IF part) are matched by the inference engine to the data elements present at that time. As rules are applied they create more data, changing the problem state, enabling other rules to become eligible. Data become available by being read from a file, entered by a user or created through the actions of rules. They are stored in a temporary memory, referred to by such terms as "working memory" or "session context." In most contemporary expert system tools, the data are object-oriented (cf. Cox and Novobilski, 1991).

One thing an expert system does not have is an explicit control structure. The inference engine determines which rules to apply at any moment. This contrasts with conventional approaches, where the programmer uses such techniques as looping, subroutines, and "Go To" statements to explicitly control the program's flow. The lack of an explicit control structure is one reason why expert systems are well suited for use in unstructured tasks, such as those for which we have developed our prototypes. This idea is illustrated in Section 17.4.2.

In principle, one could develop an expert system from scratch, building the tools to encode rules and data as well as building an inference engine. However, most expert system programmers do their work in development environments or "shells" that include rule editors, object editors, an inference engine (or sometimes both forward and backward engines) and other tools. These are specialized for expert system development and help produce an application that performs efficiently. They are derived from significant research in computer science (see, for example, Forgy, 1982, or Cooper and Wogrin, 1988, Chapter 10, for a discussion of one procedure to optimize an inference engine). Organizations that are considering this technology are advised to use an expert system shell instead of a traditional computer language to avoid reinventing the wheel. A relatively recent review of available development products can be found in Stylianou *et al.* (1992).

17.3.3   Knowledge Engineering

Expert system shells are not equipped with knowledge from any particular content area. That must be extracted from the minds of experts. This extraction process is usually carried out by a knowledge engineer who interviews the experts about their practices in particular circumstances or observes experts at work and encodes those practices as rules. It can be difficult for experts to articulate what they do because their procedures have become second nature over the years. The knowledge engineer must help the expert to specify his or her knowledge at a sufficiently detailed level so that it can be modeled as rules that successfully solve problems. Knowledge engineering is an iterative process of modeling the expert's knowledge and refining the model based on the expert's feedback. Because no rules can be encoded until the relevant expertise is elicited, knowledge engineering is considered the primary bottleneck in developing expert systems (Hoffman, 1987).

17.4   CASE STUDIES

We have chosen two classification activities at the U.S. Bureau of Labor Statistics (BLS) to explore the applicability of expert system techniques. Both rely on heuristic knowledge. The first is a data review activity in which an analyst evaluates decisions made by data collectors about the comparability of products. It is a classification task in the sense that the reviewer assigns each such decision to one of a small number of categories to indicate the adequacy of the decisions, and therefore, the usefulness of the data. The second task for which we have explored expert system techniques is a type of occupational classification done primarily during the interview. Unlike the occupational classification tasks described in Section 17.2, for which a straightforward dictionary look-up is possible, this task requires many subtle distinctions and a flexible dialogue between the interviewer and respondent.

17.4.1 Compass

The Consumer Price Index (CPI) is conducted monthly by BLS to measure the average change in price for a fixed set of goods and services. The index is updated each month by comparing the price of commodities in the current month to the price previously collected. Because of marketplace changes, a product priced in an earlier month may not currently be available. In order to maintain a continuous sample under these conditions, a comparable product may sometimes be substituted for the unavailable product. Commodity analysts (CAs) review each product substitution to determine its comparability to the original product. If the CA deems the substitute and original products to be comparable then their prices can be compared in computing the CPI.

In addition to price data, the field representatives collect data about numerous specifications of a product—for example "screen size" and "sound quality" for televisions. The specifications in each product area are listed in a separate checklist. A particular collection of specification values from a checklist describes a particular product from that area. The CAs rely on the specification values in a product description when judging comparability: if the overall pattern of specification values for the original and substitute products are judged to be sufficiently similar, then the products are considered comparable.

CAs' expertise is usually concentrated in a small number of product areas. However, their decision strategies are general across product types: analysts tend to look for evidence that products are not comparable, and only after failing to find such evidence do they accept the substitution as comparable. Such a process is efficient because it is possible to reject some substitutions when the original and substitute product differ on just one specification. In contrast, if the original and substitute products match on a particular specification, it is not possible to determine their comparability; they must essentially match on all relevant specifications in order to be judged comparable.

At certain points in the monthly production cycle, comparability decisions can constitute a substantial portion of the CA's duties. Of the 78,000 products priced each month for the CPI, about 2,700 are substitutes and must be reviewed by a staff of about 25 CAs. Automating some part of this would conserve expensive resources. In addition to being labor intensive, reviewing substitutions is prone to various mental "slips." Rare as these may be, they are a potential source of error for the CPI, one that might be reduced through computer support.

17.4.1.1 The Expert System
COMPASS (Comparability Assistant) is an experimental expert system designed to check CAs' comparability judgments about product substitutions (Conrad

et al., 1992).[1] It is intended to be used after a CA has reviewed a month of substitution data. At that time, COMPASS can evaluate the comparability of those same substitutions. The value of the procedure lies in the discrepancies that might arise between CA decisions and those made by COMPASS: because such differences are potentially due to errors by the CA, resolving them should improve data quality.

In principle, a tool such as COMPASS could be used for the routine cases, in place of the CA, or embedded in a CAPI instrument for data collection. The current architecture and knowledge engineering procedures would not change if the systems' functionality were extended in these directions. It has been implemented as a quality check because this allows us to explore the feasibility of expert system technology without affecting production activities.

To date, the COMPASS experiment has been conducted for 13 of the roughly 350 product categories surveyed for the CPI. The thought is that knowledge bases will eventually be developed for all product areas following the procedure used thus far. Four categories were chosen from the food area (Fruit, Cereal, Fish, and Dinner), three from the apparel family (Women's Swimsuits, Women's Pants and Shorts, and Women's Nightwear), three from services (Theater Admission, Automobile Finance Charges, and Hospital and Patient Services), and three that were unrelated to each other (TV, Automotive Body Work, and Ski Equipment).

Knowledge Engineering.
Expert systems are sometimes criticized because after being developed for test cases, they are not easily scaled up for production use (e.g., Creecy, *et al.*, 1992). In particular, large knowledge bases can be unwieldy and hard to maintain (Bachant and Soloway, 1989). These issues are addressed in COMPASS by building separate, relatively small knowledge bases for each product category. In order to spread the burden of developing 350 individual knowledge bases, the CAs themselves do the knowledge engineering. The basic idea is for each CA to develop a knowledge base for the product area(s) in which he or she is expert. This practice seems to have the additional benefit of enhancing COMPASS's credibility since the CA who will use COMPASS has encoded his or her own expertise.

One reason that professional knowledge engineers are usually involved in this type of activity is because they have the skills to elicit knowledge from experts—even when experts are unaware of that knowledge (Hoffman, 1987; Kelly, 1991). The substitution review task makes knowledge engineering by the CAs possible because CAs can often reach a decision on the basis of a single inference (rule), and rarely need more than two. Such inferences are easier for CAs to articulate than are the more complex inference sequences found in other tasks to which this technology has been applied.

---

[1] This should not be confused with another expert system, also called COMPASS, described by Prerau *et al.* (1985).

RULE for comparability A spec
    IF new_a OF dB3 E31011R 1 <> old_a OF dB3 E31011R 1
    THEN comparability OF quote IS no

Figure 17.1 Example rule from television knowledge base.

Even if experts can articulate their own knowledge, they are usually unable to write rules that a computer can interpret. This is not an obstacle for CAs because the expert system shell in which COMPASS is implemented uses simple, English-like syntax in its rule language. In addition, CAs are provided with a set of rule templates which present the structure of common comparability rules but are written without reference to particular product areas or specifications. In many cases, by adding only a few pieces of information to a rule template, the CA can construct a rule for his or her particular product area. Finally, CAs are given the basic computer code to run the system regardless of the product area. This enables CAs to concentrate on entering their knowledge for a product area rather than thinking about how the program will run.

An example rule is presented in Figure 17.1. It is taken from the knowledge base for Televisions ("E31011" is a code for TVs, well known to the CAs in this area). The "A" specification describes a product's color capability, where a value of 1 indicates black and white and a value of 2 indicates color ("new_a" is the value for the substitute product and "old_a" is the value for the original). The rule concludes the products are not comparable if the two products have different values for this specification (" <> " indicates "does not equal"), that is, if one TV set is black and white and the other is color.

Architecture
CAs tend to gather evidence to support one of a small set of hypotheses, for example, "not comparable," rather than constructing hypotheses from the patterns of specification values. This is mirrored in the system by its primary use of a backward chaining (i.e., hypothesis driven) inference engine. In addition, CAs search for evidence of non-comparability prior to evidence for comparability. COMPASS's decision making is structured so that it attempts to rule out comparability in several different ways before concluding two products are comparable.

17.4.2 MatchMaker

The second task that we have modeled with expert system techniques is a type of occupational classification known as job matching. The Occupational Compensation Survey Program (OCSP) publishes rates of pay for about 80 different job categories. The categories are those found in the U.S. Federal Government and are used in part to set the wages of Federal employees. The data are collected by means of face to face, establishment interviews, in which a field economist (FE) attempts to match positions found in an establishment to government job definitions. Once a job has been matched—and many are not matched—the FE can record the wages paid by the establishment to employees holding that job. The FE, therefore, is both the interviewer and the coder.

The OCSP job definitions range in length from several paragraphs to several pages, and describe the job at up to eight levels of seniority. Job matching, therefore, requires the FE to be (1) intimately familiar with the government descriptions, (2) skilled at interpreting the respondents' descriptions of establishment jobs, and (3) able to judge the similarity of one to the other. The FE must formulate questions during the course of the interview, which are appropriate to the situation and elicit informative responses.

There is no scripted questionnaire for the interview. One reason for this is to enable FEs to keep the length of the interview to a minimum by asking primarily about those jobs that will plausibly be found in the establishment—a kind of knowledge that is not explicitly documented—and asking only those questions necessary to confirm a match. This requires more flexibility than can be accommodated through a structured survey instrument.

While computer support for job matching could increase consistency between interviews and could ensure that the FE inquires about all relevant information, current CAPI tools do not provide the needed flexibility. Traditional CAPI tools primarily implement structured paper and pencil instruments, where the skip pattern among questions is specified in advance and is invariant. In principle an expert system could enable FEs to tailor the sequence of their questions to the particular interview situation while ensuring they collect all information about relevant job criteria.

17.4.2.1 *The Expert System*

MatchMaker is a prototype expert system to support job matching. It questions the FE about characteristics of an establishment job that must be present in order to match the job. The FE responds to MatchMaker on the basis of information from the respondent, so the FE must craft an appropriate question to the respondent in order to reply to MatchMaker. By using MatchMaker, any two FEs will ask respondents about the same features of a job. MatchMaker has been designed for use on a notebook computer in the interview, and could also be used to (1) prepare for an interview, (2) create a data report after an interview, and (3) train FEs to collect data.

In our research efforts to date, we have encoded knowledge for six job definitions: Accountant, Accounting Clerk, Computer Programmer, Engineer, Engineering Technician, and Personnel Specialist. These were chosen to span the range of job-matching complexity. In addition to knowledge for individual jobs, we have encoded knowledge about the jobs that are likely to be found in particular industrial sectors in establishments of various sizes. This is intended to help FEs generate hypotheses about likely jobs and confirm them by questioning the respondent.

Knowledge Engineering
The expertise in MatchMaker was encoded by analyzing publications, by interviewing and observing FEs, and by searching pertinent databases. Our analysis of the job matching task revealed that, in practice, most matches hinge on the job's duties, its required skills and education, the supervisory relations (who reports to whom) and the absence of "exclusionary evidence." Each job definition includes a list of criteria that specifically rule out a match even if other criteria would confirm the match—"exclusions." For example, a company position that otherwise matches the Computer Programmer definition but requires a bachelor's degree in a scientific field other than computer science is explicitly excluded. As a result, the knowledge engineering staff developed rules primarily about duties, skills, and education.

Once the knowledge was implemented as functioning software, several FEs reviewed it and, based on their comments, the knowledge base was refined. While there is a single knowledge base for all of the jobs in MatchMaker, in practice, the knowledge for each job is modular. As in COMPASS, this is intended to simplify the development and maintenance of the knowledge as the full set of job definitions is encoded.

Architecture
Much like the FEs, MatchMaker seeks evidence to confirm each job-match hypothesis. As a result, it is built around a backward chaining inference engine. Each hypothesis is supplied by the FE or by MatchMaker based on historical job-match data for similar establishments. The inference engine then queries the FE for the evidence required to apply appropriate rules. It first seeks to confirm the overall match, for example, an Accountant, and then at a particular level, for example Accountant IV. Under some conditions, when a match is not confirmed, MatchMaker can propose alternative jobs.

One advantage to the expert system approach in the case of job matching is that the inference engine determines the branching sequence (effectively, the skip pattern). This is helpful because the numerous job attributes under consideration combine into a complex branching structure (see Figure 17.2) that would be hard to encode explicitly for the full set of jobs, and that would make a traditional CAPI instrument difficult to maintain. Rather than focusing on the precise sequence in which rules should be applied, the knowledge engineer can focus on content knowledge that needs to be captured in rules. The inference engine determines which pieces of information to query based on what information has (and has not) been collected at any point.

To illustrate this, consider the logic depicted in Figure 17.2 to match an establishment job to the definition of personnel specialist. If the duties (upper left) match those in the definition of Personnel Specialist (the "yes" path), but none of the education requirements (middle left) are satisfied (the "no" branch pointing to the right), the inference engine applies a rule that redirects control to an alternative, less skilled job—Personnel Assistant (far right). There is nothing in the rule redirecting control to Personnel Assistant that indicates its

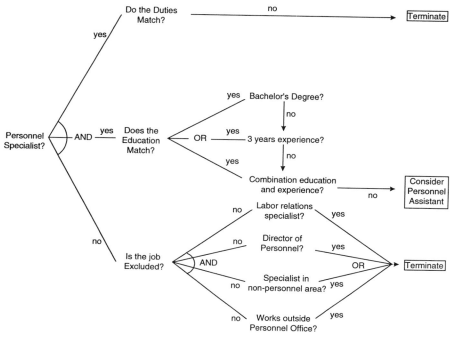

Figure 17.2 Simplified branching structure in MatchMaker. In order for a hypothesis to be confirmed, all conditions connected by an AND must be satisfied, and any one of the conditions connected by an OR must be satisfied.

sequence, simply the knowledge that an alternative job shares some of the duties of Personnel Specialist and requires less education.

By building flexibility into the design of MatchMaker, we intended to design a software system that would fit into the FE's work. This seems to have been reasonably successful. Twenty-two FEs were exposed to the prototype and were asked several questions including "How well does the system model the steps involved in job matching?" On average, they rated it 6.1 out of 9 where 1 was "not at all" and 9 was "completely."

17.5 EVALUATION AND RESOURCE ISSUES

Before advocating that survey organizations invest in expert systems we need to address two issues about the technology: "Does it work?" and "What resources are required?" To address the first of these, we present performance data from a COMPASS knowledge base. To address the second, we focus on human resources and likely reactions by personnel asked to use expert systems. It is hard to develop monetary, cost estimates based on our work to date because our systems have not been deployed in production environments.

17.5.1 Performance Evaluation

A cornerstone of the autocoding and computer-assisted coding research is quality measurement of the sort described in Section 17.2. Expert systems for classifying survey responses need to be subjected to similar evaluation. Because COMPASS has been designed as a post-production check on CA decision making, it is straightforward to compare its performance to that of the CAs.

The evaluation procedure we used compares the performance of the analyst and the system for a given knowledge base, run on three months of data. The statistics of interest are the proportion of substitutions that COMPASS was able to evaluate and the correlations between CAs and COMPASS for three comparability decisions: comparable, not comparable, and "quality adjustable," that is, comparable with a statistical adjustment (Armknecht and Weyback, 1989). Once an acceptable level of performance has been attained, the knowledge base is run again on three more months of data, one month at a time. Any additional knowledge engineering is performed at this point and statistics are calculated again.

We present results below from a representative knowledge base, Women's Pants and Shorts. COMPASS referred 19% of the decisions to the analyst, recognizing that it lacked sufficient knowledge to make these decisions. (This is equivalent to a "production rate" of 81% in the terminology of autocoding quality measurement.) Over the four data sets used to develop the knowledge base, the system and analyst agreed on 68% of the trials and disagreed on 13%. The correlation between the system's and analysts' decisions over the four iterations is .79, $p < .0001$ ($N = 300$).

A more detailed way of evaluating COMPASS's performance is to separately compare its decisions to those produced by the CA for each type of decision. Table 17.1 presents this sort of analysis for each iteration of the knowledge base. The table can be thought of as four separate 3×3 tables, one for each iteration. Within each iteration, the nine cells are created by crossing COMPASS's three decisions with the CA's three decisions. The entry in a given cell represents the proportion of times (for that iteration) that that combination of COMPASS and CA decisions occurred. For example, the uppermost left cell in the first iteration indicates that both COMPASS and the CA reached a "Yes" decision for 71% of the cases.

A high degree of correspondence between COMPASS and the CA would appear as large numbers along the main diagonal and numbers close to zero in the off-diagonal cells. Iterations 1, 3, and 4 approach this pattern, but the row for analyst *No* judgments diverges, showing small numbers in all cells, with the smallest in the diagonal cell. For the second iteration, there are simply no observations in this row.

Some of the discrepancies in the table are the result of errors by the CA that COMPASS has not made—exactly the situation for which it was designed. For example, in the second iteration, 15 out of 19 discrepancies were attributed to

Table 17.1   Proportions of Comparability Decisions for All Combinations of CA and COMPASS Decisions

Iteration	CA	COMPASS decision		
		Yes	No	Adjust
	Yes	.71	.01	.01
1	No	.05	.01	.02
	Adjust	.02	0	.16
	Yes	.84	.01	.12
2	No	—	—	—
	Adjust	.01	0	.01
	Yes	.73	.02	0
3	No	.06	0	0
	Adjust	0	0	.19
	Yes	.77	0	0
4	No	.10	0	0
	Adjust	0	0	.13

analyst error. Detecting and subsequently repairing errors like this will undoubtedly improve overall data quality.

17.5.2   Resources and Organizational Issues

17.5.2.1   Human Resources

Professional knowledge engineers are skilled in both rule-based programming and knowledge elicitation. Most production programmers in survey organizations are not skilled in either. Rule-based programming is considerably different from conventional, "procedural" programming and, in our experience, has been difficult to teach. Knowledge elicitation requires interpersonal skills and a grasp of cognitive science (Kelly, 1991). Given the unusual combination of expertise required to build expert systems, it is hard to do without knowledge engineers. In the COMPASS project, we created a situation in which the CAs could do their own knowledge engineering, but this was only possible because project members understood the technology well enough to develop tools that reduced the programming burden on the CAs and because much of the relevant knowledge was already encoded in checklists.

In addition to knowledge engineers, experts are needed to supply the knowledge that will ultimately be encoded as rules. This can involve a substantial amount of the experts' time, so their managers must reduce other obligations. A certain amount of coordination is required to enable knowledge elicitation to occur. The expert and knowledge engineer are likely to work in different offices or groups, sometimes in different regions of the country.

17.5.2.2 Consequences of Introducing Expert Systems

One of the social consequences of developing expert systems is that experts may perceive knowledge engineering as an evaluation of their knowledge or performance. In developing COMPASS and MatchMaker, we needed to reassure FEs and CAs that this was not the case. Additionally, introducing an expert system (or any automation) is likely to change the nature of the expert's work by handling the routine cases and referring the more subtle and complex cases to the expert. While this should reduce boredom, it may increase the amount of stress on the expert by requiring only difficult judgments; this, in turn, can reduce morale (Hauser and Hebert, 1992).

One of the technical consequences of introducing expert systems is that the sponsoring organization is obliged to maintain the knowledge base. In the case of MatchMaker, job descriptions change and so the knowledge base must be updated accordingly. Over the lifespan of an expert system, the most frequent and, therefore, costly activity may well be maintaining the knowledge base (Bachant and Soloway, 1989).

17.5.2.3 Advocacy

The complexity of large organizations, such as government statistical agencies, can derail the development of an expert system. Alternatively, it is possible for an expert system to be technically successful but, ultimately, unused (Hayes-Roth and Jacobstein, 1994). To overcome organizational inertia and to promote the use of the expert system, the project needs an advocate within the organization—possibly the knowledge engineer (Kelly, 1991). This person must be able to communicate the vision of the project to the skeptics and to shepherd the project around institutional obstacles.

17.6 CONCLUSIONS

Based on our experience, expert systems are a promising technology for systematizing the classification of certain survey data. Rather than widespread use of stand-alone expert systems, we may see increased embedding of the technology in other software systems. The autocoding applications discussed in Section 17.2 have begun to move in this direction. When considered in this light, expert system technology has a particularly bright future in surveys because there are numerous data processing tasks currently executed by large software systems that are sometimes simplified by introducing executable, expert knowledge.

An additional organizational benefit that results from building expert systems is documentation. Not only are previously undocumented practices made explicit and concrete, but the documentation that arises from this activity can facilitate training and technology transfer. By creating a tutoring interface to an expert system's knowledge base, an organization can make this knowledge available to its members—even if the expert from whom it is derived is unavailable.

Finally, the early AI vision of replacing workers with smart software seems to have evolved into one in which employees are supported in their decision making by this software. That is the route we have taken with COMPASS and MatchMaker, and where the value of the approach seems to be greatest. The goal is to improve the quality of the decision making—not necessarily the quantity of decisions made for a fixed cost. Expert systems are highly specialized in their knowledge and generally incapable of the common sense reasoning at which people are so good. However, expert systems do not experience memory lapses and do not get tired, ill or hungry. By simultaneously exploiting the strengths of the software and the strengths of the users, the quality of these important statistics can only be improved.

ACKNOWLEDGMENTS

The views expressed here are those of the author and do not necessarily reflect the position of the U.S. Bureau of Labor Statistics. I wish to thank the following colleagues for their invaluable contributions to the activities and ideas reported in this chapter: Marty Appel, Jelke Bethlehem, Elizabeth Brooks, Paul Carney, Cathryn Dippo, Dan Gillman, Brad Hesse, Rick Kamalich, Dan Kasprzyk, Jim Kennedy, Brian Kojetin, Leda Kydoniefs, Lars Lyberg, Jean Martin, and Dan Pratt. Of course, any errors or misrepresentations in the text are the responsibility of the author.

REFERENCES

Andersson, R., and Lyberg, L. (1983), "Automated Coding at Statistics Sweden," *Proceedings of the Section on Survey Research Methods*, American Statistical Association, pp. 41-50.

Appel, M. V., and Hellerman, E. (1983), "Census Bureau Experience with Automated Industry and Occupational Coding," *Proceedings of the Section on Survey Research Methods*, American Statistical Association, pp. 32–40.

Appel, M. V., and Scopp, T. S. (1987), "Automated Industry and Occupational Coding," paper presented at Development of Statistical Tools Seminar on Development of Statistical Expert Systems, Luxembourg.

Armknecht, P., and Weyback, D. (1989), "Adjustments for Quality Change in the U.S. Consumer Price Index," *Journal of Official Statistics*, Vol. 5, pp. 107–123.

Bachant, J., and Soloway, E. (1989), "The Engineering of XCON," *Communications of the ACM*, Vol. 32, pp. 311–317.

Bechtel, W., and Abrahamson, A. (1991), *Connectionism and the Mind: An Introduction to Parallel Processing in Networks*, Cambridge, MA: Blackwell Publishers.

Bethke, A. D., and Pratt, D. J. (1989), "Automatic Coding Methods and Practices," Research Triangle Institute, unpublished manuscript.

Bobbitt, L. G., and Carroll, C. D. (1993), "Coding Major Field of Study," *Proceedings of the Section on Survey Research Methods*, American Statistical Association, pp. 177–182.

Buchanan, B. G., and Shortliffe, E. S. (1984), *Rule-Based Expert Systems: The MYCIN Experiments of the Stanford Heuristic Programming Project*, Reading, MA: Addison-Wesley.

Cantor, D., and Esposito, J. L. (1992), "Evaluating Interviewer Style for Collecting Industry and Occupation Information," *Proceedings of the Section on Survey Research Methods*, American Statistical Association, pp. 661–665.

Ciok, R. (1993), "The Results of Automated Coding in the 1991 Canadian Census of Population," *Proceedings of the Annual Research Conference*, U.S. Bureau of the Census, pp. 747–765.

Conrad, F., Kamalich, R., Longacre, J., and Barry, D. (1993), "An Expert System for Reviewing Commodity Substitutions in the Consumer Price Index," *Proceedings of the Ninth Conference on Artificial Intelligence for Applications*, IEEE Computer Society, pp. 299–305.

Cooper, T. A., and Wogrin, N. (1988), *Rule-Based Programming with OPS5*, San Mateo, CA: Morgan-Kaufmann.

Cox, B. J., and Novobilski, A. J. (1991), *Object-Oriented Programming: An Evolutionary Approach*, Second Edition, Reading Massachusetts: Addison-Wesley.

Creecy, R. H., Masand, B. M., Smith, S., and Waltz, D. (1992), "Trading MIPS and Memory for Knowledge Engineering," *Communications of the ACM*, Vol. 35, pp. 48–63.

Forgy, C. L. (1982), "RETE: A Fast Algorithm for the Many Pattern/Many Object Pattern Matching Problem," *Artificial Intelligence*, Vol. 19, pp. 17–37.

Gale, W. A., Hand, D. J., and Kelly, A. E. (1993), "Statistical Applications of Artificial Intelligence," in R. Rao, (ed.), *Handbook of Statistics*, Vol. 9, pp. 537–576, Amsterdam: Elsevier Science Publishers.

Gillman, D., and Appel, M. (1994), "Automated Coding Research at the Census Bureau," *Statistical Research Report Series No. RR94/04*, U.S. Bureau of the Census.

Glaser, R., and Chi, M. T. H. (1988), "Overview," in M. T. H. Chi, Glaser, R., and M. J. Farr (eds.), *The Nature of Expertise*, pp. xv–xxviii, Hillsdale, NJ.: Lawrence Erlbaum, Inc.

Halfpenny, P., Parthemore, J., Taylor, J., and Wilson, I. (1992), "A Knowledge Based System to Provide Intelligent Support for Writing Questionnaires," in A. Westlake, (ed.) *Survey and Statistical Computing*, Amsterdam: Elsevier Science Publishers.

Hauser, R. D., and Hebert, F. J. (1992), "Managerial Issues in Expert System Implementation," *SAM Advanced Management Journal*, Winter, pp. 10–15.

Hayes-Roth, F., and Jacobstein, N. (1994), "The State of Knowledge-Based Systems," *Communications of the ACM*, Vol. 37, pp. 27–39.

Hoffman, R. R. (1987), "The Problem of Extracting the Knowledge of Experts from the Perspective of Experimental Psychology," *AI Magazine*, Vol. 8, pp. 53–64.

Kahn, G. (1988), "MORE: From Observing Knowledge Engineers to Automating Knowledge Acquisition," in S. Marcus (ed.), *Automating Knowledge Acquisition for Expert Systems*, pp. 7–36, Boston: Kluwer Academic Publishers.

Kelly, R. V. (1991), *Practical Knowledge Engineering*, Bedford, MA: Digital Press.

Leonard-Barton, D. (1987), "The Case for Integrative Innovation: An Expert System at Digital," *Sloan Management Review*, pp. 7–19.

Lyberg, L., and Dean, P. (1990), "International Review of Approaches to Automated Coding," paper presented at the Conference on Advanced Computing for the Social Sciences, Williamsburg, VA.

Nilson, N. (1980), *Principles of Artificial Intelligence*, Palo Alto, CA: Tioga Publishing Company.

O'Reagan, R. T. (1972), "Computer Assigned Codes from Verbal Responses," *Communications of the ACM*, Vol. 15, pp. 455–459.

Poundstone, W. (1988), *Labyrinths of Reason: Paradox, Puzzles and the Frailty of Knowledge*, New York: Doubleday.

Pratt, D. J., and Burkheimer, G. J. (1994), "On-Line Automatic Coding in Computer-Assisted Data Collection," unpublished manuscript.

Pratt, D. J., and Mays, J. W. (1989), "Automatic Coding of Transcript Data for a Survey of Recent College Graduates," *Proceedings of the Section on Survey Research Methods*, American Statistical Association, pp. 796–801.

Prerau, D. S., Gunderson, A. S., Reinke, R. E., and Goyal, S. K., (1985), "The Compass Expert System: Verification, Technology Transfer, and Expansion," *Proceedings of the Second Conference on Artificial Intelligence Applications*, IEEE Computer Society, pp. 597–602.

Raud, R., and Fallig, M. A. (1993), "Automating the Coding Process with Neural Networks," *Quirk's Marketing Research Review*, May, pp. 14–47.

Rummelhart, D. L., McClelland, J. L., and the PDP Research Group, (1986), *Parallel Distributed Processing: Explorations in the Microstructure of Cognition*, Vol. 1: *Foundations*, Cambridge, MA: MIT Press.

Stanfil, C., and Waltz, D. L. (1986), "Toward Memory-Based Reasoning," *Communications of the ACM*, Vol. 29, pp. 1213–1228.

Stylianou, A. C., Madey, G. R., and Smith, R. D. (1992), "Selection Criteria for Expert System Shells: A Socio-Technical Framework," *Communications of the ACM*, Vol. 35, pp. 30–48.

Taylor, J., Parthemore, J., Wilson, I., and Halfpenny, P. (1992), "Computer Aided Questionnaire Design," paper presented at the Conference on Computing in the Social Sciences, Ann Arbor, MI.

Van Bastelaer, A. M. L., Hofman, L. M. P. B., and Jonker, K. J. (1987), "Computer-Assisted Coding of Occupation," *Automation in Survey Processing*, pp. 77–86, Statistics Netherlands.

CHAPTER 18

Editing of Survey Data: How Much Is Enough?

Leopold Granquist
Statistics Sweden

John G. Kovar
Statistics Canada

18.1 INTRODUCTION

Data editing consumes a significant proportion of the resources of all statistical agencies responsible for data collection and dissemination. Automation of the editing process has until recently not been useful in making the process more efficient, or more cost effective. The effect of editing, especially in the case of repeated surveys that collect quantitative data, is a steeply decreasing function of both time and costs. Numerous recent studies suggest that it is possible to target the resources available for data editing more effectively without compromising data quality. This chapter underlines this point and emphasizes that editing must serve a larger function than just "correcting" the data. It should be noted that although this chapter slants towards business surveys the results related to cost savings are applicable to any survey that collects quantitative data.

While editing encompasses a wide variety of activities, ranging from interviewer field checks to subject matter validation of estimates, we concentrate here on the collection- and processing-related activities. Thus we limit the scope of the discussion to editing activities related to respondent contact, actual data collection, respondent follow-up, and manual or machine verification of rules. Of particular importance is the interplay of these operations in a fully

Survey Measurement and Process Quality, Edited by Lyberg, Biemer, Collins, de Leeuw, Dippo, Schwarz, Trewin.
ISBN 0-471-16559-X © 1997 John Wiley & Sons, Inc.

automated setting. Imputation of survey data is only of tangential importance in this chapter, while we specifically exclude from the discussion undertakings related to estimate validation, subject matter analysis and other pre-publication verification functions.

Definitions of data editing vary widely, and often encompass most of the activities named above. The Data Editing Joint Group defined editing as "an activity aimed at the acquirement of data which meet certain requirements . . ." (Data Editing Joint Group, 1982). *Survey data editing* is the procedure for detecting, by means of edit rules, and for adjusting, manually or automatically, errors resulting from data collection or data capture (Granquist, 1995). Editing aimed at ensuring validity and consistency of individual data records is generally referred to as *micro-editing*. By contrast, approaches which ensure the reasonableness of data aggregates are often referred to as *macro-editing*, but also as output or aggregate editing (Hughes *et al.*, 1990; Kovar, 1991). Procedures which target only some of the micro-data items or records for review, by prioritizing the manual work and establishing appropriate and efficient process and edit boundaries, are termed *selective editing* methods. Pierzchala (1990) provides an excellent glossary of edit-related terms and their inter-relationship.

The goals of editing are threefold (Granquist, 1984): to provide information about the quality of the data, to provide the basis for the (future) improvement of the survey vehicle, and to tidy up the data. Or, as essentially restated by Chinnappa *et al.* (1990): editing is the activity aimed at gathering intelligence related to significant differences in the data for analytical purposes, providing feedback that can lead to improvements in data collection and processing, and reducing the level of error present in the data and ensuring a degree of consistency, integrity, and coherence.

There is a good reason to believe that a disproportionate amount of resources is concentrated on the third objective of "cleaning up of the data." An explanation may be found in a brief historical background: the advent of computers was recognized by survey designers and managers as a means of reviewing all records by consistently applying sophisticated checks requiring computational power to detect most of the errors in data that could not be found by means of manual review. The focus of both the methodological work and the applications was on the possibilities of enhancing the checks. Nordbotten's (1963) work was considered as a theoretical basis for many of the automated editing systems that were constructed in the late 1960s and the 1970s. Even though automated imputation rules were often built in, these early systems produced thousands of error messages that had to be examined by subject matter personnel, by referring back to the original forms and in more complicated cases to the respondents themselves. Changes were entered in batch and edited once again by computer (see Arvas *et al.*, 1973 for an example). The set-up and maintenance costs of these rigid and cumbersome systems were high and the demands on human resources were not decreased.

Further automation streamlined the process, but the gains of the automation were used to allow the editors to process more records and to make more contacts with respondents in order to resolve encountered problems. In fact, the Subcommittee on Data Editing in Federal Statistical Agencies (Federal Committee on Statistical Methodology, 1990) has found that "about 60 percent of (U.S.) Federal survey managers reported that they refer *all* data that fail edit checks to subject matter specialists or editors for review and resolution." In the 1980s, technology made it possible to move the editing step "on line," thus allowing the editors to process more records with more checks and to make more changes, more easily, often without a trace. In other words, the new automation developments did not change the original approach in order to overcome the drawbacks of manual editing, and the opportunity to rethink the methods and make them more efficient and effective was missed.

Why did this happen? Initially, the approach seemed to be successful. Errors, even serious errors, were often detected, but it was felt that there must be more errors in the data. Furthermore, it was found that many of the problems could be solved only by re-contacting the respondents. Thus the focus continued to be on possible checks, and not on possible serious error sources. In addition, no attention was devoted to the fact that recontacting the respondents to discuss only the data that failed edits was an inefficient way of obtaining high-quality data. The underlying assumptions were that checks themselves could not harm the accuracy of the estimates, and that the editors (in complicated cases subject matter experts) always succeed in ascertaining whether a failed value was correct, and if not, in finding more accurate values. In Section 18.2, and particularly in Section 18.3, examples are presented which show that these assumptions do not hold, resulting in inefficient and even counterproductive editing methods. It is further pointed out that editing alone actually cannot find certain kinds of errors.

The purpose of editing activities is, of course, to establish internally consistent micro-data records to guarantee consistent cross tabulations at all levels of detail. While consistent micro-data are desirable, the benefits must be evaluated in light of the costs. These costs include not only financial and human resources, but also costs associated with losses in timeliness and excessive respondent burden (see Section 18.2). As well, there is a danger of distorting "true" values to fit them to preconceived models of "clean" data. Overedited surveys give users a false sense of security as far as data quality is concerned, especially since quality measures are usually not reported and often are not even available (Kovar, 1990). In fact, editing processes were hardly ever evaluated until the end of the 1980s. Furthermore, the few early evaluation studies did not capture the attention of survey managers, maybe because the results seemed to contradict their experience that more errors were detected by applying more edits and more recontacts. Thus survey designers and managers did not recognize the need for a total survey view on the collection and processing of data until a number of evaluations showed that overediting occurred in the majority of surveys.

The high cost of editing has prompted a number of statistical agencies around the world to consider alternatives to the traditional approach. While keeping the importance of editing in mind, it has been suggested on a number of occasions that many surveys are overedited. That is, the final survey estimates would not have been significantly different had the editing process been curtailed. Establishment surveys with their highly skewed quantitative data offer a particularly great potential for making the editing process more efficient. In Section 18.3 the impact of editing is evaluated on a number of processes showing that in theory, concentrating resources on the areas of high impact and selecting only some records for further editing or follow-up is a workable solution. Section 18.4 of this chapter provides a review of the currently available selective editing methods and illustrates their practicality via empirical examples from around the world. Particular attention is paid not only to the process and the tools, but also to the human and capital resources needed. However, there are limitations and quality concerns related to the selective editing approach despite its cost savings. Section 18.5 reviews these aspects of the approach.

The primary goals of editing applicable to all surveys, that is, those of providing information about the quality of the data and providing the basis for the improvement of the survey process are revisited in Section 18.6. The power and general availability of the computer allows statistical offices to shift data editing to the early stages of processing, often at the point when the respondent is still available (Lepp and Linacre, 1993). The emergence of generalized data entry software makes the task of tracking errors not only plausible, but objective. Best practices can thus be more easily identified and, as Lepp and Linacre point out, built into the software. The total quality approach to survey processing is underscored, with particular emphasis on "doing it right the first time" (Linacre, 1991; Granquist, 1995). Strategies which combine automated data entry, selective editing and automated imputation are described. The chapter concludes with some brief recommendations, general observations and a modest glance into the future.

18.2 COST OF EDITING

Even in the 1990s, editing is essentially as expensive as it was in the 1970s, although the process has been largely rationalized by continuous exploitation of technological developments. An overview of modern data editing systems is presented in Pierzchala (1995). The monetary costs of editing, expressed as a fraction of the total survey budget, have been estimated at around 20 percent in the case of household surveys and surveys of individuals, and as much as 40 percent in the case of business surveys in studies conducted all over the world as indicated in Granquist (1982), Federal Committee on Statistical Methodology (1990), and Gagnon et al. (1994) among others. Unfortunately, the efficiencies gained by rationalizing the editing process have been mostly consumed by increasing the number of questionnaires inspected by subject matter specialists or editors (Pierzchala, 1995) without necessarily leading to improved data quality.

By far the most costly editing task involves recontacting the respondents in order to resolve questionable answers. This is a high-cost activity for both the statistical office and for the respondents. The associated costs are likely of the same order for both parties. In many situations, weeks or even months after the respondents have provided the answers, the same respondents are asked to spend time with the interviewer to confirm data accuracy and in many cases to provide only marginally more accurate responses. The bad-will costs accrued in querying respondents cannot be neglected, in particular when queries leave the data unchanged.

Furthermore, there are also quality costs associated with editing that must be considered. Editing takes time, and excessive editing will cause losses in timeliness and hence decrease the data relevance. For example, Pullum *et al.* (1986) found that machine editing of the World Fertility Survey delayed the publication of the results by about one year. Another quality-related cost of editing is the opportunity cost: time and resources spent on editing might have had a higher quality pay-off if allocated to other tasks, for example, to efforts related to raising the response rates. Highly qualified subject matter experts involved in editing are better used for functions such as conducting studies of the survey data, thus augmenting their information value. The implementation of graphical editing in the U.S. Current Employment Survey (CES) increased the productivity by 50 percent thus permitting the twelve subject matter specialists involved in the editing process 14 days every month for analyzing the CES data (Esposito *et al.*, 1994). The costs incurred in essentially misusing over-qualified personnel must not be disregarded.

Finally, one must not overlook the indirect costs associated with an undue degree of confidence in data quality and respondent reporting capacity that result from overediting the survey data sets. This is of particular importance when, as is usually the case, performance measures and audit trails are scarce or non-existent. Overedited survey data will often lead analysts to rediscover the editors' models, generally with an undue degree of confidence. For example, it was not until a demographer "discovered" that wives are on average two years younger than their husbands that the edit rule which performed this exact imputation was removed from the Canadian Census system!

18.3 IMPACT OF EDITING

As indicated in the previous section, many of the data items that fail edit checks are subjected to extensive manual review. This section illustrates that the strategy of increasing the number of flagged questionnaires and recontacts cannot be justified on the grounds of data quality improvements, and, that practically unchanged survey estimates may be obtained while reducing the editing process considerably. We argue that too many minor changes are being made at a significant expense. We establish a general notion of the extent to which data should be edited in current systems. A related question addressed in this section deals with the possible accuracy that the manual review can achieve and the likelihood that the editing process will reveal all of the errors that can have a significant impact on the estimates.

Edits may be classified into two broad classes: fatal edits and query edits. Fatal edits identify data items that are certainly in error, while query edits point to data that have a high probability of being erroneous. Examples of fatal errors, that is, errors detected by fatal edits, include invalid or missing entries, as well as errors due to inconsistencies. By contrast, query edits identify data items that fall outside predominantly subjective edit bounds, items that are relatively high or low as compared to other data on the same questionnaire, and other suspicious entries. In order to maintain user confidence, particularly when micro-data are being disseminated, it is necessary that fatal errors be removed. There is no argument that editing is well suited for detecting and handling such errors, but this specific task is not responsible for the high cost of the manual review component of editing. The unacceptable costs of editing are associated with the query edits. While it is commendable that attention be paid to nonsampling errors and the inherent biases that they cause, the costs of editing operations must be defensible in light of their benefits.

It is argued in this chapter that editing should consist of detecting and correcting fatal errors, and of verifying potential outliers. Now, while in demographic surveys the notion of inconsistency is relatively "black and white," inconsistency of economic data is more of a continuous concept. For example, a sum of components can potentially not add up to the stated total by a little bit, or by a lot. Most survey managers, however, are uncomfortable releasing even mildly inconsistent data. A distinction must be made between fixing of inconsistencies by necessity, for data quality reasons on the one hand, and because of operational convenience and credibility reasons on the other hand. A judgement call must thus be made in distinguishing failures due to inconsistency that are sufficiently severe to merit respondent follow-up, and those that can be fixed by other, preferably automated means. The former errors must be addressed through careful review, the latter as expediently as possible.

Query edits flag data that look suspicious. Verifications with respondents often lead to little or no change, though in the case of true outliers such confirmations are valuable. In most surveys, too many values are flagged as suspicious, resulting in too many follow-ups. Evidently, neither the edit bounds, nor the dividing line between edit failures that must or must not be followed up is precisely established and will to some extent depend on the survey in question and the end uses to which the data are put. This, for the most part, is the subject of this chapter.

How much does the editing process change the reported data? Analysis of changes to original data have been undertaken as part of many evaluation studies. Linacre and Trewin (1989) provide an overview of three major studies of the editing process in three Australian Surveys: 1979/80 Manufacturing

Census, 1983/84 Agricultural Census, and the 1985/86 Retail Census. One major conclusion from each of the studies is that many values are being changed by insignificant amounts. For example, for the items turnover in the Manufacturing Census, number of livestock in the Agricultural Census, and turnover in the Retail Census (short forms), it was found that a large number of small, random changes were being made for small units having little effect, and that the majority of the total change of the items was due to a few changes to large units.

Boucher (1991) has undertaken an exploratory study using the Canadian Annual Survey of Manufactures and showed that there was little difference between data subjected to just a few rudimentary edits and the final data. In fact, at the global level, for the principal statistics, the difference was less than 2 percent, while at a more detailed (SIC) level the differences were less than 18 percent. The study found that a small percentage of records accounted for the majority of the total change. More detailed investigation showed that reducing the usual editing effort by about half could reduce this difference to less than one percent at the global level and less than 10 percent at the detailed level (Kozak, 1993).

In editing the 1982 U.S. Retail, Wholesale, and Services Censuses, Greenberg and Petkunas (1986) also showed that few errors are responsible for the majority of changes. Common errors in reported data are due to data capture and reporting in dollars when reporting in thousands is requested. Keying errors may be large when digits are added or when the error is committed in the first or the first two digits. It was found that all the keying and dollar reporting errors were referred to manual review, but that few contributed to a great share of the total change. This observation, that a few changes accounted for a major part of the total change, was analyzed by creating graphs and tables showing the percentage of total change by percentage of cases. For the six economic items and for five of the six kind-of-business variables, approximately 5 percent of the cases contributed to over 90 percent of the total change (sum of all absolute changes).

Similar results were found by Wahlström (1990) and Hedlin (1993), however, many more records were subjected to manual review than in the U.S. study. For example Wahlström found that 26 percent of the 3,919 reporting units to the 1989 Swedish Annual Survey of Financial Accounts had the values of the principal item "Value Added" changed. About 25 percent of the changed values contributed to 95 percent of the total change, and 7 percent of the largest changes brought the estimate within 1 percent of the final global estimate. Hedlin investigated the impact of editing for all items of the 1991 Swedish Annual Survey of Manufacturing and found for example that of all of the about 200,000 values (about 100 items for some 7,300 establishments), around 26,000 were changed. On average, of all items, 50 percent of the smallest changes contributed less than 1 percent to the final estimate, and 15 percent of the changes were less than 2 percent of the original value.

The manual review of flagged data leaves many of the suspicious values unchanged. Data on hit-rates, that is the share of the number of flags that result in changes to the original data, are rarely reported in evaluations or studies of editing processes. The following three exceptions are thus of particular interest. In the Australian Retail Census, the hit-rate of the query edits was found to be 23 percent (Linacre and Trewin, 1989). In the Swedish Quarterly Survey on Wages and Employment, the hit-rate of the current editing process was found to be 28 percent (Lindström, 1991). In his experiments, Lindström showed that the hit-rate could be raised to 62 percent by applying the aggregate method (Granquist, 1990), thus reducing the flagged records due to the query edits by 80 percent. The hit-rate of the edits (range checks with predetermined subjective bounds) of the U.S. Livestock Slaughter Data Survey was considered so low that it was decided to develop more efficient edits (Mazur, 1990). This resulted in a cost saving of approximately 75 percent.

Changes made by the editors, especially when new data are found as a result of contacts with the respondents, are often considered by survey managers as "corrections." That is, it is assumed that accurate, true data are found. Thus when new editing methods are evaluated against the outcome of the current editing process, subject matter managers are inclined to require that the effect of the changes *not* detected by the method under study should be negligible. The underlying assumption here is that current editing processes detect all important errors and manage to "correct" all flagged data. This assumption is challenged here by presenting contrasting data from various studies.

Linacre and Trewin (1989) and Corby (1984) offer results showing that editing can in fact be counterproductive. A sample of the edited forms of the 1984/1985 Australian Retail Census was analyzed by an experienced subject matter statistician with respect to the likely accuracy of the editing changes made during the processing of the census. For the short forms, edited by less experienced editors, the differences between the post-edited data (the data resulting from the expert review of the editing changes) and the originally edited data, expressed as percentages of the post-edited data, were -8.7 and $+6$ percent for the wages/salaries and turnover items, respectively. Corby (1984) provides a detailed report of a re-interview study of some principal items of the 1977 U.S. Economic Censuses. The objective was to measure the accuracy of the selected items. As the original reported data were available, the quality of the editing could be studied. Estimates of the total for each census item were calculated based on each of the re-interview, published, and original data, and compared to each other. It was found that census editing changed the original data too far in the right direction for four items, and in the wrong direction for six items. For five items editing changed the data in the right direction but not far enough. Individual respondent errors were also studied in order to find out whether reporting errors could be reduced by making simple changes to the forms or instructions. However, it turned out that there was no easy way to do so. It was also found that a large number of establishments made errors, which the respondents could not correct because their accounting systems did not allow them to separate the item components they had wrongly included or excluded. Had these records been selected for recontacts, more accurate data could not have been provided.

An additional issue is whether computer checks can identify all erroneous data that have considerable effect on the estimates. Evidently, commonly used checks cannot identify data that are affected by small but systematic errors reported consistently in repeated surveys. Such errors are likely to occur when the survey definitions and the definitions of the items in the accounting systems of the establishments do not coincide. For example, Werking *et al.* (1988) found that when an experimental group of respondents was made to apply the survey definitions, the estimate of the main item "Production Worker Earnings" became 10.7 (standard error 3.2) for the experimental group, in contrast with 1.6 for the control group.

These examples illustrate that further editing by narrower bounds or by more follow-ups will not generally yield higher-quality data, irrespective of the design of the edits or the skillfulness of the editors in coping with the respondent. The computer edits have to be targeted to current, special error types, and the respondents have to have sufficient reporting capacity to provide data free from these special errors. An example of targeting the edits to a special type of error is reported in Mazur (1990). For the Livestock Slaughter Data Survey at the U.S. Department of Agriculture, Mazur creates a statistical edit by using historical data and Tukey's biweight (see Hoaglin *et al.*, 1983) to identify "inliers," that is, units which report the same values every survey period. Furthermore, editors' mistakes may actually make the estimates worse, for example, when editors interpret their job as the task of fitting reported data to the models imposed by the edits, and are likely to change data just because the computer suggests that there is something wrong (Boucher, 1991). In discussions with the Canadian Annual Survey of Manufactures team, it was speculated that a point in time exists when just about as many new errors are introduced as are removed. This hypothesis was put forth as a result of a study where the non-critical units were reworked in the complete fashion usually afforded only to the large units (Astles, 1994). It was found that global estimates were within 1 to 2 percent of the original values, and that usually the differences were due to a difference of opinion between the editors rather than missed errors. In other words, it was felt that a lot of the errors found on reworking the editing step were actually introduced by the original editors.

In conclusion, we suggest that given the high cost of editing and its limitations, there is little gain in addressing all edit failures, since many do not significantly alter the estimates. The studies cited above indicate that for detecting errors in general, the query edits for individual items should be designed to flag only potential outliers. Going further than this will probably not result in higher data quality, because, for example, respondents may not find it worthwhile to investigate the questioned data when the data seem reasonable. According to the results of the cited evaluations it seems that for individual items, not more than 5 to 10 percent of the number of changed values in traditional editing of today should be flagged for manual review. Note that we do not suggest that only 5 percent of currently flagged *records* are worth manual review. This number will depend on the number of items of the survey, and a number of technical and practical factors. The following sections address the question of "How many records are enough?" in more detail.

18.4 SELECTIVE EDITING

Having established that the effect of changes to reported data as a consequence of editing is not linear, it remains to be shown that viable selective editing strategies can be put into practice. In other words, given that not all errors are of equal importance, can they be ordered with respect to their impact on estimates either *a priori*, or, at least in real time during survey processing, without having examined all cases?

In the last ten or so years, numerous selective editing strategies have been proposed. They use methods known as macro-editing, aggregate editing, output editing, statistical editing, top-down editing, verification of top contributors to estimates (weighted or not), graphical editing, Hidiroglou–Berthelot bounds, and others to accomplish their goals (Granquist, 1990, 1993, 1995; Chinnappa et al., 1990; Hughes et al., 1990; Kovar, 1990; Hidiroglou and Berthelot, 1986; Latouche and Berthelot, 1992). Generically, the term *selective editing* includes any approach which focuses the editor's attention on only a subset of the potentially erroneous micro-data items that would be identified by traditional editing methods.

Many of the selective editing methods are very similar, or even indistinguishable (e.g., aggregate and output editing). There is a great deal of overlap between the methods, and often only very subtle differences discern them from each other. Pursey (1994) distinguishes between macro-based and micro-based selective editing methods. The former are methods which identify micro-data items that are to be reviewed because they lead to aberrant macro-level statistics. As such, the macro-based methods can only be used when all or most of the data are available. The latter, micro-based methods require only knowledge of the values of the current record and possibly some auxiliary data. These methods can thus be used as soon as the data records are available. For example, a rudimentary application of output editing requires knowledge of all data values for a given variable in a given aggregate cell in order to establish the current data value's impact. By contrast, a simple application of Hidiroglou–Berthelot bounds (Hidiroglou and Berthelot, 1986), parameters of which have been established based on data from past survey occasions, allows for an immediate decision once the given data item has been processed. Most selective editing applications strike a balance between these two extremes, and edit the records in batches. Usually the whole survey is processed in three to five such lots, as the records arrive. The top-down approach can then be executed; Hidiroglou–Berthelot bounds can be calculated; graphical displays can be examined interactively, and so on. The individual record's effect is not related to a domain estimate, but instead to the partial aggregates as they are formed by the records in the batch. The specific application's needs would determine the appropriate balance of macro- versus micro-based selective editing. No matter how the suspicious data items are identified, the key feature of selective editing is that not all queries are followed up with the respondents, but only those of significant impact.

As noted previously, however, most survey managers are unlikely to be comfortable releasing data that may fail some basic edits, no matter how small the effect of the error. The corrections of the lesser errors are usually made for processing convenience, rather than quality reasons, and need to be made quickly and objectively. For this reason, most successful selective editing applications make use of some form of automated imputation. The existence of ready-made, re-useable, and objective imputation software often decides the viability of a selective editing approach.

Finally, as one of the objectives of implementing selective editing methods is to reduce respondent burden, care must be taken that suspicious data items are followed up intelligently, in a co-ordinated fashion. This can be accomplished by simply grouping together all high impact edit failures for a given respondent to be followed up at the same time. Alternately, it can be accomplished by the use of some form of a score function (Latouche and Berthelot, 1992; Lawrence and McDavitt, 1994). Such score functions attempt to identify data records (rather than data items) that are to be followed up based on the errors' potential effect on the estimates. This assessment is made as a function of the unit's size, the size of the potential error, the survey weight, the importance of the unit within a domain, the relative importance of the potentially erroneous variable, and operational considerations. Many formulations exist; Latouche *et al.* (1994) present two excellent examples.

Can selective editing be put into practice? Certainly! The following examples provide only a few of the possibilities.

The paper by Greenberg and Petkunas (1986) contains a description of the editing system used for the processing of the 1982 U.S. Retail, Wholesale, and Service Censuses. The strategy implemented in this system consists of automated editing and imputation followed by a manual review of large changes to reported data, and of changes or imputations to large establishments, and, finally, by manual imputation of all cases where the system failed in finding an appropriate value. Records that have their data changed are sent back to the automated editing system. The edits are mainly ratio edits with relatively wide acceptance limits determined from statistics of edited values from the preceding census. Because the rates of change to reported data are low, the changes are automated, and only a small percentage of the changes are referred for manual review, the need for manual resources is modest compared to traditional editing systems.

Boucher (1991) and Kozak (1993) detail an approach used in the editing of the Canadian Annual Survey of Manufactures. Based on past information, the survey units are separated (*a priori*) into critical and non-critical streams.

Establishments which produce significant amounts of important commodities within an industry are termed critical, the remainder are referred to as non-critical. A sufficient number of units are labeled critical so that a minimum of 80 percent of the estimated total output within an industry is accounted for. Nonetheless, less than 40 percent of all the establishments are part of the critical stream. The usual, detailed edits are applied to all critical units. The non-critical units are subjected to a minimal pre-edit treatment prior to data capture, simplified commodity coding, reduced edit set consisting mostly of validity edits, and finally automated imputation. All units are then combined and macro-editing techniques are used to identify non-critical units of high impact which are manually reviewed. The average time to edit a non-critical unit is about a quarter of the time it takes to edit a critical unit. Significant savings can thus be achieved with minimal effect on data quality as compared to the full, detailed editing approach used previously (Boucher et al., 1993). Recent estimates suggest that some 20 person years were saved over the last four years, while the volume of output was increased. This represents almost 20 percent of the operational capacity.

Latouche et al. (1994) make use of the Hidiroglou–Berthelot bounds and a score function to identify units that are to be followed up in the Canadian Survey of Employment, Payroll, and Hours. The Hidiroglou–Berthelot bounds ensure that small changes in large units are treated more seriously than large changes in small units. Using a score function, follow-up resources are concentrated on records of high impact, remaining records are imputed. Some of the records destined for imputation are actually also followed up for purposes of verification of the imputation strategy. In the simulation studies, only 20 percent of the units outside the edit bounds were followed up, resulting in estimates that for most variables were within 2 percent (at the national level) of the estimates obtained in the traditional manner. At the early stages of operationalizing this new process, however, the follow-up rate was conservatively set at 50 percent. Of particular note in this application of selective editing is that the complete process was rethought. The procedures were streamlined, the edit set reduced, and the data collection, capture, editing, and follow-up consolidated. As a result, at most one follow-up per respondent is now possible. Rationalizing the complete process at the same time allowed some functions to be moved from one processing stage to another, and often from several stages to just one. On the negative side, the process assumes that all edit failures point to errors, and either targets them for review or imputation. A more relaxed set of edits should likely be used to drive the imputation module. The authors are currently evaluating alternate options.

Van de Pol's (1994) preliminary studies demonstrate that reducing the editing effort for the Netherlands Annual Construction Survey to about 25 percent results in estimates that are within half a percent of those obtained with extensive editing. This is well within the confidence interval of these estimates. In fact, van de Pol shows that, as in other similar studies, few corrections account for most of the total change, but that even these changes do not move the estimates by statistically significant amounts. This is likely due to the fact that data entry is done by subject matter specialists who presumably remove gross errors prior to data capture. A score function, loosely based on the Hidiroglou–Berthelot approach, was used to prioritize the units that were edited. Of special note is that subsequent studies indicate that even complex econometric measures, not just sample means, were estimated sufficiently accurately based on the selectively edited data set.

Numerous other studies suggest that the current editing efforts can be safely reduced by a substantial amount. A significance editing strategy was implemented in the Australian Weekly Earnings Survey in 1992, see Lawrence and McDavitt (1994). The number of edit queries is still about 40 percent of the failed records as was found in the study preceding the implementation. The quality of the estimates has not been changed at the industry level, but the person-year requirements for editing have been reduced from seven or eight to three or four, or by about half, see McDavitt et al. (1992). The use of graphical editing can also yield significant savings: Anderson (1989) managed to reduce the need for query edits by 75 percent; Esposito et al. (1994) decreased the need for manual review by six person-years.

How much money can be saved? Most authors are cautious when it comes to citing realized monetary savings. This reluctance is in part due to the fact that some of these approaches have been implemented by necessity, for reasons of cost avoidance rather than cost savings. However, it is safe to say that the realized efficiencies can be quite substantial. The cited references target rejecting about only half as many records as the traditional methods would reject. This of course will not translate into a 50 percent saving because the rejected records likely need more than an average amount of attention at follow-up, and because the remaining records must be cleaned up by imputation which itself is not a process without its own costs. But note that not only are the monetary savings substantial, but that the process has been centralized and rationalized, resulting in a reduced respondent burden and better data quality. Furthermore, as the job of following up respondents becomes less monotonous, more challenging, and more demanding in terms of knowledge of the subject matter, the editors acquire a sense of added responsibility and accountability. Such empowerment generally results in more productive employees. In other words, the efficiency of selective editing must be measured not only in financial terms, but also in terms of improved timeliness, data quality and related output improvements, as well as in terms of reduced respondent burden and job enrichment of the editors.

How much editing is enough? Clearly this is a complex question which must be answered on a case-by-case basis. Nonetheless, as the above studies demonstrate, targeting about half of the traditional edit failures in the common domains is likely sufficient, and we suggest that any additional efforts are wasted. When the preplanned aggregate estimates are essentially left unchanged by further editing, one is likely introducing as much error as is being corrected.

18.5 LIMITATIONS

It is clear from the preceding sections that large savings can be realized by re-examining the editing process of most surveys. The use of selective editing procedures, while appealing in many instances, needs to be evaluated carefully. Clearly statistical programs whose sole goal is to produce aggregate data at relatively coarse levels need not edit every record, making selective editing very attractive. On the other hand, when small domain statistics are produced regularly, especially on an *ad hoc* basis, or when micro-data files are to be produced as part of the regular output, the need for internal consistency of all records increases. This does not necessarily mean that all records should be manually inspected. Automated imputation software can go a long way in establishing internally consistent data sets. However, the balance between manual editing and automated imputation must be examined. Such a balance would be established as a function of resources available, the quality of the automated imputation system, and a realistic (not perceived) need for absolute perfection. For example, Pullum *et al.* (1986) argue that even estimates of complex multivariate coefficients are not sensitive to whether consistency is achieved by systematic defaults or by detailed case-by-case editing. They continue by noting that while consistency is desirable mostly for processing convenience only, it should be achieved quickly and almost never by referring back to the actual questionnaires. A number of the cited references make use of automated imputation to resolve minor (i.e., lesser effect) inconsistencies, while concentrating their editing budgets on units of substantial impact.

The potential for savings can, however, be reduced by a number of mitigating factors. Before implementing any new strategy, there are initial costs, developmental costs, costs due to feasibility and parallel testing, and costs due to the need for more highly trained staff. These may offset a great deal of the initial savings, though streamlining the process makes the job of quantifying the errors and their sources much easier. This alone may well be worth the initial cost. The opportunity to re-establish the priorities and goals of the program may also be part of the unquantifiable benefits. Secondly, sparse domains such as rare industries within a small geographical area may afford little if any room for selective editing, thus reducing the possible savings. Generally, the cited authors state the need to recognize that small cells must be afforded more careful attention. Thirdly, the idea of selective editing relies predominantly on the differential effect of the errors. Clearly then, selective editing approaches are much more easily implemented when editing continuous data with varying survey weights. In other words, selective editing is likely of limited usefulness in demographic surveys as compared to economic collections. Fourthly, care must be taken that planned savings actually materialize, by ensuring that the new procedures are properly followed. Most selective applications at Statistics Canada implement some form of sample quality control in order to ensure the adherence to stated procedures, since it has been found that inconsistency in data is often the result of inconsistency in the application of the prescribed approaches. Editors need time to learn to be comfortable in letting imperfect data records go by, contrary to tradition. Quality control methods can be of great value in identifying the need for staff training and for elaborating and improving process documentation.

While the re-evaluation and rethinking of the traditional editing process is valuable in itself, not all selective methods can be implemented in their optimal form. For example, it is found that the best selective editing method for the Canadian Survey of Employment, Payroll, and Hours would involve using a score function which needs as input previous month's data at the collection unit level. Given the decentralized mode of data collection, the size of the required files, and the power of the computing environment which runs the data collection software, it was first thought that a sub-optimal method would have to be used. Such a method would have used the previous month's aggregates (not micro-data) as inputs to the score function. Ultimately, the more refined method was implemented, but the important point is that the selective editing approach has not been abandoned at the development stage!

Because selective editing methods are necessarily based on more assumptions than traditional methods, regular re-examination of the underlying, and often unspecified, postulates is essential. Verification of subjective constants that drive the system must be made on a regular basis; assumptions about the presumed size of the units confirmed. In fact, after a careful study of alternative editing methods at Statistics Canada, Chinnappa *et al.* (1990) strongly support the use of selective editing approaches, but warn explicitly against using such techniques without careful study and simulation of the methods prior to their adoption. In particular, they argue that use of new methods must not result in significant data quality losses.

Finally, there will always exist surveys, usually small, one-time surveys, for which the initial investment in selective editing software and procedures development cannot be justified. However, while developing specialized software in such instances is often more economical, the emergence of generalized and reusable systems may tip the balance of such justification in the other direction fairly soon. To wit, few of the Canadian selective editing proposals would have been possible without the existence of a generalized edit and imputation system, but are easily defended once such systems have been made available. Furthermore, while the introduction of selective editing methods to over 50 very small Current Surveys conducted by Statistics Canada's Industry Division has admittedly yielded negligible monetary savings, the improvement in data quality and timeliness, and the reduction in respondent burden has made the change-over worthwhile.

18.6 PRINCIPLES OF EDITING

We conclude this chapter with a short discussion of what we propose are the primary goals of editing. While a certain amount of editing will always be essential in order to eliminate gross errors, its more productive role lies in its ability to provide information about the quality of the collected data and thus form the basis for future improvement of the whole survey process. It has been established that editing is costly, and that its contribution to error reduction is limited. Careful reworking of current editing procedures can yield significant savings which should be redirected to other stages of the survey process. It is important to rationalize wisely the allocation of these scarce resources. While it is acknowledged that savings due to implementation of selective editing methods are more likely to be realized in the case of quantitative surveys, the general principles discussed below apply equally to all surveys.

In the quest to reduce errors in survey data, it is essential to look upstream, rather than attempting to clean up at the end. The adage "do it right the first time" is very appropriate. Editing results can be used to advantage in sharpening survey concepts and definitions and in improving the survey instrument design. More resources should be dedicated to these functions in order to help prevent errors. Statistics Canada's advancement of a policy on questionnaire design and testing, together with the established Questionnaire Design Resource Centre, is aiming in the right direction. Similar trends are being followed at the Australian Bureau of Statistics. However, we have as yet to see a report on an editing process where this principle has been applied, and which resulted in changes to, for example, the questionnaire.

Secondly, opportunities must be taken in moving some of the traditional editing functions to the early stages of the survey process, preferably while the respondent is still available (Lepp and Linacre, 1993). The emergence of computer assisted telephone or personal interviewing software (CATI/CAPI), along with the appropriate hardware, makes this task more easily achieved. In fact, the use of some of the computer assisted self-interviewing techniques (CASIC), such as touch-tone data entry or electronic questionnaires, can be used to move the editing step, at least partially, to the respondent. Appropriately constructed survey instruments that enable the respondents to check their data as they are being entered can save the statistical offices even greater resources.

Recently developed generalized, re-usable software for survey data collection, capture, editing, and imputation is usually equipped with facilities and functions which allow for better tracking and monitoring of the processes. Audit trails, performance measures, and diagnostic statistics can now be made readily available, even though they have been grossly neglected in the past. These measures can be extremely useful in identifying best practices and thus in aiding the continuous improvement of the editing process. For example, conceptual misunderstandings between the data collector and the respondent can be promptly identified.

However, care must be exercised when moving away from manual and semi-automated processes to fully automated ones. Parallel testing should usually be advocated. It is important at this stage to rethink the objectives and the scope of the undertaking, and to take the opportunity to re-engineer the whole process, rather than just converting the individual steps and adding on new ones. The edits themselves often need to be redesigned. This must be done in such a way so as to ensure that the *complete* edit set is not self-contradictory or inconsistent, and that it does not contain redundant rules. As well, the richness of the data holdings of the statistical office can often be exploited. Effective and innovative uses of auxiliary data should be encouraged.

By re-engineering the complete process, optimal combination of automated data entry, selective editing, and automated imputation can be made. This does not necessarily mean that all steps of the process must be automated. For example, there is little gain in converting reliable mail respondents to computer assisted modes of data collection. Woelfle (1993) describes an effective and practical mix of paper and pencil collection with computer assisted interviewing which is optimal with respect to timeliness, cost, and respondent burden. This strategy is made possible by the relatively sophisticated Central Management Support System within Statistics Canada's DC2 (Data Collection and Data Capture System). The incorporation of selective editing and optimal follow-up, sample follow-up of nonrespondents, and the imputation of records with less critical errors in order to maintain consistency makes this a viable approach.

Finally, some of the savings reaped can be used to expand outputs and conduct more analyses. For example, some of the savings realized in re-engineering the Canadian Annual Survey of Manufactures have been used to provide detailed information on an annual basis where before such information was only available every two years. Only one of the eleven person-years saved over a two-year period was used for this purpose. Better analyses of the results help turn the data into information.

18.7 CONCLUDING REMARKS

Many statistical offices are risking too much in their quest for perfection. Subscribing to the notion that if "a little chocolate after dinner is good, more must be better," well-intentioned individuals edit data to excruciating detail at tremendous cost. We have demonstrated that editing, while essential, can be reduced substantially. The role of editing must be re-examined, and more emphasis placed on using editing to learn about the data collection process, in order to concentrate on preventing errors rather than fixing them.

Evidently editing efforts can be reduced, but not blindly. It is likely that the relative simplicity of this specific part of the survey process, which makes it much more easily understood, contributes enormously to the overuse of editing. However, sufficient studies must be conducted before reducing the editors' tasks. The goals of such an exercise must not consist solely of the desire to save financial resources, but must include those of improving data quality and timeliness of outputs and of reducing respondent burden. While software for conducting a number of the actual tasks exists today, optimizing the process remains largely a manual task. Further integration will likely remedy the situation in the near future.

In order for the editing strategies of tomorrow to be successful, today's secondary uses of editing must gain in prominence. Learning from the results of editing must become paramount. Performance measures and other diagnostics must be studied with a view to improving both the process as well as the survey instrument. The survey takers must position themselves to be able to better quantify some of the nonsampling error sources. Editing will have to serve a greater function than just a data correction tool.

How much editing is enough? Clearly the current practice constitutes too much. On the other hand, editing top contributors without spot-checks on the other records is not sufficient, and some middle ground must be established. Given the high cost of editing and its potential to introduce, after a certain point, as much new error as is being corrected, suggests that the current allocation is far from optimal. Clearly these scarce resources can be better allocated. The above-cited studies suggest that eliminating as much as 50 percent of the editing effort may be possible without significant effect on data quality: none of the cited studies suggests less. Individual circumstances may dictate smaller reductions; the opportunity for redirection of funds may allow larger reductions. Numerous suggestions to this end have been put forth in this chapter. It is also apparent that future budgets of statistical offices are more likely to be decreasing than increasing. Thus the final question is not whether we can afford to reduce editing but rather can we afford not to?

REFERENCES

Arvas, C., Granquist, L., and Ohlsson, G. (1973), "Automatic Editing in the Yearly Survey of Manufacturing in Sweden, GPI," Conference of European Statisticians, Working Paper 9/124.

Anderson, K. (1989), "Enhancing Clerical Cost-Effectiveness in the Average Weekly Earnings," unpublished report, Belconnen: Australian Bureau of Statistics.

Astles, D. (1994), "Summary of Opportunity Analysis," unpublished report, Ottawa: Statistics Canada.

Boucher, L. (1991), "Micro-Editing for the Annual Survey of Manufactures: What is the Value Added?," *Proceedings of the Annual Research Conference*, U.S. Bureau of the Census, Washington, DC, pp. 765–781.

Boucher, L., Simard, J.-P., and Gosselin, J.-F. (1993), "Macro-Editing, a Case Study: Selective Editing for the Annual Survey of Manufactures Conducted by Statistics Canada," *Proceedings of the International Conference on Establishment Surveys*, American Statistical Association, pp. 362–367.

Chinnappa, N., Collins, R., Gosselin, J.-F., Murray, T. S., and Simard, C. (1990), "Macro Editing at Statistics Canada," unpublished report of the Statistics Canada Working Group on Strategies for Macro Editing, prepared for the Statistics Canada Advisory Committee on Statistical Methods (January), Ottawa: Statistics Canada.

Corby, C. (1984), *Content Evaluation of the 1977 Economic Censuses*, Statistical Research Division Report Series No. CENSUS/SRD/RR-84-29, Washington, DC: U.S. Bureau of the Census.

Data Editing Joint Group (1982), *Glossary of Terms*, Working Paper Version 2, July 1982, New York: United Nations.

Esposito, R., Fox, J. K., Lin, D. Y., and Tidemann, K. (1994), "ARIES—A Visual Patch in the Investigation of Statistical Data," *Journal of Computational and Graphical Statistics*, 3, pp. 113–125.

Federal Committee on Statistical Methodology (1990), *Data Editing in Federal Statistical Agencies*, Statistical Policy Working Paper 18, Washington, DC: U.S. Office of Management and Budget.

Gagnon, F., Gough, H., and Yeo, D. (1994), "Survey of Editing Practices in Statistics Canada," unpublished report, Ottawa: Statistics Canada.

Granquist, L. (1982), "On Generalized Editing Programs and the Solution of the Data Quality Problems," UNDP/ECE, Statistical Computing Project, Data Editing Joint Group, Working Paper No. 17, New York: United Nations.

Granquist, L. (1984), "On the Role of Editing," *Statistical Review*, 2, pp. 105–118.

Granquist, L. (1990), "A Review of Some Macro-editing Methods for Rationalizing the Editing Process," *Proceedings of Symposium 90: Measurement and Improvement of Data Quality*, Ottawa: Statistics Canada, pp. 225–234.

Granquist, L. (1993), "International Review of Research on Data Editing Strategies," *Bulletin of the International Statistical Institute: Proceedings of the 49th Session*, Florence, Italy, Contributed Papers Book 1, pp. 515–516.

Granquist, L. (1995), "Improving the Traditional Editing Process," in B.G. Cox, D.A. Binder, N. Chinnappa, A. Christianson, M.J. Colledge, and P.S. Kott (eds.) *Business Survey Methods*, New York: Wiley, pp. 385–401.

Greenberg, B., and Petkunas, T. (1986), *An Evaluation of Edit and Imputation Procedures Used in the 1982 Economic Censuses in Business Division*, 1982 Economic Censuses and Census of Government Evaluation Studies, Washington, DC: U.S. Bureau of the Census, pp. 85–98.

Hedlin, D. (1993), "A Comparison of Raw and Edited Data of the Manufacturing Survey," unpublished report, Stockholm: Statistics Sweden.

Hidiroglou, M. A., and Berthelot, J.-M. (1986), "Statistical Editing and Imputation for Periodic Business Surveys," *Survey Methodology*, 12, pp. 73–84.

Hoaglin, D. C., Mosteller, F., and Tukey, J. F. (1983), *Understanding Robust and Exploratory Data Analysis*, New York: Wiley.

Hughes, P. J., McDermid, I., and Linacre, S. (1990), "The Use of Graphical Methods in Editing," *Proceedings of the Annual Research Conference*, U.S. Bureau of the Census, Washington, DC, pp. 538–550.

Kovar, J. G. (1990), "Data Editing: A Discussion," *Proceedings of the Annual Research Conference*, U.S. Bureau of the Census, Washington, DC, pp. 551–554.

Kovar, J. G. (1991), "The Impact of Selective Editing on Data Quality," working paper no. 5 presented at the Conference of European Statisticians, Work Session on Statistical Data Editing, Geneva, Switzerland, October 28–31.

Kozak, R. (1993), "Selective Editing and its Impact on Data Quality for the Canadian Annual Survey of Manufactures," working paper no. 5, presented at the Conference of European Statisticians, Work Session on Statistical Data Editing, Stockholm, Sweden, October 11–15.

Latouche, M., Bureau, M., and Croal, J. (1994), "Development of a Cost-Effective Edit and Follow-Up Process: The Canadian Survey of Employment Experience," working paper no. 10, presented at the Conference of European Statisticians, Work Session on Statistical Data Editing, Cork, Ireland, October 17–20.

Latouche, M., and Berthelot, J.-M. (1992), "Use of a Score Function to Prioritize and Limit Recontacts in Editing Business Surveys," *Journal of Official Statistics*, 8, pp. 389–440.

Lawrence, D., and McDavitt, C. (1994), "Significance Editing in the Australian Survey of Average Weekly Earnings," *Journal of Official Statistics*, 10, pp. 437–447.

Lepp, H., and Linacre, S. (1993), "Improving the Efficiency and Effectiveness of Editing in a Statistical Agency," *Bulletin of the International Statistical Institute*: *Proceedings of the 49th Session*, Florence, Italy, Contributed Papers Book 2, pp. 111–112.

Linacre, S. J. (1991), "Approaches to Quality Assurance in the Australian Bureau of Statistics Business Surveys," *Bulletin of the International Statistical Institute*: *Proceedings of the 48th Session*, Cairo, Egypt, Book 2, pp. 297–321.

Linacre, S. J., and Trewin, D. J. (1989), "Evaluation of Errors and Appropriate Resource Allocation in Economic Collections," *Proceedings of the Annual Research Conference*, U.S. Bureau of the Census, Washington, DC, pp. 197–209.

Lindström, K. (1991), "A Macro-Editing Application Developed for PC-SAS," *Statistical Journal*, 8, pp. 155–165.

Mazur, C. (1990), *Statistical Edit System for Livestock Slaughter Data*, Staff Research Report No. SRB-90-01, Washington, DC: U.S. Department of Agriculture, National Agriculture Statistics Service.

McDavitt, C., Lawrence, D., and Farwell, K. (1992), "The AWE Significance Editing Study," unpublished report, Belconnen: Australian Bureau of Statistics.

Nordbotten, S. (1963), "Automatic Editing of Individual Statistical Observations," Conference of European Statisticians, Statistical Standards and Studies No. 2, New York: United Nations.

Pierzchala, M. (1990), "A Review of the State of the Art in Automated Data Editing and Imputation," *Journal of Official Statistics*, 6, pp. 355–377.

Pierzchala, M. (1995), "Editing Systems and Software," in B.G. Cox, D.A. Binder, N. Chinnappa, A. Christianson, M.J. Colledge, and P.S. Kott (eds.), *Business Survey Methods*, New York: Wiley, pp. 425–441.

Pullum, T.W., Harpham, T., and Ozsever, N. (1986), "The Machine Editing of Large-Sample Surveys: The Experience of the World Fertility Survey," *International Statistical Review*, 54, pp. 311–326.

Pursey, S. (1994), "Current and Future Approaches to Editing Canadian Trade Import Data," *Proceedings of the Survey Research Methods Section, American Statistical Association*, pp. 105–109.

Van de Pol, F. (1994), "Selective Editing in the Netherlands Annual Construction Survey," working paper no. 11, presented at the Conference of European Statisticians, Work Session on Statistical Data Editing, Cork, Ireland, October 17–20.

Wahlström, C. (1990), "The Effects of Editing—A Study on the Annual Survey of Financial Accounts in Sweden," unpublished report, Stockholm: Statistics Sweden (in Swedish).

Werking, G., Tupek, A., and Clayton, R. (1988), "CATI and Touchtone Self-Response Applications for Establishment Surveys," *Journal of Official Statistics*, 4, pp. 349–362.

Woelfle, L. (1993), "Mixed Mode Data Collection and Data Processing for Economic Surveys," paper presented at the International Conference on Establishment Surveys, Buffalo, NY.

CHAPTER 19

The Quality of Occupational Coding in the United Kingdom

Pamela Campanelli
Survey Methods Centre at SCPR, London

Katarina Thomson
SCPR, London

Nick Moon
NOP Social and Political, London

Tessa Staples
Census Division, Office for National Statistics, London

19.1 INTRODUCTION

Survey coding can be seen as the process whereby textual information is classified into mutually exclusive categories and assigned numeric values (code numbers). The code number list (coding frame) is usually supplemented by a set of coding instructions to clarify the coding decisions which need to be made. As with any process, error can occur. For example, there can be coding decision errors, misapplication of coding rules for equating code numbers to text, etc. This chapter reports on the quality of occupational coding in the United Kingdom (U.K.), including recent computer-assisted options and plans. It also illustrates factors which should be a concern in any coding operation.

Currently, the majority of survey coding in the U.K. is conducted manually, either in the office by specially trained office coders or for some surveys by

Survey Measurement and Process Quality, Edited by Lyberg, Biemer, Collins, de Leeuw, Dippo, Schwarz, Trewin.
ISBN 0-471-16559-X © 1997 John Wiley & Sons, Inc.

interviewers as part of their fieldwork duties. Coding quality is typically measured in terms of reliability, looking at the extent to which two independent coders would assign the same code. As described in Section 19.3.1 of this chapter, reliability statistics generally vary between 0 and 1, with 1 indicating perfect reliability.

Past research suggests that the level of coder reliability can be seen to depend on a number of factors, such as the type of question, the nature of the answers, the length and adequacy of the coding frame, and the training and supervision of the coders. For example, Kalton and Stowell (1979) looked at coder reliability on questions about railway noise with coding frames of 26 to 64 codes. At the start of coding, they found reliability figures to be on average 0.72 with a range from 0.68 to 0.81 for their long code frames. In another study with questions about Industrial Tribunals, Collins and Kalton (1980) found the average reliability over ten questions to be 0.70 with a range from 0.62 to 0.78. Collins and O'Brien (1981) asked seven survey organizations to perform a coding reliability experiment on three questions from recent surveys. They arrived at an overall average of 0.72, with average values for the organizations ranging from 0.58 to 0.80 and average values for questions ranging from 0.53 to 0.87. They found the highest reliability rates among those questions where the majority of respondents tended to use obvious key words and the lowest reliability rates for the often vague "Why do you say that?" type of follow-up question. Data from Durbin and Stuart (1954) suggest that the coding of factual information is more reliable than the coding of judgments and attitudes.

19.1.1 Occupational Coding

Collection of information for the coding of occupations in the U.K. usually begins with the interviewer asking a number of questions about the respondent's occupation, some precoded, such as status in employment (whether self-employed; whether a manager or a supervisor or other employee) and number of employees in the respondent's organization, and others recorded verbatim by the interviewer, such as actual job title and description of work done. The most widely used occupational coding scheme for social surveys in the U.K. is the Standard Occupational Classification (SOC). SOC was developed by the Office of Population Censuses and Surveys (OPCS) (now the Office for National Statistics, ONS), the Department of Employment and the Institute for Employment Research (IER) at the University of Warwick.

The SOC scheme (OPCS, 1990) is built up from several hundred occupational unit groups designated by three-digit codes. These can be aggregated to form meaningful two-digit and one-digit summary classifications (referred to as the Minor and the Major Group, respectively). For example, code 202 (Physicists, Geologists, and Meteorologists) is under Minor Group 20 (Natural Scientists) which is under Major Group 2 (Professional Occupations). This hierarchical system is similar to those used in other countries, as well as being similar to the International Standard Classification of Occupations. In the U.K., the three-digit SOC classification, in combination with information on status in employment and number of employees in an organization, also provides the basis for assigning individuals to a variety of socio-economic and social class variables.

19.1.2 The Reliability of Occupational Coding

Occupational coding presents a special coding situation. The information collected for coding is in a sense factual, but at the same time subject to many different ways of describing the job. The coding frame is extremely long (371 categories in SOC at the three-digit level), but well supported by documentary material. An experiment to estimate the reliability of occupational coding was carried out by ONS in the early 1980s (Butcher et al., 1981; Elliot, 1982, 1983). The study involved twelve coders: three were experienced, five had some experience, and four were total novices. They each coded 1,200 questionnaires to Operational Occupation Group—a predecessor of SOC which had 348 categories at the three-digit level. The results showed that the reliability of expert coders was 0.84. The comparable figures for intermediates and novices were both 0.77 and the overall figure was 0.78. Other studies of the reliability of occupational coding have also been conducted in the U.K. For example, looking at coding at the three-digit level, Dodd (1985) found a figure of 0.80 for office coding and a figure of 0.70 in comparing office coding to interviewer field coding, White (1983) found a figure of 0.74 in comparing office coding to interviewer field coding, and Martin et al. (1995) found a figure of 0.82 for office coding and 0.74 for interview field coding.

19.1.3 Computer-Aided Coding

One of the many strands of the computer revolution in survey data collection has been the development of computerized systems for the coding of data from open-ended questions. Some of the earliest examples of this are found in Corbett (1972) and O'Reagan (1972). MacDonald (1982) was one of the first in the U.K. to develop a working system. Computer-aided coding is normally either *computer-assisted*, where a human coder assigns code numbers while working interactively with a computer or *computer-automated*, where codes are assigned directly by the computer without human intervention. It should be noted that in most instances, computer-automated coding needs to be coupled with another method as there are always cases which cannot be coded by the automated system.

A key step in the development of a computer-aided coding system is the building of a computer-stored dictionary. As suggested by Lyberg and Dean (1992), dictionaries which are constructed from coding manuals have a number of disadvantages. Coding manuals rely on the imagination and experience of the coder and this is not easily translated to the machine. It is more efficient to build the dictionary based on respondents' actual verbatim answers.

Dictionaries, in turn, need to be continually updated, as it is seldom possible to create a very good dictionary initially.

As part of the coding process, verbatim information from respondents needs to be entered so that it can be compared and matched to the computer-stored dictionary. The data entry of the verbatim information can take place at the same time as the rest of the questionnaire is keyed, it can be keyed separately, or in some programs can be entered by the coder.

Matching can take one of two forms: exact matching (where the verbatim information is identical with the dictionary entry) and inexact matching (where the verbatim information is considered similar enough to the dictionary entry to be considered a match). Computer-automated coding systems have mainly been evaluated by their exact and inexact match rates.

The majority of national statistical offices in developed countries now use some form of computer-aided coding for their occupational coding (see Lyberg and Dean, 1992, for an international review; as well as Andersson and Lyberg, 1983; van Bastelaer *et al.*, 1987; Chen *et al.*, 1993; Embury, 1988; Hale, 1988; Lery and Stephany, 1985; Schuerhoff *et al.*, 1991; and Wenzowski, 1988 for country-specific details). Due to the size of the surveys they typically conduct, it is the national statistical offices that most feel the weight of the coding process and have been driven to develop these computerized methods.

19.1.4   Computer-Aided Occupational Coding in the U.K.

The largest potential user of computer-aided occupational coding in the U.K. is the Office for National Statistics (ONS) for use in the census of population, and the recent 1991 census marks the first use of any form of computer-aided occupational coding. This involved the development of their own in-house computerized index system, subsequently called CACOC (Computer-Assisted Census Occupational Coding) which mirrored the Australian system (Embury, 1988).

During the period immediately preceding the census, the independent development of more sophisticated occupational coding software was also taking place within the U.K. In 1986, the University of Warwick (IER) in conjunction with the University of Cambridge began development of a computerized coding program later to be known as CASOC—Computer-Assisted Standard Occupational Coding. CASOC can operate on a continuum between computer-assisted and computer-automated coding and uses a combination of exact and inexact matching. The dictionary contains over 23,000 job titles (originally taken from Volume 2 of the SOC manual—OPCS, 1990) and there are approximately 1,200 default codes for the coding of low quality occupational information. CASOC was developed to mirror SOC coding using the SOC manual and incorporates all of the standard ONS coding decision rules. The input data consist of verbatim job titles and job description text. As opposed to some systems, CASOC claims that there is no need for standardizing the descriptions in preparation for matching (e.g., through deleting unnecessary words, correcting word order, etc.). As well as acting as a computerized coding tool, CASOC has several utility files that allow the conversion of the SOC codes to other current and historical occupational and socio-economic code frames.

The use of some form of computer-aided occupational coding is growing in the U.K. Despite the introduction of CASOC, however, there is still a tendency for organizations to develop systems to fit their own specific needs. For the future, both CASOC and non-CASOC users in the U.K. are investigating ways to incorporate computer-assisted coding modules into computer-assisted data collection programs. Computerized coding systems from other countries are also having an effect on U.K. decisions.

19.1.5   General Coding Quality

Coding departments at survey organizations in the U.K. typically use inspection methods to assess and maintain an acceptable level of quality in survey coding. Generally, inspection methods take the form of selecting a sample of each coder's work and inspecting it to determine the number of codes in error. This can simply be a check of the coder's work or preferably an independent blind recoding, followed by adjudication. Appropriate feedback can then be given to coders. Ideally, feedback should also be given to interviewers about the kind of information which should be collected to facilitate the coder's task.

A contrast to traditional inspection methods is continuous quality improvement (CQI; see Imai, 1986). As described by Biemer and Caspar (1994, p. 309), CQI aims to achieve "the smallest error rate possible by continually improving the quality of the product for the duration of the operation." Biemer and Caspar (1994) describe, for example, the successful implementation of CQI for industry and occupation coding at their organization. This included steps such as weekly identification of problem codes, provision of feedback at both the group and individual levels, and a team-oriented approach where the feedback is discussed in a group setting.

19.1.6   The Possible Effect of Computer-Aided Coding

We hypothesized that computer-aided coding should affect a coding operation in a number of ways. In principle *computer-automated coding* should have complete reliability. This of course assumes that the same input file stimulus and version of the program are used. With automated coding, speed is maximized and cost is minimized because human coding intervention is removed (except for the initial keying of the job title, any other relevant job description text, and serial number). This, of course, assumes that the automated system will be able to assign a code to all cases. Also, there is a strong trade-off, between the speed and cost advantages and the possible effect of automated coding on validity. In the case of the CASOC program, for example, both exact and inexact matching are used. Thus, not all matches will lead to a valid code.

In contrast, we hypothesized that *computer-assisted coding* could make a small but meaningful improvement in coding reliability and potentially validity. For example, by suggesting particular code options, a computer-assisted program would potentially narrow the coders' frame of reference. In theory, coders will be more likely to pick from among the computer-suggested codes. (This is similar to the situation of "dependent code adjudication," where adjudicators are influenced to assign a best code from among those already chosen [see Minton, 1972].) This tendency should increase coding reliability and, in contrast to dependent code adjudication, computer-assisted coding should improve validity (to the extent that the computer-suggested codes are accurate). As CASOC in computer-assisted mode looks up the job title in the index and offers all relevant codes, it should also potentially save human coders considerable time. (There is, however, a trade-off in that the job text and serial number need to be keyed either by the coder or in advance of the coding operation.) As a record of the codings is kept by the computer, computer-assisted coding could facilitate feedback to coders as part of a CQI system (see Section 19.1.5).

19.2 DATA COLLECTION

In the remainder of this chapter, we describe the two major projects in the U.K. which were set up to examine coding quality within the context of new computer-aided coding methods. These two studies originated at different organizations, but were conducted in parallel, using a similar approach for measuring coding quality.

The first study was conducted at the Joint Centre for Survey Methods, now the Survey Methods Centre at Social and Community Planning Research (SCPR; see Thomson and Hallett, 1993). It used five ONS office coders each working independently and a subsample of job titles and descriptions collected by the ONS Omnibus survey from a monthly multitopic probability survey of 2,000 adults in Great Britain. Rather than draw a simple random sample of these cases, difficult cases (identified through a previous study) were oversampled with a disproportionate stratified sample design. This allowed for the examination of the responsiveness of computer-aided coding under "difficult" conditions. In the analyses that follow, the data from Study 1 are weighted by the inverse of their probability of selection so as to allow for generalization to the whole of the types of cases that were encountered on the Omnibus survey. This was also necessary for comparison with other studies, such as the ONS study (Butcher et al., 1981; Elliot, 1992, 1983) and Study 2.

The second study used British Household Panel Study (BHPS) data (see Campanelli and Moon, 1994) as coded by NOP Research, London. The BHPS is a ten-year household-based survey, with some 10,000 interviews carried out each year by NOP on behalf of the ESRC Research Centre on Micro-Social Change at Essex University. A random subsample of 322 descriptions of occupations had been selected at random for a blind recoding as part of standard quality control procedures. Each of the 322 job titles and descriptions was then coded twice more by the NOP coding team, again blind, this time using CASOC in computer-assisted mode. The interval between the original coding and the computer-aided coding was long enough for there to be little chance of coders remembering their previously assigned codes. A research director from NOP then coded them all both manually and with computer assistance (again with a long gap in between); a senior researcher from the Research Centre coded them manually; and a research assistant used CASOC in 100 percent computer-automated mode (Auto1 CASOC). Examination of the data suggested that although the research assistant was not making decisions about what code to assign, she *was* making decisions about how much, if any, job description text should be entered in addition to the job title. A ninth coding was therefore conducted in which solely the job title was entered (Auto2 CASOC). No weighting was necessary for Study 2 as these data represented a simple random sample of all BHPS questionnaires.

Both Study 1 and Study 2 also employed three experts on SOC coding, including two of the authors of the CASOC system, to code all cases for comparison with the standard coders' work.

19.3 MEASURES OF RELIABILITY AND VALIDITY

Before continuing, we need to clarify terminology. Terms such as reliability, validity, variance, and bias can mean different things to practitioners in different disciplines (see Groves, 1989). In this chapter, we define *reliability* as the extent to which different coders assign the same code to the same case. Unreliability can therefore be seen as increased *variance* around estimates, such as the proportion of the population possessing a particular coded characteristic. Coder variance, in turn, can be seen in terms of *simple coder variance*, in which each coder adds an element of random noise around a particular estimate, and *correlated coder variance*, in which coders are seen to deviate in their use of the code frame from each other in systematic ways (see Kalton and Stowell, 1979; Jabine and Tepping, 1973). (This latter class of error can also be seen as net *biases* for the individual coders, the effect of which is to increase the variance due to the variability among these biases.) As discussed in Section 19.3.2, simple and correlated variances have been a general survey concern for a number of years and apply to survey processes other than coding.

We use the word *validity* here to mean the extent to which coders assign the "right" code, that is, the code deemed to be the one to which the underlying occupation most properly belongs.

19.3.1 Basic Measures of Reliability

A straightforward index of reliability is the *proportion of agreement* or $\bar{P}$ (see Kalton and Stowell, 1979; Elliot, 1982). It can be estimated by computing the proportion of all paired comparisons in which two codings agree. The $\bar{P}$ ranges from 1, if all the coders are in complete agreement on all questionnaires, to 0 if none of the coders agrees with any other on any of the questionnaires. It is also possible to calculate the proportion of agreement for each coder, P_L. The P_L values can be used to identify particular coders who deviate from the norm and are thus very useful for quality control purposes.

With any comparison between raters, some degree of agreement can occur by chance alone and a preferred estimator is Kappa, κ (Fleiss, 1971). With a long coding frame like SOC, however, the level of chance agreements will be low and it is usually sufficient to use $\bar{P}$ (see Elliot, 1982). For this chapter the results of Study 1 and Study 2 are reported using $\bar{P}$. In these studies κ mirrored $\bar{P}$ but at a level between 0.002 and 0.005 below $\bar{P}$.

With the number of coders and sample sizes used in Studies 1 and 2, the standard errors for $\bar{P}$ tend to be very small (e.g., 0.002 for Study 1 and 0.001 for Study 2). Thus any two values which are at least different by one percent will generally be statistically different. One can question, however, whether a coder reliability of 0.79 is meaningfully different from one of 0.80.

It should also be pointed out that it is possible to investigate the reliability of individual codes (Kalton and Stowell, 1979; Collins and Kalton, 1980; Collins and O'Brien, 1981). Although not discussed in this chapter, this is a useful technique for identifying inherent weaknesses in a given code frame so that either the frame can be improved or coder training can be targeted at these weak points.

19.3.2 Measures of Simple and Correlated Coder Variance

In normal non-experimental situations, simple coder variance is confounded with sampling variance so that estimated standard errors of means and proportions already include simple coder variance. In contrast, as summarized by Kalton and Stowell (1979), the effect of correlated coder variance is to multiply the variance of a sample estimate by a factor, which we will call *Codeff*, as shown in (19.1).

$$Codeff = (1 + \rho_c(M - 1)(1 - \kappa_i)) \tag{19.1}$$

where ρ_c is the intra-class correlation for coders, M is the average coder workload, and κ_i is the reliability of the individual code.

Variance inflation due to correlated error applies to several other survey processes. Take, for example, the seminal work on interviewer variance by Mahalanobis (1946) or the later work of Kish (1962), or the general response variance model discussed by Hansen *et al.* (1961). (See also Bailar and Dalenius (1969) for a discussion of study designs to measure these components.) The effects of clustering on sampling variances are also well documented (e.g., Kish, 1965). Thus, we can see the similarity of the variance inflation factors: *Codeff* (for coders), *Inteff* (for interviewers) and *Deff* (for cluster sampling).

$$Inteff = (1 + \rho_i(m - 1)) \tag{19.2}$$

where ρ_i is the intraclass correlation for interviewers and m is the average interviewer workload

$$Deff = (1 + \rho_s(b - 1)) \tag{19.3}$$

where ρ_s is the intraclass correlation for clusters and b is the average cluster take.

It should be noted that Kish's ρ_i (for interviewers) is roughly equivalent to $\rho_c(1 - \kappa_i)$ (see Kalton and Stowell, 1979). Also ρ_i represents a slightly different measure under the Hansen *et al.* (1961) model than under the Kish (1962) model. Under the former, it represents a measure of the interviewer variance divided by the interviewer plus error variance. Under the latter, it represents a measure of the interviewer variance divided by the variance of the true values plus the interviewer and error variance (see Groves, 1989; Biemer and Stokes, 1991).

Typically, for most variables, values of ρ_c, ρ_i, and ρ_s tend to be very small (e.g., values greater than 0.1 are rare). Yet as can be seen from all three formulae, large inflation factors can result if the workload/cluster size is large. For example, with a ρ_c of 0.02, a κ_i of 0.8, and an average workload per coder of say 401 questionnaires, the variance multiplier is 2.6. This suggests that more coders doing fewer questionnaires is better. However, this must be balanced against the costs of having large numbers of coders who may be poorly trained and inadequately supervised. (In the past, this has often been the argument against using interviewers as coders, but see Martin *et al.*, 1995.) Ideally, an optimum balance should be determined.

19.3.3 Measures of Validity

For occupational coding, validity is a difficult concept to operationalize in a wholly satisfactory way. Perhaps the theoretically ideal criterion would be the classification arrived at by a panel of experts able to question each job-holder and to observe him/her in the work situation. In practice, surveys have to rely on respondents' answers to the occupation and industry questions and in most cases this information has also been filtered by an interviewer. Both Study 1 and Study 2 began with such pre-recorded information from the interviewer. A measure of "validity" was implemented by inviting three independent SOC coding experts to code both sets of data.

19.4 RESULTS

In this section we look at what these two projects have to say about the reliability and validity of occupational coding as well as the implications of

Table 19.1   The Reliability of Coders ($\bar{P}$)

STUDY 1		STUDY 2	
Coder	SOC 3-digit	Coder	SOC 3-digit
		Coder1	0.79
OPCS1	0.76	Coder2	0.80
OPCS2	0.79	Researcher1	0.78
OPCS3	0.80	Researcher2	0.77
OPCS4	0.75	Coder1 CASOC	0.80
OPCS5	0.78	Coder2 CASOC	0.81
		Researcher1 CASOC	0.79
		Auto1 CASOC	0.81
		Auto2 CASOC	0.69
ALL	0.78	ALL	0.78
SE	(0.002)		(0.001)

correlated coder variance on the variance of survey estimates. The contrast between manual and computer-aided coding is also continued.

19.4.1   Reliability Estimates

As shown in the first panel of Table 19.1, an overall $\bar{P}$ of 0.78 with a standard error of (0.002) was obtained for Study 1. This figure is similar to OPCS's figure of 0.77 for intermediate and novice coders (Butcher *et al.*, 1981; Elliot, 1992, 1983). The five OPCS office coders used in Study 1 can also be considered as intermediate/novice coders for although they were experienced occupational coders, they were new to the SOC frame and to using CASOC. This suggests that computer-assisted coding may not substantially improve reliability compared with manual coding.

With the design of Study 2 one can explicitly compare the reliability of manual to computer-assisted coding among the same coders. Despite the vast array of coders and methods, the similarity of values across coders is the most notable finding in the second panel of Table 19.1. It can, however, be seen that the professional coders tend to have higher reliability figures than the researchers. It can also be seen that there is only a one percent difference between the professional coders using a manual method and their use of a computer-assisted method, again suggesting that computer-assisted coding does not have a substantial effect on reliability. The findings with respect to computer-automated coding are of particular interest. When job title is supplemented by thoughtfully chosen job text (Auto1 CASOC), the values for computer-automated coding are not distinguishable from manual coding. When, however, the input is limited to *just* the job title, reliability falls substantially (Auto2 CASOC).

For both Study 1 and Study 2, amalgamating the code frame after coding yielded small increases in coding reliability. A gain of 3 percent was made in reducing the 371 categories (SOC) down to 77 categories (Minor Group) and a larger gain (5 and 8 percent, respectively) was made with the further reduction to 9 categories (Major Group). Study 1 also looked at the reliability of the related coding schemes (i.e., Socio-Economic Group (20 categories) and Social Class (6 categories)) and found these to be similar in reliability to Major Group (9 categories). This is meaningful as it is these shortened code frames, rather than the full three-digit SOC frame, which are typically used in analyses in the U.K. (Also, as suggested by Martin *et al.* (1995), differences between the quality of office coders and interviewer coders are negligible when these shorter code frames are considered.) These figures, however, suggest that at least half of coder unreliability still remains at the major group level due to disagreement in the first digit. This is an important finding as it points to the types of disagreement that are occurring between coders and has implications for SOC coder training.

19.4.2 Correlated Coder Variance

In both studies the values for ρ_c tend to be very small, with one exception less than 0.0058 (see Tables 19.2a and 19.2b). Negative values for ρ_c are also shown in the tables. Technically under the model, the true value of ρ_c should not be negative. Such estimates can occur given the estimation process and the small sample sizes. A comparison of the two tables demonstrates the idiosyncrasies of the coding teams at the two different organizations as different patterns of significant coder variance were found. (Significant coder variance was assessed by Cochran's Q test.)

These tables also allow us to discuss the actual effect of correlated coder variance on the precision of survey estimates. We know from the BHPS, upon which Study 2 is based, that the proportion of people who hold occupations in Major Group 4 is 0.20 and a 95 percent confidence interval around this estimate, assuming a simple random sample and a sample size of 10,000, is ± 0.0080. Taking account of the inflation factor, *Codeff*, based on a workload size of 322 cases as given in Table 19.2b ($\sqrt{1.09} = 1.044$), provides a revised confidence interval of ± 0.0084. As the actual coder workload for the BHPS was closer to 5,000, a *Codeff* of 2.36 should be used, yielding a revised confidence interval of ± 0.0123. Although in terms of absolute magnitude this increase in the width of the confidence interval is small, it actually reflects a reduction in effective sample size from 10,000 to 4,230. Note that simply increasing the number of coders from two to five would have increased the effective sample size from 4,230 to 6,530.

19.4.3 Validity Estimates

Tables 19.3a and 19.3b show the "validity" of each coder's work, as assessed against the codes chosen by the three expert coders. Overall, with the exception

Table 19.2a   Study 1: Examining Simple and Correlated Coder Variance

Major group	Description	F_i	Q_i	Cochran's Q	p-value	ρ_c	Codeff
1	Managers/Administrators	223	.852	4.50		−.0003	.98
2	Prof Occupations	162	.904	9.24		−.0004	.98
3	Assoc Prof & Technical	144	.835	10.67	*	.0026	1.19
4	Clerical & Secretarial	321	.892	14.03	**	.0025	1.13
5	Craft & Related	250	.861	1.42		−.0018	.88
6	Personal/Protective Services	218	.964	4.29		−.0018	.97
7	Sales	138	.841	3.11		.0006	1.04
8	Plant & Machine Operatives	246	.853	21.21	***	.0118	1.79
9	Other Occupations	216	.891	16.08	**	.0058	1.28

Table 19.2b   Study 2: Examining Simple and Correlated Coder Variance

Major group	Description	F_i	Q_i	Cochran's Q	p-value	ρ_c	Codeff
1	Managers/Administrators	333	.881	8.06		.0005	1.02
2	Prof Occupations	200	.859	2.83		−.0019	.91
3	Assoc Prof & Technical	219	.836	11.13		.0018	1.11
4	Clerical & Secretarial	493	.935	14.53	*	.0034	1.09
5	Craft & Related	312	.929	7.27		.0001	1.00
6	Personal/Protective Services	176	.950	12.65		.0025	1.04
7	Sales	187	.888	5.29		−.0008	.97
8	Plant & Machine Operatives	354	.904	6.94		.0000	1.00
9	Other Occupatiøms	300	.943	14.00		.0031	1.06

F_i = Number of times code appears in the data; Q_i = The conditional probability that the second of two coders assigns code i given that the first coder assigned code i averaged over all coders and all cases; Cochran's Q = Measure of the significance of correlated coder variance; ρ_c = Intraclass correlation coefficient for coders; *Codeff* = Variance inflation factor due to correlated coder variance.

* $p < .05$; ** $p < .01$; *** $p < .001$

Study 1 assumes a workload of 401 cases; Study 2 assumes a workload of 322 cases.

of Auto2 CASOC in Study 2, the validity level is fairly high, suggesting a high quality of coding in both of the studies. Study 2 shows that in comparison with Expert 3, professional coders tend to have slightly higher validity scores than the researchers. In comparison with Experts 1 and 2, professional coders scored slightly higher when using computer-assisted coding than they did when coding manually. It is surprising to note a validity estimate of 0.86 was obtained from the first computer-automated run (Auto1 CASOC) in comparison to Expert 1. This high value is not maintained in comparison to the other expert coders and it must be borne in mind that computer-automated CASOC is being tested

Table 19.3a   Study 1: The Validity of Each Coder as Compared to the Experts

	Expert 1 3-digit	Expert 2 3-digit	Expert 3 3-digit
Coder1	0.75	0.75	0.76
Coder2	0.75	0.73	0.75
Coder3	0.79	0.76	0.77
Coder4	0.74	0.69	0.69
Coder5	0.77	0.74	0.75

Table 19.3b   Study 2: The Validity of Each Coder as Compared to the Experts

	Expert 1 3-digit	Expert 2 3-digit	Expert 3 3-digit
Coder1	0.78	0.81	0.82
Coder2	0.82	0.81	0.84
Researcher1	0.80	0.81	0.81
Researcher2	0.81	0.83	0.81
Coder1 CASOC	0.80	0.82	0.82
Coder2 CASOC	0.83	0.85	0.84
Res1 CASOC	0.81	0.80	0.81
Auto1 CASOC	0.86	0.81	0.83
Auto2 CASOC	0.74	0.67	0.71

against two of the authors of the program. The validity figures for Auto2 CASOC are much more along expected lines.

Although not apparent from the overall reliability and validity figures, it should be noted that if computer-automated CASOC is in disagreement with the other coders, its suggested code may be much further from "truth" in terms of what we might call a plausibility scale. We suspect that this has to do with the dictionary in the version of CASOC we were using and the way inexact matching is handled for certain keywords. For example, with some occupations CASOC is very sensitive to small differences in how the title is entered. The description "bar staff" matches exactly to SOC code 622 (bar staff). If, however, the respondent has specified his/her occupation as "bar work," category 958 (Cleaners and domestics) comes up. If this is typed in as "barwork" (without a space), category 516 (Metal Working Production and Maintenance Fitters: Barman (railway shed)) comes up. If the respondent has specified "pub work," category 892 (Worker Water Board) comes up.

19.5   DISCUSSION

This chapter had the dual purpose of illustrating techniques for studying coder reliability and validity and looking at the effect of new computer-aided methods on occupation coding in the U.K.

Although the future of occupational coding in the U.K. clearly lies in the direction of computer-aided coding, the results of the two empirical studies described in this chapter are mixed. For example, with *computer-assisted* coding, we saw only modest gains in reliability and validity as compared to manual coding. In addition, although other organizations have noted time savings with the use of computer-assisted CASOC, we did not. The coders in these two studies, however, were new to the use of the program. Study 2 did, however, reveal that computer-assisted coding can be a very useful training tool. This has implications for interviewer field coding (see below). A computer-assisted coding system could also form part of a CQI system as it would serve as an easy mechanism from which group and individual coder feedback could be given. As suggested by Biemer and Caspar (1994), a CQI system can lead to significant improvements in coding quality and cost which more than offset the additional costs of implementation.

In contrast, *computer-automated* coding offered the greatest time and cost savings (particularly for large jobs), as the CASOC program assigned a code in 99 percent of cases. However, the CASOC system is sensitive to the amount and type of text which is entered and may score significantly lower than manual coders with respect to the plausibility of the code. For the future, ways to improve the reliability and validity of computer-automated coding need to be investigated. Computer-automated systems tend to standardize descriptions entered in preparation for matching by doing some type of parsing on the textual data (Lyberg and Dean, 1992). The high reliability and validity values obtained for Auto1 CASOC in comparison to Auto2 CASOC suggest that there should be an explicit investigation into the amount and type of text that should be entered for CASOC. The reliability and validity of *computer-automated* coding can also be improved by coupling it with another method. The key here is to be able to accurately identify those cases that have a low probability of a valid match and target these for coding via other methods of coding. The CASOC team at IER is currently considering ways in which this can be accomplished.

Another direction being investigated, is the possibility of incorporating a *computer-assisted* subroutine into a CAPI/CATI environment to facilitate interviewer field coding. This option offers a good solution to the trade-off between several quality issues and practical constraints. For example, the disadvantage of using interviewers as coders is that they typically do not achieve the same levels of accuracy as specialist office coders. There are also the associated training and supervision costs. On the other hand, interviewers who code occupation should have a better idea of what constitutes a good occupational description and probe accordingly. More importantly, a large number of interviewers acting as coders will mean that each has a small workload and this can play a major role in reducing the effect of correlated coder variance. However, the simple coder variance could increase if the workload is decreased too much and quality cannot be maintained among a large number of coders. A built-in computer-assisted module in a CATI/CAPI program should not only facilitate interviewer training in coding, but also clearly delineate for the interviewer which extra probes are necessary to ensure the selection of the most accurate code.

ACKNOWLEDGMENTS

The authors are particularly grateful to the work of Peter Elias, Ken Prandy, and Iain Noble who participated in this experiment and to Steve Hallett and Kathy Jones who assisted in the analysis of the data.

REFERENCES

Andersson, R., and Lyberg, L. (1983), "Automated Coding at Statistics Sweden," *Proceedings of the Section on Survey Research Methods*, Alexandria, VA: American Statistical Association, pp. 41–50.

Bailar, B.A., and Dalenius, T. (1969), "Estimating the Response Variance Components of the U.S. Bureau of the Census's Survey Model," *Sankhyā*, Ser. B, 31, pp. 341–360.

Biemer, P., and Caspar, R. (1994), "Continuous Quality Improvement for Survey Operations: Some General Principals and Applications," *Journal of Official Statistics*, 10, pp. 307–326.

Biemer, P., and Stokes, S.L. (1991), "Approaches to the Modelling of Measurement Error," in P.P. Biemer, R.M. Groves, L.E. Lyberg, N.A. Mathiowetz, and S. Sudman (eds.), *Measurement Errors in Surveys*, New York: John Wiley and Sons, Inc., pp. 487–516.

Butcher, B., Martin, J., and John, P. (1981), "Proposal for a Study of Coder Variability in Coding Occupations," *OPCS Survey Methodology Bulletin*, 12, pp. 16–19.

Campanelli, P., and Moon, N. (1994), "Computer Aided Occupational Class Coding," paper presented at the annual meeting of the World Association for Public Opinion Research, Boston, MA.

Chen, B., Creecy, R.H., and Appel, M.V. (1993), "Error Control of Automated Industry and Occupation Coding," *Journal of Official Statistics*, 9, pp. 729–745.

Corbett, J.P. (1972), "Encoding from Free Word Descriptions," unpublished report, Washington DC: U.S. Bureau of the Census.

Collins, M., and Kalton, G. (1980), "Coding Verbatim Answers to Open Questions," *Journal of the Market Research Society*, 22, 4, pp. 239–247.

Collins, M., and O'Brien, J. (1981), "How Reliable Is the Coding Process?" *The Market Research Society Annual Conference Papers*.

Dodd, T. (1985), "An Assessment of the Efficiency of the Coding of Occupation and

Industry by Interviewers," *New Methodology Series*, No. NM 14, London: Office of Population Censuses and Surveys.

Durbin, J., and Stuart, A. (1954), "An Experimental Comparison Between Coders," *Journal of Marketing*, 19, pp. 54–66.

Elliot, D. (1982), "Variability in Occupation and Social Class Coding," unpublished report, London: OPCS.

Elliot, D. (1983), "A Study of Variation in Occupation and Social Class Coding— Summary of Results," OPCS *Survey Methodology Bulletin*, 14, pp. 48–49.

Embury, B. (1988), "The Methodology of Occupation Coding in the 1991 Census," unpublished report, Australia: Australian Bureau of Statistics.

Fleiss, J. (1971), "Measuring Nominal Scale Agreement Among Many Raters," *Psychological Bulletin*, 76, pp. 378–382.

Groves, R. (1989), *Survey Errors and Survey Costs*, New York: Wiley.

Hale, A. (1988), "Computer-Assisted Industry and Occupation Coding in the Canadian Labour Force Survey," *Proceedings of the Fourth Annual Research Conference*, Washington, DC: U.S. Bureau of the Census.

Hansen, M.H., Hurwitz, W.N., and Bershad, M.A., (1961), "Measurement Errors in Censuses and Surveys," *Bulletin of the International Statistical Institute*, 38, 2, pp. 359–374.

Imai, M. (1986), "Kaizen: The Key to Japan's Competitive Success," New York: McGraw-Hill.

Jabine, T.B., and Tepping, B.J. (1973), "Controlling the Quality of Occupation and Industry Data," *Bulletin of the International Statistical Institute*, pp. 360–392.

Kalton, G., and Stowell, R. (1979), "A Study of Coder Variability," *The Journal of the Royal Statistical Society*, Series C, 28, 3, pp. 276–289.

Kish, L. (1962), "Studies of Interviewer Variance for Attitudinal Variables," *Journal of the American Statistical Association*, 57, pp. 92–115.

Kish, L. (1965), *Survey Sampling*, New York: Wiley.

Lery, A., and Stephany, A. (1985), *COLIBRI II: On-line Key Entry and Coding of 1982 Population Census Forms*, Technical Report No. 1830/RP82, France: INSEE.

Lyberg, L., and Dean, P. (1992), "Automated Coding of Survey Responses: An International Review," paper presented at the Conference of European Statisticians, Washington, DC.

MacDonald, C. (1982), "Coding Open-Ended Questions with the Help of a Computer," *Journal of the Market Research Society*, 24, 1, pp. 9–28.

Mahalanobis, P.C. (1946), "Recent Experiments in Statistical Sampling in the Indian Statistical Institute," *Journal of the Royal Statistical Society*, 109, pp. 325–378.

Martin, J., Bushnell, D., Campanelli, P., and Thomas, R. (1995), "A Comparison of Interviewer and Office Coding of Occupations," *Joint Proceedings of ASA/AAPOR: Section of Survey Research Methods*, American Statistical Association, Washington, DC.

Minton, G. (1972), "Verification Error in Single Sampling Inspection Plans for Processing Survey Data," *Journal of the American Statistical Association*, 67, pp. 46–54.

Office of Population Censuses and Surveys (1990), *Standard Occupational Classification*, Volumes 1 and 2. London: HMSO.

O'Reagan, R.T. (1972), "Computer-Assigned Codes from Verbal Responses," *Communications from the ACM*, 15, 6, pp. 455–459.

Schuerhoff, M., Roessingh, M., and Hofman, L. (1991), *Examples of Computer Assisted Coding*, Report BPA 87-M2, Netherlands: Statistics Netherlands.

Thomson, K., and Hallett, S. (1993), *Occupational Coding: CASOC Coder Variability Experiment*, Technical Report for Project 3000/19, London: Joint Centre for Survey Methods, SCPR.

van Bastelaer, A., Hofman, L., and Jonker, K. (1987), "Computer-Assisted Coding of Occupation," Chapter 6 in *Automation in Survey Processing*, Selection no. 4, Netherlands: Statistics Netherlands.

Wenzowski, M.J. (1988), "ACTR: A Generalized Automated Coding System," *Survey Methodology*, 14, pp. 299–397.

White, P. (1983), "The Continuous Manpower Survey—Feasibility Study," *OPCS Survey Methodology Bulletin*, 15, pp. 33–41.

SECTION D

Quality Assessment and Control

C H A P T E R 20

Survey Measurement and Process Improvement: Concepts and Integration

Cathryn S. Dippo
U.S. Bureau of Labor Statistics

20.1 INTRODUCTION

Survey measurement and process improvement have much in common. Theoretically, both rely on Neyman's philosophy of randomization, which went beyond that of Fisher and tied together random sampling, consistent estimators, and confidence intervals as measures of precision. Practically, both are dependent upon a production process being in a state of statistical control. Yet, despite the shared origins and assumptions, the concepts of survey measurement and process improvement are not integrated in the strict sense of the word. That is, the scientific principles of statistical measurement are not a naturally occurring part of the survey process.

In recent years, many organizations responsible for survey measurement activities have introduced some form of quality management into their operations. The chapters which follow in this section provide excellent examples of incorporating quality improvement practices into survey operations. This is an important step towards integrating process improvement into survey measurement at the implementation or operational level, but such work does not integrate the concepts or philosophies at the theoretical level. In such a synthesis, quality measurement and metadata production would receive as much attention as the production of survey measures or population estimates

Survey Measurement and Process Quality, Edited by Lyberg, Biemer, Collins, de Leeuw, Dippo, Schwarz, Trewin.
ISBN 0-471-16559-X © 1997 John Wiley & Sons, Inc.

throughout the survey measurement process, from initial discussions of concepts through the dissemination of process products.

My goal in this chapter is to share reflections on the conceptual integration of survey measurement and process improvement. Perhaps understanding the reason two concepts with the same philosophical roots have not been integrated will engender insights which will in turn promote the desired integration, and ultimately aid the new theoretical basis to find expression in actual survey operations. Thus, following the lead of many historians who believe "what is past is prologue . . .," this chapter includes a review of the past. I also focus heavily on semantics, for the meanings of words and terms provide the basis of understanding, and without understanding, we cannot hope to grasp or define a concept. I have no answers to the conundrum before us; my aim is solely to increase awareness of the need to go beyond "incorporating" or "building" process improvement "into" survey measurement.

After a brief discussion of the terms survey measurement and process improvement and quality measurement from these two perspectives in Section 20.2, I explore the meanings of the term error from these two perspectives in Sections 20.3 and 20.4. Understanding the usage of the word error and its relationship to quality is vital to grasping the theoretical constructs of survey measurement and process improvement. Next, I consider measuring error in the actual process of survey measurement from the two perspectives, and, then, in the last section, I turn to the idea of integration. Many of the issues emanate from the distinction between measures of quality for the output survey measures and for the measurement process. Understanding this distinction is a first step in defining useful measures of process quality, that is, those that can be used to improve the process. These measures help the customer determine a product's fitness for use, but in many ways, omit as much information as they contain. What they fail to convey is the information which gives a statistic meaning, that is, both the data *and* metadata about both the process *and* the product.

20.2   THE MEANING AND USAGE OF SOME TERMS

Survey measurement is a term used by many. Its meaning and interpretation are as varied as its users and the situations in which it is used. For some, the meaning of the word sample is implicit, emanating from its use, as in a measure of a population quantity based on observed data from some of the units in the population. Thus, the proportion single-parent families living in central cities of the U.S., as estimated from data collected in the Current Population Survey, is a *survey measure*. Some would say that the estimate of the same population quantity based on the decennial census is also a survey measure. For this group, the observation method—a self-administered questionnaire, rather than some form of direct observation—implicitly defines the estimate as a *survey measure*. On the other hand, a land surveyor would define the result of his or her observations with mechanical instruments rather than questions as a *survey measure*.

A common thread to all of these semantic interpretations is the act of measuring. The first definition of *measure* in the *Random House College Dictionary* (1980, p. 829) is "the act or process of ascertaining the extent, dimensions, quantity, etc., of something, especially by comparison with a standard; measurement." It is in defining measure as a *process*, "a systematic series of actions directed to some end" (Random House, 1980, p. 1055), that we get to the crux of a unified survey measurement-process improvement concept. That is, all of the activities and tasks associated with producing a survey-based statistic must be viewed as a process to which the principles of statistical measurement should be applied.

Considering the production of a statistic as a measurement process is not new; however, references in the statistical literature are rare. Eisenhart (1952, 1963) discussed measurement in the sense of calibration systems for mechanical devices; Bailar (1983) included measurement of human population characteristics using sample surveys and censuses in her broad-ranging discussion of statistical data. Her focus was on defining quality and promoting the need for statistical control of measurement processes.

However, defining survey measurement as a series of actions is not sufficient. For a measure to be useful, it must be more than just a number; it must have context. Without some knowledge about the mapping used to translate the attributes of an object to a number, the number has no meaning. For example, the information that John is 4 and Mary is 6 is not sufficient to answer the question of who is older, for John may be 4 months old and Mary may be 6 days old. Analogously in survey measurement, the information that the increase in employment is 250,000 as measured by one survey and 150,000 as measured by another is insufficient. To appropriately interpret these two statistics, the user needs *metadata* or data about the data. (For a discussion of the relationships among data, metadata, and information, see Hand, 1993.) Thus, to give meaning to the term survey measurement, attention must be paid to both the process and metadata about the process, including the inputs and outputs.

Process improvement, on the other hand, is just one of many terms used to refer to a quality philosophy. Deming's 14 points, total quality management, continuous quality or process improvement, etc., are all aimed at improving the quality of products and services through a focus on customers. Whichever philosophy you examine, the key to success is defined as a function of customers, processes, and statistical measures of quality.

Quality in the context of survey measurement is another term which has received a fair amount of attention in recent years. Some have addressed the definition of quality (e.g., Norwood, 1987); some have looked at quality management in survey organizations (e.g., Fecso, 1989; Colledge and March, 1993); and still others have discussed indicators of survey quality (e.g., Groves and Tortora, 1991). In the pursuit of quality, various approaches have been tried and discussed, including those based on standards (Freedman, 1990) and continuous quality improvement (Linacre, 1991). Colledge and March (Chapter 22) review quality policies, standards, guidelines, and recommended practices across a number of national statistical agencies. Morganstein and Marker (Chapter 21) focus on continuous quality improvement and the role of current best methods in the improvement of quality. Succeeding chapters provide some excellent examples of how quality improvements have been achieved in a variety of survey measurement situations.

The concept of quality for a survey measure is a notion; it is vague and imprecise. Measures of quality are inherent to both the concepts of survey measurement and process improvement. Both concepts also include the idea that the quality of a statistic must be defined as a function of its use, which varies by user and circumstance. Both focus on error measurement. At the same time, the meaning of error, the measures of error, and the use of error measures differ dramatically.

In the survey measurement literature, the most commonly used measure of survey quality has been mean squared error. *Error* in this context is defined in terms of mathematical expectation. The term *error* is also used to mean a mistake in process. Survey researchers have long recognized the potential for error in specific subprocesses of survey measurement such as obtaining respondent cooperation, question asking, data entry, and item coding and have devised quality control procedures to measure and monitor the implementation of defined procedures. Thus, process error measures like response rate, the rate at which interviewers substantially change a question when reading it, keying error rate, and item coding error rate are regularly computed for ongoing surveys.

To complicate the semantic problems associated with the use of the term error in a survey measurement context, one of the components of mean squared error, not the total, is often labeled measurement error. Groves (1991) discusses the meaning of the term *measurement error* as observation error across the disciplines of statistics and psychology, noting that total error is a function of both observation and nonobservation errors. The end result is that measurement error does not relate to the overall quality of a survey measure, or, said another way, the error in survey measurement is *not* the same as measurement error.

More importantly, most measures of process error do not have a direct functional relationship to mean squared error. For example, an increase in the response rate does not necessarily translate into a reduction in mean squared error. Nor does a reduction in mean squared error necessarily imply improvement in a particular process error measure.

20.3 MEAN SQUARED ERROR AS A MEASURE OF SURVEY QUALITY

The basic precept of survey sampling is to do more with less. "A sample survey can, in shorter time schedules and at lower cost, provide estimates of what might have been obtained from complete coverage using the same observational or measurement procedures. . . . A well-designed and controlled sample survey permits the achievement of higher quality . . . than can be accomplished in a complete census" (Hansen and Madow, 1976, p. 76). When in the mid-1930s, the U.S. Bureau of the Census initiated an investigation into the applicability of randomized sampling to finite populations, there was little in the way of theoretical foundation or empirical evidence to justify this statement. In particular, there was no standard measure of quality to determine whether sampling provided higher quality.

For this early investigation, a simple comparison of the estimated number of unemployed from the probability sample to that of the voluntary mail registration system sufficed. But, as survey sampling began to be used to measure additional and more conceptually complex social and economic characteristics, the need to provide additional and more sophisticated measures of survey quality and survey costs grew.

Survey quality as defined by sampling variability was first linked to survey costs by Neyman (1934). Neyman's concept of optimum allocation to minimize the variance of an estimator subject to a constraint (i.e., a fixed total sample size) provided the foundation for work that continues today on developing ever-more efficient sample designs and estimators with lower variance. Yet the situation today is as it was in 1943 when Hansen and Hurwitz (1943, p. 335) stated ". . . no theory with practical applicability has been developed which indicates a "best" system of sampling. . . ." In fact, Godambe (1955) proved that there is no uniformly best linear unbiased estimator for estimating a population total. One must make assumptions, reduce the number of options to a feasible set, and define an evaluation criterion.

Early on, Hansen and Hurwitz (1943, p. 337) argued to modify the criterion for good estimates from including only "best linear unbiased" to encompass "estimates that are nonlinear, consistent, but have a smaller mean square[d] error." While they used the terms mean squared error and variance interchangeably, neither they nor others believed sampling error equaled mean squared error or that nonsampling errors did not exist. As early as 1940, Stephan, Deming, and Hansen (Stephan et al., 1940) discussed the challenge of avoiding or reducing the biases associated with the listing procedures for households and persons within households in the 1940 census. Interest in the control and measurement of nonsampling errors eventually led to the Hansen, Hurwitz, Pritzker (1967) concept of mean squared error, which I will refer to as the HHP model.

In the HHP model, there are three factors relevant to evaluating the quality of survey estimates (x)—requirements, specifications, and operations. The requirements are reflected in the ideal goals (Z) for the survey measure. These goals which often reflect a general social or economic concept must be refined or translated into a set of defined goals (Y) through detailed specifications or working definitions. The specifications must, in turn, be translated into survey operations, which over repeated application, reflect the expected value of $x(X)$.

Consider, for example, the measurement of labor force. Conceptually, a person's labor force classification (Z) (i.e., employed, unemployed, or not in the labor force) is based upon labor market activity during the calendar week containing the 12th day of the month. The working definition of employed (or specifications) (Y) is someone who worked for pay for at least one hour *or* worked for at least one hour in his/her own business, profession, or on his/her own farm *or* worked unpaid for at least 15 hours in a family-operated farm or business *or* did not work but had a job or business from which he/she was temporarily absent because of certain specified reasons. The operating definition of employed (X) is someone for whom the response to the question "LAST WEEK, did you do ANY work for pay?" is "yes" *or* the response to this question is "no" and the response to the question "LAST WEEK, did you have a job either full or part time? Include any job from which you were temporarily absent." is "yes" *or*.... Moreover, the operating definition is supported by a set of interviewer training materials, an interviewer's manual, various memoranda supplementing and clarifying these instructions, computerized edits and imputations to resolve inconsistencies in the recorded data, etc.

In the HHP model, these three factors are used to define mean squared error as follows

$$
\begin{aligned}
\mathrm{MSE}(x) &= \mathrm{E}(x - Z)^2 \\
&= \mathrm{E}[(x - X) + (X - Y) + (Y - Z)]^2 \\
&= \mathrm{E}[(x - X)^2 + (X - Y)^2 + (Y - Z)^2 + 2(X - Y)(Y - Z)
\end{aligned}
$$

where $\mathrm{E}(x - X)^2$ is the total variance, sampling and nonsampling, $(X - Y)$ is the bias in the actual survey operations compared to the specifications, $(Y - Z)$ is the bias in the survey specifications compared to the requirements (relevance), and $2(X - Y)(Y - Z)$ is the interaction between the two bias terms.

HHP define *accuracy* as a function of the defined goals and actual performance. In other words, accuracy is defined in terms of how well operations meet specifications over repeated surveys and is equal to total variance and the first bias term. This is another instance of where, in survey measurement, terms have been given a meaning not wholly consistent with general understanding. That is, accuracy is defined in the *Random House College Dictionary* (1980, p. 10) as "condition or quality of being true, correct, or exact; precision, exactness, or correctness."

Relevance is of particular importance in the current discussion. It is through the relevance term in the HHP model that customers and uses of survey measures are incorporated into the quality measure. Morgenstern (1963) notes that "specious accuracy" occurs when accurate statistics are given, but are irrelevant for the immediate purpose—a situation he frequently encountered in economic modeling. Dalenius (1985) discusses the notion of relevant statistics, especially in the context of official statistics and their purposes. Methods for assessing or measuring relevance are extremely rare in the survey measurement literature. Moore and Shiskin (1967) try to assess the relevance of survey estimates as economic indicators using a methodology based on assigning weights to requirements and scores to specifications. In the context of the HHP model, Dalenius (1977) discusses approximating a matrix D_{ij}, where each D_{ij}

reflects the difference between the ith defined goal Y_i and the jth ideal goal Z_j for a group of potential users. But the practice is not common, and most authors simply define mean squared error as variance plus bias squared and ignore discussing relevance, in effect ignoring the customer's usage of the survey measure.

We have resorted to using this truncated definition of mean squared error as a measure of overall survey quality because truth, or Z in the HHP model, can only be defined explicitly in very rare circumstances, making it generally impossible, in practice, to measure the HHP definition of total mean squared error. However, if one is trying to reflect the customer's needs in an overall measure of survey quality and Z is truth as defined by the customer, the relevance of an estimate cannot be ignored. If the term $Y - Z$ cannot be measured, an estimate of the truncated definition of mean squared error is not sufficient, and alternative means of providing metadata about the relevance of a statistic must be provided.

20.4 THE PROCESS OF SURVEY MEASUREMENT

With such an explicit definition of survey quality, one might wonder why there was a flurry of activity in the 1980s related to defining or describing survey quality and why it is not obvious how to do better with less. The answer of course goes back to the distinction between the survey product or measure and the survey measurement process. All of the beautiful mathematics associated with estimating mean squared error or any variance or bias component assume a defined, reproducible process that is in control.

In discussing survey measurement as a process, we turn our attention to error as it is more commonly used, that is, a mistake. It is also important to differentiate between two types of errors—mistakes in specifications and mistakes in operations, in HHP terms. HHP (1967, p. 52) noted that "Good survey design calls for reasonably effective control of the accuracy of survey results through appropriate specifications for survey procedures and adequate control of the operations." Juran *et al.* (1979) make a similar distinction between quality of design and quality of conformance. When error is used to refer to a mistake, we are no longer using the framework of randomization theory and expectation over repeated events. We are dealing with a specific realization—one set of specifications for selecting a sample rather than another or one interviewer's recording of a response rather than another's of the same response.

There are a variety of process models for survey measurement, just as there are a variety of ways to disaggregate total mean squared error. Survey measurement as a process can be viewed broadly, as by Hansen and Steinberg (1956). They included errors in definition (concept, problem, universe to be studied), in specification (coverage of units, question wording, coding procedures), and in implementation as potential sources of survey error. This view, of course, corresponds to the broad interpretation of survey measure and the

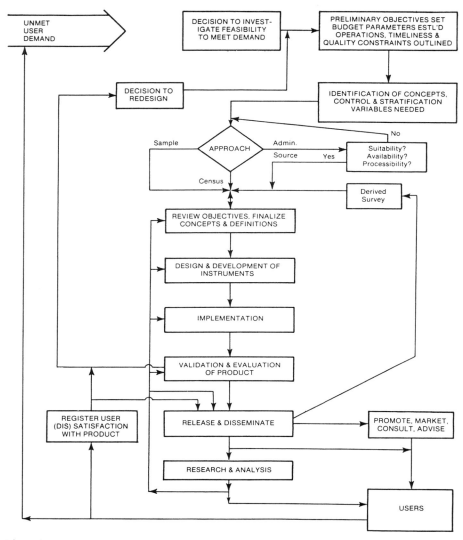

Figure 20.1 Statistics Canada's general model of the statistical survey process. Reproduced by authority of the Minister of Industry, 1996, Statistics Canada, Quality Guidelines, April 1987, Page 3.

HHP model. The process of survey measurement can also be viewed conditionally as the observation process given a design or method. This view corresponds to the truncated mean squared error model which ignores relevance.

Two versions of a broad-based model of the survey process are presented in Figures 20.1 and 20.2. Figure 20.1 is the Statistics Canada model (Statistics Canada, 1987) and Figure 20.2 is the U.S. Bureau of Labor Statistics (BLS) model (BLS, 1994), which used the Statistics Canada model as a starting point. Since the BLS model is more recent, I will discuss some of its features.

Figure 20.2 Bureau of Labor Statistics quality measurement model.

At the center of the model is the customer. As Deming (1982, p. 225) notes, "The consumer [is] the most important part of the production line." Although every survey or census begins with an expression of need by someone external to the survey-taking organization, each member of the organization and respondents will at some time also be a customer to an intermediate product of the survey process. The needs of all customers, either external or internal to the organization, have to be addressed as part of the production process. Customers often express quality criteria in qualitative terms—relevant, timely, accurate, comprehensive, etc.

The actual process of survey measurement, as outlined in the BLS model, consists of a set of six operations—conceptualization, planning, design, development, implementation, and validation. The focus of conceptualization is on defining the purpose (why) and requirements (what) of the measurement.

Defining the necessary resources (who) and timing (when) of the measurement process is the focus of planning. Design is determining the specifics of how the purpose and requirements defined during conceptualization will be met within the parameters of time and resources set during planning. Development is the preparation of the tools needed to implement the design (e.g., software, hardware, instructions, collection instruments, training materials). During validation, actual and projected performance levels are compared. In HHP terms, ideal goals are put forward during conceptualization, and defined goals are specified during planning, design, and development. It is by repeating the implementation operation that the expectations of the survey estimates are defined theoretically, and validation includes comparing survey estimates with the ideal and defined goals.

In the BLS survey measurement process model, the relationships among the operations are not sequential. One does not complete conceptualization and then move on to planning. One should not prepare a design only to discover later that the technology is not yet available for development. There is a continual flow among operations. Each is supported by the others. Each is performed to satisfy a customer need.

The process model is also recursive. Operations consist of activities which, in turn, can be viewed as having six operations. For example, the design of a questionnaire to collect consumer expenditures may involve conceptualizing alternative questionnaire formats (e.g., a diary vs scripted interview), planning who should design, develop, etc., the alternatives, designing (e.g., deciding to have separate pages for each day of the week and dividing the interview into sections based on type of purchase), developing the tools (e.g., an authoring language for computerized data collection instruments), implementing the design (e.g., using BLAISE or CASES to create both computerized self-administered questionnaires and computer assisted personal interviewing instruments), and validating that both the resulting instruments yield the information required by the customer. Each activity may also be viewed similarly, for instance, validating alternative instruments requires conceptualization of criteria for validation, etc.

The circular, recursive nature of the BLS Quality Measurement Model conforms to Shewhart's (1939) mass production process model shown in Figure 20.3. He suggests thinking about the three steps of the mass production process (specification, production, and inspection) as steps in the scientific method: making a hypothesis, carrying out an experiment, and testing the hypothesis. As he notes (p. 1), "To attain economic control and maximum quality assurance, statistical theory and techniques must enter every one of the three steps in the control of quality."

While the customer is the focus of the survey measurement process in the BLS model, the quality-cost function is the constraining influence—the set of elements which holds the pieces together to make a whole. The quality-cost function reflects the varying concepts and dimensions of quality enumerated by Rosander (1985): conformance to specifications, fitness for use, buyer

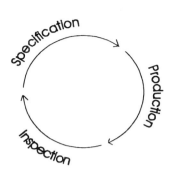

Figure 20.3 Shewhart's process model.

satisfaction, acceptability at an affordable price, and life benefits that are worth life costs. The quality-cost function also reflects the fact that resources are limited. Customer needs must be addressed within set time frames, budgets, etc. Products must meet quality criteria related to relevance, timeliness, accuracy, etc. The nonquantitative quality criteria of the customer must be translated into quantifiable requirements during conceptualization. It is through the quality-cost function that one determines whether a change in process is worth the associated costs.

However, the quality-cost function is similar to the total mean squared error in that it reflects an ideal which is rarely quantifiable. This necessitates the use of approximations and proxies—the definition of alternative quality criteria in much the same way as defined *vis-à-vis* ideal goals in the HHP model. Moreover, as in the case of the mean squared error, practitioners often revert to computing component or partial measures. On the one hand, we have response variance, response bias, nonresponse bias, etc. which are focused on the specifications or method and are measures of error in the first sense, that is, defined in terms of mathematical expectation. On the other, we have nonresponse rates, edit failure rates, undercoverage rates, etc., which are focused on controlling operations or implementation of the method and are measures of errors in the second sense, that is, mistakes. While estimates of the first set of measures are only possible in special circumstances, the second set is usually easily estimable.

20.5 MEASURING ERROR IN SURVEY MEASUREMENT

In general, the approach used in measuring and controlling errors in survey measurement parallels that used in the physical sciences. Mahalanobis (1944), in advocating the use of interpenetrating samples for controlling and studying survey error, used the term statistical engineering. Eisenhart, who actually studied with Jerzy Neyman at University College, London and eventually became the first Chief of the National Bureau of Standards' Statistical

Engineering laboratory, has been credited with the idea of measurement as a production process (Olkin, 1992). Eisenhart (1952, 1963) also makes the distinction between a measurement method and a measurement process, where process is defined as a realization of method.

Experimental studies, including many based on interpenetrating samples, have been used extensively over the years to investigate or measure the effects of alternative methods. In the United States, some of the first were in conjunction with the 1950 decennial census (Hanson and Marks, 1958). In these types of studies, as Groves (1989, p. 307) notes, "*Explanations* or *causes* of differences [in measurement] are being sought." Great care needs to be taken in the interpretation of the results of these studies, an idea I will return to later.

Sampling for controlling the quality of operations was advocated early by Deming for the 1940 U.S. census (Olkin, 1987). On the other hand, process control procedures for the Current Population Survey, which began around 1940, were minimal until the mid-1950s when a change in the number of primary sampling units from 68 to 230 resulted in significant differences in the estimates of unemployed (Hansen *et al.*, 1955). A review panel concluded the "differences arose because of the reduced supervision and training of interviewers in the original sample of 68 areas, while extensive attention was being given to training the new interviewers in the new and much larger number of primary sampling units" (Hansen, 1987, p. 185). As a consequence, process control procedures were expanded. (See Hansen and Steinberg (1956) for a detailed description.)

But, what about the assumptions behind the application of quality control procedures to survey measurement? Shewhart (1939) stated the need for reproducibility in an operation to yield a potentially infinite sequence of numbers corresponding to an infinite number of repetitions of the operation. HHP, who addressed reproducibility in the survey measurement context, defined a reproducible method as one that another investigator could follow and which would, under the same assumptions, yield the same result, at least within the range of sampling and response variance. That is, the specifications cannot be a source of variability; there is no room for differences in interpretation. In practice, this is an unobtainable goal; one can follow the recommendation of Shewhart for deciding when a process is in statistical control. His process control chart methods are based on the idea that a process is in statistical control when one can predict the level and variability of the process. While one can fairly easily determine how to use this concept in evaluating a survey operation like data keying or coding, we have not yet developed a way to determine when the processes of developing a survey question or deciding on the sample allocation among strata are in statistical control.

The application of quality control to services requires a wider approach than in manufacturing. Quality control must be applied to more than physical products; human performance, decisions, and data have to be addressed. Thus, techniques in addition to Shewhart charts are required, including "sampling for discovery, estimation, comparisons, testing effectiveness, random time sampling for work and other activity analysis, input-output analysis, learning curve analysis, written procedures and specifications, waiting or delay time analysis, and field testing and experimental design" (Rosander, 1985, p. 6).

And, what about the assumptions needed for evaluating alternative methods or specifications? Understanding the results *vis-à-vis* making a prediction or inference about the alternative methods requires one to assume there is no variability in the interpretation of the alternative specifications now or in any future implementation, and that each set of operations is and always will be in statistical control. As Deming (1987, p. 361) states "Statistical theory for analytic problems has not yet been developed. The only possible exception exists for performance or product that comes from a stable system, one that is demonstrably in statistical control." In attempting to make an inference about a method or procedure, one is using the results for analytic rather than enumerative purposes.

An *enumerative* purpose for a survey is one of description, to answer questions such as how many people are unemployed, how many gallons of oil does the U.S. have in its reserves, or how much does a family living in New York City spend on average for housing each month. An *analytic* purpose is one of cause or effect, to answer why questions such as why do not the unemployed have jobs, why are oil reserves lower or higher than last year, or why does housing cost a lot in New York City. While as Deming (1953, p. 245) states "The concepts are old," he is the one responsible for their existence in the survey measurement literature. Hahn and Meeker (1993) review the assumptions needed for making inference from both enumerative and analytic studies.

The uses of control charts are essentially analytic (Deming, 1953). The testing of alternative methodologies is analytic. The conditions that prevail when the test is conducted will never exist again. The tests should be viewed as judgment samples (Deming, 1987). Predictions about future measurements can only be expressed in terms of estimation of the parameters of a superpopulation or a stochastic process, which requires a model-dependent approach (Hansen *et al.*, 1983). The model which must be assumed is basically that the process, in our case the entire survey measurement process, is in control.

Given the circumstances under which most repetitive surveys are conducted, it is doubtful that the process of survey measurement is in control as assumed in finite population sampling theory. Survey measurement is a process in which many human beings, not machines, perform a wide variety of tasks. Over time, respondents, interviewers, data reviewers, coders, supervisors, etc., change. Even if the people performing the task remain the same, their behaviors change, since it is natural for humans to modify their behavior based upon experience and events. As a result, any estimate of mean squared error (or any subcomponent) or of any process measure is just an estimate for a particular time and circumstance. These measures must be repeatedly produced and monitored to determine whether the various processes are remaining within control limits.

These repeated measures are an integral part of the metadata needed by users to evaluate how fit a statistic is for use.

20.6   INTEGRATING PROCESS IMPROVEMENT AND SURVEY MEASUREMENT

The first of Deming's 14 points (1982, p. 343) is "Create constancy of purpose toward improvement of product and service." Since improvements in product result from improvements in process, many quality improvement advocates use the term continuous process improvement. Deming's 3rd and 4th points include the phrases "require statistical evidence that quality is built in" and "depend on meaningful measures of quality, along with price." Thus, process improvement is dependent upon having good measures of quality built into the production system. A decision to change an aspect of production should be based on facts about costs and quality, that is, the change should result in a verifiable improvement.

In the Deming philosophy, quality management is built on sound statistical and scientific principles. For any product, if you can define the production process and a measure of quality, you can implement the philosophy and improve your product. A survey measure is an easily identifiable product, several process models exist for survey measurement (as noted in Section 20.4), and there exists a whole slew of quality measures—of the survey measure and of the process. Thus, there appears to be a natural link between survey measurement and process improvement.

On a practical level, why is process improvement not built into survey measurement automatically as *the* way of doing business? That is, generally speaking, why has it been necessary for many organizations with survey measurement operations to develop and implement a separate quality management program? Or, on an even more practical level, why, for example, are measures and procedures for monitoring nonresponse not in place for most surveys? If survey measurement and process improvement were integrated, the production of process improvement measures would be synthesized into the production of the survey measures, and separate quality measurement staffs would not exist.

Some would say the problem lies in defining quality measures. It is impossible to quantify mean squared error and very expensive to quantify response bias. Improvements in indirect performance measures like response rate, while easy and inexpensive to calculate, do not necessarily result in reduced mean squared error. Others would say the problem lies in costs; defining, collecting, and evaluating data about the survey measurement process do not come free. Program managers are in what they perceive as impossible situations. Their budgets are shrinking and production costs are going up. Still others say the problem is with the primary measurement goal—the survey measures themselves. If the product is a time series statistic, some customers may be more interested in maintaining the *status quo* than in improving the measurement process. In other words, a process that is in statistical control but imprecise is better than one which is more precise and, thus, has a different expectation.

Deming (1982) addresses each of these sentiments. To meet the needs of all the customers associated with survey measurement, including data users, respondents, and survey methodologists, the focus of management needs to be on the entire survey measurement process—from conceptualization through validation. Thinking about process control for implementation activities (e.g., frame creation, sample selection, interviewing, coding, etc.) alone is not sufficient. Nor does a single measure of product quality, e.g., mean squared error, provide sufficient information for improving the survey measurement process. It is the role of management working as a team to plan for the future and invest resources in research and education (including customers). The inability to specify today a direct link between a measure of process quality and mean squared error (or some subcomponent thereof) should not preclude the computation of the measure and investigations into potential relationships (Groves, 1990). Even simple models can provide important information for making design decisions (Linacre and Trewin, 1989).

The primary tool for measuring quality is the same as it was in 1940—a sample survey. A well-designed sample survey measurement process can provide better estimates than a census, and this is true for each level in the recursive process model of survey measurement. Moreover, Mahalanobis's interpenetrating samples to evaluate alternative methodologies and data quality are still cost effective. Variance estimation using some sort of repeated subsampling is common for ongoing surveys with complex sample designs. Interpenetrating samples are used to measure interviewer and response variances. Embedded experiments can provide information useful for evaluating alternative questionnaires, procedures, etc. (Fienberg and Tanur, 1989). Subsampling is used in statistical process control for monitoring data keying and coding. Spending the money up front for a properly designed test or experiment that permits the assumption of a process in statistical control will, in the end, be more cost effective than running the risk of obtaining inadequate or unusable data.

On a theoretical level, the integration of survey measurement and process improvement is more than just focusing on the process and having the tools to measure improvements to the process. Everyone associated with the survey measurement process must be aware that the ultimate goal of producing a statistic is to present all customers, including production staff inside the organization, with meaningful data, and that, for data to provide information, it must be accompanied by metadata. Thus, for example, requirements defined during conceptualization must include both data and metadata that will meet all customers' needs. Defining the product of the survey measurement process as a set of statistics or population estimates is far from sufficient. Planning to include estimated standard errors and a source and reliability statement in dissemination products is also insufficient. Adding nonresponse rates or a quality profile to publication goals is, of course, better, but still insufficient.

What about the internal customers? What about the customer who wants to see the specific algorithm for hot-deck imputation or the reasons for using a particular question wording?

For true integration of survey measurement and process improvement to occur, the scientific principles of statistical measurement must be applied to both data *and* metadata. Measures of error whether they be from the survey measurement or process improvement perspective are in and of themselves insufficient. The process of producing and analyzing metadata must be an integral part of the entire survey measurement process from conceptualization through validation.

ACKNOWLEDGMENTS

For my appreciation of the difficulties surrounding an integration of the concepts of survey measurement and process improvement, I thank my fellow BLS Quality Measurement Team members. For my appreciation of the vagaries in the English language as used in survey research, I thank Wesley Schaible whose insight was invaluable in preparing this chapter. Thanks also to Daphne Van Buren who tirelessly searched for an endless list of references and Dan Rope for his technological feats. The views expressed in this chapter are mine alone and do not necessarily reflect those of BLS.

REFERENCES

Bailar, B. A. (1983), "The Quality of Statistical Data," *Bulletin of the International Statistical Institute*, pp. 473–495.

Colledge, M., and March, M. (1993), "Quality Management: Development of a Framework for a Statistical Agency," *Journal of Business and Economic Statistics*, 11, pp. 157–165.

Dalenius, T. (1977), "Strain at a Gnat and Swallow a Camel: Or, the Problem of Measuring Sampling and Non-sampling Errors," *Proceedings of the Social Statistics Section, American Statistical Association*, pp. 21–25.

Dalenius, T. (1985), "Relevant Official Statistics: Some Reflections on Conceptual and Operational Issues," *Journal of Official Statistics*, 1, pp. 21–33.

Deming, W. E. (1953), "On the Distinction Between Enumerative and Analytic Surveys," *Journal of the American Statistical Association*, 48, pp. 244–255.

Deming, W. E. (1982), *Quality, Productivity, and Competitive Position*, Cambridge, MA: Massachusetts Institute of Technology.

Deming, W. E. (1987), "On the Statistician's Contribution to Quality," *Bulletin of the International Statistical Institute, Proceedings of the 46th Session*, 2, pp. 355–369.

Eisenhart, C. (1952), "The Reliability of Measured Values—Part I: Fundamental Concepts," *Photogrammetric Engineering*, 18, pp. 542–561.

Eisenhart, C. (1963), "Realistic Evaluation of the Precision and Accuracy of Instrument

Calibration Systems," *Journal of Research of the National Bureau of Standards—C. Engineering and Instrumentation*, 67C, pp. 161–187.

Fecso, R. (1989), "What Is Survey Quality: Back to the Future," *Proceedings of the Section on Survey Research Methods, American Statistical Association*, pp. 88–96.

Fienberg, S., and Tanur, J. (1989), "Combining Cognitive and Statistical Approaches to Survey Design," *Science*, 243, pp. 1017–1022.

Freedman, S. R. (1990), "Quality in Federal Surveys: Do Standards Really Matter?," *Proceedings of the Section on Survey Research Methods, American Statistical Association*, pp. 11–17.

Godambe, V. P. (1955), "A Unified Theory of Sampling From Finite Populations," *Journal of the Royal Statistical Society, Series B*, 17, pp. 269–278.

Groves, R. M. (1989), *Survey Errors and Survey Costs*, New York: Wiley.

Groves, R. M. (1990), "On the Path to Quality Improvement in Social Measurement: Developing Indicators of Survey Errors and Survey Costs," *Proceedings of the Section on Survey Research Methods, American Statistical Association*, pp. 1–10.

Groves, R. M. (1991), "Measurement Error Across the Disciplines," in P. P. Biemer, R. M. Groves, L. E. Lyberg, N. A. Mathiowetz, and S. Sudman (eds.), *Measurement Errors in Surveys*, New York: Wiley, pp. 1–25.

Groves, R. M., and Tortora, R. D. (1991), "Developing a System of Indicators for Unmeasured Survey Quality Components," *Bulletin of the International Statistical Institute*, 48th Session, 2, pp. 469–486.

Hahn, G. J., and Meeker, W. Q. (1993), "Assumptions for Statistical Inference," *The American Statistician*, 47, pp. 1–11.

Hand, D. J. (1993), "Data, Metadata and Information," *Statistical Journal of the United Nations Economic Commission for Europe*, 10, pp. 143–151.

Hansen, M. H. (1987), "Some History and Reminiscences on Survey Sampling," *Statistical Science*, 2, pp. 180–190.

Hansen, M. H., and Hurwitz, W. N. (1943), "On the Theory of Sampling from Finite Populations," *Annals of Mathematical Statistics*, 14, pp. 332–362.

Hansen, M. H, Hurwitz, W. N, Nisselson, H., and Steinberg, J. (1955), "The Redesign of the Census Current Population Survey," *Journal of the American Statistical Association*, 50, pp. 701–719.

Hansen, M. H., Hurwitz, W. N., and Pritzker, L. (1967), "Standardization of Procedures for the Evaluation of Data: Measurement Errors and Statistical Standards in the Bureau of the Census," *Bulletin of the International Statistical Institute*, 36th Session, pp. 49–66.

Hansen, M. H., and Madow, W. G. (1976), "Some Important Events in the Historical Development of Sample Surveys," in D. B. Owen (ed.), *On the History of Statistics and Probability*, Marcel Dekker: New York, pp. 75–102.

Hansen, M. H., Madow, W. G., and Tepping, B. J. (1983), "An Evaluation of Model-Dependent and Probability-Sampling Inferences in Sample Surveys," *Journal of the American Statistical Association*, 78, pp. 776–793.

Hanson, R. H., and Marks, E. S. (1958), "Influence of the Interviewer on the Accuracy of Survey Results," *Journal of the American Statistical Association*, 53, pp. 635–655.

Hansen, M. H., and Steinberg, J. (1956), "Control of Errors in Surveys," *Biometrics*, 12, pp. 462–474.

Juran, J. M., Gryna, Jr., F. M., and Bingham, Jr., R. S. (1979), *Quality Control Handbook, Third Edition*, New York: McGraw-Hill.

Linacre, S. (1991), "Approaches to Quality Assurance in ABS Business Surveys," *Bulletin of the International Statistical Institute*, 48*th* Session, 2, pp. 487–511.

Linacre, S. J., and Trewin, D. J. (1989), "Evaluation of Errors and Appropriate Resource Allocation in Economic Collections," *Proceedings of the Fifth Annual Research Conference*, U.S. Bureau of the Census, Washington, DC, pp. 197–209.

Mahalanobis, P. C. (1944), "On Large-scale Sample Surveys," *Philosophical Transactions of the Royal Society of London*, Series B, 231, pp. 329–451.

Moore, G. H., and Shiskin, J. (1967), "Indicators of Business Expansions and Contractions," National Bureau of Economic Research, Occasional Paper 103.

Morgenstern, O. (1963), *On the Accuracy of Economic Observations*, New Jersey: Princeton University Press.

Neyman, J. (1934), "On the Two Different Aspects of the Representative Method: The Method of Stratified Sampling and the Method of Purposive Selection (with discussion, pp. 607–625)," *Journal of the Royal Statistical Society*, 97, pp. 558–606.

Norwood, J. L. (1987), "What is Quality?," *Proceedings of the Third Annual Research Conference*, U.S. Bureau of the Census, Washington, DC, pp. 215–219.

Olkin, I. (1987), "A Conversation with Morris Hansen," *Statistical Science*, 2, pp. 162–179.

Olkin, I. (1992), "A Conversation with Churchill Eisenhart," *Statistical Science*, 7, pp. 512–530.

Random House (1980), *The Random House College Dictionary*, Jess Stein (ed.), New York: Random House.

Rosander, A. C. (1985), *Applications of Quality Control in the Service Industries*, New York: Marcel Dekker.

Shewhart, W. A. (1939), *Statistical Method from the Viewpoint of Quality Control*, Graduate School, Washington DC: U.S. Department of Agriculture.

Statistics Canada (1987), "Statistics Canada's Policy on Informing Users of Data Quality and Methodology," *Journal of Official Statistics*, 3, pp. 83–91.

Stephan, F. F., Deming, W. E., and Hansen, M. H. (1940), "The Sampling Procedure of the 1940 Population Census," *Journal of the American Statistical Association*, 35, pp. 615–630.

U.S. Bureau of Labor Statistics (1994), *Quality Measurement Model*, Washington, DC.

CHAPTER 21

Continuous Quality Improvement in Statistical Agencies

David Morganstein and David A. Marker
Westat, Inc., Rockville Maryland

21.1 INTRODUCTION

Over the last 15 years much has been written on improving quality in industrial settings. In the last five years articles and books have appeared discussing the application of quality improvement to governments (Osborne and Gaebler, 1993; Cohen and Brand, 1993; Gore, 1993). Some articles have examined general methods of quality improvement at statistical agencies (Colledge and March, 1993; Linacre, 1991), but these have generally not reflected systematic applications. Furthermore, successful quality improvement programs should incorporate basic statistical methods as well, and this could be expected especially in statistical agencies (Simmons, 1972; Morganstein *et al.*, 1993; Morganstein and Hansen, 1990; also, see Hapuarachchi *et al.*, Chapter 26).

Clearly, statistical agencies have always been concerned with quality. The literature contains many examples describing redesigned data collection methods to improve accuracy and better ways to communicate the accuracy (and errors) of reported data (Hansen *et al.*, 1955; Groves, 1989). Nevertheless, a systematic approach to improving quality throughout an agency has been missing. Building on Dippo (Chapter 20) in which the importance of process is emphasized, this chapter discusses a systematic approach in which the survey agency itself is viewed as a system of processes that produce a final product, the survey, data registries, or other products. (This newer approach contrasts with a more traditional quality control view that focuses almost entirely on measuring the quality of the survey product with little recognition that better control of the processes will improve the result.) In this chapter the authors

Survey Measurement and Process Quality, Edited by Lyberg, Biemer, Collins, de Leeuw, Dippo, Schwarz, Trewin.
ISBN 0-471-16559-X © 1997 John Wiley & Sons, Inc.

describe how statistical agencies can systematically identify and better control processes that produce statistical products.

Section 21.2 of this chapter describes a coherent plan for continuous quality improvement, which contains several detailed steps. Examples are provided from statistical agencies for many of the steps. Section 21.3 discusses current best methods (CBMs), one of the methods included in their spectrum of standards. The Colledge and March (Chapter 22) survey of 16 agencies identified standards that address primarily objectives and definitions. Their survey revealed few instances in which descriptions of best methods for day to day operation of survey processes were used widely. Section 21.3 includes a definition and a rationale for the use of CBMs, ways to develop them, and pitfalls to be avoided.

21.2 CONTINUOUS QUALITY IMPROVEMENT

In this section specific steps are discussed for implementing a continuous improvement plan in a survey agency. We have used this method with many service and manufacturing organizations as well as in our own survey company. Several of the service organizations were government agencies. Figure 21.1, a flow chart for the continuous quality improvement process, illustrates several important aspects of the plan and their relationships, beginning with the survey customer (either external or internal). It moves quickly from focusing on the product (the survey results) to understanding and working with the survey processes and is a never-ending loop that returns to the needs of the survey client. Note that these are key elements of total quality management (Cohen and Brand, 1993).

Before examining this plan in more detail, the following definitions are offered.

External customer	The outside requester or user of the data (Parliament, the Head of State, another agency, state or local government, the research community, etc.).
Internal customer	A department or individual within the same agency receiving the product or service of another department or individual.
Measurement capability	The mean squared error of the measurement process is small relative to the overall error requirements.
Process stability	A state where the process variation consists entirely of random components.
Process capability	The random variation of a stable process is smaller than the customer's requirements or limits for that variation. A stable process is incapable if it cannot meet the customer's variation requirements.

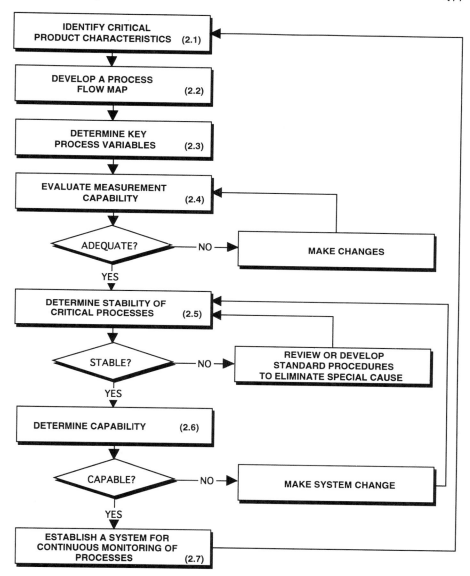

Figure 21.1 A plan for continuous quality improvement.

The first step of the plan focuses on the needs of the survey client (customer), identifying the survey characteristics most important to the client, and those aspects of the survey processes that are both important and controllable. Next, staff assess the ability to measure these processes satisfactorily. When adequate measurement capability is present, reliable data can be gathered to determine the stability of survey processes. If the processes are not stable, revised operating procedures or retraining may help to stabilize them. When the survey process is stable, the results can be compared with the customer's needs. If a process cannot consistently meet the customer's requirements, it is considered incapable, even if stable. If the process is incapable (e.g., the required response rate or level of precision cannot be consistently obtained or the data regularly reported by a specific deadline), changes to the system are required. This cycle of assessment and revision/retraining may be applied to all other activities of statistical agencies, not just the conduct of surveys. Below, each step in this plan is examined in greater detail.

21.2.1 Identify Critical Product Characteristics

One definition of quality is that a product or service needs or exceeds the customers' "*needs and expectations at a competitive price.*" In this definition, Deming (1982) suggests that the customer should be the primary focus for delivering any service or making any product. In his view, the system's objectives are defined by the needs of the customer. Defining these needs is made difficult because there are both internal and external customers. To meet all important objectives, a provider of surveys must understand the customer's requirements. A valuable initial step in identifying and understanding the customer's requirements is to distinguish carefully between broad needs that span many surveys and needs that are specific to a single effort.

Unfortunately, an individual customer's multiple needs may not be consistent. Many surveys and databases have multiple customers with divergent objectives. Whether the statistical organization is a private company, part of a university, or a government agency, understanding customers' needs and the organization's ability to resolve conflicting objectives will increase the likelihood of a successful product. In our experience, the earliest project activities are meetings with the (one or more) customers to establish priorities among conflicting survey goals that affect such critical elements as the design, time schedules and budgets.

In communicating with external customers to clarify goals for a specific survey, careful attention should be paid to the language used by survey customers and designers (Groves, 1989). For example, within one government agency the word *accuracy* was regularly used by different staff to mean both variance and mean squared error. Marker and Ryaboy (1994) point out that the term *accuracy* has been used to mean three conflicting concepts by environmental statisticians: bias, total (bias plus variance) error, and the error identified when conducting a performance evaluation audit (versus that identified by quality control samples). Without clarifying terminology, survey staff could design an "accurate" survey that may, or may not, meet the customer's goals.

A customer satisfaction survey (CSS) can provide both an overview of customers' broad needs and a review of past performance. A CSS can be used to determine the customers' definition of quality (i.e., which characteristics are truly important) and their perception of specific products and services. Another objective may be to assess customers' familiarity with the range of products and services (e.g., databases, consulting, etc.) that are available, and to identify other products and services that could appeal to customers. Surprisingly, customers may not even be aware of product or service characteristics they would value highly (e.g., 15 years ago automobile customers were not aware of the possibility of providing air bags).

21.2.2 Develop a Process Flow Map

The next activity in improving statistical products, developing a process flow map, yields a better understanding of the related processes. This step should result in far more than a superficial awareness of the activity details. Often, this step is at first rejected on the assumption that the process is already well understood. There is little doubt that those involved in the process have their own in-depth appreciation of the process details. However, based on our experience with dozens of organizations, when these impressions are compared, significant differences inevitably emerge (see Gross and Linacre, Chapter 23). All persons in the survey process view it from their own perspectives. In the absence of open dialogue, these differences will continue to exist, resulting in suboptimal processes that limit progress.

We suggest including three components in a process flow map. First, the sequence of processes is delineated, indicating decision points, the flow of the process, and the customer(s) for each step. Second, the owners of each process are identified, improving communication with the process suppliers and customers. Third, the key process variables (that is, those factors that can vary with each repetition of the process and affect critical product characteristics), decisions, or actions that can be taken by those involved in the process, are listed. Examples of process variables are questionnaire wording, staffing, sample design, and the content of an interviewer training program. Determining who owns each process and what the key process variables are requires that the staff involved in the processes participate in the development of the process flow map. The *process* by which the map is produced is as important as the final product.

As an illustration, a government statistical agency formed a quality improvement team (QIT) to attempt to reduce the time required to produce annual, national estimates of airborne emissions from two years to less than six months. The team members included those responsible for producing the estimates and those who supply the data on which the estimates are based. The team began by producing a process flow map. The process, as currently operated, included a tedious match of the names of power plants with their identification numbers found on a data tape. The team was able to eliminate this entire step when the supplier of the data, a team member, observed that the names and IDs existed on another data source. In the process of producing a process flow map, the team identified a simple improvement in existing processes that achieved their objective.

Figure 21.2 is an example of a flow chart. The goal of the team that developed

INCREASED FLEXIBILITY WITH REGIONAL STATISTICS PACKAGE

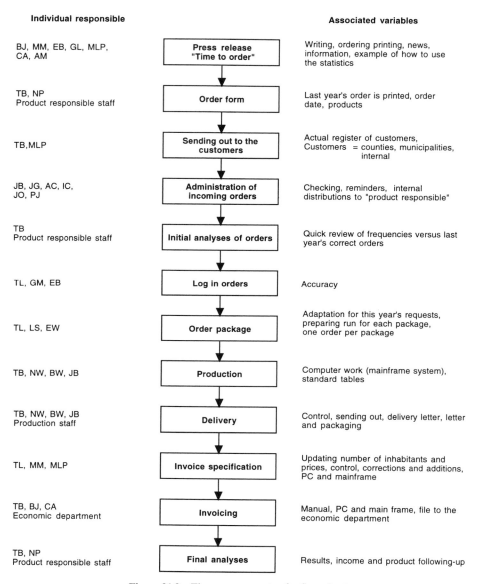

Individual responsible		Associated variables
BJ, MM, EB, GL, MLP, CA, AM	Press release "Time to order"	Writing, ordering printing, news, information, example of how to use the statistics
TB, NP Product responsible staff	Order form	Last year's order is printed, order date, products
TB,MLP	Sending out to the customers	Actual register of customers, Customers = counties, municipalities, internal
JB, JG, AC, IC, JO, PJ	Administration of incoming orders	Checking, reminders, internal distributions to "product responsible"
TB Product responsible staff	Initial analyses of orders	Quick review of frequencies versus last year's correct orders
TL, GM, EB	Log in orders	Accuracy
TL, LS, EW	Order package	Adaptation for this year's requests, preparing run for each package, one order per package
TB, NW, BW, JB	Production	Computer work (mainframe system), standard tables
TB, NW, BW, JB Production staff	Delivery	Control, sending out, delivery letter, letter and packaging
TL, MM, MLP	Invoice specification	Updating number of inhabitants and prices, control, corrections and additions, PC and mainframe
TB, BJ, CA Economic department	Invoicing	Manual, PC and main frame, file to the economic department
TB, NP Product responsible staff	Final analyses	Results, income and product following-up

Figure 21.2 Three components of a flow chart.

the flow chart was to increase the agency's flexibility in responding to customer requests. Down the center are the individual processes that are involved in producing a package of regional statistics. Down the left side of the figure is a list of the individuals responsible for each process. On the right side are variables

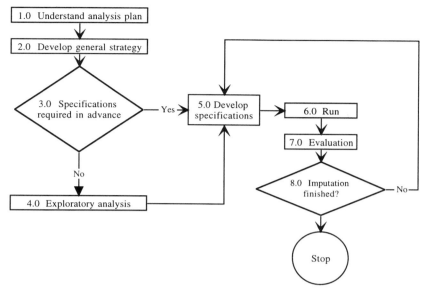

Figure 21.3   Macro-level flow chart of the imputation process.

that are associated with each of the processes using terminology understandable to those involved in the process.

In any flow chart, a balance is needed between adequate and excessive detail. Cohen and Brand (1993) describe a situation where managers were presented with a four-foot long flow chart with 97 steps. The managers' reactions were skepticism and a belief that the activity was a waste of time. Two lessons can be learned from this example: involve the managers in the development of a flow chart, and simplify the chart using macro- and micro-level charts. A macro-level chart (Figure 21.3) showing the process of imputation for non-response is useful for new staff to understand the overall process, while micro-level charts are needed to understand how the individual processes interact.

It is important to determine how part of the process is related to every other part, and how the process itself is related to other activities. Barriers are eliminated when employees understand where processes begin and end, as well as who is responsible for each component. QITs can use process flow maps to identify sources of variation in the system, and then determine whether variables reflecting those sources are currently being measured.

21.2.3   Determine Key Process Variables

We have found this step to be the most difficult. In a service activity, the processes are less tangible than in a production setting, and, therefore, more challenging to identify and to measure. The underlying processes are real, nonetheless, and they control the quality of the resulting data. Often the processes are in the form of decisions and the paths these decisions take. Survey processes include the relative amount of time and resources allocated to planning, testing, and training as compared to the actual data collection effort. Other processes include data entry, frame updating procedures, and procedures for checking publication tables.

The way in which a data collection instrument is designed is an example of a survey process. For example, instrument designers could choose to design a questionnaire entirely by themselves. Alternatively, they could choose to take the extra time to have the questionnaire reviewed by others who must use it, such as interviewers who will have to administer it. By taking this second approach, they are more likely to improve the quality of the questionnaire, and they may reduce the amount of re-work (editing, imputation, etc.) needed to identify and fix problems.

The goal of this step is to identify the critical elements that have the largest effect on process outputs, the products or service quality, and their consistency and cost. Here, staff begin to ask questions and document responses about each process element and variable. They use their collective knowledge and experience to select those processes most likely to affect quality. It is surprising how many of these critical aspects are not routinely measured, or not even measured at all.

One of the simplest, yet effective, methods for identifying key process variables is the Pareto diagram (Gitlow *et al.*, 1989). This tool is based upon the principle expressed by Vilfredo Pareto (Juran *et al.*, 1979) as sorting the "vital few versus the trivial many," that is, trying to find the relatively fewer error (or defect) types that account for the majority of all error (or defects). As an example, consider edit failure rates that some government statistical agencies track as a measure of survey quality. In the Pareto chart, Figure 21.4, error sources are arranged in descending order of failure frequency, making the identification of the *vital few* easy. The remaining *trivial many* edits are grouped together in a column entitled other. For this particular case, out of the 97 edit failures, over one-half were on question item B3. Eighty-six percent were for items B3, B2, and B5. The remaining edit checks accounted for only 14 percent of all failures. The Pareto chart enabled the staff to be more effective in deciding how to allocate resources. In this case the allowable ranges for items B3, B2, and B5 were re-evaluated so that the edit program only identified true outliers. To improve quality requires identifying the sources of edit failures and trying to minimize their future occurrence through improved question wording, revisions to allowable ranges, or some other method.

Pareto charts can be used to analyze data other than frequencies. Often the columns on the chart will represent the lost dollars or number of days lost that result from each source. When they are available, such measures of consequences are preferable to simple plotted frequencies. Unfortunately, it is often difficult to directly associate consequences with a single source.

It is also possible to identify key process variables when frequency or other types of data are not available. The cause-and-effect diagram (referred to as a

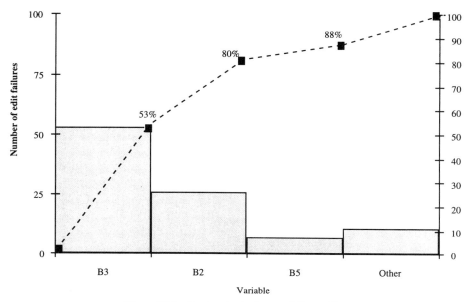

Figure 21.4   Pareto chart of range failure edits.

fishbone or an Ishikawa diagram) was designed for this purpose (Ishikawa, 1982). Figure 21.5 is a fishbone diagram describing the factors that affect development of a general strategy for imputation, developed by a QIT working on the imputation process. (Figure 21.3 was also a product of this QIT; note that "general strategy" is the second box in that figure.) From all of the factors on the fishbone, the team selects the five or six they believe to be most important. These are the factors to measure and whose variability should be reduced. Multiple factors are chosen because a knowledgeable team will generally be able to identify several important factors. However, if they rely on their best judgment to choose only one, they may select a relatively minor factor because there may be little, if any, objective basis for selecting only one. By picking several factors, they protect against such an unlucky occurrence. Once a small number of key factors have been chosen, the next step is to measure them over time. This is the topic of the next two sections.

21.2.4   Evaluate Measurement Capability

A common mistake is to collect data and reach conclusions about process stability without any knowledge of measurement error. Alternatively, researchers often select a process because it is easy to measure, rather than choosing a more important but harder-to-measure process. In our experience, measurement error is often one of the least appreciated aspects of quality improvement.

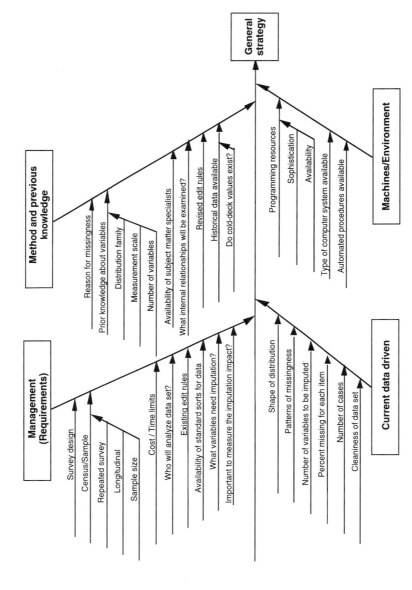

Figure 21.5 Cause-and-effect diagram of the development of a general strategy for imputation.

Customer satisfaction surveys commonly use limited (3- or 5-point) scales with the frequent result that almost all respondents select the same value (e.g., *very satisfied*). Such poor measurement systems provide little insight into correlates of satisfaction. Also, many surveys use scales that are only anchored at the extremes, thus losing information on the meaning of intermediate responses. In such cases the measurement error limits the use of the survey for the purpose of continuously improving quality.

As part of the redesign of the U.S. National Health Survey, the costs of activities related to personal interviewing were examined to help identify ways of reducing the in-person interviewing costs and to measure the effect on costs of proposed redesign methodologies. Cost data have been reported by interviewers for many years and entered into a computerized database. However, when the redesign staff conducted this examination they found that the recorded data would not support the desired analysis. Collecting the necessary data required modifying the existing interviewer procedures. Thus, the cost data were not capable of meeting the requirements of the redesign staff, and decisions on how to redesign the survey had to be made without accurate information on their effects on the cost associated with in-person data collection.

It is possible to improve the quality of processes but not be able to quantify the improvement. In the health survey example, it can be argued persuasively that the redesigned survey will improve the quality of the collected health data in terms of both accuracy and cost. However, not having adequate cost data makes it impossible to accurately measure the cost implications of the changes, and also prevents the redesign staff from using costs to assess various alternative designs. In addition, the inability to measure the extent of the improvement often makes it more difficult to gain authorization from senior management for quality improvement efforts.

21.2.5 Determine Stability of Critical Processes

Once key process variables have been identified, they are tested for stability; when survey processes are reasonably predictable (or stable), the variables can provide a basis for comparison after improvement efforts are initiated. A systematic way to acquire and analyze data is needed to bring about stability of the critical process variables. This is accomplished using, primarily, control charts, Pareto analyses, and various other statistical tools and methodologies. At this step, the sources of process and product variations and the effects on variation of changes to the process are identified.

Process variability is commonly classified into two types: *special cause* and *common cause*. Deming renamed Shewhart's original definition of *assignable cause* to *special cause* (Deming, 1982; Shewhart, 1931). In some cases, a specific, identifiable problem may be the cause of an unacceptable product or service or an unacceptable process condition. For example, a single batch of questionnaires may have been improperly coded by one individual or a defect in a diskette may cause an error in records of a file used as a sampling frame. Raw materials (e.g., paper or ink quality used to produce questionnaires or cover letters) may be out of specification. For each of these sources of variation, known as special causes, a specific one-time, local action may be enough to remedy the situation. Success in eliminating special causes of variation leads to a reduction in process variability and increased survey reliability.

In contrast, common cause variation permeates all processes. According to Deming (1993), "Common causes of variation produce points on a control chart that over a long period all fall inside the control limits. Common causes of variation stay the same, day to day, lot to lot." Poor recruiting and training practices are systemic and apply to all staff on all surveys. Out-of-date procedures and poor measurement capability affect all studies. Local problem-solving efforts cannot address common cause variation. Deming (1982) has said that "action on the system" to reduce common cause variation is the responsibility of management.

Many special cause examples can be permuted into evidence of a common cause. Poor frame data could have slipped through on one occasion (a special cause), but it could be typical of a process that had inadequate provisions for inspecting the frames (a common cause). An interviewer not following a skip pattern may have been troubled by personal problems (resulting in one-time inattention, a special cause), but if such slip-ups are common, they may indicate the need for better training or for identifying skip patterns for each new questionnaire (a common cause).

Juran *et al.* (1979) and others (Gitlow and Gitlow, 1987) estimate that only about 15 percent of process variation is due to special causes, while the remaining 85 percent (the majority of variation) is due to common causes. Juran suggests that most organizations incorrectly spend at least 50 percent of their efforts on what they consider to be special causes (e.g., reviewing memos by junior staff).

Responsibility for correcting common causes and special causes rests on different staff; therefore, an undesirable consequence is prevented by distinguishing between the two sources of variation. Identifying what went wrong at a given time is only the first step. The often overlooked second step is to distinguish a special cause from a common cause and, subsequently, to decide what staff must take what action. Making these distinctions is a primary reason for using control charts (Rosander, 1985). Simmons (1972) discusses using a control chart to compare errors for each cycle of a survey in order to determine whether a local action by operating staff is needed or whether management action to change the system is the correct response.

Figure 21.6 is a control chart from a random digit dialing telephone screening survey collecting environmental data. The chart shows the percentage of telephone numbers for which screeners were completed by county, a key process variable. Too few completed screeners would have required additional and unexpected telephone effort to identify the desired number of eligible households, delaying the conduct of the full survey in that county. A low yield also

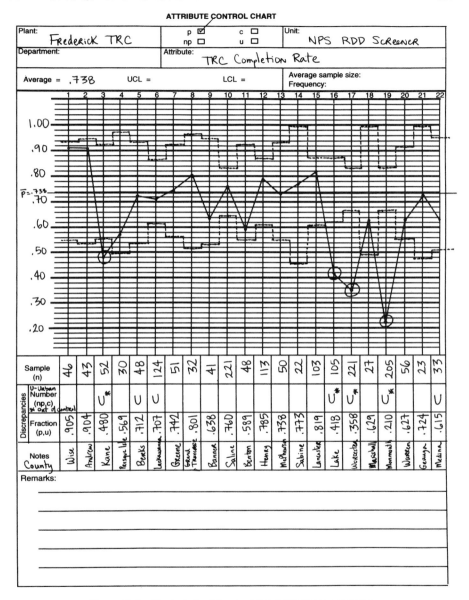

Figure 21.6   Control chart for completed telephone screeners.

could introduce a potential source of bias if the additional telephone numbers to be screened were not allocated the same amount of time for completion as were the initial telephone numbers.

The solid dark line plotted on the grid indicates the observed completion rate, while the two dotted lines indicate the bounds on the observed completion

rate assuming random variation among counties all have the same underlying rate. These are the control limits (which vary by county due to
varying sample sizes). (Using standard control chart methodology, these
upper and lower control limits are set at ± 3 standard errors from the overall
mean. Thus for a stable process, only 1 in 400 observations is expected to
exceed these limits.) As soon as any county was completed, the project staff
plotted the observed screening rate. After several dozen counties, but not all,
were completed, the control limits were determined. The team sought the reason
four counties (on the chart) had unexpectedly low completion rates. As noted
near the bottom of the chart, most of the counties in the plot were classified
as rural, with only seven classified as urban (each of these is indicated by a U).
Project staff noted that all four of the unusual counties were urban and
modified the process for all urban counties, increasing the number of telephone
numbers introduced into the telephone screener. Figure 21.7 shows the same
data just for rural counties. The process mean and control limits have been
recomputed to describe a distinct process for rural counties. Note that the
completion rates for these rural counties exhibit no unexpected variation; that
is, they are from a stable process. Use of the control chart allowed potential
problems to be anticipated and corrective actions to be planned in advance
with the customer.

21.2.6 Determine System Capability

For processes that are reasonably stable, staff can evaluate the capability of
the process to predictably meet specifications or to determine the limits of the
expected process variation. System changes to reduce variation are needed for
processes that are not capable of meeting specifications. After management
makes changes to the system, it is necessary to re-evaluate the stability of the
critical process variables before determining the new process capability.

Examples of process specifications are minimum response rates, production deadlines, and maximum coefficients of variation. The variation in key
survey processes can be reduced through adherence to standard procedures,
referred to herein as current best methods (CBMs). Section 21.3 defines
CBMs, and discusses ways of increasing the likelihood that they are understood
and used.

21.2.7 Establish a System for Continuous Monitoring of Processes

Achieving reliable and capable processes is only the beginning of the improvement process. A process found to be capable in the past should not be expected,
if left unattended, to remain capable. Technology becomes obsolete, new
methods are implemented by suppliers, supervisors and employees make
unknowing and inadvertent errors, and customer requirements change. All the
while, the objective of continuous, never-ending improvement remains. Thus,
a monitoring system is needed to keep staff informed, to provide feedback, and

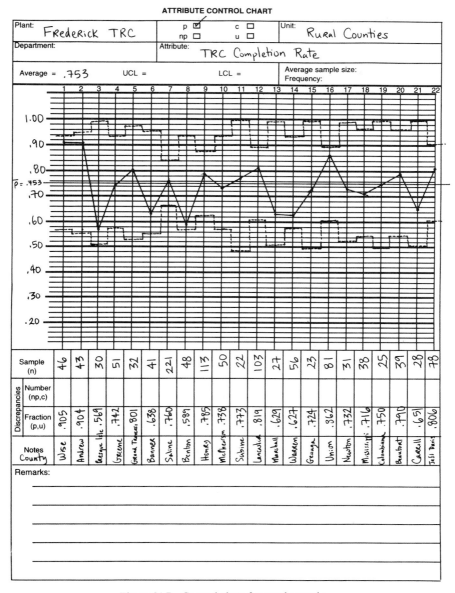

Figure 21.7   Control chart for rural counties.

to assist in controlling process variation, seeking continuous reduction in process variation through improved methods, and evaluating process changes. Current best methods (CBMs), as discussed in the next section, help minimize variation, thereby improving quality, increasing the likelihood that best practices are followed, and making monitoring processes easier.

21.3   THE ROLE OF CBMs IN THE IMPROVEMENT OF QUALITY

"It goes without saying that it will be impossible to obtain continuous improvements if best practice is not standardized and communicated to everybody concerned" (Kristensen *et al.*, 1995). Reduction of variation is central to improving quality. One of the most frequently identified sources of variation is the difference in performance among people assigned to do the same task, for example, in operations where the variation among interviewers and coders has been studied and reported on (see Batcher and Scheuren, Chapter 25). The variation among process designers developing questionnaires, sample designs, imputation and estimation procedures, etc. is less appreciated. Gross and Linacre (Chapter 23) give an example where, within the same statistical agency and for a single business register, different procedures for estimating the proportion of establishments that no longer have employees resulted in estimates used in separate surveys ranging from 31 percent to 44 percent. In this section we discuss the role of current best methods in reducing the between-person variation present in many aspects of statistical operations.

Our vision of a current best method (the reader may be familiar with the terminology *Standard Operating Procedure*, *SOP*) contains several components. One component of CBMs is documentation of procedures, discussed in more detail later. A second component of CBMs, of equal importance to the physical documentation, is management's approach (the process of developing CBMs). By conveying to the staff that continual improvement is valued, management establishes important objectives. Providing the resources needed to involve the staff in developing best practices, to document these practices, and to improve them continually develops a competitive strength in the agency. Finally, by rewarding adherence to agreed upon procedures, management reinforces actions that lead to reduced variation. Thus, the *process* of implementing CBMs may equal or exceed the value of the tangible product, the written documentation.

CBMs do not limit the creativity and innovation of employees. To the contrary, by clarifying the *current* best method, it is easy to compare it against proposed alternative procedures. If an alternative is found to be superior, then it becomes the new CBM. The place for creativity and innovation is in the development of CBMs.

Most staff want tools that help them *get it right the first time*. Re-work is a frustrating and unrewarding activity. CBMs provide a method for reducing the amount of re-work in survey operations. As reported by Batcher and Scheuren (Chapter 25), both staff and management at the U.S. Internal Revenue Service found important value in the development of and adherence to a *standard probe and response guide* for interviewers.

21.3.1   Choice of the Term CBM

The term current best method conveys a different message than the phrase standard operating procedure. The former term communicates the notion that methods are constantly improving. The ones currently documented are the best now known; however, in the future, better ones can be expected. The term CBM implies a constant effort to make improvements to the method and to operate by using the CBM, not allowing each person to decide individually how to run current operations. The acronym SOP conveys a sense of the static. It has been reported to us that the acronym SOP both raises fears of loss of individuality and fails to support the philosophy of a team approach to constant improvement.

21.3.2 Two Kinds of CBMs: Repetitive Processes and Creative Processes

CBMs can be classified into two broad categories based upon the kind of task they summarize, jobs that are repetitive and predictable or those that are unique and creative. CBMs for the former tasks can be summarized easily with a checklist (one part of a CBM) that must be followed in exact sequence for each execution of the operation. Airplane pilots, for example, follow this kind of linear CBM. Each item in the checklist is examined in the same order every time the plane is prepared for take-off or landing. In the case of survey operations, procedures for selecting (not designing) a sample from a frame fall into this class, as does creating and updating a household register, though the procedures can be quite complex. A receipt control operation is likely to be of this type, as well.

The second class of CBMs applies to less predictable processes such as those performed by physicians or scientists. Highly trained staff in a statistical agency may initially think of themselves as too creative to benefit from the use of checklists. While many such professionals may think that each task they embrace is so unique that no CBM could be relevant, they could not be farther from the truth. In some ways, such occupations need CBMs all the more because their processes are so unique, nonlinear with many decision branches, that it is very easy to forget an important element. For example, numerous critical steps are required in designing or redesigning a survey. A CBM serves as a vital reminder of factors that must be addressed. Certainly professionals who engage in short time-frame operations, such as implementing a government survey, need all the tools they can find to help prevent overlooking critical steps in complex processes.

Both kinds of processes require similar procedures to develop CBMs, as described in Section 21.3.3. Both kinds of CBMs should be regularly updated as better methods become known. CBMs for repetitive processes clearly define a sequence of steps to be followed; whereas CBMs for creative processes are more likely to define the factors that should be addressed, not how to address them.

21.3.3 Preparing CBMs

21.3.3.1 Role of a Guidance Team

Before beginning work on CBMs, it is advisable to identify a senior management guidance team who will take responsibility for the success of developing and using CBMs in the agency. The CBM guidance team may be the group with agency-wide quality improvement responsibilities or it may be a separate team that reports to the overall quality team. For a large statistical agency, it may be advisable to have a guidance team for both the entire agency and also for each department. While their roles, as outlined below, are similar, the agency-wide group is also responsible for developing and coordinating the work of the department groups.

Taking Overall Responsibility

The guidance team of senior managers and staff facilitates the CBM development because many impediments to change can only be overcome by those with authority to define the agency's mission. Most statistical organizations are capable of producing their routine product, but developing CBMs and integrating their use into the daily activity of the agency is a very different process. These activities are not likely to occur spontaneously; they require an investment of time and effort best directed by agency management.

The guidance team identifies an appropriate role for everyone in the agency who will be affected by the CBM. They address issues of communication, of solicitation of opinions; they also review the process for bringing about acceptance, increasing familiarity, and changing behavior. The guidance team considers and plans this process.

Selecting Participants

The guidance team selects and supports each CBM team, occasionally resolving conflicts of time and resources. The guidance team reviews progress and makes changes in work assignments or due dates. The team members usually are not empowered to make decisions about staff resources available to support CBM development, staff training, and compliance monitoring. Typically this can only be done by senior management representatives on the guidance team.

Planning for Change

Specific actions assist and promote changes in performance. Staff input into the selection of topics for the development of CBMs is a first step. Staff-wide review of draft documents helps accelerate their acceptance of a new procedure. Some form of training of those involved in the process is needed when the CBM is available. Again, the guidance team ensures that plans are made for this transition.

Promoting Adherence

By supporting the use of CBMs, management promotes adherence to the new methods because the availability alone of a better method is not sufficient to cause its adoption. Staff typically will use tried and true procedures unless encouraged to change. The guidance committee develops strategies for rewarding those who make the change and for identifying those who do not. Some people will change methods more quickly; others need encouragement, training, and leadership from management to adopt the improved methods.

21.3.3.2 CBM Development Process

The steps shown in Figure 21.8 suggest a careful CBM development process. The left-hand column identifies who is responsible for each step in the development of CBMs. The right-hand column identifies the key process characteristics that should be considered at each step. The flow chart displays steps for systematically developing CBMs used to ensure quality in an agency: (1) identify processes vital to the mission of the group that will benefit from widespread staff use of a CBM; (2) select the team who will write the CBM; (3) develop the actual CBM documentation; (4) communicate the details of the CBM to the staff; and, (5) put into place a plan to help ensure adherence of the procedures. These steps are now described in more detail.

Identifying Vital Processes

As discussed in Section 21.2, although there are likely to be innumerable processes in each department of a statistical agency, CBMs should be developed only for the vital few. The guidance team selects a process that is absolutely critical to completing the mission of the department. In a department charged with maintaining frames or registers, for example, critical processes include procedures for updating entries and procedures for eliminating duplicate entries. In a sampling department, vital processes are sample selection, weighting of survey results, and imputation. In a data entry department, examples might include rules for data retrieval or procedures for resolving conflicting codes. When too many processes are selected, the effort required to apply CBMs may exhaust the staff.

To some degree, the guidance team should involve the entire department in the selection process, even if staff involvement is limited to contributing suggestions. Widespread acceptance and implementation of new methods depends upon the support of almost all staff members. The entire department should be made aware that ideas for CBMs are being collected and that a team will be assembled to document the selected processes. They should know that their ideas and comments on written drafts are important. As an agency integrates continuous quality improvement into its basic way of operating, the role of CBM process selection naturally shifts towards the staff involved in the processes. Initially, management takes more of a leadership role in selecting processes, but its role shifts to ensuring adequate resources as the quality effort matures.

Selecting the CBM Team

A CBM team is often a continuation of an existing quality improvement team that has been working on identifying sources of variation, eliminating them, and thereby determining the process whose best practice is to be documented. A small team comprising both experienced and new members brings an important balance to the development process. Those staff who have dealt with the department's problem will know about challenging difficulties and some methods for dealing with them. In contrast, new members provide a fresh

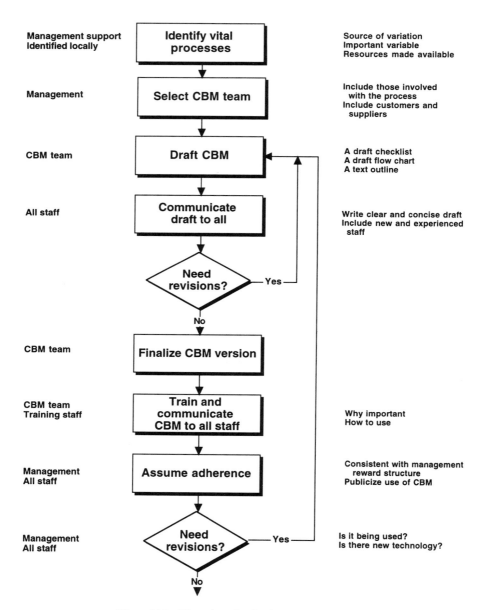

Figure 21.8 Flow chart for development of CBMs.

perspective, sometimes unencumbered by the blinders of familiarity. Documents written entirely by the most sophisticated and knowledgeable may be unreadable by those less experienced. It is also helpful to include representatives of the internal suppliers and customers of this process on the team. At the same time, a team that is too small may be overwhelmed by the task of preparing the documentation and one that is too large may be paralyzed by the diverse view points and differences.

Recognizing each staff member's role in different aspects of the CBM development is directly related to the degree of benefit the CBM will yield. Ideas for new CBMs or revisions to existing ones must flow from many members of the department. Ideally, all staff members contribute; however, in practice, this does not occur. The perception of openness, though, is related to acceptance. Staff who feel that they may contribute are more likely to change old practices and follow a new procedure than are staff who feel excluded from the process.

We are familiar with agencies that have developed CBMs quickly by giving the task to a single, technically knowledgeable individual. In some cases, the individual was a part of a technical support group whose job was to provide technical assistance to the agency. In every such case, the resulting document was carefully written, reasonably thorough and may even have passed the requirements of standards for organizations such as ISO-9000. However, in choosing one knowledgeable individual and thus a short development time, these agencies were inadvertently overlooking the degree of acceptance the document was likely to experience. In every such case, the CBM was used rarely and by few of the agency's staff. Documents developed without consensus, written with input from few staff members, or written by people outside the agency are unlikely to be used widely.

Drafting the CBM
Although the written documentation can take many forms, two important components are recommended: a summary and a detail section. The summary may be in the form of a checklist or a flow chart and provides an overview of the critical steps of the process, presented in chronological order. Figure 21.9 is a checklist for single-stage list sampling, with references to the detailed text sections. The detail sections provide the background and justification for these steps. The details may include many options and guidelines for selection among them. The summary is useful for all staff members, old and new, while the detail is of most value to new members or less experienced staff. The written documentation should provide enough detail to enable someone unfamiliar with the activity to perform the required tasks.

In Section 21.2 the value of flow-charting critical processes of the operation was discussed. Flow charts or checklists should clearly identify who participates in decision-making (e.g., the left-hand column of Figure 21.8). In many cases, this may be more than one individual; however, improvements in efficiency will result when clear lines of decision-making are agreed to and documented. Flow

SINGLE STAGE LIST SAMPLING CHECKLIST

Project number _____ Date started _____ Name of project _____
Designer _____ (List other members of the team, if any)

Text ref.	Action or decision	Chk	Notes
	SAMPLE DESIGN		
1.1	Identify frame(s)		
1.1.1	Determine adequacy of coverage		
1.1.2	Develop plan for duplicates		
1.2	Decide whether (and how) to stratify		
1.2.1	Choose stratifiers		
1.2.2	Choose number of strata		
1.2.3	Allocate sample to strata		
1.2.4	Determine effect of strat. on var. estimation		
1.3	Decide whether to sort		
1.3.1	Choose sort variables		
1.3.2	Establish class intervals (if necessary)		
1.3.3	Set sort order		
1.4	Determine if MOS other than count		
1.4.1	Identify MOS		
1.4.2	Choose minimum MOS		
1.5	Identify certainties		
2.1	Preliminary specifications to programmer		
2.2	Written instructions to programmer		
2.2.1	Define eligible records		
2.2.2	Choose way to handle missing fields		
2.2.3	Complete WESSAMP specifications		
2.2.4	Specify random starts		
2.2.5	Specify selection if MOS to be used		
2.2.6	Specify target sample size or rate		
2.2.7	Document to permit reproducing selection		
2.2.8	Select substitutes if subs are to be used		
2.3	Develop plan for QC verification		
2.3.1	No sample units below minimum MOS		
2.3.2	Frequency counts by strata and sub-domains		
2.3.3	Sum MOS by strata and overall		
2.3.4	List a portion of the frame with sample indicated		
2.3.5	Examine weighted counts		
3.	**DOCUMENTATION**		
3.1	Complete design with modifications		
3.2	Sample selection		
3.3	Memos and notes		
3.4	Worksheets		
3.5	File prepared for archives		

Figure 21.9 Single-stage list sampling checklist.

charts also identify the owners of various subprocesses. A typical example would be the owner of a report that is routinely generated. Often the work of a CBM team uncovers reports without owners (or reports are found for which there are no longer any customers) and other actions that were routine but no longer have value to the department.

Once the team has drafted a CBM, it is vital that the draft (checklists, flow charts and detailed backgrounds) be circulated to all other staff who are involved in the given process. Their input on the draft CBM is vital for two reasons: based on their knowledge of the process, they may discover preferable alternative procedures, and for the CBM to be effective, all staff must agree that the CBM indeed describes the best way to operate the process. When all staff are involved in the development, they have a feeling of ownership in the CBM that helps ensure its utility. The CBM team discusses the recommendations from other staff and revises the document, repeatedly if necessary, until a CBM is developed that is agreeable to all affected staff.

Communicating and Training Staff

Although preparing documentation for a vital process is challenging and requires a significant commitment of effort, this task alone cannot change the way staff carry out their jobs. Communicating and training staff in the objectives of developing CBMs and their importance to operations are equally important. Following document preparation, training sessions for staff can help to address questions and discuss details.

The results of these communication activities have benefit beyond the parts of the agency directly involved with the development of the CBM. They help to break down barriers and develop an agency-wide team spirit, particularly if the CBM teams include both customers and suppliers of the process. As staff learn about CBMs from other programs (e.g., for updating frames or registers), they may find assistance in resolving some of their own challenges. In this way, the quality of the entire agency improves.

Assuring Adherence

Obtaining feedback on staff use of the CBM documents ensures that the CBM is relevant and *current*. This cannot be accomplished without measuring its usefulness. For this, an informal process of *auditing* is recommended, unless legal/contractual requirements for adherence are imposed. Staff should be assured that the primary purpose of the monitoring is to assess and increase the value of the procedure to them. The objective of monitoring is not punitive.

Some examples of methods for such informal audits are: ask staff periodically whether they find the CBMs useful, have staff make internal presentations on the use and limitations of the CBMs, and encourage staff who believe the CBMs are not appropriate for a given application to explain why they propose not to follow the CBM (e.g., new technology has made the CBM out-of-date). By such methods, the use of CBMs is encouraged, and any needed revisions to the CBMs also become clarified.

Members of the CBM team are well suited to performing an adherence check. Often the team is charged with deciding when revisions are required, in which case monitoring usage is extremely helpful in the process. Managers should be involved in this monitoring as well to convey the message that using the CBM is important to the department. Managers also are responsible for examining their reward structure to ensure that it encourages the use of CBMs by their staff (e.g., is serving on a CBM team considered a positive activity or an interference with other work they are expected to complete).

21.3.4 CBMs: A Management Tool

CBMs can be powerful tools for promoting a constant improvement culture and managers may take many actions to accelerate the development and use of CBMs. Rewarding staff for that behavior is one. Management can provide many incentives (i.e., for public recognition, advancement, financial rewards, etc.) for staff to promote the use of CBMs to guide process operations. If management changes the current environment to one that encourages staff to dedicate a portion of their time to the constant evolution of procedures, and participates actively in the process, asking new questions during the process, staff will know they have support and understanding in using CBMs.

As mentioned earlier, the presence of documented CBMs is not enough to bring about improvements in quality; the development process is equally important. The right people need to be involved in the preparation and revision of CBMs. If the process owners are excluded from the CBMs' preparation process (e.g., by hiring an outside "expert" to write CBMs), then the chances of the CBM being used are greatly reduced. In addition, a system for internal auditing of adherence to the CBMs helps ensure that the written CBM matches staff practices.

One measure of the effectiveness of CBMs comes in the form of feedback from new members of a statistical department. New employees coming from other organizations that do not offer CBMs comment on the speed and ease with which they are able to participate fully in teams, to learn procedures and to prepare specifications. The indication is that the CBMs were the critical element in this process. In another case, CBMs facilitated staff agreement on a set of procedures to follow for imputation of missing data. The development of CBMs identified improvements necessary for standardized software. The use of CBM checklists has reduced the chance of accidentally omitting key steps in processes such as sampling and weighting of survey data. The detailed CBM description of why certain procedures are followed have allowed new staff to understand and appreciate the need for the recommended processes. External clients have used these detailed descriptions to better understand the processes being implemented.

21.4 CONCLUSIONS

As the survey conducted by Colledge and March (Chapter 22) indicated, statistical agencies have made quality an important element of their missions. Technological innovations and improved technical methods emerge continuously. Modern approaches to organizing and managing, however, can add new impetus to quality improvement efforts. Many agencies deviate from the suggestions presented in this chapter in two important ways: they do not use a structured plan with new tools, such as the development of written CBMs, and they do not have a *systematic* approach to continuous quality improvement guided by top management with the involvement of all staff throughout the agency. Synthesizing a new and better procedure than is currently used by any one staff member, although a time-consuming process, is one step that can yield improved efficiency for the entire agency.

Changing the structure of an agency so that it continuously strives to improve quality is not easy. It is likely to take one or two years to modify the actions of management. Only then can staff throughout the agency be expected to approach quality improvement in a systematic way. Thus, such a change involves a multi-year commitment of staff time, money, and above all, the personal involvement of management. These components are needed before an agency can incorporate the goal of continuous quality improvement into all of its basic operations.

REFERENCES

Cohen, S., and Brand, R. (1993), *Total Quality Management in Government*, San Francisco: Jossey-Bass Publishers.

Colledge, M., and March, M. (1993), "Quality Management: Development of a Framework for a Statistical Agency," *Journal of Business and Economic Statistics*, 11, pp. 157–165.

Deming, W.E. (1982), *Quality, Productivity, and Competitive Position*, Cambridge, MA: Massachusetts Institute of Technology.

Deming, W.E. (1993), *The New Economics for Industry, Government, Education*, Cambridge, MA: Massachusetts Institute of Technology, Center for Advanced Engineering Study.

Gitlow, H.S., and Gitlow, S.J. (1987), *The Deming Guide to Quality and Competitive Position*, Englewood Cliffs, NJ: Prentice-Hall, Inc.

Gitlow, H.S., Gitlow, S.J., Oppenheim, A., and Oppenheim, R. (1989), *Tools and Methods for the Improvement of Quality*, Homewood, IL: Irwin.

Gore, A. (1993), *National Performance Review*, Washington, DC: U.S. Government Printing Office.

Groves, R.M. (1989), *Survey Errors and Survey Costs*, New York: Wiley.

Hansen, M.H., Hurwitz, W.N., Nisselson, H., and Steinberg, J. (1955), "The Redesign of the Census Current Population Survey," *Journal of the American Statistical Association*, pp. 701–719.

Ishikawa, K. (1982), *Guide to Quality Control*, White Plains, New York: Asian Productivity Agency, Kraus International Publications.

Juran, J.M., Gryna, Jr., F.M., and Bingham, Jr, R.S. (1979), *Quality Control Handbook*, Third edition, New York: McGraw-Hill.

Kristensen, K., Dahlgaard, J.J., and Kanji, G.K. (1995), "TQM Leadership," in G.K. Kanji (ed.), *Total Quality Management, Proceedings of the First World Congress*, London: Chapman & Hall.

Linacre, S.J. (1991), "Approaches to Quality Assurance in ABS Business Surveys," *Proceedings of the ISI, 48th Session*, 2, pp. 487–511.

Marker, D.A., and Ryaboy, S. (1994), "The Quality of Environmental Databases," in C.R. Cothern, and N.P. Ross (eds.), *Environmental Statistics, Assessment, and Forecasting*, Boca Raton, FL: Lewis Publishers.

Morganstein, D., Brick, J.M., Burke, J., Cantor, D., and Judkins, D. (1993), "Statistical Research in a Large-Scale Private Survey Research Agency," *Journal of Official Statistics*, 9, 1, pp. 233–243.

Morganstein, D.R., and Hansen, M. (1990), "Survey Operations Processes: The Key to Quality Improvement," in G.E. Liepins, and V.R.R. Uppuluri (eds.), *Data Quality Control*, New York: Marcel Dekker Inc., pp. 91–104.

Osborne, D., and Gaebler, T. (1993), *Reinventing Government*, New York: Penguin Group.

Rosander, A.C. (1985), *Applications of Quality Control in the Service Industries*, New York and Basel: ASQC Quality Press.

Shewhart, W.A. (1931), *The Economic Control of Quality of Manufactured Product*, Van Nostrand. Reprinted in 1981 by the American Society for Quality Control.

Simmons, W.R. (1972), "Operational Control of Sample Surveys," Manual Series, No. 2, Chapel Hill, N.C.

CHAPTER 22

Quality Policies, Standards, Guidelines, and Recommended Practices at National Statistical Agencies

Michael Colledge
Australian Bureau of Statistics

Mary March
Statistics Canada

22.1 INTRODUCTION

This chapter deals with policies, standards, guidelines, and recommended practices concerning quality—"quality practices" for short—in government statistical agencies responsible for collecting, compiling, analyzing, and publishing statistical information. Based on a survey of 16 national statistical agencies around the world (see Appendix at end of chapter), this chapter summarizes existing quality practices and discusses their origins and effectiveness. In this context, "quality" refers to the relevance, accessibility, accuracy, timeliness, and cost of statistical products and services from the point of view of an agency's clients, and to the ease, cost, and confidentiality of data collection from the perspective of respondents. In other words, quality is interpreted in a broader sense than that of data accuracy alone. On the other hand, the focus is on practices that are of an essentially statistical nature, and, in some sense, unique to statistical agencies. Thus, the chapter does not attempt to cover the full spectrum of activities associated with quality management, in particular

Survey Measurement and Process Quality, Edited by Lyberg, Biemer, Collins, de Leeuw, Dippo, Schwarz, Trewin.
ISBN 0-471-16559-X © 1997 John Wiley & Sons, Inc.

excluding planning, organizational, and personnel related practices which, although figuring prominently in Total Quality Management literature, are widely applicable to statistical and non-statistical agencies alike.

Quality practices are viewed as lying on a continuum, from "policies" which are rigidly enforced through "standards" and "guidelines" to "recommended practices," which are optional. More precisely, the following terminology is used.

Policy:	A directive from agency management, which may be required by law or self-imposed by the agency, and which should be followed without exception.
Standard:	A practice that is followed almost without exception. Deviations are not recommended and require the approval of senior management. Corrective action should be taken when a standard is not being met.
Guideline:	A practice that should be followed unless there are good documented reasons for not doing so.
Recommended Practice:	A practice promoted by an agency as being consistent with its objectives, but not obligatory, and from which deviations are allowed without requiring documentation or justification.

There is considerable interest amongst statistical agencies in quality practices. They are seen as an important mechanism for improving the quality of statistical processes and products in all aspects, particularly those which require co-ordination and harmonization, and for supporting economies of scale. Quality practices can help, for example, in ensuring that products are relevant to client needs; that there are no gaps or duplication in the product range; that quality is defined in client terms; that agency data holdings are clearly visible and readily accessible in whatever format clients want; that statistics from separate survey operations are mutually consistent; and that costs are minimized by efficient internal sharing of resources and data. In addition, quality practices can enhance clients' perceptions of an agency's processes and products by providing quality assurance, thereby contributing to the agency's reputation. To cite a U.S. National Agricultural Statistical Service (NASS) (1983) report on long-range planning, "The first building block for all future activities of the statistical reporting service must be statistical standards."

There is, however, a potential downside to quality practices in so far as they imply standardization. From an individual client's perspective, quality standards may not coincide with the client's particular requirements, such as statistics for "non-standard" geographical areas. If quality is to be defined in client terms, in accordance with modern quality management theory, introduction and enforcement of quality practices must be balanced against the need to make allowance for individual clients. Viewed from within a statistical agency, enforced quality practices may be perceived by staff as involving unnecessary bureaucracy and extra work with no obvious benefits, as they support global rather than local optima and reduce the scope for individual decision making. The same NASS (1983) report noted that "one man's standard is another man's barrier," and, although made in reference to employee work quotas, one of Deming's famous 14 points referring to elimination of work standards is also indicative of the need to strike a balance between quality practices and allowing for individual initiatives.

Weighing the potential benefits against the implementation problems leads statistical agencies to be ambivalent in their approach to quality practices. For example, Statistics Canada receives numerous requests from other agencies for its *Quality Guidelines* (Statistics Canada, 1987) and, within the agency itself, the document is considered important, at least as evidenced by a 1993 survey of directors. Yet the same survey indicated that hardly anybody referred to *Quality Guidelines* on a routine basis. This example is typical of the attitude that exists towards quality practices—they are greatly sought after, but often not used.

22.2 QUALITY PRACTICES SURVEY

A nonprobability sample of statistical agencies was selected. Twenty-one questionnaires were mailed and 15 completed responses were received. There was no follow-up of the nonrespondents. Subsequently, two agencies responsible for statistical regulations were also contacted, one of whom provided additional information. For each existing, documented quality practice, respondents were asked to provide information concerning:

- name of the practice, summary of its primary objectives, and brief description;
- application area(s) (selected from pick list);
- type (policy required by law, policy, standard, guideline, recommended practice);
- date of introduction and comments on origin and development;
- degree of compliance and comments on application and enforcement;
- whether the practice is one of the agency's most significant and effective practices.

Respondents were also asked to list and describe briefly any effective but undocumented practices and practices still in development, and were invited to add any other comments.

From the outset, it was known that the questions about quality practices were not likely to be straightforward or easy to answer, and that a good deal

Table 22.1 Practices by Main Area of Application and Type

Main area application	Policy (Law)	Policy	Guideline	Standard	Recom. practice	Not specified	TOTAL
				Type			
Estab. client requirements, market research	0	2	1	4	1	1	9
Specification/use populations, data items	0	1	0	9	0	0	10
Specification/use classifications	2	3	1	11	0	1	18
Selection, testing collection instruments	0	0	0	0	0	0	0
Questionnaire design	0	4	0	1	0	1	6
Electronic forms design (CATI, CAPI)	0	0	0	0	0	0	0
Frame, business register	3	0	1	5	0	0	9
Sample design, selection	0	0	0	2	1	0	3
Survey system and procedure testing	0	0	0	0	0	0	0
Minimizing, measuring respondent burden	1	7	2	0	0	1	11
Minimizing, measuring response error	0	0	0	0	0	0	0
Confidentiality	6	1	0	2	0	0	9
Data collection, capture, follow-up	0	6	0	4	1	0	11
Editing, imputation	0	0	0	1	0	0	1

							Total
Weighting, estimation, outliers	0	0	1	0	3	0	4
Seasonal adjustment, trend calculation	0	1	2	0	1	0	4
Data analysis	0	0	1	0	0	0	1
Text content, presentation	0	8	0	4	0	1	13
Tabular presentation	0	1	0	2	0	0	3
Graphical presentation	0	2	1	1	0	0	4
Spatial presentation	0	0	0	0	0	0	0
Data dissemination	3	10	6	3	2	0	22
Data and metadata management	2	2	2	1	2	2	11
Survey and program evaluation	0	1	0	3	0	0	4
Survey design (general)	1	0	5	2	1	0	9
Systems design	0	1	0	1	1	1	4
Organization planning	0	1	0	2	3	0	6
Personnel, training	0	2	0	0	1	0	3
Program specific	2	0	1	0	1	0	4
ALL AREAS	20	53	24	58	16	8	179

of variation could be expected regarding what constituted a "quality practice" and the distinctions between "policy," "standard," "guideline," and "recommended practice." To address this problem, an example of a response was included with the questionnaire, based on practices actually existing at the Australian Bureau of Statistics and Statistics Canada.

22.3 GENERAL SUMMARY OF RESULTS

Sixteen agencies reported a total of 179 practices, 41 of which were designated as being among "their five most significant and effective" (henceforth referred to as simply "most effective"). It should not be construed that these agencies have an average of 11 quality practices each, as the questionnaire did not request exhaustive coverage.

The distribution of quality practices by type and main application area is shown in Table 22.1. The most commonly occurring type of practice is "standard" and the least common is "recommended practice." As a general rule (but with exceptions), policies tend to be more sharply focused, reflecting, perhaps, the accompanying requirement for enforcement and audit, whereas guidelines are more often general, covering a wide range of application areas.

The distributions of degree of compliance by type of practice and by main application area are shown in Tables 22.2 and 22.3, respectively. As is to be expected, compliance depends upon type—policies required by law being most highly complied with, and recommended practices least so. There is also a correlation between compliance and application area which reflects the underlying relationship between type and application area. Compliance is notably high in the areas of confidentiality and data dissemination, both of which involve contact with the public. Compliance is low in the areas of classifications, textual content, and presentation, all of which involve practices aimed at conceptual integration and reporting of errors.

The distribution of practices by date of introduction is shown in Figure 22.1. The median date is 1990. The average date is 1980; this figure is considerably lower than the median due to the inclusion by some agencies of some long-established practices. The oldest practice was reported by the U.S. National Agricultural Statistical Service; it dates back to 1909 and refers to confidentiality procedures.

22.3.1 Statistics Acts

Each agency works within the framework of a basic statistics act, often in conjunction with one or more additional acts relating specifically to the statistical program, for example, specifying the timing and content of a population census. The statistics acts can all be presumed to embody some quality practices referring to the interfaces between the statistical agency and

Table 22.2   Practices by Main Area of Application and Degree of Compliance

Main area application	Degree of compliance				
	High	Medium	Low	Not specified	TOTAL
Estab. client requirements, market research	3	1	0	5	9
Specifications/use of populations, data items	2	0	0	8	10
Specification/use classifications	3	4	0	11	18
Selection, testing collection instruments	0	0	0	0	0
Questionnaire design	4	1	0	1	6
Electronic forms design (CATI, CAPI)	0	0	0	0	0
Frame, Business Register	5	2	1	1	9
Sample design, selection	1	1	0	1	3
Survey system and procedure testing	0	0	0	0	0
Minimizing, measuring respondent burden	5	3	0	3	11
Minimizing, measuring response error	0	0	0	0	0
Confidentiality	7	0	0	2	9
Data collection, capture, follow-up	5	3	1	2	11
Editing, imputation	1	0	0	0	1
Weighting, estimation, outliers	1	2	1	0	4
Seasonal adjustment, trend calculation	1	2	0	1	4
Data analysis	1	0	0	0	1
Test content, presentation	2	6	1	4	13
Tabular presentation	1	0	0	2	3
Graphical presentation	0	1	1	2	4
Spatial presentation	0	0	0	0	0
Data dissemination	10	1	1	10	22
Data and metadata management	5	3	0	3	11
Survey and program evaluation	1	0	0	3	4
Survey design (general)	1	2	1	5	9
Systems design	2	0	0	2	4
Organization planning	3	2	0	1	6
Personnel, training	1	1	0	1	3
Program specific	1	2	0	1	4
ALL AREAS	66	37	7	69	179

the public, particularly the public as respondents. Most, if not all, acts specify the obligations of the agency with respect to the release of data, the confidentiality of individual responses, the requirement to state the purposes for which the data are being collected, and the rights of respondents and their obligations to respond and to provide accurate information.

In some countries, the statistics act has additional explicit references to quality practices. In this respect the recently revised Statistics Act (Statistics Finland, 1994) is probably one of the most comprehensive. It includes statements referring to:

Table 22.3   Practices by Degree of Compliance and Type

Degree of compliance	Type						
	Policy (law)	Policy	Guideline	Standard	Recom. practice	Not specified	TOTAL
High	14	23	7	18	3	1	66
Medium	4	7	9	6	10	1	7
Low	0	2	2	2	1	0	7
Not specified	2	21	6	32	2	6	69
ALL AREAS	20	53	24	58	16	8	179

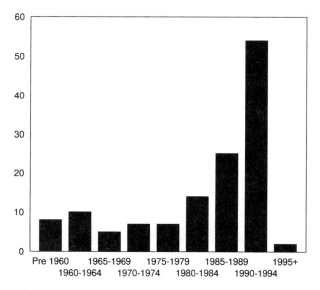

Figure 22.1   Distribution of practices by date of introduction.

- *collection instruments:* "The collection of data for statistics shall primarily draw upon data collected in other connections";
- *definitions and classifications:* "To ensure comparability of data, uniform concepts, definitions and classifications shall, where possible, be used in the compilation of statistics";
- *minimizing respondent burden:* "Before the statistical authority decides upon the data to be collected . . . it shall negotiate with the respondents . . . . The negotiations shall be arranged so early that the views of the respondents can be taken into account . . . . The authority compiling the statistics shall ensure that respondents are requested to submit only data

that are necessary for the compilation of the statistics. The data shall be collected in a manner that is economical and causes the respondents as little inconvenience as possible";

- *data content and dissemination:* "Statistics shall be as reliable as possible and they shall give the right description of the social circumstances and their development. The statistics shall be published as soon as possible after their completion";
- *systems design and data management:* "The data shall be processed in accordance with good statistical practice . . . . The authority compiling the statistics shall ensure that the data are properly secured against unauthorized processing, use, destruction and alteration as well as against theft . . . ."

Two common quality practices usually supported by legislation but not necessarily embedded in the statistics act are: (1) the requirement to submit proposals for new surveys, or major changes to ongoing ones, to a designated senior public body, and (2) the establishment of some form of national "watchdog" to review the statistical program, particularly as regards the relevance of the statistics produced, and to advise the statistical agency accordingly. Specific examples are discussed in Section 22.4.

22.3.2 Inter-Agency and International Policies and Standards

Several of the European responding agencies made reference to European Economic Council regulations produced by EUROSTAT which embody quality practices. Two regulations concerning the use of standard statistical units (EEC, 1993a) and business registers (EEC, 1993b) were explicitly cited. The Statistics Finland response noted that "the effects of the Economic Union are important for quality," and that "the quality of international comparisons . . . will be of great importance for statistics in the future." Statistics Netherlands indicated that their policy for use of a standard industrial classification originates from European Economic Community goals as well as being a long-standing national practice. In a similar fashion, the United States' statistical agencies referred to standards set by the Office of Management and Budget as quality practices.

There were no references to the International Organization for Standardization (1994) 9000 Series of standards for quality management and quality assurance. This is rather surprising as these standards specifically include guidelines for service organizations.

22.3.3 Development, Implementation, Effectiveness, and Compliance

In many agencies it appears there is no single group responsible for quality practices. With the exception of policies required by law or associated with classifications, the development and introduction of quality practices often seem to be rather haphazard—a matter of chance whether an individual or group within the agency has taken the initiative—rather than being the result of a systematic process. For example, the Australian Bureau of Statistics reported that, over a two-year period, a staff member extracted senior management decisions for the previous 10 years from meeting records and converted them into formal policies (including several quality policies) which were then accepted *en bloc*. This has probably resulted in many more policies than would have been the case had each policy been debated individually by a senior body.

The quality practices that respondents considered to be their most effective are spread across the spectrum of application areas, with confidentiality being notably well represented. The effectiveness of a quality practice may be viewed in terms of two interrelated components, namely the intrinsic utility of the practice and the degree of compliance. Rather surprisingly, respondents did not seem to regard a high degree of compliance as a requirement for designating a practice as most effective, suggesting they focused more on intrinsic utility in making their judgements. This conclusion ties in with a comment "that quality practices are often not well marketed, that they do not make the transition from conceptual development to practical implementation." In addition, it was noted that quality practices are often allowed to fall out-of-date after the initial enthusiasm for them has dissipated. (To help prevent this, the U.S. National Agricultural Statistics Service has a policy for documentation and maintenance of policies and standards.) Also, judging by the number of missing entries, many respondents had difficulty assessing the degree of compliance, even of policies. This suggests that monitoring and enforcement of quality practices is often limited or non-existent.

22.3.4   Quality Practices and Ethics

A number of quality practices refer to the ethical behavior patterns required of survey practitioners, for example, that sampled units should be selected randomly within a sampling stratum irrespective of collection costs, that interviewers should not create responses that they have not received or attempt to influence responses, that individual responses should be kept confidential, and that statistics should not be released ahead of the publication schedule or for personal gain. For example, the U.S. Bureau of Labor Statistics has a data collection integrity policy stating that "employees must not deliberately misrepresent the source of the data, the method of data collection, the data received from respondents, or entries on administrative reporting forms." On the same theme, the U.S. Bureau of the Census has a re-interview policy of which one objective is to detect interview falsification.

This coincidence of ethical requirements and quality practices occurs in part because high ethical standards in data collection and processing tend to produce unbiased, i.e., accurate statistics. It also occurs because "quality" has been defined very broadly, thereby taking into account the costs to respondents and users of unethical behavior such as confidentiality breach or premature release. In fact, Statistics Finland reported its professional ethics policy as a quality practice.

22.4 QUALITY PRACTICES BY AREA OF APPLICATION

The following paragraphs describe briefly some practices in selected application areas, beginning with those referring to client requirements. For brevity, practices which have been designated by reporting agencies as among their "most significant and effective" are referred to simply as "most effective."

22.4.1 Establishing Client Requirements

User/advisory groups
The Institut National de la Statistique et des Etudes Economiques (INSEE) in France has a recommended practice for introduction of client groups by industrial sector. Its objective is to improve relationships between businesses and the statistical agency by ensuring that agency requests match the business data available and that agency products meet business needs. Although having no formally documented practice in this respect, Statistics Canada also has a wide range of "user advisory committees," and the U.S. Bureau of the Census periodically convenes advisory groups of outside experts in particular techniques or representing special interests.

Conducting client satisfaction surveys
Several agencies reported practices for measuring client satisfaction. For example, in 1994, the U.S. Bureau of Labor Statistics (BLS) introduced a continuing, transaction-based customer service survey as part of its response to Executive Order 12862 (The White House, 1993) requiring each U.S. Federal agency to set customer service standards. On a rotating basis, each BLS office sends a service delivery questionnaire to a sample of its (telephone and mail contact) customers.

Identifying and liaising with key clients
The Statistics New Zealand standard for key client liaison, on the basis of which key clients are identified and visited monthly, is cited by the agency as most effective. The Australian Bureau of Statistics has a similar practice.

Documenting and analyzing help desk enquiries
A Statistics New Zealand practice requires that, on a rotating basis, the client service staff code and record clients' requests, using a central database system built specifically for this purpose. It is acknowledged that coding during periods of peak activity is less reliable than at quieter times. (This is a specific example of a general problem in making quality measurements.)

22.4.2 Statistical Units, Data Items, and Classifications

The majority of the practices referring to statistical units and data items that were reported relate to business statistics. It is widely acknowledged that such practices are difficult to implement effectively. For example, the Statistics Netherlands's response notes that about five years of discussion were required to get internal agreement on business statistical units, and that a standard for coordinated concepts and definitions (meeting the needs of the national accounts and other users) also took a long time to negotiate. INSEE (France) has a proactive policy, considered most effective, of being represented on the National Accounting Council. The aim of the policy is to ensure that statistical requirements are taken into consideration in revisions to the national account- ing plan, i.e., generally accepted accounting principles, and that business organizations and agency statisticians use the same terminology. Several agencies including the U.K. Office for National Statistics noted that use of standard classifications for business statistics is enforced through use of a central business register.

22.4.3 Questionnaire Design

The majority of the practices referring to questionnaire design also relate to the collection of business statistics. There is an Australian Bureau of Statistics policy, having high compliance, with the objective of "ensuring forms are easy to complete and are designed in consultation with users, reducing the need for follow-up to verify responses." It also states that the forms (i.e., questionnaires) should "project a professional and efficient agency image to respondents, and promote a unified agency approach, distinguishable from those of other government departments." The Czech Central Statistical Office has a policy for the coordination of business survey requirements so that large enterprises receive a single (annual) questionnaire covering all needs.

The absence of responses relating to electronic forms design suggests that the electronic collection methods and technology are too experimental and new for quality practices to have developed.

22.4.4 Registers and Frames

Statistics Finland has a policy required by law ensuring that the custodians, contents, intended uses, and the destruction and archiving procedures for all registers of individual persons are documented and are made known to the public. Other practices reported in this general area refer explicitly to the contents, use and maintenance of business registers. For example, INSEE (France) is subject to a very effective policy required by law which mandates the assignment of a single and invariable registration number for enterprises and local units as a common reference point for all government agencies dealing with businesses, thereby eliminating the need for redundant storage of registra- tion information and facilitating the exchange and use of administrative data. Statistics Netherlands has a business register standard with the objectives of ensuring that "statistics on the same area are based on the same populations, that statistics on different areas do not overlap or omit populations, and that statistical units are uniformly defined and classified." The U.K. Office for National Statistics (ONS) has a policy that specifies register structure and minimum content, in accordance with Economic Community regulations (EEC 1993a, 1993b).

22.4.5 Respondent Burden

Review of survey requirements
Several agencies reported policies that require proposals for new or substantially modified surveys to be reviewed and approved by a high-level body. In France, INSEE must register its surveys with the National Council for Statistical Information. This policy was recently supplemented by another which involves the assignment of a quality indicator, termed a "general interest label," to those surveys which satisfy specified conditions of relevance and conformity to statistical quality norms. In Australia, by law, the Australian Bureau of Statistics (ABS) must table, i.e., present, new surveys in Parliament. The ABS policy goes on to state that "statistical collections which for any reason have not, or could not, be tabled in Parliament must be conducted on an overtly voluntary basis." The U.K. Office for National Statistics has a policy which requires that all new surveys of businesses or local authorities, or significant changes to existing ones, be cleared with the Survey Control Unit and approved by Ministers. Information to be submitted includes a full assessment of the costs to the respondents of compliance with the survey. In New Zealand, the requirement that the Minister responsible for statistics approve new statistical activity was waived in 1993, but there is still an "independent" review.

Spreading Respondent Burden
Five agencies reported practices for spreading respondent burden evenly throughout the sampled population, and a sixth (Statistics Finland) stated that such a practice was in development. Of the practices, four relate to business surveys, and the fifth to both business and household surveys. The Australian Bureau of Statistics policy, considered most effective, involves "synchronized" sampling. Each statistical unit in the business register is assigned a random number which is used to control the overlap of samples for successive cycles of a survey and across surveys. Statistics Sweden has a similar policy. The Statistics Netherlands policy, also considered most effective, involves a quite different technical approach. Sampling for a new survey is arranged, without introducing bias, to favor the selection of business units with the least accumulated respondent burden across ongoing surveys. INSEE (France) also has a policy for the use of negative coordination in the selection of the samples which collectively comprise the annual enterprise survey. A Statistics New Zealand guideline limits to two the number of surveys to which a small business unit should be obliged to report at any given point in time, and it stipulates that there should be an interval of not less than two years between successive participations of any given household in a survey.

22.4.6   Confidentiality

Six agencies reported policies for protection of the confidentiality of individual responses. Four of these policies are rated most effective, and in all cases compliance is high. The U.S. Bureau of Labor Statistics policy is typical, stating that "data collected or maintained by, or under the auspices of, the BLS under a pledge of confidentiality shall be treated in a manner that will assure that individually identifiable data will be used only for statistical purposes and will be accessible only to authorized persons. Pre-release economic series data prepared for release to the public will not be disclosed or used in an unauthorized manner before they have been cleared for release, and will be accessible only to authorized persons." As confidentiality is such a central feature of statistical data gathering, it is highly probable that all agencies have such a policy. The fact that it was not universally reported is likely to be indicative of differences in interpretation of what constitutes a "quality practice." For example, Statistics Sweden noted "We do not specifically consider confidentiality as a quality dimension. The way we see it is that there is strict (exogenous) legislation concerning confidentiality. Thus, we do not have any options but to stick to 'the rules of the game'."

22.4.7   Collection, Follow-up, Capture, and Coding

The U.S. Internal Revenue Service Statistics of Income Division, which acquires its data by sampling and data capturing taxation documents, has a policy for document control, standards which require all data to be captured twice and edit checks to be applied, and an overall processing management standard which not only provides operational instructions but also an outline of the division's approach to quality management and a quality audit checklist. All these practices are considered most effective and have high compliance. INSEE (France) and the U.S. Bureau of the Census have data capture and coding quality control policies. Quality control is widely used at Statistics Canada, too, though there is no formal policy statement to this effect. Statistics Canada also described a policy for reporting and use of nonresponse measurements for business and social surveys. Sixteen different nonresponse rates are defined, thirteen for use in the context of summarizing data collection and three in the context of estimation.

22.4.8   Seasonal Adjustment and Trend Calculation

Several offices have, or are developing, guidelines or recommended practices in the area of seasonal adjustment and trend calculation. Most refer explicitly to one particular method, namely X11-ARIMA. The Australian response, which refers to X11 alone, typifies the objectives of all these practices as being "to assist the media in accurately reporting significant features of the main economic indicators by: emphasizing the movements of the trend rather than the seasonally adjusted estimates; putting trend estimates in historical perspective; avoiding claims about turning points until sufficient evidence has been accumulated; performing sensitivity analyses; and identifying trend breaks."

Statistics Canada has a policy dealing with forecasting which states that "estimates for future reference periods are an acceptable agency product provided that: they are generated using mathematical models whose specifications are explicitly articulated and exposed to public scrutiny; the assumptions used in the model are clearly defined and are presented with the estimates; and anyone using the model assumptions and input data would get the same results."

22.4.9 Data Output Quality

Several agencies reported practices requiring that clients be informed about data quality. For example, there is a long established Statistics Canada guideline that requires a statement of methodology and data quality to accompany all output. Consideration is presently being given to upgrading this to a standard and paying more attention to compliance. Likewise, a U.S. Bureau of the Census practice requires that standard errors be presented with estimates from sample surveys. A comprehensive Australian Bureau of Statistics policy, primarily addressing the criteria for suppression of data on the basis of unreliability, also requires that "all publications include statements about data quality and methodology and that the size of sampling and non-sampling errors are provided."

22.4.10 Textual, Tabular, Graphical, and Spatial Presentation

Most agencies have presentation policies. For example, the U.S. Internal Revenue Service Statistics of Income Division has written guidelines and a style guide (Statistics of Income Division, 1991, 1994). Likewise the U.S. National Agricultural Statistics Service and the U.S. Bureau of Labor Statistics have editorial, tabulation, and charting guidelines. Examples of these practices include the use of a style guide supported by a computerized system for all publications, and the publication of the main economic indicators in a particular format, including a key points section which assists clients and the media in accurately interpreting significant features. Rather surprisingly, no practices were reported that deal specifically with spatial presentation.

22.4.11 Data Dissemination

The area of data dissemination is rich in quality practices, perhaps reflecting the enhanced need for consistency when dealing with the public.

Accuracy
Several agencies reported practices aimed at ensuring that statistical outputs are of uniformly good quality by requiring them to be reviewed and signed off

at a senior level prior to release. For example, the U.S. Bureau of Labor Statistics has a publishing policy which requires advance clearance through the Associate Commissioner responsible for publications to ensure conformance to bureau editorial, tabulation, and charting guidelines and to Federal publication standards. Statistics Netherlands, on the other hand, is in the process of assigning this responsibility back to the publication producer.

The Australian Bureau of Statistics and Statistics Canada have similar practices referring to the release or suppression of data subject to error. The aims of the former are to ensure that "quality labels are placed on all disseminated data to assist users in interpretation," also that "individual cells are not suppressed solely due to sampling errors; and data with serious bias are not published."

Release Date
It is common practice for agencies to publish release dates in advance and to operate with an embargo policy stating that no statistics are to be released prior to the predefined date and time. Seven reported practices relate to the conditions for pre-release of data subject to embargo, and to dealing with the media in general. A typical example is the Australian Bureau of Statistics policy for pre-release and "lock ups" which defines two exceptions to a general embargo policy and states the conditions for (1) pre-release of specified publications to government departments, and (2) for government officials to view specified publications in a "lock up" situation, attendees not being able to leave the room or communicate outside until the embargo is lifted. The U.S. Bureau of Labor Statistics has a similar policy intended to provide "an objective resolution of the tension between maintaining the confidentiality of sensitive releases and encouraging informed reporting." It allows the media controlled access to the data in a lock up situation prior to the official release. An exception to the general rule regarding near simultaneity of release is U.S. Internal Revenue Service Statistics of Income Division which releases statistics from taxation records to its main clients (Office of Tax Analysis, Congressional Joint Committee on Taxation) 60 days earlier than to other users.

Access
The U.S. Bureau of Labor Statistics (1994a) reported newly formulated client service guidelines, indicating the release schedule, information availability, uses and limits of the products, and subject matter contact points. They were developed in response to Executive Order 12862 (The White House, 1993) requiring each U.S. Federal agency to establish and implement customer service standards. The provision of statistical information to international organizations is covered by a U.S. Office of Management and Budget standard. Statistics Sweden reported as most effective its integrated publication system policy which assists clients in locating quickly and efficiently the data they need. The Australian Bureau of Statistics described an ongoing data management initiative (elaborated in Section 22.5) aimed at vastly improving data visibility and

accessibility. The Australian Bureau of Statistics also has a policy dealing with conditions for the release of microdata. It states that "unidentifiable unit records ... will be released ... from most social surveys unless there is no demand or it is not possible to produce a useful file because of confidentiality restrictions ... and from other collections subject to rigorous assessment of client demand and intended use of the data." The policy goes on to specify conditions for release, including a statement that "no release ... will be authorized where the intended use includes the matching of (these) data against data holdings from other sources."

22.4.12 Data and Metadata Management

Continuity
Two practices were reported with the basic objective of ensuring that the continuity of important time series is maintained during changes in survey procedures. The Australian Bureau of Statistics policy requires (depending upon the series and the magnitude of the changes) either a "parallel run" of the survey on both old and new bases, or estimates of the sizes of the statistical discontinuities attributable to the changes.

Documentation
Among the practices relating to the general documentation of metadata is an INSEE (France) recommended practice involving the storage of its survey documentation in a shared database, aimed at rationalizing and standardizing concepts, data item definitions and processing methods. The system is presently restricted to household surveys. Statistics Canada reported a policy for use of the "Statistical Data Documentation System" with similar objectives. The U.S. Internal Revenue Service Statistics of Income Division has a policy dealing with the description and disposition of administrative and processing records, and with the provision of comprehensive records of statistical consulting activities.

Central Shared Database
The U.K. Office for National Statistics has a guideline for the use of a central shared output database designed to hold all the data published by the agency, together with some metadata. The objectives of the guideline are to ensure that outputs produced in different formats and media contain consistent information, to provide standard format table production facilities, and public access to (nonconfidential) data. Other agencies indicated their intention to put increasing focus on the use of output databases, including the Australian Bureau of Statistics and Statistics New Zealand who both reported comprehensive metadata management systems are in development.

22.4.13 Survey and Program Evaluation

Statistics Sweden has a policy (Eklöf and Lindström, 1995), which it considers most effective, requiring an annual self-assessment of the changes in overall quality of each product produced by the agency. The results are collectively reviewed and followed up by a central unit, and provide the principal basis for setting quality improvement priorities. Statistics Canada has an ongoing program evaluation process aimed at ensuring products continue to be relevant and effective by reviewing program objectives and the extent to which they meet client needs, and by assessing program design and its benefits and costs relative to potential alternatives. The U.K. Office for National Statistics survey control policy states "All surveys of businesses or local authorities conducted at regular intervals should be thoroughly reviewed periodically (at least every three or five years depending on the frequency of the survey) and approved by ministers in order to continue .... The needs for surveys should be reviewed annually."

22.4.14   General Guidelines

Many agencies, particularly including those using contractors, for example, the U.S. National Center for Education Statistics (1991, 1992), indicated that they have general survey standards/guidelines covering a wide range of survey processes.

22.5   GENERAL OBSERVATIONS AND THEMES

The survey of quality practices had a number of limitations. First, it was not based on a probability sample and cannot be assumed representative of all statistical agencies. It did, however, include a number of agencies recognized as being in the forefront of statistical data collection, processing, and dissemination. Second, there are differences in interpretation between agencies as to what constitutes a "quality practice." As quality is built in, today's quality practices become tomorrow's routine operations. For example, practically every agency in the world uses probability sampling, edits incoming data from respondents, and adjusts for nonresponse. These are such a common part of normal procedures as not to be considered quality practices *per se*. However, agencies which are at different stages relative to one another in program development have different perceptions of what is a quality practice and what is part of routine operations. Third, even within an agency, it can be assumed that quality questionnaire responses are somewhat dependent upon the particular interests of the staff members completing them. Within the framework of these limitations, the survey results support a number of general observations, as follows.

   Judged by their responses and willingness to share information, statistical agencies are very interested in the establishment and use of quality practices as a basis for statistical integration and coordination, and to bring economies of scale and quality improvements. In the absence of agency-defined practices, staff adopt implicit, local practices, usually based on "what was done last time."

   Statistical agencies' visions of what constitutes "quality" have broadened significantly over recent years, from a focus primarily on accuracy to a wider view which includes relevance, timeliness, availability, and cost factors, including cost to respondents. There is more client service orientation. Quality is "all aspects of the statistical service which influence the usage and which the users consider important" (Eklöf and Lindström, 1995). However, agencies still acknowledge their special responsibility in assuring that output data have accuracy appropriate for their use—something that clients are not usually in a position to do themselves.

The number and nature of quality practices within an agency are dependent upon its organizational structure as illustrated by the following examples.

- Some agencies such as INSEE (France) have an extensive training program through which all professional staff pass, and, as a result, the requirement for technical standards and guidelines is reduced.
- In countries with a decentralized statistical system such as the United States, there are likely to be a number of practices explicitly aimed at coordinating the activities of the separate, component statistical agencies. Such practices are developed and promulgated by a central body, for example the U.S. Office of Management and Budget, which may or may not itself also conduct surveys. In countries with a centralized statistical system like Canada, The Netherlands, and New Zealand, there is not the same motivation to define quality practices across agencies.
- The degree of geographical dispersion of an agency's offices also has an influence. An agency with widely dispersed offices is more motivated to introduce quality practices for coordination purposes.
- Agencies which are heavily dependent on other organizations such as administrative bodies for input data (like the U.S. Internal Revenue Service Statistics of Income Division), or which contract out survey-taking activities (like the U.S. Center for Education Statistics), are more likely to have data collection quality practices to support the contractual agreements.
- Relatively newly formed agencies such as the U.S. Energy Information Administration are likely to define quality practices in order to establish a common base, as outlined by Freedman (1990).

Agencies reported more quality practices referring to business statistics than to social statistics. This may be because social surveys tend to be more diverse and less subject to standardization. Alternatively, it may simply reflect the particular interests of the persons who happened to fill in the questionnaires. As previously noted, the questionnaire was sent to a small sample, and there were no explicit instructions regarding the people who were to complete it.

In terms of development and extension of quality practices, there seem to be two major themes: (1) better defined interfaces with the outside world, respondents and clients alike; and (2) integration, both at a conceptual level and physically, for data input, processing, and output. Both themes are supported by an increased focus on metadata and the advent and application of new computing and communications technology. For example, the Statistics Sweden response noted "We put very much emphasis on the development of metadata systems. According to new statistics legislation (April 1994) a comprehensive metadata system for all official statistics is to be prepared by Statistics Sweden." The Australian Bureau of Statistics reported a major ongoing data management project, with the twin goals of improved client service through better cataloged, more visible, and more accessible output data, and integration of concepts and procedures to enhance the information content and mutual coherence of output data and to reduce systems maintenance costs. "These goals are being approached through the development, loading and use of a corporate information warehouse from which most, if not all, agency data products will ultimately be generated. The warehouse will have facilities to store, catalogue, and access all the data, and corresponding metadata, produced by the agency. Implementation of the warehouse is being coupled with the introduction of data management policies and practices to promote rationalization and integration." (Colledge and Richter, 1994).

There is also a thrust expressed by several agencies for an integrated approach to quality management itself, an overall framework within which to place quality practices. The Statistics Sweden response stressed "that we see quality work in an integrated way, and not merely as a number of (more or less) independent practices." The agency recently introduced a broad quality policy (Statistics Sweden, 1993) providing an integrated framework for all quality initiatives at the agency. The objectives of the policy are "to produce and provide objective statistical information requested by users . . . in the form and at the time stipulated by users . . . fulfilling the demands for accuracy which accompany the agency's role as professional producer of statistics." Similarly, the aim of the "Quality Measurement Model" (U.S. Bureau of Labor Statistics, 1994b), currently in development, is to aid program improvement by blending three concepts in one model: the principles of total quality management with their emphasis on customer focus; a common view of the BLS "business process," i.e., the key operations used across bureau programs; and decision making based on optimizing customer satisfaction and statistical quality within resource constraints. The model provides a framework for quality guidelines and measurements. In summary, it would seem that there is a strong desire for conceptual integration, though practical implementation is difficult.

The drive for quality practices is not confined simply to organizational units within an agency, it applies across statistical agencies and other organizations. For example, INSEE (France) reported its policy of being a full time participant in revisions of accounting principles undertaken by the national accounting organization, thereby ensuring that its requirements to collect business data are fully considered. The sharing of data internationally between statistical agencies is likely to be an increasing trend (Ryten, 1995). This will translate into a need for common data models, and international quality policies will become increasingly important. The European Economic Community regulations, for example, will extend beyond their present focus on statistical units and business registers.

ACKNOWLEDGMENTS

The authors extend their thanks to the respondents and hope the survey has been an exercise of mutual benefit. The authors would also like to thank Cathy Dippo for her helpful editorial suggestions, and Alistair Hamilton of the Australian Bureau of Statistics who prepared the tables and figure and proofread the manuscript.

REFERENCES

Colledge M.J., and Richter, W. (1994), "Data Management and the Information Warehouse: Infrastructure for Re-engineering," *Proceedings of Statistics Canada Symposium* 94, *Re-engineering for Statistical Agencies*, Statistics Canada, Ottawa.

Eklöf, J.A., and Lindström, H.L. (1995), "On the Integrated Quality Programme at Statistics Sweden," Working Paper, February 1995, Statistics Sweden.

European Economic Community (1993a), "Council Regulation (EEC) No. 696/93 on the Statistical Units for the Observation and Analysis of the Production Systems in the Community," *Official Journal of the European Communities*, No. L 76/1, March 1993.

European Economic Community (1993b), "Council Regulation (EEC) No. 2186/93 on Community Coordination in Drawing up Business Registers for Statistical Purposes," *Official Journal of the European Communities*, No. L 196/1, August 1993.

Freedman, R. (1990), "Quality in Federal Surveys: Do Standards Really Matter?" *Proceedings of the Survey Research Methods Section*, American Statistical Association, pp. 11–17.

International Organization for Standardization (1994), "ISO 9000 Quality Standards," Geneva, Switzerland.

Ryten, J. (1995), "Business Surveys in Ten Years Time," in B.G. Cox, D. Binder, B.J. Chinnappa, A. Christianson, M.J. Colledge, and P.S. Kott (ed.), *Business Survey Methods*, pp. 691–709, New York: John Wiley.

Statistics Canada (1987), "Quality Guidelines," Second Edition, April 1987, Statistics Canada, Ottawa, Canada.

Statistics Finland (1994), "The Statistics Act," Version 9-3-94, Statistics Finland, Helsinki, Finland.

Statistics Sweden (1993), "Quality Policy," Policy paper, Statistics Sweden, Stockholm, Sweden.

Statistics of Income Division (1991), "Guidelines for Writing SOI Bulletin Articles," Reference Manual, Spring 1991, Statistics of Income Division, U.S. Internal Revenue Service, Washington, DC.

Statistics of Income Division (1994), "Statistics of Income Bulletin Style Guide," Statistics of Income Division, U.S. Internal Revenue Service, Washington, DC.

The White House (1993), "Setting Customer Service Standards," Executive Order 12862, *Federal Register*, Vol 56, No. 176, The White House, Washington, DC.

U.S. Bureau of Labor Statistics (1994a), "Customer Service Guide," U.S. Department of Labor, Washington, DC.

U.S. Bureau of Labor Statistics (1994b), "The Bureau of Labor Statistics Quality Measurement Model,", U.S. Department of Labor, Washington, DC.

U.S. National Agricultural Statistics Service (1983), "Framework for the Future," Report of Long Range Planning Group, U.S. Department of Agriculture, Washington, DC.

U.S. National Center for Education Statistics (1991), "Standards for Education, Data Collection and Reporting," U.S. Department of Education, Washington, DC.

U.S. National Center for Education Statistics (1992), "NCES Statistical Standards," NCES 92-021, U.S. Department of Education, Washington, DC.

APPENDIX: AGENCIES RESPONDING TO THE SURVEY

Australian Bureau of Statistics, Statistics Canada, Czech Statistical Office, Institut National de la Statistique et des Etudes Economiques (INSEE) France, Hungarian Central Statistical Office, Statistics Finland, Statistics Netherlands, Statistics New Zealand, Statistics Sweden, U.K. Office for National Statistics, U.S. Bureau of the Census, U.S. Bureau of Labor Statistics, U.S. National Center for Education Statistics, U.S. Internal Revenue Service (Statistics of Income Division), U.S. National Agricultural Statistics Service, U.S. Office of Management and Budget.

CHAPTER 23

Improving the Comparability of Estimates Across Business Surveys

Bill Gross and Susan Linacre
Australian Bureau of Statistics

Strategies for managing quality and controlling errors in business surveys are often decided in the context of the individual survey. Methodological choices such as those on frames, unit of selection, reporting unit, treatment of nonresponse, and estimation are typically made survey by survey. Ideally they provide the best quality that is achievable, within the constraints of cost and data availability, for the user requirements of that survey. However, the implications for users with a need to integrate data from across surveys to build a broader picture are not always considered, and inevitably different strategies are chosen for different surveys, sometimes with quite important consequences for the quality of the data for those users of integrated data.

23.1 INTRODUCTION

A fundamental requirement for quality management in any survey endeavor is a thorough understanding of user requirements and the identification of quality characteristics of products and services and their implications for users. A second requirement is a full knowledge of the processes actually used to produce the products and services. This knowledge enables the measurement and improvement of quality. A third aspect of quality management is to ensure an ongoing commitment to teamwork, to common goals, and to continuous improvement of products and services to meet user needs. Colledge and March (1993) also list these as the fundamental aspects of quality management.

Survey Measurement and Process Quality, Edited by Lyberg, Biemer, Collins, de Leeuw, Dippo, Schwarz, Trewin.
ISBN 0-471-16559-X © 1997 John Wiley & Sons, Inc.

In terms of business survey design and implementation, understanding user needs is not always straightforward. There will often be a variety of users interested in different aspects of the survey. One user may be predominantly interested in the subject matter of a single survey, for example, comprehensive measures of a single industry or information on wages and salaries in the economy as a whole. Another user may be interested in the same information only as part of a larger context. Typically, the first group of users will maintain close relationships with the survey, specifying their requirements directly and even indirectly by using the data in a highly visible or salient way (e.g., in media releases). In contrast, users interested in integrating information from a number of surveys have a more difficult task. They must establish relationships with a number of survey groups—sometimes even with a number of agencies. The difficulties for cross-survey users can be exacerbated as the data used in their outputs have often been further processed or manipulated—sometimes beyond the recognition of the original data producers. The producers may fail to identify the extent to which these users rely on the survey's data products.

Over the years, statistical agencies have devoted considerable effort to provide for cross-survey users, leading to the development of the concept of integrated economic statistics. This integration is supported by infrastructure such as standard classifications and definitions, and single national registers of business units. This structure is further guided by international standards on, for example, industrial classifications and units (see U.N. Statistical Office, 1990). Implementing the concept of integration has, however, met with only partial success. Tensions exist between providing solutions for users of data from the individual surveys and solutions for users who require data from a number of surveys. Moreover, the need for methodological practices that support the production of an integrated system of information is not sufficiently recognized. To manage these conflicting interests, the needs of various users must be balanced.

This chapter describes the Survey Integration Project in the Australian Bureau of Statistics (ABS). In its initial stages, this project appeared relatively straightforward. Motivated by a desire to improve the quality of survey outputs for cross-survey users, the project's goal was to review different methodologies used in ABS surveys; investigate the reasons for differences; and evaluate the effects of these differences for cross-survey users. The agency uses a variety of methods, some of which have been in place for a long time. Some methods were well justified, evaluated, and documented. Others, which might have been equally justified, were neither evaluated nor fully documented. It became clear that to achieve the required methodological changes management was as much an issue as technical concerns.

In general, it has been found that the process of reviewing methodologies provides a number of benefits. Among these are the identification and establishment of "best practice" solutions, better documentation of current practices and of their rationale, and opportunities for sharing infrastructure (documentation and training as well as system components). (The development of "best practice" is discussed in Morganstein and Marker, Chapter 21.)

While we focus on quality improvement through consistent methodologies, it is important to recognize that in some circumstances uniform procedures can impose costs or diminish quality in individual surveys. For example, the use of identical procedures to follow-up nonrespondents in all surveys is unrealistic, because surveys differ in the accuracy of their imputation rules and as a result suffer to different degrees from the effects of nonresponse. Survey integration should not be blindly applied, but rather should aim at eliminating unjustifiable methodological differences.

The Survey Integration Project has many similarities with the Business Survey Redesign Project which was undertaken in Statistics Canada from 1985 to 1991 (see Colledge, 1987; Dodds and Colledge, 1990; and Worton and Platek, 1995). The redesign project aimed to produce "economic data which can be aligned and compared between sources and over time" (Dodds and Colledge, 1990, p. 596). One of its benefits was "the standardization of concepts, definitions, classification schemes and survey procedures" (Colledge, 1987, p. 550). The ABS Survey Integration Project shares many of these goals. However, some differences can be noted. Whereas the Survey Integration Project has been relatively small scale and has concentrated on a few aspects of methodology at a time and has sought to improve consistency in those methodologies for a wide range of surveys, the Business Survey Redesign Project was on a grand scale, involving the introduction of a new business survey systems infrastructure—including the business register and administrative data access—and the simultaneous redevelopment of some business surveys using that infrastructure.

We discuss a number of possible differences in methodology and how these can lead to inconsistent estimates from business surveys (Section 23.2). We outline a process for quality improvement through consistent methodologies (Section 23.4) and illustrate the process with a case study from the ABS (Section 23.5). Other statistical agencies are reviewed in Section 23.3.

23.2 DIFFERENCES IN METHODOLOGY

The first step in improving comparability among estimates from different surveys is to understand the aspects of methodology in which inconsistencies arise. Methodologies are grouped into categories of data items and classifications, units, frames, sample and frame maintenance procedures, and editing, imputation and estimation.

23.2.1 Data Items and Classifications

Although surveys usually operate under a framework of standard definitions and classifications, the use of the framework can differ in a number of ways. The details of the data items can differ. For example, the reference period for a data item can differ: the number of employees can be specified on a particular day or during a particular pay period—with the latter yielding slightly higher counts. Items to be included or excluded can differ. For example, termination payments may, or may not, be included in wages and salaries. Different question wording to measure the same item may result in different responses, and asking for different breakdowns of an item could result in even larger differences. For example, collecting information on the number of employees, asking that part-time and full-time employees be reported separately decreases the likelihood that part-time employees will be missed in the count of all employees, and might increase the overall count of employees. The time at which data are collected can also affect the value reported. Large businesses, like statistical agencies, may have preliminary and final versions of data.

Classification codes may be assigned by different means. For example, a unit may be coded to two different industries by two different surveys. One survey might use the code given on the business register, while the other derives the code from information collected during the survey. Whether the coding process is purely clerical or is computer assisted affects the code numbers assigned.

23.2.2   Units

A frame, generally a business register, will often include a hierarchy of units, for example, location, establishment, and enterprise. The sampling unit can differ across surveys, as can the reporting unit (i.e., the level at which data are collected) (Nijhowne (1995) defines these types of units). Different reporting units can lead to differential coverage. For example, if stores in a supermarket chain are the reporting units for one survey, but the head office reports for another, identical questions can give rise to very different responses if the list of stores is out-of-date.

Even when the same sampling and reporting units are used, different contact points in a large organization can have different perceptions of the organization's activities and so provide different information. For example, the pay and personnel areas may have a more restricted view than the financial accounts area.

23.2.3   Frames

Clearly, if frames from different sources are used, then differences in coverage are expected, but even with the same source, typically a central business register, frames can differ. For the same reference period, the frames can be extracted at different times. Frames extracted from the business register can also be augmented in ways which vary from survey to survey.

23.2.4   Sample and Frame Maintenance Procedures

Sample and frame maintenance procedures are rules for identifying and treating units whose observed structure or status differs from that recorded on the frame.

Differences in these rules have been found to have significant effects on the comparability of estimates across surveys in the ABS and we discuss this more fully in Section 23.5. Differences result from variability in detecting changes and from different treatments of changes once detected. Struijs and Willeboordse (1995) discuss the handling of such changes for a business register.

The most extreme case of a unit changing its status from that recorded is when the unit has ceased operating—a death. Identifying and treating these cases is sometimes straightforward, but often, despite follow-up, it is not possible to confirm that some units are dead. How these units should be treated in estimation usually requires a best guess about their status. Different surveys invest varying degrees of effort into follow-ups to confirm deaths or use different assumptions about the proportion of unconfirmed deaths that remain alive. Either case leads to different outcomes for survey estimates.

It is also common that a unit changes industry. Although different surveys may adopt the same treatment for such units, some may be more likely to detect such a change. A survey of manufacturers which collects data on commodities could assign industry code to all units and so detect any change of industry, but a quarterly survey of employment may not collect information which would reveal a change in industry.

23.2.5 Editing, Imputation, and Estimation

While standard classifications and data item definitions are becoming more common, substantial differences in editing, imputation, and estimation methods still exist. Stringent edits lead to more changes and, hence, to different aggregate values and often to different distributions of unit values. A strategy which concentrates on units which could have a large effect on the estimates is likely to result in few, but large, changes. Similarly, different methods of imputation will lead to different unit values and, hence, different aggregates. Kovar and Whitridge (1995) discuss different methods of imputation.

Estimates can be affected by the implementation details of a particular weighting method, such as the source or currency of a benchmark. Another aspect of estimation which can lead to substantial differences among surveys is the detection and treatment of outliers. Decisions on outliers can have a subjective component, and survey areas may come to different decisions, albeit while following the same general principles.

23.2.6 Relationship to Nonsampling Error

The concept of estimate comparability is related to the concept of nonsampling error as both are defined in terms of the estimator's expected value over samples. When we say that estimates from two surveys are not comparable we mean that their expected values are not equal. While it is logically possible that different methodologies give rise to equal nonsampling errors, and hence to comparable estimates, the case where the nonsampling errors differ is more likely. Sources of incomparability are in this way related to sources of nonsampling error.

Biemer and Fecso (1995) classify the sources of nonsampling error into specification, frame, nonresponse, processing, and measurement errors and our categories fit with their classification. The discrepancies arising from different choices of data items and classifications correspond to Biemer and Fecso's specification or measurement error, while those arising from choice of unit can be nonresponse or measurement error. The frame categories in the two lists correspond and our categories of sample and frame maintenance procedures and editing, imputation and estimation relate to Biemer and Fecso's processing errors. Our categories are also similar to Griffiths and Linacre's (1995) list of processes involved in conducting a business survey.

23.3 METHODOLOGICAL DIFFERENCES WITHIN STATISTICAL AGENCIES

The examples of differences in methodology given in Section 23.2 were drawn from the ABS experience. A survey was conducted to assess the situation in other official statistical agencies and 16 agencies[1] in 15 countries replied. The survey aimed to assess how consistent agency procedures were across surveys; whether consistency was related to the structure of the agency; and the strategies an agency used to ensure estimate comparability. The survey was not based on a probability sample of government statistical agencies, so caution should be used in making inferences to other agencies.

23.3.1 Methodological Inconsistencies

Agencies were asked for summary information on major business surveys, whether specified methodological aspects were the same for these surveys within the agency, and for some examples of differences, if they existed. Table 23.1 summarizes the responses to the questions on methodologies.

All agencies collected some data items in more than one survey. The items of Employment, Wages and Salaries, Capital Expenditure, and Stocks were listed in the questionnaire and almost all agencies collected these items more than once. The exceptions were mainly agencies that were part of a decentralized statistical structure.

In most agencies, there were some inconsistencies in the way some items were collected. Most of the examples related to breakdowns of the item (e.g.,

---

[1] Australian Bureau of Statistics; Austrian Central Statistical Office; Statistics Canada; Statistics Finland; Institut National de la Statistique et des Etudes Economiques (INSEE), France; Federal Statistical Office, Germany; Census and Statistics Department, Hong Kong; Instituto Centrale di Statistica, Italy; Statistics Netherlands; Statistics New Zealand; Department of Statistics, Singapore; Central Statistical Service, South Africa; Statistics Sweden; Central Statistical Office (now Office for National Statistics), U.K.; Bureau of the Census, U.S.A.; Bureau of Labor Statistics, U.S.A.

Table 23.1   Summary of Responses on Methodology

Question	Number "YES"	Number "NO"
1. Are any major data items obtained in more than one major survey?	16	0
2. Are the common items collected identically in surveys?	4	12
3. Are the data always collected directly from the sampling unit?	3	13
4. For the major surveys which use the business register, do those with the same reference period take a frame at exactly the same time?	6	10
5. Is the business register supplemented from other sources for some major surveys?	11	5
6. Do the major surveys define, identify and treat in the same way units whose structure or status differs from that recorded on the frame?	11	5
7. For the major data items which are common to more than one survey, are the same edits applied to the same item in different surveys?	3	13
8. For the major data items which are common to more than one survey, is the same method used to impute or adjust for nonresponse?	3	13

employment reported by different types of employees) and reference periods (e.g., quarter or year). A few concerned activity included or excluded from the item (e.g., whether stocks of goods for resale were included with stocks of own production).

In most agencies there were some cases where statistical units did not report data directly. Examples were cited of smaller units reporting, but it was more common that the head office reported for its branches. This usually involved a small number of large units. Other examples concerned data for a large proportion of units being supplied by a business association or from administrative records.

Less than half of the agencies took frames at the same time for the same reference period. Frames were taken at different times to meet different survey requirements (e.g., different amount of time required to prepare the sample). No information was supplied on the extent of the differences between frames taken at different times. The majority of agencies supplemented their business register for some surveys, but generally only for one or two major surveys.

Most agencies used the same sample and frame maintenance procedures for major surveys. Where they did not, some of the differences arose from different strategies of allowing for births in the population.

Most agencies applied different edits in different surveys, generally because different combinations of items were available for comparison in the surveys. Most agencies had some differences in nonresponse imputation in different surveys. The different methods included imputing a cell mean, adjusting previous values, and calculating a value from other items or from another survey. For editing and imputing, some agencies had standardized tools which surveys used in different ways.

In summary, while most agencies reported a consistent approach to sample and frame maintenance procedures, less consistency was found in other methodological aspects. Many inconsistencies are explained by user requirements and features particular to a specific survey. Nevertheless, it is beyond the scope of this chapter to assess how great the potential for increased consistency is.

23.3.2 Structure

Agencies were also asked about their structures: whether they were functionally specialized and whether they had a central group of business survey methodologists or a central business surveys classification group. (An agency is functionally specialized for business surveys if all business surveys are processed by one part of the organization but other parts analyze and publish the data from these surveys.) Our question is whether the level of consistency in an agency is related to these aspects of its structure. We measured the level of consistency as the count of "yes" answers to questions 2 to 4 and 6 to 8 in Table 23.1 and of a "no" to question 5. Based on these counts, an agency was classified as low (count of zero to two), medium (three or four), or high (five to seven). While this is a crude measure, it does allow some analysis of the relationships. Tables 23.2 to 23.4 show agencies classified by level of consistency and structural aspects of (1) functional specialization, (2) a central group for methodology, and (3) a central group for classifications.

Keeping in mind that the small number of functionally specialized agencies weakens any conclusion, we still see that the functionally specialized agencies have a distribution of level of consistency which is not much different from that of other agencies.

There is some evidence of a relationship between the presence of a central group of methodologists and the level of consistency. Agencies with a central

Table 23.2   Number of Agencies Classified by Level of Methodological Consistency and Whether the Agency is Functionally Specialized

Functionally specialized	Level of consistency			
	Low	Medium	High	Total
Yes	2	1	0	3
No	8	1	4	13

Table 23.3   Number of Agencies Classified by Level of Methodological Consistency and the Presence of a Central Group of Business Survey Methodologists

Central group of methodologists	Level of consistency			
	Low	Medium	High	Total
Yes	3	1	2	6
No	7	1	2	10

group of methodologists appear more likely to have a high level of consistency than those without such a group.

There is no evidence of a relationship between the presence of a central classification group and the level of consistency. As with functional specialization, the small number of agencies without a central classification group weakens any conclusion.

23.3.3   Strategies

The survey also asked for information on strategies which increased estimate comparability. The responses ranged from a highly centralized approach of "We have a legal frame which defines exactly the survey and item definitions given in explanations to the definitions," to a highly decentralized one of "The strategy for each survey is to produce estimates which are as complete and accurate as possible. The goal of each survey is to provide the best possible answers to the questions the survey was developed and designed to answer."

The agencies listed strategies that relate to organizational structure, standards, and infrastructure. The organizational strategies include having: one group responsible for all respondent contacts and another for data processing; a single classification and standards group; a single group for methodology; a unit to coordinate survey data from large businesses and to maintain information

Table 23.4   Number of Agencies Classified by Level of Methodological Consistency and the Presence of a Central Classification Group

Central classification group	Level of consistency			
	Low	Medium	High	Total
Yes	8	2	4	14
No	2	0	0	2

about their structures; and a central questionnaire design group. Standards related to classification, data items, questions, coding, and sample and frame maintenance procedures. (Policies about the use of such standards are discussed in Colledge and March (Chapter 22).) Supporting infrastructure included a central business register, generalized processing systems and standard reference numbers, documentation, and training courses. Statistical processes such as confronting data from different surveys at unit and at aggregate levels were also used.

Different approaches can be taken to ensure adherence to standards and the use of common infrastructure. In the responses provided, this was reflected in the use of terminology from "promulgate" through "promote" to "strongly encourage" to describe how the use of standards and standard infrastructure was achieved.

23.4   A PROCESS FOR INCREASING CONSISTENCY

The process of increasing methodological consistency is as much a management process as it is a technical process. The management process deals with understanding user needs, both specific to the survey and across surveys. Management must also engender a common commitment to meeting these needs as prioritized. When improved consistency is a priority need, decisions previously made at local levels may need to be shifted to central or collegiate groups. Individual surveys will have less control over their processes and reduced ability to respond to changes in survey-specific user requirements. The change may involve replacing local functions, such as frame creation, with centralized ones, leaving managers with less control over their own surveys and with a greater need to rely on other members of the cross-survey team. Technical solutions must balance the quality of individual surveys against the quality goal of estimates that can be related across surveys.

The process outlined below considers managerial as well as technical aspects of improving the integration of outputs across surveys. It arises from analysis of successes and failures and the anticipated future needs of the Survey Integration Project at ABS. Five stages are discussed that are considered vital in ensuring increased consistency across surveys. The five stages are:

1. Investigation and analysis to better understand the methodologies currently in place, their relationship to user needs, any competing requirements in terms of the best methodological approach to meet cross-survey and survey-specific user needs, and the implications of alternative methodological approaches for each of these sets of needs.

2. Agreement about the priorities in terms of user needs, the implications of results of the investigations and analyses for these needs, and appropriate methodologies to best meet the various needs.

3. Implementation of agreed methodological revisions.

4. Mechanisms, including monitoring, to maintain the gains in improved comparability of survey outputs and to maintain the awareness of cross-survey users as well as the survey-specific users.

5. Experimentation and evaluation to achieve improvement in methodologies within the framework of maintaining comparability across data.

There are parallels between these steps and the development of Current Best Methods which Morganstein and Marker describe in Chapter 21. Both require the involvement of staff at all levels to develop consistent procedures, senior management involvement, and monitoring and reviewing procedures. Survey integration at ABS is aimed at achieving consistent methodologies in a range of surveys, whereas the main, but not exclusive, focus of Morganstein and Marker is on defining a process that can be applied consistently across surveys.

23.4.1   Investigation and Analysis

Investigation across a number of surveys will require analysis of documentation and data. Procedures used in surveys may not be well documented or up-to-date, and if they are documented, it will still be necessary to see how precisely the documentation is followed. Documentation may not use consistent terminology and may have no central repository. The investigation stage need not be a comprehensive and painstaking recording of existing practices, especially if it becomes clear that change is likely. Experience indicates investigation may be better directed at some measurement of major differences, discovering reasons for existing practice and their importance in terms of meeting specific user needs, and the gathering of ideas on feasible alternative procedures.

The purpose of analysis is to understand how the diversity might affect comparability of estimates across surveys, and to develop a framework for improving this comparability through consistent procedures. Analysis might identify groups of problems, as well as solutions applicable in these groups. Analysis may also provide insight into the effects of proposed procedural changes on data series, and on specific survey users.

23.4.2   Agreement

Before any attempt at implementation is made, there needs to be agreement on the priorities for cross-survey users and survey-specific users. A common understanding is important if an appropriate balance in methods is to be achieved and, more importantly, maintained.

In addition to the consultation process which the surveys maintain with their users, the process of reaching agreement involves extensive consultation between methodologists and the surveys, to ensure that needs are understood and the procedures proposed are workable and acceptable and that any diversity is justified and accepted. Endorsement by senior management should make the process of reaching agreement easier, by making it clear that consistency is a corporate goal and, possibly, by providing required resources and supporting infrastructure. However, not only the endorsement by senior management is necessary, but the commitment of people at all levels to work as a team is vital.

23.4.3 Implementation

The time it takes to implement new procedures can vary. Changes to clerical procedures can be implemented quickly, but it is often better to delay minor changes to complex systems until a number of changes are stockpiled. When clerical or software facilities are provided centrally, changes in individual surveys must wait in turn. Where changes result in a time series break, user requirements play a major role in determining the timing of the change. The process of change needs to be coordinated and managed, especially if it occurs in a number of aspects and a number of surveys over a long period.

Some central facilities may be provided to assist change. This is obviously the case where consistency is increased by replacing local services with central ones. Where procedures are made uniform, but implemented locally, there could be central support through documentation, training, and general advice and assistance on transition problems. Generalized computer systems may also be developed or modified to support the new procedures.

23.4.4 Mechanisms to Maintain Gains

To ensure ongoing understanding of and commitment to cross-survey user needs, ongoing support mechanisms are required. These may take the form of cross-survey project teams or committees responsible for ensuring that cross-cutting issues are understood and addressed. There may also be processes to improve the relationships between the individual surveys and the key cross-survey users, in particular to ensure that survey managers identify cross-survey users as key users. The relationship between the survey and the cross-survey analyst could be built by joint work to solve some of the deficiencies of the survey output from the cross-survey view.

In addition to monitoring the new procedures at implementation, the improved consistency must be maintained. If the changes are to clerical procedures, then direct external monitoring is difficult, and it is up to surveys to enforce standard procedures. If a central service is provided, then the individual survey managers will require the central group to monitor and report on quality aspects of relevance to the users. For example, if a common quarterly frame is used, then reconciliations of changes from the previous quarter could be provided.

23.4.5 Experimentation and Evaluation

Methodological uniformity can inhibit innovation and adaptation to changed circumstances. Although some innovation comes from central research groups, it also comes from experimentation by individual survey groups. With the desire to maintain consistent procedures, individual surveys may be less likely to experiment. To minimize this risk, there should be mechanisms to encourage some experimentation away from the production environment and to encourage the examination of problems from the perspective of a number of surveys.

Lead times for implementing new approaches should take account of the slower decision-making process which may accompany uniformity. However, once successful new approaches have been developed, they can then be rapidly accepted as best practice across a range of surveys and implemented across these surveys simultaneously. Finally, it will be necessary to regularly review procedures to ensure that there has not been incremental change, through a number of small decisions, to less consistent methodologies across surveys.

23.5 CASE STUDY—SAMPLE AND FRAME MAINTENANCE PROCEDURES

This section describes a subproject within the ABS Survey Integration Project: the development and implementation of standard sample and frame maintenance procedures. As mentioned in Section 23.2, sample and frame maintenance procedures are rules for identifying and treating units whose observed structure or status differs from that recorded on the frame. The important aspects of this case study are: (1) that it arose from a real problem with survey estimates; (2) that it occurred in a statistical agency with a long standing conceptual framework, infrastructure, and groups to support integrated economic statistics; and (3) that the involvement of management at various levels has been a key part of the strategy for achieving consistency.

This work started as an attempt to remedy a situation in which changes to the business register caused large and divergent changes to estimates of major economic indicators which were also important contributors to the national accounts. Investigations revealed that a major cause was the different procedures surveys used for sample and frame maintenance. Although analysis of the changes made compensatory adjustments possible, it was considered important to reduce the diversity of the procedures.

Since 1969, ABS has worked under the conceptual framework of standard unit definitions and standard classification, and the infrastructure of a business register to support integrated economic statistics. Organizationally, there has been a central group to further develop and monitor the standards for economic statistics and a central group to provide methodological advice. Nevertheless, this conceptual and organizational infrastructure did not prevent the diversity of practices which caused the problematic estimates. One explanation for the diversity of practices is that the ABS is not functionally specialized. At the ABS, a survey manager is directly responsible for the processing of a survey, as well as for survey development and the analysis and publication of the results. Although methodologists had some involvement in developing sample and frame maintenance procedures and had a common theoretical framework, their focus had been on achieving particular survey goals, rather than maintaining a consistent or uniform approach.

The involvement of management was a key part of the strategy. The input of survey managers and their staff was needed to develop a set of workable procedures and the continuing support of senior management has helped to maintain a focus on consistent procedures.

23.5.1   Investigation and Analysis

The ABS uniform sample and frame maintenance procedures developed out of a large-scale survey of units on the ABS business register. The survey investigated units with a simple structure which had not been included in recent surveys. The main purpose was to remove dead units. It was expected that, except for the effect of sampling error, there would be no change in estimates based on the business register, because the units removed would have contributed zero to the survey estimates. However, it was found that there were substantial effects on estimates from major surveys. These differences arose because the surveys and the business register used different procedures for deciding whether a unit was dead, and once identified as such, the dead units were treated in different ways. In order to prevent substantial effects on survey estimates from register maintenance in the future, the business register and all surveys should identify deaths in the same way. Further, although the experience itself related only to the treatment of deaths, it was recognized that the issue was equally applicable to most sample and frame maintenance procedures.

The first part of the investigation was the identification of the problem and the overall solution described above. The second part was the detailed investigation of all sample and frame maintenance procedures across a wide range of surveys. The complete range of procedures was discussed with staff from different surveys with a view to documenting actual practice, and to explore possibilities for standard rules. Documentation of existing procedures was collected, but varied considerably in its detail and currency. The analysis concentrated on surveys which use the business register, although other surveys were encouraged to use an appropriate subset of the procedures.

It is worth considering how such diversity arose when there had been a central group of methodologists who might have encouraged consistency. An explanation lies in the survey managers' priorities and the methodologists' relationship to their clients—the survey managers. The survey managers often saw the specific requirements and problems of their survey as more important than consistency with other surveys. In part, they were responding to the needs of survey-specific users, but, even when they had important cross-survey users, they also saw them being well served by practices that were good for the specific survey. From a client service view, the methodologists followed the survey managers' priorities. Furthermore, it is more interesting for the methodologist to develop a solution to a particular problem than to apply a standard method.

23.5.2 Agreement

While investigating and developing procedures, it became clear that a completely uniform set of procedures was not feasible. Some surveys would need variants of the standard procedures. To limit the extent of diversity, these variants were kept within the structure of the standard procedures. They went through an approval process and were fully documented in the manual described below.

The draft procedures were circulated to all surveys. An editorial group commented on drafts and the final versions. This approach is consistent with the suggestion of Morganstein and Marker (Chapter 21) that a team of four to six people develop such documents. When the procedures were near final, they were reviewed by each survey to check that they could be implemented and to discuss any need for variants on the standard procedures. An executive level committee, including management of both survey and support groups, approved the procedures and any variations. The committee reported to senior management, both on progress and to maintain their in-principle support.

The process drew wide support. The removal of deaths from the business register had caused salient problems both for individual surveys and for cross-survey users. Thus, consistency was not seen as a mere abstraction, but as something which could prevent jolts to individual survey time series.

23.5.3 Implementation

There were two aspects to the implementation of the new procedures—changes to clerical processes and changes to computer systems. Timetables for each were negotiated with the surveys. General support from the methodology group was provided in the form of a manual, a training package, and specifications for standard flags to record which procedures were used for a unit. These standard flags were implemented in a generalized input processing system. As clerical changes would occur before general systems to support them were built or individual systems fully modified, transitional arrangements were made. The effect of changes would be measured, but the methods would differ across surveys as their existing procedures vary.

Implementation in systems has been problematic. There is a wide range of processing systems of varying ages and flexibility. Even partial implementation in the older systems is expensive and risky, and any implementation may need to wait until the systems are redeveloped. These problems highlight the relationship between achieving and maintaining consistent methodologies and developing and using generalized computing systems. With general systems in place it is easy to make changes to a number of surveys, either for greater consistency or to improve procedures. On the other hand, consistent methodologies allow generalized systems to be simpler and cheaper.

23.5.4   Maintenance

The executive committee referred to above will continue to meet on a regular basis to address methodological issues cutting across individual survey areas. It is responsible for ensuring the ongoing applicability of the procedures and for the approval of changes and variations to the procedures. A manual is readily available as an electronic corporate document and the methodology group maintains it and provides ongoing training.

To monitor the use of the procedures, two groups of measurements were proposed in addition to the usual monitoring of clerical operations by line management. Firstly, sample counts of the units identified under each procedure will be monitored by survey managers over time to detect any sudden change in practice. Secondly, population estimates of the number of units identified under each procedure (e.g., known deaths) will be derived from the individual surveys and compared across surveys.

23.5.5   Experimentation and Evaluation

The experimentation and evaluation stage has not been reached, but some activity has been foreshadowed. One such activity would be an overall review in the event that business register practices change. Best guess rules for whether a unit is dead or alive were developed and tested, but should be retested periodically.

23.6   CONCLUSION

We have focused on managing quality in survey outputs for those users who relate data across surveys. Key aspects are to understand the full set of user needs for a survey, including the needs of the cross-survey analysts, to understand current procedures and methodologies and their implications for meeting the user needs, and to continually look for better methodologies to meet these needs. We note that the implementation of better solutions for the cross-survey analyst will involve not only developing appropriate technical solutions, but also finding strategies to align the organization's culture to better meet these needs. There are very real costs of change, both in terms of effort and possible breaks in time series, and barriers to change in the form of the needs of survey specific users and the desire of surveys to maintain reasonable control over important processes. Accepting the costs and overcoming the barriers require a significant commitment from corporate and survey managers. The desire to innovate and improve procedures must also be maintained, but in the context of a team approach across survey areas.

ACKNOWLEDGMENT

The views expressed in this chapter are those of the authors and do not necessarily reflect the opinion of the Australian Bureau of Statistics.

REFERENCES

Biemer, P.P., and Fecso, R.S. (1995), "Evaluating and Controlling Measurement Error in Business Surveys," in B.G. Cox, D.A. Binder, B.N. Chinnappa, A. Christianson, M.J. Colledge, and P.S. Kott (eds.), *Business Survey Methods*, New York: Wiley, pp. 257–282.

Colledge, M. (1987), "The Business Survey Redesign Project: Implementation of a New Strategy at Statistics Canada," *Proceedings of the Third Annual Research Conference*, U.S. Bureau of the Census, Washington, DC, pp. 550–576.

Colledge, M., and March, M. (1993), "Quality Management: Development of a Framework for a Statistical Agency," *Journal of Business and Economic Statistics*, 11, pp. 157–166.

Dodds, D.J., and Colledge, M. (1990), "Using the New Business Register at Statistics Canada: Problems and Solutions," *Proceedings of the Annual Research Conference*, U.S. Bureau of the Census, Washington, DC, pp. 596–614.

Griffiths, G., and Linacre, S. (1995), "Quality Assurance for Business Surveys," in B.G. Cox, D.A. Binder, B.N. Chinnappa, A. Christianson, M.J. Colledge, and P.S. Kott (eds.), *Business Survey Methods*, New York: Wiley, pp. 673–690.

Kovar, J.G., and Whitridge, P.J. (1995), "Imputation of Business Survey Data," in B.G. Cox, D.A. Binder, B.N. Chinnappa, A. Christianson, M.J. Colledge, and P.S. Kott (eds.), *Business Survey Methods*, New York: Wiley, pp. 403–424.

Nijhowne, S. (1995), "Defining and Classifying Statistical Units," in B.G. Cox, D.A. Binder, B.N. Chinnappa, A. Christianson, M.J. Colledge, and P.S. Kott (eds.), *Business Survey Methods*, New York: Wiley, pp. 49–64.

Struijs, P., and Willeboordse, A. (1995), "Changes in Populations of Statistical Units," in B.G. Cox, D.A. Binder, B.N. Chinnappa, A. Christianson, M.J. Colledge, and P.S. Kott (eds.), *Business Survey Methods*, New York: Wiley, pp. 65–84.

U.N. Statistical Office (1990, *International Standard Industrial Classification of All Economic Activities* (*ISIC*), Statistical Papers, Series M, No. 4, Revision 3, New York: United Nations.

Worton, D.A., and Platek, R., (1995), "A History of Business Surveys at Statistics Canada: From the Era of the Gifted Amateur to That of Scientific Methodology," in B.G. Cox, D.A. Binder, B.N. Chinnappa, A. Christianson, M.J. Colledge, and P.S. Kott (eds.), *Business Survey Methods*, New York: Wiley, pp. 633–653.

CHAPTER 24

Evaluating Survey Data: Making the Transition from Pretesting to Quality Assessment

James L. Esposito
U.S. Bureau of Labor Statistics

Jennifer M. Rothgeb
U.S. Bureau of Census

24.1 INTRODUCTION

The intent of this chapter, in general terms, is to describe the problem-solving behavior of survey researchers who engage themselves in efforts to detect and minimize sources of measurement error (Biemer *et al.*, 1991; Groves, 1987, 1989; Groves *et al.*, 1988; Turner and Martin, 1984). We are concerned specifically with efforts by survey researchers to obtain high-quality survey data through improvements in questionnaire evaluation and design. As noted by Dippo (Chapter 20), a first step towards implementing continuous process improvement is to define appropriate measures and build them into the survey measurement system. As survey researchers embrace the tenets of the Cognitive Aspects of Survey Measurement (CASM) movement (Jabine *et al.*, 1984; Jobe and Mingay, 1991), they have developed models of the question asking and answering process using concepts from cognitive psychology (Tourangeau, 1984) and social psychology (Cannell *et al.*, 1981; Esposito and Jobe, 1991), and have developed methods and techniques for assessing measurement error. By providing examples and an illustration of how some of these techniques have been used to evaluate survey questions, we hope to convince readers that their use leads to improve-

Survey Measurement and Process Quality, Edited by Lyberg, Biemer, Collins, de Leeuw, Dippo, Schwarz, Trewin.
ISBN 0-471-16559-X © 1997 John Wiley & Sons, Inc.

ments in data quality and that their utility extends beyond pretesting to post implementation quality assessment as well. Our emphasis is on the *process* of evaluation, and not the strengths and weaknesses of specific techniques *per se*. (For a discussion of strengths and weaknesses of these methods see Presser and Blair, 1994; Esposito *et al.*, 1992; Cannell *et al.*, 1989; Fowler and Roman, 1992.)

The chapter is organized as follows. In Section 24.1, after briefly addressing the issue of survey quality, we identify and discuss some of the techniques that have been developed to evaluate the effectiveness of survey questions and review how others have used these techniques to assess the quality of data obtained from interviewer administered questionnaires. In Section 24.2, we report on how some of these techniques were used to identify problems and assess improvements in the quality of data from the redesigned U.S. Current Population Survey (CPS), and in Section 24.3, we propose an idealized quality assessment program for major social and economic surveys, address pragmatic issues associated with such a program, and discuss some of the benefits and costs (nonmonetary) associated with quality assessment research.

24.1.1 Survey Quality

Generating a definition of survey quality that researchers from a variety of survey-related disciplines would find acceptable may well be an impossible task. The lack of an inclusive definition, however, does not mean that the concept is poorly understood. Bailar (1984) views survey quality as a multidimensional concept, one that can be viewed in terms of the interlocking steps or stages involved in producing a given data set. She specifies a number of characteristics that, if present, would do much to assure the collection of high-quality survey data (e.g., probability sampling; conceptual clarity; operational definitions that fit concepts; reporting by the most knowledgeable respondent; accurately coded and weighted data; small sampling variances; verification procedures that show little inconsistency (Bailar, 1984, p. 43)). Other researchers (Anderson *et al.*, 1979; Groves, 1989) discuss survey quality in terms of the various types of errors that detract from data accuracy. Groves (1989), for example, distinguishes between nonobservation error (i.e., coverage error, nonresponse error, and sampling error) and measurement error (i.e., error arising from the interviewer, respondent, questionnaire, and mode of data collection).

Although the approaches are different, Bailar's and Groves's conceptualizations of survey quality are helpful in making the concept less abstract and in setting limits for this chapter. With regard to the latter, we will focus on methods and techniques that enable researchers to detect and potentially reduce sources of measurement error. (We will not be addressing the effects of nonobservation errors on survey quality.) For example, cognitive interviews and respondent debriefing techniques (e.g., follow-up probe questions) can be used to determine when survey concepts are misunderstood. Interactional coding, in addition to indicating where interviewers have difficulty reading questions as worded or where respondents have difficulty providing adequate answers, can also be used to monitor interviewer performance and evaluate mode effects. When detected by such techniques, the presumption is that problems will have a negative effect on data quality; in some cases, however, the magnitude of the effect may be difficult to assess quantitatively. Nevertheless, once identified, it is assumed that questionnaire designers (i.e., teams of subject matter specialists, behavioral scientists, and survey methodologists) can attenuate problems via question modifications (e.g., changes in wording or question structure) or other strategies (e.g., interviewer training, mode changes). The net result, if assumptions prove true, would be an increase in data quality via a reduction in measurement error. (See Section 24.3.1 for more on the topic of identifying and correcting problems with survey questions.)

24.1.2 Evaluating Survey Questionnaires, Tools of the Trade

There is an expanding literature (e.g., DeMaio, *et al.*, 1993; Forsyth and Lessler, 1991; Willis, 1994) on the methods and techniques used for pretesting survey questions (see Table 24.1). These methods and techniques can be differentiated in terms of where evaluative data are usually collected (office, laboratory, or field), the purpose of the evaluation (pretesting (PT) or quality assessment (QA)), and the source (S) and target (T) of the analytical data (respondents, interviewers, others). The information in the *purpose* column highlights a point seldom made in the literature: many of the same methods used to develop and field test survey questionnaires can be used—and have been used—for the purpose of evaluating how well questionnaire items are working *after* the questionnaire has been finalized and put into production. Information contained in Table 24.1 could be used in the early stages of planning a questionnaire evaluation plan. Researchers wishing to develop a multimethod quality assessment program could use information in the *source and target* columns to identify techniques (e.g., debriefings, behavior coding, cognitive forms appraisal) that draw information from a variety of sources (i.e., interviewers, respondents, experts), and, in so doing, allow for multiple perspectives in identifying potentially problematic survey questions.

We believe that the conceptual distinction between pretesting and quality assessment research rests with the *status of the questionnaire*. If the questionnaire is in an early developmental stage or a field testing stage when evaluative research takes place, and if changes to the questionnaire can still be made after testing, then we refer to that evaluative work as *pretesting research* (see Figure 24.1). If the questionnaire is currently in use and the purpose of the evaluative research is to determine, for example, the extent to which survey questions accurately measure the concepts they are intended to measure, then we would classify that evaluative work as *quality assessment research*. (Other uses are noted in Section 24.3.) These two designations of survey research are not exhaustive. For example, not considered above is methodological research

Table 24.1 Some Methods and Techniques Used to Evaluate Survey Questions

Method/Technique	Location of data collection	Purpose	Source (S) and target (T) of analytical data			
			Interviewers	Respondents	Other S/T	Survey Qs
Cognitive/Intensive interviews						
(1) using think-aloud technique	Lab	PT		S		T
(2) using probing technique	Lab	PT		S	T (concepts)	T
(3) using other techniques	Lab	PT		S		T
Debriefings						
(1) post-survey follow-up probes	Field	PT/QA		S	T (concepts)	T
(2) debriefing questionnaires	Field	PT/QA	S	S		T
(3) focus groups	Field/Lab	PT/QA	S	S		T
Experiments (e.g., split ballot tests)	Field/Lab	PT/QA	S	S	S (context)	T
Expert reviews						
(1) expert panels	Office	PT			S (experts)	T
(2) cognitive forms appraisals	Office	PT/QA			S (experts)	T
Interaction coding						
(1) behavior coding	Field	PT/QA	S	S	S (context)	T
(2) conversation analytic coding	Field/Lab	PT/QA	S	S	S (context)	T
(3) protocol coding	Lab	PT/QA	S	S		T
Item nonresponse analysis	Field	PT		S		T
Reinterviews	Field	QA	S	S	S (context)	T
Response distribution analysis	Field	PT		S	S (design team)	T
Vignettes	Lab/Field	PT	S	S	T (concepts)	T

Note: PT refers to pretesting and QA refers to quality assessment.
Primary Sources: DeMaio et al., 1993; Forsyth and Lessler, 1991; Willis, 1994.

INITIAL DESIGN PHASE

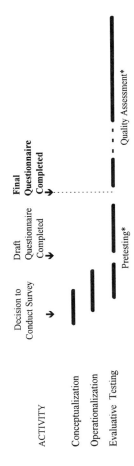

REDESIGN PHASE

* **Pretesting**, as we are using the term, refers to evaluative testing conducted prior to finalizing a particular questionnaire and includes developmental testing and field testing. **Quality assessment** (e.g., reinterview) refers to evaluative testing conducted after a particular questionnaire has been finalized. Quality assessment, when done, may or may not be a continuous process.

Figure 24.1 A simplified timeline of the survey questionnaire design and redesign process.

545

conducted primarily to demonstrate the utility of a particular technique (e.g., behavior coding) for a specific purpose (e.g., evaluating interviewer performance; see Mathiowetz and Cannell, 1980). Moreover, these two designations are not always easy to distinguish. For example, an ongoing quality monitoring program might reveal a serious flaw with a particular survey question that requires a more intrusive intervention (e.g., a change in question wording) than normally used (e.g., targeted interviewer training). Lastly, as depicted in Figure 24.1, quality assessment research and pretesting research can take place simultaneously for the same survey (e.g., during questionnaire redesign efforts). (We refer to this figure again in Section 24.3.1.)

As Table 24.1 illustrates, survey evaluation data can be collected in various locations: laboratories, offices, and field settings. In the following sections, we will consider first laboratory- and office-based methods and research, and then field-based methods and research.

24.1.3 Laboratory- and Office-Based Methods and Research

Laboratories provide investigators with a controlled environment within which to conduct questionnaire-related evaluative research (see DeMaio and Rothgeb, 1996; DeMaio et al., 1993; Dippo et al., 1993; Dippo and Norwood, 1992; Sirken 1991; Willis et al., 1991, for examples and reviews of research conducted at U.S. governmental research laboratories). Laboratory-based methods (e.g., cognitive interviews) generally focus on psychological aspects of the question-answering process (e.g., cognitive processes such as comprehension; motivational processes such as satisficing or selective reporting; see Forsyth and Lessler, 1991; Lessler et al., 1989; Royston et al., 1986; Willis, 1994). These methods have been used regularly—if not always correctly or consistently (see Blair and Presser, 1993)—by researchers to assess the understandability of survey questions and concepts. Unless there are good reasons for not doing so, researchers select and interview individuals who possess characteristics similar to those of the target population.

24.1.3.1 Cognitive and Intensive Interviews
When applied to the survey response process, the methods of cognitive and intensive interviewing rest on a multistage model of human information processing (see Willis, 1994; Willis et al., 1991). This model depicts the respondent as attempting to comprehend the target question, retrieving relevant information from memory, making decisions regarding level of effort and self-presentation, and ultimately producing a response. No presumption of strict linear processing is made; for some questions, respondents presumably switch back and forth between stages. In the *cognitive interview*, the researchers attempt to identify problems with survey questions by having respondents "think out loud" as they formulate their answers (concurrent) or shortly after they formulate their answers (retrospective). Oftentimes, researchers will incorporate probing techniques into the interview by having respondents answer a limited

number of general or item-specific probe questions. *Intensive interviews* assume the form of a cognitive interview, but differ from the latter in that they are supplemented with various other techniques—such as, paraphrasing, confidence ratings, response latency measures (Royston, 1989; Royston *et al.*, 1986; also see DeMaio and Rothgeb, 1996). Using such procedures, researchers gain insights into where respondents are experiencing cognitive difficulties and where there may be discrepancies between question intent (survey sponsor, question designer) and question interpretation (respondent).

24.1.3.2 Other Laboratory- and Office-Based Methods

In addition to cognitive and intensive interviews, there are a variety of other methods (e.g., expert evaluations, rating tasks) that are well suited to the research laboratory. Interested readers should refer to Forsyth and Lessler (1991) for a discussion of these methods. Office-based research methods are now appearing (e.g., conversation analytic coding; automated coding) that reflect the development of specialized analytical techniques (e.g., linguistic analysis, computerized coding). Other methods take advantage of the knowledge of experts who have proven expertise in designing questionnaires (e.g., cognitive forms appraisal, expert panels).

With the relatively recent dawn of the CASM movement and the establishment of three cognitive laboratories within the U.S. Federal statistical system (at the National Center for Health Statistics, the Bureau of Labor Statistics, and the Bureau of the Census), cognitive research in governmental surveys has proliferated. To give readers a sense of how these methods are used, we describe briefly some of the more innovative studies we have found in the literature.

24.1.3.3 Pretesting Research Examples

Blixt and Dykema (1993) have developed an innovative pretesting method, called *systematic intensive interviewing*, that integrates intensive cognitive interviews—involving think-aloud procedures, paraphrasing, memory probes, and other probing techniques—with behavior coding as a way of identifying problematic survey questions. In this study, a variety of different probes were asked during the cognitive interviews: definition probes ("When I use the word 'nutrition', what does that mean to you?"), frame-of-reference probes ("When I said 'other people,' what did that mean to you?"), and motivation-relevant probes ("Do you find it embarrassing to talk about how often you drink?"). Trained researchers conducted intensive interviews, which were audiotaped and later coded using: (1) standard respondent behavior codes (e.g., qualified response, request for clarification), and (2) specialized codes to reflect content generated in response to the probes mentioned above (e.g., critiques of the survey question and response options). Blixt and Dykema review data for two problematic questions and their analysis suggests that systematic intensive interviewing can be an effective method for identifying both cognitive and motivational problems with survey questions.

Bolton (1993) has developed a method of analyzing survey questions, called *automatic coding*, that integrates cognitive interviews and computerized content analysis. The coding scheme is well grounded in cognitive theory and matches instances of five verbal categories (i.e., repeat, forget, confidence, "can't say," and "don't know") and four nonverbal cue categories (e.g., pauses, broken utterances) with the contents of cognitive interview transcripts. She then used factor analysis (i.e., factor scores for comprehension, retrieval, judgment, and response) as a means for identifying specific cognitive problems with target questions. In this research, Bolton evaluated alternative customer satisfaction surveys and contrasted automatic coding with observational monitoring (i.e., a technique similar to behavior coding). She found that automatic coding was better at detecting comprehension, retrieval, and judgment problems, but that observational monitoring was better at detecting response difficulties.

24.1.3.4 *Quality Assessment Research Examples*
Forsyth *et al.* (1992) and Forsyth and Hubbard (1992) have developed a technique, called *cognitive forms appraisal*, that we believe holds great promise as a questionnaire evaluation tool—both for quality assessment and for pretesting. Items on a questionnaire are coded by experts using a coding scheme, which is grounded in a cognitive model of the survey response process. Detailed coding categories were developed to assess the demands that survey questions place on an individual's comprehension, interpretive, memory, judgment, and response generation processes. Forsyth *et al.* used this coding scheme to evaluate questions on the National Household Survey on Drug Abuse and found evidence of several types of problems (e.g., vague or ambiguous terminology, response categories with hidden definitions); later, the investigators were able to validate their findings using a small number of think-aloud interviews. (For another example of quality assessment research, see Dykema *et al.*, Chapter 12.)

24.1.4 Field-Based Methods and Research

Field-based methods draw information from two principal sources, interviewers and respondents (DeMaio *et al.*, 1993) and, as the name suggests, these techniques are conducted in the field, usually at or very near to where the survey is actually being conducted. We will consider four groupings of methods: interviewer debriefings, respondent debriefings, interaction analysis, and reinterview programs.

24.1.4.1 *Interviewer Debriefings*
The experienced field interviewer is usually one of the first persons to know how well survey questions are being understood by respondents (Converse and Schuman, 1974; DeMaio, 1983; cf. Bischoping, 1989). This expert knowledge can be tapped by a variety of methods; three of the more common techniques are to use rating forms, focus groups, and interviewer debriefing questionnaires (Fowler, 1989; Fowler and Roman, 1992; Esposito and Hess, 1992). The

debriefing questionnaire has the advantage of being more representative of the larger group of interviewers; focus groups have the advantage of providing greater depth and insight into the questions that may be causing problems for respondents and interviewers. Though useful, one must recognize the subjective nature of interviewer debriefing data and take steps to obtain input from a variety of sources (e.g., interviewers, respondents, experts, coded interaction data from actual interviews) when attempting to identify and to amend problematic survey questions.

24.1.4.2 Respondent Debriefings

Perhaps the most obvious way to gather data on how well survey questions are understood is to ask the persons answering the questions—the respondents. Belson (1981) has done some very interesting research using follow-up probes to determine when concepts and terms are being understood in a way unintended by the survey sponsor or the questionnaire designer. Other researchers have used vignettes as a way of debriefing respondents (Martin, 1986; Polivka and Martin, 1992). Another approach would be to use focus groups for the same purpose (Palmisano, 1989). The commonality that unites each of these techniques is that respondents are providing information that can be used to assess their understanding (or misunderstanding) of survey questions.

24.1.4.3 Interaction Analysis

Interaction analysis involves monitoring survey behavior (most often in a natural survey context), coding behavioral exchanges between interviewers and respondents, and tabulating the frequency of a predetermined set of behavior codes for specific survey questions. Questions with relatively high frequencies of unacceptable behavior codes are flagged as problematic (for an example, see Dykema et al., Chapter 12; for reviews, see Fowler and Cannell, 1996; Esposito et al., 1994). The most commonly used application of interaction analysis, *behavior coding*, was developed by Cannell and his colleagues (e.g., Cannell et al., 1968; Marquis, 1969; Cannel et al., 1975). Morton-Williams and Sykes (e.g., Morton-Williams, 1979; Morton-Williams and Sykes, 1984; Sykes and Morton-Williams, 1987) have also made significant contributions to the literature.

24.1.4.4 Reinterview Programs

As the name of this method suggests, reinterviews involve conducting a second interview with a given unit (Biemer and Forsman, 1992; Cantwell et al., 1992; Forsman and Schreiner, 1991). And insofar as there are a variety of purposes (e.g., to evaluate field work; to estimate response variance and response bias) and ways of conducting the second interview (e.g., same respondent vs most knowledgeable respondent; same interviewer vs different interviewer; same questions vs conceptually similar questions), one can think of reinterview programs as involving a family of closely related techniques. Depending on its purpose and scope, a reinterview program may require a greater investment of resources relative to other evaluation methodologies and may alienate some respondents (Blair and Sudman, 1993).

24.1.4.5 Pretesting Research Examples

Work by Fowler (1992) and his colleagues (Cannell et al., 1989) provides a particularly instructive example of the use of behavior coding (and response distribution analysis) to evaluate unclear survey questions. These investigators developed a 60-item questionnaire comprising questions from a variety of health surveys. Fowler's research targeted seven of the more problematic questions identified through standard behavior coding procedures (e.g., "Do you exercise or play sports regularly?"). Each of these questions contained poorly defined terms or concepts (e.g., exercise). The questions were revised and incorporated into a new questionnaire that was subsequently readministered; in one case, one very long and difficult question was replaced with several shorter questions. Behavior coding and response distribution analyses revealed significant improvements for many of the questions. Even after the modifications, however, several of the revised questions had relatively high percentages of inadequate answers and requests for clarification. This outcome underscores two important points about "fixing" survey questions: (1) sometimes an attempt to fix one problem results in the creation of one or more other problems, and (2) some questions are very difficult to repair (e.g., complex questions), and doing so could affect respondent burden by increasing the total number of questions being asked in the survey.

Another study with interesting implications was conducted by Willis (1991) on a draft health questionnaire that involved behavior coding (done while the interview was in progress), interviewer debriefings, and observer debriefings. After completing 49 field interviews, observers and interviewers were debriefed on their perceptions regarding problems with specific survey questions. Willis found that the debriefing produced more reports of problematic questions (87) than behavior coding (62). Of the 94 questions identified as problematic, the two methods agreed 59 percent of the time. Rather than be overly concerned by the relatively low level of between-method correspondence, Willis made a very provocative observation: "Under circumstances in which the behavior coding appears to be more conservative than the debriefing, it is perhaps best to view those questions found to be problematic under both methods as *clear candidates for modification*, and to consider as additional possibilities those questions identified by the debriefing alone" (p. 12, italics added). We will expand on this insight later in the chapter (see Section 24.3.1).

24.1.5.6 Quality Assessment Research Examples

Although their research was clearly done with a different purpose in mind (i.e., to demonstrate the utility of two distinct pretesting methodologies), recent research by Oksenberg, Cannell, and their colleagues (Oksenberg et al., 1991; Cannell et al., 1989) provide some very useful examples of how to do excellent quality assessment research. In the work reported by Oksenberg et al., the researchers started by generating a 60-item questionnaire covering a wide range of medical issues and question types. All questions were currently being used in major health surveys and presumably had been pretested in some manner. The research team administered the questionnaire by telephone to 164 respondents. Interviewer-respondent interactions for all interviews were behavior coded using standard procedures. The findings were startling: 60 percent of the questions tested had the inadequate answer code assigned 15 percent of the time or more; this code is given when the respondent's answer does not satisfy the question objective. Fifty percent of the questions had the request for clarification code assigned 10 percent of the time or more. In 104 interviews, four types of special follow-up probe questions (i.e., general, comprehension, retrieval and response category selection probes) were asked after target questions were asked. Of the different types of probes used, the comprehension probes proved to be the most useful. These probes uncovered problems with questions that some respondents originally answered with little difficulty; such problems go undetected by behavior coding. By recognizing and demonstrating the complementary aspects of question evaluation methods (in this case, behavior coding and follow-up probes), these researchers make a major contribution to the literature. An equally important contribution, strikingly demonstrated by the data presented above and in their report, was that conventional pretesting techniques do not always catch serious problems with survey questions.

24.2 MAKING TRANSITIONS FROM PRETESTING TO QUALITY ASSESSMENT: THE REDESIGN OF THE CURRENT POPULATION SURVEY (CPS) AS A CASE STUDY

In the mid-1980s, inspired by the increased application of cognitive psychology to survey measurement and advances in computer assisted interviewing, the U.S. Bureau of Labor Statistics (BLS) and the U.S. Bureau of the Census decided to redesign the Current Population Survey (CPS) for use in a user friendly, computer assisted interviewing environment. Both pretesting and quality assessment methodologies were used in the redesign (Campanelli *et al.*, 1991; Esposito *et al.*, 1991, 1992; Rothgeb *et al.*, 1991). During the period 1986–89, preliminary research was conducted to identify conceptual problems in the CPS that needed to be addressed in the redesign process (see Copeland and Rothgeb, 1990, for a brief review; also see Martin, 1987). A variety of techniques were used by BLS and Census Bureau researchers in conducting studies to identify problems with the CPS questionnaire:

- interviewer debriefings using focus groups with CPS interviewers (U.S. Bureau of Labor Statistics, 1988)
- respondent debriefings using focus groups, follow-up probe questions, and vignettes (Palmisano, 1989; Campanelli *et al.*, 1989; Campanelli *et al.*, 1991)

- categorical sorting tasks (Fracasso, 1989)
- field experiments (Westat/AIR, 1989a, 1989b).

Findings from these studies, and other sources (e.g., National Commission on Employment and Unemployment Statistics, 1979; U.S. Bureau of Labor Statistics, 1986, 1987, 1988), helped to set the stage for the redesign of the CPS.

24.2.1   The CPS Redesign: Pretesting Research (Phase One and Phase Two)

The CPS redesign involved three phases. The first two phases were designed to evaluate items on alternative versions of the CPS questionnaire.

24.2.1.1   Phase One

During the phase one field test, two alternative CPS questionnaires (versions B and C) were field tested with the existing CPS (version A) serving as the control. After this initial pretesting phase, the best questions from the alternative questionnaires were synthesized into a single alternative questionnaire (version D), which was tested in phase two. The following pretesting methodologies were used during this initial phase: (a) interviewer debriefings, (b) field-based respondent debriefings, (c) behavior coding, and (d) item-based response analysis.

Interviewer Debriefings
Phase one research utilized two aspects of interviewer debriefing: (1) the completion of a self-administered questionnaire, and (2) participation in focus group discussions with other interviewers. Although the two interviewer debriefing techniques utilized different formats, they sought to collect similar information and, as a result, shared a similar underlying structure. Both the questionnaire and the moderator's focus group guidelines were structured to proceed from interviewers' general preferences for a particular questionnaire version (A, B, or C) to their specific evaluations of a particular question, or series of questions. From interviewer debriefings, researchers were able to obtain interviewers' perceptions of problems with specific questionnaire items.

Respondent Debriefings
To obtain information about respondents' understanding of CPS questions, field-based respondent debriefings were conducted with household respondents after their fourth and final monthly interview. This postinterview consisted either of vignettes or a series of appropriate follow-up probe questions. For example, when given work or nonwork vignettes to evaluate (e.g., "In addition to attending her regular college classes, Bill (Jan) earned some money tending bar for a fraternity (sorority) party last week."), respondents were asked to classify the target person (Bill or Jan) as either working or not working. To assess whether a reported business in the household satisfied BLS criteria for

a business, respondents had to answer "yes" to at least one of several follow-up probe questions (e.g., "Do you advertise the products or services of the business, for example, by displaying a sign, or listing the business in the phone book or newspapers?"). Respondents' eligibility for a set of follow-up questions was determined by their responses during the main interview. From respondent debriefings, researchers were able to obtain quantitative data regarding respondent comprehension of concepts and questions, and make revisions, as appropriate.

Behavior Coding
Using a specially developed form and coding procedures (based on work by Cannell *et al.*, 1975, 1989; Shepard and Vincent, 1991), interviewer–respondent interactions were monitored and coded by BLS and Census Bureau researchers. Procedures were designed to allow the coding of interviewer–respondent exchanges *during an actual interview.* Monitors noted whether interviewers read a question exactly as worded, with a slight change in wording, or with a major change in wording. Deviations in question wording were coded as major changes if they altered the meaning or intent of the question. For the respondent, researchers distinguished among the following behaviors: gives adequate answer, gives qualified answer, gives inadequate answer, asks for clarification, interrupts, does not know, or refuses to answer. Behavior coding helped to identify items that caused problems for interviewers (e.g., manifested by deviations from the exact question wording) or that caused problems for respondents (e.g., manifested by inadequate answers, requests for clarification, interruptions, etc.).

Item-Based Response Analysis
The purpose of nonresponse and response distribution analyses was to determine the extent to which differences in question wording or question sequencing produced different patterns of responses. *Item nonresponse rates* were defined as the percent of persons eligible for a question who did not provide a substantive response; this included persons who refused to answer and persons who said they did not know the answer. Refusal rates provided data on the sensitivity of particular questions. "Don't know" rates provided an indication of respondent task difficulty. *Response distributions* were generated and tabulated for all survey questions, with special attention being paid to the response distributions of comparable questions which differed across questionnaires.

Table 24.2 summarizes how the methods described above were used during phase one to select the CPS "work" question for the version D questionnaire tested in phase two. Other CPS questions were evaluated in a similar fashion. In reviewing this table (and, later, Table 24.3), the reader should take note of two things. First, with the exception of respondent debriefing data (i.e., the follow-up probe questions), evaluative methods provide only indirect information regarding survey data quality. By comparing evaluative data (e.g., behavior coding data) across alternative work questions, analysts infer which question wording is producing higher quality data (e.g., the question with higher

Table 24.2   Selecting the *Work* Question for Version D

Alternative Questions

Version A: Did you do any work at all LAST WEEK, not counting work around the house?

Version B: LAST WEEK, did you do any work for pay or profit?

Version C: LAST WEEK, did you do any work at all? Include work for pay or other types of compensation?

Goal: To select a *work* question for the version D questionnaire that best operationalizes the concept of work and that minimizes problems for respondents and interviewers. (Criteria for the concept of work include: work for one hour or more for pay or profit, pay-in-kind, or unpaid work in a family business or farm for 15+ hours during the reference week.)

Measurement Issues: To determine effects of question wording on respondents' interpretation of "work" as a concept and the reporting of work activities.

Results of Evaluative Methodologies and Techniques

A. Behavior Coding: Data analyses provide support for selecting version B question.

- Marginally significant difference among alternative versions of the work question with respect to the percentage of time interviewer read the question exactly as worded (A = 94.3%; B = 98.8%; C = 93.9%).

- Nonsignificant difference among alternative versions of the work question with respect to the percentage of respondents who gave an adequate answer to the question (A = 90.9%; B = 95.6%; C = 91.9%).

B. Interviewer Debriefings: Debriefings suggest that interviewers (and respondents) experience some difficulties with all three versions of the work question.

   (1) Focus groups. Some interviewers report not liking the A question because it sounds demeaning to housewives and because it is confusing to some respondents (e.g., volunteer workers). The use of the term "profit" in the B question confuses some respondents—especially those who do not have a business. The use of the phrase "other types of compensation" in the C question confuses some respondents and some interviewers, too.

   (2) Debriefing questionnaire ($N = 68$ interviewers). When asked what question was most difficult for them to ask, two interviewers selected the version A work question (too wordy or awkwardly worded), three selected the version B work question (confusing, ambiguous, difficult to understand), and three selected the version C question (same reasons as B). When asked what question appeared to be most difficult for respondents to answer, five interviewers selected the version B work question (confusing, ambiguous, difficult to understand) and three selected the version C work question (same reasons as B). When asked what terms or concepts were most commonly misunderstood by respondents, six interviewers mentioned "working for pay or other types of compensation"; four mentioned "working for pay or profit" or just "profit"; and four mentioned "work" or "work vs employed."

C. Respondent Debriefings: Data analyses provide some support for all three questions.

    (1) Follow-up probe questions. All three work questions were effective at identifying employed persons. Differences in the percentage of *employed individuals missed* for all possible question pairings were not significant (A = 2.0%; B = 1.8%; C = 1.1%).

    (2) Vignettes. No one of the three work questions was better at eliciting responses that match CPS definitions (i.e., no one question clearly outperformed the other two alternatives). Some evidence to suggest that version B question wording may be less inclusive than other alternatives, in that a higher percentage of respondents say "no" to all vignette scenarios. Version B question is less successful than alternatives in correctly classifying marginal work activities (e.g., work in the home), but better at correctly classifying nonwork activities (e.g., volunteer service).

D. Item-Based Response Analysis: Data analyses suggest that no one question is better or worse than the alternatives.

    (1) Response-distribution analyses. All three work questions produced approximately the same percentage of individuals reported as working (A = 59.16%; B = 57.95%; C = 58.71%; differences in stated percentages for all possible question pairings are not significant).

    (2) Nonresponse analyses. Very little item nonresponse across versions (A = 0.18%; B = 0.18%; C = 0.22%).

Recommendation (R) and Justification (J)

R: Adopt a slightly modified version of the version B work question for the version D questionnaire: "LAST WEEK, did you do ANY work for (either) pay (or profit)?" parentheticals to be read only if respondent answers "yes" to the prior question regarding a family business or farm (i.e., "Does anyone in this household have a business or farm?"). Interviewers to emphasize the reference period "LAST WEEK" and the word "ANY."

J: Response analyses and respondent debriefings were inconclusive; that is to say, there was little or no evidence to suggest that any one of the question alternatives was better or worse than the others. Behavior coding analyses provided support for selection of the version B work question. Interviewer debriefings indicated that all three work questions have problems. Some of the confusion regarding the word "profit" in the version B question is easily rectified by having that word only appear if someone in the household has a business or a farm.

---

percentages of exact question readings and higher percentages of adequate answers). And second, analytical methods do not always produce clear cut results, especially when the concept being measured has a multifaceted definition (e.g., the concept of work). We believe the current example is more representative of the types of problems that survey researchers encounter and solve on a regular basis.

24.2.1.2 Phase Two

During the phase two field test, a newly synthesized questionnaire (version D) was tested with the existing CPS again serving as control. The primary objective of phase two was to identify problem areas in question wording in order to finalize development of the revised CPS questionnaire. Evaluation of the data was used to determine if there were any "fatal flaws" in version D (e.g., gross errors in the way questions or concepts were being understood) and make necessary modifications. As a result, some of the methodologies were utilized differently. For example, in phase two, interviewer debriefings were conducted via focus groups only, and the analytical scope was narrower. The greatest concern was with survey questions that had direct effects on labor force classification (e.g., items having to do with work and layoff from work), and the protocol reflected this concern. Behavior coding was conducted to determine whether there were any problematic items on the version D questionnaire; items on the control questionnaire (version A) were not evaluated. Respondent debriefings (follow-up probes and vignettes) and response distribution analyses were similar to those used in phase one; and when these data were available, differences between version A and version D were examined. Relying on data generated by the various pretesting techniques described above, researchers found no evidence of serious flaws in the version D questionnaire (see Table 24.3). Only minor wording changes were recommended.

24.2.2 The CPS Redesign: Quality Assessment Research (Phase Three)

Beginning in July 1992, the revised CPS questionnaire was tested in a separate survey (12,000 households per month) conducted in parallel with the existing CPS (60,000 households per month) to determine the effect of content revisions and a new data collection technology on labor force estimates. The parallel survey (phase three) differed from the prior phases in several respects. For example, the phase three sample consisted of households from an address list, and interviewing was done in person during the first- and fifth-month interviews and by telephone during the subsequent months. In addition to assessing the effects of the new questionnaire and the new data collection technology on labor force estimates, interviewer debriefings, behavior coding, respondent debriefings, and item-based response analyses were again conducted during phase three to assess the quality of data produced by the revised questionnaire. Only now, given the distinction suggested in Section 24.1.2, researchers were no longer pretesting, but rather conducting a quality assessment of the changes made to the revised CPS questionnaire.

24.2.2.1 Phase Three Methods and Results

Both focus groups and a standardized interviewer debriefing questionnaire were used to debrief interviewers during this phase. The primary objective of interviewer debriefings was to identify problems with the revised questionnaire and with procedures that required additional interviewer training or enhancements to the interviewer manual. Focus groups were conducted after interviewers had at least three to four months experience using the revised questionnaire and the new data collection technology (e.g., laptop computers). Interviewers were asked to identify questions that caused the most difficulties for them in terms of getting adequate answers from respondents. In order to have some measure of the magnitude of the problems, interviewers were asked to use rating forms to estimate the percentage of the time they had difficulty getting an adequate answer from respondents for each of the problematic items identified. The most problematic questions, as identified by interviewers participating in the focus groups, were then included in a standardized debriefing questionnaire that was sent out to the entire staff of CPS interviewers. In this self-administered debriefing questionnaire, interviewers were asked to estimate what percentage of the time they had difficulty getting an adequate answer from respondents when asked particular questions.

During phase three, behavior coding was used to assess the quality of the revised questionnaire by examining interviewer–respondent interactions for various survey questions. It was assumed that a high percentage of interviewers reading questions exactly as worded (or with a slight change) and respondents providing adequate (or qualified) answers would contribute positively to data quality. With the consent of respondents, a selected group of interviewers audiotaped 364 parallel survey interviews. Due to the large number of taped interviews, four CPS experienced supervisory staff were trained to do the behavior coding. The coding staff used a standardized coding form that was adapted for office-based coding; in terms of content, this form was very similar to the ones used in earlier phases of testing.

Respondent debriefings were conducted during phase three to assess question comprehension and response formulation. As in other phases, follow-up probe questions were asked only of household members for whom the questions were applicable. Vignettes were not used during this phase. In addition to providing an indication of how respondents interpret the "work" question, some measures of response accuracy (e.g., missed employment data) were obtained from these debriefing questions (Rothgeb, 1994).

Response distribution analyses in phase three were conducted, in part, to determine if the revised questionnaire and new data collection technology produced response patterns consistent with those obtained in phase two. (These data were also important in comparing differences in labor force classifications produced by the existing (version A) and revised CPS questionnaires.) Item nonresponse data were also produced for selected items that were suspected to be vulnerable to higher than usual refusals or "don't know" responses. So that there would be enough cases to conduct analyses for infrequently asked items on remote question-asking paths, data were cumulated over a 12-month time period. Data gathered through other methods (e.g., respondent debriefing) were useful in explaining differences in response patterns when such differences were found (Rothgeb, 1994).

It is important to realize that, as with pretesting, measures of quality assessment provide both qualitative and quantitative data. Data quality acceptance thresholds can vary among questions being evaluated. Questionnaire items that are most critical to the construct of a concept will most likely require stricter criteria for indicating acceptable data quality. Also, it should be noted here that while the quality assessment methods discussed above frequently provide complementary data, they sometimes yield incongruent data. Given the unique perspectives of the various data sources (i.e., coders, interviewers, respondents, and questionnaire designers), the existence of some incongruencies should not be viewed as an indication that the methods are unsuitable for quality assessment purposes.

To illustrate the transition from pretesting to quality assessment, we will continue with the example of how the methods described above were used in the evaluation of the CPS "work" question (revised question: "LAST WEEK, did you do ANY work for (either) pay (or profit)?"). We choose the "work" question for illustrative purposes because it is one of the few CPS questions for which analytical data are available for all of the evaluative techniques described above. It is also the CPS question asked of the greatest universe of persons.

Table 24.3 displays evaluation data for the CPS "work" question from each of the three different test phases. Results from phase three *suggest* that the survey data obtained for this question was of fairly high quality. Behavior coding data indicate that interviewers read the revised "work" question exactly, or with only slight changes, nearly all the time. Respondents interviewed via CATI provided an adequate (or qualified) response 98 percent of the time, while those interviewed in person via CAPI provided an adequate (or qualified) response 93 percent of the time.

Data from the interviewer debriefings were mixed. Focus group participants reported that respondents appear to have difficulty answering the revised "work" question; in fact, of all the questions evaluated, they rated it the most difficult question for which to obtain an adequate answer. During the focus group session, these interviewers were asked: "What percentage of the time do you have difficulty getting an adequate answer from respondents when asking this question?" Mean and median ratings of 35.5 and 34 percent, respectively, were obtained. This same question appeared on the self-administered interviewer debriefing questionnaire, which was rated by 345 of 400 field interviewers (86 percent), and it produced a mean rating of 18.2 percent and median rating of 10 percent. It is interesting that the focus group participants, on average, reported they had difficulty getting an adequate answer to the work question nearly 36 percent of the time, yet behavior coding data indicated that respondents provided "adequate answers" to the question a very high percentage of the time (98 percent for CATI and 93 percent for CAPI). The discrepancy can be resolved by listening to actual CPS interviews. Some respondents answer the "work" question by providing relevant information ("Well, just my regular job.") or by asking a question in return ("Do you mean my regular job?"). (This would appear to be an example of "reporting" behavior, i.e., answering

Table 24.3 Selected Results for the "Work" Question from the Pretest (PT) and Quality Assessment (QA) Phases of the CPS Redesign

Test/Design	Dates	Sample size	Questionnaire version	Interviewer debriefing	Respondent debriefing % Missed employment	Behavior coding INT Code (% E + mC)	Behavior coding RSP Code (% AA + qA)	Response distribution % Yes (% NR)
Phase 1 CATI/RDD (PT)	July 1990–January 1991	70,000 hhlds Cumulative All versions	A	see Table 24.2	2.0%	94% CATI	91% CATI	59.16% (0.18%)
			B	see Table 24.2	1.8%	99% CATI	96% CATI	57.95% (0.18%)
			C	see Table 24.2	1.1%	94% CATI	92% CATI	58.71% (0.22%)
Phase 2 CATI/RDD (PT)	July 1991–October 1991	32,000 hhlds Cumulative Both versions	A	—	3.8% (2.2% paid)	—	—	57.74% (0.15%)
			D	FG: "just my job"	2.6% (2.0% paid)	100% CATI	95% CATI	57.01% (0.08%)
Phase 3 CATI/CAPI address list sample (QA)	July 1992–December 1993	144,000 hhlds Cumulative (1993 only)	Revised questionnaire	FG: 35.5% IDQ: 18.2%	2.9% (1.6% paid)	100% CATI 99% CAPI	98% CATI 93% CAPI	58.58% (0.16%)

Abbreviations: FG refers to focus-group data; IDQ refers to interviewer-debriefing-questionnaire data; INT refers to interviewer and RSP to respondent; (% E + mC) refers to the percentage of exact and minor-change question readings; (% AA + qA) refers to the percentage of adequate and qualified answers; (% NR) refers to the nonresponse percentage (i.e., refusals and "don't know" responses); hhlds indicates households.

a question by providing relevant information rather than a direct answer; see Schaeffer *et al.*, 1996, for a discussion.) When they hear such responses, most interviewers either probe or check the "yes" response option. The behavior coders, based on our training, coded such responses as adequate answers, because the implication of either response is that the respondent had a job and most probably worked at that job last week. In retrospect, however, such responses (as interviewers note) should perhaps be considered as only marginally adequate. In its present context, the question does appear to confuse some respondents; but again, other data suggest that we are obtaining accurate information from this question.

The findings discussed above provide a good example of how important it is to utilize more than one evaluation method. While interviewers are an invaluable source of information regarding problems they or respondents are having with survey questions, sometimes the information they provide is colored by the most salient interview situations, which may not always be typical. If assessment of data quality for the "work" question had been based solely on the results of interviewer focus groups, the sponsor (BLS) may have been alarmed by the results. Fortunately, more objective quality assessment data were available from behavior coding and other evaluation methodologies, such as respondent debriefing and response distributions. Differences between the mean and medians obtained from the interviewer focus groups versus those from the standardized self-administered questionnaire suggest that collective behavior may influence the reporting of problems. Some focus group participants may not have originally thought the "work" question was problematic; hearing others report it as problematic, however, may have influenced their ratings of the item. It is also possible that some interviewers may have forgotten having difficulty with this question, recalling their experiences only after being cued by others who identified the question as problematic during group discussion. In contrast, the debriefing questionnaires were completed independently; in the absence of cues provided by others, interviewers may have underestimated the frequency of problems experienced with this question. We suspect the true level of difficulty with the "work" question probably lies somewhere in between the medians obtained from the debriefing questionnaire (10 percent) and the focus groups (34 percent).

Results from respondent debriefing analyses suggest that the revised "work" question is producing high-quality data. Through the years, there had been concern that the classification of marginal work activities, unpaid work in a family business or farm, and work in the underground economy may have been underreported in the CPS (Martin and Polivka, 1992). The revised questionnaire was designed to improve reporting of these types of work. To obtain estimates of missed employment, persons not reported as employed in the main survey were asked in the respondent debriefing if they had done any work at all during the reference week, even for only a few hours. This was intended to obtain estimates of persons that may not have been reported as working when questions in the main survey were asked, but who had actually performed some type of work activity during the reference week. Only 1.6 percent of persons, not reported as employed in the main survey, were reported in the respondent debriefing as having done some paid work during the reference week. (Including unpaid and paid work, 2.9 percent were reported as working.) These data indicate that missed employment is somewhere between 1.6 and 2.9 percent. It should be noted that reported missed employment was slightly higher during phase three than during pretesting (phases one and two). However, the samples for phases one and two did not include nontelephone households, whereas the sample for phase three did. It is likely that nontelephone households would have a higher proportion of persons engaged in casual, informal, or marginal work activities than telephone households; such activities may not be reported as work in the main survey and would consequently result in a higher number of reports of missed employment.

Response distribution data revealed that 58.6 percent of persons for whom this initial work question was asked were reported as working last week. These data were consistent with estimates obtained during previous phases. Item nonresponse was virtually nonexistent for this question.

Taken as a whole, analytical data collected in phase three suggest that: (a) most of the items on the revised questionnaire were being read by interviewers as worded, (b) most of the items were being understood by respondents as intended, and (c) labor force misclassification (e.g., as measured by percentage of missed employment and employment to population ratios) was relatively low for the revised questionnaire (Rothgeb, 1994; Polivka, 1994; Dippo *et al.*, 1995). We should note that other more standard measures were also used to monitor the data collection process (e.g., gross and net difference rates from reinterview; noninterview rates by questionnaire, item, and interviewer).

24.3 CONCLUDING REMARKS

In the two preceding sections, we have provided some background material on various methods that are available for evaluating survey questionnaires (i.e., via pretesting and quality assessment research). In addition, we have tried to make that material more palpable by providing a case study of how various methodologies were actually used in evaluating the redesigned CPS questionnaire. In this final section, we delineate the broad outline of an idealized quality assessment program for major social and economic surveys, address several pragmatic issues associated with such a program, and consider some of the costs and benefits associated with quality assessment research.

Before proceeding, however, we would like to mention several uses to which a quality assessment program might be put. One use might be to suggest ideas for enhancing interviewer training. A second use might be to check the reliability of pretest results using the full production sample (e.g., phase three of the CPS redesign). A third use might be to assess the degree to which *any existing questionnaire* is providing quality data; if serious data quality problems are uncovered, survey sponsors may choose: (1) to initiate a formal plan for pretesting a redesigned questionnaire, or (2) to make modifications based on quality assessment data only.

24.3.1 An Idealized Quality Assessment Program

As a means of contributing to the validity and reliability of questionnaire data, we advocate a comprehensive, ongoing, multimethodology, multitechnique quality assessment program for surveys that provide key social and economic indicators. By *comprehensive*, we mean a research program that investigates the various and interrelated sources of measurement error (i.e., interviewers, respondents, data collection contexts; see Esposito and Jobe, 1991), and one that incorporates laboratory, office, and field research. By *ongoing*, we mean a program that incorporates regular or periodic evaluations of the survey instrument; this element is consistent with the idea of continuous measurement as described by Dippo (Chapter 20). By *multimethodology*, we mean a program that utilizes analytical data from multiple methods and multiple sources: (1) interviewers, (2) respondents, (3) survey sponsors and designers, and (4) natural context interactions between interviewers and respondents. And by *multitechnique*, we mean a program that incorporates multiple techniques (e.g., focus groups, debriefing questionnaires), when resources are available, for collecting information from the sources noted above. In our view, the multimethodology criterion is critical, because it provides researchers with evaluative data from four key perspectives; the chances of making poor decisions regarding survey questions are minimized when all perspectives are considered, not just one or two. The multitechnique criterion is also important, because it enables researchers to build on the strengths of individual techniques while, at the same time, compensating for the unique weaknesses associated with each.

It is recognized that the ongoing program of research activities outlined above could be costly; but in considering the costs of producing quality data, one must also consider the costs (e.g., to users) of *not* producing quality data. Even so, we recognize that any quality assessment program, given the current fiscal environment, must be reasonably economical to implement. With that constraint in mind, we propose a flexible program that integrates laboratory and behavior coding in a symbiotic way. The core of the program would consist of three field methodologies (interviewer debriefings, respondent debriefings, and behavior coding) and various laboratory- and office-based techniques; such a program would be designed to supplement rather than replace existing quality assessment measures (e.g., a reinterview program). The field methodologies used in a quality assessment program should: (a) inform researchers as to where questionnaire-related problems exist, (b) provide explanations for those problems, and (c) suggest hypotheses regarding the causes of problems that can be tested under controlled laboratory conditions. In addition to testing the causal hypotheses, laboratory- and office-based research (e.g., cognitive interviews, cognitive forms appraisals) is likely to generate its own set of hypotheses which could be addressed in subsequent field research, if resources permit.

These recommendations in hand, we now wish to raise and address three pragmatic issues regarding the quality assessment program outlined above:

1. *Given a specific evaluation method, how does a researcher determine when a "problem" exists with a given questionnaire item?* In practice, decisions as to what constitutes a "problem" with a particular questionnaire item really start with the survey concepts and the way those concepts have been operationalized in a given question (see Figure 24.1). If substantive concepts are poorly defined or undefined, the researcher has no way of judging how well specific concepts have been operationalized in a given question and no means of accurately determining if respondents understand the question. If there are no explicit question objectives, then researchers have no basis on which to evaluate a respondent's answer. (See Schwarz (Chapter 1) and Hox (Chapter 2) for an extended discussion of these issues.) Even with clear question objectives, however, establishing criteria for what constitutes a problem with a given question tends to be a subjective process; moreover, these criteria often vary across researchers and within methods. In the case of behavior coding, for example, criteria for identifying problematic questions are suggested by researchers who have substantial experience with the technique. Cannell *et al.* (1989) have flagged questions as problematic if specific behavior codes (e.g., requests for clarification) appear 15 percent or more of the time the question is asked. Many practitioners follow this rule of thumb; however, other researchers (e.g., Esposito *et al.*, 1991) have used more stringent criteria (e.g., flagging a question as problematic if adequate answers are not obtained from respondents at least 90 percent of the time the question is asked). The analysis of data from follow-up probe questions is somewhat less subjective in that probe questions are typically designed as alternative measures of the concept under consideration, and any substantial discrepancy in response distributions between the probe question and the target question is generally taken as a sign of trouble. With other techniques, interviewers or respondents identify problems by recalling the difficulties they experienced with a given questionnaire item, and then researchers must decide how much weight to assign to such judgments. In sum, based on experience (Esposito *et al.*, 1991, 1992) and our reading of the literature (e.g., Cannell *et al.*, 1989; Presser and Blair, 1994; Willis, 1991), we believe all evaluative methods have inherent weaknesses; moreover, deciding what is or is not a problematic question tends to be a subjective process. This is why we regard single method evaluation plans to be very risky.

2. *Given the use of multiple evaluation methods (e.g., behavior coding, follow-up probe questions, interviewer debriefing questionnaire), how does the researcher identify "problematic" questionnaire items?* One strategy that might be useful— at least until other strategies are developed—is to rely on what we refer to as the *relative confidence model*. This model is an extension of an idea, suggested

Model A: One Evaluation Method

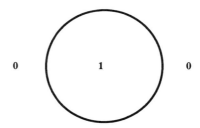

Model B: Two Evaluation Methods

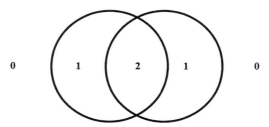

Model C: Three Evaluation Methods

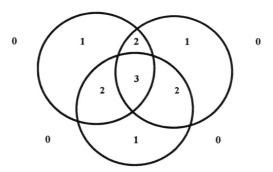

Figure 24.2 A relative confidence model for identifying problematic survey questions.

by Willis (1991), which we interpret as follows: different evaluative methods can be viewed as complementing one another; and when used in combination, multiple methods may provide a more accurate overall means of identifying problematic questions than single methods alone. We use Venn diagrams to illustrate the model (see Figure 24.2). Each circle represents a different questionnaire evaluation method drawing information from a *different source* (e.g., interviewers, respondents, experts). In Model A, a single evaluation method has identified a certain *group of questions* as problematic (area 1). Given this single method, we have no basis for viewing the questions outside the circle as problematic (area 0). In Model B, two evaluation methods are used, and three areas circumscribed. The questions falling in area 2 have been identified as problematic by both methods; the questions falling in area 1 have been identified as problematic by one method only; and the questions falling outside these areas have not been identified as problematic by either method (area 0). *In selecting questions for review and possible revision*, we would feel most confident selecting area 2 questions as problematic, and somewhat less confident in selecting area 1 questions. Model C (three evaluation methods) follows the same logic. Our confidence in correctly selecting problematic questions for review and revision would be greatest for area 3 questions, and would decrease incrementally for areas 2 and 1, respectively. (Please note that the logic supporting the relative confidence model does not generalize as well to multitechnique comparisons that draw evaluative information from a single source (e.g., using debriefing questionnaires and focus groups to gather information from interviewers only). We would expect greater overlap between circles, but our analysis would be limited to the perspectives of a single source (i.e., interviewers) and to the particular weaknesses of the techniques used.)

3. *Given accurate identification of problematic questions (and the desire to make changes in the questionnaire), how does the researcher go about revising such questions?* One strategy would be to adopt a test-fix-retest model, similar to the process described above for phase one of the CPS redesign. Here one revises a question based on the problems identified during the initial pretesting stage and then retests the question in a second evaluative stage. But, occasionally, in revising a question to solve one problem, the question designer creates another. We really need a set of guidelines, based on empirical data, for designing good survey questions. Belson (1981, p. 389) offers some useful suggestions in this regard. For example, he suggests that we *avoid*: "loading up the question with a lot of different or defining terms; ...; the use of words that are not the usual working tools of the respondent; ...; giving the respondent a difficult task to perform; giving the respondent a task that calls for a major memory effort; ...." Another possibility would be to evaluate revised questions using cognitive forms appraisals before adding them to the host questionnaire (Forsyth et al., 1992; Forsyth and Hubbard, 1992).

24.3.2 The Costs and Benefits of Quality Assessment Research

Survey quality cannot be achieved without incurring costs (Groves, 1989). The kind of quality assessment program described above for major social and economic surveys involves an ongoing commitment of staff and fiscal resources. Moreover, improvements in survey quality generally entail some change in the data collection process (e.g., question or procedural modifications). Program managers, though they welcome measures to improve quality, must also be concerned with the utility of their product to data users (e.g., time series integrity). These competing demands—to maintain quality levels and to assure the comparability of statistics across time—put program managers in a difficult position. To change or not to change, that is their dilemma, and it is not one that survey researchers should take lightly. Whenever quality-related improvements are recommended, program managers have every right to expect that survey researchers provide data on the anticipated consequences of those changes on their statistical products (e.g., the unemployment rate, consumer price index, the poverty index) and on operational aspects of the survey (e.g., data processing edits).

Of course, there are also substantial benefits to investing resources in quality assessment research. The most obvious benefits to users are more valid and reliable statistics (i.e., more accurate data) and greater confidence in the decisions that are made on the basis of those statistics. An obvious benefit for the organizations that produce these social and economic data is enhanced credibility. But there are benefits for the research community as well. There is an axiom in clinical psychology that goes something like this: if you want a clear understanding of normal human behavior, study maladaptive behavior patterns and attempt to determine their origins. A similar axiom might apply to survey response models and questionnaire design: if you want to understand how respondents answer survey questions and how to design questionnaires that produce high-quality data, identify problematic questions and attempt to determine why those problems exist. We believe that one of the main benefits of quality assessment research will be to help survey researchers to understand the processes (social and psychological) involved in responding to survey questions and, in so doing, help us all to design better questionnaires.

ACKNOWLEDGMENTS

Parts of this chapter (i.e., those pertaining to the redesign of the CPS) report on methods and research undertaken by staff at the U.S. Bureau of the Census and the U.S. Bureau of Labor Statistics (BLS). We wish to acknowledge the work of the individuals at both agencies who made the redesign a success, especially those whose work we have drawn upon and cited herein. The views expressed are attributable to the authors and do not necessarily reflect the views of the Census Bureau or the BLS. We also wish to thank Cathryn Dippo and Elizabeth Martin for very helpful comments to earlier drafts of this chapter.

REFERENCES

Anderson, R., Kaspar, J., Frankel, M.R., and Associates (1979), *Total Survey Error*, San Francisco: Jossey-Bass.

Bailar, B. (1984), "The Quality of Survey Data," *Proceedings of the Section on Survey Research Methods*, American Statistical Association, pp. 43–52.

Belson, W.R. (1981), *The Design and Understanding of Survey Questions*, Aldershot, U.K.: Gower.

Biemer, P.P., and Forsman, G. (1992), "On the Quality of Reinterview Data with Application to the Current Population Survey," *Journal of the American Statistical Association*, 87, pp. 915–923.

Biemer, P.B., Groves, R.M., Lyberg, L.E., Mathiowetz, N.A., and Sudman, S. (eds.) (1991), *Measurement Errors in Surveys*, New York: Wiley.

Bischoping, K. (1989), "An Evaluation of Interviewer Debriefing in Survey Pretests," in C. Cannell *et al.* (eds.), *New Techniques for Pretesting Survey Questions*, [Final Report], Ann Arbor, MI: Survey Research Center, University of Michigan.

Blair, J., and Presser, S. (1993), "Survey Procedures for Conducting Cognitive Interviews: A Review of Theory and Practice," *Proceedings of the Section on Survey Research Methods*, American Statistical Association, pp. 370–375.

Blair, J., and Sudman, S. (1993), "Respondent Perceptions of Reinterview," *Proceedings of the Annual Research Conference*, U.S. Bureau of the Census, Washington, DC, pp. 701–716.

Blixt, S., and Dykema, J. (1993), "Before the Pretest: Question Development Strategies," *Proceedings of the Section on Survey Research Methods*, American Statistical Association, pp. 1142–1147.

Bolton, R., (1993), "Pretesting Questionnaires: Content Analysis of Respondents' Concurrent Protocols," *Marketing Science*, 12, pp. 280–303.

Campanelli, P.C., Martin, E.A., and Creighton, K.P. (1989), "Respondents' Understanding of Labor Force Concepts: Insights from Debriefing Studies," *Proceedings of the Annual Research Conference*, U.S. Bureau of the Census, Washington, DC, pp. 361–374.

Campanelli, P.C., Martin, E.A., and Rothgeb, J.M. (1991), "The Use of Respondent and Interviewer Debriefing Studies as a Way to Study Response Error in Survey Data," *The Statistician*, 40, pp. 253–264.

Campanelli, P.C., Rothgeb, J.M., Esposito, J.L., and Polivka, A.E. (1991), "Methodologies for Evaluating Survey Questions: An Illustration from a CPS CATI/RDD Test," paper presented at the Annual Meeting at the American Association for Public Opinion Research, Phoenix, AZ.

Cannell, C.F., Fowler, F.J., and Marquis, K. (1968), *The Influence of Interviewer and Respondent Psychological and Behavioral Variables on the Reporting in Household Interviews*, Vital Health and Statistics, Series 2, Number 26, Washington, DC: Government Printing Office.

Cannell, C., Lawson, S.A., and Hausser, D.L. (1975), *A Technique for Evaluating Interviewer Performance*, Ann Arbor, MI: Survey Research Center, University of Michigan.

Cannell, C.F., Miller, P.V., and Oksenberg, L. (1981), "Research on Interviewing Techniques," in S. Leinhardt (ed.), *Sociological Methodology*, San Francisco: Jossey-Bass, pp. 389–437.

Cannell, C., and Oksenberg, L. (1988), "Observation of Behavior in Telephone Interviews", in R.M. Groves, P.P. Biemer, L.E. Lyberg, J.T. Massey, W.L. Nicholls, II, and J. Waksberg (eds.), *Telephone Survey Methodology*, New York: Wiley, pp. 475–495.

Cannell, C., Oksenberg, L., Kalton, G., Bischoping, K., and Fowler, F.J. (eds.) (1989),

New Techniques for Pretesting Survey Questions, [Final Report], Ann Arbor, MI: Survey Research Center, University of Michigan.

Cantwell, P.J., Bushery, J.M., and Biemer, P. (1992), "Toward a Quality Improvement System for Field Interviewing: Putting Content Reinterview into Perspective," *Proceedings of the Section on Survey Research Methods*, American Statistical Association, pp. 74–83.

Converse, J.M., and Schuman, H. (1974), *Conversations at Random*, New York: Wiley.

Copeland, K., and Rothgeb, J.M. (1990), "Testing Alternative Questionnaires for the Current Population Survey," *Proceedings of the Section on Survey Research Methods*, American Statistical Association, pp. 63–71.

DeMaio, T.J. (1983), "Learning from Interviewers," in T.J. DeMaio (ed.), *Approaches to Developing Questionnaires*, Statistical Policy Working Paper 10, Washington, DC: Office of Management and Budget, pp. 119–136.

DeMaio, T., Mathiowetz, N., Rothgeb, J., Beach, M.E., and Durant, S. (1993), *Protocol for Pretesting Demographic Surveys at the Census Bureau*, Census Bureau Monograph, Washington, DC: U.S. Bureau of the Census.

DeMaio, T., and Rothgeb, J. (1996), "Cognitive Interviewing Techniques: In the Lab and in the Field," in N. Schwarz, and S. Sudman (eds.), *Determining Processes Used to Answer Questions*, San Francisco: Jossey-Bass.

Dippo, C., Polivka, A., Creighton, K., Kostanich, D., and Rothgeb, J. (1995), "Redesigning a Questionnaire for Computer-Assisted Data Collection: The Current Population Survey Experience," unpublished report, Washington, DC: U.S. Bureau of Labor Statistics.

Dippo, C.S., and Norwood, J.L. (1992), "A Review of Research at the Bureau of Labor Statistics," in J.M. Tanur (ed.), *Questions about Questions*, New York: Russell Sage Foundation, pp. 271–290.

Dippo, C.S., Tucker, C., and Valliant, R. (1993), "Survey Methods Research at the U.S. Bureau of Labor Statistics," *Journal of Official Statistics*, 9, pp. 121–135.

Esposito, J.L., Campanelli, P.C., Rothgeb, J., and Polivka, A.E. (1991), "Determining Which Questions Are Best: Methodologies for Evaluating Survey Questions," *Proceedings of the Section on Survey Research Methods*, American Statistical Association, pp. 46–55.

Esposito, J.L., and Hess, J. (1992), "The Use of Interviewer Debriefings to Identify Problematic Questions on Alternate Questionnaires," paper presented at the Annual Meeting of the American Association for Public Opinion Research, St. Petersburg, FL.

Esposito, J.L., and Jobe, J.B. (1991), "A General Model of the Survey Interaction Process," *Proceedings of the Annual Research Conference*, U.S. Bureau of the Census, Washington, DC, pp. 537–560.

Esposito, J.L., Rothgeb, J.M., and Campanelli, P.C. (1994), "The Utility and Flexibility of Behavior Coding as a Method for Evaluating Questionnaires," paper presented at the American Association for Public Opinion Research, Danvers, MA.

Esposito, J.L., Rothgeb, J.M., Polivka, A.E., Hess, J., and Campanelli, P. (1992), "Methodologies for Evaluating Survey Questions: Some Lessons from the Redesign of the Current Population Survey," paper presented at the International Conference on Social Science Methodology, Trento, Italy.

Forsman, G., and Schreiner, I. (1991), "The Design and Analysis of Reinterview," in P.P. Biemer, R.M. Groves, L.E. Lyberg, N.A. Mathiowetz, and S. Sudman (eds.), *Measurement Errors in Surveys*, New York: Wiley, pp. 279–301.

Forsyth, B.H., and Hubbard, M.L. (1992), "A Method for Identifying Cognitive Properties of Survey Items," *Proceedings of the Section on Survey Research Methods*, American Statistical Association, pp. 470–475.

Forsyth, B.H., and Lessler, J.T. (1991), "Cognitive Laboratory Methods: A Taxonomy," in P.P. Biemer, R.M. Groves, L.E. Lyberg, N.A. Mathiowetz, and S. Sudman (eds.), *Measurement Errors in Surveys*, New York: Wiley, pp. 393–418.

Forsyth, B.H., Lessler, J.T., and Hubbard, M.L. (1992), "Cognitive Evaluation of the Questionnaire," in C.F. Turner, J.T. Lessler, and J.C. Gfroerer (eds.), *Survey Measurement of Drug Use: Methodological Studies*, U.S. Department of Health and Human Services, Washington, DC, pp. 13–52.

Fowler, F.J. (1989), "Evaluation of Special Training and Debriefing Procedures for Pretest Interviews," in C. Cannell *et al.* (eds.), *New Techniques for Pretesting Survey Questions*, [Final Report], Ann Arbor, MI: Survey Research Center, University of Michigan.

Fowler, F.J. (1992), "How Unclear Terms Affect Survey Data," *Public Opinion Quarterly*, 56, pp. 218–231.

Fowler, F.J., and Cannell, C.F. (1996), "Using Behavior Coding to Identify Cognitive Problems with Survey Questions," in N. Schwarz, and S. Sudman (eds.), *Methods of Determining Cognitive Processes in Answering Questions*, San Francisco: Jossey-Bass.

Fowler, F.J., and Roman, A.M. (1992), "A Study of Approaches to Survey Question Evaluation," working paper for the U.S. Bureau of the Census, Washington, DC.

Fracasso, M.P. (1989), "Categorization of Responses to the Open-Ended Labor Force Questions in the Current Population Survey (CPS)," *Proceedings of the Section on Survey Research Methods*, American Statistical Association, pp. 481–485.

Groves, R.M. (1987), "Research on Survey Data Quality," *Public Opinion Quarterly*, 51, pp. S156–S172.

Groves, R.M. (1989), *Survey Errors and Survey Costs*, New York: Wiley.

Groves, R.M., Biemer, P.P., Lyberg, L.E., Massey, J.T., Nicholls, II, W.L., and Waksberg, J. (eds.) (1988), *Telephone Survey Methodology*, New York: Wiley.

Jabine, T.B., Straf, M., Tanur, J., and Tourangeau, R. (1984), *Cognitive Aspects of Survey Methodology: Building a Bridge Between Disciplines*, Washington, DC: National Academy Press.

Jobe, J., and Mingay, D. (1991), "Cognition and Survey Measurement: History and Overview," *Applied Cognitive Psychology*, 5, pp. 175–192.

Lessler, J., Tourangeau, R., and Salter, W. (1989), *Questionnaire Design in the Cognitive Research Laboratory: Results of an Experimental Prototype*, Vital and Health Statistics, Series 6, Number 1, DHHS Publication No. (PHS) 89-1076, Washington, DC: Government Printing Office.

Marquis, K. (1969), "Interviewer-Respondent Interaction in a Household Interview," *Proceedings of the Social Statistics Section*, American Statistical Association, pp. 24–30.

Martin, E. (1986), "Report on the Development of Alternative Screening Procedures for the National Crime Survey," unpublished report, Washington, DC: Bureau of Social Science Research.

Martin, E. (1987), "Some Conceptual Problems in the Current Population Survey," *Proceedings of the Section on Survey Research Methods*, American Statistical Association, pp. 420–424.

Martin, E., and Polivka, A.E. (1992), "The Effect of Questionnaire Redesign on Conceptual Problems in the Current Population Survey," *Proceedings of the Section on Survey Research Methods*, American Statistical Association, pp. 655–660.

Mathiowetz, N.A., and Cannell, C.F. (1980), "Coding Interviewer Behavior as a Method of Evaluating Performance," *Proceedings of the Section on Survey Research Methods*, American Statistical Association, pp. 525–528.

Morton-Williams, J. (1979), "The Use of 'Verbal Interaction Coding' for Evaluating a Questionnaire," *Quality and Quantity*, 13, pp. 59–75.

Morton-Williams, J., and Sykes, W. (1984), "The Use of Interaction Coding and Follow-up Interviews to Investigate the Comprehension of Survey Questions," *Journal of the Market Research Society*, 26, pp. 109–127.

National Commission on Employment and Unemployment Statistics (1979), *Counting the Labor Force*, Washington, DC: Government Printing Office.

Oksenberg, L., Cannell, C., and Kalton, G. (1991), "New Strategies for Pretesting Questionnaires," *Journal of Official Statistics*, 7, pp. 349–365.

Palmisano, M. (1989), "Respondent Understanding of Key Labor Force Concepts Used in the CPS," unpublished paper, Washington, DC: U.S. Bureau of Labor Statistics.

Polivka, A.E. (1994), *Comparison of Labor Force Estimates from the Parallel Survey and the CPS During* 1993: *Major Labor Force Estimates*, CPS Overlap Analysis Team Technical Report 1, Washington, DC: U.S. Bureau of Labor Statistics and U.S. Bureau of the Census.

Polivka, A.E., and Martin, E.A. (1992), "The Use of Vignettes in Pretesting and Selecting Questions," paper presented at the annual meeting of the American Association of Public Opinion Research, St. Petersburg, FL.

Presser, S., and Blair, J. (1994), "Survey Pretesting: Do Different Methods Produce Different Results?," in P.V. Marsden (ed.), *Sociological Methodology*, Volume 24, Oxford: Basil Blackwell, pp. 73–104.

Rothgeb, J.M. (1994), *Revisions to the CPS Questionnaire: Effects on Data Quality*, CPS Overlap Analysis Team Technical Report 2, Washington, DC: U.S. Bureau of the Census and the U.S. Bureau of Labor Statistics.

Rothgeb, J.M., Polivka, A.E., Creighton, K.P., and Cohany, S.R. (1991), "Development of the Proposed Revised Current Population Survey Questionnaire," *Proceedings of the Section on Survey Research Methods*, American Statistical Association, pp. 56–65.

Royston, P. (1989), "Using Intensive Interviews to Evaluate Questions," *Conference Proceedings: Health Survey Research Methods*, DHHS Publication No. (PHS) 89-3447, Washington, DC: National Center for Health Services Research and Health Care Technology Assessment.

Royston, P., Bercini, D., Sirken, M., and Mingay, D. (1986), "Questionnaire Design Research Laboratory," *Proceedings of the Section on Survey Research Methods*, American Statistical Association, pp. 703–707.

Schaeffer, N.C., Maynard, D.W., and Cradock, R.M. (1996), "From Paradigm to Prototype and Back Again: Interactive Aspects of 'Cognitive Processing' in Standardized Survey Interviews," in N. Schwarz, and S. Sudman (eds.), *Determining Processes Used to Answer Questions*, San Francisco: Jossey-Bass.

Shepard, J., and Vincent, C. (1991), "Interviewer–Respondent Interactions in CATI Interviews," *Proceedings of the Annual Research Conference*, U.S. Bureau of the Census, Washington, DC, pp. 523–536.

Sirken, M. (1991), "The Role of a Cognitive Laboratory in a Statistical Agency," seminar on Quality of Federal Data (Part 2 of 3), Statistical Working Paper 20, Washington, DC: Office of Management and Budget, pp. 268–277.

Sykes, W., and Morton-Williams, J. (1987), "Evaluating Survey Questions," *Journal of Official Statistics*, 3, pp. 191–207.

Tourangeau, R. (1984), "Cognitive Science and Survey Methods," in T. Jabine *et al.* (eds.), *Cognitive Aspects of Survey Methodology: Building a Bridge Between Disciplines*, Washington, DC: National Academy Press, pp. 73–100.

Turner, C.F., and Martin, E. (eds.) (1984), *Surveying Subjective Phenomena* (Volume 1), New York: Russell Sage Foundation.

U.S. Bureau of Labor Statistics (1986), "Report of the BLS-Census Bureau Questionnaire Design Task Force," staff report, Washington DC: U.S. Department of Labor, Bureau of Labor Statistics.

U.S. Bureau of Labor Statistics (1987), "Second Report of the BLS-Census Bureau Questionnaire Design Task Force," staff report, Washington DC: U.S. Department of Labor, Bureau of Labor Statistics.

U.S. Bureau of Labor Statistics (1988), "Response Errors on Labor Force Questions Based on Consultations with Current Population Survey Interviewers in the United States," paper prepared for the OECD Working Party on Employment and Unemployment Statistics, Washington, DC: U.S. Bureau of Labor Statistics.

Westat/AIR (1989a), *Research on Hours of Work Questions in the Current Population Survey*, Final Report, Bureau of Labor Statistics Contract No. J-9-J-8-0083, Rockville, MD: Westat, Inc.

Westat/AIR (1989b), *Research on Industry and Occupation Questions in the Current Population Survey*, Final Report, Bureau of Labor Statistics Contract No. J-9-J-8-0083, Rockville, MD: Westat, Inc.

Willis, G. (1991), "The Use of Behavior Coding to Evaluate a Draft Health-Survey Questionnaire," paper presented at the Annual Meeting of the American Association for Public Opinion Research, Phoenix, AZ.

Willis, G. (1994), *Cognitive Interviewing and Questionnaire Design: A Training Manual*, Cognitive Methods Staff Working Paper Series, No. 7, Hyattsville, MD: U.S. National Center for Health Statistics, Office of Research and Methodology.

Willis, G.B., Royston, P., and Bercini, D. (1991), "The Use of Verbal Report Methods in the Development and Testing of Survey Questionnaires," *Applied Cognitive Psychology*, 5, pp. 175–192.

CHAPTER 25

CATI Site Management in a Survey of Service Quality

Mary Batcher
Internal Revenue Service, Washington, DC

Fritz Scheuren
The George Washington University, Washington, DC

25.1 INTRODUCTION

For many years, surveys were conducted either by self-administered mail collections or through face to face or telephone interviews using paper and pencil administration. There were, and continue to be, advantages and disadvantages of each mode. These have been well-documented in the literature (Groves, 1989; Dillman, 1978; Tarnai and Dillman; 1992; Bishop *et al.*, 1988; Ayidiya and McClendon, 1990; Dillman and Tarnai, 1991; De Leeuw and van der Zouwen, 1988; Sykes and Collins, 1988). Increased automation is changing the balance however—in particular the introduction of computer assisted telephone interviewing (CATI) and computer assisted personal interviewing (CAPI). These technologies have done more than simply automate a manual process. They offer significant opportunities to standardize operational procedures, provide rapid reaction to problems that might occur during data collection, and reduce the time needed to compile and disseminate survey results. For the remainder of this chapter, we will focus on CATI systems. A good discussion of the advantages and opportunities associated with CAPI can be found in Nicholls and Kindel (1993).

The survey process changes with the use of CATI systems. These systems present opportunities for monitoring interviewer performance and training

Survey Measurement and Process Quality, Edited by Lyberg, Biemer, Collins, de Leeuw, Dippo, Schwarz, Trewin.
ISBN 0-471-16559-X © 1997 John Wiley & Sons, Inc.

interviewers in short segments that are immediately applicable. The automated nature of the data collection allows rapid tabulation of results by interviewer and statistical monitoring can be done on a daily or weekly basis. This allows the quick identification of unusual interviewer results like very high or low refusal rates, unusual responses to survey questions, long or short completion times, etc. These statistics can serve to identify interviewers to be monitored more closely to identify possible causes. Successful interviewer practices can be quickly communicated to other interviewers and unsuccessful practices can be immediately addressed through retraining or clarification of procedures. In summary, CATI changes the basic survey paradigm from a single drawn-out linear process with several steps to a linked series of short cycle processes (with built-in feedbacks).

In turn, the short cycle processes, available in a CATI environment, provide a major quality opportunity. Now, it is well known that in recent years, quality management has become an important aspect of U.S. businesses and, more recently, of the federal government (Deming, 1986; Hahn and Meeker, 1993; Osborne and Gaebler, 1992). A critical component of achieving quality improvement is the reduction of cycle time. Each cycle presents an opportunity to learn something about a process. It is an opportunity to observe (i.e., collect data) and act on what is learned; it is also an opportunity to experiment with different methods, techniques, or technologies. With short cycle times, results are available much faster; consequently, organizational and individual learning (and improvement) can also occur more quickly. Short cycle times support the many small incremental improvements that have proven so successful in Japanese industry (Womack *et al.*, 1990). We can capitalize on the reduced cycle times created in a CATI environment to improve the quality of surveys as part of a continuous quality improvement process. (See Chapters 20 and 21 also for related discussions of process improvement.)

If we view one week of a CATI survey implementation as a cycle, in a three month survey collection period, we have 12 opportunities to embed experiments, improve procedures, etc.—each time learning from and improving on the previous iteration. Of course, this approach has the potential to change the measuring device (i.e., combination of instrument, interviewers, and procedures) over the course of the collection. Care must be taken to make changes and experiment on line in a rational fashion that will not bias survey estimates. However, much can be done with small embedded experiments using only a few interviewers and with approaches whose primary effect is to reduce interviewer variability.

For the remainder of the chapter, we present a case study in which we have attempted to capitalize on some of the survey quality improvement opportunities afforded by a CATI system at the U.S. Internal Revenue Service (IRS). Although we have not been able to achieve the full potential for operational management and data dissemination inherent in the system, we have used the availability of immediate results in sample and interviewer management and in rapid dissemination of survey results. As we describe more fully later in the chapter, one feature of our system is that we have replicated samples over interviewers each week. In a sense, we have fully interpenetrating subsamples. This presents a unique opportunity to study interviewer effects (Stokes and Yeh, 1988).

25.2 CASE STUDY BACKGROUND

Our survey quality improvement case study occurs within, and as part of, a larger effort to improve the quality of a major service provided by the IRS. We describe the improvement of the survey used to measure the accuracy of the telephone assistance provided to taxpayers. It is important to distinguish the survey improvement, which is the focus of this chapter, from the service improvement which is the reason for the survey and is briefly described to set context.

Worldwide, many government agencies that deal with the public directly are increasingly interested in measuring and improving the quality of the services they provide. In the United States, for example, several U.S. Federal Government agencies are establishing formal mechanisms for this. Our survey is an essential component of one such effort—an attempt to better measure telephone inquiry accuracy at the national tax collecting agency in the U.S.—the Internal Revenue Service. The Internal Revenue Service, or IRS, may, by some measures, be among the most efficient institutions of its type anywhere (best in the world in terms of administrative costs per dollar collected). Despite this, in a recent overall assessment, only 55 percent of the American people were satisfied with the agency—the worst score of any large organization in the U.S., public or private. This was not much of a surprise to us; our own polling showed similar results. Do we deserve such a low rating? Let us look at the facts about the accuracy of the telephone assistance provided to taxpayers. The best evidence we have suggests that in the late 1980s perhaps only about two-thirds (or less) of the individual tax inquiries were being answered with complete accuracy. Improvement efforts began once there was widespread acceptance of this fact and a belief we could get better. This took a while but by 1990 the accuracy rate had jumped to roughly 80 percent and has continued high since (nearing 90 percent for 1995). This improvement was a highly significant result, both subjectively and statistically—greeted with praise, even by some of our severest critics (*Money Magazine*, 1990).

25.3 SERVICE QUALITY MEASURE

The improvement process started with an attempt to develop a credible methodology to assess the quality of telephone assistance. King (1987) suggests several indicators that might be used to measure quality (see also Osborne and Gaebler, 1992). They include unsolicited complaints and compliments, solicited customer opinions, direct observation of the interchange, and "shopping" the service. Shopping the service, where the evaluator poses as a customer and seeks the service, allows us to assess quality in a more structured situation. The evaluator determines the specific services sought and can, therefore, narrow the range to some well-defined, structured set of requests and possible outcomes which can be identified in advance and provided a quality rating. While this approach, like the other options, has its limitations it is the method that seemed to work best for our task of measuring the accuracy of the IRS telephone assistance.[1]

The first step in designing the shopping survey for assessing the accuracy of our telephone information service was to identify and describe the key components of the service so they could be accurately represented in our survey. They are: customers (taxpayers seeking information), service providers (IRS telephone assistors), and the interaction between customer and provider (the telephone interchange). Each component must be considered in our survey design and in continuing survey quality improvement efforts.

25.3.1 Customers

Our customers provide the models for our callers and our survey questions; they are, overwhelmingly, individuals seeking information to help them file their own tax returns. Can we narrow this further? In a sense, anyone with a telephone can call us, but a test call survey could not hope to capture all of this variety. For our test call survey, we decided to pose as callers who are informed about their situations and who have all relevant information at hand. How realistic was this? Over the years, we have transcribed samples of actual conversations between our customers and our service providers to gain some insight into the characteristics of the people who use our service: well-prepared callers[2] turn out to be the norm.

It was not enough to model typical customer behavior in the test calls; we also needed to pick the kinds of questions that our customers regularly ask. Here we relied not only on the transcription of a random sample of calls to help us, but instituted a system in each call site where every day, on a sample basis, incoming calls were categorized by type (e.g., questions about filing status,

---

[1] The survey we designed to "shop" the toll-free information service we provide was a program of test calls. A major factor in the design of the test call program was to determine the degree of similarity each element of the test call system would possess with its corresponding element in the real world system. There were two competing objectives to balance in the program design: one was to make the test call program as realistic as possible and the other was to impose measurement control, in the sense of reducing variation due to components of the measurement system.

[2] To further strengthen our measurement and to improve service, as part of the installation of the testing program, we added an instruction to the tax package to try to increase the likelihood that people would call us after they had assembled all the relevant facts about their situations. (Incidentally, it should go without saying that IRS, of course, still attempts to provide service to those customers whose inquiries are unfocused or who are not ready with needed information—just that the test call program does not measure such service directly.)

the earned income tax credit, taxable income, etc.). This so-called "ticking" operation allowed us to keep on top of what was being asked in the real world and to create and properly weight our test calls appropriately, so that the statistics matched, to the extent possible, our real service quality.[3]

25.3.2 Telephone Assistors

The second component of our telephone service is the service provider, the telephone assistor. At IRS, there can be as many as 7,000 such assistors employed annually. Our telephone assistors receive four to six weeks of initial training in tax law and telephone interpersonal skills. They then begin to respond to telephone inquiries.

During the tax filing period each year, approximately 50 percent of the assistors are temporary workers, hired for the peak period. About 10 percent of the seasonal assistors return year after year. Among permanent employees, the telephone assistance positions have commonly been used as stepping stones for better IRS jobs. In summary, there is a high turnover in assistors and a large proportion is relatively inexperienced and has received only the initial, first year training.[4]

25.3.3 Caller and Assistor Interactions

The third component arises from the interaction of a particular customer and assistor. In obtaining information, we are inherently dealing with a complex human exchange in which the customer, the taxpayer (or interviewer/test caller), attempts to convey his or her problem to the assistor. The assistor, in turn, tries to determine the question being asked by the taxpayer and to answer it correctly. To the extent this "interaction" term is large, we are unable to generalize as readily from any observed interchange of information between customer and provider—whether in a live or test call environment.

25.3.4 Inference Setting

It might be worth stating explicitly how the test call survey data are employed. Their primary use is, during the February to April testing period, to make week to week comparisons of assistor accuracy nationally. IRS taxpayer assistance

---

[3] The actual questions employed went through a full battery of cognitive and pilot testing before use—much of this work was done with the aid of the cognitive laboratory at the U.S. Bureau of Labor Statistics (van Melis-Wright *et al.*, 1990). Final decisions on the questions involved both IRS and General Accounting Office (GAO) lawyers; in this selection and refinement process, GAO played, of course, part of its necessary oversight role.

[4] In this situation of seasonal and relatively low-paying jobs, quality problems might be expected, and, indeed as we have indicated, do arise. The test call program, once its measures were accepted, became a vehicle to use as a yardstick for experimental efforts to improve the training and management of assistors.

site comparisons are also done annually and comparisons from year to year are made as well.

Conditional inferences are made, holding the interviewers (test callers in our setting) and questions fixed, when making weekly national comparisons. For these to be effective, of course, there has to be a high degree of commonality in test callers and questions each week and the interaction term must be small. With somewhat less control, conditional inferences are still possible, again trying to hold test callers and questions fixed across sites for the whole filing season. On the other hand, there have been numerous tax law changes over time; also, from year to year, there were quite a few changes in the questions used. Finally, inevitably, staff turnover occurred among those making test calls. Under these circumstances, measuring change over the period, say, from 1989 to 1995, is more problematic but still possible, albeit harder to interpret.

In the early years of the test call program, numerous sample design and estimation features were used to minimize the sampling variance of these comparisons. These are covered elsewhere (Batcher and Scheuren, 1989); our purpose here is to look at the quality management of the survey in terms of reducing the nonsampling error contribution to the estimators. We confine our attention just to measuring and controlling the nonsampling errors in the weekly national estimates. For site to site differences the main problem is the small size of the sample but some comments will be made in passing on the nonsampling error component of these differences as well.

25.4 SURVEY QUALITY MANAGEMENT

Despite what may be an unconventional application, the test call program in many ways resembles a standard CATI operation very closely. Nicholls (1988) lists six features of CATI systems. His list may be a point of departure for us in a description of the quality management procedures being undertaken.

25.4.1 Nature of CATI Operation

Each of the features Nicholls lists has a counterpart in our testing, as is seen below:

Sample Management. Each weekly sample is managed completely by the computer. Questions are assigned to call sites and times in a way that assures overall balance nationally by week and by call site for the whole filing season.
Online call scheduling and case management. Call times are computer pre-specified to the minute, although some leeway is allowed in practice. To further increase balance, *separate identically designed subsamples are maintained for each caller.* This is an opportunity not mentioned by Nicholls and something that may not be done in other CATI operations. When callbacks are required, a record is made which allows for later attempts and follow-ups.

Online interviews. A complete set of question screens and response categories is used in the test calls, with a full audit trail available for subsequent oversight as needed. Text and precoded responses are entered by callers as the interview takes place.

Online monitoring. Online supervisory monitoring was possible from the beginning. Initially, however, the caller computer screen was not available for simultaneous viewing and concerns existed about a deterioration in line quality that might have alerted the caller to the fact that he or she was being listened to. Online independent monitoring by GAO and later by a GAO contractor was also available eventually from a remote location.

Automatic record keeping. Management statistics of all types are available from the computer and are carefully examined daily. Examples here are call times (actual vs scheduled), completion rates, caller by caller accuracy rates, site statistics, length of calls and so forth.

Preparation of data sets. Each week, on Friday afternoon, the test call survey results are moved electronically, via modem, to a mainframe where the statistical summaries required by IRS management are produced. Actual counts of the number of completed calls by category are used to weight up the test calls to be "representative" of the total volume of live calls answered around the country. (Beginning in 1995, the test question distribution is nearly identical to the taxpayer question distribution and the category accuracies are extremely close. We are therefore experimenting with dropping the category weighting for the weekly estimates.) Various edits have to be performed, as a double check, to assure consistency and to adjust for any missing data that may have arisen due to uncompleted calls. Test call survey results are available in an internal agencywide electronic bulletin board on Tuesday in the week following the calls. This timeliness makes management action possible almost immediately, should it be needed.

25.4.2 Test Call Operations

The weekly sample size (see Table 25.1) has varied over time for several reasons. Budget issues played a major role but not because of any lessening in the funding of the test call survey program. Rather, the agencywide resources available to staff and to answer taxpayer questions dropped over the years, particularly between 1991 and 1992. This affected the test call program by making it harder to complete the test calls each week—leading to a reduction in the scheduled sample, since the designs of the early years were just too large to be carried out successfully. (Incidentally, changes in the estimation were made and the loss in precision at the national level is less than the ratio of sample sizes would suggest.)

Until 1994, the number of test callers per week was stable, by design, at eight to ten each week and about the same number during the filing season. A hiring freeze during the 1994 testing period caused a jump in the number of callers

Table 25.1   Weekly Sample Sizes

Year	Weekly calls	Total callers	Average calls per caller
1989	1274	8	159
1990	1253	8	157
1991	1234	10	123
1992	886	10	89
1993	781	9	87
1994	786	15	52
1995	794	9	88

because people had to be borrowed for the telephone interviewing and could not always be spared for the entire period. Anyway, the change in the number of callers, combined with an increase in the number of questions (mentioned earlier), strained the CATI operation considerably, as we will see in the next section.

25.5   QUALITY RESULTS

In what follows we give a brief "run-through" of the quality management results over the years and provide some reasons for what was and was not achieved. Our approach is basically graphical.

First, it builds on the identical sample designs which exist within but not between years for each caller and the fact that the weekly caller sample sizes may be large enough for the central limit theorem to be operating and for the caller distributions to each be normally distributed. The 1994 season is the only case where this assumption might be questionable. The conventional rule for approximate normality is that the sample size n must be larger than

$$n \geqslant 5/\min[p, (1 - p)]$$

where p in our application is the accuracy rate. This formula means we would need sample sizes of 50 or more per caller, given an accuracy rate of $p = 90$ percent. Of course, the rule assumes simple random sampling which is not our setting. Even so, the results appear satisfactory, as discussed below.

Second, if we achieved our quality management objectives the weekly individual caller distributions should all have approximately the same mean and standard deviation for a given week because of the design replicates. Furthermore, and perhaps of even more importance, the standard errors should not depart greatly from those predicted by the sampling design, under the assumption of no caller effect or no caller interaction (i.e., that the weekly processes were under statistical control).

OK, what happened? Figure 25.1 provides a plot for each year for all caller data for that year. This approach to the analysis is a compromise partly determined by space limits and partly due to the small number of weeks for which there are caller observations (just 12). The plots provided are normal probability plots. Here we are looking at a conventional display. Departures from normality are visible, when they occur, by raggedness in the plot and the extent it departs in shape from a straight line.

What should we expect to see in these plots? Suppose, for example, that the CATI operation is running such that:

- first, the system being measured is stable from week to week (which it is not),
- second, that all callers for a given week follow the same normal distribution (which they do not).

Then, it would follow that the individual caller distributions come from the same normal distribution for any given year. Put another way, suppose the changes in the CATI measurement process and in IRS accuracy are assumed to have taken place between years and not both between and within them. This is, of course, not what happened; even so, it may be a useful reference to keep in mind as we look at the much more complicated reality of the actual weekly caller patterns over time.

What then are these year to year changes? We have already mentioned improvements in IRS accuracy, changes in question mix (and number), and the weekly caller sample size (which varied by more than a factor of three between 1989 and 1995). If the changes were between years, then the weekly data for each year would be normal with distributions for which the caller mean μ, caller standard deviation σ, and coefficients of variation could be approximated as shown in Table 25.2.

How well does this model of what happened work? If a nominal level of significance of $\alpha = .01$ is employed, then one cannot reject the null hypothesis

Table 25.2 Estimated Caller Means, Standard Deviations, and Coefficients of Variation

Test year	Estimated mean μ (in %)	Estimated standard deviation σ (in %)	Coefficient of variation
1989	64	7.0	0.11
1990	78	4.6	0.06
1991	84	5.1	0.06
1992	88	4.9	0.06
1993	90	4.3	0.05
1994	90	4.0	0.04
1995	91	4.5	0.05

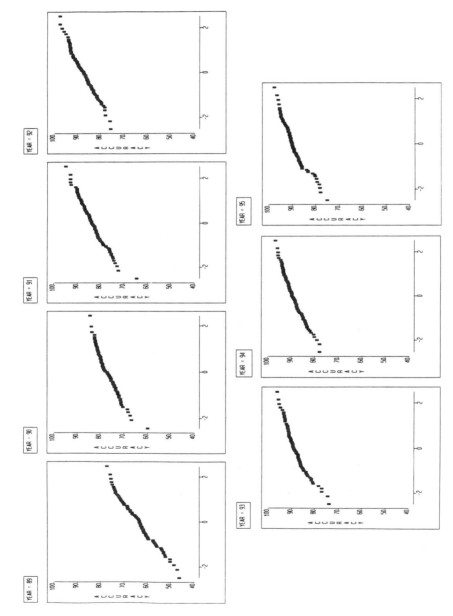

Figure 25.1 Normal probability plots of caller data, 1989–1995.

(i.e., reject the normal model) in 1989, 1991, 1992, and 1994. At $\alpha = .005$, we cannot reject in 1990 either. Only for 1993 and 1995 does this simple model fit quite badly. The visual evidence provides more insight into what is going on:

For 1993 the distribution is somewhat skewed with a few points (six in fact) well away from the rest. If these are trimmed off, the distribution becomes roughly symmetric; and, at the nominal $\alpha = .01$ level, one cannot reject the null hypothesis that all the remaining values come from the same normal distribution (the p-value for this test is .09).

For 1995 we have a similar problem, except there are more outliers this time (about nine), with over half of these coming from the first two weeks of the survey. Again if these "out of control" cases are omitted the remaining observations are roughly normal; and, at the nominal $\alpha = .01$ level one cannot reject the null hypothesis that all the remaining values come from the same normal distribution (the p-value for this test is .60).

Out of control observations appear to exist in other years as well but to such a small extent that our simple first approximation might be judged acceptable.

Of course, more is going on in these data. Differences exist between the first and second half of each year's testing period (see Table 25.3). Particularly noticeable are the within year shifts for the early years (1989 to 1992) when IRS assistor accuracy levels were improving rapidly. For the recent 1993 to 1994 period, there have been small differences in accuracy between the first and second half of the testing period. However, the 1995 testing period began with a drop in accuracy from the 1994 level and increased during the course of the year, with the greatest improvement in accuracy coming during the early weeks. This may have been the result of learning on the part of the new assistors hired after the 1994 hiring freeze ended.

Another way to visualize what is going on is to look at the interquartile ranges of the interviewer results over time. Figure 25.2 displays the median accuracy rates obtained for the callers over all the years of the test call survey. The general accuracy level rises relatively steeply in the early years and levels off by about 1992–1993. Figure 25.3 displays the interquartile ranges of

Table 25.3   Accuracy Rates for First and Second Half of the Testing Period

Test year	First half	Second half	Difference
1989	60.4	65.7	5.3
1990	75.7	79.1	3.4
1991	81.5	86.6	5.1
1992	86.7	90.1	3.4
1993	89.5	89.3	−0.2
1994	89.6	90.1	0.5
1995	88.1	92.1	4.0

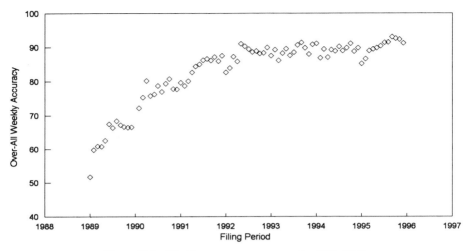

Figure 25.2   Weekly survey accuracy results, 1989–1995.

the callers' accuracy obtained by week over time. If we ignore the general level increase and focus on the length of these intervals, we can see that the interquartile ranges are fairly stable, somewhat larger in the first year but stabilizing after that. This is consistent with what is going on over time in the normal plots (with the standard deviations and coefficients of variation) but now our results are distribution free.

When we look at this stability in light of operational changes, it is somewhat

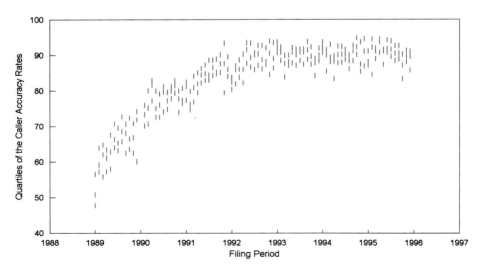

Figure 25.3   Weekly caller interquartile ranges, 1989–1995.

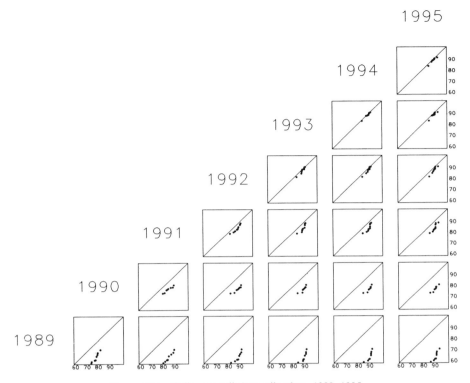

Figure 25.4   Caller quantile/quantile plots, 1989–1995.

surprising. Over the course of these years when the standard deviations and interquartile ranges showed little movement, the test call survey operation showed changes in the numbers of callers, sample sizes, number and mix of survey questions, and average number of questions per caller. We will speculate about probable reasons for this stability in our concluding comments.

Figure 25.4 is a scatterplot matrix of the caller quantile/quantile charts over time (Cleveland, 1993). These are plots of the quantiles of the caller distribution for each year plotted against the distributions for the remaining years. As with the normal probability plots, when the points fall on a straight line, the two distributions have approximately the same shape. When that line is 45 degrees, they have the same variances. Shifts in the line, up or down, indicate differences in the means of the distributions. The first comparison, 1989 to 1990, shows not only the biggest shift in overall level but also that it may be different in other ways as well. As we move to later pairs of years, the distributions look much more similar. After the first couple of years, adjacent periods, in particular, look nearly identical. Again, we see that our distributions are moving toward and maintaining stability.

25.6   CONCLUSIONS

The IRS test call system is subject to numerous outside forces which require statistical and management strategies to minimize quality costs when these changes and pressures occur. The evidence examined here suggests that our test call approach so far has weathered several of these quite well—including downsizing, personnel freezes, budget cutting, and shifts in upper management emphasis (see also Chapter 21). In our environment, many tools exist because of the CATI system that makes it possible to closely manage and improve quality. As mentioned at the outset, our CATI system provides a great advantage over traditional survey practice in the area of survey quality improvement where the short cycle times afforded by the computerized technology allow for many more iterations of improvements than are possible in a traditional survey setting.

What is at the heart of the success of the test call program is, though, a combination of the teamwork and sense of shared purpose that develops as part of the preparation for each year's testing. Clearly a success, too, are the management strategies that are used to make each test call season an operational replicate of the other years'. A great deal of organizational learning has occurred over the years of test call operation and is continuing to occur. This has strengthened the statistical design controls as well as the management strategies used to reduce the nonsampling errors contributed by the test callers.

Prior to testing each year, an intense period of operational training for the test callers using the test questions is conducted. This begins in December and peaks in January, just before the February to April measurement period. While usually there are some holdover test callers from previous years, several are new. This makes for lots of hard work and always raises concerns about year to year comparability. Such comparability problems exist, of course, in any interview setting even when the questions remain unchanged since, typically, interviewers change.

Performance checks using the same or similar questions are made each year and provide a way of directly addressing caller to caller variability during training and even overall measurement stability. Different problems arise each filing season; but, so far at least, the callers that prove capable of doing this work yield, on the whole as we have seen, reassuringly stable results. Part of the reason for this is that the number of callers is small (after weeding out individuals not suited for such work).

Perhaps the most important reason in most years for the measurement robustness of the test call survey is that there are caller meetings virtually every morning to talk through problems and misunderstandings that may arise. These meetings instill a pride in the work and keep performance from drifting over time. Efficiencies in operation also result, as has been noted in other settings (e.g., Biemer and Caspar, 1994). We suspect that our CATI operation inherently offers ways to make trade-offs visible. Because of the culture that is created by a shared sense of purpose, there is a team effort to minimize any adverse consequences that otherwise might arise. In summary, the use of modern technology and modern quality principles has led to a remarkably stable measurement in a very unstable world.

REFERENCES

Ayidiya, S.A., and McClendon, M.J. (1990), "Response Effects in Mail Surveys," *Public Opinion Quarterly*, 54, pp. 229–247.

Batcher, M., and Scheuren, F. (1989), "The IRS Test Call Program," *Proceedings of the Business and Economic Statistics Section*, American Statistical Association, pp. 648–653.

Biemer, P., and Caspar, R. (1994), "Continuous Quality Improvement for Survey Operations: Some General Principles and Applications," *Journal of Official Statistics*, 10, 3, pp. 307–326.

Bishop, G.F., Hippler, H.-J., Schwarz, N., and Strack, F. (1988), "A Comparison of Response Effects in Self-Administered and Telephone Surveys" in R.M. Groves, P.P. Biemer, L.E. Lyberg, J.T. Massey, W.L. Nicholls II, and J. Waksberg (eds.), *Telephone Survey Methodology*, New York: Wiley, pp. 321–340.

Cleveland, W.S. (1993), *Visualizing Data*, Summit: Hobart Press.

De Leeuw, E.D., and van der Zouwen, J. (1988), "Data Quality in Telephone and Face to Face Surveys: A Comparative Meta-Analysis," in R.M. Groves, P.P. Biemer, L.E. Lyberg, J.T. Massey, W.L. Nicholls II, and J. Waksberg (eds.), *Telephone Survey Methodology*, New York: Wiley, pp. 283–299.

Deming, W.E. (1986), *Out of the Crisis*, Cambridge: Massachusetts Institute of Technology.

Dillman, D.A. (1978), *Mail and Telephone Surveys: The Total Design Method*, New York: Wiley.

Dillman, D.A., and Tarnai, J. (1991), "Mode Effects of Cognitively Designed Recall Questions: A Comparison of Answers to Telephone and Mail Surveys," in P. Biemer, R. Groves, L. Lyberg, N. Mathiowetz, and S. Sudman (eds.), *Measurement Errors in Surveys*, New York: Wiley, pp. 74–93.

Groves, R.M. (1989), *Survey Errors and Survey Costs*, New York: Wiley.

Hahn, G.J., and Meeker, W.Q. (1993), "Developing a System of Indicators for Unmeasured Survey Quality Components," *The American Statistician*, 47, pp. 1–11.

King, C.A. (1987), "A Framework for a Service Quality Assurance System," *Quality Progress*, September, pp. 27–32.

Money Magazine (1990), "Surprise! The IRS Gets More Helpful."

Nicholls II, W.L. (1988), "Computer-Assisted Telephone Interviewing: A General Introduction," in R.M. Groves, P.P. Biemer, L.E. Lyberg, J.T. Massey, W.L. Nicholls II, and J. Waksberg (eds.), *Telephone Survey Methodology*, New York: Wiley, pp. 377–402.

Nicholls II, W.L., and Kindel, K. (1993), "Case Management and Communications for Computer Assisted Personal Interviewing," *Journal of Official Statistics*, 9, pp. 623–639.

Osborne, D., and Gaebler, T. (1992), *Reinventing Government: How the Entrepreneurial Spirit is Transforming the Public Sector*, Reading, MA: Addison-Wesley.

Stokes, L., and Yeh, M. (1988), "Searching for Causes of Interviewer Effects in Telephone Surveys" in R.M. Groves, P.P. Biemer, L.E. Lyberg, J.T. Massey, W.L. Nicholls II, and J. Waksberg (eds.), *Telephone Survey Methodology*, New York: Wiley, pp. 357–373.

Sykes, W., and Collins, M. (1988), "Effects of Mode of Interview: Experiments in the U.K.," in R.M. Groves, P.P. Biemer, L.E. Lyberg, J.T. Massey, W.L. Nicholls II, and J. Waksberg (eds.), *Telephone Survey Methodology*, New York: Wiley, pp. 301–320.

Tarnai, J., and Dillman, D.A. (1992), "Questionnaire Content as a Source of Response Differences in Mail and Telephone Surveys," in N. Schwarz, and S. Sudman (eds.), *Context Effects in Social and Psychological Research*, New York: Springer-Verlag.

Van Melis-Wright, M., Batcher, M., Stone, D., and Scheuren, F. (1990), "Cognitive Psychological Approaches in the Evaluation of Information Exchange Processes," *Proceedings of the Section on Survey Research Methods*, American Statistical Association, pp. 89–94.

Womack, J.P., Jones, D.T., and Roos, D. (1990), *The Machine That Changed the World*, New York: Harper.

CHAPTER 26

Using Statistical Methods Applicable to Autocorrelated Processes to Analyze Survey Process Quality Data

Piyasena Hapuarachchi, Mary March, and Adam Wronski
Statistics Canada

26.1   INTRODUCTION

Managers of monthly surveys use various measures to monitor the quality and stability of survey processes. If the process is long running, long time series of these measures may have been accumulated and analyses of these series of data, even a visual examination, can be very informative. Usually, managers compare new measures to historical values to make judgments about the quality of a process. These quality data series are often autocorrelated. An example of a series most managers watch is the survey's nonresponse rates. Frequently, a visual examination of a graph of this series reveals seasonality. Sometimes, also, there is an obvious upward or downward trend. The increasing availability of computers and analysis software has enabled meaningful and useful analyses of historical series of survey quality data to be performed quickly and inexpensively. Our purpose in this chapter is to show a method of analysing longitudinal series of survey quality measures using Shewhart control charts of the residuals resulting after fitting the series to appropriate time series models.

Control charts are widely used as a statistical tool for evaluating the stability of industrial processes. A typical control chart consists of a centre line representing the expected value of a process parameter (e.g., mean, proportion,

Survey Measurement and Process Quality, Edited by Lyberg, Biemer, Collins, de Leeuw, Dippo, Schwarz, Trewin.
ISBN 0-471-16559-X   © 1997 John Wiley & Sons, Inc.

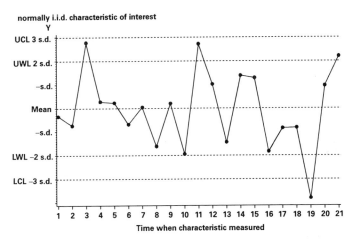

Figure 26.1 Control chart for independent observations. UCL and LCL are upper and lower control limits respectively while UWL and LWL are warning limits.

standard deviation, etc.) and two lines known as upper and lower control limits. The two lines are normally set at three standard deviations from the center line. The chart may also contain upper and lower warning limits set at two standard deviations from the center line.

The control chart generally works as follows. Random samples taken at regular intervals are used to obtain estimates of a "process average" at those points. These estimates are then plotted on the control chart. If a plotted point falls outside the control limits, the process is stopped and action is initiated to find the cause for this "out-of-control" situation and to eliminate it. This quality control method is often referred to as Statistical Process Control (SPC) since process stability is judged objectively using the statistical limits that are on the control chart. An example of a control chart is shown in Figure 26.1.

Most standard references on statistical methods for quality control such as Burr (1976), Duncan (1986), and Juran and Gryna (1988) explain the use of control charts. Since the introduction of control charts by Shewhart more than 50 years ago, they have been widely used in industry for several purposes. These include studies of process capability and measure capability, presentation of results of designed experiments, acceptance sampling, and process control (Schilling and Nelson, 1976). Construction of SPC charts depends on the quality variable under consideration. If the quality measurement is a continuous random variable, charts known as X-charts, R-charts, and s-charts are used. If the quality variable is an attribute such as the number of defective units in a sample, charts called p-charts, np-charts, and u-charts are commonly used.

For the highly correlated data typical in monthly surveys, ignoring the autocorrelation results in wider control limits and, hence, less ability to detect changes in the process. Devising control chart methods which have good properties for autocorrelated data is a challenging problem. Recently, in the process control literature, several approaches to constructing control charts when process data are autocorrelated have been developed, analyzed, compared, and criticized by Ryan (1994). These include ignoring the autocorrelation and widening the control limits' interval and fitting a time series model to the data and then using either of a variety of procedures on the residuals. The widening of the control limits interval method proposed by Barker (1993) requires a form of control charts known as *k*-charts. Barr (1993) discusses a method using residuals from an Exponentially Weighted Moving Average (EWMA) Control forecast that gives good results only if the process fits an Integrated Moving Average model called IMA(1,1). Ryan (1994) has shown that fitting an Autoregressive Integrated Moving Average (ARIMA) time series model and applying some form of control chart methodology with the residuals can give good or poor results depending on the parameters of the ARIMA model. The Average Run Length (ARL) of the residual chart, a measure of apparent randomness, can be much larger than the ARL one would expect if there were no autocorrelation.

Generally, it has been established that commonly used charts might not have desirable (ARL) properties when applied to model residuals. More sophisticated control chart methods described in the literature may produce acceptable theoretical results, but these have not been shown yet.

In this chapter we introduce a method which we have used to analyze these data and to monitor their behavior over time. Information regarding applicability and limitations of the method is also presented. A few interesting examples are shown.

26.2 MEASUREMENTS FREQUENTLY USED TO MANAGE THE QUALITY OF AN ONGOING MONTHLY SURVEY

A manager of a monthly survey must carefully monitor survey processes to ensure that estimates obtained from it are of the expected quality. The manager is interested in understanding historical behavior of the processes and their effects on the survey in order to judge whether observed changes in behavior require an intervention. Measures of possible interest to the manager because of their relationship to or effect on survey processes include nonresponse rates, coefficients of variation (CV), edit failure rates, imputation rates, weighted imputation rates, slippage (or undercoverage) rates, design effects, interviewer turnover rates, vacant dwelling rates, new construction rates, business bankruptcy rates, and even weather measures such as temperature. After some analysis to decide their influence on survey quality, a manager can select measures which are worthwhile monitoring. These will vary depending on the survey.

The Canadian Labour Force Survey (LFS) and the Monthly Survey of Employment, Payrolls, and Hours (SEPH) are important monthly surveys conducted by Statistics Canada. The primary objective of the LFS is to provide comprehensive, timely, and accurate estimates of the number, characteristics, and activities of the employed, unemployed, and persons who are not in the labor force. Its secondary objective is to serve as a general survey vehicle for the collection of a wide range of information on the Canadian population by supplementary surveys. The objective of the SEPH is to measure characteristics of employment including the total number of payroll employees, average weekly and hourly earnings, and average weekly hours.

Both surveys are large operations. The LFS currently has a sample size of approximately 63,200 households per month. These sampling units are selected using a rotating panel design. Each household remains in the sample for six months and one-sixth of the households are replaced each month. The survey involves collecting data for about 140,000 persons per month. The SEPH consists of a census of the approximately 20,000 establishments with 300 or more employees and a sample of about 9,700 establishments with 299 employees or less. Data from administrative sources are combined with data collected from the establishment survey to produce estimates for the full range of SEPH variables.

Numerous quality-related measures are computed every month for these two surveys. We have selected several LFS measures for illustrative purposes along with the difference in LFS and SEPH estimates of employment (the employed minus employees for a comparable universe). These include the nonresponse rate, slippage rate, vacancy rate, as well as the coefficient of variation for the estimated number of persons employed. The nonresponse rate is calculated each month as the ratio of the number of households for which data were not collected to the total number of eligible households in the sample. The vacancy rate is the proportion of dwellings which the interviewers classified as vacant. These measures tend to be serially correlated since households are in the sample for six consecutive months. Seasonality is also often present since the time of year influences the likelihood of a household being home and of certain types of dwellings being vacant.

As an example of a serially correlated series, consider the nonresponse rates for the LFS shown in Figure 26.2, a plot of nonresponse rates at the national level over all types of nonresponse (due to various reasons such as households being inaccessible, their being not-at-home, refusals of respondents to give information, sample units being unable to provide required information, etc.) from January 1985 to January 1995. A very prominent feature of the series is its seasonal fluctuations. The highest observed nonresponse rates are in July, the lowest in October of each year. From early 1994, there appears to have been a steady upward trend. Clearly, serial correlation, seasonality, and the trend should be taken into consideration in the analysis of this series.

26.3   USING CONTROL CHARTS FOR ANALYSIS OF MONTHLY SURVEY MEASURES

As indicated earlier, quality measures tend to be one of two types: attributes or continuous variables. If the measure is continuous, two assumptions are

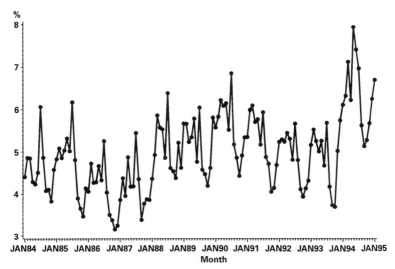

Figure 26.2   Labour Force Survey (LFS) nonresponse rates.

generally made in the construction of control charts—normality and independence. Under these assumptions, constants required to construct control charts
for various sample sizes have been tabulated in standard literature on quality
control such as Burr (1976) and Duncan (1986). These tables help quality
control practitioners avoid excessive computing in determining control limits
and allow a conclusion about the quality of a product to be reached quickly.

Lucas and Saccucci (1990 a,b) note that many processes exhibit some degree
of autocorrelation. In discussions published with this paper, Faltin *et al.*
(1990) and Hunter (1990) observe that, in today's world, there are hardly any
processes where the Shewhart chart display is strictly appropriate because of
possible autocorrelation between adjacent observations. When the measurements are serially correlated, the process mean may drift. (Such a drift may
sometimes be identified by plotting the data.) A possible symptom indicating
the presence of serial correlation might be the observation of more out-of-
control points than warranted.

Woodall and Faltin (1993) present an excellent summary on autocorrelated
data and statistical process control, discussing the effects of autocorrelation on
control charts. They note that autocorrelation often increases the variability of
the process and that positive correlation results in an increase in the number
of out-of-control signals. Misleading conclusions are drawn if one tries to
associate such out-of-control signals with assignable causes of variability or
instability. A correlation of 0.5 at lag one (i.e., of successive observations) can
cause an increase in the number of out-of-control signals by a factor of eight
times. Woodall and Faltin suggest that a first approach might be to attempt
to remove the source of serial correlation. If the source cannot be removed,

they recommend that sampling (or measuring) less frequently might remove it. If less frequent sampling is not possible, Vasilopoulis (1974) suggests modifying the control limits on X-bar charts under the assumption that the observations were following an autoregressive process said to be of order one. His procedure is extended to apply to autoregressive processes of order two by Vasilopoulis and Stamboulis (1978). Spurrier and Thombs (1987) discuss a method of constructing control charts for observations with cyclical behavior and a similar approach based on a "periodogram" is used by Beneke *et al.* (1988).

A commonly used approach suggested by Alwan and Roberts (1988) and Montgomery and Mastrangelo (1991) is to fit a time series model to the observations and then apply the standard control charts to the residuals. If the model is correct, the residuals should be independent and identically distributed. They suggest that a process level shift may show up as spikes in the residual chart rather than as a level shift. Thus, it would not be as simple to draw conclusions from residual charts as it is from standard charts. The next section proposes a method of superimposing residual charts on the time series plots of the original observations and their forecast values to enable easier detection of process level changes.

26.4 A METHODOLOGY FOR ANALYSIS OF MONTHLY SERIES OF SURVEY QUALITY DATA

We use the Box–Jenkins ARIMA modeling procedure, a widely used technique in time series analysis, to fit a time series model to the selected data series. With an ARIMA model, we assume the time-sequenced observations are statistically dependent. ARIMA models are especially suited for short term forecasting because the procedure puts heavy emphasis on the recent rather than distant past. The models are particularly useful for forecasting data series which contain seasonal (or other periodic) correlation patterns. The ARIMA modeling procedure consists of three steps: model identification, estimation, and checking. These steps may have to be repeated more than once to achieve a satisfactory time series model. The goal of the fitting process is to choose an ARIMA model which includes the smallest number of nonzero parameters needed to adequately match the correlation patterns in the available data. The goal is to achieve a model that is as simple as possible.

The Box–Jenkins modeling procedure involves computing and examining various quantities such as a sample autocorrelation function, inverse autocorrelations, a sample partial autocorrelation function, differencing data (if differencing is necessary to obtain stationarity), residuals, and others. The number of available observations should be large (generally, at least 50), otherwise, the ARIMA analysis may not be reliable. Ideally about 100 observations are needed. For a detailed discussion of this procedure see Box and Jenkins (1976) or Pankratz (1983).

Testing for "interventions" during ARIMA modeling can sometimes be

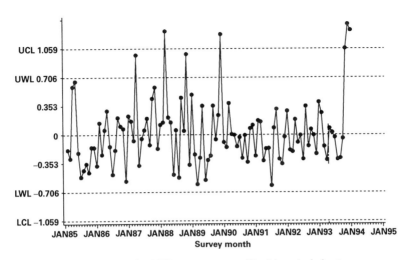

Figure 26.3   The LFS nonresponse residuals' control chart.

appropriate if at a given time period there was a significant level shift or correlation pattern change in a series. Longitudinal surveys periodically undergo various changes—either due to changes in methodology or to outside influences.

26.4.1   Control Charts

Figure 26.3 is an example of a control chart containing a plot of residuals obtained after fitting an ARIMA model to LFS response rates in Figure 26.2.

In this case, several out-of-control signals are evident. Between January 1990 and October 1993, the residuals appear to have been quite stable. Since then, we note a definite upward trend.

26.4.2   A Modified Control Chart

We have found that it is useful to look at a graphic representation of the model and the estimated trend along with the observed values. In addition, it is at the same time helpful to plot the control limits based on the model confidence intervals on the same graph. The model is fitted using data up to a given date. (We used April 1993 as the end date for the LFS nonresponse data.) Subsequent model values are forecasts based on the earlier data. We call this a "modified" control chart.

We find the modified control charts easier to interpret and explain, especially in cases where the residuals chart contains out-of-control signals, apparent clustering near the mean, and possible trends. Control limits, trend, forecasts and actual data can be represented on one graph. An example using LFS nonresponse data is given in Figure 26.4. The solid starred line represents the

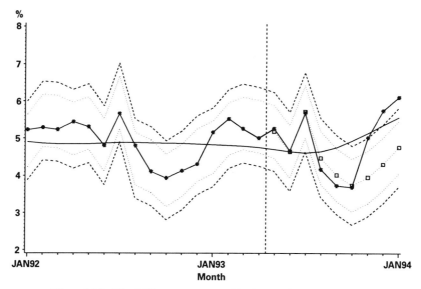

Figure 26.4 The LFS nonresponse residuals modified control chart.

original data. The smooth solid line represents the trend. The most interesting part of the graph is on the right of the vertical line dividing the original series and the forecast. The data after April 1993 were not used for fitting the model. Comparing these later values to the forecast (squares) and the control and warning limits (dashed lines) we see that there has clearly been an increase in the level of nonresponse since October 1993 that cannot be completely ascribed to the usual seasonal pattern.

Consultation with LFS statisticians who regularly monitor operations determined that this behavior might be explained by the shift to computer assisted personal interviewing which began in October 1993. There were data transmission problems in the first few months which caused a larger number of households than usual to be missed. Correction of the problem led to a more stable situation.

If the residuals are not normally distributed, a form of a modified control chart is possible. We can use ARIMA 99 and 95 percent forecast intervals to calculate control limits instead of standard deviations. Such charts should be used with caution since we have found that in examples when residuals are normally distributed, these intervals are usually wider than the control limits, that is, it is harder to detect changes to the process.

26.5 EXAMPLES

We have experimented with this analytical tool, applying it to several other series of quality data and, in cooperation with staff working on the surveys,

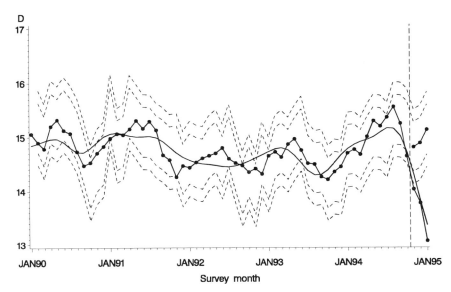

Figure 26.5 LFS vacancy rates.

looked at its ability to detect level changes and changing patterns. A few examples are illustrated.

Figure 26.5 is a modified control chart obtained as a result of analyzing LFS vacancy rates. This rate is a measure of the frequency with which interviewers classify dwellings in the survey area sample as vacant. Survey managers monitor these rates from month to month because there is a risk of undercoverage due to not-at-home households being incorrectly classified as vacant dwellings. A seasonal model was fitted to the series of data from January 1990. Recently, there has been a sudden downward trend. A likely cause is considered to be the recent reallocation of the sample to contain a higher proportion of urban dwellings.

Figures 26.6 and 26.7 show series where the models fitted contained a strong seasonal component with an increasing trend. In both cases, the models were fitted using 41 points (January 1991 to May 1994). The more recent values of LFS CVs (shown in Figure 26.6) are generally higher than the warning limits about the forecast levels. In Figure 26.7, the actual values (the shorter curve) appear to be "in control."

26.6 CONCLUSIONS

We have found the analysis technique described in this chapter very useful in helping us obtain better insight into the behavior of quality measures associated with monthly surveys. As we indicated before, many of the series derived from monthly surveys are autocorrelated (because of seasonal factors and because

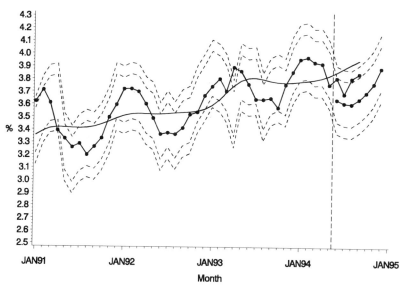

Figure 26.6 LFS CVs of estimated numbers of employed.

of rotating panel sample designs). Standard control charts are not applicable because of possible serial correlation. Therefore, to construct control charts for serially correlated data, we propose the following procedure. Identify the time series model which best describes the quality measures series. Estimate the parameters of the model. Use the resulting model to construct an appropriate

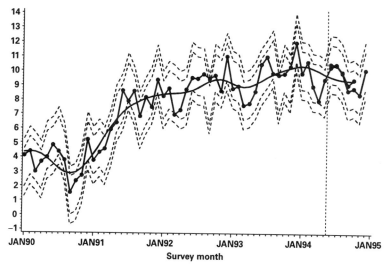

Figure 26.7 LFS estimates of the number of employed minus SEPH estimates of the number of jobs.

control chart for residuals generated from this model. As observations for new time points are available, calculate the residuals and plot them in the residuals control chart. If any of the residuals fall outside the control limits, examine why the out-of-control situation has occurred.

Interpretation of the control charts and even the modified control charts is often difficult. As a result, we do not feel that we can promote this as a general tool for use by survey managers and operational staff. However, it is a useful analysis tool for more sophisticated users.

REFERENCES

Adams, B.M., and Lin, W. (1994), "Monitoring Autocorrelated Data with a Combined EWMA-Shewhart Control Chart," *Proceedings of the Section on Quality and Productivity*, American Statistical Association, pp. 104–107.

Alwan, L.C., and Roberts, H.V. (1988), "Time Series Modelling for Statistical Control," *Journal of Business and Economic Statistics*, 6, pp. 87–95.

Barker, T. (1993), Letter to the Editor, *Quality Progress*, 26(1), p. 6.

Barr, T. (1993), "Performance of Quality Control Procedures When Monitoring Correlated Processes," unpublished Ph.D. dissertation, School of Industrial and Systems Engineering, Georgia Institute of Technology, Atlanta, Georgia.

Beneke, M., Leemis, L.M., Schlegel, R.E., and Foote, B.L. (1988), "Spectral Analysis in Quality Control: A Control Chart Based on the Periodogram," *Technometrics*, 30 (1), pp. 63–70.

Burr, I.W. (1976), *Statistical Quality Control Methods*, Marcel Dekker, New York.

Box, G.E.P., and Jenkins, G.M. (1976), *Time Series Analysis: Forecasting and Control*, Holden-Day.

Duncan, A.J. (1986), *Quality Control and Industrial Statistics*, Fifth edition, Irwin, San Francisco.

Faltin, F.W., Hahn, G.J., and Tucker, W. (1990), Discussion of Lucas and Saccucci paper, *Technometrics*, 32, pp. 19–20.

Hunter, J.S. (1990), Discussion of Lucas and Saccucci paper, Technometrics, 32, pp. 21–22.

Juran, J.M., and Gryna, F. M. (1988), *Juran's Quality Control Handbook*, Fourth Edition, McGraw-Hill, Inc.

Lucas, J.M., and Saccucci, M.S. (1990a), "Exponentially Weighted Moving Average Control Schemes: Properties and Enhancements," *Technometrics*, 32, pp. 1–29.

Lucas, J.M., and Saccucci, M.S. (1990b), "Exponentially Weighted Moving Average Control Schemes: Properties and Enhancements (Revised Version)," Drexel University Faculty Working Paper Series.

Montgomery, D.C., and Mastrangelo, C.M. (1991), "Some Statistical Process Control Methods for Autocorrelated Data," *Journal of Quality Technology*, 23, pp. 179–204.

Pankratz, A. (1983), *Forecasting with Univariate Box-Jenkins Models: Concepts and Cases*, New York: Wiley.

Ryan, T.P. (1994), "Methods for Charting Autocorrelated Process Data—What Will and Will Not Work?," unpublished draft paper, presented at the Joint Statistical Meetings, Toronto, 1994.

Schilling, E.G., and Nelson, P.R. (1976), "The Effect of Non-normality on the Theoretical Limits of $\bar{X}$-Charts," *Journal of Quality Technology*, 8, pp. 183–188.

Spurrier, J.D., and Thombs, L.A. (1987), "Control Charts for Detecting Cyclical Behaviour," *Technometrics*, 29, pp. 163–171.

Statistics Canada Household Surveys Division (Monthly), *The Labour Force*, Statistics Canada, Catalogue No. 71-001.

Statistics Canada Labour Division (Monthly), *Employment, Earnings, and Hours*, Statistics Canada, Catalogue No. 72-002.

Vasilopoulis, A.V. (1974), *Second Order Autoregressive Model Applied to Quality Control*, Ph.D. dissertation, New York University, New York.

Vasilopoulis, A.V., and Stamboulis, A.P. (1978), "Modification of Control Limits in the Presence of Serial Correlation," *Journal of Quality Technology*, 10, pp. 20–30.

Woodall, W.H., and Faltin, F.W. (1993), "Autocorrelated Data and SPC," *ASQC Statistics Division Newsletter*, 13 (4), pp. 18–21.

Error Effects on Estimation, Analyses, and Interpretation

CHAPTER 27

A Review of Measurement Error Effects on the Analysis of Survey Data

Paul P. Biemer
Research Triangle Institute, Research Triangle Park

Dennis Trewin
Australian Bureau of Statistics

27.1 INTRODUCTION

This section reviews the effects of measurement error on a number of standard univariate and multivariate analysis techniques. Much has appeared in the literature about the existence of measurement error in survey data, methods for reducing measurement error through better survey design, methods for compensating for the error effects at the analysis stage, and standards for informing users of the risks in ignoring measurement errors in the interpretation of the final results. The present chapter addresses the motivation for all this attention to measurement error.

Why are we concerned about measurement error? For example, it is well known that when independent random errors are added to survey data, the variance increases but the expected value of the mean or total remains unbiased so long as the random errors have zero expectation. Although these types of errors may increase standard errors, the standard errors can be reduced by increasing the sample sizes. Other than their potential to increase costs, why should we be concerned with these errors? This chapter responds to that

Survey Measurement and Process Quality, Edited by Lyberg, Biemer, Collins, de Leeuw, Dippo, Schwarz, Trewin.
ISBN 0-471-16559-X © 1997 John Wiley & Sons, Inc.

question and shows that even independent errors having zero mean can create that analysts may use. In fact, the population mean or total and its standard error are the only two commonly used statistics that are not biased by these errors. The chapter also examines the effects of errors that are clustered within interviewer or other operator assignments on data analytical methods.

In Section 27.2, we consider models for the analysis of continuous survey data as well as models which are more appropriate for categorical data analysis. In addition, for each of these model forms we consider the effects of measurement errors when the errors are uncorrelated between the units in the sample as well as when correlations between the units are introduced by an interviewer, coder, supervisor, or other survey operator. This section of the chapter sets the stage for the review of the error effects that follows.

Section 27.3 considers the effect of correlated and uncorrelated error on univariate analysis; in particular, descriptive statistics such as means, proportions, totals, and population quantiles. Section 27.4 deals with the effects of error on multivariate data analysis, including: regression analysis, correlation analysis, analysis of variance, analysis of covariance, and categorical data analysis (specifically, goodness-of-fit tests and tests of independence between two variables). In all these discussions, our goal is to provide insight into the nature of the error effects and a basic understanding of the potential risks involved in ignoring the errors in data analysis. We do not attempt to provide a complete review of the models that have been proposed for studying measurement error's effect on data analysis. Nor do we attempt to provide thorough coverage of the methods for compensating for their effects. To do so would lengthen the chapter unacceptably. However, in Section 27.5 we provide a summary of the most commonly used error compensation methods and references to articles in this and other volumes that deal with these issues.

27.2   MODELS FOR STUDYING THE MEASUREMENT ERROR EFFECTS

Survey error models are essential for understanding the effects of measurement errors on statistics and methods of statistical inference. Error models allow us to concisely and precisely communicate the nature of the errors that are being considered, the survey conditions that give rise to them, the ways in which errors affect the expected values of the estimates, variances and variance estimators, and the magnitudes of these effects. To facilitate our discussion, we will employ two general models for measurement error: one for continuous data and one for binary data. In addition, to facilitate the exposition of the main results, we assume simple random sampling (SRS) without replacement where the sampling fraction, n/N, is negligible.

27.2.1   Error Model for Continuous Data

In this chapter, the most general form of the continuous error model that is considered assumes that an observation, y_j, on a randomly selected unit j is the sum of two components: the true value, μ_j, for the unit and an error, d_j, which may be attributed to the measurement process, including the interviewer, the questionnaire, the respondent, the mode of interview, the interview setting, and so on.

Let $U = \{1, 2, \ldots, N\}$ denote the label set for the target population consisting of N units. Then the expected value and variance of the true value, μ_j, associated with the jth unit are given, respectively, by

$$\bar{M} = \frac{\sum_{j=1}^{N} \mu_j}{N} \tag{27.1}$$

and

$$\sigma_\mu^2 = \frac{\sum_{j=1}^{N} (\mu_j - \bar{M})^2}{N}. \tag{27.2}$$

In its most general form, the structure of the error, d_j, provides for potential correlations among the measurement errors due to *interviewers, supervisors, coders,* or other *survey operators.* Let $S = \{1, 2, \ldots, n\}$ denote the sample of $n = Im$ units and assume that S is partitioned into I assignments of $m = n/I$ units. (The assumption of equal assignment sizes is made solely to facilitate the discussion of error effects and does not limit the generalizability of the results to follow.) Denote the set of units assigned to the ith operator by S_i. Now, relabel d_j as d_{ij} where the subscript ij denotes the jth unit in S_i, for $j = 1, \ldots, m$ and $i = 1, \ldots, I$. For unit $j \in S_i$, we assume that d_{ij} is the sum of two error terms, b_i and ε_j, where b_i is an *operator error* which is assumed to be same for all units in the ith operator's assignment and ε_j is the unit-specific or *elementary error* due to the respondent as well as other sources of error, including the operator. Thus, the model for the ijth observation is

$$\begin{aligned} y_{ij} &= \mu_{ij} + d_{ij} \\ &= \mu_{ij} + b_i + \varepsilon_{ij}. \end{aligned} \tag{27.3}$$

In evaluating operator error, if our primary interest is specifically the I operators employed for the present survey, then it is appropriate to assume that the errors, b_i, are fixed constants or operator biases. However, often in planning for future surveys, we may be more interested in the population of potential operators from which the current group was chosen, reasoning that if the survey were to be repeated, the operators for the future survey would be recruited from this population. In this case, the assumption of random operator effects may be more appropriate. For random operator effects, we essentially assume that the I operator errors, b_i, constitute a random sample from an infinite population of potential operator errors having mean and variance B_b

and σ_b^2, respectively. For either case, we assume that the elementary errors are random variables with mean B_ε, variance σ_ε^2, and $\text{Cov}(\mu_{ij}, d_{ij}) = 0$. Then the covariance structure of the d_{ij} is

$$
\begin{aligned}
\text{Cov}(d_{ij}, d_{i'j'}) &= \sigma_b^2 + \sigma_\varepsilon^2 \quad \text{for } i = i'; j = j' \\
&= \sigma_b^2 \quad \text{for } i = i'; j \neq j' \\
&= 0 \quad \text{for } i \neq i'.
\end{aligned}
\tag{27.4}
$$

Under the assumption of fixed operator effects, the covariance structure is modified by setting $\sigma_b^2 = 0$. Thus, fixed operator effects is a special case of the general random operator effects model. Another special case of the general model that we will consider extensively is the case of no operator effects; i.e., $b_i \equiv 0$ for all i. This model is often referred to as the *uncorrelated error model* since then $\text{Cov}(d_{ij}, d_{ij'}) = 0$ when $j \neq j'$. Under this assumption, the covariance structure in (27.4) is also modified by setting $\sigma_b^2 = 0$.

27.2.2 Error Model for Binary Data

For binary survey responses (for example, "yes/no" or other responses which may be coded as 0 or 1), the assumptions of the continuous data model are not appropriate; for example, the assumption that $\text{Cov}(\mu_i, d_{ij}) = 0$ does not generally hold. Therefore, an alternative model has been developed for this type of data. As for the continuous data model, we assume that the population is partitioned into I operator assignments: $S_1, \ldots, S_I$. Let y_{ij} denote the observation for the jth unit in S_i $(i = 1, \ldots, I)$ where $y_{ij} = 1$ if the response is "yes" and $y_{ij} = 0$ if "no." Let μ_{ij} denote the corresponding true value which is 1 if unit (i, j) is a true "yes" and 0 if a true "no." For binary data, (27.1) and (27.2) are given by $\pi = \text{P}(\mu_{ij} = 1)$ and $\pi(1 - \pi)$, respectively, where π is the finite population proportion. Then, for $j \in S_i$, we define the following misclassification probabilities

$$
\begin{aligned}
\theta_i &= \text{P}(y_{ij} = 0 \mid \mu_{ij} = 1) \\
\phi_i &= \text{P}(y_{ij} = 1 \mid \mu_{ij} = 0)
\end{aligned}
\tag{27.5}
$$

where θ_i and ϕ_i are referred to as the *probability of a false negative* and the *probability of a false positive*, respectively. In terms of Equation (27.3), the distribution of the error, d_{ij}, is given in Table 27.1. Further assume

$$
\begin{bmatrix} \theta_i \\ \phi_i \end{bmatrix} \sim \left(\begin{bmatrix} \theta \\ \phi \end{bmatrix}, \begin{bmatrix} \sigma_{\theta\theta} & \sigma_{\theta\phi} \\ \sigma_{\phi\theta} & \sigma_{\phi\phi} \end{bmatrix} \right).
\tag{27.6}
$$

The binary data model is a generalization of Cochran's (1968) model which

Table 27.1   Error Distribution for the Binary Data Error Model

y_{ij}	μ_{ij}	d_{ij}	$P(y_{ij} \mid \mu_{ij})$
1	1	0	$1 - \theta_i$
0	1	-1	θ_i
1	0	1	ϕ_i
0	0	0	$1 - \phi_i$

specified that $\theta_i = \theta_{i'} = \theta$ and $\phi_i = \phi_{i'} = \phi$. (Note that $1 - \theta_i$ and $1 - \phi_i$ are sometimes called the *sensitivity* and *specificity* of a measure, respectively.) Here we allow the misclassification probabilities to depend upon the operator, thus introducing correlated errors into the model.

As in the case of continuous data, if operator effects are assumed to be fixed, then $\sigma_{\theta\theta} = \sigma_{\phi\phi} = \sigma_{\theta\phi} = 0$ in (27.6), although the misclassification probabilities are still indexed by i to account for the different probabilities across operators. Under the assumption of no operator effects, then, as in Cochran's model, we set $\theta_i = \theta_{i'} = \theta$ and $\phi_i = \phi_{i'} = \phi$ which implies that $\sigma_{\theta\theta} = \sigma_{\phi\phi} = \sigma_{\theta\phi} = 0$. This model is referred to as the *uncorrelated binary model* since $\text{Cov}(d_{ij}, d_{i'j'}) = 0$ for $(i, j) \neq (i'j')$. Note, however, from Table 27.1 that in the binary case, the errors are negatively correlated with the true values.

27.3   THE EFFECT OF MEASUREMENT ERROR ON UNIVARIATE ANALYSIS

The estimation of means, totals, proportions, medians, or other statistics that help to describe the populations of interest is a routine component of survey data analysis and, hence, the study of these measurement error effects on descriptive statistics is relevant to all survey work. Early papers by Hansen *et al.* (1951), Sukhatme and Seth (1952), and Kish and Lansing (1954) considered the effect of correlated and uncorrelated error on estimators of the population mean. A number of authors since then have refined the concepts and have empirically investigated the nonsampling error effects on estimators and their variances. Hansen *et al.* (1961, 1964) examined the effects of measurement error on the sample proportion using models which are more appropriate for binary data. Cochran refined the measurement error model for binary data and considered the effects of uncorrelated error on means, proportions, and a number of standard statistical techniques. In this section, we present some of the results of these early papers in the context of the models presented in Section 27.2.

27.3.1   Estimators of Means and Totals for Continuous Data and Their Standard Errors

27.3.1.1   Uncorrelated Error Effects

Assume the uncorrelated error model, i.e., model (27.3) with variance–covariance structure given by (27.4) after setting σ_b^2 to 0. For SRS, the usual estimator of $\bar{M}$ is

$$\bar{y} = \frac{1}{n} \sum_{i=1}^{I} \sum_{j=1}^{m} y_{ij}. \tag{27.7}$$

Now, since the total, M, is simply $N\bar{M}$, it suffices to consider the effects of measurement error on $\bar{M}$ only.

It is easily shown from (27.3) that

$$E(\bar{y}) = \bar{M} + B_d \tag{27.8}$$

where B_d is the bias in the sample mean. Note that $B_d = B_b + B_\varepsilon$, the sum of the bias due to operators and the bias due to other sources. If $B_d = 0$, then the sample mean is an unbiased estimator of the population mean. This condition implies that either the means of both the operator error and the elementary error distributions are zero, or that the operator and elementary error biases cancel out one another. Thus, when $B_d = 0$, the sample mean is unbiased for the population mean even in the presence of measurement error.

The variance of $\bar{y}$ is given by

$$\mathrm{Var}(\bar{y}) = \frac{\sigma_\mu^2 + \sigma_\varepsilon^2}{n}$$

$$= \frac{1}{R} \frac{\sigma_\mu^2}{n} \tag{27.9}$$

where

$$R = \frac{\sigma_\mu^2}{\sigma_\mu^2 + \sigma_\varepsilon^2}. \tag{27.10}$$

We refer to R as the reliability ratio (see, for example, Sukhatme and Seth, 1952). Thus, we see from (27.9) that as a consequence of uncorrelated error, the variance of $\bar{y}$ is increased and the "effective sample size" for the survey is $n' = Rn$. Put another way, if c is the cost of observing a single unit, then $c(n/R - n)/cn = (1 - R)/R$ is the proportionate increase in survey cost due to measurement error for a fixed level of precision in the estimator $\bar{y}$. To illustrate the effect, suppose that $R = 0.7$ for some survey questions. Then, $(1 - R)/R = 0.43$ and thus, the total survey cost is increased by 43% as a result of uncorrelated measurement error.

As we see, the reliability ratio is an ubiquitous parameter in the study of measurement error. It is the ratio of the variance of the true values to the total variance of a single observation, which under the uncorrelated model, is the sum of the true value variance and the error variance. A more general expression for R is

$$R = \frac{\text{VE}(y_{ij} \mid ij)}{\text{V}(y_{ij})} \tag{27.11}$$

The numerator in this expression is the finite population variance of the conditional expectation of observation (i, j) over the measurement error distribution. Sometimes $E(y_{ij} \mid ij)$ is called the *true score* for unit (i, j) (Lord and Novick, 1968) which is equal to the true value, μ_j, plus the measurement bias, B_d. Thus, under our assumed model, the variance of the true scores is σ_μ^2 which is identical to the variance of the true values.

Now, consider the estimator of $V(\bar{y})$ under the uncorrelated error model. The usual estimator is

$$v(\bar{y}) = \frac{s^2}{n} = \frac{\sum_{i,j}(y_{ij} - \bar{y})^2}{n(n-1)}. \tag{27.12}$$

It can be shown (see, for example, Cochran, 1968) that

$$E[v(\bar{y})] = \frac{\sigma_\mu^2 + \sigma_\varepsilon^2}{n}$$

$$= \text{Var}(\bar{y}) \tag{27.13}$$

and, thus, the usual estimator is unbiased under uncorrelated measurement error. Thus, although the variance of the sample mean is increased, the usual estimator of the variance is increased by the same amount. As we shall see, the estimator, $v(\bar{y})$, is the only estimator we study that is unbiased in the presence of uncorrelated error even when $B_d \neq 0$.

27.3.1.2 *Correlated Error Effects*

Now assume model (27.3) with variance-covariance structure given by (27.4) where $\sigma_b^2 > 0$. Under this model, the expected value of $\bar{y}$ is still given by (27.8) since we have assumed that the mean of the operator error distribution is still B_b. The variance of the estimator is changed, however.

Sukhatme and Seth (1952) first showed that

$$\text{Var}(\bar{y}) = \frac{\sigma_\mu^2 + \sigma_\varepsilon^2}{n} + \frac{\sigma_b^2}{I} \tag{27.14}$$

where the last term on the right is the increase due to variable operator error. Thus, the operator contribution to variance decreases as the number of operators, I, increases whereas the elementary error contribution decreases as the number of *sample units* increases. Alternatively, (27.14) may be written as

$$\text{Var}(\bar{y}) = \frac{\sigma_\mu^2 + \sigma_b^2 + \sigma_\varepsilon^2}{n}[1 + (m - 1)\rho_y]$$

$$= \frac{\sigma_\mu^2}{n}\frac{1}{R}[1 + (m - 1)\rho_y] \qquad (27.15)$$

where

$$\rho_y = \frac{\sigma_b^2}{\sigma_\mu^2 + \sigma_b^2 + \sigma_\varepsilon^2} \qquad (27.16)$$

and, from (27.11)

$$R = \frac{\sigma_\mu^2}{\sigma_\mu^2 + \sigma_b^2 + \sigma_\varepsilon^2}. \qquad (27.17)$$

When the operator is the interviewer, the term ρ_y is called the *intrainterviewer correlation coefficient* since it is the correlation between the errors for two observations in the same interviewer assignment. The term $[1 + (m - 1)\rho_y]$ is sometimes referred to as the interviewer *design effect* or *deff* in analogy of the corresponding term for the sampling design effect in cluster sampling. The interviewer or operator deff reflects the increase in variance as a result of operator variance. For example, if the average interviewer assignment consists of 50 units and $\rho_y = 0.01$, then $[1 + (m - 1)\rho_y] = 1 + 49(0.01)] = 1.49$ and, hence, the variance of the sample mean is increased by 49% due to correlated interviewer error. Furthermore, combined with a reliability of say, $R = 0.7$, the increase in variance due to measurement error is $1.49/0.7 = 2.13$. Thus, in this example, the effect of measurement error variance is to more than double variance! Such situations are not uncommon in the survey literature (see, for example, Groves, 1989, pp. 365–368). In Figure 27.1 we have plotted the increase in variance due to correlated interviewer error, viz., $(m - 1)\rho_y$, for $0.0 \le \rho_y \le 0.1$ and alternative values of m. Note that as the size of the interviewer assignment increases, the effect of interviewer variance is also substantially increased. This figure illustrates that even for small interviewer effects and moderate assignment sizes, interviewer variance can be several times greater than sampling variance.

The study of interviewer and other operator effects has played a very important role in the development of survey methods. As an example, at the U.S. Bureau of the Census, concern regarding the effect census enumerators were having on the 1950 census was key in the decision to adopt a self-enumeration census design for subsequent censuses (Hanson and Marks, 1958) since for self-enumeration, measurement errors are essentially uncorrelated between households. The reduction of ρ_y is also the primary motivation for standardized interviewing (Fowler and Mangione, 1990) and coding methods (Biemer and Caspar, 1994).

Groves (1989, p. 365) reports the values of ρ_y for interviewers estimated in 10

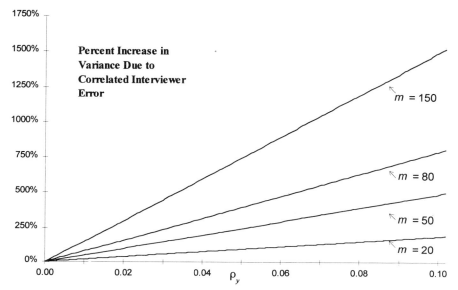

Figure 27.1 The increase in variance due to correlated interviewer error as a function of intra-interviewer correlation, ρ_y, and assignment size, m.

personal interview surveys and shows that the values of ρ_y averaged over a large number of survey characteristics measured in the studies ranged from 0.005 to 0.102. Most interviewer variance studies found in the literature report ρ_y in this range. For example, for centralized telephone interviewing, a typical value of ρ_y is 0.01. Face to face interviewing typically produces ρ_y that are somewhat higher.

Now, consider the expectation of $v(\bar{y})$ in (27.12). Hansen *et al.* (1961) showed that

$$E[v(\bar{y})] = \frac{\sigma_\mu^2 + \sigma_b^2 + \sigma_\varepsilon^2}{n}\left(1 - \frac{m-1}{n-1}\rho_y\right) \tag{27.18}$$

and, thus, the relative bias in the usual estimator of the variance is

$$\text{Relbias} = \frac{E[v(\bar{y})] - \text{Var}(\bar{y})}{\text{Var}(\bar{y})}$$

$$= -\frac{(m-1)\rho_y}{1+(m-1)\rho_y}\left(\frac{n}{n-1}\right)$$

$$\approx -\frac{(m-1)\rho_y}{1+(m-1)\rho_y} \tag{27.19}$$

for large n. From this expression we see that the relbias is a monotonically decreasing function of ρ_y and further, the amount of the underestimate is roughly equal to the increase in variance due to correlated interviewer error. Hence, the usual variance estimator does not reflect any of the increase in variance due to correlated error. For a typical value of ρ_y, say $\rho_y = 0.01$, the relative bias ranges from -16% for $m = 20$ to -50% for $m = 100$. Note from (27.19) that the largest absolute value of the relbias is approximately $(m - 1)/m \times 100\%$.

These results indicate that interviewer variance can substantially increase the variance and bias of estimators of the population mean. In fact, it is not uncommon for the magnitude of the interviewer variance to exceed that of the sampling variance. Furthermore, under simple random sampling and in the presence of interviewer error, the usual textbook formulas for estimating the sampling variance of means and totals are negatively biased, the bias being approximately equal to the magnitude of the increase in variance, *viz.* $(m - 1)\rho_y$. Wolter (1985, pp. 381–392) derives similar results under complex sampling and shows that random groups and other pseudo-replication methods for estimating standard errors can eliminate much of the bias in estimates of $\text{Var}(\bar{y})$.

27.3.2   Estimators of Quantiles

Quantiles are used quite often in social science and in the development of social policy. For example, quantiles of population income distributions are often used in setting poverty thresholds for government assistance programs. In addition, the 90%, 95%, and 99% quantiles of sampling distributions are commonly used to determine the critical values for tests of hypotheses and the margins of error for confidence intervals. The effects of uncorrelated measurement error on the estimation of quantiles have been considered by Fuller (1995) assuming the population to be sampled is normally distributed. Using Fuller's notation, the quantile function for the random variable X is

$$Q_X(a) = F_X^{-1}(a) \tag{27.20}$$

where

$$F_X(x) = P(X \le x) = a. \tag{27.21}$$

Thus, $Q_X(a)$ is the inverse of the cumulative distribution function for a variable. Some examples of quantiles are: $Q_X(0.5)$, the median of a distribution; $Q_X(0.25)$, the first quartile; and $Q_X(0.75) - Q_X(0.25)$, the interquartile range.

To illustrate the effect of measurement error on the cumulative distribution function, suppose that the true values are distributed in the population as an $N(0, 1)$ random variable and let the observations, y_{ij}, follow the uncorrelated error model with d_{ij} also distributed as an $N(0, 1)$, i.e., $B_d = 0$ and $\sigma_d^2 = 1$. Thus, y_{ij} is distributed as an $N(0, 2)$ random variable. This situation is depicted in Figure 27.2 which shows a plot of the normal $N(0, 1)$) distribution (the distribution of the μ_{ij}) superimposed on a plot of the $N(0,2)$ distribution (the distribution of y_{ij}). Also shown is the 95th percentile of each distribution. Note

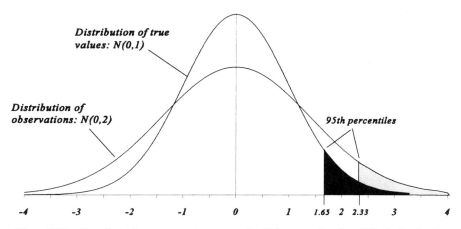

Figure 27.2   The effect of measurement error on the 95th percentile of an N(0, 1) distribution.

that the 95th percentile of the y-distribution is somewhat larger than that for the μ-distribution.

Now, suppose $Y = \mu + d$. If Y and μ are normally distributed, $P(Y \leq y_0) = a = P(\mu \leq \mu_0)$ implies that

$$\frac{y_0 - \bar{Y}}{\sigma_y} = \frac{\mu_0 - \bar{M}}{\sigma_\mu} \tag{27.22}$$

where y_0 and μ_0 are the ath quantiles of the distributions of y and μ, respectively. Thus, the quantile of the centered distribution of observations, $y - \bar{Y}$, in terms of the centered distribution of the true values, $\mu - \bar{M}$ is given by

$$Q_{y - \bar{Y}}(a) = R^{-1/2} Q_{\mu - \bar{M}}(a) \tag{27.23}$$

where $Q_{y - \bar{M}}(a)$ and $Q_{\mu - \bar{M}}(a)$ are the quantile functions for $y - \bar{M}$ and $\mu - \bar{M}$, respectively and R is the reliability ratio defined in Section 27.2. To illustrate with Figure 27.2, $Q_{\mu - 0}(0.95) = 1.65$ for the N(0, 1) distribution. Since $\sigma_d^2 = 1$, $R = 0.5$ and, thus, $Q_{\mu - 0}(0.95) = (\sqrt{0.5})^{-1} \times 1.65 = 2.33$ for the distributions of the observations, y as shown in the figure.

Furthermore,

$$\begin{aligned} Q_y(a) &= Q_{y - \bar{Y}}(a) + \bar{Y} \\ &= R^{-1/2} Q_{\mu - \bar{M}}(a) + \bar{Y} \\ &= R^{-1/2} [Q_\mu(a) - \bar{M}] + \bar{Y} \end{aligned} \tag{27.24}$$

where B_d is the measurement bias defined in Section 27.2. Thus, if $\hat{Q}_y(a)$ is an

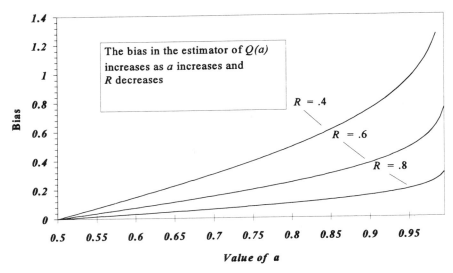

Figure 27.3   The bias in $\hat{Q}_y(a)$ for estimating the mean of an $N(0, 1)$ distribution when y is measured with reliability R.

unbiased estimator of $Q_y(a)$ in the absence of measurement error, then

$$\text{E}[\hat{Q}_y(a)] - Q_\mu(a) = (R^{-1/2} - 1)[Q_\mu(a) - \bar{M}] + B_d \qquad (27.25)$$

under the measurement error model. Figure 27.3 displays this function when μ_i is distributed as an $N(0, 1)$ distribution and y_i is observed with reliability, $R = 0.4$, 0.6, and 0.8. Note that, for $a \geq 0.5$, the bias increases as a increases and R decreases. It is easily shown that these results for the quantiles of a normal distribution under the uncorrelated error model still hold for the correlated error model with the definition of reliability given by (27.17).

Comparing the bias of the sample median ($a = 0.5$) with that of the sample mean we note from (27.25) that they are equal under this model (*viz.*, B_d) since the mean and median of the normal distribution are equal. We now consider the efficiency of the sample median relative to the sample mean for estimating the mean of the normal distribution. From a well-known result in mathematical statistics (see, for example, Wilks, 1962, pp. 364), the variance of the asymptotic distribution of $\hat{Q}_y(0.5)$ is $N(\bar{Y}, \pi\sigma_y^2/2n)$ for large n. The distribution of $\bar{y}$ for any n is $N(\bar{Y}, \sigma_y^2/n)$. Thus, the asymptotic efficiency of $\hat{Q}_y(0.5)$ relative to $\bar{y}$ is

$$\frac{\sigma_y^2}{(\pi/2)\sigma_y^2} = 0.637. \qquad (27.26)$$

For example, for a sample of size 1,000, the sample mean is as efficient as the sample median in a sample of size $1{,}000/0.637$ or $1{,}570$. This is true regardless of the magnitude of the measurement error variance or bias.

27.3.3 Estimators of Proportions and Totals for Binary Data and Their Standard Errors

At least some variables in any survey are classification variables. Some examples are nominal variables such as race, sex, and marital status and quantitative variables such as income and age that have been recoded into classes. Any classification variable can be converted into a number of binary variables, one binary variable for each category. For example, a race variable having four categories—white, African American, Asian, and other—can be represented by four binary variables—white (yes or no), African American (yes or no), Asian (yes or no), and other (yes or no). In this way, our discussion of binary data applies to multinomial variables as well.

27.3.3.1 Uncorrelated Error Effects
Assume the uncorrelated error model for binary data given in Section 27.2.2. For SRS, the usual estimator of $\bar{M} = \pi$, for binary data, is

$$p = \frac{\sum_{i=1}^{I} \sum_{j=1}^{m} y_{ij}}{n}. \tag{27.27}$$

Cochran (1968) showed that

$$\begin{aligned} E(p) &= \pi(1 - \theta) + (1 - \pi)\phi \\ &= \pi^* \end{aligned} \tag{27.28}$$

and, the bias in the sample proportion is

$$\pi^* - \pi = -\pi\theta + (1 - \pi)\phi. \tag{27.29}$$

Thus, for θ, $\phi > 0$, the bias in the estimator of π is zero if and only if $\pi\theta = (1 - \pi)\phi$ which occurs when the number of false positives in the population exactly equals the number of false negatives, regardless of the size of π. Note further that if π is small, the consequences of a relatively small value of ϕ on the bias in p can be quite severe. For example, with $\pi = 0.01$, $\theta = 0.05$, and $\phi = 0.05$, we have $E(p) = (0.01)(0.95) + (0.99)(0.05) = 0.059$ for a relative bias of $(0.059 - 0.01)/(0.01) = 5.9$ or 590%!

As Cochran observed, if the distribution of μ_{ij} is binomial with parameters n and π, then the distribution of y_{ij} is binomial with parameters n and π^*. Using the general formula for the variance of a binomial random variable, we see that the variance of y_{ij} is

$$\text{Var}(y_{ij}) = \pi^*(1 - \pi^*). \tag{27.30}$$

Thus, the variance of y_{ij} is greater than the variance of μ_{ij} only if π^* is closer to 0.5 than π is.

The usual estimator of $\text{Var}(p)$ where $p = \bar{y}$ is

$$v(p) = \frac{pq}{n-1} \tag{27.31}$$

which has expectation

$$E[\text{Var}(p)] = \frac{\pi^*(1 - \pi^*)}{n}. \tag{27.32}$$

Thus, as in the continuous variate case, the usual estimator of the variance is unbiased under the uncorrelated error model since $\text{Var}(p)$ is (27.30) over n.

27.3.3.2 Correlated Error Effects

Now, assume the correlated error model for binary data described in Section 27.2.2. It is shown in U.S. Bureau of the Census (1985) that the expected value of p under this model is still given by (27.28). Further assume that operator assignments are *interpenetrated*, i.e., units are randomly assigned to operators so that $E(\sum_j \mu_{ij}/m \mid i) = \pi$, where $E(\cdot \mid i)$ is the conditional expectation given the ith operator assignment area. Biemer and Stokes (1991) showed that

$$\text{Var}(p) = \frac{\pi^*(1 - \pi^*)}{n} [1 + (m - 1)\rho_y] \tag{27.33}$$

under these conditions which is similar in form to (27.15) for continuous data, except now

$$\rho_y = \frac{\pi^2 \sigma_{\theta\theta} + (1 - \pi)^2 \sigma_{\phi\phi}^2 - 2\pi(1 - \pi)\sigma_{\theta\phi}}{\pi^*(1 - \pi^*)}. \tag{27.34}$$

Thus, $\text{Var}(p)$ is increased by the factor $[1 + (m - 1)\rho_y]$ as a result of correlated operator error. The increase in variance is $(m - 1)\rho_y$ which is the same function plotted in Figure 27.1 and, therefore, the discussion of the effect of interviewer errors on the variance of means in the continuous case also applies to the case of binary response variables.

Note that $\rho_y = 0$ and the interviewer deff $= 1$ if the misclassification variance components are all 0. This is equivalent to $\theta_i = \theta$ and $\phi_i = \phi$, for all i which is the condition for uncorrelated operator error. It may also be shown that the interviewer deff is 1 if $(1 - \pi)\phi_i = \pi\theta_i$ for all i; i.e., if the number of false positives is equal to the number of false negatives for all operators. Note that, if this condition holds, then the expected number of positives for each operator assignment is π and, thus, the mean of each operator's assignment is unbiased for π.

Now, consider the estimator of Var(p) under the correlated error model. The usual estimator of Var(p) is

$$v(p) = \frac{pq}{n-1}. \tag{27.35}$$

Biemer and Stokes (1991) showed that

$$E[\mathrm{Var}(p)] = \frac{\pi^*(1 - \pi^*)}{n}\left[1 - \frac{m-1}{n-1}\rho_y\right]. \tag{27.36}$$

Thus, the relative bias in the usual estimator of the variance is

$$\mathrm{Relbias} = -\frac{(m-1)\rho_y}{1 + (m-1)\rho_y}\left(\frac{n}{n-1}\right)$$

$$\approx -\frac{(m-1)\rho_y}{1 + (m-1)\rho_y} \tag{27.37}$$

which is identical to (27.19) except that ρ_y in this case is given by (27.34). As such, the relationship between the relbias, ρ_y, and m is identical to that discussed previously for $\bar{y}$.

27.4 MEASUREMENT ERROR EFFECTS ON MULTIVARIATE ANALYSIS

In this section, we consider the effects of measurement errors on regression and correlation coefficients and analysis of variance and covariance. Our objective here is to provide some insight into the nature of measurement error effects on data analysis and, thus, we restrict our attention to simple models having practical importance, yet rather strong assumptions. Where we are able, we provide indications of how the error effects may change under more realistic and complex modeling assumptions.

27.4.1 Regression and Correlation Analysis

A number of authors have considered the effects of measurement error on estimates of regression coefficients; see, for example, Madansky (1959), Cochran (1968), and more recently, Fuller (1991). The simplest situation considered is the regression of a dependent variable z on explanatory variable μ where we assume that z is measured without error and y is the measurement of μ which follows the continuous data uncorrelated measurement error model. The case where z is also measured with error and where measurement errors are correlated will be considered subsequently.

Suppose we wish to estimate β_0 and β in the model

$$z_{ij} = \beta_0 + \beta\mu_{ij} + \xi_{ij} \tag{27.38}$$

where β_0 and β are the intercept and slope coefficients, respectively, and the ξ_{ij} are independent, normally distributed model errors with mean 0 and variance, σ_ξ^2. Further assume that the observed value of μ_{ij}, viz. y_{ij}, follows the uncorrelated error model in Section 27.2.1, and that the measurement error d_{ij} is normally distributed and independent of ξ_{ij}. The usual estimator of β is

$$\hat{\beta} = \frac{\sum_{i=1}^n (z_i - \bar{z})(y_i - \bar{y})}{\sum_{i=1}^n (y_i - \bar{y})^2}. \tag{27.39}$$

It is well-known that the expected value of $\hat{\beta}$ is

$$E(\hat{\beta}) = R\beta \tag{27.40}$$

where R is the reliability ratio given in (27.10). Thus, under the uncorrelated measurement error model, the estimator of β is *attenuated* or biased toward 0. Furthermore, the power associated with the usual statistical test, H_0: $\beta = 0$, is reduced as a result of this bias.

Now consider the usual estimator of β_0, viz.

$$\hat{\beta}_0 = \bar{z} - \hat{\beta}\bar{y}. \tag{27.41}$$

It can be shown from (27.38) and (27.40) that

$$E(\hat{\beta}_0) = \beta_0 + \beta\bar{M}[1 - R - B_d/\bar{M}]. \tag{27.42}$$

Since the sign of the second term on the right in (27.42) may be either positive or negative, the direction of the intercept bias may be either positive or negative.

As an illustration of these effects, consider the regressions displayed in Figure 27.4. The dark ellipse represents the cloud of points (μ_{ij}, z_{ij}) and the line through this point cloud represents the true regression of z_{ij} on μ_{ij}. The light ellipse represents the cloud of points (y_{ij}, z_{ij}). Note that when the explanatory variable is measured with error, the effect on the true point cloud is to spread the cloud right and left, corresponding to the increased variability along the abscissa. The estimated regression line, drawn through the center of the light ellipse, has a smaller slope than the true regression line, which is drawn through the center of the dark ellipse, thus demonstrating graphically the mathematical result in (27.40). In addition, the result in (27.42) can be seen in this figure. In Figure 27.4, both β and $\bar{M}$ are positive and $B_d = \bar{Y} - \bar{M} = 0$. Thus, the bracketed term in (27.42) is positive, resulting in a positive bias in the estimator of β_0 and an increase in the observed intercept. Finally, note from the figure that as B_d

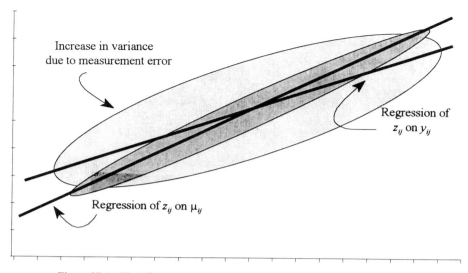

Figure 27.4 The effect of uncorrelated error on the regression of z_{ij} on μ_{ij}.

increases the light ellipse will move to the right as the observations of the explanatory variable are shifted positively along the abscissa. The corresponding shifting of the observed regression line to the right will result in a smaller and smaller intercept. Likewise, as B_d becomes increasingly negative, the regression in the figure will be shifted to the left, increasing the expected value of the estimate intercept. This can also be seen from (27.42). A large positive relative bias term, $B_d/\bar{M}$, results in a negative bracketed term in (27.42) and thus a negative bias while a large negative $B_d/\bar{M}$ results in a positive bracketed term and bias.

When the dependent variable, z_{ij}, is also measured with error, then $z_{ij} = \eta_{ij} + \delta_{ij}$ where η_{ij} is the true value component and δ_{ij} is the measurement error, in analogy to (27.3). Then, letting $\xi'_{ij} = \xi_{ij} - \delta_{ij}$, we can rewrite (27.38) as $\eta_{ij} = \beta_0 + \beta\mu_{ij} + \xi'_{ij}$. Now, the results in the preceding still hold when η_{ij} is substituted for z_{ij} and ξ'_{ij} is substituted for ξ_{ij}, provided that ξ_{ij} and d_{ij} in (27.3) are independent. Note, that the model variance is increased when $\text{Cov}(\xi_{ij}, \delta_{ij}) \geq 0$ and is otherwise reduced.

Chai (1971) considers the case where measurement errors are correlated both within variables and between variables. He shows that, in the presence of correlated error, the slope coefficient may be substantially overestimated or underestimated, according to the signs and magnitudes of the covariances between the errors. Thus, the results of the uncorrelated error model which suggests that the slope coefficient in regression analysis is always attenuated is an oversimplification of the actual effect of measurement error on both β and β_0 when errors are correlated.

Now consider the effect of uncorrelated measurement error on the estimation of the correlation between z and y when z is observed without error. Note that

$$[\text{Corr}(z, y)]^2 = \frac{[\text{Cov}(z, y)]^2}{\text{Var}(z)\,\text{Var}(y)}$$

$$= \frac{[\text{Cov}(z, \mu)]^2}{\text{Var}(z)\,\text{Var}(\mu)} \frac{\text{Var}(\mu)}{\text{Var}(y)} \qquad (27.43)$$

$$= [\text{Corr}(z, \mu)]^2 R$$

and thus

$$\text{Corr}(z, y) = \sqrt{R}\,\text{Corr}(z, \mu). \qquad (27.44)$$

Thus, under the uncorrelated error model, the correlation coefficient is attenuated by the square root of the reliability ratio. When both z and y are subject to error, we have that

$$\text{Corr}(z, y) = \sqrt{R_z R_y}\,\text{Corr}(\eta, \mu) \qquad (27.45)$$

where R_y is the reliability ratio for y, R_z is the reliability ratio for z, and η and μ are the true values associated with z and y, respectively.

27.4.2 Analysis of Variance and Covariance

Cochran (1968) considers the consequences of measurement error in the analysis of variance (ANOVA). In this section, we consider the comparison of two independent samples; however, the effects apply similarly in general ANOVA situations. For comparing the means of two groups, say Group A and Group B, the results presented in Section 27.3.1 for the estimation of means and their standard errors are applicable. Let $\bar{y}_A$ and $\bar{y}_B$ denote the means of simple random samples from Group A and Group B, respectively. From Section 27.3.1, we know that the expected value of the difference, $\bar{y}_A - \bar{y}_B$ under the null hypothesis, H_0: $\bar{M}_A = \bar{M}_B$, is not zero but is equal to $B_{dA} - B_{dB}$, where B_{dA} and B_{dB} are the measurement biases associated with the two groups. Thus, H_0 may be rejected when it is true more often than the nominal significance level. Furthermore, we know from (27.15) that the standard errors of the means are inflated by uncorrelated and correlated variance components. Thus, if the measurement errors are independent between groups, the variance of the difference is the sum of the variances and is also inflated. This effect would tend to reduce the power of the test of H_0. Without some knowledge of the measurement error bias and variance for both groups, the effects of measurement error on ANOVA are largely unpredictable.

Analysis of covariance (ANOCOV) is frequently used in the analysis of survey data to reduce the error variance in ANOVA or to remove differences due to uncontrolled variables between the groups (subpopulations or domains) of interest so that they can be compared with regard to some analysis variable.

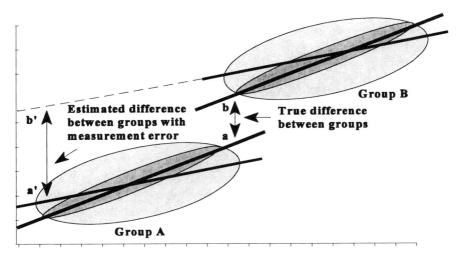

Figure 27.5 The effect of measurement errors on analysis of covariance for two groups.

The latter use of ANOCOV is particularly important in observational studies where the independent variables can neither be controlled experimentally nor randomly assigned to the survey respondents. Here we consider the consequences of measurement error in the control variable.

The effect of measurement error in ANOCOV for dependent variable z_{ij}, measured without error, can be illustrated as in Figure 27.5. The two pairs of ellipses represent the point clouds for Groups A and B. For each group the interpretation of the dark and light ellipses is identical to that described for Figure 27.4; i.e., error in the control variable (plotted on the abscissa) spreads the observed point cloud right and left of the true set of points, thus leveling or attenuating the slopes of the regression line. Although each group may have a different measurement error distribution, the slopes of the regression lines are assumed to be the same for each group.

In ANOCOV, the distance between the regression lines is the difference between the group means for z_{ij} after accounting or adjusting for the control variable. In the absence of measurement error, the lines passing through the dark ellipses define the relationship between the dependent variable and the control variable in each group. The true difference between the groups then is the distance $a - b$. When measurement error is added to the control variable, the slope's coefficient estimated for the control variable is attenuated as illustrated by the regression lines passing through the two light ellipses. Furthermore, the estimated difference between the two groups, illustrated in Figure 27.5 by the line $a' - b'$, is biased. In this case, the observed difference between the groups after accounting for the control variable is greater than the true difference. In general, however, the bias in the comparison may be either positive or negative depending upon the magnitude of the measurement error variances and the magnitude and direction of the measurement biases in both groups. This graphical illustration of the effect of measurement on ANOCOV is due to Lord (1960) and the reader is referred to that paper for a more in-depth consideration of the topic.

27.4.3 Categorical Data Analysis

In Section 27.3.3.1, we considered the effects of classification error on univariate data analysis. This section briefly discusses the consequences of classification error for polytomous response variables and multivariate analysis. In particular, we consider the standard test for goodness-of-fit and the test for independence for two-way cross-classification. As we have done previously, our goal here is to provide insight into the nature of the error effects relying primarily on simple models and illustrations. Kuha and Skinner (Chapter 28) provide an excellent review of categorical data analysis in the presence of classification error as well as an extensive list of references in this area. Readers seeking more detail are referred to their chapter. Mote and Anderson (1965) also consider this topic for SRS sampling and Rao and Thomas (1991) extend many of their results to complex sampling designs. The section draws extensively from these works.

27.4.3.1 Goodness-of-Fit Tests

Let π_i denote the true proportion in the population who belong to the ith category of a K-category classification variable, μ. Let π_{0i}, $i = 1, \ldots, K$, denote constants in $[0, 1]$ such that $\sum_i^K \pi_{0i} = 1$. The goodness-of-fit hypothesis is

$$H_0: \pi_i = \pi_{0i}, \quad i = 1, \ldots, K. \tag{27.46}$$

As before, we assume an SRS of size n is selected from the population and denote by p_i the sample proportion in the ith class and by π_i^* the expected value of p_i over both the sampling and measurement error distributions.

As an example, suppose μ is an age variable with $K = 4$ categories, *viz.* "less than 21," "21 to 35," "36 to 65," and "over 65." Further assume the previous census proportions in these categories are π_{0i}, $i = 1, 2, 3, 4$ and we wish to test whether the current target population proportions, π_i, are the same as the census proportions.

The usual test of goodness-of-fit uses the Pearson test statistic given by

$$X^2 = n \sum_{i=1}^{K} \frac{(p_i - \pi_{01})^2}{\pi_{0i}} \tag{27.47}$$

which is distributed asymptotically as a χ^2_{K-1} variable under H_0 when there are no classification errors. With classification error, Mote and Anderson (1965) show that the test leads to greater probability of Type I error than the nominal level for the test. They further propose several procedures for adjusting X^2 to compensate for misclassification which correct the test size but will reduce the limiting power of the test.

To illustrate the effect of classification error on X^2 consider a simple model for misclassification originally proposed by Mote and Anderson (1965). For ordered categories, it may be reasonable to assume that misclassification occurs only in adjacent categories. Considering the age example above, this implies that respondents might be misclassified by one category to the left or right of their correct age category, but are very unlikely to be misclassified by more than that. Let r, a constant in $[0, 1]$, denote an unknown rate of misclassification from one category to an adjacent category and further assume r is the same for all categories. Thus,

$$\pi_1^* = \pi_1(1 - r) + \pi_2 r$$

$$\pi_i^* = \pi_{i-1}r + \pi_i(1 - 2r) + \pi_{i+1}r, \quad i = 2, \ldots, K - 1 \qquad (27.48)$$

$$\pi_K^* = \pi_{K-1}r + \pi_K(1 - r).$$

In order to illustrate the effect of the increased probability of Type I error with classification error under this model, we computed X^2 in (27.47) under H_0 for $0 \leq r \leq 0.25$ with $K = 4$ and arbitrarily setting $\pi_{01} = 0.4$, $\pi_{02} = 0.3$, $\pi_{03} = 0.2$, and $\pi_{04} = 0.1$. Figure 27.6 shows the result for three values of n: 500, 2,500, and 5,000. The horizontal line in the figure is the critical value for the test with $\alpha = \Pr(\text{Type I error}) = 0.05$; i.e., $\chi^2_{3,0.95}$. Note that H_0 would be erroneously rejected (indicated by a value of X^2 above the horizontal line) when $r = 0.36$, $n = 500$ (not shown on the chart), $r = 0.16$, $n = 2,500$, and $r = 0.11$, $n = 5,000$, thus demonstrating that in the presence of classification error, the nominal

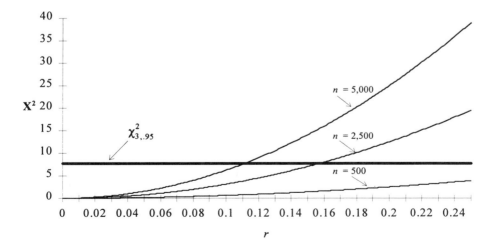

Figure 27.6   The effect of classification error on X^2 as a function of sample size (n) and the rate of misclassification to adjacent categories.

α-level of the test can be substantially smaller than the realized α-level. Furthermore, this effect increases with n, all else being equal.

To adjust X^2 for classification error, it is necessary to fit a model that represents the misclassification mechanism. This usually requires additional information from an external source. Kuha and Skinner (Chapter 28) identify two possible classes of external information which may be known for at least a sample of the population of interest:

a. A *gold standard* measurement (sometimes referred to as a *criterion* measure); and

b. repeated measurement of the misclassified variable.

A *gold standard* refers to error free measurements from an external source that can be used for validation (e.g., administrative records). Alternatively, it could be provided through a reinterview of the original survey respondents using a more accurate or *preferred* survey process; for example, a face to face reinterview using the best interviewers while the original survey data was collected by centralized telephone interviewing using less qualified interviewers. For the purposes of this analysis, the *gold standard* would be regarded as correct.

Repeated measurements can be obtained through reinterviewing, although not necessarily with improved methods. Kuha and Skinner point out that information about misclassification parameters can be obtained through the use of statistical models and the relationship between the repeated measurements of the characteristic and its actual value. The most common assumption is that the repeated measurements are independent of each other.

Mote and Anderson (1965) describe a procedure for estimating r (or θ in their notation) directly from the survey data assuming (27.48) holds. They show that the asymptotic power of the resulting test is less than that for the test without classification error.

27.4.3.2 Tests for Association

Consider a two-way cross-classification of the variables μ_A and μ_B with I and J categories, respectively. Let π_{ij} denote the population proportion in the ijth cell and let $\pi_{i\cdot} = \sum_j \pi_{ij}$ and $\pi_{\cdot j} = \sum_i \pi_{ij}$ denote the row and column marginal probabilities, respectively. The hypothesis that μ_A and μ_B are independent is

$$H_0: \pi_{ij} = \pi_{i\cdot}\pi_{\cdot j}, \quad i = 1, \ldots, I; j = 1, \ldots, J. \tag{27.49}$$

The usual χ^2 test statistic is

$$X^2 = n \sum_{i=1}^{I} \sum_{j=1}^{J} \frac{(p_{ij} - p_{i\cdot}p_{\cdot j})^2}{p_{i\cdot}p_{\cdot j}} \tag{27.50}$$

where p_{ij} is the proportion of the sample in the ijth cell and $p_{i\cdot}$ and $p_{\cdot j}$ are the row and column marginal probabilities, respectively.

If the errors of classification for μ_A are independent of those for μ_B then, as Assakul and Proctor (1967) show, H_0 is equivalent to the test $H_0^*: \pi_{ij}^* = \pi_{i\cdot}^* \pi_{\cdot j}^*$ where $\pi_{ij}^* = E(p_{ij})$, using the usual critical values for the test of $\chi^2_{(I-1)(J-1)}$. However, the asymptotic power of this test will be less than if there were no measurement errors.

To illustrate this effect, we consider an extension of the simple model for classification error used for the goodness-of-fit illustration in Figure 27.6. Thus, we assume that misclassification occurs only to adjacent categories at a rate r_A for μ_A and r_B for μ_B. To further simplify the discussion, we consider only square tables (i.e., I and J equal to K, say) and, since our goal is to illustrate the reduction of power of the test resulting from misclassification, we assume a perfect correlation between μ_A and μ_B. Thus, we assume that $\pi_{ii} > 0$ and $\pi_{ij} = 0$ for $i \neq j$ and define the following expected cell probabilities

$$\pi_{12}^* = \pi_{11}r_A + 0.5\pi_{22}r_B \quad \pi_{21}^* = 0.5\pi_{22}r_A + \pi_{11}r_B$$

$$\pi_{K,K-1}^* = \pi_{K,K}r_A + 0.5\pi_{K-1,K-1}r_B \quad \pi_{K-1,K}^* = 0.5\pi_{K-1,K-1}r_A + \pi_{K,K}r_B$$

$$\pi_{ii}^* = \pi_{ii}(1 - r_A - r_B)$$

$$\pi_{i+1,i}^* = 0.5(\pi_{i+1,i+1}r_A + \pi_{i-1,i-1}r_B)$$

$$\pi_{i,i+1}^* = 0.5(\pi_{i,i}r_A + \pi_{i+1,i+1}r_B), \quad i = 2, \ldots, K-1. \tag{27.51}$$

The rationale for these assumptions is similar to that for (27.48): misclassification for a cell occurs only to its adjacent cells. For two-way tables, this means that an individual who in fact belongs in cell (i,i) for $2 \leq i \leq K-1$ has a positive probability of being classified into cells $(i-1, i), (i+1, i), (i, i-1)$ and $(i, i+1)$. As an example, suppose $K=4$, $\pi_{ii}=0.25$ for $i=1, \ldots, 4$, $r_A=r_B=0.10$, i.e., 20% of the units in the diagonal cells are misclassified equally into adjacent row and column categories. Thus, the true and observed tables appear as in Table 27.2.

For the model in (27.51), let $X^2(r_A, r_B)$ denote the value of X^2 computed for y_A and y_B with misclassification parameters r_A and r_B. Thus, $X^2(0, 0)$ denotes

Table 27.2 Observed Cell Proportions for Two Variables with Classification Error When the True Proportions Are $\pi_{ii} = 0.25$, $\pi_{ij} = 0$

	True association μ_B					Observed association y_B			
μ_A	1	2	3	4	y_A	1	2	3	4
1	0.25	0	0	0	1	0.20	0.0375	0	0
2	0	0.25	0	0	2	0.0375	0.20	0.025	0
3	0	0	0.25	0	3	0	0.025	0.20	0.0375
4	0	0	0	0.25	4	0	0	0.0375	0.20

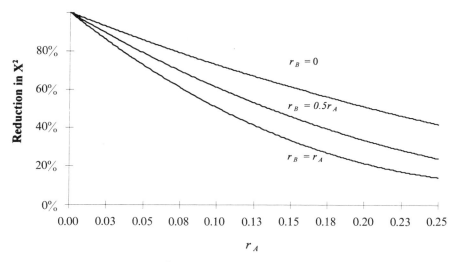

Figure 27.7 The ratio of X^2 with misclassification to X^2 without misclassification.

the value of X^2 with no misclassification (i.e., $r_A = r_B = 0$). For the example in Table 27.2, $n = 1,000$, $X^2(0, 0)$ is 2,625 (computed for the left table) and $X^2(0.10, 0.10)$ is 2,469 (computed for the right table) for a ratio of $X^2(0.10, 0.10)$ to $X^2(0, 0)$ of 0.94. Thus, the observed X^2 is 94% of the true X^2 without misclassification, illustrating the attenuation of X^2 as a result of misclassification. In Figure 27.7, the ratio $R_X^2 = X^2(r_A, r_B)/X^2(0, 0)$, which does not depend upon n, is plotted for $0 \leq r_A \leq 0.25$ and three combinations of r_A and r_B: $r_B = 0$, $r_B = 0.5r_A$, and $r_B = r_A$. In all cases, as r increases, R_X^2 decreases thus indicating attenuation of the X^2 and a loss of power in the test of association.

Mote and Anderson (1965), Rao and Thomas (1991), Kuha and Skinner (Chapter 28), and Van de Pol and Langeheine (Chapter 29) all consider more complex situations of classification error, including measurement errors that are correlated between the variables. In these situations, it is necessary to use model-based methods to obtain valid tests. It has been shown that corrections to X^2 using validation data acquired through two-phase sampling methods do not in general have a X^2 limiting distribution. However, if the model-based methods provide consistent estimators of the true values, tests can be developed which are asymptotically correct.

27.5 METHODS FOR COMPENSATING FOR THE EFFECTS OF MEASUREMENT ERROR

As the preceding discussion clearly demonstrates, the effects of measurement error analysis can be quite substantial. For univariate analysis, correlated error is more serious than uncorrelated error. However, for quantile estimation and multivariate analysis, even uncorrelated error can lead to substantial biases in the outcome measures. In the presence of measurement error, estimators will be biased, variances typically will be inflated, hypothesis testing will be invalid or have less power; and conclusions from other forms of analysis could be quite misleading. Indeed, the field of survey methodology is devoted primarily to minimize the effect of measurement error on survey data through intelligent survey design methods. These involve cognitive methods for questionnaire design; methods for reducing interviewer and other operator errors through improved recruitment, training, management, and supervision; the use of respondent-friendly interviewing methods and modes of interview, and so on (see Groves, 1989 for a review of these methods). Despite these efforts at error reduction and control, measurement errors will continue to plague data collection, and methods for compensating for error effects in the final results are needed.

The survey literature contains numerous articles proposing methods for accommodating measurement errors in the analysis of survey data. This section of the volume is devoted to these methods. In addition, Biemer *et al.* (1991, Section E) contains nine chapters on error compensation for the types of analysis we have discussed here, as well as structural equation modeling, path analysis, and event history analysis. The methods for error compensation may be characterized by two requirements: (a) a model for the error structure and (b) external, auxiliary data for estimating the error parameters indicated by the model. For modeling measurement error, the models specified in Section 27.2 can usually be modified to adequately describe most measurement processes. However, for some situations these general model formulations are inadequate and alternative models must be specified. Pullum and Stokes (Chapter 31) present just such a situation for modeling the error in the estimation of fertility rates and van de Pol and Langeheine (Chapter 29) provide an application of the use of latent class models for analyzing longitudinal categorical data.

The requirement of external data is usually met by record checks, two-phase sampling methods for obtaining more accurate measurements, replicated measurements through reinterviewing a subsample of respondents, and data from a previous survey or census. Kuha and Skinner (Chapter 28) classify external information as either validation data (i.e., administrative record data, highly accurate remeasurements, or other data that may be considered as error-free for the purpose of the analysis) and replicated measurements obtained through reinterviews with respondents. The methods proposed by Rao and Sitter (Chapter 33) assume validation data are available for adjusting the estimates of means and totals for the measurement bias. Their chapter focuses on resampling methods for estimating the standard errors of the adjusted estimates. Chapter 29 by van de Pol and Langeheine provides an application of replicated measurements and the use of latent class models for the analysis of longitudinal data. By separating measurement errors from actual movements in labor market categories, they attempt to measure gross flows as well as net change. Pullum and Stokes (Chapter 31) also provide an application of

Table 27.3 Summary of the Error Effect on Estimators and Data Analysis

Statistic/parameter	Error structure	Error effect
$\bar{y}$	Uncorrelated or correlated	$\text{Bias} = B_d$
p		$\text{Bias} = -\pi\theta + (1 - \pi)\phi$
$\text{Var}(\bar{y})$	Uncorrelated	$\text{Var}(\bar{y}) = \sigma_\mu^2/(Rn)$
$\text{Var}(p)$		$\text{Var}(p) = \pi^*(1 - \pi^*)/n$
$v(\bar{y})$	Uncorrelated	unbiased (if no fpc)
$v(p)$		unbiased (if no fpc)
$\text{Var}(\bar{y})$	Correlated	$\text{Var}(\bar{y}) = \dfrac{\sigma_\mu^2}{Rn}\left(1 + (m - 1)\rho_y\right)$
$\text{Var}(p)$		
$v(\bar{y})$	Correlated	$\text{Relbias} = -\dfrac{(m - 1)\rho_y}{1 + (m - 1)\rho_y}$
$v(p)$		

$\hat{Q}_y(a)$	Uncorrelated or correlated	bias $= (R^{-1/2}-1)[Q_\mu(a)-\bar{M}]+B_d$
β	Uncorrelated	$E(\hat{\beta})=R\beta$
β_0	Correlated	$E(\hat{\beta}_0)=\beta_0+\beta\bar{M}[1-R-B_d/\bar{M}]$
β, β_0	Uncorrelated	May be biased in either direction
$\mathrm{Corr}(z, y)$, y with error		$\mathrm{Corr}(z,y)=\sqrt{R}\,\mathrm{Corr}(z,\mu)$
$\mathrm{Corr}(z, y)$, z and y with error		$\mathrm{Corr}(z,y)=\sqrt{R_z R_y}\,\mathrm{Corr}(\eta,\mu)$
$\mathrm{Corr}(z, y)$, y with error	Correlated	May be biased in either direction
$\mathrm{Corr}(z, y)$, z and y with error		
ANOVA ANOCOV	Correlated or uncorrelated	Factor differences may be biased in either direction
X^2 for goodness-of-fit	Uncorrelated	X^2 biased; test size greater than nominal level
X^2 for association	Uncorrelated	X^2 biased; loss of power

measurement error analysis using longitudinal data. Using demographic analysis modeling techniques and general estimating equations (GEE), they provide a useful analysis of the omission and displacement of births in Pakistan. Van de Pol and Langeheine (Chapter 29) and Nusser and Fuller (Chapter 30) both incorporate replicated measures into their analysis of measurement error.

Bassi and Fabbris (Chapter 32) provide an example of the estimation of measurement error components using ANOVA techniques which do not require external data. Rather, by *interpenetrating* (or randomly distributing) interviewer and other operator assignments using experimental design methods, estimates of operator variance and the intrainterviewer correlation coefficient (ρ_y) are possible. Their chapter considers the precision of such estimates as well as the additional benefits of combining interpenetrating assignments with replicated measurements for estimating interviewer variance.

These are just some of the methods available for accommodating measurement error. For categorical data, Kuha and Skinner (Chapter 28) summarize a range of adjustment methods both with the availability of validation data and repeated measurements. Each has its own advantages and disadvantages in terms of ease of use, efficiency and validity for subsequent hypothesis testing. An important new development for repeated measurements is the increasing use of latent class models to extract information about the true values. Although their potential has been recognized for some time, it is only recently that latent class models have really been used with any frequency.

Regression and correlation analysis is another area where there are several well-known methods for adjusting for the bias due to measurement errors. The use of instrumental variables is one common method and these are summarized in most econometrics texts (see, Fuller, 1987 for a description of these methods). Maximum likelihood estimation methods have also been proposed although they require strong assumptions about the measurement process and are limited unless these assumptions are plausible for the purpose at hand. Fuller (1991) develops a more useful method when repeated measurements are available.

As a summary of the error effects discussed in this chapter, Table 27.3 is provided. This summary clearly demonstrates that measurement errors can sometimes invalidate standard methods for data analysis while at other times they result in only minor effects. To determine whether measurement errors are a problem requires consideration of the type of analysis to be conducted as well as knowledge of the magnitudes of the bias causing components of error for that analysis. Although inspecting the data for suspicious entries is a necessary step in survey data processing, little information on the biasing effects of measurement error can be gleaned by inspection alone. Rather, what is needed is well-specified models and sufficient external data to estimate the error components and adjust for the error effects during data analysis.

Fuller (1991) advocates devoting some substantial fraction (say 25 percent) of the survey budget for the reinterviewing of respondents in order to correct estimated regression coefficients for attenuation and shows that this policy would produce estimates with smaller mean squared error than a policy of no reinterviews. However, there is little evidence that such policies, as prudent as they appear, garner much support from survey sponsors. Perhaps one reason for the lack of support for using either repeated measurements or gold standard measurements to adjust the results of survey analyses is the difficulty in obtaining remeasurements that satisfy the assumptions of the adjustment methods. Our hope is that this chapter, by attempting to increase the awareness of and concern for the biasing effects of measurement error for all types of data analyses, will help generate greater attention to the needs in these areas.

ACKNOWLEDGMENTS

Our sincere thanks to James Chromy and Siu-Ming Tam who provided useful comments on a previous draft of this chapter.

REFERENCES

Assakul, K., and Proctor, C.H. (1967), "Testing Independence in Two-Way Contingency Tables with Data Subject to Misclassification," *Psychometrika*, 32, 1, pp. 67–76.

Biemer, P.P., and Caspar, R. (1994), "Continuous Quality Improvement for Survey Operations: Some General Principles and Applications," *Journal of Official Statistics*, 10, 3, pp. 307–326.

Biemer, P.B., Groves, R.M., Lyberg, L.E., Mathiowetz, N.A., and Sudman, S. (eds) (1991), *Measurement Errors in Surveys*, New York: Wiley and Sons.

Biemer, P.B., and Stokes, S.L. (1991), "Approaches to the Modeling of Measurement Errors," in P.P. Biemer, R.M. Groves, L.E. Lyberg, N.A. Mathiowetz, and S. Sudman (eds), *Measurement Errors in Surveys*, New York: Wiley and Sons, pp. 487–516.

Chai, J.J. (1971), "Correlated Measurement Errors and the Least Squares Estimator of the Regression Coefficient," *Journal of the American Statistical Association*, 66, pp. 478–483.

Cochran, W.G. (1968), "Errors of Measurement in Statistics," *Technometrics*, 10, 4, pp. 637–666.

Fowler, F.J., and Mangione, T.W. (1990), *Standardized Survey Interviewing*, Newbury Park, CA: Sage Publications.

Fuller, W.A. (1987), *Measurement Error Models*, New York: John Wiley & Sons.

Fuller, W.A. (1991), "Regression Estimation in the Presence of Measurement Error," in P.P. Biemer, R.M. Groves, L.E. Lyberg, N.A. Mathiowetz, and S. Sudman (eds), *Measurement Errors in Surveys*, New York: Wiley and Sons, pp. 617–636.

Fuller, W.A. (1995), "Estimation in the Presence of Measurement Error," *International Statistical Review*, 63, 2.

Groves, R.M. (1989), *Survey Errors and Survey Costs*, New York: John Wiley & Sons.

Hansen, M.H., Hurwitz, W.N., and Bershad, M.A. (1964), "Measurement Errors in Censuses and Surveys," *Bulletin of the International Statistical Institute*, 38, 2, pp. 359–374.

Hansen, M.H., Hurwitz, W.N., and Pritzker, L. (1964), "The Estimation and Interpretation of Gross Differences and the Simple Response Variance," in C.R. Rao (ed.), *Contributions to Statistics*, Calcutta: Pergamon Press Ltd., pp. 111–136.

Hansen, M.H., Hurwitz, W.N., Marks, E.S., and Mauldin, W.P. (1951), "Response Errors in Surveys," *Journal of the American Statistical Association*, 46, pp. 147–190.

Hanson, R.H., and Marks, E.S. (1958), "Influence of the Interviewer on the Accuracy of Survey Results," *Journal of the American Statistical Association*, 53, pp. 635–655.

Kish, L., and Lansing, J.B. (1954), "Response Errors in Estimating the Value of Homes," *Journal of the American Statistical Association*, 49, pp. 520–538.

Lord, F. (1960), "Large-sample Covariance Analysis When the Control Variables Are Subject to Error," *Journal of the American Statistical Association*, 55, pp. 307–321.

Lord, F., and Novick, M.R. (1968), *Statistical Theories of Mental Test Scores*, Reading, MA., Addison-Wesley.

Madansky, A. (1959), "The Fitting of Straight Lines When Both Variables Are Subject to Error," *Journal of the American Statistical Association*, 54, pp. 173–205.

Mote, V.L., and Anderson, R.L. (1965), "An Investigation of the Effect of Misclassification on the Properties of Chi-square Tests in the Analysis of Categorical Data," *Biometrika*, 52, 1 and 2, pp. 95–109.

Rao, J.N.K., and Thomas, D.R. (1991), "Chi-squared Tests with Complex Survey Data Subject to Misclassification Error," in P.P. Biemer, R.M. Groves, L.E. Lyberg, N.A. Mathiowetz, and S. Sudman, (eds), *Measurement Errors in Surveys*, New York: Wiley and Sons, pp. 637–664.

Sukhatme, P.V., and Seth, G.R. (1952), "Nonsampling Errors in Surveys," *Journal of Indian Society of Agricultural Statistics*, 5, pp. 5–41.

U.S. Bureau of the Census (1985), "Evaluation of Censuses of Population and Housing," STD-ISP-TR-5, Washington, D.C.: U.S. Government Printing Office.

Wilks, S.S. (1962), *Mathematical Statistics*, New York: John Wiley & Sons.

Wolter, K.M. (1985), *Introduction to Variance Estimation*, New York: Springer-Verlag.

CHAPTER 28

Categorical Data Analysis and Misclassification

Jouni Kuha
Nuffield College, Oxford
Chris Skinner
University of Southampton

28.1 INTRODUCTION

It has long been recognized that categorical variables are often subject to misclassification and that this can distort the results of data analyses. This chapter will review the literature on the effects of misclassification and on methods to adjust for these effects. In particular, we draw on a large epidemiological literature which appears not to be well-known to many survey researchers. A related review paper, with emphasis on the $2 \times 2 \times 2$ table, is Chen (1989). Dalenius (1977) gives a bibliography, now somewhat out of date, on misclassification and other nonsampling errors in surveys.

We begin in Section 28.2 with an illustrative example of misclassification in the 1991 population census of England and Wales. We then consider in Section 28.3 the possible effects of misclassification on categorical data analysis. In Section 28.4 we examine ways of collecting data to estimate misclassification parameters, and in Section 28.5 describe how these data may be used to adjust for the effects of misclassification. Finally, in Section 28.6 we make some concluding remarks and suggest some topics for further research.

28.2 AN EXAMPLE: THE 1991 CENSUS IN ENGLAND AND WALES

Following the 1991 population census in England and Wales, a Census Validation Survey (CVS) was carried out with the twin aims: (1) to assess the

Survey Measurement and Process Quality, Edited by Lyberg, Biemer, Collins, de Leeuw, Dippo, Schwarz, Trewin.
ISBN 0-471-16559-X © 1997 John Wiley & Sons, Inc.

coverage of the census and (2) to assess the quality of the census data by taking more accurate measurements of census variables for a sample of households. We shall be concerned only with the latter "Quality Check" and take no account of effects of noncoverage. In contrast to the census, in which data were collected by self-completion forms, the CVS Quality Check employed experienced interviewers to conduct face to face interviews, checking census information with each adult in the household where possible. In addition, the CVS questionnaires were more detailed than the original census forms and were designed to probe for the most accurate answer.

A comparison of the CVS responses with the original census data provides an opportunity to study misclassification in the census. Some results for single variables are presented in OPCS (1994). We extend this work to some multivariate analyses.

Around 6,000 households were selected for the Quality Check and the response rate was 89 percent. We present unweighted results for the subsample of 7,614 individuals of working age (18–64 for men, 18–59 for women) who had no missing values on the variables to be studied. See Dale and Marsh (1993) and OPCS (1994) for more details of the methods of study.

Tables 28.1 and 28.2 provide estimates of "misclassification matrices" for two variables: economic activity and ethnic group. The columns contain the percentages of individuals of a given "true" CVS status who are classified in the census into different categories.

Economic activity is a variable with a fair degree of misclassification, notably for those classified as unemployed in the CVS for whom only 85 percent are correctly classified. It is interesting that the amount of misclassification in Table 28.1 is similar to that in Table 5 of Chua and Fuller (1987), estimated from U.S. Current Population Survey data. This suggests that the results we report here for illustration may not be untypical of survey data, even though there are

Table 28.1 Economic Activity Reported in Census and in Census Validation Survey. Numbers in the Table Are Column Percentages and (in Parentheses) Cell Counts

Census	Census Validation Survey			
	Employed	Unemployed	Inactive	Total
Employed	97.64	4.26	3.34	
	(5,264)	(27)	(53)	(5,344)
Unemployed	0.54	85.33	4.34	
	(29)	(541)	(69)	(639)
Inactive	1.82	10.41	92.32	
	(98)	(66)	(1,467)	(1,631)
Total	(5,391)	(634)	(1,589)	(7,614)

Table 28.2 Ethnic Group Reported in Census and in Census Validation Survey. Numbers in the Table Are Column Percentages and (in Parentheses) Cell Counts

	Census Validation Survey		
Census	White	Other	Total
White	99.86	5.66	
	(6,968)	(36)	(7,004)
Other	0.14	94.34	
	(10)	(600)	(610)
Total	(6,978)	(636)	(7,614)

a number of reasons why the quality of census data may be expected to be inferior to that of interview surveys.

Ethnic group is subject to less misclassification than economic activity, although it is still estimated that 6 percent of those "truly" nonwhite are misclassified as white in the census.

28.3 EFFECTS OF MISCLASSIFICATION ON CATEGORICAL DATA ANALYSIS

28.3.1 Univariate Analyses

We first consider the basic question of how to represent misclassification in a finite population. We shall use the term *true variable* to refer to the hypothetical variable which the *classified variable* is intended to measure. Alternatively, if an operational definition is required, the true variable may be equated with the *preferred procedure* (see Section 28.4.1). Misclassification is represented, following Bross (1954), by treating the values A_i of the true variable as well-defined and fixed, with the values A_i^* of the classified variable determined from the A_i by means of a random process

$$\Pr(A_i^* = j \mid A_i = k) = \theta_{jk}; \quad j, k = 1, \dots, m \qquad (28.1)$$

governed by parameters θ_{jk}. Viewing the population units in true category k as drawn from an infinite superpopulation of such units, θ_{jk} may be interpreted as the proportion of these units which would be classified as category j.

Alternative approaches to representing misclassification are possible. Koch (1969) considers a very general approach in which A_i^* is treated as the outcome of just one of a series of repeated trials of the measuring instrument. Lessler and Kalsbeek (1992, sec. 10.3) consider, in contrast, a nonstochastic approach in which the θ_{jk} consist of finite population proportions.

The parameters θ_{jk} governing the misclassification mechanism may be collected into an $m \times m$ *misclassification matrix* $\Theta = [\theta_{jk}]$. The elements of this matrix are all nonnegative and the columns must sum to one. In the simplest binary case where the variable indicates presence ($A_i = 2$) or absence ($A_i = 1$) of a characteristic, Θ is a function of just two parameters

$$\begin{pmatrix} \theta_{11} & \theta_{12} \\ \theta_{21} & \theta_{22} \end{pmatrix} = \begin{pmatrix} \beta & 1 - \alpha \\ 1 - \beta & \alpha \end{pmatrix}.$$

In the medical literature (e.g., Rogan and Gladen, 1978) where the characteristic is usually some disease, α is termed the *sensitivity* of the survey instrument and β is termed the *specificity*. The measurement characteristics of different instruments may be compared in terms of these two quantities. For example, a survey instrument which classifies a person as "disabled" if any one of a long list of questions has a positive response might have a "good" sensitivity, that is, most disabled people will be "picked up," but may have poor specificity, that is, many nondisabled persons may be wrongly classified as disabled.

Some illustrative sample estimates of misclassification matrices (multiplied by 100) have been presented in Section 28.2. For example, from Table 28.2 the estimated sensitivity of the variable ethnic group (as a measure of nonwhite vs white) is 0.9434 and estimated specificity is 0.9986.

Let us now turn to the effect of misclassification on univariate categorical data analysis. Let $N_A(j)$ be the number of population units for which $A_i = j$ and let $P_A(j) = N_A(j)/N$, where $N = \sum N_A(j)$ is the population size. Suppose that the aim is to estimate the vector $\mathbf{P}_A = (P_A(1), \ldots, P_A(m))'$ of population proportions. If misclassification is ignored then $\mathbf{P}_A$ will generally be estimated by some vector $\mathbf{p}_{A*} = (p_{A*}(1), \ldots, p_{A*}(m))'$, where

$$p_{A*}(j) = \sum_{i \in s} w_i I(A_i^* = j) \tag{28.2}$$

s denotes the sample, w_i the sample weight for unit i and $I(\cdot)$ the indicator function. Letting E_m denote expectation under our misclassification model, it follows from (28.1) that

$$E_m[I(A_i^* = j)] = \sum_{k=1}^{m} \theta_{jk} I(A_i = k)$$

so that

$$E_m(\mathbf{p}_{A*}) = \Theta \mathbf{p}_A \tag{28.3}$$

where $\mathbf{p}_A = (p_A(1), \ldots, p_A(m))'$ and $p_A(k) = \sum_{i \in s} w_i I(A_i = k)$. Assuming that the w_i are defined so that $\mathbf{p}_A$ is approximately unbiased for $\mathbf{P}_A$ under the sampling scheme, the bias arising from misclassification may be expressed as

$$\text{Bias}(\mathbf{p}_{A*}) = (\mathbf{\Theta} - \mathbf{I})\mathbf{P}_A \tag{28.4}$$

where $\mathbf{I}$ is the $m \times m$ identity matrix. Note that this bias is evaluated with respect to both the misclassification model and the sampling scheme.

The nature of the bias is easiest to interpret in the binary case where $\text{Bias}[p_{A*}(1)] = -\text{Bias}[p_{A*}(2)]$ and

$$\text{Bias}[p_{A*}(2)] = (1 - \beta)P_A(1) - (1 - \alpha)P_A(2) \tag{28.5}$$

(e.g., Cochran, 1968; Schwartz, 1985). Even if the misclassification matrix differs from the identity matrix, it is possible for there to be no net bias in $p_{A*}(2)$ if the two errors of misclassification are mutually compensating, that is if $(1 - \beta)P_A(1) = (1 - \alpha)P_A(2)$, for example, if $P_A(2) = 0.6$, $\beta = 0.97$, $\alpha = 0.98$. The possibility of such compensation is discussed by Chua and Fuller (1987) and is illustrated in Table 28.1. The 27 unemployed individuals misclassified as employed are almost exactly compensated for by the 29 employed individuals who are misclassified as unemployed. Misclassification between the states unemployed and inactive is similarly compensating. Note in any case that the condition for unbiasedness is not simply a property of the misclassification parameters θ_{jk} but also of the true parameters $\mathbf{P}_A$. Thus an instrument with a given misclassification matrix may lead to biased estimates in one population but unbiased estimates in another one.

In summary, misclassification can lead to arbitrary forms of bias in the estimation of the population proportions falling into the categories of a given variable. In some circumstances the amount of misclassification out of a given category will exactly balance the amount of misclassification into that category from other categories and no net bias will result.

28.3.2 Bivariate Analyses

28.3.2.1 Estimation for 2 × 2 Tables

The simplest form of analysis we consider is the comparison of proportions between two subgroups. Letting A be the binary outcome variable defining the proportion of interest and B the binary variable defining the two subgroups, we consider first the case when A is misclassified as A^*. This case has been widely discussed in the epidemiological literature (e.g., Rubin *et al.*, 1956) for applications where A defines presence or absence of a disease and B defines two exposure groups, for example, smokers and nonsmokers. Let $P_{A|B}(j|l)$ denote the proportion of units for which $A_i = j$ amongst the subpopulation for which $B_i = l$ and let $p_{A*|B}(j|l)$ be the estimator of $P_{A|B}(j|l)$ analogous to $p_{A*}(j)$ in (28.2). We consider the bias of $p_{A*|B}(2|2) - p_{A*|B}(2|1)$ as an estimator of the difference between the population proportions $P_{A|B}(2|2) - P_{A|B}(2|1)$.

The expression for the bias simplifies under the condition that the same sensitivity α and specificity β apply in both categories of B. A more precise statement of this condition is as follows: the misclassification mechanism for A^* is said to be *nondifferential* with respect to B if the same mechanism holds for each category of B, that is

$$\Pr(A_i^* = j \mid A_i = k, B_i = l) = \theta_{jk} \quad \text{for all } l. \tag{28.6}$$

It follows from (28.3.5) that under nondifferential misclassification

$$\text{Bias}[p_{A^*\mid B}(2\mid 2) - p_{A^*\mid B}(2\mid 1)] = -[(1-\alpha) + (1-\beta)][P_{A\mid B}(2\mid 2) - P_{A\mid B}(2\mid 1)]$$

or alternatively that

$$E[p_{A^*\mid B}(2\mid 2) - p_{A^*\mid B}(2\mid 1)] \doteq [1 - (1-\alpha) - (1-\beta)][P_{A\mid B}(2\mid 2) - P_{A\mid B}(2\mid 1)].$$
$$\tag{28.7}$$

In practice, we expect the misclassification probabilities $1 - \alpha$ and $1 - \beta$ to be less than 0.5 so that the factor $[1 - (1 - \alpha) - (1 - \beta)]$ in (28.7) will be a positive constant less than one and thus the effect of misclassification (in expectation) is to *attenuate* the difference in subclass proportions (see Rubin *et al.*, 1956; Newell, 1963; Buell and Dunn, 1964; White, 1986). In other words, the "effect" of B on A is made to seem less than it actually is. Similarly, it may be shown that misclassification "attenuates" the ratio $p_{A^*\mid B}(2\mid 2)/p_{A^*\mid B}(2\mid 1)$ under nondifferential misclassification away from $P_{A\mid B}(2\mid 2)/P_{A\mid B}(2\mid 1)$ towards the null value of one (Copeland *et al.*, 1977; White, 1986). In the epidemiological case, this ratio is the "relative risk" that $A = 2$ for subgroup $B = 2$ compared to subgroup $B = 1$.

Note that the two kinds of misclassification, represented by $1 - \alpha$ and $1 - \beta$, combine with the same sign in (28.7) whereas they have opposite signs in (28.5). Thus, misclassification between pairs of categories which may be mutually compensating in univariate analyses will not be such in bivariate analyses. There may thus be some reasons to expect misclassification to have a greater relative effect in such analyses. This is illustrated numerically later in this section.

If, instead, the subgroup variable B is misclassified but the outcome variable A is correctly classified and furthermore if misclassification of B is nondifferential with respect to A then similar types of attenuation occur (Rogot, 1971; Shy *et al.*, 1978; Greenland, 1982; Flegal *et al.*, 1986).

To allow for the possibility of misclassification in both A and B, we suppose in general that the pair (A_i^*, B_i^*) are jointly determined from the pair (A_i, B_i) by the random process $\Pr(A_i^* = j^*, B_i^* = k^* \mid A_i = j, B_i = k) = \phi_{j^*k^*jk}$, say. Misclassification of A and B is said to be *independent* if

$$\Pr(A_i^* = j^*, B_i^* = k^* \mid A_i = j, B_i = k)$$
$$= \Pr(A_i^* = j^* \mid A_i = j, B_i = k) \Pr(B_i^* = k^* \mid A_i = j, B_i = k)$$

and *nondifferential* if both the misclassification of A is nondifferential with respect to B and the misclassification of B is nondifferential with respect to A, that is

$$\Pr(A_i^* = j^* \mid A_i = j, B_i = k) = \Pr(A_i^* = j^* \mid A_i = j)$$

and

$$\Pr(B_i^* = k^* \mid A_i = j, B_i = k) = \Pr(B_i^* = k^* \mid B_i = k).$$

Under the condition of independent nondifferential misclassification in A and B, Gullen *et al.* (1968) show that the estimate of $P_{A|B}(2|2) - P_{A|B}(2|1)$ is attenuated in a similar way to the earlier results (see also Keys and Kihlberg, 1963; Barron, 1977).

A special case of some interest arises in longitudinal surveys when A and B are the values of a variable at consecutive waves of the survey and the parameters $\Pr(A_i = k, B_i = l)$ represent the "gross flows" between category k and category l (Chua and Fuller, 1987). In this case misclassification can induce severe relative bias in estimates of the off-diagonal flows ($k \neq l$).

Let us now turn to the situation where the misclassification mechanism for A^* (or B^*) is not nondifferential, that is it is *differential*. In this case the bias in the estimation of $P_{A|B}(2|2) - P_{A|B}(2|1)$ can take any arbitrary form, just as in the univariate case, and need not involve attenuation. The possibility of nonattenuating effects was emphasized by Diamond and Lilienfeld (1962a, 1962b) but their discussion makes the rather unnatural assumption that the conditional distribution of A given A^* and B does not depend on B (rather than A^* given A and B as in (28.6)). Other authors (Newell, 1962; Keys and Kihlberg, 1963; Buell and Dunn, 1964) point out this source of confusion. A clear account of the possible effects of differential misclassification is presented by Goldberg (1975) (see also Copeland *et al.*, 1977; Shy *et al.*, 1978).

Changes in the categories of a misclassified variable may turn a nondifferential misclassification mechanism into a differential one. For example, suppose that a variable A with three categories is misclassified nondifferentially with respect to a binary variable B. If two of the categories of A are combined, misclassification in the resulting 2×2 table is not necessarily nondifferential (Wachholder *et al.*, 1991). Similarly, differential misclassification may arise when an erroneously measured continuous variable X is dichotomized, even if measurement error in X is nondifferential (Flegal *et al.*, 1991).

In order to illustrate the above results we first let the outcome variable A be the binary indicator of whether an individual is economically active (participates in the labor force) and the binary subgroup variable B to be age group: under or over 45.

In Table 28.3 we see that the misclassification matrices for the two groups are similar and the hypothesis of nondifferential misclassification in (28.6) is tenable. (The likelihood ratio test statistic for this hypothesis versus a saturated model, under the assumption of multinomial sampling, takes the value 4.99 on 2 d.f. with a *p*-value greater than 0.05. Under the actual complex design the *p*-value may be expected to be greater.)

Table 28.3   Economic Activity Reported in Census and in Census Validation Survey in Two Age Groups. Numbers in the Table Are Column Percentages and (in Parentheses) Cell Counts

	Age under 45			Age over 45		
	Census Validation Survey			Census Validation Survey		
Census	Active[a]	Inactive	Total	Active[a]	Inactive	Total
Active[a]	97.32 (4,038)	6.43 (60)	(4,098)	97.17 (1,823)	9.45 (62)	(1,885)
Inactive	2.68 (111)	93.57 (873)	(984)	2.83 (53)	90.55 (594)	(647)
Total	(4,149)	(933)	(5,082)	(1,876)	(656)	(2,532)

[a] Employed or unemployed.

The consequence of such nondifferential misclassification is shown in Table 28.4. The difference in activity rates between the two age groups is attenuated from 7.55 percent in the CVS to 6.19 percent in the census, a proportionate reduction of 18.0 percent. This may be compared to the expected proportionate reduction obtained from formula (28.7) as $(1 - \alpha) + (1 - \beta)$ and estimated from Table 28.3 as $7.7 + 2.4 = 10.4$ percent. The difference between 18 and 10.4 percent may be attributed to sampling error (the standard error of the estimate 18 per cent is obtained as 6.5 per cent by the δ-method under the multinomial assumption). Indeed, it seems worth noting that the estimated relative bias of 10.4 percent is of a similar order to the relative standard error of the observed

Table 28.4   Percentage of Economically Inactive in Two Age Groups, with Economic Activity Assessed by Census and by Census Validation Survey

Age	Census Validation Survey			Census	
	Active[a]	(s. e.)[b]	Total	Active[a]	(s. e.)[b]
Under 45	81.64	(0.54)	5,082	80.64	(0.55)
Over 45	74.09	(0.87)	2,532	74.45	(0.87)
Difference	7.55	(1.03)		6.19	(1.03)

[a] Employed or unemployed.
[b] Standard errors estimated under multinomial assumption.

difference, which is estimated under multinomial assumption as $1.03/6.19 = 16.6$ percent. Thus, we suggest that, at least for surveys of similar sample size, it will usually be unwise to suppose, when comparing subclass proportions, that either sampling error or measurement error bias will dominate. Both need to be taken account of.

It is also interesting to compare the relative bias here of 10.4 percent (or 18.0 percent) with that for the corresponding univariate analysis. The overall census and CVS activity rates across both age groups are obtained from Table 28.1 as 78.6 and 79.1 percent, respectively. The relative bias of the univariate analysis is thus only 0.6 percent, illustrating the earlier observation that misclassification can have more serious effects on bivariate analyses than on univariate analyses.

A second example is provided by Table 28.5, where the same outcome

Table 28.5 Economic Activity Reported in Census and in Census Validation Survey for Men and Women Separately. Numbers in the Table Are Column Percentages and (in Parentheses) Cell Counts

	Men			
	Census Validation Survey			
Census	Employed	Unemployed	Inactive	Total
Employed	98.43	4.30	4.56	
	(2,940)	(19)	(21)	(2,980)
Unemployed	0.54	91.18	6.29	
	(16)	(403)	(29)	(448)
Inactive	1.04	4.52	89.15	
	(31)	(20)	(411)	(462)
Total	(2,987)	(442)	(461)	(3,890)

	Women			
	Census Validation Survey			
Census	Employed	Unemployed	Inactive	Total
Employed	96.67	4.17	2.84	
	(2,324)	(8)	(32)	(2,364)
Unemployed	0.54	71.87	3.55	
	(13)	(138)	(40)	(191)
Inactive	2.79	23.96	93.62	
	(67)	(46)	(1,056)	(1,169)
Total	(2,404)	(192)	(1,128)	(3,724)

Table 28.6 Percentage of Unemployed Men and Women Aged 18–65, with Employment Status Assessed by Census and by Census Validation Survey

Sex	Census Validation Survey			Census	
	Unemployed	(s.e.)[a]	Total	Unemployed	(s.e.)[a]
Male	11.36	(0.51)	3,890	11.52	(0.51)
Female	5.16	(0.37)	3,724	5.13	(0.36)
Difference	6.20	(0.62)		6.39	(0.63)

[a] Standard errors estimated under multinomial assumption.

variable, econyomic activity, is compared between men and women. In this case it is clear that misclassification is differential. (The likelihood ratio test statistic for the hypothesis of nondifferential misclassification versus a saturated model takes the value 77.0 on 6 d.f. with a p-value of less than 0.001.) In particular, it is evident that the proportion of the "truly" unemployed who are misclassified as inactive is much greater for women than for men.

Table 28.6 illustrates how differential misclassification need not lead to attenuation of differences between subgroups. The true difference between the proportion of men and women unemployed (6.20 percent) is actually increased (to 6.39 percent) by misclassification.

28.3.2.2 Estimation for m × 2 Tables

Suppose we want to compare proportions, defined by a binary outcome variable A, between $m > 2$ subgroups defined by B. If A alone is misclassified and the misclassification is nondifferential with respect to B, each estimated difference $p_{A^*|B}(2\,|\,k) - p_{A^*|B}(2\,|\,l)$ ($k, l = 1, \ldots, m$) obeys a relationship similar to (28.7) and is attenuated by $1 - (1 - \alpha) - (1 - \beta)$. The ordering of the expected estimated proportions is thus the same as the ordering of the population proportions $P_{A|B}(2\,|\,1), \ldots, P_{A|B}(2\,|\,m)$.

If, instead, A is correctly classified but the subgroup variable B is nondifferentially misclassified, other effects may occur. Let the misclassification parameters for B be

$$\psi_{kl} = \Pr(B_i^* = k \,|\, B_i = l); \quad k, l = 1, \ldots, m. \tag{28.8}$$

Then in large samples the expectation of $p_{A|B^*}(2\,|\,k)$ may be approximated by

$$E[p_{A|B^*}(2\,|\,k)] \doteq \frac{\sum_l P_B(l) P_{A|B}(2\,|\,l) \psi_{kl}}{\sum_l P_B(l) \psi_{kl}}; \quad k = 1, \ldots, m.$$

Suppose that the subclasses are ordered so that $P_{A|B}(2\,|\,1) \le P_{A|B}(2\,|\,2) \le \cdots \le$

$P_{A|B}(2\,|\,m)$. Then $P_{A|B}(2\,|\,1) \leq \mathrm{E}[\,p_{A|B^*}(2\,|\,k)] \leq P_{A|B}(2\,|\,m)$ and

$$-[P_{A|B}(2\,|\,m)-P_{A|B}(2\,|\,1)] \leq \mathrm{E}[\,p_{A|B^*}(2\,|\,k)-p_{A|B^*}(2\,|\,l)] \leq P_{A|B}(2\,|\,m)-P_{A|B}(2\,|\,1)$$

for all $k, l = 1, \ldots, m$. The estimated difference in proportions between the two extreme subclasses ($B_i = 1$ and $B_i = m$) is thus again attenuated. However, estimates of other differences between subclass proportions can be biased away from or towards zero. An estimate can even have a different sign than the true difference. That is, misclassification can change the apparent ordering of the subclass proportions and distort trends in proportions across subgroups. Similar results are obtained for other measures of association, such as the relative risk $P_{A|B}(2\,|\,k)/P_{A|B}(2\,|\,l)$ and the odds ratio $\{P_{A|B}(2\,|\,k)/[1-P_{A|B}(2\,|\,k)]\}/$ $\{P_{A|B}(2\,|\,l)/[1-P_{A|B}(2\,|\,l)]\}$ used in epidemiology (Gladen and Rogan, 1979; Dosemeci et al., 1990; Birkett, 1992; see also Blettner and Wahrendorf, 1984).

Other possible biases notwithstanding, the most common effect of moderate misclassification of even a polytomous subgroup variable B seems to be that estimated subgroup proportions retain their ordering while all estimated differences are attenuated towards zero. This is the case in particular when misclassification occurs only between adjacent categories of B (Marshall et al., 1981; Birkett, 1992).

28.3.2.3 Hypothesis Testing for Two-Way Tables
An important consequence of the attenuation results is that if there is no association between A and B then there will also be no association between A^* and B^* under nondifferential misclassification in one variable (Mote and Anderson, 1975; Bross, 1954; Rubin et al., 1956; Rogot, 1961; Walsh, 1963; Gladen and Rogan, 1979) or nondifferential and independent misclassification in both variables (Assakul and Proctor, 1967; Giesbrecht, 1967). Hence, a test of no association between A and B will have the correct significance level, but, in general, the power will be reduced. Marshall et al. (1981), Walker and Blettner (1985) and Freudenheim et al. (1989) demonstrate similar results for a test of no trend in subclass proportions for tables where the outcome A is binary and a polytomous subgroup variable B is nondifferentially misclassified.

28.3.3 Multivariate Analyses

28.3.3.1 Three-Way Tables
The simplest multivariate analysis we consider is of a $2 \times 2 \times 2$ table. Let A be a binary outcome variable and let B and C be binary variables defining subgroups of the population. Let $P_{A|B,C}(j\,|\,k,l)$ be the proportion of units for which $A_i = j$ amongst the subpopulation for which $B_i = k$ and $C_i = l$. Other marginal and conditional proportions (e.g., $P_B(k)$ and $P_{C|B}(l\,|\,k)$) are denoted similarly. Suppose we wish to examine the relationship between A and B in a given category of C. Specifically, we want to estimate the differences $D_1 = P_{A|B,C}(2\,|\,2,1) - P_{A|B,C}(2\,|\,1,1)$ and $D_2 = P_{A|B,C}(2\,|\,2,2) - P_{A|B,C}(2\,|\,1,2)$. At first

assume that there is nondifferential misclassification in one of the variables only.

If C is classified correctly, we can consider the two-way tables between A and B separately in the two subgroups defined by C. The bivariate results of the previous section then apply. If A is misclassified, both D_1 and D_2 are attenuated by the same factor as in (28.7). The differences are also attenuated when B is misclassified. Because this bias depends also on the proportions $P_{B|C}(2|2)$ and $P_{B|C}(2|1)$, D_1 and D_2 are attenuated to a different degree if there is an association between B and C. Nondifferential misclassification in B can thus produce a spurious heterogeneity in estimates of D_1 and D_2 as well as mask a true heterogeneity (Greenland, 1980).

Consider now the case where C is misclassified nondifferentially as C^* but A and B are correctly classified. Let the misclassification probabilities be $\beta = \Pr(C_i^* = 1 | C_i = 1)$ and $\alpha = \Pr(C_i^* = 2 | C_i = 2)$. Using the identity

$$P_{A|B,C}(2|k, 1) = P_{A|B}(2|k) + [1 - P_{C|B}(1|k)][P_{A|B,C}(2|k, 1) - P_{A|B,C}(2|k, 2)]$$

$$(28.9)$$

it may be shown that the expected value of the estimator of $P_{A|B,C}(2|k, 1)$ is in large samples approximately

$$E[p_{A|B,C^*}(2|k, 1)] \doteq P_{A|B}(2|k) + \lambda_k[1 - P_{C|B}(1|k)]$$
$$\times [P_{A|B,C}(2|k, 1) - P_{A|B,C}(2|k, 2)] \qquad (28.10)$$

where

$$\lambda_k = \frac{\delta P_{C|B}(1|k)}{1 + \delta P_{C|B}(1|k)}$$

and $\delta = [1 - (1 - \alpha) - (1 - \beta)]/(1 - \alpha)$. In practice we may expect α and β to be greater than 0.5, in which case $0 \le \lambda_k \le 1$ and the expected value of $p_{A|B,C^*}(2|k, 1)$ lies between the true probability $P_{A|B,C}(2|k, 1)$ and $P_{A|B}(2|k)$, the proportion in the table summed over the values of C. Corresponding results may be obtained for estimates of other proportions $P_{A|B,C}(2|k, l)$, where $k, l = 1, 2$. This form of bias is known in epidemiological literature as *residual confounding* (e.g., Sabitz and Barón, 1989). It occurs because the analysis is restricted to the wrong subset of units (one for which $C_i^* = 1$ instead of $C_i = 1$) and thus the heterogeneity in the proportions due to C is not fully controlled for.

Because of residual confounding, estimates of D_1 and D_2 and measures of association between A and B controlling for C may be biased. Savitz and Barón (1989) (see also Greenland, 1980) show for the odds ratio that in a $2 \times 2 \times 2$ table the expected value of an estimated "summary" measure of the association between A and B conditional on C lies between the "crude" unconditional measure of association and the true conditional summary measure. Brenner (1993) demonstrates that this is not generally true when C has more than two categories.

The bias due to residual confounding in an estimate of D_1 or D_2 may be either away from or toward the null value. The estimate can even have a different sign than the true difference, thus indicating an association between A and B which is opposite to the true association. The estimates can also show an association where there is none ($D_1 = D_2 = 0$). Both the size and power of a test of no association between A and B are thus incorrect when C is misclassified.

The above simple example of misclassification in a three-way table can be extended in various ways. Some of the variables may be polytomous instead of dichotomous. There may be independent and nondifferential misclassification in more than one variable, in which case the effect of misclassification is a combination of attenuation and residual confounding (Fung and Howe, 1984). Finally, misclassification that is not both independent and nondifferential can produce any kinds of biases in the estimates (some examples are given by Greenland and Robins, 1985).

28.3.3.2 *General Case*

The results for three variables extend naturally to multivariate analyses with more variables. When several of the variables are misclassified, it is generally not possible to give even qualitative statements about the resulting biases. One exception is a useful result due to Korn (1981) on hypothesis testing in tables with independent and nondifferential misclassification in some of the variables. Suppose that the cell proportions in the population follow a hierarchical log-linear model. Such a model can be specified in terms of the configuration of highest-order interactions (see e.g., Bishop *et al.*, 1975, ch. 2). For example, a model for three variables where there is no association between A and B given a fixed value of C is written as $H(AC, BC)$ and a model of complete independence as $H(A, B, C)$. Korn's result states that a model is preserved by misclassification in the variable which appear only once in this model specification. A model is said to be preserved if the misclassified data also satisfy the model, that is, if the misclassification induces no spurious associations between the classified variables. Thus the model $H(AC, BC)$ is preserved by misclassification in A or B or both but not by misclassification in C. This agrees with the results obtained above for three-way tables. Suppose that B is misclassified as B^*. Since there is no association between A and B given C, that is, no AB interaction in the model, there will be no association between B^* and A given C (no AB^* interaction). Nondifferential misclassification in C, on the other hand, may produce both a spurious association between A and B given C^* (an AB interaction) and a spurious heterogeneity in these associations across levels of C^* (an ABC^* interaction).

If a model is preserved, it can be tested for goodness-of-fit using the misclassified values. The resulting test has the correct significance level but reduced power compared to a test using correctly classified values. Formulae given by Korn (1982) can be used to evaluate the power in a likelihood ratio test between two nested, preserved models.

28.4 INFERENCE ABOUT THE MISCLASSIFICATION MECHANISM

In the previous section we have seen that misclassification generally leads to biased estimation. To adjust for this bias will generally require some assumptions or information about the misclassification mechanism.

In rare cases the mechanism will be known. In particular, in the randomized response technique (Warner, 1965; Chen, 1979a) respondents operate a random device to safeguard the confidentiality of their response to sensitive questions. The survey researcher does not know the true responses but does know the misclassification matrix and can thus obtain unbiased estimates of population parameters as described in Section 28.5 (see also Press, 1968).

Much more commonly, the misclassification mechanism is unknown. In Sections 28.4.1 and 28.4.2 we examine two kinds of data which may be used to make inferences about this mechanism. Even if such data are unavailable, Tweedie *et al.* (1994, p. 22) argue that we should not "throw up our hands, note that the data are probably wrong and the conclusions biased, and assert that no claims are valid." Rather, they recommend a "what-if" sensitivity analysis in which possible biases are assessed by considering a realistic range of misclassification scenarios.

28.4.1 Validation Study with Preferred Procedure

Deming (1950) describes a *preferred procedure* as being "distinguished by the fact that it supposedly gives or would give results to what are needed for a particular end." Often the preferred procedure will be too expensive or otherwise unsuitable to use in the entire survey, thus necessitating the use of a fallible procedure. Nevertheless, it may be feasible to apply the preferred procedure for a validation sample s_v of limited size. If both the standard survey instrument A^* and the preferred procedure are administered to units in s_v and if the preferred procedure is equated with the true variable A, then the conditional probabilities θ_{jk} may be estimated directly, provided s_v is selected from the population of interest by a known probability sampling design.

28.4.1.1 Examples of Preferred Procedures

Sometimes the preferred procedure may involve the judgments of expert professionals. For example, Swires-Hennessy and Thomas (1987) describe how in some surveys of housing in Wales, the state of disrepair of a house, as assessed in a social survey, could be compared to the results of a physical inspection of the property by a professional surveyor. In the census example described in Section 28.2 the results obtained by a census enumerator are compared to the findings of an experienced interviewer. There are many similar examples in medicine where survey measurements may be compared to results of an examination by a physician (see, e.g., Cobb and Rosenbaum, 1956).

Checks against administrative records may also provide preferred procedures.

For example, Baumgarten *et al.* (1983) describe how job histories obtained by interview are compared to records kept by a pension board. Greenland (1988a) gives an example where interview data on antibiotic use were compared to medical records.

In medicine, the term "gold standard" is more commonly used. Examples include biochemical measures of smoking (Bauman and Koch, 1983) and detailed food consumption diaries as measures of diet (Willett *et al.*, 1985). Other examples are given by Copeland *et al.* (1977).

28.4.1.2 Selection of Validation Sample

Ideally the validation sample s_v should be a subsample of the main sample obtained by a known randomized *double sampling* scheme, also known as an *internal validation study*. Sometimes, however, there may be practical reasons which prevent such double sampling. For example, in a panel survey it may be desirable to avoid the additional burden of a validation study on panel members in order not to endanger future response rates. Thus in the Panel Study of Income Dynamics (Hill, 1992) survey responses were validated by comparing them to company records in a separate sample of employees of one large firm. For such *external validation samples* it will often be unreasonable to assume that the sample proportions in the categories of A are unbiased estimates of the proportions $P_A(j)$ in the population of interest. It may be more plausible to suppose that the misclassification matrix Θ can be estimated without bias from the validation sample. This would follow if the misclassification mechanism was homogenous across the validation sample and the population of interest. This homogeneity assumption may be investigated to some extent by assessing the dependence of the relationship between A_i^* and A_i on covariates such as age and sex. If, for example, it is found that the misclassification rate depends on sex then it will be necessary to allow for different Θ matrices for males and females. A validation sample of one sex only would then not be sufficient.

For a double sample it will generally be possible not only to obtain direct estimates of the misclassification probabilities θ_{jk} but also of the population proportions $\mathbf{P}_A$. Furthermore, it is possible to estimate conditional probabilities of the form

$$\Pr(A_i = k \mid A_i^* = j) = c_{kj} = \frac{P_A(k)\theta_{jk}}{\sum_{l=1}^m P_A(l)\theta_{jl}} \tag{28.11}$$

that is, probabilities that a unit classified as category j truly belongs to category k. Applying a term used by Carroll (1992) in the context of measurement error of continuous variables, we call c_{kj} $(k, j = 1, \ldots, m)$ *calibration probabilities*. When A is binary, the terms *positive predictive value* for $\Pr(A_i = 2 \mid A_i^* = 2)$ and *negative predictive value* for $\Pr(A_i = 1 \mid A_i^* = 1)$ are often used in the medical literature. Some estimation methods that adjust for misclassification make use of calibration probabilities instead of misclassification probabilities.

One option in an internal validation sample is to stratify with respect to A^*

and draw a specified proportion of units within each category of A^* (Deming, 1977; Haitovsky and Rapp, 1992). An extreme form of this design is to validate all units in some categories of A^* and none of the units in other categories. This may be very efficient if the aim is to test the hypothesis of no association between two variables (Zelen and Haitovsky, 1991). In contrast, for some external validation studies it may be possible to stratify by A, for example, if selecting a sample of individuals from a set of administrative records which will subsequently be administered a survey questionnaire to obtain the measure A^*. In such cases only θ_{jk} can be estimated but not $\mathbf{P}_A$. Stratification is particularly useful when some proportions in $\mathbf{P}_A$ are small, because it ensures that a sufficient number of units is sampled from all categories of A. Marshall (1990) compares validation study designs with different margins fixed.

There is also the question of the optimal choice of the sizes of the validation sample and the main sample. This will depend crucially on the ratio of costs of taking the basic survey measurement and the preferred procedure, as well as on the accuracy of the survey measurements. In some situations it may be most cost-effective to allocate all resources to a smaller, high-quality sample of measurements using the preferred procedure rather than a combination of a large, misclassified main sample and a small validation study (Tenenbein, 1970, 1972; Palmgren, 1987; Greenland, 1988b).

28.4.2 Repeated Measurements

Sometimes it is not feasible to use the preferred procedure even for a small validation sample. An alternative is to collect *repeated measurements* of the misclassified variables. In a survey this may mean *reinterviewing* at least once some of the subjects in the sample (Forsman and Schreiner, 1991). Although there may be misclassification errors in all of the repeated measurements, information about misclassification parameters can be extracted using statistical methods under appropriate assumptions about the relationship between the repeated measurements and the true variables.

It is shown in Section 28.5.3 that at least three repeated measurements are often required, although two replicates are sufficient in some cases. A common assumption is that these measurements are independent of each other given the true value of the underlying variable. Fortunately it is not necessary that the repeated measurements have the same distribution, and so they may be obtained from different methods. For example, repeated interviews may be conducted by different interviewers (Chua and Fuller, 1987, give an example from the U.S. Current Population Survey) or several family members may be given the same questions. Harper (1964) suggests that a questionnaire may contain several questions about a variable of interest, differently worded and separated by other questions. Other examples can be obtained from the examples of Section 28.4.1 by assuming that the preferred procedure is not accurate but contains misclassification errors of its own (Wacholder *et al.*, 1993). In these cases some of the measurements come not from the respondents but from other sources and in such cases it seems more plausible that they are conditionally independent given the true variable.

As in a validation study, the sample upon which repeated measurements are taken may be either external or internal. Similar considerations to those in Section 28.4.1 apply. Fuller (1990, Section 5) discusses the effect of the choice of n_v for two repeated measurements of a binary response.

28.5 METHODS OF ADJUSTING FOR THE EFFECTS OF MISCLASSIFICATION

Using the kinds of data described in the previous section, it is possible to adjust for the effects of misclassification and obtain estimates with reduced bias. In this section we discuss first estimation with validation data and then methods for repeated measurements.

28.5.1 Simple Matrix Methods Using Validation Data

The most straightforward and intuitively natural way to adjust for misclassification is via simple back-calculation methods. Adjusted cell counts are obtained by multiplying a vector of observed counts by an adjustment matrix and the analysis is performed on the transformed values. We refer to such methods as *matrix methods* of adjustment. We first discuss methods where the adjustment matrix is the inverse of the estimated misclassification matrix; next we consider adjustment where the elements of the adjustment matrix are the estimated calibration probabilities of the true variables given the classified variables. We also make some comparisons between these two approaches.

28.5.1.1 Adjustments Using Misclassification Probabilities

Excellent descriptions of these matrix methods are given by Greenland (1988a, appendix I) and Selén (1986). The basic ideas can be stated very briefly. Suppose we want to estimate the vector $\mathbf{P}_A = (P_A(1), \ldots, P_A(m))'$ of proportions of units falling into categories of variable A in the population. Note that here the m categories of A may represent cells in a cross-classification of several variables, ordered in some way. Let $\boldsymbol{\Theta} = [\theta_{jk}]$ be the misclassification matrix of probabilities that a unit from category k is classified as category j of variable A^* (cf. Equation 28.1). We assume that $\boldsymbol{\Theta}$ is nonsingular, that is, misclassification is not totally random. The main sample from the population consists of n_p units for which only the misclassified category is known. Let $\mathbf{p}_{A*}$ be an estimate of $\mathbf{P}_A$, defined as in (28.2), computed from this sample. In expectation, it is related to $\mathbf{P}_A$ via

$$E(\mathbf{p}_{A*}) = \boldsymbol{\Theta}\mathbf{P}_A \tag{28.12}$$

where E denotes expectation with respect to both the misclassification model and the sampling scheme. Suppose we have validation data from an internal or external validation sample of n_v units, independent of the main sample. From it we obtain a consistent estimate $\hat{\Theta}$ of Θ. For an external validation sample, a consistent estimate of $\mathbf{P}_A$ is given by

$$\hat{\mathbf{P}}_A^m = \hat{\Theta}^{-1}\mathbf{p}_{A*}. \tag{28.13}$$

If the validation sample is internal, an estimate of $\mathbf{P}_A$ from the true values in the validation sample, denoted by $\mathbf{p}_A^v$, may be combined with (28.13) to give a more efficient estimate

$$\hat{\mathbf{P}}_A^m = \frac{1}{n_p + n_v} [n_p(\hat{\Theta}^{-1}\mathbf{p}_{A*}) + n_v\mathbf{p}_A^v\}. \tag{28.14}$$

Note that here it is assumed that the probability that $\hat{\Theta}$ is singular is negligible. The requirement is hardly restrictive if the classification method is reasonable and the sample size is large enough.

Using the delta method, the asymptotic covariance matrix of $\hat{\mathbf{P}}_A^m$ defined by (28.13) may be shown to be

$$\text{cov}(\hat{\mathbf{P}}_A^m) = \hat{\Theta}^{-1} \text{cov}(\mathbf{p}_{A*})(\hat{\Theta}^{-1})' + \text{cov}[\hat{\Theta}^{-1}E(\mathbf{p}_{A*})] \tag{28.15}$$

and a similar formula may be obtained for (28.14). Another application of the delta method yields a formula for $\text{cov}[\hat{\Theta}^{-1}E(\mathbf{p}_{A*})]$ (see Greenland, 1988a). Replacing unknown quantities with their estimates, we obtain a consistent estimate of $\text{cov}(\hat{\mathbf{P}}_A^m)$ which takes into account the variability in both $\mathbf{p}_{A*}$ and $\hat{\Theta}$.

28.5.1.2 *Adjustments Using Calibration Probabilities*
In the previous section matrix adjustments were defined in terms of the inverse of the misclassification matrix. An alternative adjustment method is to multiply the observed proportions by an estimate of the calibration matrix $\mathbf{C} = [c_{kj}]$. Suppose that the validation sample is internal and that the sampling method is such that the distribution of true cell counts in each sample is multinomial. It can be shown (Tenenbein, 1970, 1972; Hochberg, 1977) that the maximum likelihood estimate of $\mathbf{P}_A$ is

$$\hat{\mathbf{P}}_A^c = \frac{1}{n_p + n_v} [n_p(\hat{\mathbf{C}}\mathbf{p}_{A*}) + n_v\mathbf{p}_A^v] \tag{28.16}$$

where $\hat{\mathbf{C}}$ is the estimate of $\mathbf{C}$ computed from the validation sample. The estimated true count in each category in the main sample is thus obtained by allocating the units from classified categories into true categories according to the estimated calibration probabilities. Suitably modified, the result holds also when the samples are stratified with respect to a variable which is not

Table 28.7   Data for Simple Matrix Adjustments with Measured Variable A^* and "True" Variable A

			A^*		
			1	2	
Validation sample	A	1	n_{11}	n_{12}	$n_{1\cdot}$
		2	n_{21}	n_{22}	$n_{2\cdot}$
			$n_{\cdot 1}$	$n_{\cdot 2}$	n_v
Main sample	A	$1+2$	m_1	m_2	n_p

misclassified and the joint distribution of the variables is thus a product of several multinomials. The asymptotic covariance matrix of $\hat{\mathbf{P}}_A^c$ is given by Tenenbein (1972).

28.5.1.3   Comparisons

We have seen that for multinomial data the estimate $\hat{\mathbf{P}}_A^c$ given by (28.16) is the maximum likelihood estimate of $\mathbf{P}_A$. Thus it is asymptotically more efficient than $\hat{\mathbf{P}}_A^m$ obtained from (28.14). But how much more efficient? In this section we compare the two estimates in a simple binomial case. The conclusion will be that $\hat{\mathbf{P}}_A^c$ is clearly superior to $\hat{\mathbf{P}}_A^m$.

Let A be a dichotomous variable with population proportions $\mathbf{P}_A = (P_A(1), P_A(2)) = (1 - P, P)$. Suppose A is misclassified as A^* with sensitivity $\alpha = \Pr(A_i^* = 2 \mid A_i = 2)$ and specificity $\beta = \Pr(A_i^* = 1 \mid A_i = 1)$ for each unit i. Assume that we have a main sample of n_p units and an independent internal validation sample of n_v units, both drawn from the population using simple random sampling. The data are thus as shown in Table 28.7. The biased estimate of P based on A^* in the main sample is $p^* = m_2/n_p$. Its expected value is $P^* = (1 - \beta)(1 - P) + \alpha P$.

Since the validation sample is internal, we can estimate P from it using the unbiased estimate $\hat{P}^v = n_{2\cdot}/n_v$. Its variance is $\text{var}(\hat{P}^v) = [P(1 - P)]/n_v$ ignoring finite population corrections. We will use $\hat{P}^v$ as a benchmark for evaluating the efficiencies of the estimates (28.14) and (28.16).

Let $f = n_v/(n_v + n_p)$ be the proportion of sampled units allocated to the validation sample. From (28.16), the maximum likelihood estimate of P is

$$\hat{P}^c = (1 - f)\tilde{P}^c + f\hat{P}^v \tag{28.17}$$

where

$$\tilde{P}^c = \frac{1}{n_p}\left(\frac{n_{21}}{n_{\cdot 1}}m_1 + \frac{n_{22}}{n_{\cdot 2}}m_2\right). \tag{28.18}$$

Its asymptotic variance is

$$\text{var}(\hat{P}^c) = \text{var}(\hat{P}^v)[1 - (1 - f)K] \tag{28.19}$$

where

$$K = \frac{P(1 - P)(\alpha + \beta - 1)^2}{P^*(1 - P^*)}. \tag{28.20}$$

The quantity K is the square of the correlation between A and A^* and is thus a simple scalar summary of the quality of the measurement (Tenenbein, 1970). Since $0 \le K \le 1$, $\hat{P}^c$ is never less efficient asymptotically than $\hat{P}^v$. If n_p increases while n_v is held fixed, the asymptotic relative efficiency $\text{var}(\hat{P}^v)/\text{var}(\hat{P}^c)$ tends to $1/(1 - K)$.

The estimate of P corresponding to (28.14) is

$$\hat{P}^m = (1 - f)\tilde{P}^m + f\hat{P}^v \tag{28.21}$$

where

$$\tilde{P}^m = \frac{p^* + \hat{\beta} - 1}{\hat{\alpha} + \hat{\beta} - 1} \tag{28.22}$$

and $\hat{\alpha} = n_{22}/n_{2.}$ and $\hat{\beta} = n_{11}/n_{1.}$ are the estimates of α and β computed from the validation sample. Its asymptotic variance (Marshall, 1990) is

$$\text{var}(\hat{P}^m) = \text{var}(\hat{P}^v)\left[1 - (1 - f)\left(\frac{2K - 1)}{K}\right)\right]$$

$$= \text{var}(\hat{P}^c) + \text{var}(\hat{P}^v)\left\{\left[\frac{(K - 1)^2}{K}\right](1 - f)\right\}. \tag{28.23}$$

Clearly $\text{var}(\hat{P}^m) \ge \text{var}(\hat{P}^c)$ for all values of the parameters. More remarkably, we see that $\text{var}(\hat{P}^m) > \text{var}(\hat{P}^v)$ when $K < 1/2$. Thus incorporating the n_p observations of the main sample through (28.14) gives *worse* estimates than using the validation sample alone. Furthermore, for a fixed n_v the variance of $\hat{P}^m$ actually *increases* as n_p increases.

The value $K = 0.5$ corresponds to a correlation of 0.71, which is not exceptionally low. Table 28.8 shows some parameter combinations (P, α, β) which gives values of K around 0.5. As expected, K is an increasing function of both α and β, with $K = 1$ at $\alpha = \beta = 1$. However, for any values of α and β both less than one, $K < 0.5$ when P or $1 - P$ is close enough to zero. This is illustrated in Figure 28.1, which shows the asymptotic relative efficiencies $\text{var}(\hat{P}^v)/\text{var}(\hat{P}^c)$ and $\text{var}(\hat{P}^v)/\text{var}(\hat{P}^m)$ for a range of values of P when $\alpha = 0.9$, $\beta = 0.85$ and $f = 1/6$.

A further problem with the estimate $\hat{P}^m$ is that it is not constrained to lie between zero and one. We have conducted some simulation studies to examine this problem. For smaller values of K a nonnegligible proportion of the

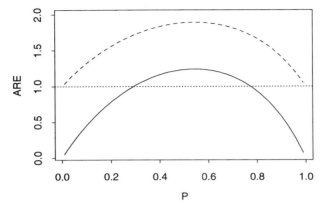

Figure 28.1 Asymptotic relative efficiencies of $\hat{P}^m$ (solid line) and $\hat{P}^c$ (dashed line) relative to $\hat{P}^v$ for $P \in [0.01, 0.99]$, $\alpha = 0.9$, $\beta = 0.85$, and $f = 1/6$.

simulated samples yields estimates which are negative or larger than one. Largely due to these outlying values we have found that the asymptotic formula (28.23) can quite badly underestimate the actual variance of $\hat{P}^m$. In contrast, we found that the corresponding formula for $\hat{P}^c$ in (28.19) seems to track the actual variance very well even in moderate size samples.

In the simple example of this section the maximum likelihood estimate $\hat{P}^c$ is clearly superior to $\hat{P}^m$. However, $\hat{P}^c$ is generally only appropriate when the validation sample is internal, since for external samples it is usually more plausible that the misclassification parameters θ_{jk} are "transportable" to the main population than the calibration probabilities c_{kj}. When only external validation data are available, the poor performance of $\hat{P}^m$ is thus a major problem.

Marshall (1990) compares the estimates $\hat{P}^m$ and $\hat{P}^c$ in the same binomial example as in this section. He also argues strongly in favor of the maximum likelihood estimate $\hat{P}^c$, although he does not note the potentially poor performance of $\hat{P}^m$ relative to $\hat{P}^v$. Selén (1986) gives some simulation results

Table 28.8 Values of the Squared Correlation K and the Asymptotic Relative Efficiencies $\text{ARE}(\hat{P}^m) = \text{var}(\hat{P}^v)/\text{var}(\hat{P}^m)$ and $\text{ARE}(\hat{P}^c) = \text{var}(\hat{P}^v)/\text{var}(\hat{P}^c)$ for $f = 0$ and Some Parameter Combinations

P	α	β	K	$\text{ARE}(\hat{P}^m)$	$\text{ARE}(\hat{P}^c)$
0.5	0.67	1	0.50	1.02	2.02
0.5	0.85	0.85	0.49	0.96	1.96
0.3	0.80	0.90	0.48	0.93	1.93
0.1	0.93	0.93	0.51	1.02	2.02
0.7	0.80	0.95	0.48	0.94	1.94
0.9	0.95	0.85	0.51	1.04	2.04

comparing a matrix estimator and the maximum likelihood estimator for a 2×2 table where one variable is misclassified. In most of his examples, however, misclassification probabilities are considered known. It would be of interest to examine whether the conclusions reached above for estimators of simple proportions hold also for bivariate and multivariate analyses.

28.5.1.4 *Literature*

There is a large literature on applications of the matrix method to various misclassification problems. Rogan and Gladen (1978) and Quade *et al.* (1980) describe its use in estimating a single proportion. Tenenbein (1970) considers the same problem, but using calibration probabilities instead of misclassification probabilities. Tenenbein (1972) extends the formulae to problems where proportions falling into more than two categories are estimated. Tenenbein (1971), Hochberg and Tenenbein (1983) and Haitovsky and Rapp (1992) describe various ways of sharpening the analysis by refining the double sampling methodology.

Matrix adjustments for two-way tables have often been obtained as a by-product of studying the effects of misclassification in such tables (Copeland *et al.*, 1977; Shy *et al.*, 1978; Barron, 1977; Greenland, 1982). These authors consider misclassification in one or both of the variables, as well as both differential and nondifferential misclassification. They, however, assume that the misclassification probabilities are known and fixed and do not give variance estimates for the adjusted tables. Greenland and Kleinbaum (1983) give a clear account of matrix adjustments (with fixed Θ and no variance formulae) in two-way tables, possibly involving matching of units with respect to one of the variables, with differential or nondifferential misclassification in one or both of the variables. They also point out that the matrix adjustments need to be modified slightly when the main sample is stratified with respect to a misclassified variable, that is, when the sample is drawn so that the number of units in each category of a misclassified variable is fixed. In this case the sensitivity and specificity *in the sample* depend on the sampling fractions (see also Chen, 1989).

Matrix adjustments for two-way tables in various special cases are considered by Elton and Duffy (1983), Green (1983; see also Begg, 1984) and Greenland (1989). Greenland (1988a) presents variance estimates for some functions of two-way tables. He also gives a general description of the matrix adjustment methods using misclassification probabilities. Finally, several general epidemiological textbooks (e.g., Fleiss, 1981; Kleinbaum *et al.*, 1982) discuss simple examples of the matrix method, although typically for known misclassification matrices only.

28.5.2 Model-Based Methods Using Validation Data

In the previous section we described matrix adjustment methods which give maximum likelihood estimates of cell proportions when an internal validation sample is available. These estimates are simple and easy to compute, but they also have several disadvantages. First, it is not possible to test hypotheses about or fit models to the resulting estimated true tables using standard methods, since these take no account of the extra uncertainty arising from misclassification adjustment. Second, for large tables, especially those based on several variables, these estimates of Θ will often be very imprecise because of sparseness and it is desirable to fit models to the misclassification mechanism to obtain more precise adjustments. In this section we discuss adjustment methods which yield maximum likelihood estimates for a wide class of models and data structures. Inevitably, the methods are computationally more complicated than the matrix methods, generally requiring iterative techniques.

To keep the notation simple we illustrate model-based methods for two variables A and B, with corresponding misclassified variables A^* and B^*. The methods may be applied quite generally, however. We assume multinomial sampling. Some extensions to complex designs are discussed by Rao and Thomas (1991). We suppose that the aim is to model the association between the true variables.

Let $\mathbf{P}_{AB}$ be the proportions of population units belonging to categories defined by the possible values of the true variables, and let $\theta_{A^*B^*|AB}$ denote the misclassification probabilities. In the rare cases where the misclassification probabilities are known, estimation of $\mathbf{P}_{AB}$ involves a straightforward application of the EM algorithm (Dempster et al., 1977). The method is described by Chen (1978, 1979a, 1989). (Whittemore and Grosser, 1986 consider the case where the calibration probabilities are known.) Let $\mathbf{n}_{A^*B^*}$ be the observed numbers of sample units classified into the possible combinations of the classes of A^* and B^*. Suppose we wish to fit a model (e.g., one of independence) between A and B using $\mathbf{n}_{A^*B^*}$ and the known misclassification probabilities $\theta_{A^*B^*|AB}$. At the E-step of the EM algorithm an expected value of the complete table between A, B and A^*, B^* is computed as

$$\mathbf{n}_{ABA^*B^*}^{(j)} = \frac{\mathbf{p}_{AB}^{(j)}\theta_{A^*B^*|AB}}{\sum_{A,B}\mathbf{p}_{AB}^{(j)}\theta_{A^*B^*|AB}} \cdot \mathbf{n}_{A^*B^*}$$

where $\mathbf{p}_{AB}^{(j)}$ is the current estimate of $\mathbf{P}_{AB}$ and division and multiplication on the right-hand side of the equation denote elementwise division and multiplication of the relevant vectors or matrices. An expected table of the true variables is obtained by collapsing over the classified variables

$$\mathbf{n}_{AB}^{(j)} = \sum_{A^*,B^*} \mathbf{n}_{ABA^*B^*}^{(j)}.$$

At the M-step a new estimate $\mathbf{p}_{AB}^{(j+1)}$ of $\mathbf{P}_{AB}$ is obtained by fitting the required model to $n_{AB}^{(j)}$ in a standard way, and the process is iterated until convergence.

When the misclassification probabilities are not known but are estimated from a validation sample, the above procedure needs to be extended. It is convenient to define a new variable L, which identifies the sample from which a unit comes. The variable L is binary when there is one main sample and one validation sample, but other study designs are also possible. For example, Chen *et al.* (1984) consider a "triple sampling" scheme where a third sample with only accurately classified values is available.

In the following we assume that the models discussed are all log-linear and hierarchical. They can thus be specified through the highest-order interactions in the model (see Section 27.3.3). To represent the association structure between the variables, we consider models at two levels (cf. Espeland and Odoroff, 1985).

1. The model for the table between all the variables $(A, B, A^*, B^*$ and $L)$, including full interaction terms between the true variables (AB). The model describes two kinds of relationships.

   a. The differences between the main sample and the validation sample, indicated by interactions between L and some of the true variables $(A$ and $B)$. If the validation sample is internal, both samples are random samples from the same population and the sample indicator L is independent of the other variables [model $H(ABA^*B^*, L)$]. Often the first step of the analysis is to test whether the data are consistent with this model.

   b. The relationship between the true and the classified variables. This model is often known as the misclassification model. A minimum requirement is that the model contains an interaction term between each true variable and the corresponding classified variable, since otherwise the misclassification would be totally random. With this restriction, various misclassification models can be considered. For example, in the model $H(AB, AA^*, BB^*, L)$ misclassification of both A and B is nondifferential and independent; in $H(ABA^*, BB^*, L)$ it is independent and nondifferential in B but differential with respect to B in A; and in $H(ABA^*B^*, L)$ misclassification is neither independent nor nondifferential in either variable.

2. The model for the relationships between the true variables. This does not contain terms involving the classified variables. Here, L can also be omitted if it is independent of the true variables. In the bivariate case the two possible models are then the saturated model $H(AB)$ and the model of independence $H(A, B)$.

While the two submodels are both log-linear, the joint model generated by them need not be. The joint model can be fitted using the EM algorithm; alternative versions have been proposed by Chen (1979b), Chen *et al.* (1984) and Espeland and Odoroff (1985). These authors use earlier, more general results concerning maximum likelihood estimation in contingency tables with some observations only partially classified (Hocking and Oxspring, 1971, 1974; Chen and Fienberg, 1974, 1976) and sequential fitting of log-linear models

(Goodman, 1973). Espeland and Odoroff (1985) give estimated variances for the parameter estimates from the model.

Other ways of modeling contingency tables with misclassification have also been proposed. Hochberg (1977) uses the least squares method of Grizzle *et al.* (1969). Chiacchierini and Arnold (1977) apply straightforward maximization to fit a model of independence to a two-way table with misclassification. More recently, Espeland and Hui (1987) (see also Ekholm, 1991) propose a Fisher scoring algorithm for fitting log-linear misclassification models using data from different kinds of validation samples. Ekholm and Palmgren (1987) and Palmgren and Ekholm (1987) describe an alternative approach where the cell probabilities in the samples are specified directly in terms of marginal cell proportions of the true variables and the misclassification probabilities. The parameters can be estimated using a modification of the iteratively reweighted least squares algorithm.

Many of the authors discussing model-based adjustments have chosen the same set of data to illustrate their methods. Hochberg (1977) introduced an example from highway safety research involving four variables, two of which are misclassified. These data have been analyzed also by Chen *et al.* (1984; see also Chen, 1989), Espeland and Odoroff (1985) and Ekholm and Palmgren (1987) (see also Ekholm, 1992). Different authors have selected different misclassification models, each of which fits the data well, yet may lead to quite different estimates of the joint distribution of the true variables. Chen (1992) summarizes and extends these results.

We now illustrate the use of modeling with the CVS data. We shall not attempt to carry out a complete adjustment of the main census estimates. Rather, we shall just consider models for the variables in the CVS sample. The variable L above will therefore not feature in our analysis, since we only consider data for one value of L.

We take the "true" variables as:

- ACT: economic activity in the CVS (employed/unemployed/inactive)
- ETH: ethnic group in the CVS (white/other)
- AGE: age in the census (18–29/30–44/45–64 [men] or 45–59 [women])
- SEX: sex in the census

and the misclassified variables as:

- ACT*: economic activity in the census (employed/unemployed/inactive)
- ETH*: ethnic group in the census (white/other).

We thus effectively assume that age and sex are measured without error. Following the modeling approach discussed above we begin by assuming the presence of all interactions between the true variables, represented by AGE.SEX.ACT.ETH. We also assume the presence of the main effects of

Table 28.9 Log-Linear Misclassification Models Fitted to the CVS Data, Assuming Full Interactions Between the True Variables. Definitions of the Variables Are Given in the Text

Baseline model: H(AGE.SEX.ACT.ETH, ACT*, ETH*)

Term added	Deviance difference	d.f.
+ACT*.ACT	8,969	4
+ETH*.ETH	3,822	1
+ACT*.SEX	67	2
+ACT*.ACT.SEX	14	4

Final model: H(AGE.SEX.ACT.ETH, ACT*.ACT.SEX, ETH*.ETH)

ACT* and ETH*. Table 28.9 shows the effect on the deviance ($-2 \times$ log-likelihood under the multinomial assumption) of adding various interaction terms in turn. Using a forward selection approach, including only terms significant at the 95 percent level, leads to the choice of the final model indicated at the bottom of Table 28.9. A graph representing this model is displayed in Figure 28.2. The "graphical model" convention (Whittaker, 1990) is adopted, where edges between pairs of variables are absent if the variables are independent, conditional on all the remaining variables in the model. Thus misclassification of a variable A is nondifferential with respect to a variable B if the edge between $A*$ and B is absent and misclassification of two variables A and B is independent if the edge between $A*$ and $B*$ is absent. Three important results may therefore be noted from the graph.

1. Misclassification of ethnic group is nondifferential with respect to age, sex, and economic activity.
2. Misclassification of economic activity is differential with respect to sex, but not with respect to age or ethnic group (cf. earlier analysis in Section 28.3.2).
3. Misclassification of ethnic group is independent of misclassification of economic activity.

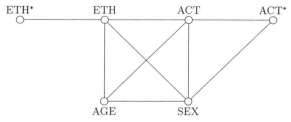

Figure 28.2 Graph representing conditional independence relationships between classified variables (ETH* and ACT*) and "true" variables (ETH, ACT, AGE and SEX).

The result of this modeling is to reduce the number of parameters in the misclassification model and thus, one may hope, improve the precision of estimation of these parameters. Thus in a saturated model there would be 180 $(= 3 \times 2 \times 3 \times 2 \times (3 \times 2 - 1))$ parameters representing the conditional distribution of the two measured variables given the four true variables, whereas in the fitted model there are only 14 $(= 2 \times 3 \times 2 + 2)$ parameters.

If we turn to the joint distribution of the true variables we find that this model can also be simplified without significant reduction in goodness-of-fit by assuming conditional independence between SEX and ETH given AGE and ACT.

28.5.3 Adjustment Methods Using Repeated Measurements

Instead of validation studies, information about misclassification parameters may come from repeated measurements of the misclassified variables. A number of estimation methods have been developed for such situations. The choice of method depends partly on the available data, for example, the number of replicates per unit, whether the measurements are obtained using one or several measurement methods and whether repeated measurements are available for all or only some of the units. Another important aspect is the complexity of the problem, for example, the number of variables and their categories, what the parameters of interest are and what kinds of models are to be fitted and what hypotheses tested.

Data from repeated measurements may be analyzed using latent class models (see, e.g., Goodman, 1978). The unobserved latent variables are the true values of the misclassified variables. Conditional on the latent values, the repeated measurements are assumed independent of each other and of other variables. Estimates are typically computed using the EM algorithm, which lends itself naturally to latent class problems. Direct Newton–Raphson maximization of the likelihood may also be feasible, and for simple models even closed-form parameter estimates are available.

28.5.3.1 *Identifiability of the Models*
As in all latent class models, identifiability is a key issue in misclassification models with repeated measurements. Not all models are identifiable without additional constraints. Here we give some identifiability results, based mainly on an excellent review article by Walter and Irwig (1988). General conditions for identifiability in misclassification problems are given by Liu and Liang (1991).

Consider first the problem of estimating the proportions $P^{(j)} = P_A^{(j)}(2)$ $(j = 1, \ldots, S)$ of units belonging to a category of a binary variable A in S subgroups of a population. Suppose R repeated measurements are observed for some or all of the units in a sample. Let β_k be the specificity and α_k the sensitivity of the kth measurement $(k = 1, \ldots, R)$. With three or more replicates $(R \geq 3)$ all parameters are identifiable for any number of populations. With $R = 2$, the parameters are not always identifiable without further constraints. When $S = 1$, a common assumption is that all misclassification probabilities are the same, that is, $\alpha_1 = \alpha_2 = \beta_1 = \beta_2 = \delta$. The estimates of $P^{(1)}$ and δ can then be computed in a closed form. Note that the less restrictive constraints $\alpha_1 = \alpha_2$ and $\beta_1 = \beta_2$ are not sufficient in this case because the symmetry of the repeated measurements implies that there are only two degrees of freedom to estimate the three unknown parameters. When $S = 2$ subgroups are compared, it is only necessary to assume that the misclassification probabilities are the same across subgroups. All the parameters $(P^{(1)}, P^{(2)}, \alpha_1, \alpha_2, \beta_1, \beta_2)$ are then identifiable.

Even in otherwise identifiable models the labeling of the latent classes is not uniquely determined (except when all probabilities are equal to 0.5). This problem is inherent in latent class models. It implies, for example, that when $(\hat{P}^{(1)}, \hat{\alpha}, \hat{\beta})$ is a maximum likelihood estimate of $(P^{(1)}, \alpha, \beta)$, then so is $(\hat{P}^{(1)\prime}, \hat{\alpha}', \hat{\beta}') = (1 - \hat{P}^{(1)}, 1 - \hat{\beta}, 1 - \hat{\alpha})$. In this case it is natural to select the set of estimates which satisfies the condition $\hat{\alpha} + \hat{\beta} - 1 \geq 0$.

Identifiability conditions are more complicated when the variable of interest has more than two categories or when there are several misclassified variables (Liu and Liang, 1991). However, it seems that three replicates are quite generally sufficient, while models with two replicates are not always identifiable. In all but the simplest problems there are many overidentified models. Some degrees of freedom are then available for testing of hypotheses and more parsimonious modeling of the variables.

28.5.3.2 *Literature*

The simplest application of repeated measurements is in the estimation of proportions in one population ($S = 1$). Harper (1964) gives estimates of a single binary proportion in two cases (with $R = 2$ and $R = 3$ measurements) where the estimates are available in a closed form. Quade *et al.* (1980) apply the EM algorithm to the second of Harper's examples. Dawid and Skene (1979) describe estimation of polytomous proportions from $R \geq 3$ repeated measurements, and Nathan and Eliav (1988) an application to a multiround survey with different modes of measurement.

Hui and Walter (1980) give closed-form estimates and information matrices for binary proportions and misclassification probabilities when $S = 2$, $R = 2$ and the misclassification probabilities differ across measurements. Differences between subgroup proportions and other measures of association can then be computed from the estimated values. Clayton (1985) studies a similar problem but with sensitivity and specificity constant across measurements and replicates available for only some of the subjects. He also explores in detail interval estimation of the parameters using profile likelihood methods. Ekholm and Palmgren (1982) consider a case where $R = 1$ but the model is identifiable because a strong structure is imposed on the proportions $P^{(j)}$ and the misclassification probabilities are the same in each subgroup.

An example of a more complicated model is given by Palmgren and Ekholm (1987) who fit models with two misclassified and repeatedly measured ($R = 2$) variables using a direct Newton–Raphson method. Duffy *et al.* (1989) consider a three-variable case with two variables misclassified. They first obtain estimates of the misclassification probabilities from a set of repeated measurements. These estimates are used to compute an estimate of the table between the true variables, which is then used in the analysis. The method is thus essentially the same as the matrix adjustments of Section 28.5.1, but with misclassification probabilities estimated from repeated measurements instead of validation data.

Estimation and modeling of any number of variables with repeated measurements is explored in two papers. Liu and Liang (1991) discuss an approach which is related to matrix adjustments, although more flexible and general. They identify one accurately classified variable as the response, associated with the other variables through a generalized linear model. The first step is to estimate the misclassification parameters from a set of repeated measures using latent class methods. Given these estimates, the parameters of the regression model are then estimated from the rest of the data using quasi-likelihood techniques. An excellent account of the full latent class approach is provided by Kaldor and Clayton (1985). They assume that repeated measurements are available for all subjects. A log-linear model is specified between all latent and observed variables and its parameters are estimated using the EM algorithm. Confidence intervals can be obtained from a profile likelihood function.

Sometimes quicker and simpler methods than maximum likelihood estimation may be used. McClish and Quade (1985) propose two sequential designs for estimating proportions. In both of these a subject is classified into one of two categories according to the majority of a number of repeated measurements. In case of a tie, an additional measurement is observed. Marshall and Graham (1984) estimate parameters of association using only the subjects for which two repeated observations are concordant. These methods are only applicable in special cases and are generally less efficient than maximum likelihood estimations (see a discussion in Walter, 1984). They may, however, be useful as initial values for iterative estimation algorithms.

Finally, we note that repeated measurements are also used in analyzing *interrater agreement*. The aim then is not so much to estimate misclassification probabilities as to examine the degree of agreement between different measurements (raters). This is summarized by indices of agreement such as the kappa coefficient (see, e.g., reviews by Landis and Koch, 1975; Fleiss, 1981; MacLure and Willett, 1987). A more sophisticated approach is to use log-linear models to model the agreement (Tanner and Young, 1985). Often the difference between these models and the ones discussed above is one of emphasis rather than content, since suitable chosen models can be used to estimate both agreement patterns and misclassification probabilities simultaneously (Espeland and Handelman, 1989).

28.5.4 Comparison of Methods

The choice between adjustments using validation information and those using repeated measurements is usually dictated by the available data, although in some studies both approaches may be feasible. Further research is needed on study design and optimal allocation of resources in cases where a choice is possible. One relevant study is Duffy *et al.* (1992). They compare estimates for a 2 × 2 table with misclassification in the explanatory variable under two study designs: an internal study with two repeated measurements and an external validation study. In all of their examples, repeated measurements give better precision than a validation study.

Marshall (1989) compares the two designs in more general terms. He, too, comes out in favor of repeated measurements. However, Marshall's main criticism of the validation methods is the sensitivity of "direct" matrix adjustments (see Section 28.5.1) to error in the estimation of the misclassification probabilities. As we have seen, however, this sensitivity may be reduced for adjustments based on calibration probabilities.

28.6 CONCLUSIONS

There is considerable evidence that some degree of misclassification occurs for most categorical variables in most surveys and that the biases caused by misclassification may often be at least as important as sampling errors. The bias effects of misclassification may be expected to be more serious for bivariate and multivariate analyses than for univariate analyses and we have illustrated this empirically. We have reviewed a number of theoretical results which provide guidance on how the nature of the bias might be predicted. A key condition, required to achieve many theoretical results, is that misclassification of a variable is nondifferential, that is, it does not depend on the values of other variables. We have seen empirically, however, that misclassification need not be nondifferential and hence its bias effect in categorical data analysis will often be difficult to predict.

It follows from this conclusion that it is important to consider ways of estimating the bias and, as the natural next step, adjusting estimates for this bias. On the whole, reliable adjustment requires additional information about the magnitude of misclassification, as discussed in Section 28.4. According to much current survey practice, this information is often not available and only sensitivity analyses are feasible (Tweedie *et al.*, 1994). However, given the potential serious bias effects that can occur, there are strong reasons why survey commissioners should invest the resources required to collect such information.

For basic estimates of univariate proportions, simple matrix adjustments may be used. For bivariate and multivariate analyses, modeling methods are usually required. A variety of modern statistical modeling procedures may be employed. We have found that the class of graphical models (Whittaker, 1990) is easily interpretable and sufficiently general for this purpose.

We end with some suggestions for areas needing further research:

- the choice of design for the collection of information about the misclassification mechanism;
- the choice of model selection procedure for model-based misclassification adjustment and the sensitivity of the resulting analysis to this choice;
- the combined treatment of both misclassification and nonresponse;
- the combined treatment of misclassification and complex sample designs;
- the combined treatment of measurement error in both continuous and discrete variables.

All of these topics for research would benefit not only from theoretical investigation but also from further empirical evaluation using real survey data.

One final suggestion, flowing naturally from our example, is that adjustments for misclassification could be made to census estimates. We are not aware that the kinds of methods described in Section 28.5 have been applied to census data but it seems plausible that, for the multivariate analysis of categorical census data, such adjustments might be more important than the widely discussed adjustments for undercoverage.

ACKNOWLEDGMENTS

We are very grateful to the Office of Population Censuses and Surveys (now the Office for National Statistics) for providing us with tabulations of CVS data and for comments. Research by C.J. Skinner was supported by Economic and Social Research Council grant number H51925500T, under its Analysis of Large and Complex Data Sets Programme. Research by J. Kuha was supported by grant number 6949 from the Research Council for the Social Sciences, the Academy of Finland.

REFERENCES

Assakul, K., and Proctor, C.H. (1967), "Testing Independence in Two-way Contingency Tables with Data Subject to Misclassification," *Psychometrica*, 32, pp. 67–76.

Barron, B.A. (1977), "The Effects of Misclassification on the Estimation of Relative Risk," *Biometrics*, 33, pp. 414–418.

Bauman, K.E., and Koch, G.C. (1983), "Validity of Self-reports and Descriptive and Analytical Conclusions: The Case of Cigarette Smoking by Adolescents and Their Mothers," *American Journal of Epidemiology*, 118, pp. 90–98.

Baumgarten, M., Siemiatycki, J., and Gibbs, G.W. (1983), "Validity of Work Histories Obtained by Interview for Epidemiologic Purposes," *American Journal of Epidemiology*, 118, pp. 583–591.

Begg, C.B. (1984), "Estimation of Risks When Verification of Disease Status Is Obtained in a Selected Group of Subjects," *American Journal of Epidemiology*, 120, pp. 328–330.

Birkett, N.J. (1992), "Effects of Nondifferential Misclassification on Estimates of Odds Ratios with Multiple Levels of Exposure," *American Journal of Epidemiology*, 136, pp. 356–362.

Bishop, Y.M.M., Fienberg, S.E., and Holland, P.W. (1975), *Discrete Multivariate Analysis: Theory and Practice*, Cambridge: The MIT Press.

Blettner, M., and Wahrendorf, J. (1984), "What Does an Observed Relative Risk Convey about Possible Misclassification," *Methods of Information in Medicine*, 23, pp. 37–40.

Brenner, H. (1993), "Bias Due to Non-differential Misclassification of Polytomous Confounders," *Journal of Clinical Epidemiology*, 46, pp. 57–63.

Bross, I. (1954), "Misclassification in 2 × 2 Tables," *Biometrics*, 10, pp. 488–495.

Buell, P., and Dunn, J.E. (1964), "The Dilution Effect of Misclassification," *American Journal of Public Health*, 54, pp. 598–602.

Carroll, R.J. (1992), "Approaches to Estimation with Error in Predictors," in L. Fahrmeir, B. Francis, R. Gilchrist, and G. Tutz (eds.), *Advances in GLIM and Statistical Modelling* (Proceedings of the GLIM92 Conference and the 7th International Workshop on Statistical Modelling, München, July 1992), New York: Springer-Verlag, pp. 40–47.

Chen, T.T. (1978), "Log-linear Models for the Categorical Data Obtained from Randomized Response Techniques," *Proceedings of the Section on Social Statistics*, American Statistical Association, pp. 284–288.

Chen, T.T. (1979a), "Analysis of Randomized Response as Purposively Misclassified Data," *Proceedings of the Section on Survey Research Methods*, American Statistical Association, pp. 158–163.

Chen, T.T. (1979b), "Log-linear Models for Categorical Data with Misclassification and Double Sampling," *Journal of the American Statistical Association*, 74, pp. 481–488.

Chen, T.T. (1989), "A Review of Methods for Misclassified Categorical Data in Epidemiology," *Statistics in Medicine*, 8, pp. 1095–1106.

Chen, T.T. (1992), Reply to Ekholm, *Statistics in Medicine*, 11, pp. 272–275.

Chen, T., and Fienberg, S.E. (1974), "Two-dimensional Contingency Tables with Both Completely and Partially Cross-classified Data," *Biometrics*, 30, pp. 629–642.

Chen, T., and Fienberg, S.E. (1976), "The Analysis of Contingency Tables with Incompletely Classified Data," *Biometrics*, 32, pp. 133–144.

Chen, T.T., Hochberg, Y., and Tenenbein, A. (1984), "Analysis of Multivariate Categorical Data with Misclassification Errors by Triple Sampling Schemes," *Journal of Statistical Planning and Inference*, 9, pp. 177–184.

Chiacchierini, R. P., and Arnold, J.C. (1977), "A Two-sample Test for Independence in 2 × 2 Contingency Tables with Both Margins Subject to Misclassification," *Journal of the American Statistical Association*, 72, pp. 170–174.

Chua, T., and Fuller, W.A. (1987), "A Model for Multinomial Response Error Applied to Labor Flows," *Journal of the American Statistical Association*, 82, pp. 46–51.

Clayton, D. (1985), "Using Test–Retest Reliability Data to Improve Estimates of Relative Risk: An Application of Latent Class Analysis," *Statistics in Medicine*, 4, pp. 445–455.

Cobb, S., and Rosenbaum, J. (1956), "A Comparison of Specific Symptom Data Obtained by Nonmedical Interviews and by Physicians," *Journal of Chronic Diseases*, 4, pp. 245–252.

Cochran, W.G. (1968), "Errors of Measurement in Statistics," *Technometrics*, 10, pp. 637–666.

Copeland, K.T., Checkoway, H., McMichael, A.J., and Holbrook, R.H. (1977), "Bias Due to Misclassification in the Estimation of Relative Risk," *American Journal of Epidemiology*, 105, pp. 488–495.

Dale, A., and Marsh, C. (eds.) (1993), *The 1991 Census User's Guide*, London: Her Majesty's Stationary Office.

Dalenius, T. (1977), "Bibliography of Non-sampling Errors in Surveys," *International Statistical Review*, 45, pp. 71–89, 181–197 and 303–317.

Dawid, A.P., and Skene, A.M. (1979), "Maximum Likelihood Estimation of Observer Error-rates Using the EM Algorithm," *Applied Statistics*, 28, pp. 20–28.

Deming, W.E. (1950), *Some Theory of Sampling*, New York: Wiley.

Deming, W.E. (1977), "An Essay on Screening, or on Two-phase Sampling, Applied to Surveys of a Community," *International Statistical Review*, 45, pp. 29–37.

Dempster, A.P., Laird, N.W., and Rubin, D.B. (1977), "Maximum Likelihood from Incomplete Data via the EM Algorithm" (with discussion), *Journal of the Royal Statistical Society*, Series B, 39, pp. 1–38.

Diamond, E.L., and Lilienfeld, A.M. (1962a), "Effects of Errors in Classification and Diagnosis in Various Types of Epidemiological Studies," *American Journal of Public Health*, 52, pp. 1137–1144.

Diamond, E.L., and Lilienfeld, A.M. (1962b), "Misclassification Errors in 2×2 Tables with One Margin Fixed: Some Further Comments," *American Journal of Public Health*, 52, pp. 2106–2110.

Dosemeci, M., Wacholder, S., and Lubin, J.H. (1990), "Does Nondifferential Misclassification of Exposure Always Bias a True Effect Toward the Null Value?" *American Journal of Epidemiology*, 132, pp. 746–748.

Duffy, S.W., Maximowitch, D.M., and Day, N.E. (1992), "External Validation, Repeat Determination, and Precision of Risk Estimation in Misclassified Exposure Data in Epidemiology," *Journal of Epidemiology and Community Health*, 46, pp. 620–624.

Duffy, S.W., Rohan, T.E., and Day, N.E. (1989), "Misclassification in More Than One Factor in a Case–Control Study: A Combination of Mantel–Haenszel and Maximum Likelihood Approaches," *Statistics in Medicine*, 8, pp. 1529–1536.

Ekholm, A. (1991), "Algorithms Versus Models for Analyzing Data that Contain Misclassification Errors," *Biometrics*, 47, pp. 1171–1182.

Ekholm, A. (1992), Letter to the Editor concerning the paper by Chen, *Statistics in Medicine*, 11, pp. 271–272.

Ekholm, A., and Palmgren, J. (1982), "A Model for a Binary Response with Misclassifications," in R. Gilchrist (ed.), *GLIM 82: Proceedings of the International Conference on Generalized Linear Models*, Heidelberg: Springer, pp. 128–143.

Ekholm, A., and Palmgren, J. (1987), "Correction for Misclassification Using Doubly Sampled Data," *Journal of Official Statistics*, 3, pp. 419–429.

Elton, R.A., and Duffy, S.W. (1983), "Correcting for the Effect of Misclassification Bias in a Case-Control Study Using Data from Two Different Questionnaires", *Biometrics*, 39, pp. 659–665.

Espeland, M.A., and Handelman, S.L. (1989), "Using Latent Class Models to Characterize and Assess Relative Error in Discrete Measurements," *Biometrics*, 45, pp. 587–599.

Espeland, M.A., and Hui, S.L. (1987), "A General Approach to Analyzing Epidemiologic Data that Contain Misclassification Errors," *Biometrics*, 43, pp. 1001–1012.

Espeland, M.A., and Odoroff, C.L. (1985), "Log-linear Models for Doubly Sampled Categorical Data Fitted by the EM Algorithm," *Journal of the American Statistical Association*, 80, pp. 663–670.

Flegal, K.M., Brownie, C., and Haas, J.D. (1986), "The Effects of Exposure Misclassification on Estimates of Relative Risk," *American Journal of Epidemiology*, 123, pp. 736–751.

Flegal, K.M., Keyl, P.M., and Nieto, F.J. (1991), "Differential Misclassification Arising from Nondifferential Errors in Exposure Measurement," *American Journal of Epidemiology*, 134, pp. 1233–1244.

Fleiss, J.L. (1981), *Statistical Methods for Rates and Proportions*, New York: Wiley.

Forsman, G., and Schreiner, I. (1991), "The Design and Analysis of Reinterview: An Overview," in P.P. Biemer, R.M. Groves, L.E. Lyberg, N.A. Mathiowetz, and S. Sudman (eds.), *Measurement Errors in Surveys*, New York: Wiley.

Freudenheim, J.L., Johnson, N.E., and Wardrop, R.L. (1989), "Nutrient Misclassification: Bias in the Odds Ratio and Loss of Power in the Mantel Test for Trend," *International Journal of Epidemiology*, 18, pp. 232–238.

Fuller, W.A. (1990), "Analysis of Repeated Surveys," *Survey Methodology*, 16, pp. 167–180.

Fung, K.Y., and Howe, G.R. (1984), "Methodological Issues in Case-Control Studies III: The Effect of Joint Misclassification of Risk Factors and Confounding Factors upon Estimation and Power," *International Journal of Epidemiology*, 13, pp. 366–370.

Giesbrecht, F.G. (1967), "Classification Errors and Measures of Association in Contingency Tables," *Proceedings of the Social Statistics Section*, American Statistical Association, pp. 271–273.

Gladen B., and Rogan, W.J. (1979). "Misclassification and the Design of Environmental Studies," *American Journal of Epidemiology*, 109, pp. 607–616.

Goldberg, J.D. (1975), "The Effects of Misclassification on the Bias in the Difference Between Two Proportions and the Relative Odds in the Fourfold Table," *Journal of the American Statistical Association*, 70, pp. 561–567.

Goodman, L.A. (1973), "The Analysis of Multidimensional Contingency Tables When Some Variables Are Posterior to Others: A Modified Path Analysis Approach," *Biometrika*, 60, pp. 179–192.

Goodman, L.A. (1978), *Analyzing Qualitative Categorical Data: Log-Linear Models and Latent-Structure Analysis*, Reading, MA: Addison-Wesley.

Green, M.S. (1983), "Use of Predictive Value to Adjust Relative Risk Estimates Biased by Misclassification of Outcome Status," *American Journal of Epidemiology*, 117, pp. 98–105.

Greenland, S. (1980), "The Effect of Misclassification in the Presence of Covariates," *American Journal of Epidemiology*, 112, pp. 564–569.

Greenland, S. (1982), "The Effect of Misclassification in Matched-Pair Case-Control Studies," *American Journal of Epidemiology*, 120, pp. 643–648.

Greenland, S. (1988a), "Variance Estimation for Epidemiologic Effect Estimates Under Misclassification," *Statistics in Medicine*, 7, pp. 745–757.

Greenland, S. (1988b), "Statistical Uncertainty Due to Misclassification: Implications for Validation Substudies," *Journal of Clinical Epidemiology*, 41, pp. 1167–1174.

Greenland, S. (1989), "On Correcting for Misclassification in Twin Studies and Other Matched-Pair Studies," *Statistics in Medicine*, 8, pp. 825–829.

Greenland, S., and Kleinbaum, D.G. (1983), "Correcting for Misclassification in Two-Way Tables and Matched-Pair Studies," *International Journal of Epidemiology*, 12, pp. 93–97.

Greenland, S., and Robins, J.M. (1985), "Confounding and Misclassification," *American Journal of Epidemiology*, 122, pp. 495–506.

Grizzle, J.E., Starmer, C.F., and Koch, G.G. (1969), "Analysis of Categorical Data by Linear Models," *Biometrics*, 25, pp. 489–504.

Gullen, W. H., Bearman, J. E., and Johnson, E.A. (1968), "Effects of Misclassification in Epidemiological Studies," *Public Health Reports*, 83, pp. 914–918.

Haitovsky, Y., and Rapp, J. (1992), "Conditional Resampling for Misclassified Multinomial Data with Applications to Sampling Inspection," *Technometrics*, 34, pp. 473–483.

Harper, D. (1964), "Misclassification in Epidemiological Surveys," *American Journal of Public Health*, 54, pp. 1882–1886.

Hill, M.S. (1992), *The Panel Study of Income Dynamics. A User's Guide*, Guides to Major Science Data Bases 2, Newbury Park, CA: Sage Publications.

Hochberg, Y. (1977), "On the Use of Double Sampling Schemes in Analyzing Categorical Data with Misclassification Errors," *Journal of the American Statistical Association*, 72, pp. 914–921.

Hochberg, Y., and Tenenbein, A. (1983), "On Triple Sampling Schemes for Estimating from Binomial Data with Misclassification Errors," *Communications in Statistics— Theory and Methods*, 12(13), pp. 1523–1533.

Hocking, R.R., and Oxspring, H.H. (1971), "Maximum Likelihood Estimation with Incomplete Multinomial Data," *Journal of the American Statistical Association*, 66, pp. 65–70.

Hocking, R.R., and Oxspring, H.H. (1974), "The Analysis of Partially Categorized Contingency Data," *Biometrics*, 30, pp. 469–483.

Hui, S.L., and Walter, S.D. (1980), "Estimating the Error Rates of Diagnostic Tests," *Biometrics*, 36, pp. 167–171.

Kaldor, J., and Clayton, D. (1985), "Latent Class Analysis in Chronic Disease Epidemiology," *Statistics in Medicine*, 4, pp. 327–335.

Keys, A., and Kihlberg, J. (1963), "Effect of Misclassification on Estimated Relative Prevalence of a Characteristic: Part I. Two Populations Infallibly Distinguished. Part II. Errors in Two Variables," *American Journal of Public Health*, 53, pp. 1656–1665.

Kleinbaum, D.G., Kupper, L.L., and Morgenstern, H. (1982), *Epidemiologic Research: Principles and Quantitative Methods*, Belmont: Lifetime Learning Publications.

Koch, G.G. (1969), "The Effect of Non-sampling Errors on Measures of Association in 2×2 Contingency Tables," *Journal of the American Statistical Association*, 64, pp. 852–863.

Korn, E.L. (1981), "Hierarchical Log-linear Models not Preserved by Classification Error," *Journal of the American Statistical Association*, 76, pp. 110–113.

Korn, E.L. (1982), "The Asymptotic Efficiency of Tests Using Misclassified Data in Contingency Tables," *Biometrics*, 38, pp. 445–450.

Landis, J.R., and Koch, G.G. (1975), "A Review of Statistical Methods in the Analysis of Data Arising from Observer Reliability Studies (Parts I and II)", *Statistical Neederlandica*, 29, pp. 101–123 and 151–161.

Lessler, J.T., and Kalsbeek, W.D. (1992), *Nonsampling Errors in Surveys*, New York: Wiley.

Liu, X., and Liang, K-Y. (1991), "Adjustment for Non-differential Misclassification Error in the Generalized Linear Model," *Statistics in Medicine*, 10, pp. 1197–1211.

McClish, D., and Quade, D. (1985), "Improving Estimates of Prevalence by Repeated Testing," *Biometrics*, 41, pp. 81–89.

MacClure, M., and Willett, W.C. (1987), "Misinterpretation and Misuse of the Kappa Statistic," *American Journal of Epidemiology*, 126, pp. 161–169.

Marshall, J.R. (1989), "The Use of Dual or Multiple Reports in Epidemiologic Studies," *Statistics in Medicine*, 8, pp. 1041–1049.

Marshall, J.R., and Graham, S. (1984), "Use of Dual Responses to Increase Validity of Case-Control Studies," *Journal of Chronic Studies*, 37, pp. 125–136.

Marshall, J.R., Priore, R., Graham, S., and Brasure, J. (1981), "On the Distortion of Risk Estimates in Multiple Exposure Level Case-Control Studies," *American Journal of Epidemiology*, 113, pp. 464–473.

Marshall, R.J. (1990), "Validation Study Methods for Estimating Exposure Proportions and Odds Ratios with Misclassified Data," *Journal of Clinical Epidemiology*, 43, pp. 941–947.

Mote, V.L., and Anderson, R.L. (1965), "An Investigation of the Effect of Misclassification on the Properties of χ^2-tests in the Analysis of Categorical Data," *Biometrika*, 52, pp. 95–109.

Nathan, G., and Eliav, T. (1988), "Comparison of Measurement Errors for Telephone Interviewing and House Visits by Misclassification Models," *Journal of Official Statistics*, 4, pp. 363–374.

Newell, D.J. (1962), "Errors in Interpretation of Errors in Epidemiology," *American Journal of Public Health*, 52, pp. 1925–1928.

Newell, D.J. (1963), "Note: Misclassification in 2×2 Tables," *Biometrics*, 19, pp. 187–188.

OPCS (1994), *First Results from the Quality Check Element of the 1991 Census Validation Survey*, Monitor SS 94/2, London: Office of Population Censuses and Surveys.

Palmgren, J. (1987), "Precision of Double Sampling Estimators for Comparing Two Probabilities," *Biometrika*, 74, pp. 687–694.

Palmgren, J., and Ekholm, A. (1987), "Exponential Family Non-Linear Models for Categorical Data with Errors of Observation," *Applied Stochastic Models and Data Analysis*, 3, pp. 111–124.

Press, S.J. (1968), "Estimating from Misclassified Data," *Journal of the American Statistical Association*, 63, pp. 123–133.

Quade, D., Lachenbruch, P.A., Whaley, F.S., McClish, D.K., and Haley, R.W. (1980), "Effects of Misclassifications on Statistical Inference in Epidemiology," *American Journal of Epidemiology*, 111, pp. 503–515.

Rao, J.N.K., and Thomas, D.R. (1991), "Chi-squared Tests with Complex Survey Data Subject to Misclassification Error," in P.P. Biemer, R.M. Groves, L.E. Lyberg, N.A. Mathiowetz, and S. Sudman (eds.), *Measurement Errors in Surveys*, New York: Wiley.

Rogan, W.J., and Gladen, B. (1978), "Estimating Prevalence from the Results of a Screening Test," *American Journal of Epidemiology*, 107, pp. 71–76.

Rogot, E. (1961), "A Note on Measurement Error and Detecting Real Differences," *Journal of the American Statistical Association*, 56, pp. 314–319.

Rubin, T., Rosenbaum, A.B., and Cobb, S. (1956), "The Use of Interview Data for the Detection of Association in Field Studies," *Journal of Chronic Diseases*, 4, pp. 253–266.

Savitz, D.A., and Barón, A.E. (1989), "Estimating and Correcting for Confounder Misclassification," *American Journal of Epidemiology*, 129, pp. 1062–1071.

Schwartz, J.E. (1985), "The Neglected Problem of Measurement Error in Categorical Data," *Sociological Methods and Research*, 13, pp. 435–466.

Selén, J. (1986), "Adjusting for Errors in Classification and Measurement in the Analysis of Partly and Purely Categorical Data," *Journal of the American Statistical Association*, 81, pp. 75–81.

Shy, C.M., Kleinbaum, D.G., and Morgenstern, H. (1978), "The Effect of Misclassification of Exposure Status in Epidemiological Studies of Air Pollution Health Effects," *Bulletin of the New York Academy of Medicine*, 54, pp. 1155–1165.

Swires-Hennessy, E., and Thomas, G.W. (1987), "The Good, the Bad and the Ugly: Multiple Stratified Sampling in the 1986 Welsh House Condition Survey," *Statistical News*, 79, pp. 24–26.

Tanner, M.A., and Young, M.A. (1985), "Modeling Agreement among Raters," *Journal of the American Statistical Association*, 80, pp. 175–180.

Tenenbein, A. (1970), "A Double Sampling Scheme for Estimating from Binomial Data with Misclassifications," *Journal of the American Statistical Association*, 65, pp. 1350–1361.

Tenenbein, A. (1971), "A Double Sampling Scheme for Estimating from Binomial Data with Misclassifications: Sample Size Determination," *Biometrics*, 27, pp. 935–944.

Tenenbein, A. (1972), "A Double Sampling Scheme for Estimating from Misclassified Multinomial Data with Applications to Sampling Inspection," *Technometrics*, 14, pp. 187–202.

Tweedie, R. L., Mengersen, K.L., and Eccleston, J.A. (1994), "Garbage in, Garbage out: Can Statistics Quantify the Effects of Poor Data?," *Chance*, 7, pp. 20–27.

Wacholder, S., Armstrong, B., and Hartge, P. (1993), "Validation Studies Using an Alloyed Gold Standard," *American Journal of Epidemiology*, 137, pp. 1251–1258.

Wacholder, S., Dosemeci, M., and Lubin, J.H. (1991), "Blind Assignment of Exposure Does Not Always Prevent Differential Misclassification," *American Journal of Epidemiology*, 134, pp. 433–437.

Walker, A.M., and Blettner, M. (1985), "Comparing Imperfect Measures of Exposure," *American Journal of Epidemiology*, 121, pp. 783–790.

Walsh, J.E. (1963), "Loss in Test Efficiency Due to Misclassification for 2 × 2 Tables," *Biometrics*, 19, pp. 158–162.

Walter, S.D. (1984), Comment to paper by Marshall and Graham (with author's reply and a further comment), *Journal of Chronic Diseases*, 37, pp. 137–139.

Walter, S.D., and Irwig, L.M. (1988), "Estimation of Test Error Rates, Disease Prevalence and Relative Risk from Misclassified Data: A Review," *Journal of Clinical Epidemiology*, 41, pp. 923–937.

Warner, S.L. (1965), "Randomized Response: A Survey Technique for Eliminating Evasive Answer Bias," *Journal of the American Statistical Association*, 60, pp. 63–69.

White, E. (1986), "The Effects of Misclassification of Disease Status in Follow-up Studies: Implication for Selecting Disease Classification Criteria," *American Journal of Epidemiology*, 124, pp. 816–825.

Whittaker, J.W. (1990), *Graphical Models in Applied Multivariate Statistics*, Chichester: Wiley.

Whittemore, A.S., and Grosser, S. (1986), "Regression Methods for Data with Incomplete Covariates," in S.H. Moolgavkar, and R.L. Prentice (eds.), *Modern Statistical Methods in Chronic Disease Epidemiology*, New York: Wiley.

Willett, W.C., Sampson, L., Stampfer, M.J., Rosner, B., Bain, C., Witschi, J., Hennekens, C.H., and Speizer, F.E. (1985), "Reproducibility and Validity of a Semiquantitative Food Frequency Questionnaire," *American Journal of Epidemiology*, 122, pp. 51–65.

Zelen, M., and Haitovsky, Y. (1991), "Testing Hypotheses with Binary Data Subject to Misclassification Errors: Analysis and Experimental Design," *Biometrika*, 78, pp. 857–865.

CHAPTER 29

Separating Change and Measurement Error in Panel Surveys With an Application to Labor Market Data

Frank van de Pol
Statistics Netherlands

Rolf Langeheine
University of Kiel

29.1 INTRODUCTION

Latent class models have been used to correct for measurement error. For panel data such models have also been used to investigate true change, controlling for the attenuating effects of measurement error. In this chapter we first use a standard method of distinguishing between measurement error and actual change. More precisely, we apply a multiple indicator latent class model to labor market data, assuming that measurement error is independent of change.

We then add a new feature to this standard model. We investigate whether people who change labor market status give less reliable answers than those who do not. In periods of transition, one's status can be ambiguous as one moves from, for instance, employment to retirement or unemployment. Our application is novel for labor market data in that we use bootstrap methods to assess model fit.

Because of some data peculiarities, we had to take into account differences

Survey Measurement and Process Quality, Edited by Lyberg, Biemer, Collins, de Leeuw, Dippo, Schwarz, Trewin.
ISBN 0-471-16559-X   © 1997 John Wiley & Sons, Inc.

in interview situation, such as face to face versus telephone interviews, and proxy versus self-response interviews. First we explain the main model and then we treat the issues in greater detail.

This chapter can be read as an introduction to the use of latent class models for distinguishing measurement error from change in categorical data. We also discuss the treatment of disturbing factors in these models and report on the effect on measurement error of labor market turnover.

Latent class models can be used in various ways. Best known is their use as a data reduction technique. Suppose we have a questionnaire with a large number of similar questions on the same subject and answers have been obtained from several hundred respondents. The multiway table of responses to these questions can be reduced to a limited number of respondent types, the latent classes. For example, one wants to assess how many youngsters are involved in juvenile delinquency, and for those who are, the type of delinquency. For this, a large number of questions can be asked, phrased like "Did you . . . in the past year" ('t Hart, 1995). Next, models with $2, 3, \ldots, A$ latent classes are fitted and a choice is made on the basis of model fit and interpretability of the outcomes. A frequency distribution of the latent classes is obtained with proportions that we denote $\delta_1, \delta_2, \ldots, \delta_A$. The analysis also produces information on the relationship between the latent variable and each question: for latent class a the probability of answer i to question 1, $\rho_{i|a}$, answer j to question 2, $\rho_{j|a}$, etc. Latent classes can be named after the questions they are most strongly related to. Under certain assumptions it is valid to estimate crosstables of latent classes with other categorical variables, like sex or religion.

For those familiar with factor analysis, which requires numerical data, it may be useful to note that latent class models, as we use them here, are built on the same assumption as factor analysis. Both models assume local independence of observed variables, given a small number of indirectly observed (latent) variables. So the associations among the observed variables should be completely explained by the associations of the latent variable(s) with observed variables (Bartholomew, 1987). Thus knowing someone's latent class to be a is enough information to give the probability of answer i on a specific indicator variable, $\rho_{i|a}$. Other questions, with answers j, k, l are not relevant for the value of this response probability, once latent class membership is given: $\rho_{i|ajkl} = \rho_{i|a}$. In the case of four questions, the proportion of the population that answers i, j, k, l is given by $P_{ijkl} = \delta_a \rho_{i|a} \rho_{j|a} \rho_{k|a} \rho_{l|a}$. Proportions and conditional probabilities sum to 1: $\sum_{ijkl} P_{ijkl} = 1$, $\sum_a \delta_a = 1$, $\sum_i \rho_{i|a} = 1, \ldots$. For a more detailed introduction, see McCutcheon (1987).

More advanced latent class models also include the relationships between several latent variables (Langeheine, 1988; Magenaars, 1993). In the present contribution there is only one latent relationship, a 3×3 turnover table (also known as gross flows table) of labor market status at two points in time. The data consist of repeated measurements of two fallible categorical indicators of the same latent variable, which is labor market status: (1) employed (em), (2) unemployed labor force (un), and (3) not in the labor force (ni). Several variables were used to assign people to a specific labor market status (Bierings *et al.*, 1991). In our application we do not aim at data reduction, but rather we want to assess the amount of measurement error involved and use this information to estimate the latent turnover table. Therefore, the latent variables should have as many latent classes, or true states, as the observed indicator variables have categories, which here are three. With this restriction, the crosstable of a latent variable and an observed indicator variable is square. When all observations are on the diagonal of such a table, measurement reliability is perfect. (For this type of application, latent classes should be ordered in such a way that the number of observations on the diagonal is maximized.)

Now let us focus on the fact that measurements were obtained at different points in time: change in labor market status may have occurred between measurements. Previous studies on this subject suggest that this change is overestimated in the presence of measurement error. We add an argument that points in the opposite direction, that is underestimation of change. We use latent class models to separate change from measurement error. Identification of latent class parameters (conditional probabilities or proportions) is possible if we have three or more measurements and change is absent (Lazarsfeld and Henry, 1968). With latent change, identification is ensured by using multiple indicators at each point in time (Langeheine and Van de Pol, 1994). With only one indicator at each time point, latent class parameters can be estimated if we make an assumption about the nature of change, for instance assume Markovian change (Wiggins, 1973). If certain time-homogeneity assumptions are made, the latent Markov model is identified for three or more measurements (Van de Pol and De Leeuw, 1986).

The data concern four variables, two indicator variables for one latent variable, observed indirectly at two points in time. The data come from a (quasi) experiment on two factors that we discuss more extensively later. The two factors are use of proxy repondents (the medium effect) and the difference between telephone and face to face interviews (the mode effect). The fact that two indicators for labor market status were available at each point in time is a unique feature of these data. The first indicator (I) was obtained in a regular interview of the Statistics Netherlands Labor Market Survey, in February–March 1990. The second indicator (II) was obtained in a short reinterview conducted about 18 days after the first interview. In the reinterview, the labor market status at the time of the regular interview was assessed again, this time by retrospective questions, referring to the indicator I time point. The experiment involved only two points in time. For the second point in time, at about May 1990, there was another regular interview, eliciting indicator I, again followed shortly by a reinterview, which captured the second measurement of indicator II. Now we have only one latent transition table and therefore we have not made any assumption (Markovian or other) about the nature of change.

Not many changes were observed, that is, most people were on the diagonal of the indirectly observed (latent) transition table. Due to measurement error some of these people are observed as contributing to gross flow: they constitute false flows. Several authors have proposed latent class models that can separate measurement unreliability from change, using the local independence assumption. Poterba and Summers (1986) and Chua and Fuller (1987) proposed employment status models for one or two regular interview waves and a reinterview which is assumed to have an unbiased margin. Abowd and Zellner (1985) developed a model for many points in time. They estimated adjusted monthly gross flows for the period 1977 to 1982. Singh and Rao (1995) give an overview of this literature and extend it with respect to heterogeneity of response probabilities. Skinner and Humphreys (1994) use instrumental variables for data at two points in time. Other related literature is Shockey (1988), Skinner and Torelli (1993) and Humphreys (1996).

The models in the literature assume that the probability of an (in)correct response depends on the current latent class only. If these models are correct, false flows are only to a small extent compensated by false stability of people who in fact did change labor market status, but who were erroneously observed to remain in the same category. Compensation is partial; false stability occurs less often as only a small proportion actually changes labor market status. Assuming change-independent measurement reliability, the net effect of measurement error correction is a sizeable downward adjustment of observed flows.

We are fortunate to have *two* indicators at each point in time. This enables us to relax the assumption of change-independent reliability. We propose a model that takes into account a possibly lower probability of a correct response for people who recently changed employment status. This means that people who are not on the diagonal of the latent transition table have a higher probability to be erroneously observed on the diagonal than with the older models mentioned earlier. For the data set we analyzed, this false stability is large enough to compensate most of the false flows, that is the downward adjustment of observed flows is almost absent.

In Section 29.2 the model is formalized. Section 29.3 describes the data. In Section 29.4, the model is refined and results are presented, both with and without change-dependent reliability. Section 29.5 concludes.

29.2 THE MODEL

The data are a crosstable of four variables, two indicator variables for one latent, indirectly observed, variable at two points in time. We assume that at a given time point people belong to one of three indirectly observed states (latent classes), (1) employed (em), (2) unemployed labor force (un) and (3) not in the labor force (ni). The proportion in state a is denoted δ_a for time point 1.

Two indicator variables contain fallible answers concerning labor market status. Table 29.1 shows the crosstable of the latent variable at time point 1 and the first indicator, I, in terms of one margin and row proportions. The left column contains the margin, the initial proportions, with, for instance, the

Table 29.1 The Relationship Between a Latent Variable and an Indicator, Written in Terms of One Margin and Conditional Probabilities

		Initial proportions	Response probabilities indicator I			
			em	un	ni	Total
Latent variable	Employed	δ_{em}	$\rho_{em\mid em}$	$\rho_{un\mid em}$	$\rho_{ni\mid em}$	1
at time 1	Unemployed	δ_{un}	$\rho_{em\mid un}$	$\rho_{un\mid un}$	$\rho_{ni\mid un}$	1
	Not in lab. force	δ_{ni}	$\rho_{em\mid ni}$	$\rho_{un\mid ni}$	$\rho_{ni\mid ni}$	1
	Total	1				

proportion in state a, δ_a. The right side of the table contains the row proportions, with the probability to answer i, given state a, $\rho_{i\mid a}$, as a typical element. The conditional probability $\rho_{i\mid a}$ represents a response probability. Latent classes are ordered such that if $i = a$ then $\rho_{i\mid a}$ is the probability of a correct response, or the reliability for short. If $i \neq a$, $\rho_{i\mid a}$ is the probability of an erroneous response. The proportion with characteristics a and i can then be written as $P_{ai} \equiv \delta_a \, \rho_{i\mid a}$. This is not restrictive; it is simply notation for the crosstable.

About 18 days later answer j was obtained in the retrospective reinterview. This is the other time point 1 indicator, denoted II. We first describe the threeway table of the latent variable at time point 1 and the time point 1 indicators I and II without restrictions. We denote the proportion with characteristics a, i, j as $P_{aij} \equiv \delta_a \rho_{i\mid a} \rho_{j\mid ai}$.

Local independence involves assuming that indicator II is independent of indicator I, given latent state a,

$$\rho_{j\mid ai} = \rho_{j\mid a}. \tag{29.1}$$

In numerical data terminology, the measurement error of indicator I is uncorrelated with the measurement error of indicator II, given the latent variable. The response probabilities in our model are an average. Some respondents have a somewhat higher probability to give a correct response (with both indicators), and some have a somewhat smaller response probability.

Although this assumption helps identify the latent labor market status proportions (δ's) and the response probabilities (ρ's), it is not enough to get an identified model. The observed crosstable of two indicators contains fewer independent proportions than the model has latent proportions (δ's) and response probabilities (ρ's). More variables must be included in the analysis to obtain an identified model.

Now we consider a second point in time. Latent labor market status may change between both points in time. The proportion of those in state a at time 1 who make a transition to state b at time 2 is denoted $\tau_{b\mid a}$. The transition probability is assumed to be the same for all people in state a, irrespective of whether correct answers, $i = a, j = a$, were given or not: $\tau_{b|aij} = \tau_{b|a}$. Then the proportion with characteristics a, i, j, b is

$$P_{aijb} = \delta_a \rho_{i|a} \rho_{j|a} \tau_{b|a}. \tag{29.2}$$

For a person in class b at time point 2 we have responses k and l on indicators I and II. We always use the same index for one specific indicator, time combination; i for indicator I, time 1, j for indicator II, time 1, k for indicator I, time 2 and l for indicator II, time 2. As for time 1, local independence is assumed,

$$\rho_{k|aijb} = \rho_{k|ab} \quad \text{and} \quad \rho_{l|aijbk} = \rho_{l|ab}. \tag{29.3}$$

Thus the complete table proportion with characteristics a, i, j, b, k, l is

$$P_{aijbkl} = \delta_a \rho_{i|a} \rho_{j|a} \tau_{b|a} \rho_{k|ab} \rho_{l|ab}. \tag{29.4a}$$

A proportion in the observable table, P_{ijkl}, is obtained by summing over the nonobservable classes a and b, $\sum_{ab} P_{aijbkl}$.

An interesting point we believe merits attention is whether reliability at time 2 can be assumed to be independent of the true state at time 1,

$$\rho_{k|ab} = \rho_{k|b} \quad \text{and} \quad \rho_{l|ab} = \rho_{l|b} \tag{29.5}$$

in short: whether reliability can be assumed to be independent of change. In fact, taking time 1 labor market status into account with respect to time 2 response probabilities is a way of dealing with correlated response errors. We return to this point of view in the last section.

Chua and Fuller (1987), Abowd and Zellner (1985), and others were forced to assume change-independent reliability, because their data had only one indicator for each point in time. Actual cases may be different, however. Employees who lose their jobs or retire from the labor force (em → un, em → ni) will generally use the holidays they are still entitled to. This causes ambiguity as to the exact time of the event. Also the exact moment when people give up looking for a job (un → ni) is hard to fix, as is the moment of entering the labor force. Does a job start with the first day at work, after a medical examination, or when one was hired (un → em, ni → em)? Do people start looking for a job when their education is finished, before that time, or after a holiday (ni → un)? Moreover social desirability may influence the answers that are given, especially when unemployment is involved (em → un, un → ni, ni → un).

In the model with change-independent reliability, an indicator's response probabilities can be arranged in a 3×3 matrix, as in Table 29.1. When reliability is allowed to be change-dependent, we have reliability matrices with nine rows, one row for each combination of latent classes at times 1 and 2 (see the Appendix). This means that change-dependence involves $(9 - 3) \times 3 = 18$ additional parameters for each indicator. Only two out of three parameters are independent, because probabilities sum to 1 over the first index: $\sum_k \rho_{k|ab} = \sum_l \rho_{l|ab} = 1$. So we have 24 additional independent parameters, 12 for indicator I and 12 for indicator II.

This amounts to 56 independent parameters (2 δ's, 6 τ's and 48 ρ's) for the change-dependent reliability model, which have to be estimated from a multiway table with 3^4 cells, holding 80 independent observed proportions. It turns out that this is too many parameters for the data set we use, which has 26 empty cells. So many parameters for so few respondents who change labor market status would result in unacceptably large confidence intervals, even if indicator I reliability is set equal to indicator II reliability, $\rho_{k|ab} = \rho_{l|ab}$ ($k = l$, $a \neq b$; not in a table). For this reason we used only four additional independent parameters. One parameter is for the probability of a correct answer (the reliability), $\rho_{k|ab}$ with $k = b$ or $\rho_{l|ab}$ with $l = b$:

$$\rho_{2|12(\text{em} \to \text{un})} = \rho_{3|13(\text{em} \to \text{ni})} = \rho_{3|23(\text{un} \to \text{ni})} = \rho_{1|31(\text{ni} \to \text{em})} = \rho_{2|32(\text{ni} \to \text{un})}$$

$$(29.6a)$$

with the exception of the transition from unemployment to employment. Another parameter represents the probability of giving an erroneous answer which would have been correct at the previous point in time, $\rho_{k|ab}$ with $k = a$ or $\rho_{l|ab}$ with $l = a$,

$$\rho_{1|12(\text{em} \to \text{un})} = \rho_{1|13(\text{em} \to \text{ni})} = \rho_{2|23(\text{un} \to \text{ni})} = \rho_{3|31(\text{ni} \to \text{em})} = \rho_{3|32(\text{ni} \to \text{un})}$$

$$(29.6b)$$

again omitting the transition from unemployment to employment. Because only three answers are possible (probabilities sum to 1 over the first index), the probability of giving an answer which is false at both points in time is also equal, $\rho_{k|ab}$ with $k \neq a$, $k \neq b$ or $\rho_{l|ab}$ with $l \neq a$, $l \neq b$,

$$\rho_{3|12(\text{em} \to \text{un})} = \rho_{2|13(\text{em} \to \text{ni})} = \rho_{1|23(\text{un} \to \text{ni})} = \rho_{2|31(\text{ni} \to \text{em})} = \rho_{1|32(\text{ni} \to \text{un})}.$$

$$(29.6c)$$

These parameters are assumed to be the same for indicators I and II. However, the reliability of answers for people who remain in the same labor market status ($a = b$: em $\to$ em, un $\to$ un, ni $\to$ ni) may differ for indicators I and II.

We decided to use two distinct parameters for the transition from unemployment to employment ($\rho_{1|21}$, $\rho_{2|21}$ and $\rho_{3|21}$; un $\to$ em), because our analyses suggest that unemployed who just got a job give more reliable answers than other changers. One explanation is that this is such an important event that both interviewer and respondent are less likely to make mistakes. Moreover, the event is so desirable that everybody may agree a job starts the moment one is hired, not at a later time like the first day at work. More research needs to be done, both to corroborate and to explain our results.

29.3 DATA AND ESTIMATION

To test our hypothesis that people with changing labor market status answer with lower reliability than others, we have data on 3,305 respondents in the

15–64 age group. Panel data were collected from February to June 1990 by Statistics Netherlands. Four interviews were obtained referring to two points in time which were three months apart. For both points in time two indicators were obtained, indicator I from an interview on the current situation and indicator II from a retrospective reinterview, which took place two to three weeks later. At each interview several questions were asked from which labor market status was established, including questions on employment, hours worked, willingness to accept a job, and job search activities (Bierings *et al.*, 1991).

The cross-tabulation of two indicators (3×3 cells) at two points in time yields a $(3 \times 3)^2 = 81$ cell table with typical proportion p_{ijkl}. Given starting values for the parameters, a first estimate of the population table with the typical cell $\hat{P}_{ijkl} = \sum_{ab} \hat{P}_{aijbkl}$ can be computed according to formula (29.4a). For the (kernel of the) log-likelihood for multinomially distributed data from a random sample of size n,

$$\ln(L) = n \sum_i \sum_j \sum_k \sum_l p_{ijkl} \ln \hat{P}_{ijkl} \qquad (29.7)$$

the EM algorithm can be used to find maximum likelihood parameter estimates (Van de Pol and De Leeuw, 1986). Equality restrictions were treated in Van de Pol and Langeheine (1990) and Mooijaart and Van der Heijden (1992). With the program PanMark (Van de Pol *et al.*, 1991), we used 100 sets of starting values to avoid local maxima. For this model, no evidence of local maxima was found.

We ignored panel members who responded in some waves only, mostly because the change-dependent reliability model is not identified when an indicator is missing. One could include these respondents into the analysis, using equality assumptions for reliability and turnover parameters (ρ's and τ's). Then estimates of indirectly observed margins (δ's) and directly observed margins would be different (Van de Pol, 1992). The conclusions about differences between turnover tables in this chapter (Table 29.3) would nevertheless be the same. For this reason we choose not to introduce a further complication. See also Stasny (1986).

While choosing a model, it is important that all parameters can be identified for the data set at hand. This was checked by inspecting the eigenvalues of the information matrix (the inverse of the matrix of second-order derivatives of the likelihood with respect to all independent parameters). When an eigenvalue is zero, the model is not identified. When an eigenvalue is close to zero, the model is not desirable either, because one or more parameters will have very large standard errors. Inspection of the correlation matrix of the parameters shows which parameters might be set equal.

Moreover, we are not interested in results from a model which does not hold for the population. We use the likelihood ratio statistic,

$$G^2 = 2n \sum_i \sum_j \sum_k \sum_l p_{ijkl} \ln(p_{ijkl}/\hat{P}_{ijkl}) \qquad (29.8)$$

as a measure of lack of fit between the model and the data. To describe whether the lack of fit for a given model is so large that it cannot be attributed to sampling inaccuracy alone, one usually uses the chi-square probability $p*$ of getting a value of G^2 larger than the G^2 obtained. To select the proper chi-square distribution one has to compute df = the number of independent proportions in the observed table—here 80—minus the number of independent parameters. Dependence between proportions is due to the fact that they sum to 1, $\sum_{ijk} p_{hijk} = 1$, $\sum_a \delta_a = 1$, $\sum_i \rho_{i|a} = 1, \ldots, \sum_b \tau_{b|a} = 1$. Also, the equality restrictions should be taken into account.

The usual decision rule is that if $p*$ is smaller than 0.05, we should reject the hypothesis that the model holds for the population. (In the hypothetical case that only one model is tested, this means we are willing to take a risk of 5 percent to reject a model which actually holds, that is, of making a type I error.) However, the theoretical chi-square distribution is only valid if all observed frequencies approach infinity, that is, if they are large, say larger than 10. When we look at the 81-cell table, which is the sum of the four "tables" in the Appendix, we see that only 18 cells contain more than 10 observations. Due to this lack of observations, the G^2 for model fit cannot be evaluated using the theoretical chi-square distribution.

An unfeasible, but illustrative, alternative to using the probability $p*$ from the theoretical chi-square distribution is to sample repeatedly from the same population. Then, in the long run a proportion p of all G^2's would be larger than our original G^2. Instead, we used a Monte Carlo bootstrap procedure (Aitkin et al., 1981). The Monte Carlo bootstrap repeatedly draws a sample from all crosstable cells (i, j, k, l) with estimated population proportion $\hat{P}_{ijkl}$ (Davis, 1993) obtained under the assumption that the model is true (equation 29.4a). For each of, say, 10,000 bootstrap samples, parameters are estimated, and a G^2 is obtained. The bootstrap G^2's form a distribution for the original G^2, under the assumption that the model is true. The proportion of bootstrap G^2's that are larger than the original G^2 is an estimate of p (Langeheine et al., 1995). Bootstrap methods can also be applied to estimate standard errors of model parameters (Skinner and Humphreys, 1994).

When parameters are estimated with the panel data, it turns out that the model fits badly. The bootstrap p is smaller than 0.001, whether we assume change-dependent reliability or not. An explanation for this may be heterogeneity. Some subpopulations have deviant parameter values. Fortunately we can divide our sample into subsamples which are different with respect to labor market status and changes therein. As mentioned in Section 29.1, the quasi-experiment was intended to test medium and mode effects. The respondent usually is the person drawn in the sample, but with indicator I a proxy was allowed to give the answers for the sample person. With indicator II the interviewer would come back if the sample person was not at home. Sample persons for whom data are obtained by proxy are more often employed and are absorbed more readily into the labor force when unemployed. Moreover it was found, that for this proxy group the answers obtained are somewhat less reliable. The interview mode was always face to face interviews, except for indicator I at the second point in time, which was a telephone interview. Results on these effects, which were not large, have been reported in Kempkens and Van den Hurk (1993) and Van de Pol (1992).

We can use the medium variable to obtain a more realistic, better fitting model by allowing different parameter estimates for self-response and proxy interviews. The experiment was designed to have a high percentage of proxy interviews (about 50 percent) with indicator I, and no proxy interviews with indicator II (reinterviews). In this way four subsamples ($h = 1, \ldots, 4$) can be distinguished, self-response at time 1 and self-response at time 2 (self–self), self–proxy, proxy–self and proxy–proxy. These data are found in the Appendix. The latent turnover table is allowed to be different for each subsample h, that is $\delta_{a|h}$ and $\tau_{b|ah}$ depend on h. Response probabilities are allowed to depend on the medium (self-response versus proxy) at the relevant point in time only: $\rho_{i|af}$ (time 1) depends only on f—with two values, self-response ($h = 1, 2$) or proxy ($h = 3, 4$)—and $\rho_{k|abg}$ (time 2) depends only on g—also with two values, self-reponse ($h = 1, 3$) or proxy ($h = 2, 4$). For indicator II a reinterviewer effect is taken into account, which is only present when indicator I was measured in a self-interview (indices f and g): $\rho_{j|af}$ and $\rho_{l|abg}$. Change-dependent response probabilities remain restricted as previously (Equations 29.6a, 29.6b, 29.6c). With these modifications, the complete data model (formula 29.4a) becomes

$$P_{haijbkl} = \gamma_h \delta_{a|h} \rho_{i|af} \rho_{j|af} \tau_{b|ah} \rho_{k|abg} \rho_{l|abg} \qquad (29.4b)$$

with γ_h the proportion in subsample h. This is a model for simultaneous analysis of a set of multidimensional contingency tables (Clogg and Goodman, 1984; Van de Pol and Langeheine, 1990).

29.4 DO CHANGERS ANSWER WITH A LOWER RELIABILITY?

Before we arrive at the question we want to answer, we have to trim our model some more. A critical reader might argue that testing several models bears the risk of capitalizing on chance. However, we will not just look at the fit of several models to the data (the saturated model), but also test the fit of a more restricted model, given a less restricted model. For the former we use bootstrap methods, for the latter we rely on the asymptotic chi-square distribution (Agresti and Yang, 1987). Moreover we note that the authors of several outstanding articles on labor market dynamics do not report on the overall fit of their models at all (Poterba and Summers, 1986; Chua and Fuller, 1987; Singh and Rao, 1995).

As already stated, the model that does not distinguish between self-response and proxy interviews does not fit, regardless of whether reliability is allowed to depend on change or not (model A). In model B we make this distinction for the indirectly observed (latent) table, but not for the response probabilities. Table 29.2 shows that model B does not fit either. The alternative model

Table 29.2   Model Fit

Model	Reliability does not depend on change			Reliability depends on change		
	# parameters	G^2	Bootstrap p	# parameters	G^2	Bootstrap p
B	59	266	0.00	63	253	0.00
C	65	248	0.01	**69**	**217**	**0.08**
D	**77**	**210**	**0.04**	81	203	0.06
E	83	203	0.02	87	196	0.03

As many bootstrap samples were drawn as necessary to obtain a <0.0015 standard error of p.
B: Reliability of self-response interviews is equal to reliability of proxy interviews.
C: As D, but reliability of indicator II is time-homogeneous. (See text for change-dependent version.)
D: As E, but reliability of indicator I is equal to reliability of indicator II if the former was a proxy interview (only for time 1).
E: Reliability of self-response interviews is not equal to reliability of proxy interviews.

described in the previous section, allowing different parameters for self-response and proxy interviews, has more parameters than necessary (model E). To obtain more accurate parameter estimates we try to make model E more parsimonious by introducing additional equality constraints.

For model D we make use of the fact that a self-response reinterview which follows a proxy interview is in fact the first interview with that panel member. Therefore a self-response interview ρ of indicator I is set equal to a ρ (for the same category, class combination) of indicator II, if this interview followed a proxy interview,

$$\rho_{i|af} = \rho_{j|a\bar{f}} \ (j = i, f = \text{self-response}, \ \bar{f} = \text{proxy; time 1}). \quad (29.9)$$

This restriction is not applied for time 2, $\rho_{k|abg} \neq \rho_{l|ab\bar{g}}$ ($l = k$; $g = $ self-response), because of the fact that at time 2 indicator I is a telephone interview, whereas indicator II was a face to face interview. (At time point 1 both indicators were obtained in a face to face interview.) The latter assumption would result in a worse fit because the reliability of the telephone interview turns out to be somewhat different from the reliability of the other interviews, which are all face to face.

Model D has an acceptable fit ($p = 0.06$) when reliability is allowed to depend on change. For this model we see only weak evidence that people who change labor market status answer with a lower reliability. When the restriction of change-independent reliability is added to the model, the fit is only slightly worse, $p = 0.04$. Table 29.2 shows that the G^2 statistic is seven higher (210–203) when this restriction is applied, whereas four fewer parameters are estimated. If asymptotics are approximately valid here, and we have reason to believe so (Agresti and Yang, 1987), this is not a significant difference: the theoretical chi-square table gives $p^* = 0.14$.

Before discussing model C, a model for which we do find a significant difference in favor of our hypothesis, we will first have a look at the point estimates in Table 29.3. The observed proportions p_{hijkl} (frequencies are found in the Appendix) are condensed to the turnover table of indicator II, $\sum_{hik} p_{hijkl}$, which is displayed in the second row of Table 29.3. Although less suitable as a benchmark, we also display the turnover table of indicator I, $\sum_{hjl} p_{hijkl}$ (the first row in Table 29.3). The indicator I turnover table displays more flows, which are probably false flows due to the lower reliability of proxy interviews and of telephone interviews, the interview mode of indicator I at time 2. In the other rows of Table 29.3 we have indirectly observed, that is, model-corrected, estimates for the same table which were obtained by summing the complete table , $\hat{P}_{haijbkl}$ (Formula 29.4b), over nonrelevant dimensions, $\sum_{hijkl} \hat{P}_{haijbkl}$.

The last row of Table 29.3 clearly shows that ignoring change-dependent reliability leads to a large downward turnover correction. Compared with indicator II, we see half as much turnover as before correction, 4.2 percent instead of 8.1 percent. The change-independent version of model D finds 1 percent more remaining employed (em $\rightarrow$ em), almost 1 percent more remaining unemployed (un $\rightarrow$ un), and 2 percent more remaining out of the labor force (ni $\rightarrow$ ni).

When we allow the reliability to be lower for people who change labor market status, corrections are small. For model D the corrected table shows 7.3 percent turnover, which is not far from the 8.1 percent in the observed indicator II table. With the more restricted model C, which will be explained below, one finds 8.6 percent estimated turnover. According to this model, indicator II does not overestimate change, as the error correction literature would suggest, but underestimates change. Indicator I, which due to lower reliability shows 9.3 percent turn-over, still slightly overestimates change according to model C (with change-dependent reliability). Our interpretation is that, due to the very high reliability of nonchangers' answers, false flows are relatively rare. They occur about as often as false stability.

We now discuss model C. It is like model D, with the additional idea that reliability might be the same at both points in time for indicator II. (This is not true for indicator I due to the interview mode, face to face or telephone, which is not the same at both points in time.) As with model D a reinterview effect is taken into account; reinterviews that follow a self-response interview have another (higher) reliability than reinterviews following a proxy interview. For the change-independent version of model C

$$\rho_{j|af} = \rho_{l|bg} \quad (j = l, a = b, f = g). \tag{29.10a}$$

In the change-dependent version we have distinct parameters for changers and nonchangers. For this reason we cannot make the same equality restrictions, but only something similar,

$$\rho_{j|af} = \rho_{l|abg} \quad (j = l, a = b, f = g) \tag{29.10b}$$

Table 29.3 Observed and Model-Based Turnover Tables

				Time 1 × Time 2					
	em→em	em→un	em→ni	un→em	un→un	un→ni	ni→em	ni→un	ni→ni	Total
Indicator I	0.631	0.008	0.027	0.018	0.041	0.015	0.011	0.015	0.235	1
Indicator II	0.640	0.008	0.015	0.017	0.045	0.008	0.014	0.019	0.234	1
Change-dep. C	**0.633**	**0.009**	**0.023**	**0.017**	**0.047**	**0.007**	**0.014**	**0.017**	**0.234**	**1**
Change-dep. D	0.636	0.008	0.022	0.015	0.049	0.006	0.009	0.013	0.242	1
Change-indep. D	**0.651**	**0.005**	**0.009**	**0.016**	**0.052**	**0.003**	**0.004**	**0.007**	**0.254**	**1**

Key: em, employed; un, unemployed labor force; ni, not in the labor force.

which states that the reliability of nonchangers' answers at time 2 is equal to the reliability of anybody's, including changers, at time 1. In fact we assume that the reliability of nonchangers' answers to be a little lower at time 2; but only a little because the vast majority consists of nonchangers (Table 29.3). We experimented with artificial data to assess the sensitivity of the model to these assumptions (Van de Pol, 1992). It turned out that assuming the reliability of nonchangers' answers at time 2 to be a little lower than it is, does not favor our hypothesis.

Conclusions based on model C are clearly in favor of our hypothesis that people who change labor market status give less reliable answers. This model fits well if we allow for our hypothesis ($p = 0.08$), and fits a lot worse if we do not ($p = 0.01$). The difference in fit between models is clearly significant ($G^2 = 248 - 217 = 31$; $p^* \ll 0.001$ according to the theoretical chi-square table).

Finally the present data set enables us to test an assumption which has been made by other authors (work on data without multiple indicators) namely, symmetry in the matrices of response probabilities,

$$\rho_{i|a} = \rho_{i'|a'} \quad (i' = a, a' = i), \ldots, \rho_{l|b} = \rho_{l'b'} \quad (l' = b, b' = l). \quad (29.11)$$

If the margins of both indicators are equal at time t, this assumption is equal to the one made by Chua and Fuller (1987). Model D with change-independent reliability has 56 parameters with this assumption, 21 less than without it. This model does not hold here, $G^2 = 287$ with bootstrap $p = 0.00$. This illustrates that all models, also the ones presented previously, represent reality in a simplified manner. Models that take more variables into account will generally outperform simpler models.

29.5   DISCUSSION

This chapter shows a way to test the hypothesis of change-dependent measurement reliability for categorical data. Multiple indicators are helpful to obtain an identified model. In that case one does not need measurements on more than two points in time, and hence assumptions on the process of change need not be made. If only one indicator is available, measurements are needed for more than two occasions and restrictive (Markovian) assumptions are needed about the nature of the indirectly observed process of change.

These models can also be developed further. The present model can be viewed as a way to take autocorrelated measurement error into account, as far as it is caused by the difference between changers and nonchangers. Singh and Rao (1995) tackled the problem of change-dependent response probabilities in a different way. They assumed that change-dependent response probabilities arise as a mixture of perfect reliability and independent classification errors. We doubt, however, that this covers all types of autocorrelations.

Humphreys (1996) takes a very general approach by allowing response probabilities to be heterogeneous because of a random component, which may be correlated in time. Moreover, autocorrelated measurement error is also modeled by regressing these random components on respondent characteristics, such as education, region, and year of birth. This approach requires distributional assumptions with respect to the random terms, though.

In this chapter we arrived at model D, which suggests, however weakly, that transitional states in labor market status give rise to less reliable responses. We see that even insignificant deviations in response reliability almost eliminate the effect of independent error correction as modeled by Abowd and Zellner (1985), Poterba and Summers (1986), Chua and Fuller (1987) and Skinner and Humphreys (1994). The weak indication turns into strong evidence if we are willing to assume that the second reinterview (indicator II at time 2) was almost as reliable as the first reinterview.

What our results showed with certainty is that taking this type of autocorrelated measurement error into account will produce less dramatic corrections to the turnover table. Also the results of Humphreys (1996) show that a model with autocorrelated measurement error increases labor market stability much less, or not at all, compared to an independent error correction model. Moreover, his autocorrelated error model fits much better than an independent error model. Singh and Rao (1995), on the other hand, conclude that the independent error classification model is, in most cases, fairly robust, but their results cover autocorrelated measurement error in a different way.

All in all, our results shed doubt on the use of change-independent error correction models. Such models may boost labor market stability to an unrealistically high level.

ACKNOWLEDGMENTS

The views expressed in this chapter are those of the authors and do not necessarily reflect the policies of Statistics Netherlands.

The authors thank P. Biemer, A. Boomsma, C. Dippo, A. Israëls, L. Kempkens, J. Kuha, L. Lyberg, C. Skinner, D. Trewin, G. J. van den Hurk and P. van der Heijden for comments and help.

REFERENCES

Abowd, J.M., and Zellner, A. (1985), "Estimating Gross Labor Force Flows," *Journal of Business and Economic Statistics*, 3, pp. 254–283.

Agresti, A., and Yang, M.C. (1987), "An Empirical Investigation of Some Effects of Sparseness in Contingency Tables," *Computational Statistics and Data Analysis*, 5, pp. 9–21.

Aitkin, M., Anderson, D., and Hinde, J. (1981), "Statistical Modelling of Data on Teaching Styles," *Journal of the Royal Statistical Society*, A, 144, pp. 419–461.

Bartholomew, D.J. (1987), *Latent Variable Models and Factor Analysis*, London: Griffin.

Bierings, H.B.A., Imbens, J.C.M., and van Bochove, C.A. (1991), "The Labour Force Definition," in C.A. van Bochove, W.J. Keller, H.K. van Tuinen, and W.F.M. de Vries (eds.), *CBS-select 7, Statistical Integration*, The Hague: Staatsuitgeverij, pp. 41–62.

Chua, T.C., and Fuller, W.A. (1987), "A Model for Multinomial Response Error Applied to Labor Flows," *Journal of the American Statistical Association*, 82, pp. 46–51.

Clogg, C.C., and Goodman, L.A. (1984), "Latent Structure Analysis of a Set of Multidimensional Contingency Tables," *Journal of the American Statistical Association*, 79, pp. 762–771.

Davis, C.S. (1993), "The Computer Generation of Multinomial Random Variates," *Computational Statistics & Data Analysis*, 16, pp. 205–217.

Hagenaars, J.A.P. (1993), *Loglinear Models with Latent Variables*, Newbury Park, CA: Sage.

't Hart, H. (1995), "Criminaliteit in latente klassen; een voorlopige analyse van de structuur van het crimineel gedrag van jongeren," Report 5 to the Ministry of Justice, Utrecht: Utrecht University (in Dutch).

Humphreys, K. (1996), "The Latent Markov Chain with Multivariate Random Effects: An Evaluation of Instruments Measuring Labour Market Status in the British Household Panel Study," Southampton University.

Kempkens, L., and Van den Hurk, D.J. (1993), "Gegevens over de beroepsbevolking verkregen uit panel-onderzoek; een vooronderzoek naar proxi- en medium-effecten," *Supplement bij de Sociaal-Economische Maandstatistiek*, 6, pp. 4–14 (in Dutch).

Langeheine, R. (1988), "New Developments in Latent Class Theory," in R. Langeheine, and J. Rost (eds.), *Latent Trait and Latent Class Models*, New York: Plenum Press, pp. 77–108.

Langeheine, R., Pannekoek, J., and Van de Pol, F. (1995), "Bootstrapping Goodness-of-Fit Measures in Categorical Data Analysis," to appear in *Sociological Methods & Research*.

Langeheine, R., and Van de Pol, F. (1994), "Discrete Time Mixed Markov Latent Class Models," in A. Dale, and R. Davies (eds.), *Analyzing Social and Political Change: A Casebook of Methods*, New York: Sage, pp. 170–197.

Lazarsfeld, P.F., and Henry, N.W. (1986), *Latent Structure Analysis*, Boston: Houghton-Mifflin.

Magenaars, J.A. (1993), *Loglinear Models with Latent Variables*, Sage University Papers 94, Newbury Park, CA: Sage.

McCutcheon, A.L. (1987), *Latent Class Analysis*, Sage University Paper Series 64, Newbury Park CA: Sage.

Mooijaart, A., and Van der Heijden, P.G.M., (1992), "The EM Algorithm for Latent Class Analysis with Equality Constraints," *Psychometrika*, 57, pp. 261–269.

Poterba, J.M., and Summers, L.H. (1986), "Reporting Errors and Labour Market Dynamics," *Econometrica*, 54, pp. 1319–1338.

Read, T.R.C., and Cressie, N.A.C. (1988), *Goodness-of-Fit Statistics for Discrete Multivariate Data*, New York: Springer.

Shockey, J.W. (1988), "Adjusting for Response Error in Panel Surveys: A Latent Class Approach," *Sociological Methods & Research*, 17, pp. 65–92.

Singh, A.C., and Rao, J.N.K. (1995), "On the Adjustment of Gross Flow Estimates for Classification Error with Application to Data from the Canadian Labour Force Survey," *Journal of the American Statistical Association*, 90, pp. 478–488.

Skinner, C.J., and Torelli, N. (1993), "Measurement Error and the Estimation of Gross Flows from Longitudinal Economic Data," *Statistica*, 53, pp. 391–405.

Skinner, C.J., and Humphreys, K. (1994), "Instrumental Variable Estimation of Gross Flows in the Presence of Measurement Error," University of Southampton, paper submitted for publication.

Stasny, E.A. (1986), "Estimating Gross Flows Using Panel Data with Nonresponse: An Example from the Canadian Labour Force Survey," *Journal of the American Statistical Association*, 81, pp. 42–47.

Van de Pol, F.J.R. (1992), "Proxy- en telefooneffecten in het EBB-panelexperiment," Voorburg: Statistics Netherlands.

Van de Pol, F.J.R., and De Leeuw, J. (1986), "A Latent Markov Model to Correct for Measurement Error," *Sociological Methods & Research*, 15, pp. 118–141.

Van de Pol, F.J.R., and Langeheine, R. (1990), "Mixed Markov Latent Class Models," in C.C. Clogg (ed.), *Sociological Methodology 1990*, Oxford: Blackwell, pp. 213–247.

Van de Pol, F.J.R., Langeheine, R., and De Jong, W. (1991), "PANMARK User Manual. PANel Analysis Using MARKov Chains, version 2.2," Voorburg: Statistics Netherlands

Wiggins, L.M. (1973), *Panel Analysis: Latent Probability Models for Attitude and Behavior Processes*, New York: Elsevier.

APPENDIX (*overleaf*)

APPENDIX

Table A.29.1   Observed Frequencies in Four Subsamples
(Indicator I Medium at Time 1 × Indicator I Medium at Time 2)
Self–Self (Upper Left),
Self–Proxy (Upper Right), Proxy–Self (Bottom Left) and
Proxy–Proxy (Bottom Right)

Indicator II, time 1 × time 2 (reinterviews)

Indicator I, time 1 ↑ × time 2	em ↑ em	un ↑ em	ni ↑ em	em ↑ un	un ↑ un	nu ↑ un	em ↑ ni	un ↑ ni	ni ↑ ni	em ↑ em	un ↑ em	ni ↑ em	em ↑ un	un ↑ un	ni ↑ un	em ↑ ni	un ↑ ni	ni ↑ ni
em → em	508	2	7	1	1	–	5	–	1	282	2	3	–	–	–	1	–	1
em → un	5	7	–	–	–	1	–	–	–	2	2	–	1	–	–	–	–	–
em → ni	15	3	13	–	–	–	–	–	6	8	–	5	–	–	–	–	–	1
un → em	1	–	–	17	1	–	1	–	–	–	–	–	12	1	–	1	–	–
un → un	–	–	–	1	71	4	–	2	–	–	–	–	1	24	4	–	1	–
un → ni	–	–	–	–	4	5	–	–	6	–	–	–	–	3	–	2	2	7
ni → em	–	–	–	3	–	–	4	–	2	–	–	–	–	–	–	2	1	5
ni → un	–	–	–	–	4	–	–	12	6	–	–	–	–	1	–	–	2	5
ni → ni	1	–	1	–	1	–	3	7	336	–	–	–	–	–	3	1	7	114
em → em	320	1	1	–	–	–	1	–	–	930	3	8	3	–	–	4	–	1
em → un	–	1	–	–	–	–	–	–	–	4	1	–	–	1	–	–	–	1
em → ni	1	1	3	–	–	–	–	–	1	17	2	6	–	–	–	1	–	3
un → em	1	–	–	2	–	–	1	–	–	3	1	–	12	1	–	1	1	1
un → un	–	–	–	–	6	–	–	–	–	–	1	–	1	17	1	–	2	1
un → ni	–	–	–	–	2	2	–	–	2	1	–	–	1	3	2	–	1	7
ni → em	2	–	–	–	–	–	–	–	1	8	–	–	–	–	–	6	1	2
ni → un	–	–	–	–	1	1	–	2	–	–	–	–	–	3	–	–	4	9
ni → ni	1	–	1	1	1	2	1	2	56	4	–	–	1	3	2	12	15	199

Key: em, employed; un, unemployed labor force; ni, not in the labor force.

Estimating Usual Dietary Intake Distributions: Adjusting for Measurement Error and Nonnormality in 24-Hour Food Intake Data

Sarah M. Nusser, Wayne A. Fuller
Iowa State University

Patricia M. Guenther
U.S. Department of Agriculture

30.1 INTRODUCTION

Food consumption data are regularly collected by the U.S. Department of Agriculture (USDA) for the purposes of evaluating dietary status and formulating policy related to dietary consumption. When chronic phenomena such as inadequate dietary intake or long-term exposure to food contaminants are considered, researchers focus on the concept of an individual's *usual* intake. The usual intake is defined as the long-run average daily intake of a dietary component for the individual. The emphasis on usual intake focuses on long-term patterns of consumption rather than consumption levels on any given day.

Usual intake distributions for dietary components can be used to produce estimates of risk at the population level. For example, the proportion of a population whose usual intakes are less or greater than a specified value (dietary

Survey Measurement and Process Quality, Edited by Lyberg, Biemer, Collins, de Leeuw, Dippo, Schwarz, Trewin.
ISBN 0-471-16559-X © 1997 John Wiley & Sons, Inc.

inadequacy or excess) can be determined from a usual intake distribution. Alternatively, given a known concentration of a contaminant in a food and a toxic intake level, the proportion of individuals whose long-run consumption is above the toxic threshold can be estimated.

The usual intake of a food cannot be directly observed. Direct observation would require respondents to complete dietary intake questionnaires over a long period of time (e.g., a year) without altering their consumption patterns. The average of a respondent's daily intakes would then be the usual intake. A more realistic method of obtaining information on usual intakes involves asking each respondent to report daily food intakes on a few randomly selected days. This reduces respondent burden and increases the chances of obtaining accurate food intake data. However, because daily food intakes measure the usual intake with error and there is large day to day variability in consumption, a measurement error framework is needed to estimate usual intake distributions for foods and other dietary components.

Daily intake variation consists of two components. The first component is measurement or response error, the failure of the respondent to correctly report the amounts of food actually consumed. The second source of variability is the individual's day to day variation in food consumption. The sum of these two components produces a large within-person variance component that tends to be heterogeneous across subjects (Hegsted, 1972, 1982; Beaton et al., 1979; National Research Council (NRC), 1986; Nusser et al., 1996). In addition, the data often display systematic variation associated with day of week and day of interview. It is well known that food intake is higher on the weekend than during the week. The day of interview identifies the position in a sequence of interviews. Thus, the second day of interview is the second in a sequence of interviews. For studies in which 24-hour food intakes are recorded, intakes reported on the first interview day are believed to be more accurate than data collected on subsequent days.

When the observed data are approximately normally distributed, fixed and random effects can be easily estimated using simple measurement error models. However, researchers have shown that intake distributions for most dietary components are right skewed (Sempos et al., 1985; NRC, 1986; Emrich et al., 1989; Aickin and Ritenbaugh, 1991; Carriquiry et al., 1993). For foods that are not consumed daily, the distribution typically has a spike at zero corresponding to nonconsumers of the food, and a unimodal or J-shaped distribution of usual intakes for the consumers in the population (Nusser, 1995). Thus, methodology based on a measurement error model must account for the different distributional shapes inherent in daily and usual intakes.

A report developed by the National Research Council (1986) represents one of the first attempts to develop a method of estimating usual intake distributions that recognizes the presence of within-person variance and nonnormality in daily intake data. They proposed log transforming the data, shrinking the log mean intakes so that the shrunken means have variance equal to the estimated among-individuals component, and then transforming back the shrunken means. The estimated distribution of usual intakes is the estimated distribution of the back transformed shrunken means.

In cooperation with USDA, researchers at Iowa State University (ISU) have extended these ideas for estimating usual intake distributions for dietary components that are consumed on a nearly daily basis (Nusser et al., 1996). Daily intake data are adjusted for nuisance effects, and the intake data on each sample day are adjusted to have a mean and variance equal to that of the first sample day because the first day is believed to be the most accurate. The ISU approach then uses a semiparametric procedure to transform the adjusted daily intake data to normality. The transformed observed intake data are assumed to follow a measurement error model, and normal distribution methods are used to estimate the parameters of the model. Finally, a transformation that carries the normal usual intake distribution back to the original scale is estimated. The back transformation of the fitted normal distribution adjusts for the bias associated with applying the inverse of the nonlinear forward transformation to a mean distribution. The back transformation is used to define the distribution of usual intakes in the original scale. The approach was developed to produce an algorithm suitable for computer implementation and applicable to a large number of dietary components.

Neither the NRC nor the ISU method is well-suited for estimating distributions of usual food intakes because many food items are consumed on only a fraction of the sample days for a portion of the population. Daily intake data for infrequently consumed foods contain a substantial number of zero intakes that arise from individuals who never consume the food, or from persons who are consumers, but did not consume the food on sample days. Thus, the measurement error approach must be augmented to account for the mixture of the consumer and nonconsumer distributions that arise with food intake data. Estimation is complicated by the fact that subpopulation membership (consumer or nonconsumer) may not be identifiable for each subject.

In this chapter, we describe a procedure for estimating usual intake distributions for foods and other dietary components that are not consumed on a daily basis. The procedure is an extension of the measurement error approach of Nusser et al. (1996) to settings in which the data arise from a mixture of a single-valued nonconsumer distribution and a continuous, but not necessarily normal, consumer distribution. We concentrate here on the case where an individual's usual intake is unrelated to the individual's probability of consumption. The usual intake for individual i is modeled as the individual's usual intake on days that the food is consumed multiplied by the individual's probability of consuming the food on any given day. The ISU method for estimating usual nutrient intake distributions is applied to the positive intakes to estimate a consumption day usual intake distribution for the population. The distribution of the probability of consumption is estimated, and used to construct the joint distribution of consumption day usual intakes and consumption probabilities. This joint distribution is used to derive the usual intake distribution for all days.

We begin by describing characteristics of food intake data. The proposed methodology is presented and illustrated with data from the USDA's 1985 Continuing Survey of Food Intakes by Individuals (CSFII).

30.2 CHARACTERISTICS OF FOOD INTAKE DATA

Nusser *et al.* (1996) provide a description of characteristics of daily intake data for dietary components that are consumed daily (or very nearly daily) from the USDA's 1985 CSFII. They found that distributions of daily intakes were generally skewed to the right, that the within-person variability was sizable relative to the among-individual variability, that the within-person variances were related to the mean, and that day of week and sequence (day of interview) effects were significant.

To investigate the patterns of food intake data, daily intakes were examined for several foods from the 1985 CSFII four-day data set (USDA, 1987). These data were used because they contain four daily intake observations on each individual, and thus contain considerable information regarding the underlying patterns of food consumption for individuals. Food groups were selected that provide a wide range of consumption patterns: dark green vegetables, apples, alcoholic beverages, diet soda, eggs, beef, fruit, and milk products. We used the same set of respondents used by Nusser *et al.* (1996) consisting of 743 women aged 25–50 who were meal planners/preparers and were not pregnant or lactating. For each food, two data sets were considered: (1) a data set containing intakes for all interview days, and (2) the set of positive intakes from the days on which the respondents consumed the food. These data sets were examined to consider patterns relevant to the usual intake distribution for all days and the usual intake distribution for consumption days only.

Table 30.1 provides information on the frequency of consumption for these food groups. The columns of the table contain the percentage of women who consumed the food group on k out of the four days of recorded intake, where $k = 0, 1, 2, 3$, or 4. Note that for the more specific and less commonly consumed food groups (dark green vegetables, apples, alcoholic beverages, and diet soda), a substantial proportion of respondents did not consume the food group on any of the sample days. When broader or more commonly consumed classes of foods are considered (e.g., beef, eggs, fruit, and milk products), the percentage of respondents consuming the food group on at least one day is much larger. Except for fruit and milk products, the percentage of women consuming the food group for a consumption frequency class is inversely related to the frequency of consumption. For fruit, however, the percentages across consumption frequency classes are quite similar. For milk products, the percentage of individuals in the consumption frequency class increases with the frequency of consumption. In the procedure we propose, at least some women

Table 30.1 Percentage of Women Who Consumed from Each Food Group on 0, 1, 2, 3, or 4 of the Sample Days

Food group	Number of sample days on which food group was consumed				
	0	1	2	3	4
Dark green vegetables	68[a]	25	6	1	0
Apples	69	21	7	2	1
Alcoholic beverages	70	15	7	3	4
Diet soda	60	15	10	8	7
Eggs	40	32	19	7	2
Beef	36	38	20	6	0
Fruit	19	22	22	18	19
Milk products	4	9	16	27	44

[a] Percentage of women consuming the specified food group on the indicated number of sample days.

Source: USDA 1985 Continuing Survey of Food Intakes by Individuals (unweighted).

must have more than one day of intake data in order to estimate variance components. For these data, the number of women with two or more positive intakes varies from 48 for dark green vegetables to 646 for milk products.

The distribution of individual mean daily intakes is expected to reflect some of the characteristics that appear in the usual intake distribution for all days. Relative frequency histograms for individual daily intake means are generally J-shaped, although the histogram for apples appears to be bimodal (Figure 30.1). The distributions for more frequently consumed foods such as eggs, beef, fruit, and milk products are less skewed than for other foods.

When all days of zero intake are removed, the distribution of individual mean daily intakes for consumption days provides information on the shape of the consumption day usual intake distribution. These distributions are generally unimodal, and exhibit a high degree of skewness (Figure 30.2). The distributions for a few of the food groups (dark green vegetables, alcoholic beverages, and milk products) are more J-shaped.

As expected, tests based on the positive intake data indicated that the null hypothesis of normality is rejected for all of the food groups (Schaller, 1993). In addition, we were unable to find power transformations that produced normal distributions for any of the food groups, indicating that the semiparametric transformation proposed by Nusser *et al.* (1996) is needed to transform consumption data to normality.

Within-individual standard deviations calculated from positive intake data

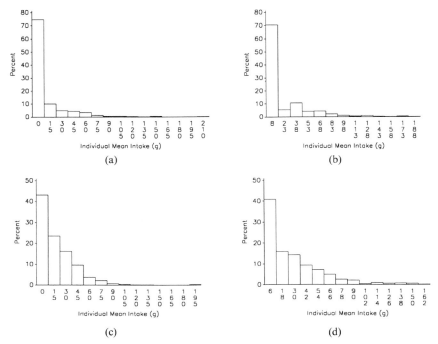

Figure 30.1   Relative frequency histograms for individual daily intake means calculated using all four days of intake per respondent for (a) dark green vegetables, (b) apples, (c) eggs, and (d) beef.

plotted against individual means on consumption days indicate that for many food groups, there is a positive relationship between within-person variances and individual means (Schaller, 1993). Some of the relationships are not as strong as those observed in Nusser *et al.* (1996) for nutrient intakes, but a pattern of heterogeneous variances is still evident.

Considerable within-person variability is present in the positive food intake data. Ratios of within-person to among-person variation range from 1.4 to 8.2, with several foods exhibiting ratios in the 3–6 range.

Table 30.2 presents information on the observed intake distributions for each of the four sample days. For some food groups, the mean positive intake changes substantially across interview days. Mean positive intakes for dark green vegetables and alcoholic beverages are approximately 20 percent higher on the first interview day than on subsequent days. Intake of beef on the first day is also higher than the mean for the remaining three days. For the other food groups, mean positive intakes are roughly constant across interview days. The standard deviations showed the same patterns across interview days as the means (data not shown). The number of individuals reporting consumption is higher on the first sample day for alcoholic beverages and diet sodas, but is relatively constant across interview days for the other food groups.

Exploratory analyses indicate that for some food groups, intakes on

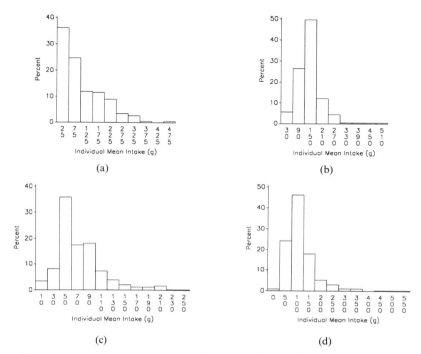

Figure 30.2   Relative frequency histograms for individual daily intake means calculated using only the positive intake values for each respondent for (a) dark green vegetables, (b) apples, (c) eggs, and (d) beef.

consumption days are correlated with the probability of consuming the food. Table 30.3 presents the mean and standard deviation for intakes on consumption days for women who consumed the food on 1, 2, 3, or 4 out of the 4 sample days. No significant correlation was detected between an individual's mean consumption day intake and the number of sample days on which the food was consumed by the individual for dark green vegetables, apples, beef, and eggs. However, statistically significant ($p < 0.01$) positive correlations exist for alcoholic beverages ($r = 0.19$), diet soda ($r = 0.43$), fruit ($r = 0.25$), and milk ($r = 0.31$). The standard deviation also appears to increase with consumption frequency for diet soda. Results from a test of homogeneity of means across consumption classes confirmed these observations.

30.3   ESTIMATING USUAL INTAKE DISTRIBUTIONS FOR INFREQUENTLY CONSUMED DIETARY COMPONENTS

30.3.1   Model

For most foods or food groups, it is assumed that a portion of the population never consumes the food. Thus, the usual intake distribution for infrequently

Table 30.2   Mean Positive Intake (and the Number of Respondents with Positive Intakes) for Each Sample Day and for Each of Eight Foods

Food group	Sample day			
	1	2	3	4
Dark green vegetables	122[a]	84	103	107
	(64)[b]	(63)	(76)	(84)
Apples	130	135	137	153
	(83)	(60)	(98)	(92)
Alcoholic beverages	504	425	392	406
	(121)	(108)	(93)	(90)
Diet soda	534	534	537	524
	(183)	(172)	(143)	(158)
Eggs	73	74	73	68
	(178)	(189)	(176)	(196)
Beef	118	112	109	103
	(177)	(191)	(162)	(180)
Fruit	227	249	240	238
	(350)	(365)	(376)	(359)
Milk products	245	234	231	246
	(563)	(537)	(562)	(550)

[a] Mean positive intake (g) on the sample day.
[b] Number of respondents (out of 743) with positive intakes on the sample day.
Source: USDA 1985 Continuing Survey of Food Intakes by Individuals (unweighted).

consumed dietary components generally consists of a spike at zero corresponding to the nonconsumers of the component, and a continuous distribution of positive usual intakes for consumers. The usual food intake on all days for an individual can be modeled as the individual's usual intake on consumption days (usual intake conditional on positive intake) multiplied by the individual's probability of consuming the food on any day. The proposed approach is to set aside the zero intakes and transform the positive intakes to normality using a modification of the measurement error approach described in Nusser et al. (1996). The conditional distribution of usual intake for consumption days is estimated in the normal scale using a measurement error model, and transformed back to the original scale. Next, a distribution of individual consumption probabilities (i.e., the probability that an individual consumes the dietary component on any given day) is estimated. The unconditional usual intake distribution for all days is then estimated from the joint distribution of conditional usual intakes and individual consumption probabilities.

In what follows, let Y_{ij} be the observed intake for individual i on day j, y_i

Table 30.3   Mean (and Standard Deviation) of Individual Mean Intakes Calculated from Positive Intakes Only, for Each Consumption Frequency, and for Each of Eight Foods

Food group	Number of sample days on which food was consumed by respondent			
	1	2	3	4
Dark green vegetables	102[a] (89)[b]	106 (79)	122 (120)	—
Apples	141 (71)	133 (46)	148 (45)	137 (51)
Alcoholic beverages	303 (359)	441 (525)	587 (537)	465 (323)
Diet soda	358 (177)	442 (200)	512 (282)	705 (383)
Eggs	70 (43)	72 (35)	70 (28)	81 (48)
Beef	111 (72)	112 (54)	106 (39)	—
Fruit	181 (130)	220 (125)	230 (108)	272 (118)
Milk products	126 (144)	165 (125)	193 (129)	279 (199)

[a] Mean of individual mean intakes (g) calculated using positive intake values only.
[b] Standard deviation of individual mean intakes (g) calculated using positive intakes.
Source: USDA 1985 Continuing Survey of Food Intakes by Individuals (unweighted).

be the usual intake of individual i, p_i be the probability that individual i consumes the dietary component on any given day, Y_{ij}^* be the observed intake when intake is positive, and y_i^* be usual intake when intake is positive. Note that $y_i^* = E\{Y_{ij} | i \text{ and } Y_{ij} > 0\}$ is the expected value of the positive intakes for individual i. Our model is

$$y_i = y_i^* p_i \tag{30.1}$$

$$p_i \sim g(p; \boldsymbol{\theta}) \tag{30.2}$$

$$X_{ij}^* = T(Y_{ij}^*; \mathbf{a}_{ij}, \boldsymbol{\alpha}) \tag{30.3}$$

$$X_{ij}^* = x_i^* + u_{ij} \tag{30.4}$$

$$x_i^* \sim NI(\mu_{x*}, \sigma_{x*}^2) \tag{30.5}$$

$$u_{ij} \sim NI(0, \sigma_u^2) \tag{30.6}$$

$$y_i^* = \eta(x_i^*, \boldsymbol{\beta}) \tag{30.7}$$

where T is a transformation of Y_{ij}^* that is a function of characteristics of the observations, such as day of week and day of interview, denoted by $\mathbf{a}_{ij}$, and of a parameter vector $\boldsymbol{\alpha}$; p_i has a distribution g that depends on the parameter vector $\boldsymbol{\theta}$; and η is the back transformation depending on parameter vector $\boldsymbol{\beta}$ that carries normal conditional usual intakes to the original scale. The distribution of usual intakes is determined by the joint distribution of p_i and y_i^*. We assume that the u_{ij} are independent given i, and that p_i is independent of y_i^*.

Because of the complexity of the model, components of the model are estimated separately. First, the transformation T that carries the original Y_{ij}^* into X_{ij}^* is estimated. The steps that define T are presented in Sections 30.3.2.1, 30.3.2.2 and 30.3.2.3, and are summarized in Table 30.4. Using the X_{ij}^*, the parameters $(\mu_{x*}, \sigma_{x*}^2, \sigma_u^2)$ and the transformation η are estimated (Section 30.3.2.4). The parameters of the distribution of p_i are estimated using information on the number of days individuals consume the food (Section 30.3.3). The distribution of y^* is combined with the distribution of p to obtain the distribution of usual intakes, y (Section 30.3.4). Measurement error is important at two points in the estimation. A part of the transformation T adjusts for systematic measurement error by transforming reported consumption on the second, third, and fourth days to the level observed on the first day. The variable u_{ij} represents the day to day variability due to variation in consumption and to errors in reporting.

In the work of Nusser et al. (1996) the error variance σ_u^2 is permitted to vary from individual to individual. In our analysis of food intakes, we adopt the simpler model of common variance in the normal scale. Extensions to the more complicated model in Nusser et al. (1996) are straightforward. Empirical studies indicate adjustments that account for the heteroscedasticity in the normal scale have little influence on the results.

30.3.2 Estimating the Consumption Day Usual Intake Distribution

30.3.2.1 Data Adjustments

The positive food intake data are first adjusted for nuisance effects. The nuisance variables will vary with the study. For the 1985 CSFII, adjustments are made for day of week and day of interview effects. The distribution of daily intakes on the first sample day is taken to be the reference standard because first day data are considered to be the most accurate information available in the sample. The methods described below can be modified to incorporate other reference standards.

The adjustment for nuisance effects is similar to a standard regression approach. The data are regressed on variables representing nuisance effects

Table 30.4 Summary of Adjustments and Transformations Applied to Positive Intakes That Define the Transformation T in (30.1), and the Purpose of Each Step

Section	Adjustment/Transformation	Purpose
30.3.2.1	Regress positive intakes (Y_{ij}^*) on nuisance variables, generate predicted intakes without nuisance effects ($\hat{Y}_{ij}^*$).	Removes effects of nuisance variables such as day of week from data.
	Ratio-adjust positive intakes using predicted values to day 1 mean, (Y_{aij}^*).	Accounts for under reporting observed after first sample day.
	Power transform data (V_{ij}^*), scale transformed data to day 1 variance ($\tilde{V}_{ij}^*$), back transform to original scale ($\tilde{Y}_{ij}^*$).	Adjusts for differences in variance observed across sample days by creating data with homogeneous day 1 variances across sample days. Power transformed scale improves estimation of adjustment factors for variances.
30.3.2.2	Obtain equal weight sample ($\ddot{Y}_{ij}^*$).	Incorporates sampling weights. Simplifies models used to estimate usual intake distribution by removing need to incorporate sampling weights.
30.3.2.3	Power transformation of equal weight data, and cubic spline fit of normal scores to power-transformed data, X_{ij}^*.	Transforms data to normality so that normal measurement error methods can be used. Spline provides flexibility for dietary components which cannot be transformed to normality by a simple power function. Two-stage process reduces number of parameters required to define transformation.

using a linear model. For example, dummy variables for day of week are included in the regression. The estimated model is used to adjust the data to the first day mean (rather than the grand mean) using a procedure that is the multiplicative analog of the standard linear adjustment to increase the chances that adjusted intakes are nonnegative. Let $\{Y_{ij}^*: i = 1, 2, \ldots, n^*$ individuals, $j = 1, 2, \ldots, r_i$ days$\}$ be the set of unadjusted positive observed intakes for a dietary component, where n^* is the number of individuals with at least one positive intake and r_i is the number of positive intakes for individual i. Let $\bar{Y}_{\cdot 1}^*$ be the (weighted) mean of the day one positive intakes, and $\hat{Y}_{ij}$ be the predicted values from the (weighted) multiple regression of Y_{ij}^* on the nuisance effect variables. The data adjusted for nuisance effects are

$$Y_{aij}^* = \hat{Y}_{ij}^{*-1} \bar{Y}_{\cdot 1}^* Y_{ij}^*. \tag{30.8}$$

A second transformation is applied to these data to produce approximately homogeneous distributions across days in the normal scale. The positive data are transformed to approximate normality using a power transformation. A grid search is used to determine the power transformation that brings the data closest to normality. A segmented linear transformation is used to center and scale the data on day j ($j = 2, 3, \ldots, r$) such that the data on day j have the mean and variance of day one. Let γ be the power that best transforms the positive data Y^*_{aij} to normality, and let

$$V^*_{ij} = (Y^*_{aij})^{\gamma}.$$

The sample mean and variance of the transformed positive intakes on day j are denoted by $\hat{\mu}_j$ and $\hat{\sigma}^2_j$, respectively. The data for day j are adjusted to the day one mean and variance as follows:

$$\tilde{V}^*_{ij} = \begin{cases} \hat{\mu}_1 + \hat{\sigma}^{-1}_j \hat{\sigma}_1 (V^*_{ij} - \hat{\mu}_j) & \text{if } V^*_{ij} > 2|a_j| \\ \hat{\mu}_1 + \hat{\sigma}^{-1}_j \hat{\sigma}_1 (V^*_{ij} - \hat{\mu}_j) - b_j[1 - (2|a_j|)^{-1}V^*_{ij}] & \text{otherwise} \end{cases} \quad (30.9)$$

where

$$a_j = \hat{\mu}_j - \hat{\sigma}^{-1}_1 \hat{\sigma}_j \hat{\mu}_1$$

and

$$b_j = \hat{\mu}_1 - \hat{\sigma}^{-1}_j \hat{\sigma}_1 \hat{\mu}_j.$$

The constants a_j and b_j are the points of intersection between the line defined in the first component of equation (30.9) and the V^* and $\tilde{V}^*$ axes, respectively. The second line of equation (30.9) is a modification to the linear transformation in the first line to ensure that adjusted transformed intakes are positive and that zero intakes are transformed into zero intakes (empirical results indicate that very few, if any, observations fall into the $[0, 2|a_j|]$ interval). Adjusted original scale intakes are defined by

$$\tilde{Y}^*_{ij} = \tilde{V}^{*1/\gamma}_{ij}.$$

30.3.2.2   *Obtaining an Equal Weight Sample*
Our procedure is designed for complex samples in which individuals have different sample weights. For ease of analysis and to increase computational efficiency, a new set of sample values is constructed in which each individual has the same weight. The sample distribution function of the new values is a close approximation to that of the adjusted data.

The new values are generated using a smooth estimate of the cumulative distribution function for the adjusted intakes. To estimate the cumulative distribution function, the weight for a positive intake on day j for individual i

is defined by

$$w_{ij} = r_i^{-1} w_i$$

where w_i is the original sample weight for individual i, and r_i is the number of positive intakes recorded for individual i, $i = 1, 2, \ldots, n^*$. A piecewise linear estimator, $\hat{F}$, of the distribution function for observed intakes, F, is developed by connecting the midpoints of the rises in the empirical cumulative distribution function using procedures outlined in Nusser *et al.* (1996).

An equal weight sample is constructed from $\hat{F}$ by calculating n^* intake values at equal probability intervals. These n^* equal weight values replace the ranked set of the first observed positive intakes recorded for each of the n^* individuals. Using $\ddot{Y}^*_{s(1)}$ to denote the equal weight sample value for the first observed positive intake for an individual whose adjusted positive intake has rank s, we have

$$\ddot{Y}^*_{s(1)} = \hat{F}^{-1}\left(\frac{s - 0.5}{n^*}\right)$$

where $s = 1, 2, \ldots, n^*$. The first positive reported intake value of individual i, which is of rank s among the $\tilde{Y}^*_{i1}$, is replaced by $\ddot{Y}^*_{s(1)}$ and denoted by $\ddot{Y}^*_{i(1)}$. Positive intakes from subsequent interviews are adjusted to maintain the individual's day to day structure. For the tth subsequent day of observed positive intake for individual i, where $t = 2, \ldots, r_i$, the intake for individual i is defined by

$$\ddot{Y}^*_{i(t)} = \tilde{Y}^{*-1}_{i(1)} \ddot{Y}^*_{i(1)} \tilde{Y}^*_{i(t)}.$$

We let $\ddot{Y}^*_{ij}$ denote the adjusted equal weight values, where j denotes the original sample day.

30.3.2.3 *Transformation to Normality*

The adjusted equal weight sample values, $\ddot{Y}^*_{ij}$, are transformed to normality using the two-step semiparametric procedure defined in Nusser *et al.* (1996). A grid search is used to determine the power transformation that minimizes the squared deviation between the normal score and the power transformed data. Then a smooth cubic spline is used to estimate the function that takes the power transformed data to normality. This procedure can be viewed as a semiparametric version of the Lin and Vonesh (1989) procedure. The function created by this two-step process is defined to be the function carrying the adjusted data to normality. The transformed positive intakes are defined by

$$X^*_{ij} = \omega(\ddot{Y}^*_{ij})$$

where ω is used to denote the transformation composed of the spline transformation applied to the power of the adjusted equal weight observations.

30.3.2.4 Estimating the Conditional Usual Intake Distribution in Normal Scale

The normal data are used to estimate a distribution of usual intakes based on the measurement error model proposed by Nusser *et al.* (1996). The transformed positive intakes, X_{ij}^*, are assumed to satisfy (30.4), (30.5), and (30.6).

Under this model, the mean μ_x of the normal usual intake distribution is estimated by

$$\hat{\mu}_x = n^{*-1} \sum_{i=1}^{n^*} \bar{X}_{i\cdot}^*$$

where

$$\bar{X}_{i\cdot}^* = r_i^{-1} \sum_{j=1}^{r_i} X_{ij}^*.$$

The variance $\sigma_{x^*}^2$ is estimated using Henderson's method III (Graybill, 1976), which is a method of estimating variance components for unbalanced data.

30.3.2.5 Distribution of Positive Usual Intakes

Given estimates of $\sigma_{x^*}^2$ and σ_u^2, the transformation η of (30.7) is estimated. This requires two steps. A set of estimated x^*-values, denoted by $\ddot{x}^*$, is created with the property that the mean and variance of the set is equal to the estimated mean and variance of x^*. The estimated x^* values are defined by

$$\ddot{x}_i^* = (\hat{\sigma}_{x^*}^2 + r_i^{-1}\hat{\sigma}_u^2)^{-1/2}\hat{\sigma}_{x^*}\bar{X}_{i\cdot}^*.$$

Then the estimated usual intake that would be generated by an individual with usual intake $\ddot{x}_i^*$ is calulated as

$$\ddot{y}_i^* = \sum_{i=-4}^{4} b_i \omega^{-1}(\ddot{x}_i^* + c_i)$$

where ω^{-1} is the inverse of the function ω and (b_i, c_i), $i = -4, -3, \ldots, 4$ is such that $\sum_{i=-4}^{4} b_i c_i = 0$, $c_i = -c_{-i}$, $\sum_{i=-4}^{4} b_i c_i^2 = \sigma_u^2$, and $\sum_{i=-4}^{4} b_i c_i^4 = 3\sigma_u^4$.

The function η is then estimated by the spline regression of $\ddot{x}_i^*$ on the power of the $\ddot{y}_i^*$. The power used in ω is used in η, and the spline procedure is that defined in the semiparametric transformation ω. The distribution of positive usual intakes is defined as the distribution of $y^* = \eta(x^*)$, where $x^* \sim N(\mu_{x^*}, \sigma_{x^*}^2)$.

30.3.3 Estimating the Distribution of Individual Consumption Probabilities

To estimate the unconditional distribution of usual intakes, an estimate of the distribution of the individual probabilities of consumption, $g(p)$, is required. The information available to support estimation is the proportion of sample days on which the food is consumed by individual i. Attempts to model the distribution of consumption probabilities with logistic regression and with a beta distribution resulted in estimated distributions that were not consistent with the observed data.

Therefore, it was decided to model the consumption probability distribution as a discrete distribution with K probability values, p_k, each with probability mass θ_k. In the examples below, we use $K = 51$ equally spaced mass points, $\{p_k\} = \{0.0, 0.02, 0.04, \ldots, 1.0\}$. Let $\hat{\Psi}_l$ denote the observed (weighted) relative frequency of individuals who consume the food on l out of r days, where $l = 0, 1, \ldots, r$. The $\hat{\Psi}_l$ are assumed to arise from a mixture of the K binomial probabilities of consumption on l out of r days, with binomial parameters of r and p_k, and mixture parameters $\boldsymbol{\theta} = (\theta_1, \theta_2, \ldots, \theta_K)$, where $\theta_k = [0, 1]$ and $\sum_{k=1}^{K} \theta_k = 1$. Hence, the expected value for $\hat{\Psi}_l$ is equal to

$$\Psi_l(\boldsymbol{\theta}) = \sum_{k \in A_l} \theta_k \binom{r}{l} p_k^l (1 - p_k)^{r-l}$$

where

$$A_l = \begin{cases} \{1, 2, \ldots, K-1\} & \text{if } l < r \\ \{2, 3, \ldots, K\} & \text{if } l = r. \end{cases}$$

Using the notation above, the minimum chi-squared estimator for this problem is defined as the value of $\boldsymbol{\theta}$ that minimizes

$$n \sum_{l=0}^{r} [\hat{\Psi}_l - \Psi_l(\boldsymbol{\theta})]^2 [\Psi_l(\boldsymbol{\theta})]^{-1}$$

(Agresti, 1990, p. 471). However, in our problem, the number of parameters, K, exceeds the number of terms in the chi-squared objective function, $r + 1$. Thus, we include an entropy term in the objective function to smooth the observed distribution over the K mass points of the distribution.

Entropy, as a measure of uncertainty, was introduced by Shannon (1948), and its use as a principle in statistical estimation was discussed by Jaynes (1957). Maximum entropy estimation is often used when the number of parameters to be estimated exceeds the amount of data available for estimation. The K probabilities, θ_k, of a discrete distribution with K mass points are obtained by maximizing

$$\Gamma = - \sum_{k=1}^{K} \theta_k \ln \theta_k$$

subject to $\sum_{k=1}^{K} \theta_k = 1$ and constraints that represent the known information regarding the θ_k, where $\theta_k \in [0, 1]$ and $\theta \ln \theta$ is zero for $\theta = 0$. In the absence of any prior information, Γ is maximized when $\theta_k = K^{-1}$ for all k; that is, when there is complete uncertainty about the probability of the K events. The function Γ reaches a global minimum when θ_k is one for some k and zero for all other values of k.

The modified minimum chi-squared estimator for our problem is defined as the value of θ that minimizes

$$n \sum_{l=0}^{r} [\hat{\Psi}_l - \Psi_l(\theta)]^2 \tilde{\Psi}_l^{-1} + \sum_{k=2}^{K} \frac{\theta_k}{1 - \theta_1} \ln \left(\frac{\theta_k}{1 - \theta_1} \right)$$

where $\sum_{k=1}^{K} \theta_k = 1$, $\theta_k \in [0, 1]$,

$$\tilde{\Psi}_l = \begin{cases} \max\{\hat{\Psi}_0, (1 - \bar{\Psi})^r\} & \text{if } l = 0 \\ \max\{\hat{\Psi}_r, \bar{\Psi}^r\} & \text{if } l = r \\ \hat{\Psi}_l(1 - \tilde{\Psi}_0 - \tilde{\Psi}_r)(1 - \hat{\Psi}_0 - \hat{\Psi}_r)^{-1} & \text{if } l = 1, 2, \ldots, r - 1 \end{cases}$$

and $\bar{\Psi} = n^{-1} \sum_{i=1}^{n} (r_i/r)$. The chi-squared term contains the sample information. The modified denominator in the chi-squared term prevents numerical difficulties that can arise when $\hat{\Psi}_l = 0$. The resulting estimator is closely related to the modified chi-squared estimator (Agresti, 1990, p. 472). The maximum entropy term smooths the mass across all possible values of p_k given the sample information in the chi-squared term. Note that θ_1, which is the proportion of the population that never consumes the food, is not included in the entropy term. The $\theta_2, \theta_3, \ldots, \theta_K$ are the parameters associated with consumers and θ_1 is the fraction of nonconsumers in the population.

A FORTRAN program using an IMSL subroutine was written to produce estimates of the consumption probability distribution parameters θ. Details of this procedure are presented in Zheng (1995). Let $g(p, \hat{\theta})$ denote the estimated consumption probability distribution.

30.3.4 Estimating the Unconditional Usual Intake Distribution

The unconditional usual intake for individual i, denoted by y_i, is

$$y_i = y_i^* p_i$$

where y_i^* is the usual intake given that the food is consumed and p_i is the probability of positive consumption. The distribution of usual intakes for consumers is the distribution of $y_i^* p_i$, which can be derived from the joint distribution of y_i^* and p_i. If y_i^* is independent of p_i, then the joint distribution of y_i^* and p_i is

$$f(y_i^*)g(p_i).$$

The cumulative distribution function of the unconditional usual intakes is then

$$H(y) = \Pr(y^*p \leq y) = \theta_1 + \sum_{k=2}^{K} \theta_k \int_0^{y/p_k} f(y^*)\,dy^*. \qquad (30.10)$$

Numerical integration is required to estimate the unconditional usual intake distribution H. An adaptive quadrature algorithm was developed to perform these calculations.

30.4 APPLICATION TO 1985 CSFII DATA

Our estimation methods were applied to dark green vegetable, apple, egg, and beef consumption data for the 743 nonpregnant, nonlactating, 25–50 year-old women from the 1985 Continuing Survey of Food Intakes by Individuals (CSFII) described in Section 30.2. On the basis of the analyses associated with Table 30.3, it is assumed that the conditional usual intake for these foods is independent of the probability of consumption. First, adjusted equal weight positive intakes were used to estimate the consumption day usual intake distribution using the semiparametric transformation approach described in Section 30.3.2. The semiparametric transformation provided a substantial improvement over the power transformation initially selected as the best power for transforming the data to normality for all foods. The deviation of the power-transformed data from normality was more severe than that found for nutrients in Nusser *et al.* (1996). The ratio of the estimated measurement error variance to the usual intake variance in normal scale ($\hat{\sigma}_{x^*}^{-2}\hat{\sigma}_u^2$) ranged from 1.1 to 5.6, indicating large within-person variability relative to among-person variability for these foods. A plot of the estimated consumption day usual intake distribution for each of the four foods is presented in Figure 30.3. The estimated mean, standard deviation, and skewness coefficient for the conditional usual intake distribution are presented in Table 30.5. The mean consumption day intakes are roughly one medium serving for dark green vegetables and apples, and about 1 to 1.5 servings for eggs and beef (Pao *et al.*, 1982).

The consumption probability distribution for each food was estimated using the observed relative frequencies for consumption classes listed in Table 30.1. The distributions are presented in Figure 30.4, and the estimated proportions of nonconsumers are listed in Table 30.5. For apples, the mean of the distribution is 0.11, indicating that, on average, apples are consumed about once every nine days by this population. Dark green vegetable and beef consumption tends to have sharply defined peaks, indicating modal consumption patterns. Our procedure for estimating consumption probability distributions proved to be very flexible. Estimated distributions for fruit and milk products

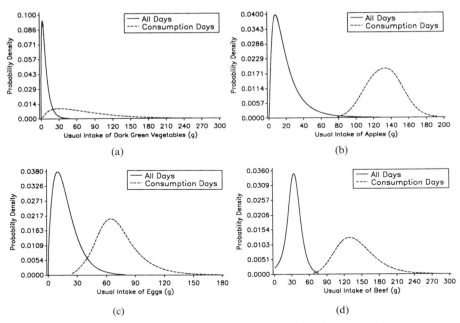

Figure 30.3 Estimated usual intake distributions on consumption days (dashed line) and on all days (solid line) for (a) dark green vegetables, (b) apples, (c) eggs, and (d) beef.

(not presented here) reflected the flat and increasing shapes, respectively, expected for these foods based on the consumption patterns noted in Table 30.1.

Under the assumption that intake levels and the probability of consumption are independent, the consumer usual intake distribution for all days was

Table 30.5 **Estimated Mean (g), Standard Deviation (SD), and Skewness Coefficient for Usual Intake Distributions on Consumption Days Only and on All Days for Consumers (in Grams), and Estimated Proportion of Nonconsumers**

	Consumption day usual intakes			Percentage nonconsumers	Usual intakes on all days for consumers		
Food group	Mean	SD	Skewness		Mean	SD	Skewness
Dark green vegetables	72	52	1.3	14.5	8.2	7.4	2.2
Apples	132	20	0.1	29.7	21.1	17.8	1.9
Eggs	73	24	1.1	3.5	18.7	13.3	1.4
Beef	141	34	0.8	0.003	33.4	12.5	0.6

Source: USDA 1985 Continuing Survey of Food Intakes by Individuals (unweighted).

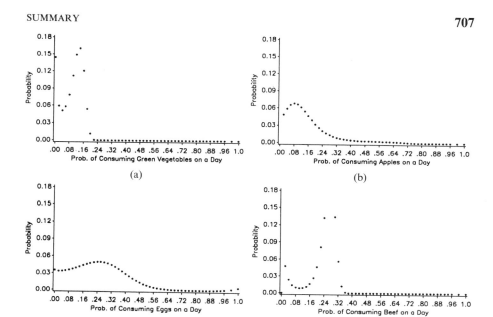

Figure 30.4 Estimated distribution of consumption probabilities for (a) dark green vegetables, (b) apples, (c) eggs, and (d) beef.

estimated using equation (30.10). The estimated distributions for consumers are presented in Figure 30.3. The estimated mean, standard deviation, and skewness coefficient for the consumer usual intake distributions are listed in Table 30.5. The results indicate that a variety of shapes can be estimated using this procedure.

30.5   SUMMARY

We have developed a method for estimating the distribution of an unobservable random variable from data that are subject to considerable measurement error and that arise from a mixture of two populations, one of which has a single-valued distribution and the other having a continuous unimodal distribution. Although we motivate the methodology development with a specific problem in dietary assessment, the method is more broadly applicable. Mixture populations arise frequently in health and reliability studies in which data are nonnormal and subject to measurement error.

The method requires that at least two positive intakes be recorded for a subset of the subjects in order to estimate the variance components for the measurement error model. Thus, this procedure may not be applicable to foods that are so rarely consumed by the population under study that very few individuals report two consumption days.

The specific approach presented here is appropriate for food intakes when the probability that a subject consumes a food is unrelated to the usual intake on consumption days. Further work is planned to develop models that account for dependence between the probability of consumption and conditional usual intakes. Such an extension will permit the methods to be applied to a broader set of foods. In addition, research is under way to develop variance estimators for the estimated parameters of the usual intake distribution, and software is being written so that our methods can be readily implemented.

ACKNOWLEDGMENTS

This research was partly supported by Cooperative Agreement 58-3198-2-006 between the Agricultural Research Service, U.S. Department of Agriculture and the Center for Agricultural and Rural Development, Iowa State University. We thank S. B. Schaller, K. W. Dodd, and Z. Zheng, who developed the software and analyzed food consumption data for this chapter.

REFERENCES

Agresti, A. (1990), *Categorical Data Analysis*, New York: John Wiley & Sons.

Aickin, M., and Ritenbaugh, C. (1991), "Estimation of the True Distribution of Vitamin A Intake by the Unmixing Algorithm," *Communications in Statistics-Simulations*, 20, pp. 255–280.

Beaton, G.H., Milner, J., Corey, P., *et al.* (1979), "Sources of Variance in 24-hour Dietary Recall Data: Implications for Nutrition Study Design and Interpretation," *American Journal of Clinical Nutrition*, 32, pp. 2546–2559.

Carriquiry, A.L., Jensen, H.H., Dodd, K.W., Nusser, S.M., Guenther, P.M., and Fuller, W.A. (1993), "Estimating Usual Intake Distributions," unpublished manuscript, Ames, IA: Department of Statistics, Iowa State University.

Emrich, L.J., Dennison, D., and Dennison, K.F. (1989), "Distributional Shape of Nutrition Data," *Journal of American Dietetics Association*, 89, pp. 665–670.

Graybill, F.A. (1976), *Theory and Application of the Linear Model*, Pacific Grove, CA: Wadsworth & Brooks/Cole Advanced Books & Software.

Hegsted, D.M. (1972), "Problems in the Use and Interpretation of the Recommended Daily Allowances," *Ecology of Food and Nutrition*, 1, pp. 255–265.

Hegsted, D.M. (1982), "The Classic Approach—The USDA Nationwide Food Consumption Survey," *American Journal of Clinical Nutrition*, 35, pp. 1302–1305.

Jaynes, E.T. (1957), "Information Theory and Statistical Mechanics," *Physics Review*, 106, pp. 620–630.

Lin, L.I.-K., and Vonesh, E.F. (1989), "An Empirical Nonlinear Data-Fitting Approach for Transforming Data to Normality," *The American Statistician*, 43, pp. 237–243.

National Research Council (1986), *Nutrient Adequacy*, Washington, DC: National Academy Press.

Nusser, S.M. (1995), "Estimating Usual Intake Distributions from 24-Hour Dietary Intake Data," *Proceedings of the Section on Epidemiology*, American Statistical Association, pp. 49–58.

Nusser, S.M., Carriquiry, A.L., Dodd, K.W., and Fuller, W.A. (1996), "A Semiparametric Transformation Approach to Estimating Usual Daily Intake Distributions," *Journal of the American Statistical Association*, 91, pp. 1440–1449.

Pao, E.M., Fleming, K.H., Guenther, P.M., and Mickle, S.J. (1982), *Foods Commonly Eaten by Individuals: Amount Per Day and Per Eating Occasion*, HERR No. 44, U.S. Department of Agriculture, Human Nutrition Information Service, Hyattsville, MD.

Sempos, C.T., Johnson, N.E., Smith, E.L., and Gilligan, C. (1985), "Effects of Intra-individual and Interindividual Variation in Repeated Dietary Records," *American Journal of Epidemiology*, 121, pp. 120–130.

Schaller, S. (1993), "Estimating Usual Intake Distributions for Dietary Components with Many Zero Daily Intakes," unpublished Creative Component for M.S. degree, Ames, IA: Department of Statistics, Iowa State University.

Shannon, C.E. (1948), "The Mathematical Theory of Communication," *Bell System Technical Journal*, 27, pp. 379–423.

U.S. Department of Agriculture, Human Nutrition Information Service (1989), *Continuing Survey of Food Intakes by Individuals, Women 19–50 Years and their Children 1–5 Years, 4 Days, 1985*, CSFII Report No. 85–4, p. 182.

Zheng, Z. (1995), "Modified Minimum Chi-Squared Estimation of Food Consumption Probability Distributions," unpublished Creative Component for M.S. degree, Ames, IA: Department of Statistics, Iowa State University.

CHAPTER 31

Identifying and Adjusting for Recall Error with Application to Fertility Surveys

Thomas W. Pullum and S. Lynne Stokes
The University of Texas at Austin

31.1 INTRODUCTION

Since the 1950s, surveys have been conducted periodically in developing countries to estimate current levels of fertility and contraceptive use. Such estimates are crucial for population projections and for the evaluation of family planning programs. The vital statistics systems in most countries are unable to produce reliable estimates of fertility, and in any case would not provide information about contraception and the relationship of these characteristics to a range of socio-economic covariates.

The most successful of these surveys have been those conducted within the framework of two successive programs: the World Fertility Survey (WFS), which extended from 1973 to 1984 and included 42 surveys in developing countries, and the Demographic and Health Surveys (DHS), which have been in operation since 1984 and had conducted more than 70 surveys by the end of 1994. These surveys typically include 4,000 to 8,000 respondents, with fertility estimates derived from the birth histories obtained from each respondent. Women (respondents and interviewers in these surveys are female) are asked to list all of their births, from earliest to most recent, and questions are then asked about the sex and survivorship of these children. Probes are used in an effort to avoid omission of births, particularly births in the relatively distant

Survey Measurement and Process Quality, Edited by Lyberg, Biemer, Collins, de Leeuw, Dippo, Schwarz, Trewin.
ISBN 0-471-16559-X © 1997 John Wiley & Sons, Inc.

past that resulted in a child death, which the respondent might not initially mention.

Fertility rates are calculated from the birth histories by aggregating the numbers of births that occurred in specified intervals of time and age or marital duration, aggregating the woman-years lived in such windows (interpreted as the collective exposure to the risk of childbearing), and then dividing the number of births by the number of woman-years. Many of these surveys, especially more recently, include all women (between the ages of 15 and 49) without respect to marital status, so all-woman age-specific fertility rates can be calculated directly. The sum of the age-specific fertility rates is the Total Fertility Rate (TFR), a cross-sectional measure of the eventual number of children a woman will have if she survives the childbearing years.

Some surveys—including the one from Pakistan that will be the focus of this chapter—are limited to women who have been married at least once. Fertility rates from such surveys are described as marital fertility rates. They can be shifted to all-woman rates only by referring to an accompanying household survey, which has a listing of all the members of the sample households, with their ages and marital statuses.

Marital fertility rates will be calculated here for intervals of marital duration, that is, elapsed years since first marriage (usually 0–4, 5–9, . . . , 30–34). The sum of the marital duration-specific rates will be referred to as the Total Marital Fertility Rate (TMFR), a cross-sectional measure of the eventual number of children a woman will have *if she marries* and survives the childbearing years.

All DHS have included questions about the health of children. Those conducted from approximately 1988 to 1992, as part of DHS-II (the second cycle of DHS), contained more extensive questions about the health of children under five years of age, including childhood illnesses, vaccinations, and measurements of height and weight.

In some of the surveys with extensive questions on child health, discontinuities have been observed in the reporting of births near the threshold of eligibility for these questions, suggesting that some birthdates were changed or some children were even omitted. Interviewers could shorten the interview by dropping some children who would have required the extensive questions or by shifting them beyond the five-year threshold. It is particularly easy to shift a birthdate beyond the threshold in contexts where the respondent may not know the true birthdates of her children. If omission and displacement occur to a substantial degree, the spurious conclusion may be reached that fertility has declined in recent years.

The purpose of the research described here is to suggest strategies to identify this type of systematic omission and displacement and to adjust for its effects. We will focus on the DHS conducted during 1990–91 in Pakistan, the Pakistan Demographic and Health Survey (PDHS), carried out by Pakistan's National Institute of Population Studies (NIPS). The Pakistan survey is the one with the strongest evidence of this type of misreporting.

This chapter works with the numbers of births and amounts of marital exposure that were reported for two five-year windows of time before the survey. The first analysis will use the individual-level data on the numbers of births and the amount of exposure to each interval of marital duration in each window. The second analysis will work from aggregated data, the numerators and denominators of the observed duration- and period-specific fertility rates. Both formats will be used to estimate simultaneously the levels of misreporting and the true levels of fertility.

31.2 BACKGROUND

31.2.1 Fertility Surveys in Pakistan

With an estimated population of 115 million in 1991, Pakistan is one of the largest countries in the world that has not yet shown convincing evidence of fertility decline. The reasons why fertility remains high there, despite substantial inputs into the national family planning program, include such factors as low female literacy and the seclusion of women, high infant mortality, a strong preference for sons, religious opposition, and the absence of strong political support for the family planning program.

There is a chronic pattern of displacement in Pakistan, which shows up in census data as well as surveys, such that children tend to be reported as somewhat older than they actually are. This pattern has been described in detail by Shah et al. (1986), Retherford et al. (1987), and Pullum (1990). Because of this displacement of children to higher ages, earlier surveys (Alam and Dinesen, 1984) were initially misinterpreted as showing evidence of recent fertility decline. The displacement discussed in this chapter will necessarily consist of a combination of that induced by the questionnaire structure and that which would have occurred even without any threshold.

The authors of the main report on the PDHS concluded that the TFR for ages 15–49 was 5.4 during the six years before the survey (National Institute of Population Studies, 1992, p. 42). The figure of 5.4, if true, would constitute a dramatic decline from earlier levels. A six-year window or time interval was used for the PDHS estimate, even though it is standard practice at DHS to use a five-year window. The estimate for a standard five-year window was so low, because of the threshold effect, that the authors decided to incorporate an additional year.

At the time of the PDHS, only 11.8 percent of currently married women were currently using a contraceptive method. This is one of the lowest levels of contraceptive prevalence outside sub-Saharan Africa. By comparison, 40 percent of couples in Bangladesh are estimated to be current users, 43 percent in India, and 62 percent in Sri Lanka. International comparisons show that contraceptive prevalence must be around 70 percent for fertility to be at replacement level—a TFR of about 2.1. In such comparisons, a prevalence of

Table 31.1   Unweighted Counts of Births from the PDHS Birth Histories, Reported for the Months January 1981 through December 1990. There are 8,418 Births in Window 1, 6,309 Births in Window 2, and a Total of 14,727 Births

	Window 1					Window 2				
	1981	1982	1983	1984	1985	1986	1987	1988	1989	1990
Jan.	67	88	92	91	107	55	67	105	92	142
Feb.	77	96	101	102	121	73	66	92	65	124
March	76	111	107	128	128	74	89	109	81	95
April	104	136	149	134	169	103	121	102	89	116
May	127	155	145	173	157	106	117	97	77	76
June	132	208	207	190	215	117	171	132	102	122
July	121	176	174	173	175	98	119	111	86	112
August	119	186	140	190	193	95	116	121	109	131
Sept.	115	160	153	181	202	97	117	106	115	129
Oct.	89	130	115	146	136	91	99	114	122	142
Nov.	104	114	128	132	156	73	102	104	110	133
Dec.	114	162	149	163	229	106	97	134	130	113
Totals	1,245	1,722	1,660	1,803	1,988	1,088	1,281	1,327	1,178	1,435

11.8 percent would correspond approximately with a TFR of 6.5 children (Weinberger, 1991).

Even before the PDHS field work was carried out, plans were made for methodological research based on reinterviews of a subsample of respondents. The Reinterview Survey was analyzed by Curtis and Arnold (1994). Although NIPS kindly gave us access to the PDHS data files, we have not been allowed to use the Reinterview Survey itself, so only limited comparisons with it can be made in this chapter.

The PDHS included 6,910 ever-married women. Our analysis excludes 299 women whose date of marriage was missing and is therefore limited to 6,611 women.

31.2.2   Reporting Errors in the Pakistan Demographic and Health Survey

The most striking evidence of misreporting in the PDHS comes from a simple listing of the numbers of births reported for each month before the survey, provided in Table 31.1.

The births in Table 31.1 and in all subsequent analyses are unweighted. The PDHS had a multistage cluster sampling design, with a rather wide range of sampling weights. All estimates quoted above from the PDHS report were weighted to compensate for the sampling design. However, for the purpose of modeling the misreporting process, all design effects will be ignored. Some comparisons between estimates will be confounded by this difference.

Two five-year intervals prior to the survey are of interest. Window 1 extends from January 1981 through December 1985, and window 2 extends from January 1986 through December 1990. The more extensive questioning about child health was limited to births reported for window 2. Births occurring after December 1990 were also eligible for the extra questions but will be ignored because there were few of them and it is simpler to use windows of observation and intervals of marital duration that are consistently of the same length, e.g., five years.

Table 31.1 shows that substantially fewer births were reported for window 2 than for window 1. If fertility *rates* are constant over an interval of time, then a retrospective survey of ever-married women should actually show a steady increase in the annual counts of births as one moves closer to the survey. Eligibility for inclusion in the survey is defined at the date of interview, and a substantial proportion of the women were not even married five or ten years before the survey. As they age into eligibility, so to speak, the reported numbers of births should increase. However, the PDHS shows 25 percent fewer births in window 2 than in window 1. This might at first seem to be evidence of genuine fertility decline, but the annual numbers of births *within* window 2, from 1986 to 1990, tended to increase.

Figure 31.1 gives a graphical representation of the monthly birth counts. It also indicates the numbers of births that would have been expected each month if the duration-specific rates were constant throughout the ten-year interval. (Such rates, produced by pooling the reported births and exposure without respect to the calendar year in which they occurred, are shown later in the last column of Table 31.2.) The discontinuities in Table 31.1 and Figure 31.1 give clear evidence of some combination of omission and displacement of births.

A more subtle manifestation of misreporting appears in the marital fertility rates calculated from the PDHS. Indeed, it is not common demographic practice to prepare a table such as Table 31.1, although the authors of the PDHS report did provide a weighted version of this table in an appendix (National Institute of Population Studies, 1992, p. 221). Table 31.2, prepared in order to describe levels and trends in fertility, could well be the first evidence of problems that a demographer would encounter.

The marital fertility rates in Table 31.2 are based on the numbers of births reported within the respective intervals of time, 1981–85 and 1986–90 (windows 1 and 2), and within durations 0–4 years, 5–9 years, . . . , 30–34 years (durations 1, . . . , 7) after the date of first marriage. The denominators of the rates are woman-years lived after marriage by the sample women, whether or not the women were *continuously* married. The sum of these duration-specific rates, when multiplied by five (because the rates are measured as births per year, but the duration intervals are five years long) is the Total Marital Fertility Rate (TMFR), described above.

The two five-year windows described in Table 31.2 imply a pattern which is demographically implausible. First, the TMFR for 1981–85, 9.415, is much too high to be credible. Such a figure is far above any estimate of completed

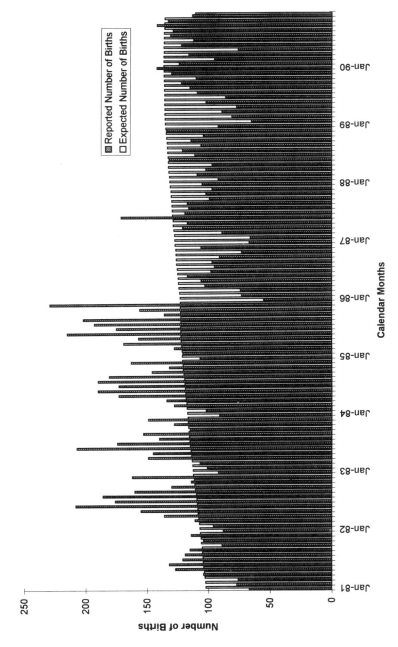

Figure 31.1 Reported and estimated numbers of births by calendar months, Pakistan DHS 1990–91. Estimated numbers are based on an assumption of constant duration-specific fertility rates throughout 1981–90.

Table 31.2 Observed Marital Fertility Rates from the PDHS, for 1981–85, 1986–90, and 1981–90

Years since first marriage	(1) 1981–1985	(2) 1986–1990	(3) Combined 1981–1990
0–4	0.426	0.314	0.371
5–9	0.414	0.274	0.339
10–14	0.352	0.211	0.275
15–19	0.277	0.161	0.209
20–24	0.182	0.083	0.119
25–29	0.114	0.045	0.060
30–34	0.117	0.021	0.027
TMFR	9.415	5.545	7.000

marital fertility that has ever been put forward for Pakistan. Second, the TMFR for 1986–90, 5.545, is obviously too low. If taken at face value, these estimates would imply a decline of more than 40 percent across five years, despite a very low level of contraceptive use.

A third and certainly spurious implication of Table 31.2 is that the decline in the rates included the earliest marital durations. In a country at the early stages of fertility decline, such as Pakistan, fertility regulation always takes the form of limitation—that is, the prevention of higher-order births. Indeed, in Pakistan, the most important effective method, female sterilization, is rarely adopted before the woman has had four or more births. In Pakistan, there is a strong desire to conceive as soon as possible after marriage, and it is virtually unimaginable that a couple would deliberately delay the first birth. Spacing of early births is rarely a goal, despite the increasing emphasis of the family program on spacing because it improves the chances of child survival. Given the emphasis on fecundity early in marriage and on having sons, any real fertility decline would probably not show up before five or even ten years of marriage.

31.2.3 The Pakistan Reinterview Survey

The reinterview survey was carried out during September–November 1991, about eight months after the main PDHS interviews. It did not include the extensive questions about child health or any other kind of calendar threshold, and was presumably more accurate, although certain background distortions—such as the tendency to exaggerate the ages of young children, were probably present in both surveys. Forty-four clusters, about 10 percent of the clusters in the main survey, were selected, and all of the women in those clusters who had participated in the main survey were subject to reinterview. There were 709 such women; 528 of them were actually reinterviewed; and 505 could be confidently linked to original cases. We are limited by the fact that the latter survey was small and had a rather high non-interview rate, as well as by the fact that the report on the reinterview survey (Curtis and Arnold, 1994) contains few numbers that can be compared directly with our results.

Comparing the main survey and reinterview survey in terms of the women who appeared in both of them, the authors (Curtis and Arnold, 1994, p. 34) distinguished between children who were reported as survivors or child deaths as of the date of the main survey. The following findings are most relevant:

1. The main survey reported 2,016 children ever-born and the reinterview survey reported 2,055, implying an omission of just under 2 percent in the main survey. However, these numbers refer to children born *since* 1960, a much longer reference period than the 10 years (1981–90) we used.

2. Among *surviving* children, the main survey reported 1,797 children ever-born and the reinterview survey reported 1,774 children (since 1960). That is, there is no evidence of omission among survivors; on the contrary, if there was omission, it appeared more in the reinterview survey.

3. Among *surviving* children, the main survey reported 335 during 1986–89 and the reinterview reported 459, implying that 27 percent of the children in this four-year interval were displaced. Most appear to have been displaced into 1984 and 1985, although the reinterview survey also indicated slightly more births in 1990 than were originally reported. (It is not clear how a shift into 1990 could be related to the structure of the interview.)

4. Among children *who died*, 219 were reported in the main survey and 281 in the reinterview survey (since 1960). It thus appears that there was some omission of dead children, although spread out over a long period of time. The reinterview survey reported more children who died, compared to the main survey, for every calendar year after 1980 except for 1982.

Curtis and Arnold (1994, p. 37) estimated that the TFR for the period 0–5 years before the survey was underestimated by between 8 percent and 16 percent, placing the true value in the range of 6.0 to 6.4.

31.3 INDIVIDUAL-LEVEL ESTIMATION OF TRUE RATES AND MISREPORTING PARAMETERS

31.3.1 A Model for Misreporting

In order to calculate the duration-specific numbers of births or amounts of exposure, we must assume that the dates of marriage are not displaced. This assumption is probably only questionable for younger women, because interviewers otherwise would have been limited in their ability to backdate a birth if it occurred soon after marriage. It seems unlikely that the interviewer would intentionally misstate the date of a birth to make it appear premarital. In Pakistan, there is negligible evidence of premarital fertility in all surveys, including the PDHS.

Accepting the date of marriage as given, the data set allows us to calculate each woman's births and exposure to each marital duration in each window. Following a common demographic practice for the analysis of marital fertility, no deductions from exposure will be made for widowhood, divorce, or other kinds of separation. Moreover, the calculation of exposure will be simplified by not going below the level of the calendar year as an interval of time.

Intervals of marital duration will be five years long, with the exception that the first interval will be separated into year 0 and years 1–4. The reason for doing this is that generally, in Pakistan, there is no cohabitation before marriage, and very little fertility is observed in the first eight months or so of marriage. Heterogeneity within the first five years is neatly captured by breaking out the first year. Year 0 will be referred to as duration 0, and years 1–4 will be referred to as duration 1.

For woman j, let a_{tj} and b_{tj} refer to the actual and observed numbers of births, respectively, in calendar year t ($t = 1981, \ldots, 1990$). Let d_{tj} refer to the true number of births that were displaced out of calendar year t in window 2 $(1986, \ldots, 1990)$ and into any calendar year in window 1 $(1981, \ldots, 1985)$. Assume that no other kinds of displacement occur. Let w_{tj} refer to the true number of births that occurred in calendar year t in window 2 but were completely omitted from the birth history. For woman j, the subtotals of the actual numbers of births in windows 1 and 2 will be $a_{1 \cdot j}$ and $a_{2 \cdot j}$; the subtotals of her observed numbers of births will be $b_{1 \cdot j}$ and $b_{2 \cdot j}$; her total number of displaced births will be d_j, and her total number of omitted births will be w_j. Thus, the general misreporting model states that

$$
\begin{aligned}
b_{1 \cdot j} &= a_{1 \cdot j} + d_j & &\text{in window 1} \\
b_{2 \cdot j} &= a_{2 \cdot j} - d_j - w_j & &\text{in window 2.}
\end{aligned}
\tag{31.1}
$$

It would be desirable to be able to use the model to estimate omission and displacement simultaneously, but the proposed model confounds these two types of misreporting with each other. Because the reinterview survey did not find evidence of omission of recent births in Pakistan, our model will not include the possibility of that type of misreporting in window 2. That is, we set w_{tj} in (31.1) identically equal to zero for all women j.

We cannot observe the actual numbers of births, of course, but we assume that their expected values are determined by the woman's marital duration during each year of the window. That is, the expected number of births for woman j in year t is given by

$$
\alpha_{tj} \equiv \mathrm{E}(a_{tj}) = \sum_{k=0}^{7} I_{tjk}\,\varphi_{kt}
\tag{31.2}
$$

where k identifies an interval of marital duration, corresponding to years $0, 1-4, 5-9, \ldots, 30-34$; φ_{kt} is the marital fertility rate for duration k in year t; and I_{tjk} is an indicator variable that is 1 if woman j was exposed to duration interval k in year t and 0 if she was not.

We also cannot observe the number of displaced births, but will assume that an actual birth in year t in window 2 has a probability δ_{tj} of being displaced into window 1. That is, the expected number of displaced births in year t will be $E(d_{tj} \mid a_{tj}) = \delta_{tj}a_{tj}$ and the expected total number of displaced births for woman j will be

$$E(d_j) = E\left[\sum_{t=1986}^{1990} \delta_{tj}a_{tj} \right] = \sum_{t=1986}^{1990} \delta_{tj}\alpha_{tj}. \tag{31.3}$$

Taking the expected values of (31.1), we have

$$
\begin{aligned}
E(b_{1 \cdot j}) &= \sum_{t=1981}^{1985} \alpha_{tj} + \sum_{t=1986}^{1990} \delta_{tj}\alpha_{tj} \text{ in window 1} \\
E(b_{2 \cdot j}) &= \sum_{t=1986}^{1990} \alpha_{tj} - \sum_{t=1986}^{1990} \delta_{tj}\alpha_{tj} \text{ in window 2.}
\end{aligned}
\tag{31.4}
$$

The parameters on the right hand side of the model defined by (31.2)–(31.4) are the probabilities of displacement out of the years $1986, \ldots, 1990$, namely δ_{tj}, and the duration-specific fertility rates for duration k in year t, namely φ_{kt}. It is necessary to reduce the number of these parameters in order to achieve identifiability.

A parameterization of the φ_{kt}'s will be based on a model for marital fertility developed by Coale and Trussell (1974). Their model links the marital fertility rates for contracepting and noncontracepting populations by $\varphi_k = Mn_k \exp(mv_k)$, where n_k is the rate for a natural or noncontracepting reference population, v_k describes the typical pattern of contraceptive effect, M is a scale factor for the overall level of fertility, and m is a scale factor for the level of contraceptive effect. The first v_k coefficient is set at zero (otherwise the v_k's would be confounded with m), but the remaining v_k's and the n_k's were estimated by Coale and Trussell from observed marital fertility schedules. This model has been widely used and validated by subsequent researchers.

Coale and Trussell presented their model in terms of marital fertility rates for the six *age* intervals $20-25, \ldots, 45-49$. However, we shall apply it directly to the eight marital durations $0, 1-4, \ldots, 30-34$, by equating the successive intervals of duration to the corresponding age intervals and equating the terms for the first two intervals and the final two intervals. The values of the v_k coefficients will be drawn from the more recent and complete empirical validation of the Coale–Trussell model by Xie and Pimentel (1992): $v_0 = v_1 = 0.0$, $v_2 = -0.335$, $v_3 = -0.717$, $v_4 = -1.186$, $v_5 = -1.671$, and $v_6 = v_7 = -1.115$. This pattern for the effect of contraceptive use corresponds with the observation that, early in a fertility transition, the main use of fertility regulation is to limit the higher order births rather than to limit or even to space the births early in marriage. The v_k terms increase in magnitude with age (or duration or parity) because of the increasing motivation to use contraception to prevent high order births, but they diminish somewhat in magnitude at the highest ages (or durations or parities) because the oldest cohorts of women see themselves as having a low risk of conceiving. The v_k estimates from Xie and Pimentel (1992) were estimated from a pooling of data from 41 national surveys and represent a synthesis of settings with data of relatively good quality.

Attaching subscripts 1 and 2 to the Coale–Trussell model for years t_1 and t_2, we have $\varphi_{k1} = M_1 n_k \exp(m_1 v_k)$ and $\varphi_{k2} = M_2 n_k \exp(m_2 v_k)$. We can then show that $\varphi_{k2} \cong \varphi_{k1}(1 + \Delta m v_k)$, where $\Delta m = m_2 - m_1$, since $\exp(\Delta m v_k) \cong 1 + \Delta m v_k$. Okun (1994) has shown that a change in m is an excellent indicator of change in fertility control, even though the magnitude of m is a less reliable indicator of the level of control in a given population.

Applying this relationship to the calendar years 1981 to 1990, we have

$$\varphi_{kt} \cong \varphi_k(1 + c_t v_k \Delta m) \tag{31.5}$$

where $\varphi_k = \varphi_{k,1981}$ gives the estimated rates in 1981 for $k = 0, \ldots, 7$ and

$$c_t = (t - 1981)/10 \tag{31.6}$$

describes the linear change in m. No change in contraceptive use would imply $\Delta m = 0$, and therefore no change in the true rates.

We examined a variety of simple models for δ_{tj}, including (a) $\delta_{tj} = 0$; (b) $\delta_{tj} = \delta$ if woman j was married at any time in window 1, and 0 otherwise; (c) $\delta_{tj} = \delta(e_{1j}/5)$ if $t \leq 1985 + r$, where $r = 2, 3, 4$, or 5, and $\delta_{tj} = 0$ otherwise, and where e_{1j} is the number of years of marital exposure that woman j had in window 1. Several models appeared approximately equally good as measured by their deviance and the smoothness of their estimated φ_k's. Among these best models was (c) with $r = 5$. Detailed results and discussion will be restricted to this model. That is, we assume

$$\delta_{tj} = \delta(e_{1j}/5). \tag{31.7}$$

With this model, the probability that any birth in window 2 will be displaced into window 1 is proportional to the amount of marital exposure of the woman to window 1. For most women, $e_{1j} = 5$ years, that is, $\delta_{tj} = \delta$. If a woman's marriage occurred in window 2, she will have had no exposure to window 1, and no possibility of a birth being displaced into window 1. If her marriage occurred in window 1, then δ_{tj} will be between zero and δ.

Our goal is to estimate nine parameters, δ and $\varphi_0, \ldots, \varphi_7$, appearing in the model defined above. The problem remains that displacement is confounded with actual fertility decline, so that we cannot estimate δ and Δm simultaneously.

Our strategy will be to estimate δ and the φ_k's for three hypothetical values of Δm which we believe represent a lower bound, a likely or most plausible value, and an upper bound for fertility change over the decade. We justify these choices by the following argument.

Roughly speaking, with the linearized model, Δm values of 0.1 and 0.2 will correspond to declines of 7 and 13 percent, respectively, across the ten-year interval 1981–91, in the Total Marital Fertility Rate. A very liberal interpretation of the contraceptive use information in the PDHS suggests that 0.2 could be taken as an upper bound for Δm, although it is more likely that this coefficient is 0.1 or less. As a rough calibration based on Weinberger (1991), an increase in prevalence of 5 percent across ten years would lead to an actual decline of about $(5)(0.074) = 0.37$ children in the TFR, and with a baseline TFR of about six children, this would translate into a percentage decline of about 6 percent, approximately corresponding with $\Delta m = 0.1$. Estimates of the eight rates in 1981 and the displacement probability δ will be developed for hypothetical values $\Delta m = 0.0$, 0.1, and 0.2.

31.3.2 Estimation

The model described in the previous section can be summarized as follows

$$E(b_{1,j}) = \sum_{k=0}^{7} \varphi_k \left[\sum_{t=1981}^{1985} I_{tjk}(1 + c_t v_k \Delta m) + \sum_{t=1986}^{1990} \delta(e_{1j}/5)I_{tjk}(1 + c_t v_k \Delta m) \right]$$

(31.8)

$$E(b_{2,j}) = \sum_{k=0}^{7} \varphi_k \left[\sum_{t=1986}^{1990} (1 - \delta e_{1j}/5)I_{tjk}(1 + c_t v_k \Delta m) \right].$$

To estimate this model, we must assume further structure. First we assume that

$$\mathrm{Var}(b_{w,j}) = \gamma E(b_{w,j})$$

$$\mathrm{Cov}(b_{w,j}, b_{w',j'}) = 0 \text{ for } j \neq j', \text{ any } w \text{ and } w', \text{ and} \quad (31.9)$$

$$\mathrm{Cov}(b_{w \cdot j}, b_{w' \cdot j}) = \rho \sqrt{\mathrm{Var}(b_{w \cdot j}) \, \mathrm{Var}(b_{w' \cdot j})} \text{ for } w \neq w'.$$

Here ρ denotes the correlation between the number of births to the same woman in the two windows. A positive correlation is plausible in a context of low contraceptive use, because fertility will be largely determined by fecundability, or biological capacity, which varies from couple to couple.

If ρ were known to be zero, the parameters of this model could be estimated using the usual methods of generalized linear models, assuming a Poisson mean-variance structure and an identity link. Since the expectation of $b_{w \cdot j}$ is not linear in all nine parameters, an iterative procedure could be used for estimation as described, for example, in McCullagh and Nelder (1989, ch. 11). This would consist of iteratively estimating the φ_k's and then δ until convergence is obtained.

Liang and Zeger (1986) proposed an extension to the class of generalized linear models which can be used to estimate such models when not all observations are independent. Their method, known as the method of generalized estimating equations or GEE, was developed for the analysis of longitudinal data such as ours. Their estimates of the regression parameters are obtained by solving a system of estimating equations which reduce to the likelihood equations of the generalized linear model if $\rho = 0$. Their estimators are consistent under mild assumptions about the mean-variance relationship and the correlation structure among the time-dependent observations on a single individual, and are asymptotically normally distributed.

We used the GEE procedure to fit the model defined in (31.8) and (31.9). The SAS macro described in Karim and Zeger (1988) was used for implementation. (A SAS procedure to solve GEE models is under development by SAS Institute (Dunlop, 1995).) Iteration between estimation of the φ_k's and δ was required, using GEE at each step. Convergence was generally obtained in about six steps after using starting values from a nonlinear regression procedure.

31.3.3   Results

Table 31.3 gives the results of fitting of the individual-level model (31.8) and (31.9) for three values of the change parameter Δm: 0.0, 0.1, 0.2. It is very unlikely that fertility actually fell by an amount as large as would be implied by $\Delta m = 0.2$—if it fell at all—because of the low level of current contraceptive use observed in the PDHS. However, the value of assuming such an outer limit is that it measures the sensitivity of components of the model to an assumed trend.

The model indicates high levels of displacement for all assumed amounts of change. The estimates of the probability of displacement, for mothers who were married during all of window 1, range from 23.9 to 26.5 percent, and the volume of displacement ranges from 1,405 to 1,586 births, or 18.3 to 20.1 percent of all actual births. The fitted TMFR in window 2 ranges from 6.8 to 7.0. In Pakistan, the TFR can be estimated from the TMFR by multiplying by 0.88. (The PDHS report gave a six-year TMFR of 6.105 births, compared with a standard all-woman TFR of 5.355 births (National Institute of Population Studies, 1992, pp. 42–43). Assuming that both of these rates are subject to the same reporting errors for births, an approximate conversion from a TMFR to a TFR can be made in Pakistan by multiplying the TMFR by $5.355/6.105 = 0.88$.) Hence we estimate a range in the TFR from 6.0 to 6.2.

For each of the three models, ρ was estimated to be approximately 0.18, confirming the expected positive correlation between a woman's number of births in the two windows. The scale parameter γ was estimated to be about 0.75 for each of the models, suggesting that the number of births is underdispersed when compared with a Poisson random variable. This is expected, because of the intervals of pregnancy and post partum amenorrhoea associated with each birth. Finally, the error sums of squares of these fitted models are

Table 31.3 Results of Misreporting Model for Individual-Level Data: Optimized Parameter Estimates for $\Delta m = 0.0$, $\Delta m = 0.1$, and $\Delta m = 0.2$

True duration-specific marital fertility rates and standard errors

Duration	$\Delta m = 0.0$ Rate	s.e.	$\Delta m = 0.1$ Rate	s.e.	$\Delta m = 0.2$ Rate	s.e.
			Window 1 (1981–85)			
0	0.018	0.010	0.019	0.010	0.019	0.010
1	0.364	0.006	0.367	0.006	0.371	0.006
2	0.356	0.006	0.359	0.006	0.362	0.006
3	0.291	0.006	0.297	0.006	0.301	0.007
4	0.226	0.007	0.232	0.007	0.238	0.007
5	0.143	0.008	0.148	0.008	0.153	0.008
6	0.062	0.008	0.063	0.008	0.064	0.009
7	0.035	0.015	0.036	0.015	0.038	0.016
TMFR	7.04	0.09	7.16	0.10	7.28	0.10
			Window 2 (1986–90)			
0	0.018	0.010	0.019	0.010	0.019	0.010
1	0.364	0.006	0.367	0.006	0.371	0.006
2	0.356	0.006	0.353	0.006	0.350	0.006
3	0.291	0.006	0.286	0.006	0.279	0.006
4	0.226	0.007	0.218	0.007	0.209	0.006
5	0.143	0.008	0.135	0.007	0.126	0.007
6	0.062	0.008	0.059	0.008	0.057	0.008
7	0.035	0.015	0.034	0.014	0.033	0.014
TMFR	7.04	0.09	6.91	0.09	6.77	0.10
			Estimate of displacement parameter and standard error			
δ	0.265	0.007	0.252	0.007	0.239	0.007
			Volume of displacement			
Δ	1,586.1		1,497.2		1,405.4	
			Estimated true numbers of births			
Window 1	6,831.9		6,920.8		7,012.6	
Window 2	7,895.1		7,806.2		7,714.4	
			Proportions of births displaced			
	0.201		0.193		0.183	

similar, providing little reason to prefer one value of Δm to another within this range.

31.3.4 Sensitivity to Alternative Assumptions

Although they will not be described in detail, several modifications to this procedure have been attempted, with results that lend confidence to the estimates.

As discussed in Section 31.3.1, a variety of models for the displacement probability δ_{tj} were examined. The poorest fitting model was the one which assumed $\delta_{tj} = 0$. The best fitting models for δ_{tj} that we tested were those in which $\delta_{tj} = \delta(e_{1j}/5)$ if $t - 1985 \leq 3$, 4, or 5, and $\delta_{tj} = 0$ otherwise. The resulting ranges of the error sum of squares for these models were small and there was little variation in the resulting estimates of the TMFR.

Stability in the estimates of displacement for alternative hypothetical values of Δm also implies little sensitivity to the magnitude of the v_k coefficients. These coefficients are approximately linear in marital duration, so there is a simple trade-off between multiplying them by some factor c and multiplying Δm by $1/c$.

We attempted to validate the model by applying it to data which had been generated with preset values of the φ_k's and δ and a Poisson distribution of the numbers of actual births. The model was able to retrieve these preset values, generally within the accuracy of twice their estimated standard errors. However, when applied to the general model in (31.1), including omission as well as displacement, the estimation procedure could not successfully retrieve the pre-set values, and that is the main reason we cannot advocate the estimation of omission and displacement simultaneously.

31.4 AGGREGATE-LEVEL ESTIMATION OF TRUE RATES AND MISREPORTING PARAMETERS

31.4.1 A Model for Aggregated Data

We now propose a simple modification of the individual-level model in order to suggest how misreporting can be estimated in the absence of access to the individual data file. In a sense this is a naive model because it makes several simplifying assumptions, but it is presented because it readily confirms the estimates of rates and displacement in the better justified individual-level model and it only requires information that would typically be available for a secondary analysis. The earlier definitions and assumptions will be retained, with the principal modification that all variables will be aggregated across individuals and the outcome variable, the number of births, will be *dis*-aggregated across intervals of marital duration. Duration 0 (the first year of marriage) will now be combined with duration 1, years 1–4 since first marriage, because tabulations of births by marital duration do not normally subdivide years 0–4. This model has fourteen equations, one for each combination of the two time windows and the seven marital durations.

The observed number of births to the sample in duration k and window t ($t = 1, 2$) will be b_{kt} and the years of marital exposure to this combination will be e_{kt}. The uncorrected marital fertility rate for this combination will simply be b_{kt}/e_{kt}. Many survey reports publish such numerators and denominators.

The windows and durations have the same lengths, namely five years. If an interval of marital duration k straddles the boundary between windows 1 and 2, and a birth in the window 2 portion is displaced back into window 1, then it will either remain in that same interval of marital duration (k) or be shifted to the preceding interval ($k - 1$). As an exception, if a birth in duration $k = 1$ is displaced, it must remain in duration 1, because we have stipulated that no birth will be misreported to be premarital. Otherwise, we assume that the probabilities of omission and displacement do not depend on k, nor on the specific location of a birth within duration k or within window 2.

Except for duration $k = 1$, we assume that the allocations of shifted births into durations $k - 1$ and k will be proportional to the reported woman-years of marital exposure to durations $k - 1$ and k in window 1. For example, of the births which actually occurred in duration 2 in window 2, but are transferred back into window 1, a proportion $e_{11}/(e_{11} + e_{21})$ will be misreported in duration 1 and a proportion $e_{21}/(e_{11} + e_{21})$ will be misreported in duration 2. We arbitrarily assume that the probability of displacement out of duration $k = 1$ is half of the probability for other durations. Therefore, the expected number of births reported in window 1, for each of the seven marital durations, will be

$$E(b_{11}) = e_{11}\varphi_{11} + \delta e_{12}\varphi_{12}/2 + \delta e_{22}\varphi_{22}[e_{11}/(e_{11} + e_{21})]$$

$$E(b_{21}) = e_{21}\varphi_{21} + \delta e_{22}\varphi_{22}[e_{21}/(e_{11} + e_{21})] + \delta e_{32}\varphi_{32}[e_{21}/(e_{21} + e_{31})]$$

$$E(b_{31}) = e_{31}\varphi_{31} + \delta e_{32}\varphi_{32}[e_{31}/(e_{21} + e_{31})] + \delta e_{42}\varphi_{42}[e_{31}/(e_{31} + e_{41})]$$

$$E(b_{41}) = e_{41}\varphi_{41} + \delta e_{42}\varphi_{42}[e_{41}/(e_{31} + e_{41})] + \delta e_{52}\varphi_{52}[e_{41}/(e_{41} + e_{51})]$$

$$E(b_{51}) = e_{51}\varphi_{51} + \delta e_{52}\varphi_{52}[e_{51}/(e_{41} + e_{51})] + \delta e_{62}\varphi_{62}[e_{51}/(e_{51} + e_{61})]$$

$$E(b_{61}) = e_{61}\varphi_{61} + \delta e_{62}\varphi_{62}[e_{61}/(e_{51} + e_{61})] + \delta e_{72}\varphi_{72}[e_{61}/(e_{61} + e_{71})]$$

$$E(b_{71}) = e_{71}\varphi_{71} + \delta e_{72}\varphi_{72}[e_{71}/(e_{61} + e_{71})].$$

In window 2, the expected number of reported births will have deficits from both omission and displacement:

$$E(b_{12}) = (1 - \omega - \delta/2)e_{12}\varphi_{12}$$

$$E(b_{22}) = (1 - \omega - \delta)e_{22}\varphi_{22}$$

$$E(b_{32}) = (1 - \omega - \delta)e_{32}\varphi_{32} \tag{31.10}$$

$$E(b_{42}) = (1 - \omega - \delta)e_{42}\varphi_{42}$$
$$E(b_{52}) = (1 - \omega - \delta)e_{52}\varphi_{52}$$
$$E(b_{62}) = (1 - \omega - \delta)e_{62}\varphi_{62}$$
$$E(b_{72}) = (1 - \omega - \delta)e_{72}\varphi_{72}. \tag{31.10}$$

Here ω is the conditional probability that a birth will be omitted, given that it has not been displaced. The 14 equations in (31.10) correspond to the two individual-level equations in (31.1). To reduce the number of parameters to be estimated, the rates in the two windows will be linked by

$$\varphi_{k2} = \varphi_{k1}[1 + (\Delta m/2)v_k].$$

Here, as before, Δm is the hypothesized ten-year change in the Coale–Trussell m parameter; it must be divided by two because the midpoints of the five-year rates for 1981–85 and 1986–90 are five years apart. This model implies that the probabilities of omission or displacement out of window 2 are uniform with respect to the time (within window 2) when the birth actually occurred. The numerators and denominators of the rates in Table 31.2, the quantities b_{kt} and e_{kt}, are given in Table 31.4. (Note that the exposures in Table 31.4 are in months, rather than years.)

The degree of omission and displacement is described by the probabilities ω and δ, that apply to the births that actually occurred in window 2. With estimates of the actual numbers of births to each woman, we can estimate the total *numbers* of births that were omitted and displaced, referred to with the symbols Ω and Δ, respectively. Let B_1 and B_2 be the observed total numbers of births in windows 1 and 2, respectively, and let A_1 and A_2 be the estimated actual numbers. Also define the totals $A = A_1 + A_2$ and $B = B_1 + B_2$. Then

Table 31.4 Numerators and Denominators of Observed Duration-Specific Marital Fertility Rates, PDHS. Exposure is Given in Months and Must Be Divided by Twelve to Give Annual Rates

Marital duration	Window 1 (1981–85)		Window 2 (1986–90)	
	Births	Exposure	Births	Exposure
1	2,888	81,440	2,049	78,288
2	2,458	71,238	1,854	81,192
3	1,708	58,166	1,248	71,008
4	943	40,898	777	57,973
5	356	23,411	282	40,737
6	61	6,395	88	23,320
7	4	409	11	6,334

$A - B = \Omega$ and $A_2 - B_2 = \Omega + \Delta$ lead after algebraic manipulation to $\Omega = A - B$ and $\Delta = B_1 - A_1$.

31.4.2   Estimation

For the same reasons that omission was dropped from model (31.1), namely, the evidence from the Reinterview Survey that omission was negligible in the PDHS and the instability of simultaneous estimates of omission and displacement, we set $\omega = 0$ and only use (31.10) to estimate displacement. The 14 observations, consisting of the intervals of marital duration k $(k = 1, \dots, 7)$ and time t $(t = 1, 2)$, are assumed to be statistically independent of each other. Obviously, they are not, but no correction is made for this fact. For each of them, there are three distinguishable numbers of births: (a) the reported number, b_{kt}; (b) the expected or "fitted" value of the reported number under the misreporting model, $E(b_{kt})$, given above; and (c) the expected value of the "true" number of births, $\alpha_{kt} = e_{kt}\varphi_{kt}$. We estimate the misreporting parameter δ and the true rates in window 1, $\varphi_{11}, \dots, \varphi_{71}$, by optimizing the fit of the estimated numbers (b) to the observed numbers (a). This includes requiring that the totals for the "fitted" birth counts for windows 1 and 2 match the corresponding totals of observed birth counts. We assume that the observed numbers have a Poisson distribution around the "fitted" numbers; the log of the likelihood ratio is

$$ l = \sum_{kt} \left[(b_{kt} - \beta_{kt}) - b_{kt} \log(b_{kt}/\beta_{kt}) \right]. $$

Almost exactly the same results can be obtained by alternatively assuming a normal distribution in place of a Poisson. The estimates were obtained using GLIM.

As before, the change parameter Δm is not estimated but is preset at three possible values: 0.0, 0.1, and 0.2. To estimate δ and $\varphi_1, \dots, \varphi_7$ we iterate across two Poisson regressions with linear link functions. In the first iteration, δ is set at a trial value and the rates $\varphi_{11}, \dots, \varphi_{71}$ are calculated as the coefficients in a regression with no constant term. Using the fitted values of the seven rates and the preset value of Δm, the estimate of δ is then updated with another regression, and so on. The quality of the fit is indicated by the deviance, which would have a chi-square distribution with $14 - 7 - 2 - 1 = 4$ degrees of freedom if the observations were independent.

31.4.3   Results

Application of the three scenarios with $\Delta m = 0.0$, 0.1, and 0.2 is described in Table 31.5. The estimated proportions of cases displaced are high and stable in a range of 18.4 to 20.3 percent, with a volume of 1,422 to 1,607 true births, or 18.4 to 20.3 percent of the actual numbers of births. The estimated true TMFR in window 2 is in a range from 6.93 to 7.13, leading (after multiplication by the

Table 31.5   Results of Misreporting Model for Aggregated Data: Optimized Parameter Estimates for $\Delta m = 0.0$, $\Delta m = 0.1$, and $\Delta m = 0.2$

True duration-specific marital fertility rates and standard errors*

Duration	$\Delta m = 0.0$ Rate	s.e.	$\Delta m = 0.1$ Rate	s.e.	$\Delta m = 0.2$ Rate	s.e.
			Window 1 (1981–85)			
1	0.349	0.005	0.350	0.005	0.351	0.005
2	0.346	0.006	0.349	0.006	0.352	0.006
3	0.280	0.005	0.286	0.006	0.292	0.006
4	0.219	0.005	0.226	0.006	0.234	0.006
5	0.126	0.005	0.133	0.005	0.140	0.006
6	0.070	0.006	0.072	0.006	0.074	0.006
7	0.036	0.009	0.038	0.009	0.039	0.010
TMFR	7.13		7.27		7.41	
			Window 2 (1986–90)			
1	0.349	0.005	0.350	0.005	0.351	0.005
2	0.346	0.006	0.343	0.005	0.340	0.005
3	0.280	0.005	0.276	0.005	0.271	0.005
4	0.219	0.005	0.213	0.005	0.206	0.005
5	0.126	0.005	0.122	0.005	0.117	0.005
6	0.070	0.006	0.068	0.006	0.066	0.006
7	0.036	0.009	0.035	0.009	0.035	0.009
TMFR	7.13		7.03		6.93	
		Estimate of displacement parameter and standard error*				
δ	0.240	0.009	0.228	0.009	0.216	0.009
			Volume of displacement			
Δ	1,607.4		1,517.3		1,422.3	
			Estimated true numbers of births			
Window 1	6,810.6		6,995.7		6,995.7	
Window 2	7,916.4		7,826.3		7,731.3	
			Proportions of births displaced			
	0.203		0.194		0.184	
			Fit (4 degrees of freedom)			
Deviance	32.5		22.8		15.8	

* Nominal standard errors under crude assumption of independent observations.

conversion factor 0.88) to an estimated true TFR in a range from 6.1 to 6.3. There is a very close correspondence with the results of the individual-level model, particularly in the volume of misreported (displaced) births.

The deviance or chi-square statistic is low for these three hypothesized values of Δm, considering that the sample includes 14,727 reported births, but is never below the nominal critical value. It can be mentioned that the deviance could be further reduced by allowing larger (but implausible) values of Δm. The bulk of the deviance in the fits described here is concentrated in the last two or three marital durations, in which the numbers of births are relatively small. The model fits very well within the duration intervals where most of the births occurred.

31.5 DISCUSSION

This chapter has proposed two quite similar models, one for the kind of aggregated data that might be published in a survey report, and the other for individual-level data that could only be obtained from access to the detailed birth histories. Our preference is for the individual-level model because it uses more information and makes more plausible assumptions. The two models are very close in terms of the estimated "true" TFRs in 1986–90 and in the estimated volume of displacement. The models cannot provide stable estimates of omission and displacement simultaneously, but for the Pakistan survey there is a basis for setting omission at zero.

If we are willing to grant that contraceptive use may have increased in Pakistan by as much as 5 percent across the ten years before the survey—and this would be a substantial increase, given that the (stated) prevalence in 1990–91 was only 11.8 percent, including traditional methods—then the contraceptive change parameter, Δm, would be approximately our middle value, 0.1. This level of change would imply (in the individual-level model) that 19 percent of the births that actually occurred in window 2—that is, 1,497 births—were displaced back into window 1. It would imply that the TMFRs in the two windows were 7.2 and 6.9; or, converting to approximate TFRs with a multiplier of 0.88, that the TFRs were 6.3 and 6.1, respectively. A shift to the most extreme scenarios, $\Delta m = 0.0$ or $\Delta m = 0.2$, would only alter these estimates by a small amount. It is clear that the PDHS report, despite extending the reference period from five years to six years, underestimated the true TFR by at least half a child.

It is reassuring that our estimates of displacement and the true TFR in 1986–90 are quite similar to those found in the reinterview survey. The estimated changes in actual fertility are also quite similar to those estimated with a completely different approach (parity progression) by Brass and Juarez (1983).

Although the interviewers—with an incentive provided by the questionnaire design—are the alleged source of displacement across the 1985–86 threshold, this model has not been disaggregated by interviewers. It is plausible that there was some heterogeneity across interviewers. The data allow identification of interviewers, as well as team supervisors, time of day of the interview, etc. The model also has not allowed for heterogeneity by type of place of residence, region of residence, education of respondent, and other characteristics of the respondents which might well be related to their knowledge of birthdates and the amount of misreporting. A fully comprehensive misreporting model in the context of Pakistan would also include parameters for the progressive background displacement of young children to older ages and parameters to represent digit preference, but the task of estimating all such effects simultaneously is indeed formidable.

The models described here, and particularly the method of generalized estimating equations, could be useful for modeling measurement error in any retrospective survey. GEE models allow specification of a nonzero correlation structure among observations in the sample, such as those made on the same sample unit at different points in time. Thus, for example, models that describe the extent of telescoping in a crime survey, or the extent of heaping of employment duration in a retrospective labor force survey, might use this estimation methodology.

ACKNOWLEDGMENTS

This chapter is a revision of a paper prepared for the Annual Meetings of the Population Association of America, April 6–8, 1995, San Francisco, CA.

This research was supported in part by research grant R01HD31063 from the National Institute of Child Health and Human Development. We gratefully acknowledge the helpful comments of Fred Arnold.

REFERENCES

Alam, I., and Dinesen, B. (1984), *Fertility in Pakistan: A Review of Findings from the Pakistan Fertility Survey*, Voorburg, Netherlands: International Statistical Institute.

Brass, W., and Juarez, F. (1983), "Censored Cohort Parity Progression Ratios from Birth Histories," *Asian and Pacific Census Forum*, 10, pp. 5–13.

Coale, A.J., and Trussell, T.J. (1974), "Model Fertility Schedules: Variations in the Age Structure of Childbearing in Human Populations," *Population Index*, 40, pp. 185–258.

Curtis, S.L., and Arnold, F. (1994), "An Evaluation of the Pakistan DHS Survey Based on the Reinterview Survey," *Occasional Papers*, No. 1. Calverton, MD: Macro International, Inc.

Dunlop, D.D. (1995), "Regression for Longitudinal Data: A Bridge from Least Squares Regression," *The American Statistician*, 48, pp. 299–303.

Karim, M.R., and Zeger, S.L. (1988), "GEE: A SAS Macro for Longitudinal Data

Analysis," Technical Report 674, Department of Biostatistics, The Johns Hopkins University.

Liang, K.-L., and Zeger, S.L. (1986), "Longitudinal Data Analysis Using Generalized Linear Models," *Biometrika* 73, pp. 13–22.

McCullagh, P., and Nelder, J.A. (1989), *Generalized Linear Models*, second edition, London: Chapman & Hall.

National Institute of Population Studies (1992), *Report on the Pakistan Demographic and Health Survey 1990–91*, Islamabad, Pakistan.

Okun, B.S. (1994), "Evaluating Methods for Detecting Fertility Control: Coale and Trussell's Model and Cohort Parity Analysis," *Population Studies*, 48, pp. 193–222.

Pullum, T.W. (1990), "Statistical Methods to Adjust for Date and Age Misreporting to Improve Estimates of Vital Rates in Pakistan," *Statistics in Medicine*, 10, pp. 191–200.

Retherford, R.D., Mirza, G.M., Irfan, M., and Alam, I. (1987), "Fertility Trends in Pakistan—the Decline that Wasn't," *Asian and Pacific Population Forum*, 1, pp. 1–10.

Shah, I.H., Pullum, T.W., and Irfan, M. (1986), "Fertility in Pakistan During the 1970s", *Journal of Biosocial Science*, 18, pp. 215–29.

Weinberger, M.B. (1991), "Recent Trends in Contraceptive Behavior," *Proceedings of the Demographic and Health Surveys World Conference*, Vol. 1, pp. 555–573.

Xie, Y., and Pimentel, E.E. (1992), "Age Patterns of Marital Fertility: Revising the Coale–Trussell Method," *Journal of the American Statistical Association*, 87, pp. 977–984.

CHAPTER 32

Estimators of Nonsampling Errors in Interview–Reinterview Supervised Surveys with Interpenetrated Assignments

Francesca Bassi and Luigi Fabbris
University of Padua

32.1 INTRODUCTION

In surveys where data are collected by interviewers, the major sources of nonsampling error are respondents, interviewers, supervisors and other field staff trainers, coders, and keyers. Respondents may give incorrect responses because of recall failure, social conditioning, confusion, and so on. If the responses deviate from a true value in a completely random fashion, so-called uncorrelated response error results. However, if the errors tend to be systematic, bias is introduced.

Interviewer error may be related to the interviewer's personal characteristics, attitudes, and behavior during the interview. The error is a bias if it is referred to a single interviewer. On the other hand, it is a correlated random error if it is referred to the sampled population. The errors of coders and other operators who influence the data are statistically similar to the errors of interviewers.

Supervisor error is another potential source of bias. Through training and communication, supervisors directly influence the interviewers and, consequently, indirectly influence responses collected by interviewers. The systematic errors of a particular group of supervisors may be modeled as fixed bias while the

Survey Measurement and Process Quality, Edited by Lyberg, Biemer, Collins, de Leeuw, Dippo, Schwarz, Trewin.
ISBN 0-471-16559-X   © 1997 John Wiley & Sons, Inc.

errors for a large population of supervisors is more appropriately modeled as correlated random error. In the former case, the particular group of supervisors available for the survey is considered a random sample from a large population of potential supervisors.

The training chain continues as instructed by the survey researchers, who may inadvertently pass on information, attitudes, or mindsets that will later emerge as error somewhere in the survey process. We call this type of error instruction error.

In this chapter, we propose a model for the study of nonsampling errors in an interviewer-assisted survey. Our model incorporates the effects of both interviewers and supervisors. Estimators of the various components of the variance of the sample mean and their standard errors are proposed and their efficiencies compared. In addition, the mean squared error (MSE) of estimators of the correlated response variance based upon single interviews are compared with estimates that combine the interview data with data from a reinterview survey (sometimes referred to as double sample estimators). Even though the interview–reinterview data set contains more information on the respondents, this larger data set may not improve the efficiency of the estimator of correlated response variance since the reinterview data may be correlated with the data from the parent survey. In this chapter, we provide sufficient conditions that the double sample estimator can be an improvement over the single sample estimator.

32.2 THE MODEL

Consider the situation in which a sample survey of a population is conducted using an interviewer-assisted mode of interview. Later a reinterview survey is conducted of a subsample of the survey respondents for evaluation purposes. Suppose the population, of size N^*, is divided into H strata and from each stratum a sample of units is drawn without replacement. We assume that each stratum is of equal size (N) and each stratum sample is of size $n = km$ where k denotes the number of interviewers in the evaluation and m is the size of each interviewer's assignment.

Further assume that the assignment of units to interviewers within a stratum is done randomly. This random assignment of units to interviewers is known as interpenetration (Mahalanobis, 1946) and is required for the estimation of correlated error variance in a single sample survey design. Finally, we assume that each of the H strata is assigned to one supervisor. In the following the simple assumptions of equal supervisor and interviewer assignments do not limit the generalizability of our conclusions to more complex survey situations.

We assume the same design of interpenetrated assignments used in the parent survey is also used in the reinterview survey. Further, without loss of generality, assume that all units in the parent survey are in the reinterview survey.

Table 32.1   Interviewers' and Supervisors' Assignments in the Parent and Reinterview Surveys

	Parent survey				Reinterview			
Supervisor	1		2		1		2	
Interviewer	1	2	3	4	1	2	3	4
Subsample	A	B	C	D	B	A	D	C

However, in the reinterview survey, the interviewer assignments are swapped with another interviewer within the same stratum so that each reinterviewer interviews an assignment other than his/her own. Note, the interviewers have the same supervisors in both the parent and the reinterview surveys.

To illustrate, let us consider the case with two supervisors ($H = 2$) and four interviewers ($k = 2$ in each stratum). In Table 32.1 under Parent survey, the letters A, B, C and D represent interviewer assignments of size m for the parent survey and under Reinterview the assignments of these interviewers in the reinterview survey.

Our assumptions extend those of Fellegi (1964) by including effects of one additional hierarchy of error, namely that of the supervisors. It can be easily shown that our interview–reinterview design with double interpenetration is identical to Fellegi (1964) when $H = 1$.

Under our design, the response model for a unit in stratum h is:

$$y_{hij} = \mu_{hij} + b_h + a_{hi} + r_{hij} \tag{32.1}$$

where

y_{hij} $(h = 1, \ldots, H; i = 1, \ldots, k, j = 1, \ldots, m)$ denotes the observed value of the normally distributed random variable Y,

μ_{hij} is the corresponding "true value" or error-free value for unit j, within interviewer i's assignment within stratum h,

a_{hi} is the error due to the hith interviewer,

b_h is the error of the hth supervisor, and

r_{hij} is the deviation of y_{hij} from $\mu_{hij} + a_{hi} + b_h$, which will be referred to as *respondent error*.

Thus we assume that a response is the result of the individual true value μ_{hij} distorted by the errors of the respondent, interviewer, or supervisor.

For component estimation, the following assumptions are made:

1. A sample of size $n = km$ is randomly drawn without replacement from each stratum of size N. A sample of Hk interviewers is randomly selected

from a large population of potential interviewers, each of whom is assigned a sample of m units. Each group of k interviewers operates solely within its own stratum.

2. The distributions of the error components of model (32.1) are normal with zero mean and known variance. The respondent errors r_{hij} are i.i.d. values from a $N(0, \sigma_r^2)$, the interviewer errors, a_{hi}, are i.i.d. from a $N(0, \sigma_a^2)$, and the supervisor errors, b_h, are i.d. (i.e., identically distributed but not independently) from a $N(0, \sigma_b^2)$. This implies that the joint distribution of observations, given the sample of units, is multivariate normal.

3. Interviewers' and supervisors' errors are independent of each other, and of the respondent error. For the sake of simplicity, independence is also assumed between sample deviations and the response errors.

The observed value of the jth unit in the ith assignment in stratum h, y_{hij}, may be equivalently written as

$$y_{hij} = \mu_h + d_{hij} + c_{hij} \tag{32.2}$$

(see Hansen *et al.*, 1961) where

$\mu_h = \sum_{i=1}^k \sum_{j=1}^m \mu_{hij}/n$ is the stratum mean of variable Y

$d_{hij} = r_{hij} + b_h + a_{hi}$ is the response deviation of unit j, enumerated by interviewer i, supervised by h

$c_{hij} = \mu_{hij} - \mu_h$ is the sample deviation of the same unit.

We refer to model (32.2) to define the nonsampling error components in the manner of Fellegi (1964). Let E( ) and V( ) denote expectation and variance, respectively, taken with respect to nonsampling error distributions as well as the sample design. Under assumptions (1)–(3), the variance of the observed mean of stratum h, denoted by

$$\bar{y}_h = \frac{1}{n} \sum_i^k \sum_j^m y_{hij} = \frac{1}{k} \sum_i^k \bar{y}_{hi}$$

is given by

$$V(\bar{y}_h) = \frac{N-n}{N-1} \frac{\sigma_{sh}^2}{n} + \frac{\sigma_{rh}^2}{n} [1 + (m-1)\delta_{2h} + m(k-1)\delta_{3h}] \tag{32.3}$$

where σ_{rh}^2 is the uncorrelated response variance defined as

$$\sigma_{rh}^2 = E\left\{ \frac{1}{km} \sum_{i=1}^k \sum_{j=1}^m d_{hij}^2 \right\}.$$

Here, σ_{sh}^2 is the sampling variance of variable Y in stratum h (i.e., the variance of the true values within the stratum) defined by

$$\sigma_{sh}^2 = E\left\{\frac{1}{km}\sum_{i=1}^{k}\sum_{j=1}^{m} c_{hij}^2\right\}.$$

We see that δ_{2h} is the correlation coefficient between response deviations within interviewers' assignments defined by

$$\delta_{2h} = \frac{1}{\sigma_{rh}^2}E\left\{\frac{1}{n(m-1)}\sum_{i}^{k}\sum_{j\neq j'}^{m} d_{hij}d_{hij'}\right\}$$

and $\delta_{2h}\sigma_{rh}^2$ is the (*correlated*) *interviewer variance*. Finally, we define δ_{3h}, the correlation coefficient between response deviations within supervisors' assignments, as

$$\delta_{3h} = \frac{1}{\sigma_{rh}^2}\left\{\frac{1}{nm(k-1)}\sum_{i\neq i'}^{k}\sum_{j,j'}^{m} d_{hij}d_{hi'j'}\right\}$$

where $\delta_{3h}\sigma_{rh}^2$ is the (*correlated*) *supervisor variance*.

An estimator of the overall population mean is

$$\bar{y} = \frac{1}{H}\sum_{h=1}^{H}\bar{y}_h$$

with variance

$$V(\bar{y}) = \frac{N-n}{N-1}\frac{1}{H^2n}\sum_{h=1}^{H}\sigma_{sh}^2 + \frac{1}{H^2n}\sum_{h=1}^{H}\sigma_{rh}^2[1 + (m-1)\delta_{2h} + m(k-1)\delta_{3h}]$$

$$+ \frac{1}{H^2}\sum_{h\neq h'}^{H}\delta_{4hh'}\sigma_{rh}^2\sigma_{rh'}^2 \tag{32.4}$$

where the term $\delta_{4hh'}$ is the *instructor effect*, i.e., the correlation coefficient among response deviations obtained in two different strata

$$\delta_{4hh'} = \frac{1}{\sigma_{rh}\sigma_{rh'}}E\left\{\frac{1}{k^2m^2}\sum_{i,i'}^{k}\sum_{j,j'}^{m} d_{hij}d_{h'i'j'}\right\}.$$

Note that the first terms on the right-hand sides of both (32.3) and (32.4) represent the variance due purely to sampling, while the remaining terms are all due to response errors.

Let us denote by μ and σ^2, the mean and the variance, respectively, of Y in the target population. Thus

$$\sigma^2 = E\left\{\frac{1}{Hn}\sum_{h=1}^{H}\sum_{i=1}^{k}\sum_{j=1}^{m}(y_{hij}-\mu)^2\right\} = \sigma_s^2 + \sigma_r^2$$

where $\sigma_s^2 = 1/H \sum_{h=1}^{H} \sigma_{sh}^2 + 1/H \sum_{h=1}^{H} (\mu_h - \mu)^2$ and $\sigma_r^2 = 1/H \sum_{h=1}^{H} \sigma_{rh}^2$. Furthermore, the response variance for all strata combined is the arithmetic mean across strata and

$$\delta_4 \sigma_r^2 = \frac{1}{H(H-1)} \sum_{h \neq h'}^{H} \delta_{4hh'} \sigma_{rh}^2 \sigma_{rh'}^2$$

and hence, (32.4) can be written as

$$V(\bar{y}) = \frac{N-n}{N-1} \frac{1}{Hn} \left[\sigma_s^2 - \frac{1}{H} \sum_{h=1}^{H} (\mu_h - \mu)^2 \right]$$

$$+ \frac{\sigma_r^2}{Hn} [1 + (m-1)\delta_2 + m(k-1)\delta_3 + n(H-1)\delta_4] \qquad (32.5)$$

where the terms without subscript h refer to population parameters.

Let us now consider the response error associated with the reinterview survey. We assume that both models (32.1) and (32.2) apply to reinterview data. The variance components are indexed by 1 if associated with the parent survey, by 2 if associated with the reinterview survey.

In addition to the variance components associated with reinterview, we also need to define the correlations of the variance model errors between surveys. Thus, we define the following parameters:

β_{1h} is the correlation between response errors of the same unit in different surveys, referred to as the *recall effect*, since this correlation is large if respondents remember erroneous responses given in the first survey and repeat them in the reinterview. Thus

$$\beta_{1h} = \frac{1}{\sigma_{rh1} \sigma_{rh2}} E \left\{ \frac{1}{n} \sum_i^k \sum_j^m d_{hij1} d_{hij2} \right\}$$

β_{2h} is the correlation coefficient between the response errors of two different units interviewed by the same interviewer but in different surveys. It is given by

$$\beta_{2h} = \frac{1}{\sigma_{rh1} \sigma_{rh2}} E \left\{ \frac{1}{nm} \sum_i^k \sum_{j \neq j'}^m d_{hij1} d_{hij'2} \right\}$$

β_{3h} is the correlation coefficient between response errors of units belonging to the same subsample in the two surveys. It is given by

$$\beta_{3h} = \frac{1}{\sigma_{rh1} \sigma_{rh2}} E \left\{ \frac{1}{n(m-1)} \sum_i^k \sum_{j \neq j'}^m d_{hij1} d_{hi^\circ j'2} \right\}$$

where i° denotes the interviewer in the reinterview survey and i denotes the interviewer in the parent survey.

β_{4h} is the correlation between response errors of different units assigned to different interviewers in different surveys and is given by

$$\beta_{4h} = \frac{1}{\sigma_{rh1}\sigma_{rh2}} \operatorname{E}\left\{\frac{1}{k(k-2)m^2} \sum_{i' \neq i, i^{\circ}}^{k} \sum_{j \neq j'}^{m} d_{hij1} d_{hi'j'2}\right\}$$

Note that β_{4h} is meaningless if $k = 2$ since the interviewers swap their assignments in the second interview.

Finally, $\beta_{5hh'}$ is the correlation coefficient between response deviations of units assigned to different supervisors in different surveys and is given by

$$\beta_{5hh'} = \frac{1}{\sigma_{rh1}\sigma_{rh'2}} \operatorname{E}\left\{\frac{1}{nkm} \sum_{i,i'}^{k} \sum_{j,j'}^{m} d_{hij1} d_{h'i'j'2}\right\}.$$

Note that β_{4h} and $\beta_{5hh'}$ are approximately of the same magnitude as δ_{3h} and $\delta_{4hh'}$, respectively. We see that β_{4h} is the correlation between units belonging to the same stratum and $\beta_{5hh'}$ is that between units interviewed by field personnel receiving the same instruction.

As with formula (32.5), these correlation coefficients can be defined for the entire population by averaging them across strata. In what follows, these population level correlation coefficients will be written using the same symbols but without subscript h.

32.3 ESTIMATORS OF ERROR COMPONENTS

In this section, the estimators of the uncorrelated variance (Section 32.3.1), interviewer variance (Section 32.3.2), and supervisor variance (Section 32.3.3) are obtained using analysis of variance (ANOVA) methods. Estimators suitable to data collection designs with and without reinterviewing are presented.

32.3.1 Uncorrelated Variance

An estimator for the uncorrelated response variance of stratum h, σ_{rh}^2, suggested by Fellegi (1964) is the following

$$s_{rh}^2 = \frac{1}{2k(m-1)} \sum_{i}^{k} \sum_{j}^{m} (y_{hij1} - y_{hi^{\circ}j2} - \bar{y}_{hi1} + \bar{y}_{hi^{\circ}2})^2. \tag{32.6}$$

Fellegi showed that without a second measurement for a subsample of parent survey respondents, the uncorrelated response variance is confounded with the sampling variance and is therefore not separately estimable. An estimator for the sum of sampling and uncorrelated response variance, i.e., σ_s^2 and σ_r^2, is given by Kish (1962)

$$s_{sh}^2 + s_{rh}^2 = \frac{1}{k(m-1)} \sum_{i}^{k} \sum_{j}^{m} (y_{hij} - \bar{y}_{hi})^2. \tag{32.7}$$

32.3.2 Interviewer Variance

When reinterview data are available, the estimator of the interviewer variance in stratum h, $\delta_{2h}\sigma_{rh}^2$, is given by Fellegi (1964)

$$_\mathrm{I}\hat{\delta}_{2h}s_{rh}^2 = \frac{1}{2(k-1)} \sum_i^k (\bar{y}_{hi1} - \bar{y}_{hi\circ 2} - \bar{y}_{h1} + \bar{y}_{h2})^2$$

$$-\frac{1}{2km(m-1)} \sum_i^k \sum_j^m (y_{hij1} - y_{hi\circ j2} - \bar{y}_{hi1} + \bar{y}_{hi\circ 2})^2 \qquad (32.8)$$

and, without reinterview, it is given by Kish (1962)

$$_\mathrm{II}\hat{\delta}_{2h}s_{rh}^2 = \frac{1}{k-1} \sum_i^k (\bar{y}_{hi} - \bar{y}_h)^2 - \frac{1}{km(m-1)} \sum_i^k \sum_j^m (y_{hij} - \bar{y}_{hi})^2. \qquad (32.9)$$

32.3.3 Supervisor Variance

Finally, an estimator of the supervisor variance which combines information from all strata in both surveys is given by Bassi (1991)

$$_\mathrm{I}\hat{\delta}_3 s_r^2 = \frac{1}{2(H-1)} \sum_h^H (\bar{y}_{h1} - \bar{y}_{h2} - \bar{y}_1 + \bar{y}_2)^2$$

$$-\frac{1}{2Hk(k-1)} \sum_h^H \sum_i^k (\bar{y}_{hi1} - \bar{y}_{hi\circ 2} - \bar{y}_{h1} + \bar{y}_{h2})^2. \qquad (32.10)$$

If reinterview data are not used then the estimator is given by Fabbris (1991)

$$_\mathrm{II}\hat{\delta}_3 s_r^2 = \frac{1}{H-1} \sum_h^H (\bar{y}_h - \bar{y})^2 - \frac{1}{Hk(k-1)} \sum_h^H \sum_i^k (\bar{y}_{hi} - \bar{y}_h)^2. \qquad (32.11)$$

32.4 ESTIMATOR EFFICIENCY

To assess the statistical distribution of estimators (32.6)–(32.11) the following results are needed. The first result is due to Bhat (1962) and states that under the assumption of normally distributed errors, the sample sums of squares and the sample variances are distributed as chi-squared variates. Furthermore, apart from a constant multiplier, this holds even when the observations are correlated, as long as the observations have a common variance and covariance. The second result is due to Matérn (1949), and states that two nonnegative uncorrelated quadratic forms consisting of normally distributed and correlated variables are independently distributed. These two results are applicable to model (32.1) since interviewer and supervisor sample means are assumed to be normally distributed with equal variance and common covariances.

Since the components that comprise the estimators proposed in Section 32.3 are mutually independent and distributed as chi-squared random variables apart from a constant, estimators (32.6) to (32.11) are also chi-square distributed. The simple relationship between mean and variance in the chi-squared distribution allows us to calculate the variance and, therefore, the mean squared error of each estimator.

It can be shown that the relationship between the mean and the variance of the quadratic forms holds even when observations are derived from nonnormal, continuous distributions under the condition that the fourth moment of the distribution is three times the square of the second moment. Thus, it is possible to relax the assumption of normality somewhat.

The expected values and variances of estimators (32.6)–(32.11) are given in the Appendix and the relative efficiency of alternative estimators of the correlated response variance is examined in Section 32.5.

32.5 COMPARISON OF THE ESTIMATORS

In this section, our goal is to compare the efficiency of the estimators based on the reinterview survey data with those based only on the parent survey data, under various assumptions about the magnitudes of the correlation coefficients β_i, for $i = 1, 2, 3, 4$, and 5. In what follows we assume a hierarchical relation among these correlations; namely that $\beta_i \geq \beta_{i+1}$, for $i = 2, 3, 4$, which, by the nature of the correlations, makes intuitive sense and greatly simplifies the discussion as well. This assumption is compatible with the hypothesis that supervisors are, on average, more skilled than interviewers, and researchers possess an even greater command of the survey. This hierarchy of skills is reflected again in the reinterview survey.

Before proceeding with the comparison, the following additional assumptions are needed.

1. The reinterview survey is carried out under the same essential conditions as the parent survey, ensuring identical reliability for both measurements. Then, we assume that $\sigma_{r1}^2 = \sigma_{r2}^2 = \sigma_r^2$; $\sigma_{s1}^2 = \sigma_{s2}^2 = \sigma_s^2$; $\delta_{21} = \delta_{22} = \delta_2$; $\delta_{31} = \delta_{32} = \delta_3$; and $\delta_{41} = \delta_{42} = \delta_4$. These simplifications hold if interviewer, supervisor and instruction effects do not differ between trials.

2 Interviewer error correlations are the same for both between the within occasions $\delta_2 = \beta_2$. The same holds for the supervisor errors $\delta_3 = \beta_4$ and the school effect, $\delta_4 = \beta_5$ (see also Fellegi, 1964).

3. $\beta_2 = \beta_3$, which means that the correlation between response deviations within the same interviewer's assignments in both surveys is the same as

the correlation between units belonging to the same subsample in each survey. This assumption further reduces the parametric space of our evaluation.

4. Because of the nested structure of assignments to interviewers, supervisors and instruction, we may further assume that $\delta_2 \geq \delta_3 \geq \delta_4$ and $\beta_2 = \beta_3 \geq \beta_4 \geq \beta_5$.

Assuming the correlation coefficients are nonnegative (see Fellegi, 1964) the differences between the MSEs of estimators may be precisely evaluated and suggestions for their use in empirical contexts may be derived. Using the formulas for the MSEs given in the Appendix, the estimators have been evaluated and compared in a numerical study, for a wide range of parameter values.

Under assumptions 1 through 4 above, we compare the MSEs of estimators (32.8) and (32.9) by varying only three parameters, namely β_2, β_4 and σ_r^2. For comparing the MSEs of estimators (32.10) and (32.11), β_5 must also be considered. Without loss of generality we set $\sigma_s^2 = 1$.

Estimators (32.8) and (32.9) and (32.10) and (32.11) have been compared for numerous combinations of parameters. Because of limited space, we include only a few examples that are representative of the results of this larger study. In particular, the following parameter values are considered: $\beta_1 = 0.3$ and 0.8, $\beta_2 = 0.05$, 0.1 and 0.2, $\sigma_r^2 = 0.3$, 1, 3 and 10 for estimators (32.8) and (32.9), and for estimators (32.10) and (32.11), we added the values $\beta_4 = 0.075$ and 0.15. These results are displayed in Figures 32.1 through 32.5.

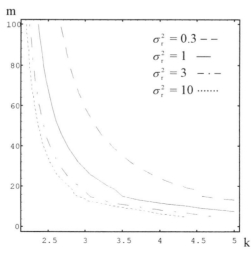

Figure 32.1 Plot of curves under which MSE(8) $\leq$ MSE(9) for different values of σ_r^2, $\beta_2 = 0.1$, $\beta_4 = 0.075$ and $\beta_1 = 0.3$.

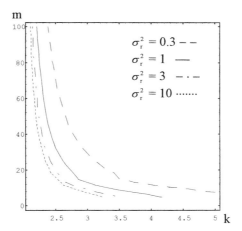

Figure 32.2 Plot of curves under which $MSE(8) \leq MSE(9)$ for different values of σ_r^2, $\beta_2 = 0.2$, $\beta_4 = 0.15$ and $\beta_1 = 0.3$.

32.6 RESULTS OF THE NUMERICAL STUDY

In this section, we describe the results of our simulation study to compare the double sample estimator of the interviewer variance with the single sample estimator under a number of configurations of parameter values. These values span the range of interest for most evaluation survey applications. In discussing our results, we make frequent reference to Figures 32.1 through 32.5 which graphically represent our results.

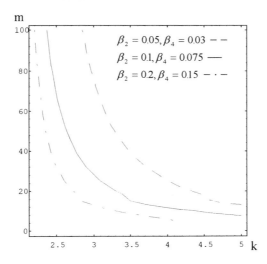

Figure 32.3 Plot of curves under which $MSE(8) \leq MSE(9)$ for different values of β_2, β_4 and $\sigma_r^2 = 1$, $\beta_1 = 0.3$.

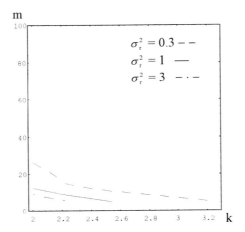

Figure 32.4 Plot of curves under which MSE(10) ≤ MSE(11) for different values of σ_r^2, $\beta_2 = 0.1$, $\beta_4 = 0.075$, $\beta_5 = 0.05$ and $\beta_1 = 0.3$.

In Figures 32.1, 32.2 and 32.3, the MSE of the interviewer variance estimator in (32.8), based on the interview–reinterview design and denoted by MSE(8), is compared with the MSE of the corresponding estimator in (32.9), based on a single survey interpenetrated assignment design and denoted by MSE(9). For alternative values of the number of interviewers per stratum (k), interviewer workloads (m), uncorrelated response variance σ_r^2, and combinations of the β coefficients as indicated in the figures, we computed the point at which MSE(9)

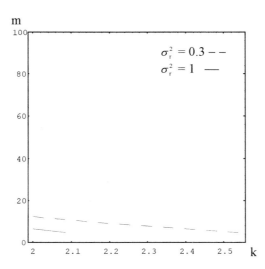

Figure 32.5 Plot of curves under which MSE(10) ≤ MSE(11) for different values of σ_r^2, $\beta_2 = 0.2$, $\beta_4 = 0.15$, $\beta_5 = 0.1$ and $\beta_1 = 0.3$.

exceeds MSE(8); that is, where the difference, MSE(8) − MSE(9), is negative and plotted this point for each combination of m and k. Thus, below the curve corresponding to a particular value of σ_r^2, the double sample estimator has smaller MSE than the single survey estimator while the opposite is true (i.e., MSE(8) exceeds MSE(9)) above the curve. It is clear from these figures that, for large values of the simple response variance, it is almost always more efficient to use the estimators based upon interpenetrated interviewer assignments alone. That is, the additional information provided by the reinterview survey does not offset the extra variance added by incorporating this information in the estimator.

Figures 32.4 and 32.5 present analogous curves for comparing estimators (32.10) and (32.11) for the supervisor effect with and without, respectively, the reinterview survey data. Again, below the curves, MSE(10) is smaller than MSE(11) indicating greater efficiency in the estimator which uses the reinterview data. Above the curves, it is more efficient to ignore the reinterview data and estimate the supervisor effect using the interpenetrated assignment information alone.

It can be seen that, for $k = 2$, when the average interviewer workload exceeds 30 (or 20 for larger values of the error correlation coefficients), it is no longer efficient to conduct a reinterview survey in order to increase the efficiency of response error estimators. In general, the interviewer and supervisor workloads are the primary determinant of the relative efficiency of the two types of estimators. For estimators based on an interview–reinterview design to be more efficient than those based solely on interpenetrated assignments, a relatively large interviewer workload as well as a large number of interviewers are needed in the evaluation survey. This is difficult to achieve in practice due to the cost of such large evaluation surveys.

The shapes of the curves are just slightly changed if β_1 varies from its value set in the tables; only if the recall effect is small (for example, less than 0.3), an increase in the relative efficiency of estimators based on the interview–reinterview design is detected.

Corollaries to these primary results are outlined in Table 32.2 where a more extensive analysis of the two components of the MSE (bias and variance) is performed under the following assumptions.

Case (1): $\beta_1 > 0$ and $\beta_i = 0$ ($i = 2, 3, 4, 5$), i.e., we assume that the respondent recall effect is positive and that the reinterviewers are different from those in the parent survey.

Case (2): $\beta_i > 0$, ($i = 1, 2, 3$) and $\beta_i = 0$ ($i = 4, 5$); this is the design considered by Fellegi (1964) in which correlation between different interviewers and different supervisors in different surveys is assumed to be 0.

Case (3): $\beta_i > 0$ ($i = 1, 2, 3, 4$) and $\beta_5 = 0$, i.e., only the correlation between different supervisors in different surveys is assumed to be zero.

Case (4): $\beta_i > 0$ ($i = 1, 2, 3, 4, 5$); the most general case of all positive correlations.

Table 32.2 Study of Bias and Variance of Estimators Under Cases (1)–(4)

	Cases			
	1	2	3	4
$\lvert\text{BIAS}(8)\rvert - \lvert\text{BIAS}(9)\rvert$	$=0$	$\geq 0\ (=0, k=2)$	$\geq 0\ (=0, k=2)$	
$\text{VAR}(8) - \text{VAR}(9)$ Sufficient conditions	<0 $\beta_1\sigma_r^2 \leq \sigma_s^2$	<0 $\delta_2\sigma_r^2 \leq \sigma_s^2$ $2\delta_3(k-1) \geq k\delta_2$	<0 $\delta_2\sigma_r^2 \leq \sigma_s^2$	
$\lvert\text{BIAS}(10)\rvert - \lvert\text{BIAS}(11)\rvert$	$=0$	>0	>0	>0
$\text{VAR}(10) - \text{VAR}(11)$ Sufficient conditions	<0 $\beta_1\sigma_r^2 \leq \sigma_s^2$	<0 $\delta_2\sigma_r^2 \leq \sigma_s^2$ $\delta_3(k-1) \geq k\delta_4$ $k\delta_2 \geq 2(k-1)/\delta_3$	<0 $\delta_2\sigma_r^2 \leq \sigma_s^2$ $\delta_3 \geq 2\delta_4$	<0 $\delta_2\sigma_r^2 \leq \sigma_s^2$

The results of the comparisons between the estimators under these four cases are summarized in Table 32.2. In the first row of Table 32.2, the direction of the difference between the absolute value of the bias of estimator (32.8), $\lvert\text{BIAS}(8)\rvert$, and the absolute value of the bias of estimator (32.9), $\lvert\text{BIAS}(9)\rvert$, is reported. The second row gives the direction of the difference between variances of the two estimators when the sufficient conditions given in the third row are satisfied. The fourth and fifth rows along with the sufficient conditions in the seventh row provide the analogous results for (32.10) and (32.11).

Under case (1), both estimators based on repeated measures are more efficient than the ones based solely on the interpenetrated parent survey data when the coefficient condition, $\beta_1\sigma_r^2 \leq \sigma_s^2$, is satisfied; i.e., when the recall error is less than or equal to the ratio between sampling and uncorrelated response variance.

Estimator (32.8) has smaller variance but is more biased than estimator (32.9) under the sufficient condition, $\delta_2\sigma_r^2 \leq \sigma_s^2$, and $2\delta_3(k-1) \geq k\delta_2$ of case (2), and the condition $\delta_2\sigma_r^2 \leq \sigma_s^2$ of case (3). The two estimators are equally biased if there are just two interviewers per stratum.

In all cases but (1), the estimator of supervisor variance with repeated measurements, (32.10), has lower variance than the corresponding estimator without repeated measurements, (32.11), under the sufficient conditions: i.e., for case (2), when $\delta_2\sigma_r^2 \leq \sigma_s^2$, $\delta_3(k-1) \geq k\delta_4$, and $k\delta_2 \geq 2(k-1)/\delta_3$; for case (3), when $\delta_2\sigma_r^2 \leq \sigma_s^2$ and $\delta_3 \geq 2\delta_4$; and, for case (4), when $\delta_2\sigma_r^2 \leq \sigma_s^2$. However, under these same conditions, estimator (32.10) is more biased than estimator (32.11).

Both estimators based on the interview–reinterview design then have lower variance than those based on a single survey when the sufficient conditions are satisfied. Furthermore, these conditions are very likely to be satisfied in practice.

Note that the regions of the parameter space considered in Table 32.2 for estimators (32.8) and (32.9) all satisfy the condition $\delta_2\sigma_r^2 \leq \sigma_s^2$ which, under case (3), is sufficient for the variance of estimator (32.8) to be smaller than the variance of estimator (32.9). Likewise, the regions of the parameter space for estimators (32.10) and (32.11) include the sufficient condition of case (4) which implies that the variance of estimator (32.10) is smaller than that of estimator (32.11). Therefore, the failure of estimators based on double sampling to be more efficient that those based upon a single survey must be due to biases in the double sample estimators that increase as the simple response variance for both surveys increases and the correlation of between-survey response errors increases.

32.7 SUMMARY AND CONCLUSIONS

In this study, a model for the evaluation of respondent, interviewer and supervisor errors is proposed. Under this model, a number of alternative estimators of the components of the correlated response variance are considered. In particular, an estimator for the evaluation of supervisor error with a doubly interpenetrated interview–reinterview design is proposed. Formulas for the variances of the proposed estimators are derived and their mean squared errors are compared with estimators proposed by other authors.

Regions of the parameter space and sufficient conditions under which the interview–reinterview estimators are more efficient than estimators that use only data from the parent survey are investigated both analytically and through a simulation study.

Essentially, the interview–reinterview estimators are more efficient when response deviations of the parent and the reinterview survey are independent. However, the single survey estimators may be more efficient if the simple response variance is large and the between-survey response error correlations are positive and large. The latter condition is very likely to occur when interviewers or supervisors are the same for the two occasions. Thus, if response errors tend to be large and correlated between surveys, using the results of a reinterview survey to estimate the supervisor or interviewer effects in the parent survey is not efficient.

For large interviewer assignments and when there are more than two interviewers per stratum, an estimator based on a single interpenetrated sample design will often yield better estimates of interviewer effect than an estimator based on a double interpenetrated reinterview design. Likewise, under these same conditions, a single interpenetrated survey may be more efficient than a double interpenetrated interview–reinterview survey design for supervisor error evaluation.

However, there are situations in which the double sample estimator is more efficient than the single interpenetrated design for estimating correlated errors. Here, we identified some combinations of m, the average interviewer assignments size, and k, the average number of interviewers per supervisor, for which this is true. Precisely, the double sampling estimator is more efficient for combinations of low values of k and m; moreover, the higher the simple response variance, the lower should be k and m for the above to happen.

In our analysis, we did not consider data collection costs under each design. Of course, the extra cost of conducting a reinterview survey would provide an even greater advantage for the single survey estimator for those cases where the single survey estimator has smaller MSE than the double sample estimator. However, for cases where the double sample estimator is more efficient, careful consideration of the costs of the reinterview survey relative to its gain in efficiency is warranted.

Our results also suggest that the topic of estimator efficiency for complex surveys deserves further attention. Composite estimators, i.e., estimators based on combinations of biased concurrent estimators, similar to those suggested by Fellegi (1964), may reduce bias and produce efficient estimates.

ACKNOWLEDGMENTS

The research work documented in this chapter has been carried out jointly by the two authors. Nevertheless, Sections 32.1 and 32.2 have been written by L. Fabbris and the following ones by F. Bassi. The authors are indebted to several colleagues for ideas and critical evaluation. Among them are Claudio Agostinelli, Paul Biemer, Gad Nathan, Lynne Stokes, and Nicola Torelli.

REFERENCES

Bassi, F. (1991), "Efficienza di stimatori dell'errore extracampionario," ("Efficiency of Estimators of Nonsampling Errors"), Ph.D. dissertation, Statistics Department, University of Padua.

Bhat, B.R. (1962), "On the Distribution of Certain Quadratic Forms in Normal Variates," *Journal of the Royal Statistical Society*, Series B, 24, pp. 148–151.

Fabbris, L. (1991), "Abbinamento tra fonti di errore nella formazione dei dati e misure dell'effetto degli errori sulle stime," ("The Pairing of Error Sources and Measures of Estimates Variance"), *SIS Bulletin*, 22, pp. 19–54.

Fellegi, I.P. (1964), "Response Variance and Its Estimation," *Journal of the American Statistical Association*, 69, pp. 496–501.

Hansen, M.H., Hurwitz, W.N., and Bershad, M.A. (1961), "Measurement Errors in Censuses and Surveys," 38 (2), pp. 359–374.

Kish, K. (1962), "Studies of Interviewer Variance for Attitudinal Variables," *Journal of the American Statistical Association*, 57, pp. 92–115.

Mahalanobis, P.C. (1946), "Recent Experiments in Statistical Sampling in the Indian Statistical Institute," *Journal of the Royal Statistical Society*, 109, pp. 325–370.

Matérn, B. (1949), "Independence of Non-Negative Quadratic Forms in Normally Correlated Variables," *Annals of Mathematical Statistics*, 20, pp. 119–120.

APPENDIX

Expected values and variances of the estimators under model (32.2) and assumption of normally distributed error components.

$$E[s_{rh}^2] = \frac{\sigma_{rh1}^2 + \sigma_{rh2}^2}{2} + \frac{N^*}{N^* - 1} \frac{\sigma_{sh1}^2 + \sigma_{sh2}^2 - 2\sigma_{sh1}\sigma_{sh2}}{2} - \frac{\delta_{2h1}\sigma_{rh1}^2 + \delta_{2h2}\sigma_{rh2}^2}{2}$$
$$- (\beta_{1h} - \beta_{3h})\sigma_{rh1}\sigma_{rh2}$$

$$V[s_{rh}^2] = \frac{2}{k(m-1)}\left[\frac{N^*}{N^*-1} \frac{\sigma_{sh1}^2 + \sigma_{sh2}^2 - 2\sigma_{sh1}\sigma_{sh2}}{2} \right.$$
$$\left. + \frac{(1-\delta_{2h1})\sigma_{rh1}^2 + (1-\delta_{2h2})\sigma_{rh2}^2}{2} - (\beta_{1h} - \beta_{3h})\sigma_{rh1}\sigma_{rh2} \right]^2$$

$$E[s_{rh}^2 + s_{sh}^2] = \frac{N^*}{N^* - 1}\sigma_{sh}^2 + (1 - \delta_{2h})\sigma_{rh}^2$$

$$V[s_{rh}^2 + s_{sh}^2] = \frac{2}{k(m-1)}\left[\frac{N^*}{N^* - 1}\sigma_{sh}^2 + (1 - \delta_{2h})\sigma_{rh}^2 \right]^2$$

$$E[_l\hat{\delta}_{2h}s_{rh}^2] = \frac{\delta_{2h1}\sigma_{rh1}^2 + \delta_{2h2}\sigma_{rh2}^2}{2} - \frac{\delta_{3h1}\sigma_{rh1}^2 + \delta_{3h2}\sigma_{rh2}^2}{2}$$
$$+ \sigma_{rh1}\sigma_{rh2}\frac{\beta_{2h} - (k-1)\beta_{3h} + (k-2)\beta_{4h}}{k-1}$$

$$V[_l\hat{\delta}_2 s_{rh}^2] = \frac{2}{m^2(k-1)}\left[\frac{N^*}{N^* - 1} \frac{\sigma_{sh1}^2 + \sigma_{sh2}^2 - 2\sigma_{sh1\,sh2}}{2} \right.$$
$$+ \frac{(1-\delta_{2h1})\sigma_{rh1}^2 + (1-\delta_{2h2})\sigma_{rh2}^2}{2}$$
$$\left. + m\frac{(\delta_{2h1} - \delta_{3h1})\sigma_{rh1}^2 + (\delta_{2h2} - \delta_{3h2})\sigma_{rh2}^2}{2} \right.$$

$$
- \sigma_{rh1}\sigma_{rh2} \frac{(k-1)\beta_{1h} - m\beta_{2h} + (m-1)(k-1)\beta_{3h} - m(k-2)\beta_{4h}}{k-1} \Bigg]^2
$$

$$
+ \frac{2}{km^2(m-1)} \Bigg[\frac{N^*}{N^*-1} \frac{\sigma_{sh1}^2 + \sigma_{sh2}^2 - 2\sigma_{sh1}\sigma_{sh2}}{2}
$$

$$
+ \frac{(1-\delta_{2h1})\sigma_{rh1}^2 + (1-\delta_{2h2})\sigma_{rh2}^2}{2} - (\beta_{1h} - \beta_{3h})\sigma_{rh1}\sigma_{rh2} \Bigg]^2
$$

$$
E[_{II}\hat{\delta}_{2h}s_{rh}^2] = \delta_{2h}\sigma_{rh}^2 - \delta_{3h}\sigma_{rh}^2
$$

$$
V[_{II}\hat{\delta}_{2h}s_{rh}^2] = \frac{2}{m^2(k-1)} \Bigg[\frac{N^*}{N^*-1}\sigma_{sh}^2 + (1-\delta_{2h})\sigma_{rh}^2 + m(\delta_{2h} - \delta_{3h})\sigma_{rh}^2 \Bigg]^2
$$

$$
+ \frac{2}{km^2(m-1)} \Bigg[\frac{N^*}{N^*-1}\sigma_{sh}^2 + (1-\delta_{2h})\sigma_{rh}^2 \Bigg]^2
$$

$$
E[_{I}\hat{\delta}_{3}s_{r}^2] = \frac{\delta_{31}\sigma_{r1}^2 + \delta_{32}\sigma_{r2}^2}{2} - \frac{\delta_{41}\sigma_{r1}^2 + \delta_{42}\sigma_{r2}^2}{2}
$$

$$
- \sigma_{r1}\sigma_{r2} \frac{(k-1)\beta_2 - \beta_4 + (k-1)\beta_5}{k-1}
$$

$$
V[_{I}\hat{\delta}_{3}s_{r'}^2] = \frac{2}{(H-1)m^2k^2} \Bigg[\frac{N^*}{N^*-1} \frac{\sigma_{s1}^2 + \sigma_{s2}^2 - 2\sigma_{s1}\sigma_{s2}}{2}
$$

$$
+ \frac{(1-\delta_{21})\sigma_{r1}^2 + (1-\delta_{22})\sigma_{r2}^2}{2}
$$

$$
+ m\frac{(\delta_{21} - \delta_{31})\sigma_{r1}^2 + (\delta_{22} - \delta_{32})\sigma_{r2}^2}{2}
$$

$$
+ mk\frac{(\delta_{31} - \delta_{41})\sigma_{r1}^2 + (\delta_{32} - \delta_{42})\sigma_{r2}^2}{2}
$$

$$
- \sigma_{rh1}\sigma_{rh2}(\beta_1 + \beta_2 + (n-1)\beta_3 + m(k-2)\beta_4 - mk\beta_5) \Bigg]^2
$$

$$
+ \frac{2}{Hk^2m^2(k-1)} \Bigg[\frac{N^*}{N^*-1} \frac{\sigma_{s1}^2 + \sigma_{s2}^2 - 2\sigma_{s1}\sigma_{s2}}{2}
$$

$$
+ \frac{(1-\delta_{21})\sigma_{r1}^2 + (1-\delta_{22})\sigma_{r2}^2}{2} + m\frac{(\delta_{21} - \delta_{31})\sigma_{r1}^2 + (\delta_{22} - \delta_{32})\sigma_{r2}^2}{2}
$$

$$
- \sigma_{r1}\sigma_{r2} \frac{(k-1)\beta_1 - m\beta_2 + (n-1)(k-1)\beta_3 - m(k-2)\beta_4}{k-1} \Bigg]^2
$$

$$E[{}_{\text{II}}\hat{\delta}_3 s_r^2] = \delta_3 \sigma_r^2 - \delta_4 \sigma_r^2$$

$$V[{}_{\text{II}}\hat{\delta}_3 s_r^2] = \frac{2}{(H-1)k^2 m^2} \left[\frac{N^*}{N^*-1} \sigma_s^2 + (1-\delta_2)\sigma_r^2 + m(\delta_2 - \delta_3)\sigma_r^2 \right.$$

$$\left. + mk(\delta_3 - \delta_4)\sigma_r^2 \right]^2 + \frac{2}{Hk^2 m^2 (k-1)}$$

$$\times \left[\frac{N^*}{N^*-1} \sigma_s^2 + (1-\delta_2)\sigma_r^2 + m(\delta_2 - \delta_3)\sigma_r^2 \right]^2$$

Variance Estimation Under Stratified Two-Phase Sampling with Applications to Measurement Bias

J. N. K. Rao and R. R. Sitter

Department of Mathematics and Statistics, Carleton University

33.1 INTRODUCTION

Two-phase sampling or double sampling is widely used in sample surveys and experimental studies. In the classic application (see, e.g., Cochran, 1977, Chapter 12), a large first-phase sample is taken in which an auxiliary variable, correlated with a characteristic of interest and relatively inexpensive to obtain, alone is measured. The second-phase subsample in which the characteristic of interest is measured is then employed to get efficient estimators of the population mean through ratio or regression estimation, using the first-phase auxiliary information. We focus on stratified two-phase sampling in this chapter.

It is now widely recognized that response or measurement errors can have a substantial effect on the accuracy of a survey estimate, as measured by its mean squared error (MSE). The general survey model developed by Hansen *et al.* (1961) has enabled us to identify the contributions to total MSE of a survey estimate from different error sources and to measure this effect. The contributions due to response errors include response variance and squared measurement bias. Extensive literature exists on the estimation of response variance (see, e.g., Biemer and Stokes, 1985; Kleffe *et al.*, 1991). In this chapter we are mainly concerned with the estimation of measurement bias, employing

Survey Measurement and Process Quality, Edited by Lyberg, Biemer, Collins, de Leeuw, Dippo, Schwarz, Trewin.
ISBN 0-471-16559-X   © 1997 John Wiley & Sons, Inc.

stratified two-phase sampling in which a subsample of survey respondents is selected in each stratum and the true values of the characteristics of interest are ascertained through some means such as reconciled reinterviews. The subsample size is typically small relative to the original sample because of the high costs involved in obtaining the true values. Madow (1965) used two-phase sampling to correct the survey estimate for measurement bias through difference estimation, whereas Brackstone et al. (1975) used the subsample data to estimate the mean squared error of the survey estimate.

Biemer and Atkinson (1993) used stratified two-phase sampling to estimate the measurement bias as opposed to eliminating the measurement bias. They proposed several estimators of the measurement bias including ratio and generalized regression estimators using the values of an auxiliary variable known for each unit in the original sample and its known population mean. Extending the single-phase bootstrap procedure of Gross (1980) (see also Bickel and Freedman, 1984; Sitter, 1992) they also obtained bootstrap variance estimators which were used to evaluate the estimators of measurement bias. For the latter purpose, data from the U.S. National Agriculture Service reinterview program was used. The bootstrap variance estimators of Biemer and Atkinson yield consistent variance estimators provided the within-stratum sampling fractions are negligible. In the case of nonnegligible first-phase sampling fractions it is, however, difficult to see a correction to their method which leads to consistent variance estimators.

In this chapter we introduce a general class of estimators which includes the estimators of Biemer and Atkinson as special cases. This class also covers classical applications of two-phase sampling. We obtain Taylor linearization and jackknife variance estimators and spell out the formulae for the specific estimators of interest. These variance estimators will remain valid even when the within-strata sampling fractions are not negligible. We also introduce a modification of the rescaling bootstrap proposed by Rao and Wu (1988) which yields consistent variance estimators in the presence of nonnegligible within-strata sampling fractions. Finally, the proposed variance estimators are evaluated through a limited simulation study.

33.2 ESTIMATION OF MEASUREMENT BIAS

Suppose we have a finite population of N units stratified into L strata. Let μ_{hi} be the true value of a characteristic of interest for the ith unit in the hth stratum $(i = 1, \ldots, N_h; \; h = 1, \ldots, L; \; \sum_h N_h = N)$, and $\bar{M} = \sum_h \sum_i \mu_{hi}/N$ be the corresponding population mean. Also, let μ'_{hi} be the average response obtained from conceptual repetitions of the measuring process on unit (hi), and $\bar{M}' = \sum_h \sum_i \mu'_{hi}/N$ be the corresponding population mean. The measurement bias is then defined as $\bar{B} = \bar{M}' - \bar{M}$.

Following Biemer and Atkinson (1993) we consider two-phase sampling in each stratum h using simple random sampling without replacement at each phase. A sample, s_{1h}, of n_{1h} units is selected in the first phase and the outcome, y_{hi}, subject to measurement error and an auxiliary value, x_{hi}, are observed for all $i \in s_{1h}$. In the second phase, a subsample, s_{2h}, of n_{2h} units is selected and the true values, μ_{hi}, are observed for all $i \in s_{2h}$. We assume that the population mean, $\bar{X} = \sum_h \sum_i x_{hi}/N$, is known. We denote the first-phase sample means as $(\bar{y}_{1h}, \bar{x}_{1h})$ and the second-phase sample means as $(\bar{y}_{2h}, \bar{x}_{2h}, \bar{\mu}_{2h})$.

The usual unbiased estimator of $\bar{M}$ is $\bar{\mu}_{2st} = \sum_h W_h \bar{\mu}_{2h}$ where $W_h = N_h/N$, i.e., $E_p(\bar{\mu}_{2st}) = \bar{M}$ where E_p denotes the expectation with respect to the sampling design. The customary estimator of measurement bias, $\bar{B}$, is the net difference rate (NDR) given by

$$\bar{b}_{2st} = \text{NDR} = \bar{y}_{2st} - \bar{\mu}_{2st} \tag{33.1}$$

where $\bar{y}_{2st} = \sum_{h=1}^{L} W_h \bar{y}_{2h}$. This estimator is unbiased in the sense $E_m E_p(\bar{b}_{2st}) = \bar{B}$, where E_m stands for the expectation with respect to the measurement process. Note that $\bar{b}_{2st}$ uses only the second-phase means $(\bar{y}_{2h}, \bar{\mu}_{2h})$.

Biemer and Atkinson (1993) suggest a number of possible alternate estimators of $\bar{B}$ which make use of the outcome values, y_{hi}, available on $s_{1\sim2h} = s_{1h} - s_{2h}$ or the auxiliary values, x_{hi}, available on s_{1h}, or both. The simplest of these is obtained from (33.1) by replacing $\bar{y}_{2st}$ by $\bar{y}_{1st} = \sum_{h=1}^{L} W_h \bar{y}_{1h}$

$$\bar{b}_{12st} = \bar{y}_{1st} - \bar{\mu}_{2st}. \tag{33.2}$$

This estimator is also unbiased for $\bar{B}$. Next $\bar{\mu}_{2st}$ in (33.1) or (33.2) can be replaced by a two-phase combined ratio estimator, $\bar{\mu}_{2stR} = (\bar{\mu}_{2st}/\bar{y}_{2st})\bar{y}_{1st}$, to get

$$\bar{b}_{2stR} = \bar{y}_{2st} - \bar{\mu}_{2stR} \tag{33.3}$$

and

$$\bar{b}_{12stR} = \bar{y}_{1st} - \bar{\mu}_{2stR}. \tag{33.4}$$

One can also use the auxiliary information, x_{hi} and $\bar{X}$, in various ways to improve the estimation of $\bar{B}$. First, using a combined ratio estimator of $\bar{M}'$, $\bar{y}_{xstR} = (\bar{y}_{1st}/\bar{x}_{1st})\bar{X}$, instead of $\bar{y}_{1st}$ in (33.4) we get

$$\bar{b}_{x2stR} = \bar{y}_{xstR} - \bar{\mu}_{2stR}. \tag{33.5}$$

Secondly, one could apply a generalized regression estimator of $\bar{M}$ (see Särndal et al., 1992, p. 360)

$$\bar{\mu}_{SSW} = \bar{\mu}_{2stR} + \frac{\bar{\mu}_{2st}}{\bar{x}_{2st}}(\bar{X} - \bar{x}_{1st})$$

in place of $\bar{\mu}_{2stR}$ in (33.5) to get

$$\bar{b}_{SSW} = \bar{y}_{xstR} - \bar{\mu}_{SSW}. \tag{33.6}$$

The alternative estimators (33.3)–(33.6) are approximately unbiased for $\bar{B}$ provided the second-phase sample size, $n_2 = \sum_h n_{2h}$, is large.

In developing variance estimators for the above estimators of measurement bias, it is convenient to view them as special cases of a general class of estimators, $\hat{\theta} = g(\bar{w}_2, \bar{v}_1)$, of a population parameter $\theta = g(\bar{W}, \bar{V})$. Here $\bar{W} = \sum_{h=1}^{L} \sum_{i \in U_h} w_{hi}/N$ is the population mean of $w = (u^T, v^T)^T$ and $\bar{V} = \sum_{h=1}^{L} \sum_{i \in U_h} v_{hi}/N$ the population mean of v, where v is observed for the entire first-phase sample while u is observed only on the second-phase sample. The sample estimators $\bar{w}_2$ and $\bar{v}_1$ are assumed to be unbiased for $\bar{W}$ and $\bar{V}$, respectively. The above set-up also covers classical applications of two-phase sampling (Cochran, 1977, Chapter 12).

Writing $\hat{\theta} = g(\bar{W} + \Delta\bar{w}_2, \bar{V} + \Delta\bar{v}_2) = h(\Delta\bar{w}_2, \Delta\bar{v}_1)$ with $\Delta\bar{w}_2 = \bar{w}_2 - \bar{W}$ and $\Delta\bar{v}_1 = \bar{v}_1 - \bar{V}$, we see that $\theta = h(0, 0)$. For example, consider the estimator $\bar{b}_{12st}$ given in (33.2). In this case $\bar{w}_2 = (\bar{u}_2, \bar{v}_2)^T = (\bar{\mu}_{2st}, \bar{y}_{2st})^T$, $\bar{v}_1 = \bar{y}_{1st}$, $\bar{U} = \bar{M}$, $\bar{V} = \bar{Y}, \hat{\theta} = \bar{v}_1 - \bar{u}_2, \theta = \bar{V} - \bar{U}$ and $h(\Delta\bar{w}_2, \Delta\bar{v}_1) = \bar{V} + \Delta\bar{v}_1 - (\bar{U} + \Delta\bar{u}_2)$. Similarly for the estimator $\bar{b}_{2stR}$ given by (33.3) we have the same $\bar{w}_2$ and $\bar{v}_1$ but now $\hat{\theta} = \bar{v}_2 - (\bar{u}_2/\bar{v}_2)\bar{v}_1$, $\theta = \bar{V} - (\bar{U}/\bar{V})\bar{V} = \bar{V} - \bar{U}$ and $h(\Delta\bar{w}_2, \Delta\bar{v}_1) = \bar{V} + \Delta\bar{v}_2 - [(\bar{U} + \Delta\bar{u}_2)/(\bar{V} + \Delta\bar{v}_2)](\bar{V} + \Delta\bar{v}_1)$. The remaining estimators can also be expressed as $\hat{\theta} = g(\bar{w}_2, \bar{v}_1)$ for suitably defined $\bar{w}_2$ and $\bar{v}_1$. In particular, for the estimator $\bar{b}_{12stR}$ given by (33.4) we have the same $\bar{w}_2$ and $\bar{v}_1$ but $\hat{\theta}$ is changed to $\hat{\theta} = \bar{v}_1 - (\bar{u}_2/\bar{v}_2)\bar{v}_1$. Turning to the estimator $\bar{b}_{x2stR}$ given by (33.5) we have $\bar{w}_2 = (\bar{u}_2, \bar{v}_2^T)^T$ with $\bar{u}_2 = \bar{\mu}_{2st}$, $\bar{v}_2 = (\bar{v}_{21}, \bar{v}_{22})^T = (\bar{y}_{2st}, \bar{x}_{2st})^T$, and $\bar{v}_1 = (\bar{v}_{11}, \bar{v}_{12})^T = (\bar{y}_{1st}, \bar{x}_{1st})^T$. Further, $\hat{\theta} = (\bar{v}_{11}/\bar{v}_{12})\bar{V}_2 - (\bar{u}_2/\bar{v}_{21})\bar{v}_{11}$ with $\bar{V}_2 = \bar{X}$. Finally, the estimator (33.6) can also be expressed in terms of the same $\bar{w}_2$ and $\bar{v}_1$ but $\hat{\theta}$ changed to $\hat{\theta} = (\bar{v}_{11}/\bar{v}_{12})\bar{V}_2 - (\bar{u}_2/\bar{v}_{21})\bar{v}_{11} - (\bar{u}_2/\bar{v}_{22})(\bar{V}_2 - \bar{v}_{12})$.

33.3   LINEARIZATION VARIANCE ESTIMATORS

The mean squared error of an estimator of $\bar{B}$, say $\bar{b}$, may be written as $\text{MSE}(\bar{b}) = \text{E}_m \text{MSE}_p(\bar{b}) + \text{E}_m(\bar{Y} - \bar{M}')^2$ assuming that $\bar{b}$ is unbiased (or approximately unbiased), where $\text{MSE}_p(\bar{b}) = \text{E}_p(\bar{b} - \bar{B}_y)^2$ with $\bar{B}_y = \bar{Y} - \bar{M}$ and $\bar{Y} = \sum_h \sum_i y_{hi}/N$. We can obtain consistent estimators of $\text{MSE}_p(\bar{b})$ and hence of the first component $\text{E}_m \text{MSE}_p(\bar{b})$. The second component, $\text{E}_m(\bar{Y} - \bar{M}')^2$, however, depends on the measurement process. For example, if the y_{hi} are independent with $\text{E}_m(y_{hi}) = \mu'_{hi}$ and $\text{V}_m(y_{hi}) = \sigma^2$, then $\text{E}_m(\bar{Y} - \bar{M}')^2 = \text{V}_m(\bar{Y}) = \sigma^2/N$. Biemer and Atkinson (1993) proposed an *ad hoc* estimator of $\text{MSE}(\bar{b})$ based on their bootstrap method, but its rationale in the context of the measurement process is not clear to us. In this chapter we confine ourselves to the estimation of design MSE, *viz.*, $\text{MSE}_p(\bar{b})$, of the six estimators given in Section 33.2.

We first consider Taylor linearization variance estimators for the general class $\hat{\theta} = g(\bar{w}_2, \bar{v}_1)$. The variance estimators for the special cases can be obtained from these general formulae, but it is easier to obtain them directly using the customary ratio approximation. The general treatment nevertheless allows us to introduce bootstrap and jackknife variance estimators more easily and to prove their consistency.

By a Taylor expansion of $\hat{\theta} = h(\Delta\bar{w}_2, \Delta\bar{v}_1)$ around the point $(\mathbf{0}, \mathbf{0})$, we obtain

$$\hat{\theta} - \theta = (\Delta\bar{w}_2)^T h^{(w)} + (\Delta\bar{v}_1)^T h^{(v)} + o_p(n_2^{-1/2}) \tag{33.7}$$

where $h^{(w)} = h^{(w)}(\mathbf{0}, \mathbf{0})$ and $h^{(v)} = h^{(v)}(\mathbf{0}, \mathbf{0})$, respectively, denote the vector of derivatives with respect to $\Delta\bar{w}_2$ and $\Delta\bar{v}_1$ evaluated at $(\mathbf{0}, \mathbf{0})$. The remainder term of the approximation is of lower order, $o_p(n_2^{-1/2})$, provided $\max_h W_h/n_{2h} = O(n_2^{-1})$ for large n_2. It now follows from (33.7) that

$$\mathrm{MSE}_p(\hat{\theta}) \doteq \sum_{\alpha=1}^{k+l} \sum_{\alpha'=1}^{k+l} h_\alpha^{(w)} h_{\alpha'}^{(w)} \mathrm{E}\{\Delta\bar{w}_{2\alpha}\Delta\bar{w}_{2\alpha'}\} + \sum_{\beta=1}^{l} \sum_{\beta'=1}^{l} h_\beta^{(v)} h_{\beta'}^{(v)} \mathrm{E}\{\Delta\bar{v}_{1\beta}\Delta\bar{v}_{1\beta'}\}$$
$$+ 2 \sum_{\alpha=1}^{k+l} \sum_{\beta=1}^{l} h_\alpha^{(w)} h_\beta^{(v)} \mathrm{E}\{\Delta\bar{w}_{2\alpha}\Delta\bar{v}_{1\beta}\} \tag{33.8}$$

where $\Delta\bar{w}_2$ is a $(k + l)$-vector with components $\Delta\bar{w}_{2\alpha}$ and $\Delta\bar{v}_1$ is an l-vector with components $\Delta\bar{v}_{1\beta}$. Under two-phase sampling within each stratum, (33.8) reduces to

$$\mathrm{MSE}_p(\hat{\theta}) \doteq \sum_{\alpha=1}^{k+l} \sum_{\alpha'=1}^{k+l} h_\alpha^{(w)} h_{\alpha'}^{(w)} S_{w\alpha\alpha'} + 2 \sum_{\alpha=1}^{k+l} \sum_{\beta=1}^{l} h_\alpha^{(w)} h_\beta^{(v)} S_{w\alpha,v\beta} + \sum_{\beta=1}^{l} \sum_{\beta'=1}^{l} h_\beta^{(v)} h_{\beta'}^{(v)} S_{v\beta\beta'}$$
$$\tag{33.9}$$

where $S_{w\alpha\alpha'} = \sum_h A_{2h} S_{w\alpha\alpha',h}$, $S_{v\beta\beta'} = \sum_h A_{1h} S_{v\beta\beta',h}$, $S_{w\alpha,v\beta} = \sum_h A_{1h} S_{w\alpha,v\beta,h}$, $A_{2h} = (1 - f_{2h}) W_h^2/n_{2h}$ and $A_{1h} = (1 - f_{1h}) W_h^2/n_{1h}$, noting that s_{2h} is a simple random sample of size n_{2h} from stratum h. Here $S_{w\alpha\alpha',h}$, $S_{v\beta\beta',h}$ and $S_{w\alpha,v\beta,h}$ are the hth stratum covariances of the characteristics w_α and $w_{\alpha'}$, v_β and $v_{\beta'}$, and w_α and v_β, respectively, and $f_{1h} = n_{1h}/N_h$ and $f_{2h} = n_{2h}/N_h$ the within-strata sampling fractions.

The formula (33.9) suggests two possible linearization variance estimators

$$v_{L0}(\hat{\theta}) = \sum_{\alpha=1}^{k+l} \sum_{\alpha'=1}^{k+l} \hat{h}_\alpha^{(w)} \hat{h}_{\alpha'}^{(w)} s_{w\alpha\alpha'} + \sum_{\beta=1}^{l} \sum_{\beta'=1}^{l} \hat{h}_\beta^{(v)} \hat{h}_{\beta'}^{(v)} s_{v\beta\beta'} + 2 \sum_{\alpha=1}^{k+l} \sum_{\beta=1}^{l} \hat{h}_\alpha^{(w)} \hat{h}_\beta^{(v)} s_{w\alpha,v\beta}$$
$$\tag{33.10}$$

obtained by replacing the strata covariances by their sample analogues $s_{w\alpha\alpha',h}$, $s_{v\beta\beta',h}$ and $s_{w\alpha,\beta,h}$ based on the second-phase sample only, and $v_{L1}(\hat{\theta})$ obtained from (33.10) by replacing $s_{w\alpha\alpha',h}$ for $\alpha, \alpha' > k$ by $s_{1w\alpha\alpha',h}$, $s_{w\alpha,v\beta,h}$ for $\alpha > k$ by $s_{1w\alpha,v\beta,h}$, and $s_{v\beta\beta',h}$ by $s_{1v\beta\beta',h}$, where $s_{1w\alpha\alpha',h}$, $s_{1w\alpha,v\beta,h}$, and $s_{1v\beta\beta',h}$ are based on the entire first-phase sample, s_{1h}. Also, the population values $\bar{W}$ and $\bar{V}$ in $h_\alpha^{(w)}$ and $h_\beta^{(v)}$ are replaced by the sample estimators $\bar{w}_2$ and $\bar{v}_1$ to get $\hat{h}_\alpha^{(w)}$ and $\hat{h}_\beta^{(v)}$. The variance estimator $v_{L1}(\hat{\theta})$ makes fuller use of the data than $v_{L0}(\hat{\theta})$ and therefore it is likely to be more efficient (Rao and Sitter, 1995).

We now illustrate the calculation of (33.4) for the estimator $\bar{b}_{2stR}$. Using the equation for $h(\Delta\bar{w}_2, \Delta\bar{v}_1)$ given in Section 33.2, we get $h_1^{(w)} = -1$, $h_2^{(w)} = 1 + R_{\mu y}$ and $h_1^{(v)} = -R_{\mu y}$, where $R_{\mu y} = \bar{M}/\bar{Y}$. Using $\hat{R}_{\mu y} = \bar{\mu}_{2st}/\bar{y}_{2st}$ as an estimator of $R_{\mu y}$ we get $\hat{h}_1^{(w)} = -1$, $\hat{h}_2^{(w)} = 1 + \hat{R}_{\mu y}$ and $\hat{h}_1^{(v)} = -\hat{R}_{\mu y}$. We also note the following relationships among the second-phase sample variances and covariances: $s_{\mu h}^2 = s_{dh}^2 + 2\hat{R}_{\mu y}s_{dyh} + \hat{R}_{\mu y}^2 s_{yh}^2$ and $s_{\mu yh} = s_{dyh} + \hat{R}_{\mu y}s_{yh}^2$ with $d_{hi} = \mu_{hi} - \hat{R}_{\mu y}y_{hi}$, $i \in s_{2h}$. Using the above partial derivatives and relationships we get from (33.10), after simplification

$$v_{L0}(\bar{b}_{2stR}) = \sum_{h=1}^{L} W_h^2 \left(\frac{1}{n_{2h}} - \frac{1}{N_h}\right)(s_{dh}^2 - 2s_{dyh} + s_{yh}^2)$$

$$+ \sum_{h=1}^{L} W_h^2 \left(\frac{1}{n_{1h}} - \frac{1}{N_h}\right)[2\hat{R}_{\mu y}\{s_{dyh} - s_{yh}^2\} + \hat{R}_{\mu y}^2 s_{yh}^2]. \quad (33.11)$$

The alternative estimator $v_{L1}(\bar{b}_{2stR})$ is obtained from (33.5) by changing s_{yh}^2 to s_{1yh}^2.

It may be noted that the direct method, using the usual ratio approximation (Cochran, 1977, p. 344)

$$\frac{\bar{\mu}_{2st}}{\bar{y}_{2st}} \bar{y}_{1st} - \bar{M} \doteq (\bar{\mu}_{2st} - R_{\mu y}\bar{y}_{2st}) + R_{\mu y}(\bar{y}_{1st} - \bar{Y})$$

gives a variance estimator which agrees precisely with (33.11).

We now spell out the linearization variance estimators v_{L0} and v_{L1} for the remaining four estimators of $\bar{B}$, omitting the technical details.

We have

$$v_{L0}(\bar{b}_{2st}) = \sum_{h=1}^{L} W_h^2 \left(\frac{1}{n_{2h}} - \frac{1}{N_h}\right)s_{bh}^2 = v_{L1}(\bar{b}_{2st}) \quad (33.12)$$

where s_{bh}^2 is the second-phase sample variance of $b_{hi} = y_{hi} - \mu_{hi}$. Turning to $\bar{b}_{12st}$, we get

$$v_{L0}(\bar{b}_{12st}) = \sum_{h=1}^{L} W_h^2 \left(\frac{1}{n_{2h}} - \frac{1}{N_h}\right)(s_{dh}^2 + 2\hat{R}_{\mu y}s_{dyh} + \hat{R}_{\mu y}^2 s_{yh}^2)$$

$$+ \sum_{h=1}^{L} W_h^2 \left(\frac{1}{n_{1h}} - \frac{1}{N_h}\right)[(1 - 2\hat{R}_{\mu y})s_{yh}^2 - 2s_{dyh}]. \quad (33.13)$$

Further, $v_{L1}(\bar{b}_{12st})$ is obtained from (33.13) by changing s_{yh}^2 to s_{1yh}^2. We next consider $\bar{b}_{12stR}$. We get

$$v_{L0}(\bar{b}_{12stR}) = \sum_{h=1}^{L} W_h^2 \left(\frac{1}{n_{2h}} - \frac{1}{N_h} \right) s_{dh}^2$$

$$+ \sum_{h=1}^{L} W_h^2 \left(\frac{1}{n_{1h}} - \frac{1}{N_h} \right) [(1 - \hat{R}_{\mu y})^2 s_{yh}^2 - 2(1 - \hat{R}_{\mu y}) s_{dyh}]. \quad (33.14)$$

Further $v_{L1}(\bar{b}_{12stR})$ is obtained from (33.14) by changing s_{yh}^2 to s_{1yh}^2.

We next consider the estimators $\bar{b}_{x2stR}$ and $\bar{b}_{SSW}$ that make use of the supplementary information, x_{hi} and $\bar{X}$. After considerable simplification, we get

$$v_{L0}(\bar{b}_{x2stR}) = \sum_{h=1}^{L} W_h^2 \left(\frac{1}{n_{2h}} - \frac{1}{N_h} \right) s_{dh}^2$$

$$+ \sum_{h=1}^{L} W_h^2 \left(\frac{1}{n_{1h}} - \frac{1}{N_h} \right) [s_{eh}^2 - 2s_{edh} + 2\hat{R}_{\mu y} s_{dyh} - 2\hat{R}_{\mu y} s_{eyh} + \hat{R}_{\mu y}^2 s_{yh}^2]$$

$$(33.15)$$

where s_{eh}^2, s_{edh} and s_{eyh} are the second-phase sample variances and covariances for $e_{hi} = y_{hi} - \hat{R}_{yx} x_{hi}$, d_{hi} and y_{hi}. Further, $v_{L1}(\bar{b}_{x2stR})$ is obtained from (33.15) by changing s_{yh}^2, s_{eh}^2 and s_{eyh} to first-phase values s_{1yh}^2, s_{1eh}^2, and s_{1eyh}. Finally we turn to $\bar{b}_{SSW}$. In this case we have

$$v_{L0}(\bar{b}_{SSW}) = v_{L0}(\bar{b}_{x2stR}) + \sum_{h=1}^{L} W_h^2 \left(\frac{1}{n_{1h}} - \frac{1}{N_h} \right)$$

$$\times [\hat{R}_{\mu x}^2 s_{xh}^2 - 2\hat{R}_{\mu x} s_{dxh} - 2\hat{R}_{\mu y} \hat{R}_{\mu x} s_{yxh} + 2\hat{R}_{\mu x} s_{exh}] \quad (33.16)$$

in obvious notation. Further, $v_{L1}(\bar{b}_{SSW})$ is obtained by changing $v_{L0}(\bar{b}_{x2stR})$, s_{xh}^2, s_{yxh} and s_{exh} to $v_{L1}(\bar{b}_{x2stR})$, s_{1xh}^2, s_{1yxh} and s_{1exh}, respectively.

33.4 BOOTSTRAP VARIANCE ESTIMATORS

Biemer and Atkinson (1993) propose an extension of the single-phase without replacement bootstrap (see Gross, 1980; Bickel and Freedman, 1984; Sitter, 1992) to the case of a stratified two-phase design. Their basic idea is as follows. Let $s_{1\sim 2h}$ denote the label set $s_{1h} - s_{2h}$ for each $h = 1, \ldots, L$, and let $s_{1\sim 2} = \bigcup_h s_{1\sim 2h}$ and $s_2 = \bigcup_h s_{2h}$. Also, consider the estimator of interest, $\hat{\theta}$, as a function of $s_{1\sim 2}$ and s_2, i.e., $\hat{\theta} = \hat{\theta}(s_{1\sim 2}, s_2)$. Assuming $k_h = N_h/n_{1h}$ to be an integer, form the label sets $U_{A(2)h}^*$ and $U_{A(1\sim 2)h}^*$ for each h consisting of k_h copies of the units in s_{2h} and $s_{1\sim 2h}$, respectively. The union of these sets forms a pseudo-population for each h. A bootstrap sample then consists of a simple random sample without replacement (SRSWOR), s_{2h}^*, of size n_{2h} from $U_{A(2)h}^*$ together with an SRSWOR, $s_{1\sim 2h}^*$, of size $n_{1\sim 2h}$ from $U_{A(1\sim 2)h}^*$, for $h = 1, \ldots, L$.

A bootstrap variance estimator $\hat{\theta}$ is given by $v_{BSS}(\hat{\theta}) = V_*(\hat{\theta}^*)$, where $\hat{\theta}^* = \hat{\theta}(s_1^*{}_{\sim 2}, s_2^*)$ with $s_1^*{}_{\sim 2} = \bigcup_h s_1^*{}_{\sim 2h}$ and $s_2^* = \bigcup_h s_{2h}^*$ and V_* denotes variance under the resampling. In practice, we use a Monte Carlo estimate of $V_*(\hat{\theta}^*)$ given by

$$v_{BA}(\hat{\theta}) = \frac{1}{Q} \sum_{q=1}^{Q} (\hat{\theta}_q^* - \hat{\theta}_\cdot^*)^2 \tag{33.17}$$

where $\hat{\theta}_1^*, \ldots, \hat{\theta}_Q^*$ are obtained by repeating the resampling procedure a large number, Q, of times and $\hat{\theta}_\cdot^* = \sum_q \hat{\theta}_q^*/Q$. The above bootstrap procedure can be extended to the general case $N_h = n_{1h}k_h + r_h$ with $0 < r_h < n_{1h}$ (see Biemer and Atkinson, 1993) but for simplicity we study here only the case $N_h/n_{1h} = k_h$, an integer for each h.

An advantage of the bootstrap estimator (33.17) and other resampling variance estimators is that the same formula can be applied to estimators, $\hat{\theta}$, of any complexity, unlike the linearization variance estimators. The above bootstrap procedure yields consistent variance estimators for the general class of estimators $\hat{\theta} = g(\bar{w}_2, \bar{v}_1)$ introduced in Section 33.2, provided the sampling fractions $f_{1h} = n_{1h}/N_h$ and $f_{2h} = n_{2h}/N_h$ are negligible, as in the application of Biemer and Atkinson (1993). However, it runs into problems if the sampling fractions f_{1h} and f_{2h} are not negligible for $h = 1, \ldots, L$. There are two reasons for this difficulty: (1) both s_{2h} and $s_{1 \sim 2h}$ are replicated k_h times to form the pseudo-population so that the resampling fractions differ from the original sampling fractions; (2) if f_{1h} and f_{2h} are negligible, then $\bar{v}_2 = \sum_{h=1}^{L} W_h \bar{v}_{2h}$ and $\bar{v}_{1 \sim 2} = \sum_{h=1}^{L} W_h \bar{v}_{1 \sim 2h}$ are uncorrelated, not so otherwise. On the other hand, the bootstrap counterparts, $\bar{v}_2^*$ and $\bar{v}_1^*{}_{\sim 2}$, are uncorrelated under resampling in either case.

To overcome this difficulty, we propose using a modification of the rescaling bootstrap (Rao and Wu, 1988) by which the correlation between $\bar{v}_2$ and $\bar{v}_{1 \sim 2}$ can be induced into the resampling procedure. The proposed rescaling bootstrap procedure is as follows. (1) Take a simple random sample with replacement (SRSWR) of size n_{2h}^* from s_{2h} to get the label set s_{2h}^* for each h. (2) Independently, select an SRSWR of size $n_1^*{}_{\sim 2h}$ from $s_{1 \sim 2h}$ to get the label set $s_1^*{}_{\sim 2h}$ for each h. (3) Let

$$\tilde{w}_{2h}^* = \bar{w}_{2h} + \lambda_h(\bar{w}_{2h}^* - \bar{w}_{2h})$$

$$\tilde{v}_1^*{}_{\sim 2h} = \bar{v}_{1 \sim 2h} + \gamma_h(\bar{v}_1^*{}_{\sim 2h} - \bar{v}_{1 \sim 2h}) + \beta_h(\bar{v}_{2h}^* - \bar{v}_{2h})$$

and

$$\tilde{v}_{1h}^* = \frac{n_{2h}}{n_{1h}} \tilde{v}_{2h}^* + \left(1 - \frac{n_{2h}}{n_{1h}}\right) \tilde{v}_1^*{}_{\sim 2h}$$

where $\lambda_h^2 = n_{2h}^*(n_{2h} - 1)^{-1}(1 - f_{2h})$,

$$\gamma_h^2 = n_1^*{}_{\sim 2h}(n_{1 \sim 2h} - 1)^{-1}(1 - f_{1h})(1 - f_{2h})^{-1}$$

and $\beta_h = -[n^*_{2h}(n_{2h} - 1)^{-1}f^2_{2h}(1 - f_{2h})^{-1}]^{1/2}$. (4) Let $\tilde{w}^*_2 = \sum^L_{h=1} W_h\tilde{w}^*_{2h}$, $\tilde{v}^*_1 = \sum^L_{h=1} W_h\tilde{v}^*_{1h}$ and $\tilde{\theta}^* = g(\tilde{w}^*_2, \tilde{v}^*_1)$. Common choices of the bootstrap sample sizes $(n^*_{2h}, n^*_{1 \sim 2h})$ are $(n_{2h}, n_{1 \sim 2h})$ and $(n_{2h} - 1, n_{1 \sim 2h} - 1)$. In the latter case, rescaling disappears if f_{1h} and f_{2h} are negligible and the method amounts to taking an SRSWR of size $n_{2h} - 1$ from s_{2h} and another SRSWR of size $n_{1 \sim 2h} - 1$ from $s_{1 \sim 2h}$ and then applying the estimator to the resampled data as in the case of Biemer and Atkinson (1993).

A bootstrap variance estimator of $\hat{\theta}$ under the above procedure is given by $V_*(\tilde{\theta}^*)$, which denotes as before the variance under resampling. A Monte Carlo estimate of $V_*(\tilde{\theta}^*)$ is obtained as

$$v_B(\hat{\theta}) = \frac{1}{Q} \sum^Q_{q=1} (\tilde{\theta}^*_q - \tilde{\theta}^*_{\cdot})^2 \tag{33.18}$$

where $\tilde{\theta}^*_1, \ldots, \tilde{\theta}^*_Q$ are obtained by repeating steps (1)–(4) a large number, Q, of times and $\tilde{\theta}^*_{\cdot} = \sum^Q_{q=1} \tilde{\theta}^*_q/Q$.

Our method ensures consistency of the bootstrap variance estimator (33.18) even if the sampling fractions are not negligible. This is achieved by the introduction of the rescaling term $\beta_h(\tilde{v}^*_{2h} - \bar{v}_{2h})$ in the definition of $\tilde{v}^*_{1 \sim 2h}$ which ensures that the bootstrap covariance of any two components $\tilde{v}^*_{2\beta h}$ and $\tilde{v}^*_{1 \sim 2\beta'h}$ of $\tilde{v}^*_{2h}$ and $\tilde{v}^*_{1 \sim 2h}$ matches the sample covariance of $\bar{v}_{2\beta h}$ and $\bar{v}_{1 \sim 2\beta'h}$. A heuristic proof of consistency is obtained by linearizing the bootstrap variance estimator (33.18) and showing its asymptotic equivalence to the linearization variance estimator (33.10).

By letting $\Delta\tilde{w}^*_2 = \tilde{w}^*_2 - \bar{w}_2$, $\Delta\tilde{v}^*_1 = \tilde{v}^*_1 - \bar{v}_1$,

$$\tilde{\theta}^* = g(\tilde{w}^*_2, \tilde{v}^*_1) = g(\bar{w}_2 + \Delta\tilde{w}^*_2, \bar{v}_1 + \Delta\tilde{v}^*_1) = \tilde{h}(\Delta\tilde{w}^*_2, \Delta\tilde{v}^*_1),$$

and noting that $\hat{\theta} = \hat{h}(0, 0)$, it is shown in Appendix 1 that for large n_2

$$v_B(\hat{\theta}) \doteq \tilde{v}_L(\hat{\theta}) = \sum^{k+l}_{\alpha=1} \sum^{k+l}_{\alpha'=1} \tilde{h}^{(w)}_\alpha \tilde{h}^{(w)}_{\alpha'} s_{w\alpha\alpha'} + \sum^l_{\beta=1} \sum^l_{\beta'=1} \tilde{h}^{(v)}_\beta \tilde{h}^{(v)}_{\beta'} \tilde{s}_{v\beta\beta'}$$
$$+ 2 \sum^{k+l}_{\alpha=1} \sum^l_{\beta=1} \tilde{h}^{(w)}_\alpha \tilde{h}^{(v)}_\beta s_{w\alpha, v\beta}. \tag{33.19}$$

Here $\tilde{h}^{(w)}_\alpha$ and $\tilde{h}^{(v)}_\beta$ are the derivatives of $\hat{h}$ with respect to the components of $\Delta\tilde{w}^*_2$ and $\Delta\tilde{v}^*_1$ evaluated at $(0, 0)$, and

$$\tilde{s}_{v\beta\beta'} = \sum^L_{h=1} W^2_h \left(\frac{1}{n_{1h}} - \frac{1}{N_h}\right)\{a_h s_{hv\beta\beta'} + (1 - a_h)s_{1 \sim 2hv\beta\beta'}\}$$

with $0 < a_h = n_{2h}(1 - f_{1h})/[n_{1h}(1 - f_{2h})] < 1$,

$$(n_{2h} - 1)s_{hv\beta\beta'} = \sum_{i \in s_{2h}} (v_{\beta hi} - \bar{v}_{2\beta h})(v_{\beta'hi} - \bar{v}_{2\beta'h})$$

and $(n_{1 \sim 2h} - 1)s_{1 \sim 2h\beta\beta'} = \sum_{i \in s_{1 \sim 2h}}(v_{\beta hi} - \bar{v}_{1 \sim 2\beta h})(v_{\beta' hi} - \bar{v}_{1 \sim 2\beta' h})$. Note that $\tilde{h}_\alpha^{(w)}$ and $\tilde{h}_\beta^{(v)}$ are not identical to $\hat{h}_\alpha^{(w)}$ and $\hat{h}_\beta^{(v)}$ in (33.10), but are approximately equal to them as $n_2 \to \infty$. Comparing (33.19) and (33.10) we see that the bootstrap variance estimator, $v_B(\hat\theta)$, is consistent.

It is clear that $\tilde{v}_L(\hat\theta)$ could be used as an alternative linearization variance estimator. One could also replace $\hat{h}_\alpha^{(w)}$ and $\hat{h}_\beta^{(v)}$ by $\tilde{h}_\alpha^{(w)}$ and $\tilde{h}_\beta^{(v)}$ in the formulae for $v_{L0}(\hat\theta)$ or $v_{L1}(\hat\theta)$ to get alternative linearization variance estimators $\tilde{v}_{L0}(\hat\theta)$ or $\tilde{v}_{L1}(\hat\theta)$. In the special case of a ratio estimator under two-phase simple random sampling, Rao and Sitter (1995) have proposed a variance estimator analogous to $\tilde{v}_{L1}(\hat\theta)$. They have shown in a limited simulation study that this variance estimator has better conditional properties than the analog to $v_{L1}(\hat\theta)$.

33.5 JACKKNIFE VARIANCE ESTIMATORS

Jackknife variance estimation is commonly used in sample surveys because of the advantage of applying the same formula to all estimators $\hat\theta$, as in the case of the bootstrap. Moreover, limited empirical results (e.g., Rao and Wu, 1988; Sitter, 1992) have indicated that the jackknife leads to more stable variance estimators as compared to the bootstrap. It is therefore worthwhile to extend the jackknife to the present case of stratified two-phase sampling.

We mimic the rescaling bootstrap of Section 33.4, using a delete one-unit jackknife. Our method is as follows. (1) For each h, let $\bar{\boldsymbol{v}}_{1h}(j) = (n_{1h}\bar{\boldsymbol{v}}_{1h} - \boldsymbol{v}_{hj})/(n_{1h} - 1)$ for all $j \in s_{1h}$; $\bar{\boldsymbol{w}}_{2h}(j) = (n_{2h}\bar{\boldsymbol{w}}_{2h} - \boldsymbol{w}_{hj})/(n_{2h} - 1)$ if $j \in s_{2h}$ and $\bar{\boldsymbol{w}}_{2h}(j) = \bar{\boldsymbol{w}}_{2h}$ if $j \in s_{1 \sim 2h}$; $\bar{\boldsymbol{v}}_{2h}(j) = (n_{2h}\bar{\boldsymbol{v}}_{2h} - \boldsymbol{v}_{hj})/(n_{2h} - 1)$ if $j \in s_{2h}$ and $\bar{\boldsymbol{v}}_{2h}(j) = \bar{\boldsymbol{v}}_{2h}$ if $j \in s_{1 \sim 2h}$; $\bar{\boldsymbol{v}}_{1 \sim 2h}(j) = \bar{\boldsymbol{v}}_{1 \sim 2h}$ if $j \in s_{2h}$ and

$$\bar{\boldsymbol{v}}_{1 \sim 2h}(j) = (n_{1 \sim 2h}\bar{\boldsymbol{v}}_{1 \sim 2h} - \boldsymbol{v}_{hj})/(n_{1 \sim 2h} - 1)$$

if $j \in s_{1 \sim 2h}$. (2) Define the rescaled means as follows

$$\tilde{\boldsymbol{w}}_{2h}(j) = \bar{\boldsymbol{w}}_{2h} + \tilde\lambda_h(\bar{\boldsymbol{w}}_{2h}(j) - \bar{\boldsymbol{w}}_{2h})$$

$$\tilde{\boldsymbol{v}}_{1 \sim 2h}(j) = \bar{\boldsymbol{v}}_{1 \sim 2h} + \tilde\gamma_h(\bar{\boldsymbol{v}}_{1 \sim 2h}(j) - \bar{\boldsymbol{v}}_{1 \sim 2h}) + \tilde\beta_h(\bar{\boldsymbol{v}}_{2h}(j) - \bar{\boldsymbol{v}}_{2h})$$

and

$$\tilde{\boldsymbol{v}}_{1h}(j) = \frac{n_{2h}}{n_{1h}}\tilde{\boldsymbol{v}}_{2h}(j) + \left(1 - \frac{n_{2h}}{n_{1h}}\right)\tilde{\boldsymbol{v}}_{1 \sim 2h}(j)$$

where $\tilde\lambda_h$, $\tilde\gamma_h$ and $\tilde\beta_h$ are obtained from the scale factors λ_h, γ_h and β_h for the bootstrap by replacing $n_{2h}^*(n_{2h} - 1)^{-1}$ and $n_{1 \sim 2h}^*(n_{1 \sim 2h} - 1)^{-1}$ by $(n_{2h} - 1)n_{2h}^{-1}$ and $(n_{1 \sim 2h} - 1)n_{1 \sim 2h}^{-1}$, respectively. (3) Let $\tilde{w}_2(hj) = \sum_{h' \neq h} W_{h'}\bar{w}_{2h'} + W_h\tilde{w}_{2h}(j)$, $\tilde{v}_1(hj) = \sum_{h' \neq h} W_{h'}\bar{v}_{1h'} + W_h\tilde{v}_{1h}(j)$, and $\hat\theta(hj) = g[\tilde{w}_2(hj), \tilde{v}_2(hj)]$. (4) A jackknife variance estimator for $\hat\theta = g(\bar{w}_2, \bar{v}_1)$ is then given by

$$v_J(\hat\theta) = \sum_{h=1}^{L} \sum_{j \in s_{1h}} (\hat\theta(hj) - \hat\theta)^2. \tag{33.20}$$

It is shown in Appendix 2 that $v_J(\hat{\theta}) \doteq \tilde{v}_L(\hat{\theta})$ given by (33.19). Thus the internally rescaled jackknife and bootstrap yield approximately the same variance estimator for large n_2.

33.6 SIMULATION STUDY

To study the finite sample properties of the proposed variance estimators, we conducted a limited simulation study. For this purpose, we created a stratified finite population with two strata of sizes N_1 and N_2. The characteristics y, x, and μ for the ith unit in the hth stratum were generated using the following model

$$x_{hi} \sim \text{gamma}(a_h, b_h), \quad \mu_{hi} = \alpha_h x_{hi} + \sqrt{x_{hi}} e_{hi}$$

$$\text{E}(y_{hi} \mid x_{hi}) = \beta_h x_{hi} \quad \text{and} \quad \text{V}(y_{hi} \mid x_{hi}) = x_{hi} \sigma_{\varepsilon h}^2$$

for specified values of a_h, b_h, α_h, β_h and $(\sigma_{eh}^2, \sigma_{\varepsilon h}^2)$, where $e_{hi} \sim \text{N}(0, \sigma_{eh}^2)$ and, for each x_{hi}, y_{hi} is generated from a gamma distribution with parameters chosen to give the above mean and variance, and x_{hi} and e_{hi} are independently distributed. Thus, the mean, variance and coefficient of variation of x_{hi} are given by $\mu_{xh} = a_h b_h$, $\sigma_{xh}^2 = a_h b_h^2$ and $C_{xh} = \sigma_{xh}/\mu_{xh} = 1/\sqrt{a_h}$, respectively. Further, the mean and variance of y_{hi} are $\mu_{yh} = \beta_h \mu_{xh}$ and $\sigma_{yh}^2 = \beta_h^2 \sigma_{xh}^2 + \mu_{xh} \sigma_{\varepsilon h}^2$, and $\text{corr}(x_{hi}, y_{hi}) = \rho_{xyh} = \beta_h \sigma_{xh}/\sigma_{yh}$, while the mean and variance of μ_{hi} are $\mu_{\mu h} = \alpha_h \mu_{xh}$ and $\sigma_{\mu h}^2 = \alpha_h^2 \sigma_{xh}^2 + \mu_{xh} \sigma_{eh}^2$, and $\text{corr}(x_{hi}, \mu_{hi}) = \rho_{x\mu h} = \alpha_h \sigma_{xh}/\sigma_{\mu h}$.

For a given (N_1, N_2) combination, we first generated a finite population of values $\{y_{hi}, \mu_{hi}, x_{hi}\}$ using the above model with $a_1 = 0.5$, $a_2 = 2.0$, $\beta_1 = 1.5$, $\beta_2 = 1.0$, $\alpha_1 = 1.25$, $\alpha_2 = 0.75$, $\mu_{x1} = 48$ and $\mu_{x2} = 96$ and choosing σ_{eh} and $\sigma_{\varepsilon h}$ so that $\rho_{x\mu h} = 0.8$ and $\rho_{xyh} = 0.8$.

From a given finite population we then generated $K = 10,000$ independent two-phase random samples with first-phase stratum sample sizes n_{1h} and second-phase stratum sample sizes n_{2h} for a given (n_{1h}, n_{2h}) combination. For each simulation run each of the six estimators of measurement bias given in Section 33.2 was calculated. The simulated design MSE of a specified estimator $\bar{b}$, was calculated as

$$\text{MSE}_p(\bar{b}) = \frac{1}{K} \sum_{k=1}^{K} \{\bar{b}_k - \bar{B}_y\}^2$$

where $\bar{B}_y = \bar{Y} - \bar{M}$ and $\bar{b}_k$ is the value of $\bar{b}$ for the kth simulation run. The simulated percentage relative bias and MSE of a specified variance estimator, v, were calculated as

$$\text{RB}(v) = 100\{\tilde{v} - \text{MSE}_p(\bar{b})\}/\text{MSE}_p(\bar{b})$$

Table 33.1 **Comparing the % Relative Bias and Mean Squared Error of v_{L0}, v_{L1}, v_J, v_B and v_{BA} When Applied to $\bar{b}_{2st}$, $\bar{b}_{12st}$, $\bar{b}_{2stR}$, $\bar{b}_{12stR}$, $\bar{b}_{x2stR}$ and $\bar{b}_{SSW}$**

$\bar{b}$	% RB(v)					MSE(v)/MSE(v_{L1})				
	v_{L0}	v_{L1}	v_J	v_B	v_{BA}	v_{L0}	v_{L1}	v_J	v_B	v_{BA}
Case A. $N_1 = 800$, $N_2 = 600$, $n_{11} = 200$, $n_{21} = 40$, $n_{12} = 150$, $n_{22} = 30$										
$\bar{b}_{2st}$	1.70	1.70	1.70	2.44	−21.07	1.00	1.00	1.00	1.10	0.87
$\bar{b}_{12st}$	0.20	3.77	0.22	0.66	−21.47	0.88	1.00	0.65	0.72	0.80
$\bar{b}_{2stR}$	−2.35	−2.37	1.08	0.30	−24.77	2.28	1.00	1.79	1.73	1.78
$\bar{b}_{12stR}$	−3.36	−3.57	4.27	3.53	−18.65	1.03	1.00	1.36	1.30	1.08
$\bar{b}_{xstR}$	−4.40	−4.27	2.76	2.41	−18.02	1.05	1.00	1.44	1.44	1.21
$\bar{b}_{SSW}$	−4.14	−4.29	4.01	3.16	−19.17	1.01	1.00	1.36	1.30	1.09
Case B. $N_1 = 400$, $N_2 = 600$, $n_{11} = 100$, $n_{21} = 50$, $n_{12} = 150$, $n_{22} = 75$										
$\bar{b}_{2st}$	−1.04	−1.04	−1.04	−0.58	−15.78	1.00	1.00	1.00	1.16	1.15
$\bar{b}_{12st}$	−1.33	−0.97	−1.35	−0.83	−15.38	1.45	1.00	1.09	1.41	1.77
$\bar{b}_{2stR}$	−1.36	−1.42	−0.28	−0.20	−21.23	1.54	1.00	1.11	1.31	1.80
$\bar{b}_{12stR}$	−2.65	−2.80	−1.07	−1.37	−13.77	1.05	1.00	0.93	1.06	1.15
$\bar{b}_{xstR}$	−2.01	−2.00	−0.48	−0.64	−11.50	1.18	1.00	0.98	1.15	1.19
$\bar{b}_{SSW}$	−2.90	−3.01	−0.71	−1.42	−14.23	1.02	1.00	0.90	1.01	1.12

and

$$\text{MSE}(v) = \frac{1}{K} \sum_{k=1}^{K} \{v^{(k)} - \text{MSE}_p(\bar{b})\}^2$$

where $v^{(k)}$ is the value of v for the kth simulation run and $\bar{v} = \sum_k v^{(k)}/K$.

Table 33.1 summarizes the results. It considers two cases: case A—$N_1 = 800$, $N_2 = 600$, $n_{11} = 200$, $n_{12} = 150$, $n_{21} = 40$ and $n_{22} = 30$; case B—$N_1 = 400$, $N_2 = 600$, $n_{11} = 100$, $n_{12} = 150$, $n_{21} = 50$ and $n_{22} = 75$.

Table 33.1 reports the values of percentage relative bias of variance estimators, RB$\{v(\hat{\theta})\}$, for the linearization, v_{L0} and v_{L1}, the jackknife, v_J, the rescaled bootstrap, v_B, and the Biemer–Atkinson bootstrap, v_{BA}, for each of the six estimators of response bias. It also gives the efficiency of variance estimators, $v(\hat{\theta})$, relative to $v_{L1}(\hat{\theta})$, as measured by MSE$\{v(\hat{\theta})\}$/MSE$\{v_{L1}(\hat{\theta})\}$.

For the sake of space, we do not present the resulting MSE$_p(\bar{b})$ for the six estimators. However, in both cases A and B, $\bar{b}_{12stR}$ and $\bar{b}_{SSW}$ were significantly more efficient than the other four estimators, followed by $\bar{b}_{x2stR}$, $\bar{b}_{12st}$, $\bar{b}_{2st}$ and $\bar{b}_{2stR}$. It is clear from Table 33.1 that v_{L0}, v_{L1}, v_J and v_B all perform well in terms of relative bias with absolute percentage relative bias less than 5 percent in both cases A and B. The Biemer and Atkinson bootstrap variance estimator, v_{BA}, however, performs poorly with a negative percentage relative bias as high as -25 percent. This poor performance is due to the presence of significant within-strata sampling fractions. By its construction, Biemer and Atkinson's bootstrap implicitly amounts to correcting for the within-strata sampling fractions with $(1 - f_{1h})$, but it is clear from the development of linearization variance estimators in Section 33.3 that there are components of the variance which have within-strata finite population corrections of $(1 - f_{2h})$ (see (33.11) for example). Since $(1 - f_{1h}) \leq (1 - f_{2h})$ this results in a negative bias which becomes more pronounced when f_{1h} and f_{2h} differ greatly. Unfortunately, in two-phase sampling situations this is most often the case. This is illustrated in Table 33.1 by comparing the relative bias of v_{BA} in case A to case B. In case A, $1 - f_{1h} = 0.75$ and $1 - f_{2h} = 0.95$, while in case B, $1 - f_{1h} = 0.75$ and $1 - f_{2h} = 0.875$ for $h = 1, 2$. So $(1 - f_{1h})/(1 - f_{2h}) \doteq 0.8$ and 0.86 in cases A and B, respectively, and the relative bias is smaller in absolute value in case B than in case A as expected. We should note that in the situation considered by Biemer and Atkinson (1993) where v_{BA} was proposed, the sampling fractions were negligible and thus v_{BA} would behave well.

Turning to the efficiency of the variance estimators, we first note that v_{L0}, v_{L1}, and v_J are algebraically equivalent for the linear estimator $\bar{b}_{2st}$ and thus perform equivalently. Looking further, we see that on average v_{L1} is most efficient, having the smallest MSE in 8 of the 12 cases presented. Beyond this, it is difficult to see too much which can be said to hold generally for all six estimators. However, in view of the fact that our results on $\text{MSE}_p(\bar{b})$ for these six estimators indicated that $\bar{b}_{12stR}$ and $\bar{b}_{SSW}$ are significantly more efficient, it may be more relevant to look at $\bar{b}_{12stR}$ and $\bar{b}_{SSW}$ separately. For these two estimators there is a distinct difference in the relative performances of the various estimators between cases A and B, likely due to the larger second-phase sample size of case B. In case A, v_{L1} is most efficient closely followed by v_{L0} and then v_{BA}, while v_J and v_B are much less efficient. In case B, v_J is most efficient followed closely by v_{L1}, v_{L0} and v_B, while v_{BA} is a bit less efficient. The fact that v_{BA} has smaller MSE than v_J and v_B in case A is likely due to the same property that resulted in its large absolute relative bias. The fact that v_{BA} uses $1 - f_{1h}$ instead of the larger $1 - f_{2h}$ will not only cause a negative bias but also a smaller variance.

33.7 CONCLUSIONS

We have studied the estimation of response bias under stratified two-phase sampling in which a subsample of survey respondents is selected in each stratum and the true values are ascertained through some costly means such as reconciled reinterviews. We have introduced a general class of estimators which included the estimators of Biemer and Atkinson (1993) as special cases. We have also obtained Taylor linearization, jackknife and rescaling bootstrap variance estimators. These variance estimators remain valid even when the within-strata sampling fractions are not negligible. Our simulation results indicated that the proposed variance estimators perform well in terms of relative bias, while the bootstrap variance estimator of Biemer and Atkinson (1993) leads to significant underestimation in the presence of significant within-strata sampling fractions. Generally the Taylor and jackknife variance estimators are more efficient.

ACKNOWLEDGMENTS

This research was supported by grants from the Natural Sciences and Engineering Council of Canada.

REFERENCES

Bickel, P., and Freedman, D.A. (1984), "Asymptotic Normality and the Bootstrap in Stratified Sampling," *Annals of Statistics*, 12, pp. 470–482.

Biemer, P.P., and Atkinson, D. (1993), "Estimation of Measurement Bias Using a Model Prediction Approach," *Survey Methodology*, 19, pp. 127–136.

Biemer, P.P., and Stokes, S.L. (1985), "Optimal Design of Interviewer Variance Experiments in Complex Surveys," *Journal of the American Statistical Association*, 80, pp. 158–166.

Brackstone, G.J., Gosselin, J.F., and Garton, B.E. (1975), "Measurement of Response Errors in Censuses and Surveys," *Survey Methodology*, 1, 1975, pp. 144–157.

Cochran, W.G. (1977), *Sampling Techniques* (3rd edn.), New York: Wiley.

Gross, S., (1980), "Median Estimation in Sample Surveys," *Proceedings of the Section on Survey Research Methods*, American Statistical Association, pp. 181–184.

Hansen, M.H., Hurwitz, W.N., and Bershad, M.A. (1961), "Measurement Errors in Censuses and Surveys," *Bulletin of the International Statistical Institute*, 38, pp. 359–374.

Kleffe, J., Prasad, N.G.N., and Rao, J.N.K. (1991), "'Optimal' Estimation of Correlated Response Variance Under Additive Models," *Journal of the American Statistical Association*, 86, pp. 144–150.

Madow, W.G. (1965), "On Some Aspects of Response Error Measurement," *Proceedings of the Social Statistics Section*, American Statistical Association, pp. 182–192.

Rao, J.N.K., and Sitter, R.R. (1995), "Variance Estimation Under Two-Phase Sampling with Application to Imputation for Missing Data," *Biometrika*, 82, pp. 453–460.

Rao, J.N.K., and Wu, C.F.J. (1988), "Resampling Inference with Complex Survey Data," *Journal of the American Statistical Association*, 83, pp. 231–241.

Sitter, R.R. (1992), "Comparing Three Bootstrap Methods for Survey Data," *Canadian Journal of Statistics*, 20, pp. 135–154.

Särndal, C.E., Swensson, B., and Wretman, J. (1992), *Model Assisted Survey Sampling*, New York: Springer-Verlag.

APPENDIX 1: PROOF OF (33.19)

Let $\Delta \tilde{w}_2^* = \tilde{w}_2^* - \bar{w}_2$, $\Delta \tilde{v}_2^* = \tilde{v}_2^* - \bar{v}_2$ and $\Delta \tilde{v}_1^* = \tilde{v}_1^* - \bar{v}_1$. Then, conditional on the sample, $\Delta \tilde{w}_2^*$, $\Delta \tilde{v}_2^*$ and $\Delta \tilde{v}_1^*$ are all $O_p(n_2^{-1/2})$ under the assumptions that $\max_h W_h/n_{2h}$ and $\max_h W_h/n_{1 \sim 2h}$ are of order $O(n_2^{-1})$. For discussion of these assumptions see Rao and Wu (1988). Also, note that $\tilde{\theta}^* = g(\tilde{w}_2^*, \tilde{v}_1^*) = g(\bar{w}_2 + \Delta \tilde{w}_2^*, \bar{v}_1 + \Delta \tilde{v}_1^*) = \hat{h}(\Delta \tilde{w}_2^*, \Delta \tilde{v}_1^*)$ and takes the value $\hat{\theta}$ at $(\mathbf{0}, \mathbf{0})$. Thus

$$\tilde{\theta}^* - \hat{\theta} = (\Delta \tilde{w}_2^*)^T \tilde{h}^{(w)} + (\Delta \tilde{v}_1^*)^T \tilde{h}^{(v)} + o_p(n_2^{-1/2}) \tag{33.A.1}$$

where $\tilde{h}^{(w)} = \tilde{h}^{(w)}(\mathbf{0}, \mathbf{0}) = (\tilde{h}_1^{(w)}, \ldots, \tilde{h}_{k+l}^{(w)})^T$ is the vector of derivatives of $\hat{h}$ with respect to the components of $\Delta \tilde{w}_2^*$ and $\tilde{h}^{(v)} = \tilde{h}^{(v)}(\mathbf{0}, \mathbf{0}) = (\tilde{h}_1^{(v)}, \ldots, \tilde{h}_l^{(v)})^T$ is the vector of derivatives of $\hat{h}$ with respect to the components of $\Delta \tilde{v}_1^*$. If we now square both sides of (33.A.1) and take expectations with respect to the bootstrap procedure we get

$$V_*(\hat{\theta}^*) \doteq \sum_{\alpha=1}^{k+l} \sum_{\alpha'=1}^{k+l} \tilde{h}_\alpha^{(w)} \tilde{h}_{\alpha'}^{(w)} E_* \{\Delta \tilde{w}_{2\alpha}^* \Delta \tilde{w}_{2\alpha'}^*\} + \sum_{\beta=1}^{l} \sum_{\beta'=1}^{l} \tilde{h}_\beta^{(v)} \tilde{h}_{\beta'}^{(v)} E_* \{\Delta \tilde{v}_{1\beta}^* \Delta \tilde{v}_{1\beta'}^*\}$$

$$+ 2 \sum_{\alpha=1}^{k+l} \sum_{\beta=1}^{l} \tilde{h}_\alpha^{(w)} \tilde{h}_\beta^{(v)} E_* \{\Delta \tilde{w}_{2\alpha}^* \Delta \tilde{v}_{1\beta}^*\}$$

$$= \tilde{v}_L(\hat{\theta}) \tag{33.A.2}$$

as given in (33.19).

APPENDIX 2: PROOF OF $v_J(\hat{\theta}) \doteq \tilde{v}_L(\hat{\theta})$

To compare the jackknife variance estimator to the linearization and bootstrap variance estimators of the previous sections, we will linearize the jackknife in a similar fashion to the linearization of the bootstrap in the previous section. To do this, note that

$$\Delta \tilde{w}_2(hj) = \tilde{w}_2(hj) - \bar{w}_2 = \begin{cases} -\tilde{\lambda}_h W_h \left(\dfrac{w_{hj} - \bar{w}_{2h}}{n_{2h} - 1} \right) & \text{if } j \in s_{2h} \\ \\ 0 & \text{if } j \in s_{1 \sim 2h} \end{cases}$$

and

$$\Delta \tilde{v}_1(hj) = \tilde{v}_1(hj) - \bar{v}_1 = \begin{cases} -W_h \left(\dfrac{n_{2h}}{n_{1h}} \tilde{\lambda}_h + \dfrac{n_{1 \sim 2h}}{n_{1h}} \tilde{\beta}_h \right) \left(\dfrac{v_{hj} - \bar{v}_{2h}}{n_{2h} - 1} \right) & \text{if } j \in s_{2h} \\ \\ -W_h \left(\dfrac{n_{1 \sim 2h}}{n_{1h}} \tilde{\gamma}_h \right) \left(\dfrac{v_{hj} - \bar{v}_{1 \sim 2h}}{n_{1 \sim 2h} - 1} \right) & \text{if } j \in s_{1 \sim 2h} \end{cases}$$

are $O(n_{2h}^{-1/2})$ under some conditions on the distributions of w and v. Thus

$$
\begin{aligned}
\hat{\theta}(hj) - \hat{\theta} &= g(\bar{w}_2 + \Delta\tilde{w}_2(hj), \bar{v}_1 + \Delta\tilde{v}_1(hj)) \\
&= \hat{h}(\Delta\tilde{w}_2(hj), \Delta\tilde{v}_1(hj)) \\
&\doteq (\Delta\tilde{w}_2(hj))^T \tilde{h}^{(w)} + (\Delta\tilde{v}_1(hj))^T \tilde{h}^{(v)} + o_p(n_2^{-1/2})
\end{aligned}
$$

and

$$
\begin{aligned}
v_J(\hat{\theta}) &= \sum_{h=1}^{L} \sum_{j \in s_{1h}} (\hat{\theta}(hj) - \hat{\theta})^2 \\
&\doteq \sum_{\alpha=1}^{k+l} \sum_{\alpha'=1}^{k+l} \tilde{h}_{\alpha}^{(w)} \tilde{h}_{\alpha'}^{(w)} \sum_{h} \sum_{j \sim s_{1h}} \Delta\tilde{w}_{2\alpha}(hj) \Delta\tilde{w}_{2\alpha'}(hj) \\
&\quad + \sum_{\beta=1}^{l} \sum_{\beta'=1}^{l} \tilde{h}_{\beta}^{(v)} \tilde{h}_{\beta'}^{(v)} \sum_{h} \sum_{j \in s_{1h}} \Delta\tilde{v}_{1\beta}(hj) \Delta\tilde{v}_{1\beta'}(hj) \\
&\quad + 2 \sum_{\alpha=1}^{k+l} \sum_{\beta=1}^{l} \tilde{h}_{\alpha}^{(w)} \tilde{h}_{\beta}^{(v)} \sum_{h} \sum_{j \in s_{1h}} \Delta\tilde{w}_{2\alpha}(hj) \Delta\tilde{v}_{1\beta}(hj) \\
&= v_L(\hat{\theta})
\end{aligned}
$$

as given in (33.19).

Index

WILEY SERIES IN PROBABILITY AND STATISTICS

ESTABLISHED BY WALTER A. SHEWHART AND SAMUEL S. WILKS

Editors
Vic Barnett, Ralph A. Bradley, Nicholas I. Fisher, J. Stuart Hunter,
J. B. Kadane, David G. Kendall, David W. Scott, Adrian F. M. Smith,
Jozef L. Teugels, Geoffrey S. Watson

*Now available in a lower priced paperback edition in the Wiley Classics Library.

*Now available in a lower priced paperback edition in the Wiley Classics Library.

*Now available in a lower priced paperback edition in the Wiley Classics Library.

*Now available in a lower priced paperback edition in the Wiley Classics Library.

*Now available in a lower priced paperback edition in the Wiley Classics Library.

*Now available in a lower priced paperback edition in the Wiley Classics Library.